Electromagnetic Analysis and Condition Monitoring of Synchronous Generators

Electromagnetic Analysis and Condition Monitoring of Synchronous Generators

Hossein Ehya and Jawad Faiz

IEEE Press Series on Power and Energy Systems
Ganesh Kumar Venayagamoorthy, Series Editor

IEEE PRESS

WILEY

Published by John Wiley & Sons, Inc., Hoboken, New Jersey.

Published simultaneously in Canada.

For general information on our other products and services or for technical support, please contact our Customer Care Department within the United States at (800) 762-2974, outside the United States at (317) 572-3993 or fax (317) 572-4002.

Wiley also publishes its books in a variety of electronic formats. Some content that appears in print may not be available in electronic formats. For more information about Wiley products, visit our web site at www.wiley.com.

Library of Congress Cataloging-in-Publication Data:

Names: Ehya, Hossein, author. | Faiz, Jawad, author. | John Wiley & Sons, publisher.
Title: Electromagnetic analysis and condition monitoring of synchronous generators / Hossein Ehya, Jawad Faiz.
Description: Hoboken, New Jersey : Wiley, [2023] | Publication date from ECIP data view. | Includes bibliographical references and index.
Identifiers: LCCN 2022043335 (print) | LCCN 2022043336 (ebook) | ISBN 9781119636076 (hardback) | ISBN 9781119636137 (adobe pdf) | ISBN 9781119636168 (epub)
Subjects: LCSH: Electric machinery–Monitoring. | Synchronous generators.
Classification: LCC TK2313 .F35 2023 (print) | LCC TK2313 (ebook) | DDC 621.31/34–dc23/eng/20221017
LC record available at https://lccn.loc.gov/2022043335
LC ebook record available at https://lccn.loc.gov/2022043336

Cover Design: Wiley
Cover Image: © Maciej Bledowski/Shutterstock

Set in 9.5/12.5pt STIXTwoText by Straive, Chennai, India

Contents

Author Biographies

Hossein Ehya, PhD, is a Research Scientist in the Department of Electrical Power Engineering at the Norwegian University of Science and Technology (NTNU). He has been working on the health monitoring of electric machines since 2010. From 2013 to 2018, he also worked as an electrical machine design engineer who designed over 30 industrial electric machines. Dr. Ehya was the recipient of several innovation awards from NTNU and the Research Council of Norway. His current research activities include the development of an automated health monitoring system for electric aviation and renewable energies.

Jawad Faiz, PhD, is a Professor in the School of Electrical and Computer Engineering at the University of Tehran, Iran, where his research interests are the design, modeling, and fault diagnosis of electrical machines and transformers. He is a senior member of the IEEE, a Fellow of the Iran Academy of Sciences, and a member of the Euro-Med Academy of Sciences and Arts. He has published more than 300 journal papers and presented about the same number of conference papers. He published two books by Springer and IET in the above-mentioned fields. Professor Faiz was the recipient of several international and national awards for his research activities.

Preface

Because of technological advancements, a smaller global labor force is required, but the demand for electrical energy is increasing tremendously. Electrical power, as a very efficient and clean energy, is the major type of energy used in the world. Most of the electrical power used in civilian applications is the AC type; therefore, the AC generators are the most important equipment for generating electrical power. AC power has several advantages, including it is easy and very efficient to transmit over long distances using electrical transformers, which allows its delivery to consumers in appropriate places. The loads supplied by synchronous generators dictate the size of the generator. Industrial generators generate many megawatts at high voltage, whereas the very small generators typically used in vehicles are low power ones.

Synchronous generators are one of the most expensive and important components of power systems. They are also subject to various types of failures, so early detection of incipient faults and rapid maintenance may prevent costly consequences. Many faults in synchronous generators can prevent their proper operation. Initially, at the beginning of a fault occurrence, a small decline in function from the healthy state is observed in the generator. If the fault is not diagnosed, it can gradually lead to catastrophic failure. Therefore, fault protection is vital. Some faults that occur in generators are more dangerous than others, and their quick diagnosis prevents serious damage to the generator. The operators of synchronous generators try to reduce maintenance and repair costs and to prevent unscheduled downtimes. Condition monitoring, in parallel with conventional planned maintenance schemes, is necessary. Continuous online monitoring can prevent catastrophic failure of the generators.

During the past few years, a number of condition monitoring techniques have been introduced. In addition, new and alternative online diagnostic methods have been developed and reported in the relevant literature. The operator must therefore select the most appropriate and effective monitoring systems suited to their particular generator. Despite intensive research efforts directed at fault diagnosis in induction machines, the research work pertinent to the failure of synchronous generators is very limited. This book is concerned with the in-depth study of failure and its traceable symptoms in different generator signals and parameters.

This book originated from the experience of the authors in fault diagnosis of synchronous generators at the University of Tehran, Iran, and at the Norwegian University of Science and Technology (NTNU) in Norway. This research has led to several publications in fault diagnosis of synchronous generators. The purpose of this book is to help electrical engineers in power system industries and researchers doing work in this field. An attempt has been made to collect recent work in fault diagnosis in synchronous generators in such a way that potential readers will be able to apply various types of fault diagnosis techniques. Step-by-step theoretical, analytical, numerical, and experimental implementations are provided in 16 chapters, and the contents of these chapters are explained in Chapter 1 of the book.

Several postgraduate/graduate students have helped with the preparation of the present book, and we cannot properly acknowledge them all. This book would never have been possible without their excellent research over the last 10 years. We especially thank the Department of Electrical Engineering at the Norwegian University of Science and Technology, the Norwegian Research Centre for Hydropower Technology (HydroCen), and the University of Tehran for their financial support of the projects whose outcome has been gathered here as a book.

Special thanks go to Professor Arne Nysveen at the Norwegian University of Science and Technology (NTNU) for his relentless support for the project of fault detection in synchronous machines. We also extend our thanks to Dr. Mansour Ojaghi, the previous PhD student at the University of Tehran and now Associate Professor at the University of Zanjan, Iran; Dr. Mojtaba Babaei, the previous PhD student at the Islamic Azad University of Tehran, Iran, now Assistant Professor at the Islamic Azad University, Iran; Bodil Vuttudal Wold, Head of Administration at the Department of Electric Power Engineering at NTNU; and the previous talented MSc students Ingrid Linnea Groth, Johan Henrik Holm Ebbing, Tarjei Nesbø Skreien, and Gaute Lyng Rødal at the Norwegian University of Science and Technology, and Vahid Bahari, Ali Mahmoudi, and Saeed Sedigh at the University of Tehran and also Hassan Smaeili from Karaj Art House, Karaj, Iran.

Hossein Ehya and Jawad Faiz

1

Introduction

CHAPTER MENU

1.1 Introduction to Condition Monitoring of Electric Machines

Condition monitoring in engineering systems increases the importance of automatic processing, while decreasing maintenance and repair costs. From a traditional point of view, electric machines have safe operation, and from a system point of view, they are reliable components. Therefore, for a long time industries have relied on periodic maintenance to keep their electric machines running. However, these users are now recognizing the need for condition-based maintenance for the following reasons:

1. According to reference [1], the maintenance costs for electrical machines in the US are around 2 billion USD per year, and 70% of that amount is spent on electric machines that do not have a condition monitoring system. This indicates that adopting a condition monitoring system would significantly lower maintenance costs.
2. Faults are unavoidable due to aging, environmental factors, or manufacturing problems, and they can stop an electric machine without giving any notice. Unexpected stoppage, both in power plants and in industrial settings, can result in significant economic losses. Not only do the power companies and industries have to pay for repairs and maintenance, they also cannot generate electricity or run their production lines because of unplanned stoppages. Although having a condition monitoring system cannot avoid a failure entirely, it can give the maintenance team some insight into impending issues so that they are ready with the required repair components and can stop the machine at the proper time. An unplanned outage of the machine is critical in some applications, such as hydropower plants that rely on run-off rivers during the high season of production and cannot "store" the water stream.
3. The application of condition monitoring becomes more crucial with the introduction of off-shore and on-shore wind parks. These parks are located in areas with harsh environments, where periodic inspection and maintenance are challenging and costly. Condition monitoring of electric machines in remote areas has attracted significant attention.

Electromagnetic Analysis and Condition Monitoring of Synchronous Generators, First Edition. Hossein Ehya and Jawad Faiz.
© 2023 The Institute of Electrical and Electronics Engineers, Inc. Published 2023 by John Wiley & Sons, Inc.

4. The green shift that emphasizes the application of electricity to reduce carbon emissions is now pushing industries and societies toward electrified transportation systems. These transportation systems consist of electric vehicles, electric aviation, electric ships, and hyperloops that utilize electric machines. All these systems require condition monitoring to ensure the safety of passengers.

Clearly, condition monitoring of electric machines has become crucial from different standpoints. In some applications, such as electricity generation, the cost is important, while safety is a key point in electric transportation.

Condition monitoring is similar to the check-ups that physicians perform on their patients. Here, the "patient" is the electrical machine. Several measured signals are analyzed using signal processing methods, and some features are extracted that can be used to evaluate the health status of the electric machine. If the extracted patterns are similar to any previously introduced features, the machine is deemed to have a feature-related fault that requires attention. These faults in synchronous machines are divided into the following categories:

1. Insulation and electrical faults in the stator winding.
2. Faults in the rotor field winding.
3. Faults in bearings and their seals.
4. Mechanical faults in the stator winding, such as insulation deformations and broken wedges.
5. Rotor mechanical faults, such as bending.
6. Generator core faults, such as eddy current circulation in the lamination or stator core deformation.
7. Faults in generator accessory systems, such as the lubrication, cooling, and excitation systems.

Condition monitoring of electric machines consists of the following approaches, which are covered in different chapters of this book:

1. Analytical or numerical modeling of electric machines.
2. Data acquisition in laboratory or field tests.
3. Signal processing and pattern recognition.
4. Automated fault detection using artificial intelligence.

1.2 Importance of Synchronous Generators

Synchronous generators in power stations, refineries, petrochemical industries, and industrial centers are used as the main electric power generating systems. Therefore, synchronous generators are the spinal cords of electrical power systems. Normally, these generators are designed to have robust and durable structures. However, with the passage of time and constant electromechanical energy conversion, these machines experience stress, erosion, and mechanical fatigue that leads to faults. Any severe fault and interruption in the operation of synchronous generators can create irreparable damage. Protection of faulty generators is the most important aspect of the reliability of a power system. Faulty generators cause budget planning disruptions, repair cost outlays, reduction in generation capacity, and interruptions in consumer services [2]. However, despite the importance of synchronous generators, the current protection systems are often activated only after a fault occurs, and they operate just to prevent expansion and progression of the damage [3].

The frequency and consequent speed of synchronous generators is constant. This factor requires precise design of the generator, while also significantly increasing their cost. Synchronous generators cost many times more than induction machines with the same power rating (although this comparison may not hold for very high-power generators). When a generator fault causes damage, utility companies have no expert repair personnel, so either the generator must be sent to the manufacturer for repair or the manufacturing company must send personnel to repair the generator on-site. Either scenario involves an extra cost to the power company. In addition, generator faults that do not appear immediately damaging can produce voltage harmonics that can pass into the system and interfere with power system harmonics, leading to maloperation of protection relays and/or reducing the power quality of the network. With today's emphasis on distributed generation and privatization of electrical energy generation, fault diagnosis is being recognized as essential, and its application is becoming more common.

1.3 Economic Aspects and Advantages

The economic well-being of society, in the long run, is clearly related to its capability to produce demanded goods and, to some extent, to provide needed services. This capability is directly related to its ability to use energy effectively. One measure of the quantity of energy required is the quantity of electrical energy generated over a given period. This energy is used for residential, commercial, and industrial purposes and is directly employed in some industrial processes, such as electric welding, electrolysis, industrial space heating, and lighting. Much of the generated electrical energy must be converted to mechanical energy in the form of rotating shafts or linear movements [4].

The demand for electrical energy is increasing as the availability of fossil fuels (coal, diesel, and gasoline) decreases [3]. The evolution of the synchronous generator is the result of a continuing process of maximizing and optimizing electric machine performance. These generators are almost ideal converters of energy and are characterized by a technically and economically high-quality level. This has been achieved principally through a multitude of innovations and improvements in manufacturing, testing, and operation.

Today, a number of activities are considered vital for synchronous generators. Capacity, economy, and reliability are three important factors that are taken into account. The key ideas are the extension of technology, reduction in costs, shortening delivery times, assuring quality, and operating reliably. Maintenance of a synchronous generator is required to achieve trouble-free operation, and its condition must be regularly checked to extend its lifespan. The modernization and up-rating of existing generators are performed using the products of the latest technologies, which is deemed profitable in view of the high financial values attached to the technical equipment and processes [5, 6].

A great deal of attention has been paid to introducing new concepts in synchronous generators. Recently, attempts have been made to use superconductors in the magnetic fields of rotor windings, particularly for large power generators. High-temperature superconductors have promising applications in both rotor and stator windings. Their currently uninterrupted evolution is predicted to continue to increase their importance in electric machines in the future.

1.4 Intention of the Book

The condition monitoring of synchronous generators is a very important task because these generators are the main generation equipment of power systems. Therefore, interest is increasing in technologies to improve synchronous generator protection and maintenance schemes. The techniques involved are both invasive and non-invasive and can be applied both offline and online.

Condition monitoring has the potential to reduce operating costs, enhance the reliability of operation, and improve power supply and service to customers. Therefore, the importance of developing condition monitoring for synchronous generators is a key aspect of this book. After introducing the concepts and functions of condition monitoring, current popular condition monitoring methods and the research status of condition monitoring for synchronous generators will be described. The book also points out the potential benefits of utilizing advanced signal processing and artificial intelligence techniques in developing novel condition monitoring schemes.

Most industries previously relied on reactive maintenance, which is essentially repair operations after a fault causes equipment breakdown. Because of the high cost of repairs (or the cost of replacing a new machine) and, more importantly, the costs associated with slowdowns or lack of production due to out-of-service electrical machines, these reactive maintenance systems have been replaced by periodic maintenance systems. This use of repair and maintenance systems has received the attention of industries and power stations aiming to enhance the reliability, programmed outage, and reduction of maintenance and repair costs. Therefore, most industries now follow some type of periodic maintenance for repair and maintenance of medium- and high-voltage synchronous generators and motors. However, the newest approach in developed countries is to use net systems based on predictive maintenance. In this approach, the online condition of the machine is continuously monitored and, at the first signs of any problem, the operator of the machine is informed to take appropriate action and prevent fault progression to a more severe state.

Two types of faults can trip an electric machine from the power grid: external faults and internal faults. External faults arise due to a failure in the power system, such as different types of short circuits in the power lines or inrush currents in the transformers that cause the protection system to cut off the synchronous generator from the power grid. Internal faults, by contrast, are those due to failure in the components of the synchronous generators themselves, such as windings, bearings, shafts, etc. Unlike external faults, in which the protection system takes quick action to avoid further damage to the generators, internal faults have no early detection systems for shutting down the generator at an early stage of the internal fault. Having a condition monitoring system fixed on the synchronous machine would allow the diagnosis of early-stage faults and enable acute actions to be taken to prevent progression to severe faults. This would, in turn, enhance the reliability of the generator and its related system. The following methods have been proposed so far for fault diagnosis in large electrical generators and are covered in this book:

- Magnetic field-based detection (analysis of air gap magnetic field and stray magnetic field).
- Electrical-based detection (current and voltage of the stator and rotor windings).
- Vibration-based detection.
- Acoustic-based detection.
- Chemical-based methods consisting of cooling gas analysis.
- Temperature-based detection.
- Partial discharge-based detection.

This book covers fundamental and advanced topics on synchronous generators. The book is useful for researchers in academics and for professional engineers in the industry because the selected methods are based on experimental and field test results. Four main topics are covered:

1. Principle, operation, and design of synchronous machines.
2. Electromagnetic, thermal, and mechanical modeling of synchronous machines.

3. Signal processing and machine learning.
4. Feature extraction for the fault detection of synchronous machines.

These topics are described in the following 15 chapters:

Chapter 2 gives the principles of operation of synchronous generators and different types of generators. The design procedure, analysis, and practical considerations of the generator are discussed.

Chapter 3 discusses the transformed models of multi-phase synchronous generators based on Park equations. In this chapter, parameter identification of the synchronous machine and the impact of faults on these parameters are explained.

Chapter 4 provides a preliminary discussion on various types of faults in synchronous generators. The necessity of condition monitoring and the reasons for the damage occurring in the generators are highlighted. The faults in different parts of the synchronous generator, including electrical faults, such as short circuits in the stator winding and rotor field winding, are described. The mechanical faults, including eccentricity, misalignment, and broken damper bar faults, are explained and their root causes are illustrated.

Chapter 5 illustrates the equipment required for experiments in the laboratory, since the results obtained by simulations must be verified by experimental results. This chapter illustrates how to emulate various mechanical and electrical faults in synchronous generators. Different types of signals are utilized in fault detection, indicating that various sensors can be introduced to measure signals, such as magnetic field, voltage, current, and vibration. Some approaches require a pure signal, so some methods are introduced for noise rejection purposes. The data acquisition system and its requirements for signal measurement are explained in detail.

Chapter 6 describes analytical modeling based on the wave and permanence method. The wave and permanence method provides a deep understanding of how the frequency contents of the various signals, including the air gap magnetic field, current, voltage, and vibration, are influenced by short circuit and eccentricity faults. The impact of saturation and slotting effects are studied in this chapter and the ways that saturation and slotting harmonics can interact with fault harmonics are demonstrated. The frequency harmonics analysis is used later, in Chapters 11 and 12, for pattern recognition purposes.

Chapter 7 employs the winding function method to model and analyze synchronous generators. The winding function method is a complicated modeling approach that considers the detailed specifications of the generator. This chapter discusses the detailed formulation and implementation of the synchronous generator using the winding function method and different fault implementations.

Chapter 8 explains the finite element method (FEM) formulation for generators. Two types of synchronous generators are modeled: salient pole and round rotor. The equivalent electric circuit is introduced to model the stator and rotor windings and load. Various fault implementations in FEM, such as short circuit faults in the stator and rotor field windings, static, dynamic, and mixed eccentricity, and broken damper bars, are introduced.

Chapter 9 is devoted to the general thermal analysis of electrical generators as a way to enhance their efficiency. The impact of faults on the thermal behavior and thermal aging of the synchronous machine is considered an important issue for maintenance and replacement planning. Different thermal modeling methods, including analytical and numerical methods, are introduced.

Chapter 10 gives a concise description of "signal" prior to signal processing. Signal processing is the act of applying mathematical tools to raw data to obtain a deep insight into its components. Advanced signal processing tools are introduced for fault diagnosis of synchronous generators. The signal processing tools introduced in this chapter are limited to time domain, frequency domain, and time-frequency domain signal processing tools. In addition, various types of noise are introduced and their impact on the results of signal processing tools are explained.

Chapter 11 focuses on electrical fault detection methods based on electromagnetic analysis. A brief introduction to some detection methods not based on electromagnetic analysis is also provided. The electrical faults covered in this chapter include short circuit faults in the stator and rotor field windings. Electromagnetic analysis based on both invasive and non-invasive methods is introduced and the pros and cons of these approaches are discussed in detail.

Chapter 12 discusses the electromagnetic signature analysis of mechanical faults, such as eccentricity, stator core, and broken damper bar faults. Both invasive and non-invasive fault detection methods are proposed.

Chapter 13 investigates condition monitoring based on the vibration signals in synchronous generators. First, general details on vibration in synchronous generators are presented. The vibration of the generator body is generated by unevenly distributed forces on the stator core due to the non-uniform magnetic field in the air gap. Therefore, the impact of various faults on the air gap magnetic field, and consequently on the force and vibration, are investigated for a better understanding of the root source of vibration in synchronous generators. The impact of faults on the frequency pattern of the vibration signal is also discussed.

Chapter 14 deals with the application of various machine-learning tools for fault detection in synchronous generators. Fault detection in electrical machines is tending to become a mature science that can be utilized in the industry; however, data analysis and pattern recognition need expert knowledge for data interpretation. This makes condition monitoring more costly for the end-users, indicating a need to apply machine-learning tools to automate the fault detection process. This chapter explains in detail the prerequisites for automating fault detection, including data pre-processing, selection of machine-learning algorithms, parameter selection, training, and optimization of the utilized generators.

Chapter 15 presents the historical background of electrical machine insulation and explains in detail the insulation system of the stator and rotor of the synchronous generators. Different types of partial discharges (PDs) and their risk assessment are explained. Chapter also explains in detail the online approaches that have been developed so far for PD measurement, including electrical, acoustic, chemical, and visual methods. The required sensors and practical application of the methods are discussed.

Chapter 16 deals with noise rejection methods and the interpretation of PD data. The data interpretation begins by analyzing the amplitude of the current pulses, followed by addition of the phase in which the PD appeared in this analysis. Methods for using various techniques for PD interpretation are also provided.

References

1 Ehya, H., Sadeghi, I., and Faiz, J. (2017). Online condition monitoring of large synchronous generator under eccentricity fault. In: *2017 12th IEEE Conference on Industrial Electronics and Applications (ICIEA)*, 19–24. https://doi.org/10.1109/ICIEA.2017.8282807.

2 M. A. S. K. Khan, O. Ozgonenel, and M. Azizur Rahman, Diagnosis and detection of stator faults in synchronous generators using wavelet transform, in *IEEE International Electric Machines and Drives Conference (IEMDC 2007)*, Vol. 1, pp. 184–189, May 2007.

3 Peregrinus, S.P. (1981). Power system protection. In: *Principle and Component*, vol. 1. Electrical Training Association.

4 Ministry of Economic Affairs, Agriculture and Innovation http://www.government.nl/documents-and-publications/reports/2011/11/01/energyreport-2011.html (Accessed March 08, 2011).

5 Paul, L. (1989). Cochran. In: *Poly-phase Induction Motors, Analysis: Design, and Application*. New York and Basel: Marcel Dekker Inc.

6 Vickers, V.J. (1874). Recent trends in turbo-generators. *Proc. IEE* 121 (11): 1273–1306.

2

Operation Principles, Structure, and Design of Synchronous Generators

2.1 Introduction

A synchronous generator is an alternating current (AC) machine in which an alternating electro-motive force is induced in its armature. It is generally employed to generate AC power at electrical power stations. Large AC power systems operating at fixed frequency rely almost exclusively on synchronous generators for the provision of electrical energy [1]. The frequency of the generator voltage is as follows:

$$f = pn/120 \tag{2.1}$$

where n is the rotor speed in rpm, p is the number of poles, and f is the frequency terminal voltage in Hz. The generator winding system consists of two parts: the field winding and an armature winding. The field winding is excited by direct current. This current is fed to the field winding usually on the rotor and produces a stationary magneto-motive force (MMF) wave with respect to the rotor. This excitation is usually provided by an exciter mounted on the same shaft of the generator and is supplied through two slip-rings. This exciter is an auxiliary generator that is mechanically coupled to the shaft of the main generator. In a small generator, the excitation field can be provided by permanent magnets (PMs). Another synchronous generator without any type of excitation is the synchronous reluctance generator, which is generally a low-power generator. The acceptable construction for a very small low-voltage generator is the rotating magnetic field and fixed armature type. This structure allows easy insulation of the armature coils for high-voltage (HV) parts. The load of the generator is directly supplied from the winding without involving the brushes.

Electromagnetic Analysis and Condition Monitoring of Synchronous Generators, First Edition. Hossein Ehya and Jawad Faiz.
© 2023 The Institute of Electrical and Electronics Engineers, Inc. Published 2023 by John Wiley & Sons, Inc.

The armature of synchronous generators can be on either the stator or the rotor. Therefore, the generators are constructed with a rotating armature and fixed field. In HV generators, passing a relatively high current across moving contacts is difficult, which is why the stator-wound armature is a common choice. The rotating direct current (DC) magnetic field system on the rotor permits the AC winding to be located on the stator. In this case, HV generators are more easily braced against the electromagnetic force, and the high voltage winding can be better insulated.

2.1.1 Rotating Magnetic Field of a Three-Phase Synchronous Machine

A synchronous machine operates based on the rotating magnetic field; this field has a constant magnitude, but its axis of direction rotates in space. Consider a two-pole three-phase synchronous machine: the stator winding has three coils with 120° angular spacing, as shown in Figure 2.1. Each coil of phases a, b, and c has two conductors (aa′, bb′, and cc′) connected to a balanced three-phase supply; the load is also a three-phase balanced load. Suppose, at $t = 0$ s, the current of phase a is zero. The current vectors in Figure 2.1 present the peak phase currents. The vector diagram at $t = 0$ is oriented such that the vector I_1 points along $0x$. Projecting the vectors I_2 and I_3 along $0y$, the instantaneous current is $i_1 = 0$. Since $i_1 + i_2 + i_3 = 0$, then $i_2 = -i_3$. Figure 2.1 shows the directions of the flowing current. At $t = 0$, bc and b′c′ (dotted lines) can generate MMF, which is vertically downward from a to a′. The flux leaves the stator at a (north pole) and enters the stator at a′ (south pole). After one-third of the cycle, the vector diagram rotates 120°, so $i_2 = 0$ and $i_1 = -i_3$, and, similar to the above-mentioned case, here ac and a′c′ produce MMF from b to b′. After another one-third period, the MMF acts from c to c′. Therefore, during one complete cycle, the two-pole magnetic field rotates one complete revolution. The speed of the two-pole magnetic field is $120f/2$ rpm. This description can be extended to a four-pole synchronous machine and it can be concluded that the speed of the p-pole magnetic field is $120f/p$ [2]. Figure 2.2 presents a schematic of a synchronous generator structure and an induced three-phase voltage in the stator windings.

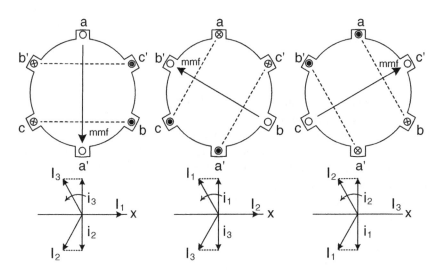

Figure 2.1 Generation of a rotating MMF by three-phase stator currents.

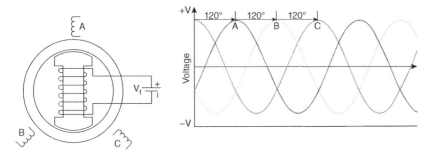

Figure 2.2 Synchronous generator structure and induced three-phase voltage in the stator windings.

2.2 History of Synchronous Generators

All synchronous generators, including diesel, gas, and steam, are utilized in thermal power stations, large hydroelectric turbines in hydro-power stations, and wind turbines in wind power stations [3, 4]. Wind turbines are coupled to the synchronous generator via a mechanical gearbox or power converters to operate over different speeds. Conversely, PM synchronous generators have become more popular due to their high-power density and less requirement for repair and maintenance [5, 6].

2.2.1 Advancement History of Synchronous Generators

2.2.1.1 Up to the Year 1970

The history of synchronous generators dates back more than 100 years. The first changes in synchronous generators occurred in the 1880s [7]. The primitive samples, similar to the DC machine, had one or two pairs of windings on the rotating armature, their ends were connected to slip rings, and fixed poles on the stator provided the excitation field. This was the so-called external-pole type generator [8]. Later, another type was considered, in which the location of the field and armature had been displaced. This sample was a primitive configuration of a synchronous generator of the so-called internal-pole type and it found a suitable position in the electrical power industry. Different shapes of magnetic poles and field windings were used on the rotor, while the stator winding was single- or three-phase. Later, the optimal case was recognized as a combination of three AC currents with phase angle differences. The stator consisted of three pairs of windings, where one side was connected to the neutral point (star) and the other side to the transmission line [9].

In fact, we are indebted to the efforts of many outstanding people, such as those in Nikola Tesla and Galileo Ferraris, and Charles Bradley, Michail Dolivo-Dobrovolsky, and Friedrich August Haselwander, who introduced the idea of the three-phase AC machine. In 1887, Haselwander designed and built the first 2.8 kW three-phase synchronous generator that operated at a speed of 960 rpm (frequency of 32 Hz). This machine had a fixed three-phase armature and a four-pole wound rotor to supply the required excitation field. This generator was used to supply the local loads (Figure 2.3) [3].

In 1891, for the first time, a combination of a generator and a long transmission line for supplying distant loads was tested successfully. In fact, the electrical power of the generator was transferred from Laphan (175 km distance) to the International Electrotechnical Exhibition in Frankfurt by a three-phase transmission line in 1891 at a phase-to-phase voltage of 95 V, phase current of 400 A,

Figure 2.3 The first three-phase generator was built by Friedrich August Haselwander in 1887. Bigbratze / Wikimedia Commons / CC BY-SA 4.0.

Figure 2.4 The first commercial generator was tested in 1891.

and frequency of 40 Hz [10]. The rotor speed of this generator was 150 rpm and it had 32 poles. Its diameter was 1752 mm and its effective length was 380 mm. The excitation current was supplied by a DC machine. Its stator had 96 slots and each slot contained a copper bar of 29 mm diameter [11]. Note that the skin effect was not realized at that time. The stator winding consisted of one bar per pole per phase. The efficiency of the generator was 96.5%, which was excellent. This generator, shown in Figure 2.4, was designed and built by Charles Brown.

In the beginning, most synchronous generators were designed for coupling to hydro-turbines. Then, powerful steam turbines were designed that were compatible with high-speed generators [12]. In this case, the insulation system became an important problem in the generator. The primitive insulation materials used were natural materials, such as cellulose, cotton, wool, and other natural fibers [13]. Natural resins extracted from plants and crude oil compounds were also used to manufacture insulation materials. The research into synthetic electrical insulations began in 1908. For the first time during the First World War, the asphalt resins, called bitumen, with mica pieces, were used as slot insulation in the stator windings of turbo-generators [14, 15]. Both sides of these pieces were surrounded by cellulose paper. In this method, the stator windings were first covered with cellulose tapes and then with two-layer tape. The winding was heated in a compartment and then subjected to a vacuum. After a few hours, a dried and porous insulation material resulted. A huge volume of the hot bitumen was then poured on to the windings under vacuum. The compartment was subsequently pressed with 550 kP pressure dry nitrogen. After a few hours, the nitrogen gas was evacuated and the windings were dried and tightened at room temperature [16].

Later, in the 1940s, the General Electric Company chose epoxy compounds to improve their stator winding insulation system. As a result, a system called rich resin was introduced in which the resin, as tapes and/or varnish, was placed between the layers [17]. From the 1940s until the 1960s, the number of insulation faults increased significantly with the increasing capacity of the generators and the resultant increase in thermal stresses. The main reason for these faults was tape separation or cracking. The reason for these faults was the uncoordinated expansion and contraction of the copper conductor and the iron core. After the Second World War, to solve this problem, the Westinghouse Company started experimental work on new polyesters and offered a system with the commercial name of thermoplastic [18, 19].

The next generation of insulation was fiberglass papers, which were used in the first half of the 1950s. Subsequently, a type of insulation that resisted partial discharge was obtained: it was a combination of 50% fiberglass strands and 50% PET strands that covered the conductor. Heating this insulation in special ovens melted the PET, and it covered the fiberglass. This insulation was used as single or multiple layers, as needed. It was a so-called poly-class with the commercial name of Daglass [20].

The most important stress exerted on the insulation was thermal stress; therefore, insulation systems were always connected to cooling systems [21]. The primitive generators were cooled only by air and the best-obtained results were at 1800 rpm in a 200 MVA generator installed in the Brooklyn borough of New York City in 1932. However, increasing the power of the generator demanded a more effective cooling system.

The idea of cooling with hydrogen was first proposed by Max Sholler in 1915 [22]. His efforts in building this type of system began in 1928 and the first 3600 rpm prototype generator was introduced in 1936. General Electric had the first hydrogen-cooled turbo-generator on the market in 1937. This technology was used in Europe after 1945. In the 1950s and 1960s, different direct cooling systems for the stator windings appeared, run by gas, oil, and water, and by the middle of the 1960s, most large generators were cooled with cool water. The rise in direct cooling technology allowed the capacity upgrading of generators up to 1500 MVA [23]. One of the prominent changes, occurring in 1960, was the production of niobium–titanium as the first commercial superconductor material, which was very well recognized in the next decades [24].

2.2.1.2 Changes in the 1970s

In the 1970s, an important change occurred in insulation processing. Before 1975, insulations were saturated with solvent resins in organic compounds [25]. In this process, the compound evaporates and is released into the air. Environmental laws and the beginning of the green movement in the early 1970s imposed serious limitations on the release of these materials, which led to their elimination from this process. As a result, the use of materials more compatible with the environment was considered in the production and repair of electrical machine insulation systems. The use of water-based resins was one of the first suggestions for this purpose, but a more comprehensive solution that remains common today was the application of solid adhesives. The production of non-solvent rich resin mica tapes has been considered in this regard [26].

One important step in the 1970s was the introduction of superconductor generators. Generally, a superconductor generator consists of a superconductor field winding and a copper armature winding. The rotor core is not normally iron, because the field winding generates a high-density magnetic field that leads to core saturation. A ferromagnetic core was used only in the stator yoke to act as a shield and to transfer the field between the poles. Removing the iron caused a reduction in the synchronous reactance of about 0.3 to 0.5 pu, which clearly led to better dynamic stability of the generator. As already pointed out, the first commercial superconductor was niobium–titanium, which had a superconducting feature at 5 °K. Of course, in the next decades, the progress of this industry led to the introduction of superconducting materials with an operating temperature of 110 °K. Based on this, superconducting materials are classified into low-temperature materials, such as niobium–titanium, and high-temperature materials, such as bismuth strontium calcium copper oxide (BSCCO-2223). In the early 1970s, researchers studying superconducting generators began using low-temperature superconductors. The Westinghouse Company began some research into a prototype two-pole generator. The outcome of this project was the building and testing of a 5 MVA generator in 1972.

In 1970, the General Electric Company began to design and build a superconducting generator using low-temperature superconductors with the aim of installing it in a power system network [27]. Building and testing this two-pole, 3000 rpm, 20 MVA generator was completed in 1979. This machine was fabricated using an air-core design method, and its field winding was cooled by liquid helium. At that time (1979), this generator was the largest tested superconducting generator [28]. Also in 1979, Westinghouse and the Electric Power Research Institute (EPRI) began to build a 300 MVA superconducting generator. However, by mutual agreement between both parties, this project was canceled in 1983 due to the world market conditions. In this context, the Siemens Company began to build low-temperature generators in the early 1970s and manufactured a sample rotor and stator with an iron core with an 850 MVA with 3000 rpm speed. However, no real performance test was ever carried out due to some problems. A superconducting rotor in a 250 MW synchronous turbo-generator was also built by Alstom.

Considering the importance of cooling for the appropriate operation of superconducting generators, new cooling methods were introduced to keep up with the developments in this industry. In 1977, Lascaris offered a two-phase cooling system (fluid-gas) for superconducting generators. In this design, a part of the winding was placed in liquid helium and cooled by helium boiling at a temperature of 4.2 °K. The centrifugal force due to rotor rotation separated the liquid from the gas [2]. In the PM, powerformer, and superconducting generators, cooling systems, excitation systems, insulation systems, and mechanical considerations were recognized as the most important factors [29].

2.2.1.3 Developments in the 1980s

In this decade, the insulation systems remained an important research topic. Alstom offered a new class of epoxy formulas with no solvent, in combination with a fabric class and a special type of mica paper, commercially called Dortenax. This insulating system had more mechanical strength, higher insulation strength, lower dielectric losses, and lower thermal resistance compared to the previous samples. The industrial technology development and renewable energy organizations in Japan had just begun a 12-year super-GM national project on superconductor projects in 1988, and the results were achieved in the next decades.

The cooling systems of superconductor generators were still under investigation. For example, a project on cooling systems under pressure was proposed in Japan. This project was offered in 1985, and it had an advanced heat exchanger and a helium fluidizing device with a capacity of 350 l/s [30]. At this point, some research on PM materials was also realized. The use of Nd-Fe-Br PMs in the 1980s created a huge change in PM generator design. The most important feature of these

PMs is their high magnetic energy (maximum flux density vs. magnetizing force, or BH), which led to a reduction in the magnetic energy unit. The high energy also allows the use of smaller PMs. Therefore, the size of other parts of the generators, such as windings and iron pieces, also decreases. Consequently, the total cost was reduced. Notably, huge research projects were devoted to self-excited and brushless generators for special applications, but they were not extended to the power station generator industry; therefore, their descriptions are ignored here [31].

A survey of different topics related to synchronous generators led to the following results by the beginning of the 1990s:

1. Previous insulating methods were improved to reduce the heat resistance of insulation.
2. Wide studies were done on brushless synchronous generators.
3. Projects on superconductor generators continued.
4. A new cooling system was offered for superconductor generators.
5. The finite element method (FEM) was widely used in the design and analysis of synchronous generators and particularly PM generators.

2.2.1.4 Developments in the 1990s

Many efforts were made to improve insulation systems in this decade. In 1990, General Electric Company (GE) offered the Micapal insulation systems, which were compounds of alkide types (which are compounds in which a metal is combined with alkyl radicals) and epoxies. The Westinghouse Electric Company proposed a new class F rotor winding insulation in 1992. This system consisted of an epoxy class layer pasted on to the copper conductor with a polyamide-epoxy glue. Its advantages included electrical and mechanical robustness and a decrease in thermal deterioration. In 1998, a Nanjing industrial group fabricating electrical machines and turbines introduced a new rotor winding insulation of saturated Nomex with a sticky varnish. The most important advantages of this material were its flexibility, insulation strength, saturation improvement with varnish, easy cleaning, and non-absorption of moisture. Some attempts were made to increase the thermal conduction of insulations in the late 1990s. For example, Miller from the Siemens-Westinghouse Company introduced a method for substituting the filling layer used in previous designs with special resins. The major advantage of this method was that it filled the air gap between the filling layer and the stator wall, thereby significantly increasing the thermal conduction of the stator insulation [32].

In the 1990s, mechanical problems in synchronous generators received more attention. A method was introduced in 1993 to reduce vibration in PM generators. This vibration is generated by the absorption of the forces applied to the stator by the rotating PMs. Vibrations were investigated using Maxwell tensors, FEM, and Fourier transform, and this ultimately led to the suggestion of a new geometrical dimension for PMs. Of course, the key condition was that the performance of the generator would not decline. Simultaneously, with these advancements, increasing the speed and memory of computers and the appearance of powerful software led to a greater use of computers for the analysis and design of synchronous generators.

Cooling system implementation procedures were also considered. Aydear studied the impact of the location of the ventilation ducts on the magnetic field of the synchronous generator using FEM and showed that the choice of appropriate ducts must be ducts is very important for preventing rises in the magnetization current and the saturation effect. The location of the ducts had a considerable impact on the yoke flux [32].

One of the most important developments in superconductor generators arose from the results of the Super GM project, which had started in the previous decade in Japan. The results of this project were the construction and testing of three superconductor rotor models for the generator stator. The first model combined with the stator gave a 79 MW output. The third model, which had a quick response excitation system, was tested in 1999 and used in a power system network.

The utilization of high-temperature superconductor materials in the 1990s marked the entry of the technology of superconducting synchronous generators into a new stage. The General Electric Company completed designing, manufacturing, and testing a high-temperature winding in the middle of this decade. A collaboration between GE and the American Westinghouse and Superconducting Company led to the design of a high-temperature four-pole, 1800 rpm, 60 Hz superconducting generator [33].

The 1990s witnessed important progress in excitation systems, such as the appearance of electronic static excitation systems. The use of these systems led to flexibility in the design of the excitation systems and resolution of the problems of brush maintenance in rotating excitations. One of the first samples of these systems was introduced by Shafer, from Basler Electric in Germany, in 1997. By that point, the application of digital systems in the generator's excitation had begun. One of the first examples of a digital excitation system was introduced by Arsag, from Zagreb University of Croatia, in 1999.

In their efforts to improve the cooling system, Siemens and the Westinghouse Companies proposed a large generator with air cooling in 1999. This design proposal was the beginning of changing the cooling designs from hydrogen to air. The use of high-temperature thin insulations for the stator and the application of computer computations of dynamic flow economized this design compared with the hydrogen-cooled generators [34].

The end of the 1990s coincided with the appearance of powerformer technology. In the early spring of 1998, Mats Leijon from the ABB Company in Sweden introduced the idea of high-voltage electrical energy generation. The most important feature of this plan was the use of the usual crossover polyethylene HV cables as transmission and distribution systems in the stator winding [35]. In this case, very high-voltage cylindrical cables were used to eliminate partial discharge and corona discharge. The first prototype powerformer was installed in the Porjus power station in northern Sweden in 1998. This powerformer had a rating voltage of 45 kV, a rating power of 11 MVA, and a rated speed of 600 rpm [36]. One of the important problems in powerformer technology was the precise fixing of the cables in slots to prevent damage to the cable outer semi-conductor layer due to vibration effects. To do this, the cables were fixed using triangular pieces of silicone rubber.

Since the stator winding current of the powerformer has negligible copper losses, the use of a water-cooling circuit is sufficient. The cooling system maintains the operating temperature around 70 °C, while the insulation of the cable has been designed for a rated temperature of 90 °C. Therefore, the powerformer can be overloaded without incurring any special problems [36].

In the 1990s, the following topics were considered in synchronous generators:

1. Activities began on high-temperature synchronous generators.
2. Application of digital and static excitation systems was extended.
3. Vibration reduction methods during generator operation were proposed.
4. In the early period of the decade, designers aimed to improve hydrogen cooling systems; however, by the end of the decade, cooling systems by air were reconsidered.

2.2.1.5 Developments After 2000

The trend of using numerical methods, and particularly FEM, continued to increase. A new FEM using curve-shaped elements in cylindrical coordinates was introduced. The advantages of this method were its high accuracy and simple formulation. This method was very appropriate for analysis of the magnetic field in cylindrical shapes, such as electrical generators [37].

A novel method for designing electrical generators was introduced in 2004. This method was a combination of FEM and analytical methods. An analytical method was used for the initial design based on the rated torque, current, and speed, while FEM was utilized for precise analysis of the magnetic fields to evolve the initial design. In this way, the required time was shortened and the cost was reduced.

Concerning insulation, some efforts were also made that led to improvements in thermal conduction, and a high thermal conduction system was presented by Toshiba and Von Roll in 2001. The impact of thermal conduction improvement in this system was evident compared with the usual system.

The General Electric Company proposed a project to build and test a 100 MVA superconductor generator in 2002. The rotor and stator cores of the generator were similar to those of the conventional generators. The aim was to adapt a conventional rotor to pass the superconductor winding magnetic field with no saturation. The most important parts of this project were the high-temperature field winding and cooling system [38]. After 2000, extensive activities ensued to build and fix powerformers. The outcomes were the installation of a number of powerformers in different power stations. These powerformers have been utilized up to 136 kV and 75 MVA.

The following topics related to the synchronous generator can be found after 2000:

1. Many attempts were made to improve the thermal conduction of stator winding insulation cooled by air, with the aim of approaching a higher power generator.
2. Powerformers were installed in different power stations.
3. The projects on high-temperature superconductor generators initiated in the 1990s continued.
4. Application of digital excitation systems, and particularly systems with several microprocessors, was extended.
5. The use of numerical methods in the design and analysis of synchronous generators, and particularly the cooling systems, was widely extended.

2.3 Types and Constructions of Synchronous Machines

The DC current of a rotor winding generates the rotor magnetic field in a synchronous generator. A prime mover rotates the rotor; therefore, the magnetic field also rotates and induces three phase voltages in the stator windings of the generator. Fundamentally, the rotor of a synchronous generator is a large electromagnet. The magnetic poles on the rotor can be salient or non-salient. In the salient-pole generator, the air gap is not uniform, which means that the air gap under the poles is short and the air gap between the poles is long. This type of rotor is generally used in synchronous generators with pole numbers of four or higher [39].

2.3.1 Non-salient or Round Rotor

These types of rotor are utilized for high-power and high-speed generators coupled to steam turbines and generally have a core of solid steel. However, the core of the low-power generator is laminated and has a cylindrical shape. The solid core can easily tolerate very large centrifugal forces. By decreasing the diameter of the rotor and increasing the axial length, the peripheral speed of the generator is reduced. The cylindrical rotor core and shaft of the generator are forged as a single piece. The high strength is the result of a complicated process of heat treating and machining operations. Along the entire axial length of the rotor, a central hole is opened for testing the forging material in the central region and relieving dangerous internal stresses in the

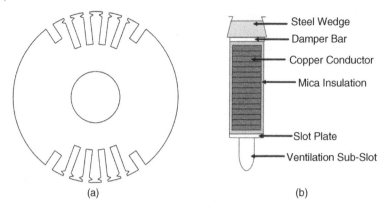

      (a)   (b)

Figure 2.5 (a) Forming pole faces on the core and (b) thick insulated copper strips used in slots.

forging. Generally, a radial slot machine is preferred [40]. The field winding is placed in radial slots. As shown in Figure 2.5, the slots on some parts of the core are omitted to form pole faces. Thick insulated copper strips, shown in Figure 2.5b, are laid in the slots of the core. A steel wedge is placed at the top of the bars to retain the winding inside the slot against the centrifugal force. The slots have a ventilating hole at the bottom. The field winding projecting ends are securely needed for the rotating rings of non-magnetic steel. Since the solid poles act as a damper bar, no damper winding is required in the round cylindrical rotor. The manufacturing cost of a round rotor generator is lower than that of the salient-pole generator due to the reduction in turbine and generator size. The low number of poles also improves the efficiency of the generator.

The rotor is usually connected directly to a prime mover, such as a steam turbine in a steam unit or a hydraulic turbine in a hydroelectric unit. The power output of the generator therefore depends on the speed of the rotor at which the turbine rotates by the prime mover as input energy. Two pole generators are typically used in coal-fired steam units due to the high pressures and temperatures associated with the steam. One winding is present in the field, which acts as one large electromagnet with two poles, as illustrated in Figure 2.6. Nuclear power plants typically use four-pole rotors, because of the lower temperature and pressure of the nuclear steam unit. Steam

Figure 2.6 The rotor of the synchronous generator.

with a lower energy content requires a larger turbine. At a frequency of 50 Hz, rotating large steam turbines at 3000 rpm would not be safe.

Gas turbines typically have a two-pole rotor, although the turbine itself may rotate at a higher speed and use some methods of gear reduction to connect to the rotor of the generator. Most of today's gas turbines or combustion turbines turn at 3000 rpm at a supply frequency of 50 Hz. The energy in the water used to turn hydroelectric generators is low; therefore, hydraulic turbines rotate at much slower speeds. Most hydroelectric units rotate below 300 rpm, resulting in a high number of poles. For example, a hydroelectric generator with a rotating speed of 100 rpm has 72 poles (for $f = 60$ Hz). The number of poles and the rotating speed used for a hydroelectric unit vary greatly, depending on the characteristics (i.e., the difference between incoming versus outgoing water elevation, etc.) of the particular water resource where a unit is located.

Collector rings are provided at the generator end of the rotor. The field windings are connected to these collector rings, as illustrated in Figure 2.7. In a direct current generator, the current passes through brushes that ride against these rings, and that current magnetizes the field. Figure 2.8 presents a cross-section through one slot, with its windings, illustrating the flow of cooling gas as well as the insulation.

In this type of rotor, the poles are not clearly visible. The generator air gap is uniform and the rotor is built as a complete cylinder. This type of rotor is usually used in two- or four-pole generators. High-speed generators are run by a steam turbine, which has high efficiency. These generators are

Figure 2.7   Collector brushes.

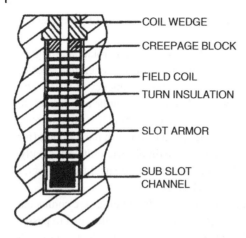

Figure 2.8 Cross-section of a rotor winding slot.

COIL WEDGE

CREEPAGE BLOCK

FIELD COIL

TURN INSULATION

SLOT ARMOR

SUB SLOT
CHANNEL

normally called non-salient pole turbo-generators [41] and those with two poles rotate at 3000 rpm and a frequency of 50 Hz. These generators are usually designed with horizontal axes and small armature diameters to restrict centrifugal forces. The peripheral speed in turbo-generators is high (175 m/s), and this generates high mechanical stresses on the rotating part of the generators. To restrict the mechanical stresses, the rotor diameter must be kept small. For this peripheral speed (3000 rpm) of the rotor, the rotor diameter is restricted to 1.2 m. To enhance the rated power of the turbo-generator, its stack length must be increased. For example, for a 500 MW turbo-generator having a rotor of 1.2 m diameter, the core length is approximately 5.5 m, the shaft length is 12 m, and the stator outer core diameter is around 3 m. Therefore, a turbo-generator is characterized by a small diameter and a long axial length. Cooling these machines, and particularly their central part, is difficult due to their long axial length and small diameter [39].

2.3.2 Salient-Pole Rotor

In a generator having a stationary field, the salient poles are similar to the excitation poles of the direct current machine, as shown in Figure 2.9. For a low number of poles, the pole width is relatively wide and the coils are often copper strip types. The width of the strips is proportional to the pole width. Obviously, synchronous machines with a high number of poles have a narrower pole width and the rectangular wire is used in the winding. These coils are insulated from each

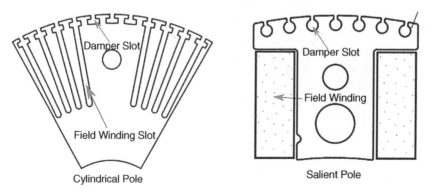

Cylindrical Pole

Salient Pole

Figure 2.9 Rotor poles of a cylindrical and salient-pole synchronous generators.

Figure 2.10   A 14-pole salient-pole rotor synchronous generator.

other and then fixed on the poles. The salient-pole rotor has been abandoned in high-speed generators in favor of the cylindrical type. The reason is that the protruding pole structure gives rise to dangerously high mechanical stresses. The air gap length between the rotor and stator in the salient-pole generator is non-uniform, leading to a non-sinusoidal armature magnetic flux distribution, which induces a non-sinusoidal electromotive force. By considering only the fundamental wave and based on the Blondel two-reaction theory, this is resolved into two components: the d-axis and the q-axis [40]. Figure 2.10 presents a 14-pole salient-pole rotor synchronous generator.

The air gap length is shorter under the poles and longer between the poles. This type of rotor is usually applicable to generators with four or more poles. These generators are usually low-speed types in the range of 150 to 600 rpm. The reason for using low-speed generators is to maintain the stability of the rotor distributed windings and the thicker rotor diameter compared with round rotor generators.

These generators are coupled to hydro-turbines and are called hydro-generators. The speed of these machines mainly depends on the water falling height, with a greater height corresponding to a greater speed. Generally, the following hydro-turbines are utilized depending on the height of water:

1. Pelton turbine: for heights higher than 400 m.
2. Francis turbine: for medium heights up to 300 m.
3. Kaplan turbine: for low heights 50 m and under.

Usually, the gateway speed (the speed arising from the main drive under sudden no-load conditions) of the hydro-turbines is higher than the rated speed:

1. Pelton turbine: 1 to 1.8 times the rated speed.
2. Francis turbine: 2 to 2.3 times the rated speed.
3. Kaplan turbine: 2.4 to 2.7 times the rated speed.

Therefore, the peripheral speed of the generator at the getaway speed is higher than that of the normal speed of non-salient pole generators. A salient-pole generator must also be designed so that it can cope with the mechanical stresses based on the getaway speed [41].

Structural details of the salient-pole generator completely differ from those of the turbo-generators (non-salient pole generators). The main difference is in the manufacturing and design of these two generators, regarding their substantial differences in the speed of the rotating parts. The higher number of poles in the salient-pole generator indirectly leads to an increased rotor diameter. The outer diameter of the stator frame of hydro-generators usually exceeds 3 m and may reach 15 m. The stator core length is very much smaller than the stator diameter. The ratio of the stator core length and the stator inner diameter for these generators is 0.15 to 0.20. Therefore, hydro-generators are characterized by a large diameter and a short axial length.

2.3.3 Synchronous Generators with Different Field Locations

All electrical generators consist of two main parts: (1) armature and (2) magnetic field (excitation) systems. Depending on the location of the field winding and armature, they are divided into two types [42]:

1. Fixed excitation windings on the stator and rotating windings on the rotor (armature), as shown in Figure 2.11a.
2. Rotating excitation windings on the rotor and armature fixed windings on the stator, as shown in Figure 2.11b.

Some low-power synchronous generators (first type) are similar to DC generators and a field excitation winding is placed on the stator. This is not recommended in medium- and high-power generators. In fact, most generators are of the second type. The reason is that the fixed armature and rotating magnetic field system is the most appropriate and most economical structure.

The following are the advantages of the different fixed rotating field and fixed armature systems:

1. **Armature voltage**: Modern large synchronous generators are generally designed for high voltages, such as 11 kV, 13.8 kV, and 25 kV. In the fixed armature system, the voltage of the armature winding can be directly connected to the terminals; therefore, the use of slip rings, which are unreliable, is avoided.

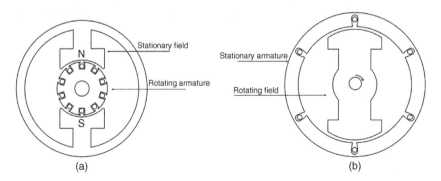

Figure 2.11 (a) Fixed magnetic field and (b) rotating magnetic field type.

2. **Magnetic field system voltage**: The rated DC voltage is very low, typically 400 V, and can be delivered to the rotating field system by slip rings.
3. **Size and number of slip rings**: Slip rings with a rotating magnetic field have low power; consequently, their sizes are considerably reduced. The number of slip rings on the rotor in the rotating field system is also significantly decreased compared with the rotating armature system.
4. **Performance of slip rings**: When the rotating field system is replaced, for high current and voltage, the performance of slip rings is unreliable.
5. **Armature insulation**: In a fixed armature system and rotating field systems, the stator windings and insulation are exposed to lower mechanical vibrations.
6. **Design and winding**: The three-phase armature on the rotor creates a heavier rotor and generates mechanical and thermal problems at high speeds.

For large modern synchronous generators, the fixed armature and rotating magnetic field are recommended.

2.3.4 Different Schemes of Excitation Systems for Synchronous Generators

Different schemes for the DC excitation source of the field winding of the large synchronous generator are introduced in the following subsections [43].

2.3.4.1 DC Excitation

This field winding of synchronous generators is a conventional excitation method. In this case, three machines, called the auxiliary exciter, the main exciter, and the main three-phase generator, are connected, and they rotate coaxially. The auxiliary exciter is a DC shunt generator that supplies the main exciter. The main exciter is a separate excited DC generator. Referring to Figure 2.12, the DC output of the main exciter supplies the exciter of the generator through brushes and slip rings [45].

2.3.4.2 Static Excitation System

In this method, the generator excitation power is taken from the output of the main three-phase generator. At this end, a three-phase step-down transformer reduces the alternating voltage to a desirable level, and then it is supplied to a full-wave three-phase thyristor rectifier bridge. A regulator, as shown in Figure 2.13, controls the firing angles of these thyristors by taking the signals from the generator output using a potential transformer (PT) and current transformer (CT). The controlled output power from the thyristors is delivered to the main generator exciter winding via the slip rings and brushes [41].

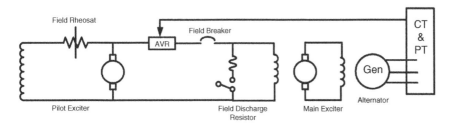

Figure 2.12   DC excitation system of synchronous generator [44].

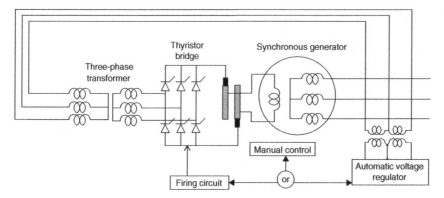

Figure 2.13 Static excitation system of a synchronous generator.

2.3.4.3 Brushless Excitation System

Figure 2.14 presents a brushless excitation method in which the silicon rectifiers are fixed on the shaft [45]. The auxiliary exciter of the generator is a PM fixed on the rotor, while the three-phase armature is fixed on the stator. The three-phase voltage from the auxiliary exciter applies the thyristor bridge outside the generator and, after rectifying, the controlled DC output is supplied to the static excitation winding of the main exciter. The three-phase power generated in the main exciter rotor is transferred to the silicon rectifier fixed on the shaft through the hollow shaft along the main shaft. The DC power from the rectifier bridge is delivered to the main generator field through the hollow shaft without slip rings and brushes. The brushless excitation system is the only hopeful solution for supplying the field winding of large turbo-generators. This excitation system is used in almost all designed and manufactured turbo-generators due to its advantages.

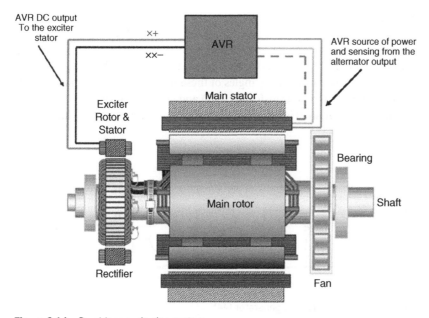

Figure 2.14 Brushless excitation system.

2.4 Voltage Equation and Rated Power of the Synchronous Generator

The phase voltage V_φ can be expressed as

$$V_\varphi = E_A - I_A R_A - jX_S I_A \tag{2.2}$$

where E_A is the induced electromotive force at the terminal of the generator, I_A is the phase A current, R_A is the phase A resistance, and X_S is the synchronous reactance. For a reactive load (I_A lagging phase angle), E_A is larger than V_φ, and for the capacitive load (leading phase angle), E_A is generally lower than V_φ. Steam turbines, hydro-turbines, or similar prime movers coupled to a synchronous generator must have a fixed speed, regardless of the demanded electrical power. The reason is that the frequency of the power system must be kept constant. The input mechanical power on the shaft is [1]

$$P_{in} = \tau_{app}\omega_m \tag{2.3}$$

where τ_{app} is the applied torque and ω_m is the prime mover mechanical speed. The converted power P_{conv} converted from mechanical energy to electrical energy and transferred to the rotor is

$$P_{conv} = \tau_{ind}\omega_m = 3E_A I_A \cos \gamma \tag{2.4}$$

where τ_{ind} is the induced torque and γ is the angle between the phasor E_A and I_A. The actual output electrical power P_{out} versus the line quantities is as follows:

$$P_{out} = \sqrt{3}V_T I_L \cos \theta = 3V_\varphi I_A \cos \theta \tag{2.5}$$

Similarly, the output reactive power Q_{out} versus line quantities is as follows:

$$Q_{out} = \sqrt{3}V_T I_L \sin \theta = 3V_\varphi I_A \sin \theta \tag{2.6}$$

If R_A is neglected, a simplified phasor diagram of a generator is produced, as shown in Figure 2.15. The vertical part of bc is expressed as $E_A \sin \delta$ or $X_S I_A \cos \theta$. Therefore,

$$I_A \cos \theta = \frac{E_A \sin \delta}{X_S} \tag{2.7}$$

By substituting Equation (2.7) in (2.5):

$$P_{out} = \frac{3V_\varphi E_A \sin \delta}{X_S} \tag{2.8}$$

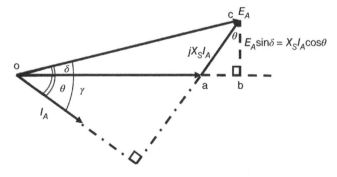

Figure 2.15 A simplified phasor diagram of a generator.

All resistances in the generator are ignored and no electrical losses are assumed. Therefore, this equation is true for both P_{out} and P_{conv} and is maximized at $\delta = 90°$:

The induced torque τ_{ind} in these generators is as follows:

$$\tau_{ind} = kB_R \times B_S = kB_R \times B_{net} \tag{2.9}$$

and the final form of Equation (2.9) is

$$\tau_{ind} = kB_R B_{net} \sin \delta \tag{2.10}$$

where δ is the torque angle (the angle between the rotor and resultant magnetic fields). The rotor flux density B_R induces E_A voltage and the resultant flux density B_{net} induces V_ϕ voltage. The angle δ is between B_R and E_A is the same as the angle between B_R and B_{net}. An alternative equation for the induced torque in the synchronous generator is as follows:

$$\tau_{ind} = \frac{3V_\varphi E_A \sin \delta}{\omega_m X_S} \tag{2.11}$$

This gives the induced torque versus electrical quantities, while Equation (2.10) indicates the torque versus magnetic quantities.

2.5 Synchronous Generator Model Parameters

Three quantities must be known in the equivalent circuit of a synchronous generator to predict the full behavior of the generator. These quantities include the relationship between the field current and the flux, synchronous reluctance, and armature resistance. A simple method is briefly proposed in this section to determine these quantities in a synchronous generator. The first step in this process is the open circuit test of the synchronous generator. In this test, the no-load generator is rotated with a rated speed and excitation current set at zero. The field current is then gradually increased and the terminal voltage of the generator is measured at each excitation current. This allows plotting of the open circuit characteristic (OCC) of the generator. The second step is the short circuit test of the generator. Again, the field current is set to zero, and the terminals are short-circuited. By increasing the excitation current, the armature current or line current is measured. This curve is called the short circuit characteristic (SCC) of the generator.

For $V_\phi = 0$, the internal impedance of the generator Z_S is obtained as follows:

$$Z_S = \sqrt{R_A^2 + X_S^2} = \frac{E_A}{I_A} \tag{2.12}$$

Since $X_S \gg R_A$, Equation (2.12) can be simplified as follows:

$$X_S \approx \frac{E_A}{I_A} = \frac{V_{\varphi.OC}}{I_A} \tag{2.13}$$

2.6 Different Operating Modes of Synchronous Machines

Synchronous machines can operate in motor or generator modes. When the synchronous machine operates in a no-load mode and the excitation field is under control, it is called a synchronous condenser or a synchronous compensator. In this mode, depending on over-excitation or under-excitation, the machine injects the reactive power into the network or absorbs it from the network. When no torque is applied to the rotor shaft and the excitation current increases quietly, the generator moves to the under-excitation mode and absorbs reactive power from the network. The advantage of a synchronous compensator compared with a static compensator (capacitor) is that it does not generate harmonics, as it switches transient modes and resonance in the network and enhances the network stability against transient modes, such as short circuit faults [42].

In the motoring mode, the interaction between the generated resultant flux by the stator winding and the rotor excitation field develops a torque on the rotor. In this case, depending on the armature current and excitation current, it operates in two under-excitation or over-excitation modes. Large machines merely operate in the motoring mode. In the generating mode, the stator flux vector lags with the rotor flux vector and follows it. The angle between these two vectors is called the load angle. In this case, the synchronous machine can also operate in under-excitation (leading) and over-excitation (lagging) modes. Generally, all synchronous generators (except in no-load or light-load cases) operate in the over-excitation mode in the normal case in order to be able to control the terminal voltage by controlling the reactive power of the network (see Figure 2.12 or 2.13).

2.7   Damper Bars in Synchronous Generators

The reasons for using damper bars in the synchronous machines are as follows:

1. **Damping oscillations**: The main duty of damper bars in synchronous generators is to damp the rotor oscillations. Reaching that goal requires the design of low-resistance damper bars. When the rotor rotates with synchronous speed, no damping torque develops from the damper bars. When the rotor speed differs from the synchronous speed (lower or higher), the rotor oscillates and the damper bars engage. For speeds lower than the synchronous speed, the damper bars develop torque in the rotating direction. This additional torque helps the rotor reach the synchronous speed. When the rotor rotates at a speed higher than the synchronous speed, the damper system develops a torque that rotates counter to the rotor, leading to a slowing down of the rotor and reaching synchronous speed [46].
2. Damper bars help to decrease the over-voltage under abnormal conditions.
3. Damper bars reduce the rotor oscillations in cases of system disorder.
4. Damper bars help to start the synchronous motor, similar to the asynchronous motor, before reaching synchronous speed.
5. Damper bars also protect the rotor field winding in the case of a short circuit fault in the stator winding.

The number of damper bars per pole is

$$N = \frac{b_p}{\tau_r} - 1 \qquad\qquad (2.14)$$

To have a good connection with damper bars, the thickness of the end rings of damper bars must be approximately 10 mm [47].

2.8   Losses and Efficiency in Synchronous Generators

Different losses in synchronous generators include ohmic losses (copper losses), core losses (iron losses), mechanical losses, and excitation losses. Ohmic losses consist of the total losses of the stator winding and rotor winding (excitation) losses. Core losses consist of stator hysteresis losses, eddy current losses, and stray losses of the core. The field winding flux (even at no-load) generates stray losses of the core. These losses are produced on the rotor surface and stator sheets. The losses of synchronous generators are divided into no-load and short circuit losses. Today, the losses of synchronous generators, including winding and lamination losses, can be obtained in detail by FEM [48].

No-load losses in the stator core are as follows:

$$P_{\text{Fe10}} = p_{\text{h}} + p_{\text{eddy}} = \left[K_{\text{h}} p_{\text{h0}} \left(\frac{f}{50} \right) + K_{\text{ed}} p_{\text{ed0}} \left(\frac{f}{50} \right)^2 \right] (B)^2 G \tag{2.15}$$

where p_{h0} and p_{ed0} are the hysteresis losses and eddy current losses per kg at a magnetic flux density of 1 T at the rated frequency, respectively, and G is the lamination weight and K_{h} and K_{ed} are the hysteresis coefficient and eddy current coefficient respectively. The short circuit losses in the stator windings are as follows:

$$P_{\text{co}_{\text{sc}}} = 3(R_{\text{s}})_{\text{DC}} K_{\text{R}} I_1{}^2 \tag{2.16}$$

where K_{R} is the skin effect of the stator windings. The excitation losses are as follows:

$$P_{exsh} = \frac{R_f I_f^2 + 2\Delta V I_f}{\eta_{ex}} \tag{2.17}$$

where ΔV is the voltage drop between the terminals of the brushes and η_{ex} is the efficiency of the excitation system R_f and I_f are the resistance and current of the excitation winding respectively. The mechanical losses also consist of the friction losses and windage with hydrogen cooling, which is usually 1% of the total losses [49].

The efficiency of the synchronous generator is:

$$\eta_{SG} = \frac{(P_2)_{electric}}{(P_1)_{mechanical}} = \frac{(P_2)_{electric}}{(P_2)_{electric} + \sum losses} \tag{2.18}$$

where η_{SG} is the synchronous generator efficiency, $(P_2)_{electric}$ is the output electrical power, and $(P_1)_{mechanical}$ is the input mechanical power of the synchronous generator. In industry, at a constant speed and voltage, the efficiency is estimated for 25%, 50%, 100%, and 125% full load. In these cases, the no-load losses are constant, and only short circuit losses vary with current variations. Therefore, the powers in the synchronous generator are as follows:

$$p_{sc} = p_{scn} \left\langle \frac{I_1}{I_n} \right\rangle^2 \tag{2.19}$$

$$K_{load} = \frac{I_1}{I_n} \tag{2.20}$$

$$P_{excsh} = P_{exchn} \left\langle \frac{I_1}{I_{fn}} \right\rangle^2 + P_{brush} \left\langle \frac{I_1}{I_{fn}} \right\rangle \tag{2.21}$$

$$p_{sc} = S_n \frac{I_1}{I_{1n}} \cos \varphi \tag{2.22}$$

where p_{sc} is the no-load short circuit losses at the stator current I_1, p_{scn} is the no-load short circuit losses at the stator rated current I_{n}, K_{Load} is the load factor, p_{excsh} is the total excitation losses at excitation current I_{f}, p_{exchn} is the no-load short circuit losses at the stator rated current I_{n}, I_{fn} is the rated excitation current, p_{brush} is the brush losses, S_{n} is the rated VA, and $\cos \varphi$ is the power factor. Therefore, the efficiency is as follows:

$$\eta_{SG} = \frac{K_{Load} \cdot S_n \cdot \cos \phi}{K_{Load} \cdot S_n \cdot \cos \phi + p_{Feo} + p_{mec} + p_{scn} + \left(\frac{I_f}{I_{fn}} \right)^2 + p_{brush} \left(\frac{I_f}{I_{fn}} \right) + p_{stray}} \tag{2.23}$$

Estimation of the losses using the above equations gives an efficiency of the generator with overall loads for a suitable design. Note that the winding losses predominate over the other losses in a synchronous generator with a direct cooling system, while in a generator with no direct cooling system, the non-winding losses are generally predominant [50].

2.9 High-Voltage Synchronous Generators

The high-voltage synchronous generators, called powerformers, have been seriously considered since the 1990s. These generators are directly connected to the power system without a step-up transformer. The new idea used in this design was the utilization of cable as the stator winding. The resulting high voltage generator is suitable for use in thermal and hydropower plants. The high efficiency, reduction in repair and maintenance costs, lower losses, and negative effects on the environment (considering the materials used) are some features of this type of generator. The high-voltage generator operates with a lower current compared with conventional generators. The maximum output voltage of this generator is restricted by the cable technology. Today, the high technology of cable manufacturing allows the design of voltage generators with 400 kV voltage. The conductor used in the high-voltage generator is a circular type, while the conductors used in conventional generators are rectangular-shaped. Consequently, the electric field in the high-voltage generators is more uniform. The winding dimensions are determined based on the system voltage and the maximum power of the generator.

In the high-voltage generators, the external layer of the cable is earthed. This restricts the electric field along the cable and, unlike the case for conventional generators, the field does not require controlling at the end winding region. The leads and connections are usually placed in an accessible empty space; therefore, the connection placement in one power station can differ from that in other

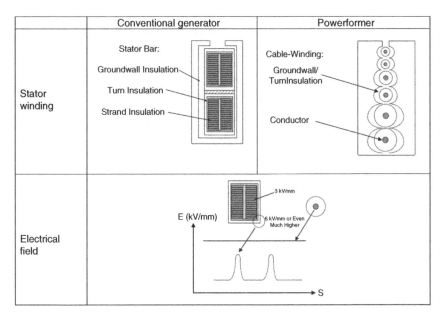

Figure 2.16 Differences in stator winding in conventional synchronous generators versus powerformers [52] / with permission of IEEE.

power stations. However, these connections and leads mean that no problems arise in this generator from vibrations and oscillations, as occurs in conventional machines [51].

Two types of cooling systems are used in the present high-voltage generators. Rotor and end windings are cooled by cold water. The water-cooled system consists of a stator core containing XLPE (a cross-linked polyethylene-insulated aluminum conductor armored cable) tubes through which water flows to cool the stator core. The disadvantages are the high impact of the network on the generator due to the elimination of the step-up transformer, and a heavy weight and fixing problems [52]. Figure 2.16 presents the differences in the stator winding structures of these generators compared with conventional generators.

2.10 Preliminary Design Considerations

To achieve the required data for different parts of the synchronous generator and to offer them to the manufacturer, the following items must be considered for the design [40]:

1. Main dimensions of the stator.
2. Complete details of the stator winding.
3. Details of the rotor and its winding.
4. Performance of designed parts.

A designer needs the following items to design the synchronous generator:

1. Detailed specifications and the needs of the customer.
2. Design equations.
3. The values of different parameters, such as specific magnetic loading and specific electrical loading.
4. Knowledge of the materials used, such as magnetization characteristics, insulation features, conductivity, and their other normal behaviors, such as the expansion coefficient and the contraction.
5. Restrictions of different performance parameters of the generator, such as core losses, efficiency, and short circuit current.

An attempt must be made to design the generator to achieve a minimal cost, small size, and low weight. Achieving all these goals simultaneously is not possible. The design of a synchronous generator with better performance will be more expensive, but a cheaper generator will have a poorer performance.

The main specifications (design input data) of the three-phase generator to initiate the required design include [40]:

1. Rating output in MVA or kVA.
2. Rated voltage in kV.
3. Speed.
4. Frequency.
5. Generator type (salient pole or round rotor).
6. Stator winding connection (the star in generators).
7. Temperature rise limits.

2.10.1 Output Equations

The output equation of the stator is the most basic tool for beginning the design of a synchronous generator. This equation indicates the relationship between the output of the generator and its stator dimensions. The output of the generator is as follows:

$$P = 3V_{ph}I_{ph} \times 10^{-3} \text{ kVA} \tag{2.24}$$

The phase-induced electromotive force is as follows:

$$E_{ph} = 4.44fT_1K_w\varphi \tag{2.25}$$

where frequency f was given by Equation (2.1). The air gap magnetic flux per pole is

$$\varphi = B_{av} \times \frac{\pi DL}{P} \tag{2.26}$$

In addition, specific electrical loading is as follows:

$$q = 3\frac{I_{ph}Z_1}{\pi D} \tag{2.27}$$

By substituting the above equations in Equation (2.24) we have

$$P = 3\left(2.22 \times \frac{Pn_s}{2} \times Z_1 \times K_w \times B_{av}\frac{\pi DL}{P}\right)I_{ph} \times 10^{-3}$$

$$= (1.11 \times \pi DL \times B_{av} \times K_w \times n_s)(3I_{ph}Z_1) \times 10^{-3}$$

$$= 1.11 \times \pi^2 D^2 LB_{av}qK_wn_s \times 10^{-3}$$

$$= KD^2Ln_s \text{ (kVA)} \tag{2.28}$$

where

$$K = 11B_{av}qK_w \times 10^{-3} \tag{2.29}$$

K is the synchronous generator output coefficient, B_{av} is the average magnetic flux density in the air gap, T, K_w is the stator winding coefficient, n_s is the speed in rps, Z_1 is the number of stator conductors per phase, D is the stator inner diameter in m, and L is the stator length in m. The output equation indicates that:

1. If a high-speed machine is designed, the D^2L (proportional with the volume) and consequently the size and cost of the machine are reduced.
2. If the output factor is higher, the size of the machine is smaller.
3. If specific electrical and magnetic loadings are large, the output factor is higher and it finally leads to reductions in the machine size and cost.

2.10.2 Selecting Specific Magnetic Loading

The cost and size of the synchronous generator clearly drop by selecting a higher magnetic flux density. However, the higher effective magnetic flux in the stator tooth and core leads to higher core losses, lower efficiency, and a higher temperature. The stability margin is an important index in the performance of a synchronous generator, as it shows the maximum power delivered at the rated conditions by the generator ($P_{max} = (V + E)/X_s$). Therefore, steady-state stability has an inverse relationship with X_s. This reactance is directly proportional to the square of the number of turns of the stator phase winding. The higher air gap magnetic flux density means a higher air gap flux. Therefore, a higher magnetic flux density reduces the synchronous reactance and increases the stability margin of the generator.

Generally, synchronous generators are used in parallel to supply the loads. The synchronizing power shows the capability of the system to keep the generators in a synchronous mode; this power has an inverse relation with X_s. Therefore, to obtain a reasonable performance of the parallel operation of these generators, a higher synchronizing power is needed, which is achieved by designing a machine with a high air gap magnetic flux density [53]. This higher flux density reduces the synchronous reactance, and this in itself causes an increase in the transient current under short circuit conditions. However, a machine designed with a low flux density can restrict electromagnetic forces in short circuit conditions [54]. The choice of a high magnetic flux density in designing synchronous machines therefore leads to some advantages and disadvantages. The advantages include reduction in the generator size, a higher stability range, reasonable performance in parallel operation, and reductions in the generator cost. The disadvantages consist of higher core losses, a rising transient short circuit current, reductions in efficiency, and higher temperature rises. Therefore, a trade-off must be considered based on the end-user needs when selecting the magnetic flux density of the machine.

With the advancement of magnetic materials, choosing a higher air gap magnetic flux density is now possible in modern synchronous machines. Its value is restricted due to the permissible temperature rise of the insulating materials; however, much progress has been made to improve the insulating materials as well. Today, insulating materials with higher temperature tolerances have been employed in modern synchronous machines.

2.10.3 Selecting Specific Electric Loading

The output equation shows that choosing higher values of specific electric loading reduces the size of the generator, leading to a lower cost of the generator for the same output. In addition, the leakage reactance and armature reaction reluctance are larger for a higher specific electric loading, which increases the synchronous reactance of the generator. Therefore, a higher value of the specific electric loading causes a lower transient current in the short circuit case, thereby allowing the generator to withstand a high sudden short circuit fault. Nevertheless, the higher specific electric loading causes a higher copper loss, lower efficiency, and a higher temperature rise. In addition, this choice increases the load stray losses. A higher specific electric loading also leads to an inherent voltage regulation, a reduction in the stability limit in the steady state, and a decrease in the synchronizing power due to the higher synchronous reactance. Higher specific electric loading can still be used for generators with good cooling systems and for generators with low voltage [55]. The choice of a higher specific electric loading leads to the following advantages and disadvantages:

Advantages

1. Reduction in the size of the synchronous machine.
2. Lowering of the transient current under short circuit conditions.
3. Cost reduction of the synchronous machine.

Disadvantages

1. Rising copper losses.
2. Higher stray load losses.
3. Higher temperature rises.

4. Poor voltage regulation.
5. Lower stability limit in steady-state conditions.
6. Lower synchronizing power.

2.10.4 Relationship between *L* and *D*

The inner diameter and stator apparent length can be estimated from D^2L in Equation 2.28. This equation shows an appropriate relation between L and D.

2.10.4.1 Salient-Pole Generators
The stator inner diameter of the salient-pole generator is much larger than that of the axial length of the stator core. The rotor diameter must also be justifiable, considering the rotor peripheral speed limit. The reason is that the rotor must withstand the centrifugal forces. For appropriate design of the peripheral speed, it is normally kept equal to 30 m/s.

2.10.4.2 Turbo-generators
The stator inner diameter is generally lower than that of the stator core axial length. These generators are normally designed for a speed of 3000 rpm. The rotor diameter (approximately equal to the stator inner diameter) is limited due to the permissible peripheral speed. Therefore, the stator inner diameter is estimated based on the peripheral speed. The peripheral speed in the turbo-generator is much higher than that of the salient pole generator [40]. The peripheral speed of these generators must not exceed 175 m/s. For an appropriate design, this speed may be kept between 130 and 150 m/s.

2.10.4.3 Short Circuit Ratio
If saturation is ignored, the short circuit ratio (SCR) is the inverse of the synchronous reactance X_d. The SCR is a very important parameter and influences several performance indexes of the generator, such as the stability limit, voltage regulation, short circuit current, synchronizing power, and parallel operation in the power system. The following advantages and disadvantages clearly indicate the impact of the SCR on the performance and cost of the generator [56]:

- **Advantages of the low value of the SCR:**

1. Short circuit current: The short circuit current directly depends on the SCR. The low values of the SCR mean less short circuit current. Consequently, control equipment costs are sharply reduced.
2. Size of generator: The designed machine with a low SCR has a shorter air gap length, which reduces the size of the generator. In addition, the required excitation will be much lower and the overall size of the generator decreases.
3. Cost of generator: As mentioned above, the cost of the generator is reduced with a small SCR. As a result, designing a synchronous generator with a small SCR is economical.
4. Active materials: A small SCR reduces active iron and copper.

- **Disadvantages of the low value of the SCR:**

1. Stability limit: The maximum accessible power is directly proportional to the SCR, so a small SCR will weaken the stability limit.
2. Voltage regulation: A synchronous generator with a small SCR will have higher voltage variations when coupled to an oscillatory load; this will weaken the inherent voltage regulation of the generator.

3. Parallel operation: Parallel operation of generators with small SCR is not very reasonable because of the low synchronizing power; these generators have more sensitivity to voltage disturbance.

These advantages and disadvantages indicate that a larger SCR leads to better performance of the machine, but a small SCR machine is recommended for economic reasons. Therefore, a correct choice must be made between a small SCR and these factors. The present procedure is to design synchronous generators with a small SCR, based on recent progress in rapid control system excitation. A small SCR is used in a less expensive generator. The SCR for turbo-generators may vary between 0.7 and 1.1, while the salient-pole synchronous generators can be designed with a larger SCR between 0.9 and 1.3.

2.10.5 Air Gap Length

The air gap length of the synchronous generator is an important design parameter that influences the performance of the generator. The air gap length makes a considerable contribution to the no-load MMF. If a synchronous generator is designed with a wide air gap length, it will have a higher reluctance due to the armature flux. This is true where the synchronous d-axis reactance is low and the SCR is high. Therefore, the correct choice of the air gap length in the synchronous generator is essential because it affects stability, which is the most important performance parameter (stability directly depends on SCR). The following are the advantages and disadvantages of a large air gap length in a synchronous generator [54]:

- **Advantages of a large air gap length**

1. Stability: Extended stability range.
2. Regulation: Smaller voltage regulation.
3. Synchronizing power: Higher synchronizing power that lowers the sensitivity of the generator to load oscillations.
4. Cooling: Better cooling on the air gap surface.
5. Noise: Reduction of the noise level.
6. Magnetic pull: Lower magnetic pull.

- **Disadvantages of a large air gap length**

1. Field MMF: More field MMF is required.
2. Size: Higher stator diameter leads to a higher generator size.
3. Magnetic leakage: Increasing magnetic leakage.
4. Copper weight: Higher weight of field windings.
5. Cost: Higher overall cost.

Therefore, the air gap length must be chosen based on the cost, performance, and the other factors.

2.11 Stator Design Considerations

The stator core consists of laminated sheets to reduce the eddy current losses of the generator in the stator frame. The stator core sheet thickness is 0.5 mm and its specific loss is 1.5 W/kg. These laminations are isolated from each other by an electric insulation material on both sides of each sheet. The stator winding is placed in the slots of the stator and on the inner edges of the sheets. A series of holes made in the stator sheets considerably improves the cooling system of the generator [57]. Specific lamination sheets made of non-magnetic steel and isolated straps keep the stator core sheets on top of each other. The stator of a small turbo-generator is made as a complete part, but that of a large hydro-generator is constructed from two and/or more parts and is wound in the place where the generator is assembled. The three-phase AC windings in the stator slots are usually wound in two layers and with a 60° phase displacement. The winding pitch is chosen to reduce the harmonics of the electromotive force waveform [57]. Figure 2.17a shows the stator core and the winding of the large synchronous generator.

2.11.1 Stator Core Outer Diameter

The stator core depth can be calculated for both types of generators, assuming the effective magnetic flux density in the stator core to be between 1.1 T and 1.3 T. The magnetic flux carried by the stator core is equal to half of the air gap flux per pole (φ). The stator core magnetic flux density B_c is

$$B_c = \frac{\varphi}{2 \times A_c} \tag{2.30}$$

the stator core cross-section A_c is

$$A_c = d_c \times L_i \tag{2.31}$$

where L_i is the stator core length and the stator core depth d_c is

$$d_c = \frac{\varphi}{2 \times B_c L_i} \tag{2.32}$$

The outer diameter of the stator core D_0 is

$$D_0 = D + 2(d_c + h_s) \tag{2.33}$$

where D is the stator inner diameter.

2.11.2 Leakage Reactance

Precise estimation of magnetic leakage is a complicated matter. The following simplified equations may be used to determine the leakage flux for voltage regulation estimation. The total magnetic flux leakage is equal to $\varphi_1 = \varphi_s + \varphi_0$, where the slot magnetic leakage flux φ_s is

$$\varphi_s = 2\sqrt{2}\mu_0 Z_s I_{ph} L_s \lambda_s \tag{2.34}$$

and the slot length L_s is equal to $L - n_v b_v$ (n_v = number of stator ducts and b_v = stator duct width), λ_s is the specific flux-linkage of the slot, and I_{ph} is the phase current. The slot leakage of forehead connections φ_0 is:

$$\varphi_0 = 2\sqrt{2}\mu_0 Z_s I_{ph} L_0 \lambda_0 \tag{2.35}$$

where Z_s is the number of conductors in the slot and λ_0 is the permeance of forehead connections. Referring to Figure 2.17b, as a typical cross-section of the winding slot, the permeance of magnetic flux leakage is:

$$\lambda_s = \frac{h_1}{3b_s} + \frac{h_2}{b_s} + \frac{2h_3}{b_s + b_0} + \frac{h_4}{b_0} \tag{2.36}$$

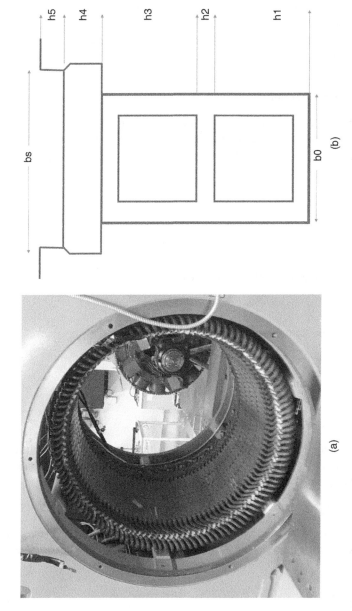

Figure 2.17 (a) The stator of a synchronous generator (b) slot dimensions.

$$L_0 \lambda_0 = \frac{\tau_p{}^2 K_s}{\pi \tau_s} \tag{2.37}$$

where K_s is the slot opening and pole pitch ratio, τ_p is the pole pitch in m, τ_s is the slot pitch, h_1, h_2, h_3, h_4, b_0, b_s are the dimensions of the slot as shown in Figure 17.b.

2.11.3 Stator Winding

The stator winding consists of one or more coils placed properly in slots and connected to each other to obtain the required phase group. The winding must be arranged such that the amplitude and frequency of the induced electromotive force in all phases are the same. The electromotive force of all phases must have identical waveforms with a 120° phase angle difference. Single- or double-layer windings may be used in the stator of the synchronous generators, depending on the applications. The double-layer winding is more common because it has some advantages over the single-layer winding. The stator windings are usually star-connected with a neutral wire to produce an isolated phase voltage. This connection also eliminates the 3rd harmonic of the electromotive force in the winding [50].

2.11.3.1 Double-Layer Winding

The most commonly applied winding type in the synchronous generator is the double-layer type with a short-pitch wound coil. The double-layer winding can be a fractional winding slot, which results in a better flux waveform. The full pitch or chord winding can be employed in the stator double layer. The chord winding has some advantages, such as harmonic reduction and shorter end winding connections. The double-layer winding has the following advantages compared to the single-layer machine [2]:

1. A better electromotive force waveform: This can be found by using a short-pitch wound winding.
2. Savings in copper: Decreasing the length of end connections in a short-pitch winding leads to a considerable saving in the consumed copper.
3. Lower winding cost: If all windings are identical, fewer tools are required, leading to a lower cost of stator winding production.
4. The convenience of winding production: If all windings are identical, they can be easily produced.
5. Fractional winding: Fractional slot winding can be used to improve the electromotive force waveform.

The double-layer winding has the following features compared with the single-layer winding:

1. Repair problem: Repairing the lower part of the damaged winding is difficult.
2. The problem of placement of last windings: Some problems arise in the placement of the stator's last winding leg above the embedded bottom leg of the winding.
3. Wider slot opening: For precise placement of windings, wider slot openings are preferred, but this causes more noise.
4. Weak slot space optimization: This is caused by insulation between the layers.
5. Winding length in the slot: Every turn has twice the stator core gross length (2L).
6. Length of end connections: This depends on the generator pole pitch; this part of the turn length is approximately 2.5 times the length of the pole pitch.
7. Length of the straight part along with core: This mainly depends on the terminal voltage of the generator; therefore, this turn length can be 0.05 times the terminal voltage in kV.

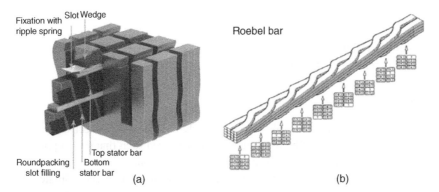

Figure 2.18 Schematic of the Roebel bar in the stator slot (a) and the transposition of the Roebel bar conductors along the axial direction of a synchronous machine (b).

8. The end winding length: The length of this part consists of a multi-turn winding or a single-turn bar, which are equal to 0.15 m.
9. Mean length in the stator in m: This is as follows [42]:

$$L_{\mathrm{mt}} = (2L + 2.5\tau_{\mathrm{P}} + 0.05 \times KV + 0.15) \tag{2.38}$$

2.11.3.2 Stator Winding Resistance

A stator with a mean turn length L_{mt}, stator conductor cross-section A_{s}, and phase turn number T_{ph} has a stator winding resistance given by $\rho L_{\mathrm{mt}} T_{\mathrm{ph}}/A_{\mathrm{s}}$, where ρ is the conductivity and the stator conductor cross-section is in mm^2.

2.11.3.3 Eddy Current Losses in Conductors

Stator windings of high-power generators carry high current and inevitably have large cross-section conductors and deep stator slots. The inductance of the central part of the conductor in the slot is high, while the inductance of the outer part is low. The difference between the central inductance and outer inductance generates eddy currents. These eddy currents generate additional I^2R losses, known as eddy current losses. If a solid conductor is used for stator winding, the eddy current will be quite high because these losses depend on the conductor depth. The use of strands as stator conductors reduces the eddy current losses. These strands are correctly isolated from each other. In addition, by displacing the strands such that every strand can continuously move in all existing positions in the core depth (called a Roebel bar), the eddy current losses are minimized [58]. Figure 2.18 shows the Roebel bar used in synchronous generators.

2.11.3.4 Eddy Current Loss Estimation

The eddy current losses in the conductors can be estimated based on the following stator slot design data:

$$K_{\mathrm{dav}} = 1 + (\alpha h_{\mathrm{c}})^4 \frac{m^2}{9} \tag{2.39}$$

where h_{c} is the depth of the strand and m is the depth of the conductor numbers in the slot. The loss mean factor α is:

$$\alpha = \sqrt{\frac{\text{Copper width in slot}}{\text{Slot width}}} \tag{2.40}$$

and

$$\text{Total eddy current losses = stator winding losses } (K_{dav} - 1) \tag{2.41}$$

The eddy current losses of the generator are considered to be almost 15% of the total copper losses and eddy current losses. The effective resistance of the stator winding can be obtained as follows:

1. Resistance drop *IR* is

$$IR = I_{ph}.R_s.K_{dav} \tag{2.42}$$

2. The effective resistance of the stator winding in per-unit (pu) is

$$\text{The effective resistance} = \frac{IR}{V_{ph}} \text{ (pu)} \tag{2.43}$$

2.11.3.5 Number of Slots

Figure 2.19 presents a typical slot and tooth shapes in synchronous generators. The appropriate number of stator slots can be chosen based on the conditions in A, B, and C.

A. The slot number must be properly chosen because it affects the cost and synchronous machine performance. No specific rule exists regarding the choice of the number of slots, but the following can be useful for choosing a proper number of slots [40]. The following are the advantages and disadvantages of choosing a large number of stator slots in synchronous generators:

Advantages:

1. Reduced leakage reactance: The leakage reactance has an inverse relationship with the number of slots per pole.
2. Lower internal temperature: Due to existing space between the conductors for air circulation.
3. Reduced tooth ripples: Due to the large number of slots.

Disadvantages:

1. High cost: High numbers of windings turns, insulation, and displacement between slots increase the cost of the generator.
2. Teeth mechanically weaken: Due to thinner teeth and the large number of slots.
3. Higher flux density in tooth: Due to thin teeth and a high number of slots.

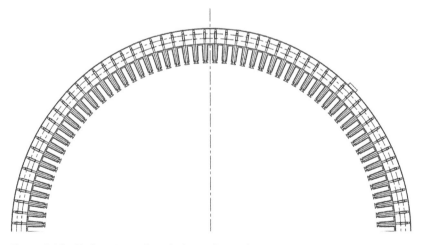

Figure 2.19 Typical slot and tooth shapes in synchronous generators.

B. Ampere conductors (slot-loading) in every slot exceed 1500.

C. Slot pitch: This depends on the slot number and must have the following approximate values in different cases:

1. In LV machines: Lower than or equal to 3.5 cm.
2. In machines up to 6 kV: Lower than or equal to 5.5 cm.
3. In machines up to 15 kV: Lower than or equal to 7.5 cm.

Generally, the number of poles in the salient-pole synchronous generators is high; therefore, the slot number in every pole of the phase is taken to be 3. In the turbo-generator, the number of slots per poles is possibly much higher (7–9) than that of the salient pole synchronous generators [59]. The number of slots in every pole per phase may be chosen as a fractional number such as $3\frac{1}{2}$ to reduce the amplitude of the harmonics in the electromotive force waveform. Therefore, the number of slots must satisfy the conditions B and C above.

2.11.3.6 Number of Turns per Phase

The number of turns per phase can be estimated using the following electromotive force equation of the synchronous generator:

$$E_{ph} = 4.44 f K_w \varphi T_{ph} \tag{2.44}$$

Therefore, the number of turns per phase is

$$T_{ph} = \frac{E_{ph}}{4.44 f \varphi K_w} \tag{2.45}$$

where K_w is the stator winding factor. The airgap flux per pole is

$$\varphi = B_{av} \times \frac{\pi D L}{P} \tag{2.46}$$

The estimated value of the turns must be rounded to an integer number. This approximated value gives the number of turns per phase. The final value of turns per phase is obtained after proper location of the conductors in the stator slots. The airgap flux must be modified based on the final number of turns per phase.

2.11.3.7 Conductor Cross-section

The conductor cross-section for a stator winding can be calculated based on the stator phase current and effective current density of the stator winding, as follows:

$$a_s = \frac{I_s}{\delta_s} \tag{2.47}$$

The current density δ_s must be chosen based on the cooling system. For a better cooling system, a higher value of δ_s is used, which reduces the cross-section of the conductor at a fixed stator current. This provides savings in copper and its cost for winding, and the total cost of the machine drops. Based on the economic considerations, a higher δ_s is chosen, even though this increases the stator winding resistance and stator copper losses and ultimately decreases the efficiency of the machine [3]. The current density must be properly chosen considering these points. A usual value for current density in the stator winding is between 3 and 5 A/mm^2.

2.11.3.8 Single-turn Bar Winding

The cross-section of the conductor in a high-power generator is large; therefore, the stator conductor is split up into several wires to reduce the eddy current losses. Each stator slot contains two conductors [50], as shown in Figure 2.20. Each conductor of the stator winding consists of two vertical stacks of copper strands properly isolated by asbestos. The strands can also be covered with glass, which has a better space factor because the thickness of the insulation layers on the strands is lower. The dimensions of a single strand are chosen based on electrical considerations and required factors of production. The width of a single strand must not exceed 7 mm, as the insulation material may be damaged due to bending. Normally, the width of each strand can vary between 4 and 7 mm. The maximum thickness of the layer must not exceed 3 mm due to eddy current losses in the conductors. Generally, the thickness of the strands in the winding depth for a given bar current is estimated based on the permissible temperature rise. The number of top bar strands is usually compared with that of the bottom bar strands. This can balance the existing temperature difference between the top and bottom (the temperature rise due to eddy current losses of the top coil) as different strands are moved to allow each strand to move continuously along all winding places to minimize circulating current losses.

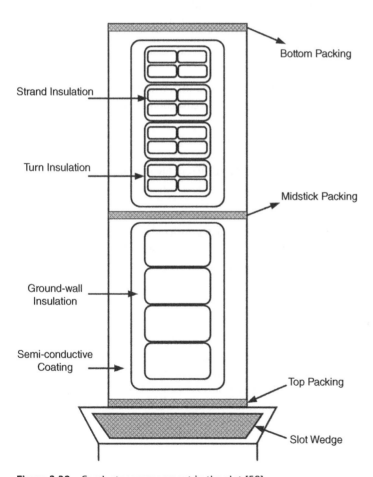

Figure 2.20 Conductor arrangement in the slot [50].

2.11.3.9 Multi-turn Windings

The multi-turn windings are wound by machinery in a complete loop, with two or three turns per loop. In multi-turn winding, each turn, as well as the individual strands, must be separately isolated. Mica or Novolac epoxy sheets are used for many parts; the mica sheets are used for forehead connections. The thickness of the insulation of each turn is determined by the generator line voltage. At this end, the insulation must be chosen so that it withstands sudden voltages equal to 1.5 times the line voltage due to lightning on the lines for a few microseconds [50].

2.11.3.10 Stator Winding Type Comparison

Stators have two types of winding: (1) single-turn bar and (2) multi-turn winding [4]. A detailed comparison of the two types of windings is as follows:

1. Multi-turn winding has more flexibility in the choice of the stator slot number compared with the single-turn bar.
2. The physical dimensions of the core are limited to 7.5 cm × 2.5 cm, an approximate slot length of 3 m, and a pole pitch of 80 cm. These limitations do not apply to the single-turn bars because these bars are hand-made.
3. The bend of the upper part of the winding that links with the core multi-turn winding needs highly flexible insulation materials. Single-turn bars have no need for a high bending feature.
4. Some problems arise in substituting faulty cores in a multi-turn winding.
5. The multi-turn winding is built by machine and is less expensive than the single-turn hand-made winding.
6. To generate a series connection, different windings in a single-turn winding need a solder clamp at the rear and front of every pair of poles, which adds to the cost compared with the multi-turn winding.
7. A full transpose in a single-turn winding with a tight connection is essential to reduce the eddy current losses.
8. Every turn of the multi-turn winding must be isolated by different layers of mica to have enough dielectric power to compensate for impulses on the wires due to lightning, but this reduces the allowable space for slot depth.

This comparison indicates that the multi-turn winding must be utilized to reduce costs. However, the multi-turn type must not be used if the stator current exceeds 1500 A. In large generators, the current exceeds 1500 A. Therefore, technically, single-turn bars must be used, although this incurs a slightly higher cost.

2.11.3.11 Winding and Slot Insulation

Establishing insulation systems for LV machines is not particularly difficult because insulation with sufficient mechanical strength is appropriate for electrical purposes as well. Breakdown of the insulation system is prevented by having a sufficient insulation thickness in HV generators. The ability of insulation materials to withstand the voltage level depends on the following factors:

1. Insulation thickness.
2. Dielectric strength of insulation material.
3. Voltage gradient in insulation material.
4. Time duration of applied voltage.

HV insulation problems in generators caused by rising voltage usually cannot be resolved by increasing the insulation thickness. The reason is the increase in the machine cost or the production of a low-output capacity machine. These problems are resolved by a proper choice of insulation system. Generally, a generator insulation structure consists of two or more types of insulation materials (composite insulation system). The stator winding insulation used in generators usually consists of the following basic materials:

1. Mica for a high strength dielectric.
2. Resin for bending and for filling blank parts to achieve higher robustness.
3. Mechanical carriers, such as dilute cellulose and film sheets, to fix the mica in position.

Therefore, the basic insulation used in generators is mica sheets and paper lined with bitumen (a substance produced by the distillation of crude oil which is known for its waterproofing and adhesive properties). This material is used in a part of the stator conductor inside the slot. Mica is used on the end part of the winding conductors. Since the insulation of the end winding must withstand the rated voltage of the adjacent windings, the total thickness of the insulation material is normally equal to 1.5 times the slot insulation. This system has been used for a long time in 200 MW generators. The demand for higher output power for the generators has led to the introduction of better insulation systems for the stator winding, leading to reductions in insulation thickness at the same rated voltage compared with the bitumen insulation [59]. Substituting a better insulation system provides a larger space for copper inside the slot and gives the designer greater freedom to design a winding for a higher temperature rise. In large generators, the bitumen insulation is replaced by a kinetics thermostat. In this insulation system, the basic insulation of the mica sheet is made from recovered pieces of mica. A glass cover with a thickness of 0.04 m is placed on the mica sheet, and the sheet is then saturated with Novolac epoxy resin to produce what is usually called the Novolac epoxy sheet system. Enough mechanical strength is obtained by the glass cover, while desirable electrical strength is provided by the resin and mica. This new insulation system has the ability to withstand heavy blows due to sudden short circuits at the generator terminals for the rated voltage. The epoxy mica insulation system is superior to the bitumen system in the following ways:

1. The dielectric strength is about 40% higher.
2. The insulation thickness of the slot is reduced by 20%.
3. The generator can operate under higher temperature rises.
4. The dielectric losses factor (δ) is smaller.
5. Heat transfer is better due to the reduction in insulation thickness.
6. Reduction of the frame size of the generator.

2.11.3.12 Stator Slot Dimensions

The number of stator slots, winding type, and copper cross-section have already been discussed. Here, the width and depth of the slot are estimated. The limit on the width depends strongly on the terminal rated voltage of the generator, because the basic insulation thickness in the winding and inside the slot can change. For the rated voltage of 11 kV up to 16 kV, the stator slot width is generally between 1.7 and 2.4 cm. The most economical and largest width for stator slots in the higher voltage generators is about 2.4 cm. The stator slot depth for the rated voltages between 11 and 16 kV is generally between 12.5 and 16.5 cm. For low voltages, it is between 7 and 12 cm [4]. The slot width is equal to the slot pitch minus the tooth width.

Flux density in the stator tooth must not exceed 1.8 T over the air gap surface in the generators; otherwise, the tooth losses increase. The flux density of the stator tooth B_t over the air gap surface can be estimated as follows [4]:

$$B_t = \frac{\varphi}{b_t L_i N_t} \tag{2.48}$$

where N_t is the tooth number in the pole arc of the generator. Therefore, the tooth width b_t is

$$b_t = \frac{\varphi}{B_t L_i N_t} \tag{2.49}$$

and the slot width b_s is

$$b_s = \tau_s - b_t \tag{2.50}$$

The available space for the conductor (effective width) can be estimated by subtracting the slot width from the total insulation width. Based on the conductor size and available space, the conductor depth may be calculated. An appropriate combination of conductors in the slot can then be chosen based on the above data. Finally, the stator slot width is estimated based on the following factors:

1. Space occupied by isolated conductors.
2. Turn insulation.
3. Slot insulation.
4. The separator of two layers.
5. Wedge.
6. Edge.

Generally, the slot depth is between 4 and 5 times the slot width.

2.11.4 Rotor Design of a Salient Pole Synchronous Generator

The armature developed MMF of operating synchronous generators is distorted by the magnetic field system. This leads to a reduction in the induced MMF in the on-load case. Therefore, the magnetic field system of synchronous machines must be sufficiently enhanced by excitation windings compared with the armature to improve the MMF at full load. A larger ratio of the field (F_f) and armature MMF (F_a) improves the regulation of the generator, although the larger ratio increases the required copper volume for winding and winding space and leads to a more expensive machine. In the initial design of the generator, the F_f/F_a ratio may vary between 1.7 and 2. When the pole height is estimated once, the no-load field MMF can be estimated and used to determine the field MMF at full load. The field winding is then designed based on the accurate electromotive force must be MMF of the magnetic field must be field at full load [4].

2.11.4.1 Pole Shape

To ensure reasonable operation for the connected load, the MMF waveform of the generator must be very close to a sinusoidal waveform. All electrical equipment is designed assuming a sinusoidal supply waveform; therefore, any deviation has a negative impact on performance. For salient-pole synchronous generators, the pole and its shoe must be shaped such that a sinusoidal electromotive force waveform is generated. To achieve this goal, the air gap must be formed so that it varies sinusoidally around the generator. If the air gap length gradually increases from the pole center to the pole tip, a reasonable waveform is obtained. The air gap length on the pole tips must be 1.5 to

2.25 times its value in the pole center to lead to a distributed sinusoidal waveform. This requires a precise sinusoidal distribution of the air gap along the x axis, l_{gx}, from pole center, as follows [4]:

$$l_{gx} = \frac{l_g}{\cos \frac{\pi x}{\tau_p}} \tag{2.51}$$

where τ_p is the pole shoe length and l_g is the air gap length.

The ratio of the pole arc and pole pitch for salient-pole generators must be properly chosen. If this ratio exceeds 0.75, the magnetic flux leakage between the two poles will cause a large magnetic flux density in the pole body and improper distribution of the magnetic flux in the stator. If this ratio is lower than 0.67, the end connections of the pole shoe will be insufficient to support the field winding. The optimum ratio between these two cases is 0.7.

2.11.4.2 Pole Dimensions

1. Axial length: The axial length of the pole may be 1 to 1.5 cm shorter than the stator core length; therefore, a rotor can freely rotate.
2. Pole width: Round or rectangular poles are used for salient-pole generators. The cross-section of the pole body can be obtained based on the pole body magnetic flux φ_p and flux density of the pole B_p [4]:

$$A_p = \frac{\varphi_p}{B_p} \tag{2.52}$$

The magnetic flux leakage factor for the pole is between 1.1 and 1.5. Therefore,

$$\varphi_p = (1.1 \text{ to } 1.5)\varphi \tag{2.53}$$

The B_p is between 1.4 and 1.6 T. The pole body b_p width is as follows:

$$b_p = \frac{A_p}{0.95 L_p} \tag{2.54}$$

For round poles the pole body cross-section A_p is

$$A_p = \frac{\pi}{4} b_p^2 \tag{2.55}$$

Therefore, the diameter of the round pole b_p can be determined.

Pole height: To be able to insert the total ampere-turn at full-load armature on the pole of excitation winding, an appropriate pole height must be chosen. The total full-load ampere-turn of every phase required by the field system for generating the useful magnetic flux ϕ and pole height are interdependent. The total ampere-turn of the pole at the full-load armature required by the field system can be approximated based on the results of this section. The ampere-turn of every pole of the armature field at the full load can be assumed to be between 1.7 and 2 times the ampere-turn per armature pole. The armature ampere-turn (AT_a) in every pole is as follows:

$$AT_a = \frac{1.35 I_{ph} T_{ph} K_w}{P} \tag{2.56}$$

where T_{ph} is the number of turns of phase winding.

Therefore, an approximate value of the ampere-turn of the full load field AT_f is

$$AT_f = (1.7 \text{ to } 2)AT_a \tag{2.57}$$

Every pole height can be estimated based on the height of the required field winding and insulation. The field winding height may be estimated approximately from the ampere-turn of the full-load field.

2.11.4.3 Copper Losses of Field Windings

The copper losses in the field winding are

$$I_f^2 \left[\frac{\rho l_{\mathrm{mtf}} T_f}{a_f} \right] = I_f^2 R_f \tag{2.58}$$

where l_{mtf} is the mean turn length, at is the cross section of the field winding.

The total area of copper in the field winding is

$$(d_f \times h_f) s_f = a_f T_f \tag{2.59}$$

Therefore,

$$a_f = \frac{d_f h_f s_f}{T_f} \tag{2.60}$$

The inner and outer surfaces of the field winding influence heat dissipation. The heat dissipated by the bottom and top of the field winding is negligible. Therefore, the field winding height is obtained as follows:

$$h_f = (I_f T_f) \left[\frac{\rho}{2 d_f h_f S_f} \right]^{\frac{1}{2}} = \frac{I_f T_f}{10^4 \sqrt{d_f R_f S_f}} \tag{2.61}$$

where

I_f: field winding current in A
T_f: total number of turns in every field winding
ρ: copper conductivity in Ω^{-1}
d_f: field winding depth in m
h_f: field winding height in m
S_f: field winding conductor cross-section in mm^2
R_f: field winding resistance in Ω

The approximate values of the quantities in the above equations are assumed as follows:

1. The full-load ATs of the field winding $I_f T_f$ can be between 1.7 and 2 times ATs of each armature pole.
2. The d_f varies from 0.03 to 0.05 m; the large depth of the winding may cause a temperature difference between the hotspot and outer surface of the field winding.
3. The permissible losses of the cooling surface in pu may be between 700 and 750 W/m^2 if the temperature rise does not exceed the permissible limit.

The pole height can be determined by adding the pole shoe, the missed space due to bending the yoke, and insulation between the winding and the frame. The pole shoe height at the center may be between 0.04 and 0.06 m. The overall space for insulation and the missed space due to bending can be approximately 0.11 to 0.15 times the pole pitch [4].

2.11.4.4 Rotor Core Depth

The rotor core depth is obtained based on the magnetic flux carried by the core section and core magnetic flux density. The magnetic flux in the core section φ_y is almost half the pole flux [4]:

$$\varphi_y = \frac{1}{2}\varphi_P \tag{2.62}$$

The core magnetic flux φ_c can be assumed to be between 1 and 1.2 T. The core cross-section A_c is as follows:

$$A_c = \frac{\varphi_c}{B_c} \tag{2.63}$$

Therefore, the core depth d_c is

$$d_c = \frac{A_c}{L_c} \tag{2.64}$$

where L_c is the rotor axial length, which may be considered equal to the stator core gross length.

2.11.4.5 Ampere-Turns of the No-Load Field

The magnetic circuit of the synchronous generator consists of the rotor core, pole, air gap, stator tooth, and stator core [4]. The design process for the calculation of the overall dimensions of these parts has already been proposed. Therefore, in this stage, the ampere-turn of the field winding at no-load can be estimated to establish the effective flux in the magnetic circuit.

The ampere-turn estimation of different parts of the magnetic circuit, based on the dimensions and design data of these parts, is briefly given in the following.

a. Air gap ampere-turn

1. The air gap magnetic flux is φ_g.
2. The effective air gap cross-section is

$$\acute{A}_g = k_f \tau_P L \tag{2.65}$$

where k_f is the pole arc per pole pitch ratio.
3. The air gap flux density B_g is

$$B_g = \frac{\varphi_g}{\acute{A}_g} \tag{2.66}$$

4. The air gap coefficient K_{gv} for the slot is

$$K_{gv} = \frac{L}{L - \eta_v b_v (1 - \delta_v)} \tag{2.67}$$

5. The air gap total coefficient K_g is

$$K_g = K_{gs} \times K_{gv} \tag{2.68}$$

b. Stator tooth ampere-turn

1. Stator tooth magnetic flux is φ.
2. The diameter at one-third of the height from the thin tooth end (D_1) is:

$$D_1 = D - 2\left(\frac{1}{3}h_s\right) \tag{2.69}$$

3. The slot pitch τ'_s versus diameter D_1 is

$$\tau'_s = \frac{\pi D_1}{\text{No. of slot}} \tag{2.70}$$

4. The tooth width b'_t at one-third of the height from the thin tooth end is

$$b'_t = \acute{\tau}_s - b_s \tag{2.71}$$

5. The stator gross length of tooth iron L_i is

$$L_i = K_i(L - n_v b_v) \tag{2.72}$$

K_i can be assumed to be between 0.92 and 0.93.

6. The effective area of all teeth under a pole pitch A_t is

$$A_t = K_f \times b'_t \times L_i \times \text{No. of teeth under a pole pitch} \tag{2.73}$$

7. The apparent magnetic flux density B_{apt} in a tooth is

$$B_{apt} = \frac{\varphi}{\acute{A}_t} \tag{2.74}$$

8. The K_s factor is

$$K_s = \frac{L \times \acute{\tau}_s}{L_i \acute{b}_t} \tag{2.75}$$

9. The total AT_t for a tooth is

$$AT_t = H_t \times h_S \tag{2.76}$$

2.11.5 Design of the Rotor of Round-Rotor Synchronous Generators

No concentrating rotor winding is present in the round-rotor generators. Instead, the winding is distributed in the rotor slots. Generally, 70% of the rotor slots are filled by the winding and the remaining 30% have no windings and form the pole arc [4]. The calculated number of rotor slots must also satisfy the following conditions to prevent undesirable harmonics in the magnetic flux density waveform:

1. A common factor between the rotor slot pitch and the number of stator slot pitches.
2. The number of rotor wound slots for two-pole synchronous generators must be divisible by 4.
3. The rotor slot width is limited to prevent stresses to the rotor tooth root.

2.11.6 Rotor Winding Design

The following items can be considered for designing rotor windings for round-rotor synchronous generators:

1. Full-load ATs of each pole can be assumed to be twice that of each armature pole AT:

$$2\left(1.35\frac{I_{ph}T_{ph}K_w}{p}\right) \tag{2.77}$$

2. The standard excitation voltage of 220 V with 15–20% tolerance is assumed:

$$V_c = \frac{(0.8 \text{ to } 0.85)}{p} \tag{2.78}$$

3. The mean turn length L_{mt} in m is estimated by the following approximated equation:

$$L_{mt} = 2L + 1.8\tau_P + 0.25 \tag{2.79}$$

4. The field conductor cross-section a_f is calculated as follows:

$$a_f = \frac{\rho l_{\mathrm{mtf}} I_f T_f}{V_C} \tag{2.80}$$

5. Assume a suitable value for the field winding current density. For machines cooled by a conventional method, a lower value of the current density, in the range of 2.5 and 3 A/mm^2, is considered. In generators with high rated values and direct cooling, a higher current density, in the range of 8 and 12 A/mm^2, is chosen [4].
6. The area of all conductors per rotor pole is estimated by $2I_f T_f / \delta_f$.
7. The number of conductors of each rotor pole is calculated by $2I_f T_f / \delta_f a_f$.
8. The resistance of the field winding R_f of each pole is

$$R_f = \frac{\rho l_{\mathrm{mtf}} T_f}{a_f} \tag{2.81}$$

9. The field winding current is $I_f = V_c / R_f$.

Usually, the ratio of slot depth and slot width is between 4 and 5. Insulation in the slot must withstand high mechanical stresses and forces due to the slot extension for different temperatures.

Normally, the following insulation materials are used for field winding:

1. Hard mica sheet (0.5 mm) around all sides of the field winding.
2. A flexible 1.5 mm mica sheet is placed on the winding field above the hard mica sheet.

Finally, a 0.6 mm steel sheet is used for all field windings. In addition, different turns in the slot are separated by a 0.3 mm separator consisting of a compressed mica sheet.

2.11.7 Synchronous Generator Excitation System Design Issues

The synchronous generator excitation system supplies the necessary excitation current for the generator operation. Two key problems define an excitation system. One is the amplification level of transient instance and the other is the maximum excitation ratio (ratio of the peak and rated voltage at the terminal of excitation windings) [60]. The level of the instantaneous amplification is a quantity that has a direct impact on the small-signal analysis and system dynamic stability. In fact, this depends on the generator oscillation frequency in the range of about one-tenth of 1 Hz and several Hz.

In the past, excitation systems consisted of large steady-state amplifiers, which stabilized the system against oscillations. In modern systems, proportional integral (PI) and proportional integral differential (PID) controllers and Fuzzy controllers are used. The maximum excitation system voltage also influences the transient stability in stationary excitation systems. In general, 160–200% of the rated excitation voltage is considered for design, and after removing the three-phase fault in the HV side of the step-up transformer connected to the generator, the stability is preserved [61].

The brushless excitation systems are less under the influence of power network faults, so the maximum factor ratio has less excitation voltage. The relation between the AC voltage of the motor drive in each phase (V_s) and the excitation system voltage (V_f) is as follows:

$$V_s \approx \frac{V_f}{2.34} \tag{2.82}$$

$$V_s = 0.78 \times I_f \tag{2.83}$$

These equations describe a loss-less excitation system. Overall, an important design problem of an excitation system is its efficiency and high reliability, which can improve the efficiency and reliability of the whole system. The long design stages of a synchronous generator and the need for many trial-and-error tests make giving a numerical example impossible [62]. Table 2.1 shows the different parameters of a typically designed salient-pole generator.

Table 2.1 Different parameters of a typically designed salient-pole generator [4].

Quantity	Symbol	Value
Full-load output (kVA)	Q	750
Line voltage (V)	V	2200
Phase	------	3
Frequency (Hz)	F	50
Speed (rpm)	N_s	375
Stator inner diameter (m)	D	1.45
Stator core gross length (m)	L	0.35
Pole pitch (m)	τ_P	0.285
Peripheral speed (m/s)	V	28.5
Pole flux (Wb)	φ	0.0535
Phase turn number	T_{ph}	112
Slot number	------	168
Conductor of each slot	------	4
Slot pitch (cm)	τ_S	2.7
Conductor size (mm^2)	------	7×7.5
Slot width (cm)	b_S	1.3
Conductor section (mm^2)	a_s	51.7
Slot depth (cm)	h_s	4.7
Winding resistance (Ω)	R_s	0.0765
Leakage reactance (pu)	L_{leak}	0.077
Short circuit ratio (pu)	SCR	1.1
Air gap length (mm)	l_g	0.38
Pole axial length (m)	l_p	0.335
Pole width (m)	b_p	0.12
Pole height (m)	h_p	0.21
Total no-load turns	AT	3928
Field full-load ATs	AT$_f$	7060
Cross section of field conductor (mm^2)	a_f	28.2
Conductor size (mm × mm)	------	15×1.9
Field current (A)	I_f	93
Turns field of each winding	T_f	76
Field winding resistance (Ω)	R_f	0.059
Stator winding copper losses (kW)	P_{cus}	11.74
Eddy current losses in conductors (kW)	P_{eddy}	1.41
Stray load losses (kW)	P_{cus}	1.53
Iron losses (kW)	P_{Fe}	1.53
Rotor copper losses (kW)	P_{cur}	8.35
Exciter losses (kW)	P_{cus}	1.15
Windage and friction losses (kW)	P_{mech}	6.45
Total losses full-load (kW)	P_t	44.49
Efficiency (%)	η	93.5

2.12 Summary

The synchronous generator is the most essential equipment in electrical energy generation. Knowledge of its structure and design procedure is necessary for both users and manufacturers of the generator. In addition, identification of new problems and updated knowledge are required.

This chapter provided a history of the synchronous generator from the beginning until the present and analysis of the generator structures and their different parameters were proposed. Finally, the design of these generators was considered. For this, information from different books and papers was considered. To increase the voltage, the invention of the powerformer at the end of the twentieth century resulted in an increase in the generated voltage up to the transmission voltage level. Some now believe that there will no longer be a use for step-up transformers in power plants a few years from now.

The technology of superconductor generators, rather than PM generators, has recently been receiving more attention. Expanding this technology in the future is expected to result in higher capacities with lower volume. Better designs and analysis methods with more accuracy are also possible due to advancements in computer processors and the use of FEM analysis [63, 64].

References

1 Mittle, V. and Mittal, A. (2009). *Design of Electrical Machines, Delhi*. India: NC Jain, Standard Publishers Distributors.

2 Boldea, I. (2015). *Synchronous Generators, Florida*. USA: CRC Press.

3 Sawhney, A. and Chakrabarti, A. (2010). *Course in Electrical Machine Design*. New Delhi, India: Dhanpat Rai & Co.

4 Upadhyay, K. (2011). *Design of Electrical Machines, New Delhi*. India: New Age International.

5 Kalsi, S., Weeber, K., Takesue, H., and Lewis, C. (2004). Development status of rotating machines employing superconducting field windings. *IEEE Proceedings* 92 (10): 1688–1704.

6 www.hitachi.com.

7 Hamdi, E.S. (1998). *Design of Small Electrical Machines*. USA: Wiley.

8 Eren, L. and Devaney, M.J. (2004). Bearing damage detection via wavelet packet decomposition of the stator current. *IEEE Trans. Instrum. Meas.* 53: 431–436.

9 Neidhofer, G. (2007). Early three-phase power [History]. *IEEE Power and Energy Magazine* 5 (5): 88–100.

10 Neidhofer, G. (1992). The evolution of the synchronous machine. *Engineering Science and Education Journal* 1 (6): 239–248.

11 Laskaris, T. (1977). A two-phase cooling system for superconducting AC generator rotors. *IEEE Trans. on Magnetics* 13 (1): 759–762.

12 Berry, P.J. and Hamdi, E.S. (2015). An investigation into damper winding failure in a large synchronous motor. In: *Proceedings of 50th International Universities' Power Engineering Conference (UPEC 2015)*, 1–4. Stafford, England.

13 Stone, G.C., Culbert, I., Dhirani, H., and Boulter, A. (2004). *Electrical Insulation for Rotating Machines*. USA: Wiley-IEEE Press.

14 Boulter, E. (2004). Historical development of rotor and stator winding insulation materials and systems. *IEEE Electrical Insulation Magazine* 20 (3): 25–39.

15 Gott, B. (1996). Advances in turbogenerator technology. *IEEE Electrical Insulation Magazine* 12 (4): 28–38.

16 Vogelsang, R., Fruth, B., and Ducry, O. (2005). Performance testing of high voltage generator- and motor insulation systems. In: *Proceedings of the 5th WSEAS/IASME International Conference on Electric Power Systems, High Voltages, Electric Machines*, 136–142. Tenerife, Spain.

17 *IEEE Recommended Practice for Testing Insulation Resistance of Electric Machinery*, IEEE Std. 43-2013, 2014.

18 David, E. and Lamarre, L. (2010). Progress in DC testing of generator stator windings: Theoretical considerations and laboratory tests. *IEEE Trans Energy Conversion* 25 (1): 49–58.

19 David, E., Lamarre, L., and Nguyen, D.N. (2007). Measurements of polarization/depolarization currents for modern epoxy mica bars in different conditions. In: *Proceedings of the IEEE Electrical Insulation Conference*, 202–206.

20 Pinto, C. (1991). An improved method of detecting contamination of HV stator windings in the field. In: *Proceedings of the IEEE Electrical Insulation Conference*, 55–59.

21 Pinto, C. (1991). Variations in the capacitance of delaminated HV stator insulation due to electrostatically generated forces. In: *Proceedings of the IEEE Electrical Insulation Conference*, 65.

22 Pinto, C. (1998). A generalized approach for study of the non-linear behavior of stator winding insulation. In: *Proceedings of the IEEE International Conference on Conduction and Breakdown in Solid Dielectrics*, 528.

23 Aglen, O. and Andersson, A. (2003). Thermal analysis of a high-speed generator. In: *38th IEEE IAS Annual Meeting on Conference Record of the Industry Applications Conference*, vol. 1, 547–554.

24 Al-Mosawi, M.K., Beduz, C., and Yang, Y. (2005). Construction of a 100 kVA high temperature superconducting synchronous generator. *IEEE Trans. on Applied Superconductivity* 15 (2): 2182–2185.

25 Bakie, E., Johnson, B., Hess, H., and Law, J. (2005). Analysis of synchronous generator internal insulation failures. In: *IEEE International Conference on Electric Machines and Drives*, 106–110.

26 Ballard, W. R. (1982). Directly cooled, rotating rectifier assembly for a synchronous machine, Google Patents.

27 Baojun, G., Yanping, L., Chuiyou, Z. et al. (2004). Powerformer – A nascent electric power generation equipment for the 21st century. *Automation of Electric Power Systems* 28 (7): 1–4.

28 Berry, P.J. and Hamdi, E.S. (2016). Monitoring the stator winding insulation condition of a large synchronous motor. In: *2016 51st International Universities Power Engineering Conference (UPEC)*, 1–6.

29 Demello, F.P. and Concordia, C. (1969). Concepts of synchronous machine stability as affected by excitation control. *IEEE Trans. on Power Apparatus and Systems* PAS-88 (4): 316–329.

30 Di Tommaso, A.O., Miceli, R., Galluzzo, G.R., and Trapanese, M. (2007, 2007). Efficiency maximization of permanent magnet synchronous generators coupled coupled to wind turbines. *IEEE Power Electronics Specialists Conference* 1267–1272.

31 Ghahremani, E. and Kamwa, I. (2011). Online state estimation of a synchronous generator using unscented Kalman filter from phasor measurements units. *IEEE Trans. on Energy Conversion* 26 (4): 1099–1108.

32 Ghahremani, E., Karrari, M., and Malik, O. (2008). Synchronous generator third-order model parameter estimation using online experimental data. *IET Generation, Transmission & Distribution* 2 (5): 708–719.

33 Ghomi, M. and Sarem, Y.N. (2007). Review of synchronous generator parameters estimation and model identification. In: *2007 42nd International Universities Power Engineering Conference*, 228–235.

34 Grauers, A. (1994). Synchronous generator and frequency converter in wind turbine applications: System design and efficiency. In: *Project for a degree of Licentiate of Engineering, Department of Electrical Machines and Power Electronics, Chalmers University of Technology*. Göteborg, Sweden.

35 Grauers, A. (1996). Efficiency of three wind energy generator systems. *IEEE Trans. on Energy Conversion* 11 (3): 650–657.

36 Han, Y., Xie, X., and Oui, W. (2001). Status quo and future trend in research on synchronous generator excitation control. *Journal of Tsinghum University* 41 (4/5): 142–146.

37 Yang, Y.-M. and Chen, X.-J. (2014). Partial discharge ultrasonic analysis for generator stator windings. *Journal of Electrical Engineering & Technology* 9 (2): 670–676.

38 Karrari, M. and Malik, O. (2004). Identification of physical parameters of a synchronous generator from online measurements. *IEEE Trans. on Energy Conversion* 19 (2): 407–415.

39 Leijon, M. (1999). Novel concept in high voltage generation: Powerformer. In: *IEE Conference Publication*, vol. 5, no. 467, 5.379. London, UK: IET.

40 Kostenko, M. and Piotrovsky, L. (1969). *Electrical Machines – Alternating Current Machines*. II, Moscow: Mir Publishers.

41 Ames, R.L. (1990). *AC generators: Design and Application*. New York: Wiley.

42 Leijon, M., Owman, F., Sorqvist, T. et al. (1999). Powerformer (R): A giant step in power plant engineering. In: *IEEE International Electric Machines and Drives Conference*, IEMDC'99, Proceedings (Cat. No. 99EX272), 830–832.

43 Rahim, A. and Mohammad, A. (1994). Improvement of synchronous generator damping through superconducting magnetic energy storage systems. *IEEE Trans. on Energy Conversion* 9 (4): 736–742.

44 Ye, L., Lin, L., and Juengst, K.-P. (2002). Application studies of superconducting fault current limiters in electric power systems. *IEEE Trans. on Applied Superconductivity* 12 (1): 900–903.

45 Saban, D.M. and Hoobler, J. (2007). Rotor cooling system for synchronous machines with conductive sleeve. *Google Patents*.

46 Miller, G. (1998). Trends in insulation materials and processes for rotating machines. *IEEE Electrical Insulation Magazine* 14 (5): 7–11.

47 Nilsson, N. and Mercurio, J. (1994). Synchronous generator capability curve testing and evaluation. *IEEE Trans. on Power Delivery* 9 (1): 414–424.

48 Satoh, Y. and Kawamura, H. (2000). Synchronous generator, Google Patents.

49 Seguchi, M., Yoneda, S., and Imani, Y. (2002). Compact and reliable structure of multi-rotor synchronous machine. *Google Patents*.

50 Lin, X.-N. and Tian, Q. (2005). Review of the design and operation of Powerformer. *Electrical Power Systems* 29 (5): 1–5.

51 Shafaie, R., Kalantar, M., and Gholami, A. (2014). Thermal analysis of a 10-MW-class wind turbine HTS synchronous generator. *IEEE Trans. on Applied Superconductivity* 24 (2): 90–98.

52 Smith, J. (1983). Overview of the development of superconducting synchronous generators. *IEEE Trans. on Magnetics* 19 (3): 522–528.

53 Stone, G.C., Boulter, E.A., Culbert, I., and Dhirani, H. (2004). *Electrical Insulation for Rotating Machines: Design Evaluation, Aging, Testing, and Repair*. USA: Wiley.

54 Subramaniam, P. and Malik, O. (1971). Digital simulation of a synchronous generator in direct-phase quantities. *Proceedings of the Institution of Electrical Engineers* 118 (1): 153–160.

55 Timperley, J. and Michalec, J. (1994). Estimating the remaining service life of asphalt-mica stator insulation. *IEEE Trans. on Energy Conversion* 9 (4): 686–694.

56 Lipo, T.A. (2012). *Analysis of Synchronous Machines, Boca Raton*. USA: CRC Press.

57 Traxler-Samek, G., Zickermann, R., and Schwery, A. (2009). Cooling airflow, losses, and temperatures in large air-cooled synchronous machines. *IEEE Trans. on Industrial Electronics* 57 (1): 172–180.

58 Wallace, R.R., Lipo, T.A., Morán, L.A., and Tapia, J.A. (1997). Design and construction of a permanent magnet axial flux synchronous generator. In: *1997 IEEE International Electric Machines and Drives Conference RecordMA1/4.3*, MA1/4.1.

59 Wang, T. and Wang, Q. (2012). Optimization design of a permanent magnet synchronous generator for a potential energy recovery system. *IEEE Trans. on Energy Conversion* 27 (4): 856–863.

60 Westlake, A., Bumby, J., and Spooner, E. (1996). Damping the power-angle oscillations of a permanent-magnet synchronous generator with particular reference to wind turbine applications. *IEE Proceedings – Electric Power Applications* 143 (3): 269–280.

61 Wu, C.J. and Lee, Y.-S. (1991). Application of superconducting magnetic energy storage unit to improve the damping of synchronous generator. *IEEE Trans. on Energy Conversion* 6 (4): 573–578.

62 Zhang, Z., Yan, Y., Yang, S., and Bo, Z. (2008). Principle of operation and feature investigation of a new topology of hybrid excitation synchronous machine. *IEEE Trans. on Magnetics* 44 (9): 2174–2180.

63 Machowski, J., Bialek, J., Robak, S., and Bumby, J. (1998). Excitation control system for use with synchronous generators. *IEE Proceedings-Generation, Transmission and Distribution* 145 (5): 537–546.

64 Hooshyar, H., Savaghebi, M., and Vahedi, A. (2007). Synchronous generator: past, present and future. In: *AFRICON Conference*, 1–7. Windhoek, South Africa.

3

Transformed Models and Parameter Identification of Synchronous Generators

3.1 Introduction

A precise description of any power system dynamic performance requires accurate modeling of synchronous generators. An accurate model requires that the parameters of the generator should be known, but manufacturers of synchronous generators do not share enough information with the designers and users to achieve high performance. Generally, some information, including the rated output power, rated/maximum speed, rated output voltage, rated current, insulation class, and operating temperature, is given, but some data, such as dimensions, weight, and variations of parameters under different operation conditions are not shared.

The available parameters of the synchronous generators vary non-linearly, depending on the environment temperature and operating conditions. The high temperature leads to a higher stator resistance, while leading to a lower output current. The magnetic field flux-linkage change is reflected in the values of the dq-axis inductances, which decrease under high load conditions and increase the magnetic saturation level. The generator parameters also change due to aging because they depend on the material properties of the generator windings, which undergo changes in their physical characteristics as they age. Major changes in machine parameters also often occur after a repair. Rewinding the rotor of a synchronous generator would alter the field resistance from its originally designed value. Thus, accurate knowledge of these parameters is required to achieve precise and robust control of the working generator.

For this reason, parameter estimation in a synchronous generator is essential to ensure the accuracy of those parameters and to enhance confidence in the interpretation of performance results. The high-performance control of synchronous generators also requires accurate knowledge of their parameters, including direct d-axis inductance (L_d), quadrature q-axis inductance (L_q), stator resistance, rotor resistance, core-loss equivalent resistance, rotor inertia, and a viscous friction

Electromagnetic Analysis and Condition Monitoring of Synchronous Generators, First Edition. Hossein Ehya and Jawad Faiz.
© 2023 The Institute of Electrical and Electronics Engineers, Inc. Published 2023 by John Wiley & Sons, Inc.

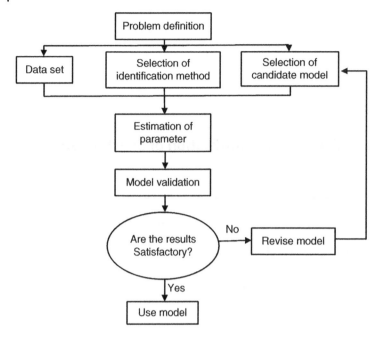

Figure 3.1 Parameter identification procedure for a synchronous machine.

coefficient. With this information, the generators can be precisely analyzed in both healthy and faulty cases, and the analytical results can be used in fault diagnosis and condition monitoring. Figure 3.1 shows the identification procedure for parameter estimation after determining the generator model [1].

Generally, basic equations for the synchronous generator can be employed to develop its detailed model. The most important equations for developing synchronous generator models are Park's equations, *dq*-axis equivalent circuits, and phase variables, including voltage, current, and flux-linkage [2, 3].

In the modeling stage, some approximations are considered, and some secondary effects are ignored. The symmetrical three-phase sinusoidally distributed stator winding with a Δ/Y connection is frequently modeled with approaches that may neglect core losses, ignore winding stray capacitances, represent the distributed windings as a concentrated winding, consider sinusoidal variations of the stator winding inductances with the rotor angular position (no harmonics), and examine linear magnetization characteristics as part of the modeling. In the synchronous generator model, the differential equations of multi-phase balanced voltage windings, inductance matrices, torque equations, and motion equations are considered. Multi-phase synchronous generators can have 1, 2, 3, 5, 6, 7, ... phases. Three-phase synchronous generators are widely used in electrical power systems, as are single-phase, two-phase, and six-phase generators, while other versions are seldom employed.

The angle between any phase magnetic axis and the field axis of the synchronous generator is time dependent. The *d*-axis and *q*-axis rotate, leading to a periodic variation in the air gap permeance. Consequently, the phase self-inductance and the mutual inductance between the phases vary periodically. Despite this variance, only the fundamental component is considered for modeling.

3.2 Multi-Phase Synchronous Generator Modeling Based on Park Equations

Multi-phase synchronous generator modeling has received less attention in the relevant literature and only a few models can be found. Generally, the number of phases of generators can be between 3 and 18 and are considered in a symmetric manner [4, 5]. However, three-phase synchronous generators are employed worldwide for the transmission and distribution of electrical energy. The generators are analyzed by a number of software programs developed using different models. Notably, the models based on the phase domain are very complicated and their solution takes a substantial amount of computation time.

Figure 3.2 shows an ideal symmetrical multi-phase synchronous generator. The coupled electric circuit approach can be employed to model a more general analysis of multi-phase synchronous generators. To this end, an idealized Park transformation or a *dq* reference frame, particularly in the rotor frame, has been used in which the magneto-motive force (MMF) and permeance have a sinusoidal distribution [2, 3]. A multi-phase synchronous generator consists of stator windings (armature) and a field winding on the rotor. Damping of the speed oscillations in the generators is accomplished with a set of dampers (copper or brass bars). In a multi-phase synchronous generator with the N phase, the stator winding magnetic axes are displaced by the electrical angle of $\delta = 2\pi/N$. The stator N-phase windings with δ displacement are: $as, bs, \ldots, ns$. The dampers are also represented on the d-axes and q-axes of the rotor (kdr, kqr) and in the field winding on the d-axis of the rotor.

The following assumptions are made to develop the voltage and flux-linkage equations.

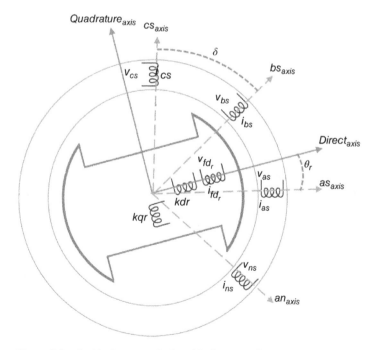

Figure 3.2 An ideal symmetrical multi-phase synchronous generator.

- A multi-pole synchronous generator is replaced by an equivalent two-pole generator.
- Stator windings generate a sinusoidal MMF and permeance is sinusoidally distributed, neglecting space harmonics.
- Stator windings resistances and leakage inductances of phase are equal.

The voltages, currents, and flux-linkage matrices (bold letters are matrices) of the synchronous generator can be written as follows:

$$\begin{bmatrix} v_{abc...n_s} \\ v_{dq_r} \end{bmatrix} = \begin{bmatrix} r_s & 0 \\ 0 & r_r \end{bmatrix} \begin{bmatrix} i_{abc...n_s} \\ v_{dq_r} \end{bmatrix} + \frac{d}{dt} \begin{bmatrix} \psi_{abc...n_s} \\ \psi_{dq_r} \end{bmatrix} \tag{3.1}$$

$$\begin{bmatrix} \psi_{abc...n_s} \\ \psi_{dq_r} \end{bmatrix} = \begin{bmatrix} L_{ss} & L_{sr} \\ L_{sr}^T & L_{rr} \end{bmatrix} \begin{bmatrix} i_{abc...n_s} \\ v_{dq_r} \end{bmatrix} \tag{3.2}$$

where

$$\boldsymbol{f}_{abc...n_s}^T = (f_{as} \ f_{bs} \ f_{cs} \cdots f_{ns}) \tag{3.3}$$

$$\boldsymbol{f}_{dq_r}^T = (f_{dr_1} \ f_{dr_2} \cdots f_{qr_1} \ f_{qr_2} \cdots) \tag{3.4}$$

where f represents v, I, Ψ, and L_{ss}, and L_{rr} are the stator self-inductance matrix and rotor self-inductance matrix, respectively, and L_{sr} is the mutual inductance between the stator and rotor windings. All depend on the saturation and rotor angular position θ_r. Synchronous generator data in the form of equivalent circuits for rotor d-axis and q-axis are available. Therefore, the generator can be analyzed using the dq frame reference. The stator inductance matrix is as follows:

$$L_{ss} = L_{ssl} + L_{ssoh} + L_{ss2h} \tag{3.5}$$

where $[L_{ssoh}]$ and $[L_{ss2h}]$ are the symmetrical component transformations to the magnetizing part of the stator self-inductance matrix, as follows:

$$\left[L_{ssl} \right]_{ik} = \begin{Bmatrix} L_{ls} & i = k \\ 0 & i \neq k \end{Bmatrix} \tag{3.6}$$

$$\left[L_{ssoh} \right]_{ik} = \left(3 \frac{2}{N} \right) M_{0s} \cos \left[(k - i)\delta \right] \tag{3.7}$$

$$\left[L_{ss2h} \right]_{ik} = \left(\frac{2}{N} \right) L_{2s} \cos \left[2\theta_r - (i + k - 2)\delta \right] \tag{3.8}$$

$$i, k \in 1, 2, 3, \ldots, N$$

By applying $[L_{ssoh}]$ and $[L_{ss2h}]$, diagonal matrices with non-zero elements only on the positions corresponding to positive and negative sequences are obtained for symmetrical multi-phase synchronous generators. From the sequence domain, differential equations of the generator are brought up to the two rotor dq components.

The $dq0$ equivalent circuits with zero-sequence circuits of w, u, and v are shown in Figure 3.3. The $dq0$ components correspond to the $(+)$, $(−)$, and (0) sequences in the symmetrical component frame. The w, u, and v components correspond to other sequences in the symmetrical component frame of the multi-phase (>3) generator. The w-axis only exists when the number of phases is even. Depending on the number of phases, additional u- and v-zero-sequence circuits may exist. The u and v components always appear in pairs.

Symmetrical multi-phase generators can now be modeled with an arbitrary number of phases, as the differential equations of the zero-sequence circuits can be solved independently. The dq equivalent circuits are also identical to the ones for three-phase machines. The generator parameters and zero sequences can be determined by appropriate tests.

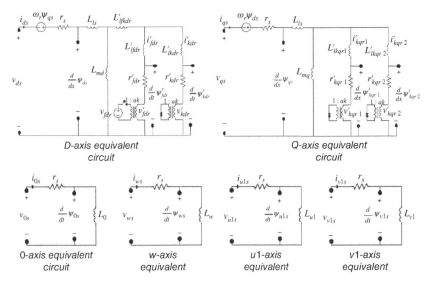

Figure 3.3 Equivalent circuit of an idealized multi-phase synchronous generator (with opposite current direction) [6] / A. B. Dehkordi (RTDS Technologies Inc).

3.2.1 Two-Phase Synchronous Generators

Generally, a single-phase synchronous generator is seldom available on the market. To supply a single-phase load, two phases of a three-phase Y-connected synchronous generator can be used. Note that a single-phase load generates magnetic flux pulsations in generators. This leads to many harmonic components of magnetic flux and causes rising core losses, while generating many harmonics-induced EMFs. If two similar windings with 90 electrical degrees are fixed on the armature (stator), a two-phase synchronous generator can be built, as shown in Figure 3.4. The rotor consists of the field winding fd and short-circuited damper bars of kd and kq. When the generator has a high number of poles and damper bars, a laminated salient-pole rotor or solid rotor can be used. Solid rotor synchronous generators normally have two or four poles. The dampers are assumed to be sinusoidally distributed windings with 90° displacements. The magnetic axes of fd and kd are identical. The kq damper is 90° ahead of the fd magnetic axis. More damper bars

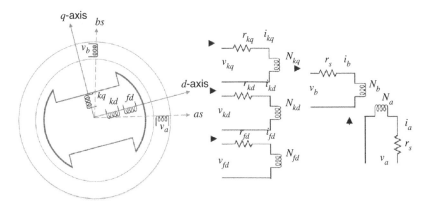

Figure 3.4 The dq model of a two-phase synchronous generator.

in the d and q axes can be considered to achieve a more precise solution. The φ_s and φ_r are the angular positions of the stator and rotor, respectively. The θ_r is the rotor electrical angular speed with respect to the q axis and $\omega_r = d\theta_r/dt$ is the rotor angular speed and $\varphi_s = \varphi_r + \theta_r$. In the salient-pole synchronous generator (SPSG), the stator is symmetrical, the rotor is asymmetrical, and the air gap is non-uniform. Therefore $L_d \neq L_q$.

The balanced steady-state stator currents cause MMF in the air gap (MMF_s) rotating at ω_e about the air gap. A direct current i_{fd} passes the fd winding, causing MMF_r fixed with respect to the rotor. However, MMF_r must rotate at ω_e, which means that $\omega_e = -\omega_r$. The voltage equations of the two-phase synchronous generator, as and bs are as follows:

$$v_{as} = r_s i_{as} + \frac{d\lambda_{as}}{dt} \tag{3.9}$$

$$v_{bs} = r_s i_{bs} + \frac{d\lambda_{bs}}{dt} \tag{3.10}$$

Their matrix form is:

$$\boldsymbol{v}_{abs} + \boldsymbol{r}_s \boldsymbol{i}_{abs} + \boldsymbol{p}\boldsymbol{\lambda}_{abs} \tag{3.11}$$

The voltage equations of the two damper bars are

$$v_{kq} = r_{kq} i_{kq} + \frac{d\lambda_{bkqs}}{dt} \tag{3.12}$$

$$v_{kd} = r_s i_{kd} + \frac{d\lambda_{kd}}{dt} \tag{3.13}$$

Their matric form is

$$\boldsymbol{v}_{qdr} + \boldsymbol{r}_r \boldsymbol{i}_{aqdrbs} + \boldsymbol{p}\boldsymbol{\lambda}_{qdr} \tag{3.14}$$

The voltage equation of the field winding is as follows:

$$v_{fd} = r_s i_{fd} + \frac{d\lambda_{fd}}{dt} \tag{3.15}$$

The resistance matrices are:

$$\boldsymbol{r}_s = \begin{bmatrix} r_s & 0 \\ 0 & r_s \end{bmatrix} \tag{3.16}$$

$$\boldsymbol{r}_r = \begin{bmatrix} r_{kq} & 0 & 0 \\ 0 & r_{fd} & 0 \\ 0 & 0 & r_{kd} \end{bmatrix} \tag{3.17}$$

The flux-linkage equations for two phases, damper bars, and field winding are as follows:

$$\lambda_{as} = L_{asas} i_{as} + L_{asbs} i_{bs} + L_{askq} i_{kq} + L_{asfd} i_{fd} + L_{askd} i_{kd} \tag{3.18}$$

$$\lambda_{bs} = L_{bsas} i_{as} + L_{bsbs} i_{bs} + L_{askq} i_{kq} + L_{bsfd} i_{fd} + L_{bskd} i_{kd} \tag{3.19}$$

$$\lambda_{kq} = L_{kqas} i_{as} + L_{kqbs} i_{bs} + L_{kqkq} i_{kq} + L_{kqfd} i_{fd} + L_{kqkd} i_{kd} \tag{3.20}$$

$$\lambda_{fd} = L_{fdas} i_{as} + L_{fdbs} i_{bs} + L_{fdkq} i_{kq} + L_{fdfd} i_{fd} + L_{fdkd} i_{kd} \tag{3.21}$$

$$\lambda_{kd} = L_{kdas} i_{as} + L_{kdbs} i_{bs} + L_{kdkq} i_{kq} + L_{kdfd} i_{qf} + L_{kdkd} i_{kd} \tag{3.22}$$

In a salient-pole generator, the stator winding self-inductances (L_{asas}) and mutual inductance between the stator windings depend on θ_r. At $\theta_r = 0$ and $180°$, $L_{asas} = L_{ls} + L_{mq}$, where L_{ls} is the

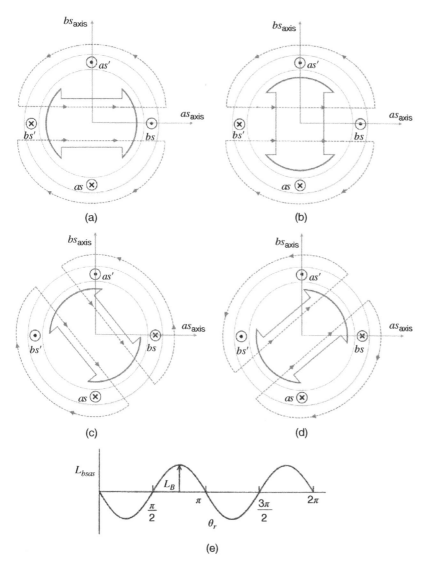

Figure 3.5 Winding *as* flux path indicating mutual coupling between stator windings to obtain L_{asas} and L_{asbs}.

stator winding leakage inductance and $L_{mq} = N_s^2 / R_{mq}$ (R_{mq} is the *q*-axis equivalent reluctance of the magnetic flux path). Figure 3.5 shows an flux path to indicate mutual coupling between stator windings. At $\theta_r = 90°$, the *d*-axis and the magnetic axis of the winding are aligned and $L_{asas} = L_{ls} + L_{md}$ (R_{md} is the *d*-axis equivalent reluctance of the magnetic flux path). L_{asas} in minimized at $\theta_r = 0$ and maximized at $\theta_r = 90$ and $270°$. The inductance varies about the positive mean value and for sinusoidal variation, it varies with $2\theta_r$. Suppose L_A is the mean value and L_B is the sinusoidal variation about the mean value; then $L_{mq} = L_A - L_B$ and $L_{md} = L_A + L_B$. By combining the above equations L_A and L_B can be obtained. For sinusoidal variation:

$$L_{asas} = L_{ls} + L_A - L_B \cos 2\theta_r \tag{3.23}$$

The mutual-inductance between stator winding a and winding b, L_{abas}, is

$$L_{asbs} = L_{bsas} = -L_B \sin 2\theta_r \tag{3.24}$$

The matrix form of the flux-linkage equations are as follows:

$$\begin{bmatrix} \lambda_{abs} \\ \lambda_{qdr} \end{bmatrix} = \begin{bmatrix} L_s & L_{sr} \\ (L_{sr})^T & L_r \end{bmatrix} \begin{bmatrix} i_{abs} \\ i_{dqr} \end{bmatrix} \tag{3.25}$$

where L_s, L_{sr}, and L_r are

$$L_s = \begin{bmatrix} L_{asas} & L_{asbs} \\ L_{bsas} & L_{bsbs} \end{bmatrix} = \begin{bmatrix} L_{ls} + L_A - L_B \cos 2\theta_r & -L_B \sin 2\theta_r \\ -L_B \sin 2\theta_r & L_{ls} + L_A - L_B \cos 2\theta_r \end{bmatrix} \tag{3.26}$$

$$L_{sr} = \begin{bmatrix} L_{askq} & L_{asfd} & L_{askd} \\ L_{bskq} & L_{bsfd} & L_{bskd} \end{bmatrix} = \begin{bmatrix} L_{skq} \cos\theta_r & L_{sfd} \sin\theta_r & L_{skd} \sin\theta_r \\ L_{skq} \sin\theta_r & -L_{sfd} \sin\theta_r & -L_{skd} \cos\theta_r \end{bmatrix} \tag{3.27}$$

$$L_r = \begin{bmatrix} L_{kqkq} & L_{kqfd} & L_{kqkd} \\ L_{fdkq} & L_{fdfd} & L_{fdkd} \\ L_{kdkq} & L_{kdfd} & L_{kdkd} \end{bmatrix} = \begin{bmatrix} L_{lkq} + L_{mkq} & 0 & 0 \\ 0 & L_{lfd} + L_{mfd} & L_{fdkd} \\ 0 & L_{fdkd} & L_{lkd} + L_{mkd} \end{bmatrix} \tag{3.28}$$

The inductances mentioned in the above matrices are as follows:

$$L_{skq} = \frac{N_{kq}}{N_s} L_{mq}$$

$$L_{sfd} = \frac{N_{fd}}{N_s} L_{md}$$

$$L_{skd} = \frac{N_{kd}}{N_s} L_{md}$$

$$L_{mkq} = \left(\frac{N_{kq}}{N_s} \right)^2 L_{mq}$$

$$L_{mfd} = \left(\frac{N_{fd}}{N_s} \right)^2 L_{md}$$

$$L_{mkd} = \left(\frac{N_{kd}}{N_s} \right)^2 L_{md}$$

$$L_{fdkd} = \frac{N_{kd}}{N_{fd}} L_{mfd} = \frac{N_{fd}}{N_{kd}} L_{mkd} \tag{3.29}$$

By referring the rotor variables to a N_s turn winding:

$$i'_j = \frac{N_j}{N_s} i_j, \quad v'_j = \frac{N_j}{N_s} v_j, \quad \lambda'_j = \frac{N_j}{N_s} \lambda_j, \qquad j = kq, fd \text{ or } kd \tag{3.30}$$

and the flux-linkage equations become

$$\begin{bmatrix} \lambda_{abs} \\ \lambda'_{qdr} \end{bmatrix} = \begin{bmatrix} L_s & L'_{sr} \\ (3L'_{sr})^T & L'_r \end{bmatrix} \begin{bmatrix} i_{abs} \\ i'_{dqr} \end{bmatrix} \tag{3.31}$$

where the matrices L'_{sr} and L'_r are as follows:

$$L'_{sr} = \begin{bmatrix} L_{mq} \cos\theta_r & L_{md} \sin\theta_r & L_{md} \sin\theta_r \\ L_{mq} \sin\theta_r & -L_{md} \cos\theta_r & -L_{md} \cos\theta_r \end{bmatrix} \tag{3.32}$$

$$\mathbf{L}_r' = \begin{bmatrix} L_{lkq}' + L_{mq} & 0 & 0 \\ 0 & L_{lfd}' + L_{md} & L_{md} \\ 0 & L_{md} & L_{lkd}' + L_{md} \end{bmatrix} \tag{3.33}$$

The voltage equations are

$$\mathbf{v}_{abs} = \mathbf{r}_s \mathbf{i}_{abs} + p\boldsymbol{\lambda}_{abs} \tag{3.34}$$

$$\mathbf{v}_{qdr}' = \mathbf{r}_s' \mathbf{i}_{qdr}' + p\boldsymbol{\lambda}_{qdr}' \tag{3.35}$$

$$\begin{bmatrix} \mathbf{v}_{abs} \\ \mathbf{v}_{qdr}' \end{bmatrix} = \begin{bmatrix} \mathbf{r}_s + p\mathbf{L}_s & p\mathbf{L}_{sr}' \\ p(\mathbf{L}_{sr}')^T & \mathbf{r}_r' + p\mathbf{L}_r' \end{bmatrix} \begin{bmatrix} \mathbf{i}_{abs} \\ \mathbf{i}_{dqr}' \end{bmatrix} \tag{3.36}$$

$$r_j' = \left(\frac{N_s}{N_j}\right)^2 r_j \tag{3.37}$$

$$L_{lj}' = \left(\frac{N_s}{N_j}\right)^2 L_{lj} \tag{3.38}$$

After some mathematical manipulations, the voltage equations in terms of currents in the rotor reference frame can be obtained as follows [7]:

$$\begin{bmatrix} v_{qs}^r \\ v_{ds}^r \\ v_{kq}'^r \\ v_{fd}'^r \\ v_{kd}'^r \end{bmatrix} = \begin{bmatrix} r_s + pL_q & \omega_r L_d & pL_{mq} & \omega_r L_{md} & \omega_r L_{md} \\ -\omega_r L_q & r_s + pL_d & -\omega_r L_{mq} & pL_{md} & pL_{md} \\ pL_{mq} & 0 & r_{kq}' + pL_{kq}' & 0 & 0 \\ 0 & pL_{md} & 0 & r_{fd}' + pL_{fd}' & pL_{md} \\ 0 & pL_{md} & 0 & pL_{md} & r_{kd}' + pL_{kd}' \end{bmatrix} \begin{bmatrix} i_{qs}^r \\ i_{ds}^r \\ i_{kq}'^r \\ i_{fd}'^r \\ i_{kd}'^r \end{bmatrix} \tag{3.39}$$

3.2.2 Three-Phase Synchronous Generators

The three-phase synchronous generator is a widely used generator, which is normally used in power plants. Three lap-wound (two-layer windings) of the generator are displaced 120 in electrical degrees. The voltage of the generator is between a low voltage of 400 V and a high voltage (HV) of 20 kV. Obviously, very high-quality insulation is required for these HV three-phase windings. Usually, Y-connected three-phase synchronous generators are preferred, because third-harmonic currents and their multiples may not circulate in the Δ-connected winding. In addition, its neutral point can be used for grounding. Multi-phase synchronous generators are recommended in some cases, such as lower per-phase current ratings, current ripples, torque pulsations, DC-link voltage ripples, and higher reliability applications [8].

Figure 3.6 shows the structure of a three-phase synchronous generator. It consists of a salient-pole rotor with field or excitation winding (fd), three-phase identical and symmetrical stator, or armature windings (a, b and c). Two lumped windings of the Q and D models have short-circuited damper bars in the DQ axes.

According to Park's transformation, the voltage (V), current (I), and flux-linkage (Ψ) on the three abc axes are projected on the dq-axes rotating at synchronous speed. If the rotor itself also rotates at the synchronous speed, Park's transformation can be applied to the three-phase synchronous

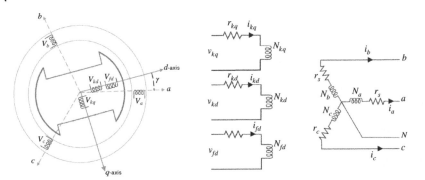

Figure 3.6 Schematic cross-section of a three-phase synchronous generator and its different windings and axes.

generator equations and reduce them to static phases of the synchronous generator [3]. As seen in Figure 3.6, the excitation winding axis is taken as the *d*-axis and the inter-polar axis as the *q*-axis. The angle between the phase a-axis and the *d*-axis is equal to γ. At time *t*, this angle is ωt, where ω is the angular speed of the rotor.

However, if the rotor of the synchronous generator does not rotate at the synchronous speed, the use of Park's transformation has no benefit. Since this transformation is standard, the parameters of the synchronous generator are given in the *dq*-axis by manufacturers.

Magnetic flux-linkages of different windings of synchronous generators depend on the self-inductances, mutual-inductances, and currents of the three-phase windings, and can be expressed in a matrix form, as follows:

$$
\begin{bmatrix} \psi_a \\ \psi_b \\ \psi_c \\ \psi_f \\ \psi_d \\ \psi_q \end{bmatrix} = \begin{bmatrix} L_{aa} & L_{ab} & L_{ac} & L_{af} & L_{ad} & L_{aq} \\ L_{ba} & L_{bb} & L_{bc} & L_{bf} & L_{bd} & L_{bq} \\ L_{ca} & L_{cb} & L_{cc} & L_{cf} & L_{cd} & L_{cq} \\ L_{fa} & L_{fb} & L_{fc} & L_{ff} & L_{fd} & L_{fq} \\ L_{da} & L_{db} & L_{dc} & L_{df} & L_{dd} & L_{dq} \\ L_{qa} & L_{qb} & L_{qc} & L_{qf} & L_{qd} & L_{qq} \end{bmatrix} \begin{bmatrix} i_a \\ i_b \\ i_c \\ i_f \\ i_d \\ i_q \end{bmatrix}
\tag{3.40}
$$

where ψ_x, i_x, and L_{xy} are the flux-linkage, the current, and the mutual-inductance between winding *x* and *y* (*x* and *y* = a, b, v, f, d, q). When *x* = *y*, it will be self-inductance. Most inductances in Equation (3.40) depend on ω. The stator self-inductances of the three phases are as follows:

$$L_{aa} = L_S + \Delta L_S \cos 2\gamma$$

$$L_{bb} = L_S + \Delta L_S \cos\left(2\gamma - \frac{2\pi}{3}\right) \tag{3.41}$$

$$L_{cc} = L_S + \Delta L_S \cos\left(2\gamma - \frac{4\pi}{3}\right)$$

where L_S is a constant value and ΔL_S is the second-order harmonic (this is zero for the round-rotor generator). The stator mutual-inductances are as follows:

$$L_{ab} = L_{ba} = -M_s - \Delta L_S \cos(2\gamma + \pi/3)$$

$$L_{bc} = L_{cb} = -M_s - \Delta L_S \cos(2\gamma + \pi) \tag{3.42}$$

$$L_{ca} = L_{ac} = -M_s - \Delta L_S \cos(2\gamma + 5\pi/3)$$

where M_s is the constant component. The mutual inductances between the stator and rotor are as follows:

$$L_{ai} = L_{ia} = M_i \cos(\gamma) \tag{3.43}$$

$$L_{bi} = L_{ib} = M_i \cos(\gamma - 2\pi/3) \tag{3.44}$$

$$L_{ci} = L_{ic} = M_i \cos(\gamma - 4\pi/3) \tag{3.45}$$

where $i = f,\ d,\ q$. Generally, the mutual inductances of the rotor in the d-axis and q-axis are $L_{fq} = L_{qf} = L_{dq} = L_{qd} = 0$. However, the self-inductances of the rotor are constant: $L_{dd} = L_d$, $L_{qq} = L_q$, and $L_{fd} = L_{df}$. By applying a linear transformation, the phasor in the stator reference frame (a, b, c) is transferred into the dq-axis of the rotor. Transforming the quantities from the dq frame of the rotor into the abc frame of the stator is possible if a zero-sequence is considered in the dq frame. In this case, it is called the $dq0$ transformation [2]:

$$\begin{bmatrix} i_d \\ i_q \\ i_0 \end{bmatrix} = \begin{bmatrix} \beta_d \cos\gamma & \beta_d \cos\left(\gamma - \dfrac{2\pi}{3}\right) & \beta_d \cos\left(\gamma - \dfrac{4\pi}{3}\right) \\ \beta_q \sin\gamma & \beta_q \sin\left(\gamma - \dfrac{2\pi}{3}\right) & \beta_q \sin\left(\gamma - \dfrac{3\pi}{3}\right) \\ \beta_0 & \beta_0 & \beta_0 \end{bmatrix} \begin{bmatrix} i_a \\ i_b \\ i_c \end{bmatrix} \tag{3.46}$$

If the reference frame is changed, β_d, β_q, and β_0 are non-zero arbitrary values. The relationship between i_{abc} and i_{dq0} is $i_{abc} = C^{-1} i_{dq0}$, where the matrix C is as follows:

$$C = \sqrt{\dfrac{2}{3}} \begin{bmatrix} -\cos\gamma & -\cos\left(\gamma - \dfrac{2\pi}{3}\right) & -\cos\left(\gamma - \dfrac{4\pi}{3}\right) \\ \sin\gamma & \sin\left(\gamma - \dfrac{2\pi}{3}\right) & \sin\left(\gamma - \dfrac{4\pi}{3}\right) \\ \sqrt{\dfrac{1}{2}} & \sqrt{\dfrac{1}{2}} & \sqrt{\dfrac{1}{2}} \end{bmatrix} \tag{3.47}$$

As already mentioned, if the rotating speed of the dq frame and the speed of the rotor are equal, the transformation is called the Park transformation. Considering that rotor currents, voltages, and flux-linkages are already in the (d, q) reference frame, no transformation is necessary for them. The flux-linkage and currents are related as follows:

$$\begin{bmatrix} \Psi_{dq0} \\ \Psi_{fdq} \end{bmatrix} = \begin{bmatrix} C & 0 \\ 0 & 1 \end{bmatrix} \begin{bmatrix} \Psi_{abc} \\ \Psi_{fdq} \end{bmatrix} \tag{3.48}$$

$$\begin{bmatrix} i_{dq0} \\ i_{fdq} \end{bmatrix} = \begin{bmatrix} C^{-1} & 0 \\ 0 & 1 \end{bmatrix} \begin{bmatrix} i_{abc} \\ i_{fdq} \end{bmatrix} \tag{3.49}$$

This can also be expressed in the following matrix form:

$$\begin{bmatrix} \psi_d \\ \psi_q \\ \psi_0 \\ \psi_f \\ \psi_D \\ \psi_Q \end{bmatrix} = \begin{bmatrix} L_d & 0 & 0 & kM_f & kM_d & 0 \\ 0 & L_q & 0 & 0 & 0 & kM_q \\ 0 & 0 & L_0 & 0 & 0 & 0 \\ kM_f & 0 & 0 & L_f & L_{fd} & 0 \\ kM_d & 0 & 0 & L_{fd} & L_d & 0 \\ 0 & kM_q & 0 & 0 & 0 & L_q \end{bmatrix} \begin{bmatrix} i_a \\ i_b \\ i_c \\ i_f \\ i_d \\ i_q \end{bmatrix} \tag{3.50}$$

The independent set of equations can be extracted from the matrix. This indicates the flux-linkages Ψ_d, Ψ_f, Ψ_D versus i_d, i_f, i_D and also Ψ_q, Ψ_Q versus i_q, i_Q. We have the following simple equation:

$$\Psi_0 = L_0 i_0 \tag{3.51}$$

Note that coils d, D, and f are placed on the d-axis and coils Q and q are on the quadrature axis q (90° shift). No coupling is evident between the coils. Note that the rotation speed of the dq reference axis in the synchronous generator is the same as the rotation speed of the rotor.

The matrix form of the voltage equation of the windings in Figure 3.6 can be determined as follows:

$$[v_a\, v_b\, v_c - v_f\, v_D\, v_Q]^T = [\mathbf{R}]\,[i_d\, i_q\, i_0 \quad i_f\, i_D\, i_Q]^T - d/dt\,[\Psi_d\, \Psi_q\, \Psi_0 \quad \Psi_f\, \Psi_D\, \Psi_Q]^T \tag{3.52}$$

where the matrix $[\mathbf{R}]$ is a diagonal matrix consisting of elements: R_a, R_b, R_c, R_f, R_D, R_Q. Also, shorted D and Q windings lead to $v_Q = v_D = 0$. If we add the speed-dependent matrix, then Equation (3.52) is completed as follows:

$$[v_a\, v_b\, v_c - v_f\, v_D\, v_Q]^T = [\mathbf{R}]\,[i_d\, i_q\, i_0\, i_f\, i_D\, i_Q]^T - d/dt\,[\Psi_d\, \Psi_q\, \Psi_0 \quad \Psi_f\, \Psi_D\, \Psi_Q]^T + \omega[-\Psi_q\, \Psi_q\ \ 0\ 0\ 0\ 0]^T \tag{3.53}$$

If the flux-linkage terms are replaced by the current terms, a more convenient form of Equation (3.52) can be obtained and easily used in synchronous generators:

$$[v_a\, v_b\, v_c - v_f\, 0\, 0]^T = -[\mathbf{A}]\,[i_d\, i_q\, i_0\, i_f\, i_D\, i_Q]^T - [\mathbf{B}]d/dt\,[i_d\, i_q\, i_0\, i_f\, i_D\, i_Q]^T \tag{3.54}$$

where

$$[A] = \begin{bmatrix} R_a & \omega L_q & 0 & 0 & 0 & \omega k M_Q \\ -\omega L_q & R_a & 0 & \omega k M_f & \omega k M_Q & 0 \\ 0 & 0 & R_c & 0 & 0 & 0 \\ 0 & 0 & 0 & R_f & 0 & 0 \\ 0 & 0 & 0 & 0 & R_D & 0 \\ 0 & 0 & 0 & 0 & 0 & R_Q \end{bmatrix} \tag{3.55}$$

$$[B] = \begin{bmatrix} L_d & 0 & 0 & k M_f & k M_D & 0 \\ 0 & L_q & 0 & 0 & 0 & k M_Q \\ 0 & 0 & L_0 & 0 & 0 & 0 \\ k M_f & 0 & 0 & L_f & L_{fD} & 0 \\ k M_D & 0 & 0 & L_{fD} & L_D & 0 \\ 0 & k M_Q & 0 & 0 & 0 & L_Q \end{bmatrix} \tag{3.56}$$

3.2.3 Six-Phase Synchronous Generators

Using more than three phases in synchronous generators is technically and economically viable. Six-phase synchronous generators were considered in the 1970s and 1980s, with applications predominantly related to uninterruptible power supply (UPS) systems. The six-phase synchronous generator has been employed to provide a single machine UPS. Six-phase synchronous generators may be used as a standalone electric energy supply in conjunction with the hydropower plant. These generators can operate with a single three-phase winding set so that a fault at one winding does not lead to system shutdown. The generator can also supply two separate three-phase loads, which represents an additional advantage. A comparative study indicated that the six-phase generator is able to deliver more power (higher power density) in the same frame. The voltage regulation is also better than its three-phase counterpart.

The behavior of a synchronous generator with two three-phase stator windings is displaced by an arbitrary angle employing an orthogonal transformation of the phase variables into a new set of *dq* variables [9]. These generators can be used in marine ships, thermal power plants (to drive induced draft fans), electric vehicles, nuclear power plants, etc. Compared to conventional three-phase generators, the six-phase synchronous generator shown in Figure 3.7 has many advantages, such as a reduced amplitude and increased frequency of torque pulsations, reduced rotor harmonic currents, reduced phase current (without increasing the phase voltage), lower DC-link current harmonics, increased power density, and improved reliability. The general feasibility of a six-phase synchronous generator has been reported in references [4] and [10]. The time-dependence coefficients of the differential equations of the generator model can be removed by *dq* transformation. Schiferl and Ong [11] have presented a six-phase synchronous generator mathematical model that can be obtained considering the mutual leakage couplings between the two sets of three-phase stator windings. The relationships of the mutual leakage inductances with a winding displacement angle and pitch for a six-phase winding configuration have been given in reference [12].

A six-phase alternator with its two three-phase stator windings displaced by an arbitrary angle was analyzed by Fuchs and Rosenberg. Orthogonal transformation of the phase variables into a new set of *dq* variables was used for this purpose. Examination of a six-phase synchronous machine (SM) operation showing its redundancy characteristics, behavior during fault conditions, the sensitivity of design parameters together with its operation when fed with a non-sinusoidal voltage source is briefly addressed [13].

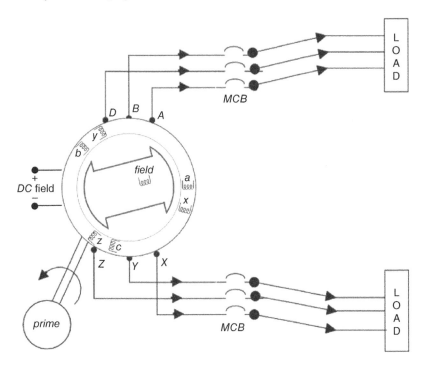

Figure 3.7 Schematic diagram of the six-phase synchronous generator system.

Three-phase synchronous generators can be replaced by six-phase AC generators to achieve better efficiency at partial load and high reliability under fault conditions, such as the most common open circuit and short circuit faults. The symmetrical component method is commonly used to analyze the unbalanced operation of synchronous generators under fault conditions. Although this method can be used in steady-state analysis of sinusoidal excitation, it loses its usefulness due to a lack of interaction between the disconnected phases and the remaining machine windings, which drastically alters the dynamic behavior of the generator [10]. An alternative approach is to use *dq*0 variables to simulate the open phase six-phase synchronous generator [14–19]. This approach eliminates the time-varying inductances in voltage equations, computational stability problems, and the need for matrix inversion [17]. Therefore, a detailed theoretical and experimental analysis of SPSG under the open circuit condition for different configurations is presented. A mathematical model can be developed for analyzing the synchronous generator behavior during the open circuit condition with single and double phases, using the *dq*0 approach.

Note that the short circuit is also a serious problem in the generator and leads to a very high current and generator outage if not cleared in a timely manner. Therefore, generator modeling for these conditions is essential, but it is a complex task [20, 21].

3.3 Mathematical Modeling

In a six-phase synchronous generator, two three-phase sets can supply separate loads. Two winding sets *abc* and *xyz* with 30° angular displacements result in an symmetrical winding, which substantially reduces low-order time harmonics (such as 5th, 7th, 17th, 19th, …) and space harmonics (no harmonics less than 11th order). The rotor has the field winding (*FR*) and the shorted damper bars (*KD*) along the *d*-axis and (*KQ*) along the *q*-axis. Figure 3.8a presents the stator and rotor axes schematically. Before deriving the generator equations, the following simplifying assumptions are applied [22]:

1. The *abc* and *xyz* are assumed symmetrical leading to air gap sinusoidal distribution.
2. The flux and MMFs are sinusoidal in space.
3. Saturation, hysteresis, and skin effects are ignored.

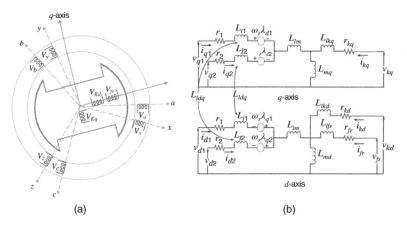

(a)   (b)

Figure 3.8 Schematic representation of a six-phase synchronous motor: (a) stator and rotor axes of a six-phase motor and (b) equivalent circuit of a six-phase synchronous motor.

The voltage and electromagnetic torque equations are written in generator variables, resulting in a set of nonlinear differential equations; this nonlinearity is attributed to the dependency of inductance on rotor position and time. All sets of generator equations are expressed in ($dq0$) variables or rotor reference frames [22, 23].

Voltage equations:

$$v_{Q1} = r_1 i_{Q1} + \frac{\omega_R}{\omega_B} \psi_{D1} + \frac{p}{\omega_B} \psi_{Q1} \tag{3.57}$$

$$v_{D1} = r_1 i_{D1} - \frac{\omega_R}{\omega_B} \psi_{Q1} + \frac{p}{\omega_B} \psi_{D1} \tag{3.58}$$

$$v_{Q2} = r_2 i_{Q2} + \frac{\omega_R}{\omega_B} \psi_{D2} + \frac{p}{\omega_B} \psi_{Q2} \tag{3.59}$$

$$v_{D2} = r_2 i_{D2} - \frac{\omega_R}{\omega_B} \psi_{Q2} + \frac{p}{\omega_B} \psi_{D2} \tag{3.60}$$

$$v_{KQ} = r_{KQ} i_{KQ} + \frac{p}{\omega_B} \psi_{KQ} \tag{3.61}$$

$$v_{KD} = r_{KD} i_{KD} + \frac{p}{\omega_B} \psi_{KD} \tag{3.62}$$

$$v_{FR} = \frac{x_{MD}}{r_{FR}} \left[r_{FR} i_{FR} + \frac{p}{\omega_B} \psi_{FR} \right] \tag{3.63}$$

where v, i, and Ψ are voltage, current, and flux-linkage, respectively. Their different indexes are for various windings, as defined above, where ω_B is the base electrical angular speed and p is the differentiation function with respect to time.

The voltage equations can be expressed in flux-linkage per second, Ψ, as follows:

$$\psi_{Q1} = x_{L1} i_{Q1} + x_{LM}(i_{Q1} + i_{Q2}) - x_{LDQ} i_{D2} + \psi_{MQ} \tag{3.64}$$

$$\psi_{D1} = x_{L1} i_{D1} + x_{LM}(i_{D1} + i_{D2}) - x_{LDQ} i_{Q2} + \psi_{MD} \tag{3.65}$$

$$\psi_{Q2} = x_{L2} i_{Q2} + x_{LM}(i_{Q1} + i_{Q2}) - x_{LDQ} i_{D1} + \psi_{MQ} \tag{3.66}$$

$$\psi_{D2} = x_{L2} i_{D2} + x_{LM}(i_{D1} + i_{D2}) - x_{LDQ} i_{DQ1} + \psi_{MD} \tag{3.67}$$

$$\psi_{KQ} = x_{LKQ} i_{KQ} + \psi_{MQ} \tag{3.68}$$

$$\psi_{KD} = x_{LKD} i_{KD} + \psi_{MD} \tag{3.69}$$

$$\psi_{FR} = x_{LFR} i_{FR} + \psi_{MD} \tag{3.70}$$

where

$$\psi_{MQ} = x_{MQ}(i_{Q1} + i_{Q2} + i_{KQ}) \tag{3.71}$$

$$\psi_{MD} = x_{MD}(i_{D1} + i_{D2} + i_{KD} + i_{FR}) \tag{3.72}$$

An equivalent circuit of the generator can be drawn using the equations where leakage reactance x_{LM} and cross mutual coupling reactance x_{LDQ} between the d- and q-axis of the stator, respectively, are given by

$$x_{LM} = x_{Lax} \cos(\xi) + x_{Lay} \cos\left(\xi + \frac{2\pi}{3}\right) + x_{Laz} \cos\left(\xi - \frac{2\pi}{3}\right) \tag{3.73}$$

$$x_{LDQ} = x_{Lax} \sin(\xi) + x_{Lay} \sin\left(\xi + \frac{2\pi}{3}\right) + x_{Laz} \sin\left(\xi - \frac{2\pi}{3}\right) \tag{3.74}$$

The leakage flux leads to mutual coupling between stator windings *abc* and *xyz*. This flux occupies the same stator slots and its related inductance is denoted by x_{LM}. It affects the harmonic coupling of the *abc* and *xyz* sets of windings. The x_{LM} depends on the winding pitch and displacement angle between the windings *abc* and *xyz*. An elaborated technique to find the slot reactance is given in reference [24]. Standard test procedures for determining generator parameters are available in references [24] and [25]. It is often convenient to express the voltage Equations (3.57) to (3.63) and flux linkages Equations (3.64) to (3.72) in terms of reactance rather than inductance. Equations are first solved for currents, and then back-substituted into the voltage equations. These mathematical manipulations yield the following integral equations:

$$\psi_{Q1} = \frac{\omega_B}{p}\left\{v_{Q1} - \frac{\omega_R}{\omega_B}\psi_{D1} - \frac{r_1}{x_B}[(x_{L2} + x_{LM})\psi_{Q1} - x_{LM}\psi_{Q2} - x_{L2}\psi_{MQ} + x_{LDQ}(\psi_{D2} - \psi_{MD})]\right\}$$

(3.75)

$$\psi_{D1} = \frac{\omega_B}{p}\left\{v_{D1} - \frac{\omega_R}{\omega_B}\psi_{Q1} - \frac{r_1}{x_B}[(x_{L2} + x_{LM})\psi_{D1} - x_{LM}\psi_{D2} - x_{L2}\psi_{MD} + x_{LDQ}(\psi_{Q2} - \psi_{MQ})]\right\}$$

(3.76)

$$\psi_{Q2} = \frac{\omega_B}{p}\left\{v_{Q2} - \frac{\omega_R}{\omega_B}\psi_{D2} - \frac{r_1}{x_B}[(x_{L2} + x_{LM})\psi_{Q2} - x_{LM}\psi_{Q1} - x_{L1}\psi_{MQ} + x_{LDQ}(\psi_{D1} - \psi_{MD})]\right\}$$

(3.77)

$$\psi_{D2} = \frac{\omega_B}{p}\left\{v_{D2} - \frac{\omega_R}{\omega_B}\psi_{Q2} - \frac{r_1}{x_B}[(x_{L1} + x_{LM})\psi_{D2} - x_{LM}\psi_{D1} - x_{L1}\psi_{MD} + x_{LDQ}(\psi_{Q1} - \psi_{MQ})]\right\}$$

(3.78)

$$\psi_{KQ} = \left[V_{KQ} - \frac{r_{KQ}}{X_{LKQ}}(\psi_{KQ} - \psi_{MQ})\right]$$

(3.79)

$$\psi_{KD} = \left[V_{KD} - \frac{r_{KD}}{X_{LKD}}(\psi_{KD} - \psi_{MD})\right]$$

(3.80)

$$\psi_{FR} = \left[\frac{v_{FR}x_{MD}}{x_{ER}} - \frac{r_{FR}}{X_{LFR}}(\psi_{FR} - \psi_{MD})\right]$$

(3.81)

Current can be expressed in terms of flux as

$$i_{Q1} = \frac{1}{x_B}[(x_{L2} + x_{LM})\psi_{Q1} - x_{LM}\psi_{Q2} - x_{L2}\psi_{MQ} + x_{LDQ}(\psi_{D2} - \psi_{MD})]$$

(3.82)

$$i_{D1} = \frac{1}{x_B}[(x_{L2} + x_{LM})\psi_{D1} - x_{LM}\psi_{D2} - x_{L2}\psi_{MD} + x_{LDQ}(\psi_{Q2} - \psi_{MD})]$$

(3.83)

$$i_{Q2} = \frac{1}{x_B}[(x_{L1} + x_{LM})\psi_{Q2} - x_{LM}\psi_{Q1} - x_{L1}\psi_{MQ} + x_{LDQ}(\psi_{D1} - \psi_{MD})]$$

(3.84)

$$i_{D2} = \frac{1}{x_B}[(x_{L1} + x_{LM})\psi_{D2} - x_{LM}\psi_{D1} - x_{L1}\psi_{MD} + x_{LDQ}(\psi_{Q1} - \psi_{MD})]$$

(3.85)

$$i_{KQ} = \frac{1}{x_{LKQ}}(\psi_{KQ} - \psi_{MQ})$$

(3.86)

$$i_{KD} = \frac{1}{x_{LKQ}}(\psi_{KD} - \psi_{MD})$$

(3.87)

$$i_{FR} = \frac{1}{x_{LFR}}(\psi_{FR} - \psi_{MD})$$

(3.88)

$$\psi_{MQ} = X_{AQ} \left[\frac{x_{L2}\psi_{Q1} + x_{L1}\psi_{Q2} + x_{LDQ}(\psi_{D2} - \psi_{D1})}{x_b} + \frac{\psi_{KQ}}{x_{LKQ}} \right] \tag{3.89}$$

$$\psi_{MD} = X_{AD} \left[\frac{x_{L2}\psi_{D1} + x_{L1}\psi_{D2} + x_{LDQ}(\psi_{Q1} - \psi_{Q2})}{x_b} + \frac{\psi_{KD}}{x_{LKQ}} + \frac{\psi_{FR}}{x_{LFR}} \right] \tag{3.90}$$

where

$$x_{AQ} = \frac{1}{\left[\dfrac{1}{x_{MQ}} + \dfrac{1}{x_{LKQ}} + \dfrac{x_{L1} + x_{L2}}{x_B} \right]} \tag{3.91}$$

$$x_{AD} = \frac{1}{\left[\dfrac{1}{x_{MD}} + \dfrac{1}{x_{LKD}} + \dfrac{1}{x_{LFR}} + \dfrac{x_{L1} + x_{L2}}{x_B} \right]} \tag{3.92}$$

$$x_B = x_{L1}x_{L2} + (x_{L1} + x_{L2})x_{LM} - x_{LDQ}^2 \tag{3.93}$$

3.3.1 Optimal Observer with Kalman Filters

Steady–state estimation (SSE) methods assume that the electrical power network operates in a quasi-steady state [26–29]. The voltage amplitude and phase angles of the buses can then be estimated by synchro-phasor measurements. SSE provides inputs of the energy management system (EMS) applications, such as automatic generation control and optimal power for monitoring purposes. In most energy resources, the dynamic state estimation (DSE) using high sampling rate synchronized measurements is required for wide-area monitoring, protection, and power system control.

Practical application of SSE and DSE is not easy. The reasons include:

1. Inaccuracy of the estimation model and parameters.
2. Possible cyberattacks on measurements necessary for estimation [30].

However, an attempt has been made to validate the dynamic model of the generator and its parameter calibration [31, 32]. If the estimator is made more robust to model uncertainty, the inaccuracy may be improved. False data injection attacks against SSE have been suggested in [33]. Mitigating the attack and securing monitoring and control of power networks have been proposed in references [34–36].

The approaches for performing DSE fall into two classes of methods that have been proposed:

1. Stochastic estimators: Given a discrete-time representation of a dynamic system, the observed measurements, and the statistical information on process noise and measurement noise, the Kalman filter (KF) and its many derivatives have been proposed that calculate the Kalman gain as a function of the relative certainty of the current state estimate and the measurements.
2. Deterministic observers: Given a continuous-time or discrete-time dynamic system depicted by state-space matrices, a combination of matrix equalities and inequalities are solved while guaranteeing asymptotic (or bounded) estimation error. The solution to these equations is often matrices that are used in an observer to estimate states and other dynamic quantities.

3.4 Parameter Estimation Algorithms

A valid model for synchronous generators is essential for a reliable analysis of stability and dynamic performance. To have a valid model, the parameters of the machine must be specified.

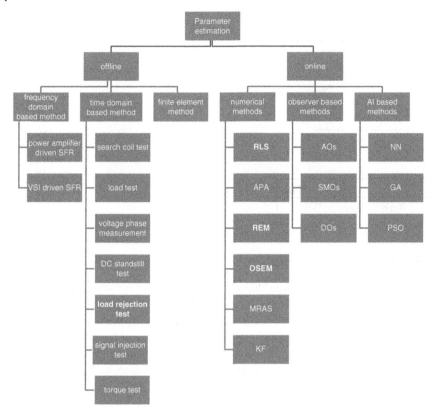

Figure 3.9 Classification of parameter estimation techniques.

Numerous parameter estimation techniques have been proposed and these are typically offline or online parameter estimation methods. The offline estimation methods are based on the calculation and formulation of frequency-domain and time-domain methods. These techniques include experimental tests, conducted in the laboratory either at standstill [37] or at a constant speed, to characterize the parameters. The numerical simulation methods, particularly finite element analysis (FEA) [38], are considered offline parameter estimation techniques. The online estimation methods are based on formulas as observer-based methods and artificial intelligence (AI)-based methods. These techniques are applied while the generator is in operation, so the parameters are estimated without disturbing machine operation [39]. In offline cases, the collected input/output data are used for parameter estimation. Offline parameter estimation is generally conducted either by disconnecting the generator from its load or at standstill conditions. Conversely, the online parameter estimation is performed when the generator is operating with a load and the new input/output data are available during real-time operation. Figure 3.9 categorizes the offline and online parameter estimation methods based on the operating conditions.

3.4.1 Offline Parameter Estimation Techniques

The offline parameter identification methods are divided into frequency domain-based, time domain-based, and finite element-based methods.

3.4.1.1 Frequency Domain-Based Methods

The frequency domain-based methods predominantly utilize the standstill frequency response (SFR) test to identify parameters in the standstill state. Generally, the frequency domain-based methods are classified as the power amplifier-driven SFR and voltage source inverter (VSI)-driven SFR. The power amplifier-driven SFR method provides accurate equivalent circuits with a low risk to the generators during the test compared to the short circuit test and DC decay test. Traditionally, the SFR is conducted by sequential evaluation of the response of a test signal with a single frequency from the predefined set of signal frequencies. The expensive high-power linear amplifier can be utilized as a waveform generator for the excitation signal of the power amplifier-driven SFR. The VSI-driven SFR may also be employed as a waveform generator. VSI is able to generate excitation signals by chopping the DC bus voltage. The accuracy, measuring time, and sensitivity of the measurement process to the disturbing noise are the key factors to be considered when selecting the excitation signal. Generally, three types of applied signals are used: periodic signals, transient signals, and aperiodic signals. Two other excitation classes are deterministic and stochastic excitation signals.

The frequency domain-based methods require further exploration to minimize the power amplifier cost, reduce the nonlinearity effects of VSI, decrease the digital-time delay effect, estimate the accurate magnetic saturation, utilize the higher-order models, and apply AI-based estimation methods to the data collected from the SFR tests. The power amplifier-driven SFR and VSI-driven SFR are commonly used for the self-commissioning and condition monitoring of generators.

3.4.1.2 Time Domain-Based Methods

Time domain-based parameter estimation methods utilize the time responses of an imposed perturbation or excitation signal on the generators for parameter identification. Note that these methods are conducted at a standstill or constant speed under special laboratory conditions. The methods include the search coil test, load test, phase voltage measurement test, AC standstill test, load rejection test, signal injection test, and torque test. The time domain-based methods are generally applied for self-commissioning, condition monitoring, and fault diagnosis [40].

Search Coil Test: The search coil is fixed in the stator and the passing magnetic flux induces a voltage in it. Sometimes, many search coils are wound around the generator during the test to allow the maximum degrees of freedom [41].

Load Test: Compared with the short circuit and open circuit tests, the load test provides better parameter identification by supplying accurate torque angle measurements. In the load test, the generator runs at a constant speed and the load varies; the currents and voltages are measured to estimate the flux linkage. The rotor speed should be sufficiently high to minimize the effects of a resistive voltage drop but not too high to distort the stator currents due to the high EMF of the generator. Position error, rotor heating, and core loss are key factors that pollute the parameter estimation. For a high-speed flux that weakens the operating range, the generator is loaded by coupling it with DC machines as a separately excited generator [41].

Phase Voltage Measurement Test: The dq-axis inductances are determined by phase voltage measurement tests, which require an additional synchronous machine, a prime mover, and a brake [40].

DC Standstill Test: The DC-step voltage is a standstill experiment needed to generate the required data to estimate the fundamental parameters of the dq-axis equivalent circuits. This test is performed by applying DC voltage to the stator phases with a short circuit field winding and a fixed rotor. The parameters are estimated by considering an equivalent circuit and comparing its response with the response of the generator. If a discrepancy is observed, the parameters of the equivalent circuit are reset. Consequently, a suitable algorithm is needed to optimize the parameters [42].

Load Rejection Test: This test is similar to the sudden short circuit test [1]. In the latter test, the time responses of the generator variables following a sudden disturbance are measured to identify the generator characteristics. After certain initial conditions are set up, the response of the generator to a load rejection is recorded. The initial conditions determine the axis on which the parameters are derived. In this case, the operating conditions are arranged so that the current flows only in the direct axis ($i_q = 0$). The unit is tripped and the resulting terminal voltage and field current decay are used to extract parameters for the d-axis model. A similar test is performed with current flowing only in the quadrature axis ($i_d = 0$) to obtain q-axis data. These tests provide data for the direct and quadrature axes. However, they are somewhat difficult to conduct. For most generators, attaining unsaturated conditions is difficult or impossible, which certainly complicates the testing and analysis of the results [1].

Signal Injection Test: The phase inductance varies at different rotor positions due to the saliency present in the rotor. The position is extracted from the inductance profile by feeding a high-frequency voltage signal to the generator phases. This method is reliable at zero speed. However, the accuracy of this method is highly influenced by the geometry of the rotor; therefore, the method is not suitable for surface-mounted PM synchronous generators. The main drawbacks of this method are the adverse effect of the injecting signal on the generator dynamics and the extra hardware requirements. This method is used for sensorless schemes at standstill [43].

Torque Test: The torque test is conducted using the DC supply to estimate the torque constant, the d-axis stator inductances L_d, and the q-axis stator inductances L_q, and by measuring the machine torque with a torque sensor. To identify the whole map of the dq-axis flux-linkages, two sets of machine data at two different rotor speeds per each load point are collected and minimized by an immune clonal-based quantum genetic algorithm (GA) [42].

3.4.1.3 Finite Element Methods

FEM is a powerful simulation tool to determine and analyze the generator parameters. Generally, FEM requires accurate knowledge of the structure, geometry, and design to estimate the parameters. The popular FEM methods for parameter estimation include the stored magnetic energy, energy dual, vector magnetic potential, energy perturbation, and air gap flux density methods. Note that the estimated parameters consider the effects of mutual inductances and changes in magnetic flux due to magnetic saturation. The FEMs are helpful for generator designers to study the geometric complexity and high local magnetic saturation. The extracted parameters can be used to design the control system. FEM can also be used to estimate the generator parameters under fault conditions [42].

3.4.1.4 An Example of Offline Parameter Estimation Using the DC Standstill Test

The parameter estimation of a synchronous generator using the DC-step voltage is presented and a model is adopted that also takes into account the unequal mutual coupling between the d-axis damper and field windings for the q-axis [19]. Their equivalent circuits are indicated in Figure 3.10.

The two-axis voltage equations that describe the synchronous generator are given by

$$v_d = \frac{\mathrm{d}}{\mathrm{d}t}\psi_d + \omega\psi_q + r_a i_a$$

$$v_f = \frac{\mathrm{d}}{\mathrm{d}t}\psi_f + r_f i_f \tag{3.94}$$

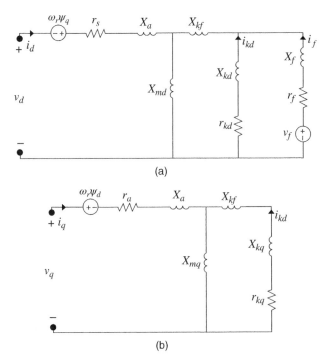

Figure 3.10 (a) The d-axis equivalent circuit and (b) q-axis equivalent circuit.

$$0 = \frac{d}{dt}\psi_{kd} + r_{kd}i_{kd}$$

$$v_q = \frac{d}{dt}\psi_q - \omega\psi_q + r_a i_q$$

$$0 = \frac{d}{dt}\psi_{kd} + r_{kq}i_{kq}$$

where v, i, r, ω, and ψ represent the voltage, current, resistance, angular speed, and flux linkage; a, f, and k denote armature, field, and damper windings, respectively, and the subscripts d and q represent the direct and quadrature axes, respectively.

The per-unit flux linkage equations that are related to the five generator windings can be expressed by

$$\psi_d = (L_{md} + L_a)i_d + L_{mq}i_f + L_{md}i_{kd} \tag{3.95}$$

$$\psi_f = L_{md}i_d + (L_{md} + L_f + L_{kf})i_{kd} \tag{3.96}$$

$$\psi_{kd} = L_{md}i_d + (L_{md} + L_{kf})i_f + (L_{md} + L_{kd} + L_{kf})i_{kd} \tag{3.97}$$

$$\psi_q = (L_{mq} + L_a)i_q + L_{mq}i_q \tag{3.98}$$

$$\psi_{kq} = L_{mq}i_q + (L_{mq} + L_{kq})i_{kq} \tag{3.99}$$

where subscript m represents mutual effect, and subscript kf indicates a leakage flux that links f and kd (but not d-axis) windings.

The DC-step voltage is a standstill test, so the rotor speed is zero and the rotational voltages that appear in Equation (3.94) vanish. Actual winding currents of the generator are measurable quantities; therefore, they offer a choice as state variables for the equations that govern the d–q machine axes. The state-space model for the d- and q-axes is given by the following equations:

$$\dot{I}_d = L_d^{-1}V_d - L_d^{-1}R_dI_d$$
$$\dot{I}_q = L_d^{-1}V_q - L_q^{-1}R_qI_q \tag{3.100}$$

where

$$\begin{cases} R_d = \text{diag}(r_a, r_f, r_{kd}) \\ R_q = \text{diag}(r_a, r_{kq}) \\ V_d = [v_d, v_f, 0]^T \\ V_q = [v_q, 0]^T \\ I_d = [i_d, i_f, i_{kd}]^T \\ I_q = [i_q, i_{kq}]^T \end{cases} \tag{3.101}$$

A genetic algorithm (GA) is used to find the fundamental parameters of the dq-axis equivalent circuits using measurements from the DC-step voltage test at each magnetic axis. A schematic diagram of the estimation procedure is shown in Figure 3.11, where the input can be either the d- or q-axis voltage and the outputs are the stator and field winding currents. The identification method is based on the minimization of a cost function. The performance of each individual is evaluated by employing the least-squares method, as given by the following, which corresponds to the dq-axis cost functions:

$$\min F_d(\theta_d) = \frac{1}{2}\sum_{i=1}^{N}\left(\left(i_{di}^e - i_{di}^t\right)^2 + \left(i_{fi}^e - i_{fi}^t\right)^2\right)$$
$$\min F_q(\theta_q) = \frac{1}{2}\sum_{i=1}^{N}\left(i_{qi}^e - i_{qi}^t\right)^2 \tag{3.102}$$

where the subscript i denotes the ith data, N is the number of data points, and superscripts e and t mean estimated and test responses. The sets of dq-axis parameters, θ_d and θ_q, are defined by

$$\theta_d = [X_d \ X_{md} \ X_f \ X_{kd} \ X_{kf} \ r_{kd} \ r_a \ r_f]$$
$$\theta_q = [X_q \ X_{mq} \ X_{kq} \ r_{kd} \ r_a] \tag{3.103}$$

where the reactance is defined as $X = \omega L$.

3.4.2 Online Parameter Estimation Techniques

The online estimation techniques can estimate the parameters during real-time operation without affecting generator operation. If problems related to persistent excitation and rank deficiency are encountered, these should be taken care of with the online parameter estimation techniques. According to the estimation types, the online parameter estimation techniques are categorized as numerical methods, observer-based methods, and AI-based methods.

Figure 3.11 Schematic diagram of the estimation process.

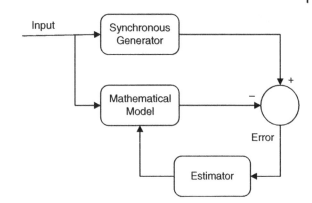

3.4.2.1 Numerical Methods

The numerical methods are commonly used for online parameter identification, such as the least squares optimization method, affine projection algorithm (APA), extended KF, model reference adaptive system (MRAS), resistance estimation methods (REMs), and orthogonal series expansion method (OSEM). The numerical-based methods are a good choice for identifying parameters for applications, such as fluctuating load operation, condition monitoring, sensor less control, and robust control [42].

Recursive Least Square (RLS): The objective of the RLS is to fit a mathematical model to a sequence of observed data by minimizing the sum of squares of the difference between the observed and computed data. The RLS is widely used to estimate some parameters, such as the stator resistance, d-axis and q-axis stator inductances, and VSI nonlinearity for the robust sensorless control algorithm, and the nominal values are set for the unidentified parameters. This method estimates the internal parameters of the generator by measuring generator output voltages, output currents, and speed. In the identification stage, the model excitation signals are the values for the current and speed of the generator, while the response signals are the output voltages. After building the model, the output currents and other quantities of the generator can be calculated by knowing the output voltages and the speed of the generator. If the model-based current estimations are sufficiently reliable to replace the current sensor information, then the generator can be controlled without having current sensors. This is more complex than for the system having non-linear equations; hence, it requires more calculations [42].

APA: In this method, the initial values of the parameters, mathematical model, output, and input of the generator are used to estimate new values of the parameters at each sample time. The APA is designed using the two-time scale approach (i.e. one APA is designed to estimate the dq-axis inductances, while the other APA is designed to estimate the stator resistance, magnetic flux-linkage, and load). The parameters estimated via the APA are also used for robust sensor-less control [44].

REMs: The REM for the stator resistance method is an online method for parameter estimation based on the extended electromotive force (EMF) model. This method is applicable to all types of synchronous generators. The stator resistance is sensitive to temperature and frequency variations. The rotor REM is based on the generator model, which takes magnetic saturation into account. This method can estimate both mechanical and electrical parameters, such as rotor resistance and angular position of the rotor. REM is applicable to all types of SMs, permanent-magnet SMs, and switched reluctance machines (SRMs). It requires continuous input and output operating data [45].

OSEM: The OSEM is an online method for estimating parameters of large-sized generators. This method estimates the armature circuit parameters of the generator using real-time operating data. This method can be easily applied when the operating data, synthetic input, and output data are available. It is an alternative method for parameter estimation. OSEM has a problem with system identification, including linear time-invariant, lumped, and distributed systems, as well as non-linear systems. The advantage of this method is the detection of turn-to-turn short circuits in the field windings of synchronous generators. Although the accuracy of this method is high, it requires time-to-time operating data of the system [1].

KF: In 1960, R. E. Kalman introduced a recursive solution of linear filtering discrete data that was employed in real time. The process was completely random, and the filter was immediately identified by Kalman. First, the recursive KF equations were extracted. Note that the KF is an extension of RLS and that it covers a wider range of estimation problems [43]. The KF can be extended to a non-linear system as an extended KF (EKF). Its main feature is the recursive processing of the noise measurement risk, which consists of the following two steps:

1. The prediction step.
2. The correction/innovation step.

This method is used to estimate the stator winding resistance, rotor flux linkages of the SM, and temperature rise in the stator winding. The KF is an estimator and can be used to track the dynamics of the generator. This approach requires an assumption of the model of the process being tracked. This method estimates the generator parameters with reasonable accuracy when the system is connected to a large disturbance (i.e. a grid or source). This method is sensitive to sampling time. The advantages of KF are the following:

1. KF has a recursive computational algorithm and can be easily simulated by a computer.
2. It has a feedback structure; therefore, it has robust performance, particularly when the statistics for noise specifications and the system model are not very accurate.
3. When the noise specifications and system model are time-varying, KF is quick to adapt to the new position.

Although KF is very heavy from a computational point of view, it has many practical advantages.

MRAS: The MRAS-based method is used for online identification of the generator parameters. MRAS computes the desired state using two different models: the reference model and the adjustable model. The error between the models is used to estimate an unknown parameter. A condition to form the MRAS is that the adjustable model should only depend on the unknown parameter. Here, the reference model is independent of the rotor speed, whereas the adjustable model depends on the rotor speed. The error signal is fed to the adaptation mechanism. The output of the adaptation mechanism is the estimated quantity, which is used for tuning the adjustable model and for feedback. It is a simple method and requires less computation, but it does not work well with low torque and at standstill conditions [43].

3.4.2.2 Observer-Based Methods

The observer-based methods, such as adaptive observers (AOs), sliding-mode observers (SMOs), and disturbance observers (DOs), are appropriate for online parameter estimation owing to their simple implementation. These methods are used for robust control applications; however, extracting the value of individual parameters from the estimated lumped disturbances is difficult.

AOs: An adaptive interconnected observer is designed in the stationary reference frame to estimate the stator resistance R_s, stator inductances L_s, magnetic flux linkage λ_m, and load torque T_L for sensor less control applications of the generators. The AO augmented with a high-frequency signal injection technique is used to estimate the R_s at the low-speed range and the λ_m at medium- and high-speed ranges. A reduced-order observer has been designed in [46] to estimate the R_s and λ_m for the stable sensorless control of the SM.

SMOs: The sliding mode control technique has shown promising results for estimating the parameters of a non-linear system. In this technique, the robustness of the electric machine parameter estimation can be guaranteed. The technique is applicable to both linear time-invariant and non-linear systems and can easily estimate the position and speed of the machine. The observer gains have a great influence on the performance of the observer and the method is mechanically unreliable [45].

DO-based methods: The DO-based methods include parameter estimation and attenuation techniques to achieve robustness against parameter uncertainties. All the disturbances related to the parameter variations are lumped together and the overall effects of parameter variations are estimated and compensated accordingly [47].

3.4.2.3 Artificial Intelligence (AI)-Based Methods

AI-based methods, such as neural network (NN), GA, and particle swarm optimization (PSO), can be used for parameter estimation. These methods are applied to self-commissioning and condition monitoring and much scope exists for their future implementation in other applications, such as sensor-less control and robust control.

NN: An online self-tuning artificial NN is developed based on the inverse dynamics of the generators to estimate the parameters lumped together for precise speed control. However, in terms of the separate estimation of parameters, the NN-based estimator is designed for the online estimation of the torque constant (or rotor flux-linkage) and stator resistance. The Elman NN is also designed and trained to simultaneously identify the stator resistance and dq-axis stator flux-linkages [46].

GA: The GA has been developed for the separate online identification of the electrical and mechanical parameters of the generator. An extended GA (i.e., an immune clonal different evolution algorithm) is designed to identify the stator resistance and the d-axis and q-axis stator inductances L_d, L_q, R_s, and λ_m using the two sets of generator equations corresponding to $i_d = 0$ and $i_d \neq 0$ [42].

PSO: PSO proposes an intelligent model parameter identification. PSO is a versatile and efficient tool technique for the design of an advanced controller, taking into account the importance of obtaining an appropriate model of the machine. This method can identify resistance and load torque disturbance. PSO is a population-based stochastic search algorithm that has been widely used to solve a broad range of optimization problems. It is capable of tracking time-varying parameters with good accuracy. However, PSO-based identification methods are not applicable to complicated non-linear models [42].

3.4.2.4 An Example of Online Parameter Estimation Using the Affine Projection Algorithm

An affine projection technique is defined as follows:

$$y(k + 1) = \phi^T(k + 1)\rho(k + 1)$$

$$\hat{\rho}(k + 1) = \hat{\rho}(k) + \gamma\phi(k + 1)[\eta I + \phi^T(k + 1)\phi(k + 1)]^{-1}[y(k + 1) - \phi^T(k + 1)\hat{\rho}(k)] \tag{3.104}$$

where y is an output signal vector, ϕ is an input signal matrix, ρ and $\hat{\rho}$ are unknown and estimated parameter vectors, and γ and η are gain matrices.

The first estimation algorithm with the sampling time (T_1) is designed to accurately estimate the dq-axis stator inductances (L_d, L_q) as follows:

$$\phi \in R^{2\times2}, y = [y_1\ y_2]^T, \rho = [L_d\ L_q]^T, \hat{\rho} = [\hat{L}_d\ \hat{L}_q]^T \tag{3.105}$$

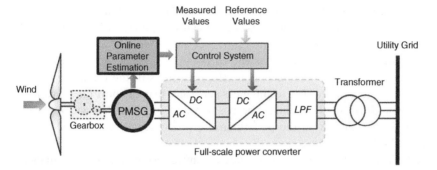

Figure 3.12 Overall block diagram of a prototype IPMSM drive system [48].

where

$$\phi^T(k+1) = \begin{bmatrix} i_d(k+1) - i_d(k) & -T_1\omega(k)i_q(k) \\ T_1\omega(k)i_d(k) & i_q(k+1) - i_q(k) \end{bmatrix} \tag{3.106}$$

$$y_1(k+1) = -T_1 R_s i_d(k) + T_1 V_d(k)$$
$$y_2(k+1) = -T_1 R_s i_q(k) + T_1 V_q(k) - T_1 \lambda_m \omega(k) \tag{3.107}$$

Next, a parameter estimation algorithm with the sampling time (T_2) can be easily established for precise estimation of the stator resistance, magnetic flux linkage, and load torque, in the following:

$$\phi \in R^{3\times3}, y = [y_1 \ y_2 \ y_3]^T, \rho = [T_L \ R_s \ \lambda_m]^T, \hat{\rho} = [\hat{T}_L \ \hat{R}_s \ \hat{\lambda}_m]^T \tag{3.108}$$

where

$$\phi^T(k+1) = \begin{bmatrix} T_2 P & 0 & 1.5T_2 P^2 i_q(k) \\ 0 & -T_2 i_d(k) & 0 \\ 0 & -T_2 i_q(k) & -T_2 \omega(k) \end{bmatrix} \tag{3.109}$$

$$y_1(k+1) = J(\omega(k+1) - \omega(k)) + T_2 B\omega(k) - 1.5T_2 P^2 (L_d - L_q) i_d(k) i_q(k)$$
$$y_2(k+1) = L_d(i_d(k+1) - i_d(k)) - T_2 V_d(k) + T_2 L_q \omega(k) i_q(k)$$
$$y_3(k+1) = L_q(i_q(k+1) - i_q(k)) - T_2 V_q(k) + T_2 L_d \omega(k) i_d(k) \tag{3.110}$$

The parameters are estimated by these equations according to the electric machine operating point, gain matrices (γ and η), and electric machine input and output. Note that each independent algorithm of the proposed online parameter estimation scheme in Equations. (3.108) and (3.110) is stable because the rank of each matrix in the equation is equal to the number of parameters identified [44]. Figure 3.12 shows an overall block diagram of a prototype interior permanent-magnet SM drive using online parameter estimation.

3.5 Parameter Accuracy Increments by Considering Saturation

The operating parameters of machines change according to the operating conditions of the generator. This is mainly due to the magnetic saturation experienced by the generator inductances. Saturation is apparent when the current through an inductor exceeds a certain limit. In effect, a saturation of an inductor occurs when the inductor core can no longer store magnetic energy.

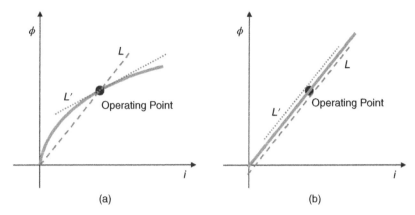

Figure 3.13 (a) Flux-linkage versus current characteristics with magnetic saturation. (b) Flux-linkage versus current characteristics with no magnetic saturation.

If magnetic saturation is neglected, the inductances are assumed constant. Under full loads, almost all generators are magnetically saturated; therefore, some discrepancies will arise between the model and the actual generator. Hence, the unsaturated model is often used with a table of inductances. However, this is inadequate to represent the magnetic saturation [49].

Figure 3.13 shows the flux-linkages versus current without and with magnetic saturation, respectively. Two kinds of inductances are shown in these figures. One is a static inductance L that represents the slope of the line from the origin to an operating point and the other is a dynamic inductance L' that represents the slope of a tangent line at the operating point. The difference between the two inductances becomes clear under magnetic saturation, and the two inductances cannot be included in the model without saturation using the table of inductances. Therefore, two kinds of inductances must be included in the model under magnetic saturation.

The effect of generator saturation inductances cannot be accounted for in offline studies. Saturation is a critical concept in generator operation; to consider it in the estimation process, one has to account for the operating level at the particular estimation interval. Contrary to the offline methods, online methods are very attractive to utility companies for identifying generator parameters because of their minimal interference in the normal power plant operation.

Saturation representation has been identified in reference [50] as the most important factor in the effort for improved synchronous generator models. The q-axis reactance is the dominant factor in determining the rotor angle for a particular operating condition. The d-axis reactance is the main factor in obtaining the excitation voltage and the power/load angle characteristics. The margins of the steady-state stability are calculated through knowledge of the power/load angle characteristics. Further, saturation plays an important role in the determination of the short circuit current. Since both are affected by saturation, it is necessary to model saturation to improve the reliability of the estimated parameters. The benefit of this estimation is improved planning of the operation of synchronous generators. The main effect of saturation in a synchronous generator is the decrease of its mutual inductances depending on the operating level of the generator. This decrease may become considerable as the generator is driven higher into saturation. Therefore, the effect of saturation must be modeled on the parameter estimation procedure.

Online estimation of the stator resistances of an SM considering magnetic saturation has been presented in reference [51]. By neglecting the saturation, the generator model is only valid for small phase currents. Thus, the estimation results are correct only for low torque, and the estimation error increases for higher torque compared to the nonlinear model with saturation and cross-magnetization. Obviously, the estimated stator resistances remain almost constant at all operating points. The conclusion drawn is that resistances can be estimated precisely at all operating points if the nonlinear model is used. By contrast, if saturation and cross-magnetization are neglected, the use of the resulting model error leads to estimation errors.

Torque ripple minimization using online estimation of the stator resistances considering the magnetic saturation has been presented in reference [52]. Considering the magnetic saturation in the RLS estimator does not significantly improve torque ripple reduction. In other words, the torque ripples can be significantly reduced using inaccurate stator resistances estimated by RLS based on a linear motor model. This is because only the *absolute values* of the resistances are estimated incorrectly if magnetic saturation is neglected.

To verify the importance of using the model with saturation, a parameter identification method based on the model without saturation has been applied in reference [49]. The results confirm the necessity of using the model with saturation in the estimation system under high magnetic saturation. A saturation model was incorporated in the estimation process to account for the change in generator inductances at different operating points in reference [50]. The saturated mutual inductances calculated at every operating point depict a decreasing trend with increasing saturation levels. The decreasing trend agrees with the theoretical expectations about the behavior of generator inductances.

3.6 Fault Detection Based on Parameter Deviation

The fundamental assumption to monitor (or supervise) a system by parameter estimation is that a fault results in the variation of one (or several) characteristic parameter (s) of the system, thus constituting the signature of this fault. According to this assumption, supervising a system involves monitoring its parameters using an identification algorithm with either offline (or by parts of samples) or online methods. In fact, this assumption can easily be invalidated by the fact that this methodology is not able to distinguish a normal parametric variation (possibly foreseeable) from that corresponding to a fault occurring randomly. This is because a model should initially be defined when estimating parameters, especially as the first reflex is to use the model of the normal operation of the system. However, a fault tends to modify the model and, in some cases, to change its structure. In most cases, a modeling error is introduced. Thus, a method is proposed again based on the parameter estimation, but that combines the two following characteristics:

1. The general model will include a model of safe functioning (or a nominal model) and a fault model (specific to each considered fault),
2. Parameters will be estimated with a priori information, which corresponds to the expertise (or knowledge) of the user on the safe functioning of the system.

3.6.1 Principle of the Method

The principle of the method is depicted here in the case of linear systems governed by differential equations with constant parameters, although this methodology is general.

Let $H(s)$ be a system of nominal transfer function, characterized by a vector θ_n. When a fault occurs, a modeling error $\delta H_i(s)$ signing the fault also appears, where $\delta H_i(s)$ is characterized by a vector θ_n. Thus, the input /output transfer function is as follows:

$$H(s) = H_n(s) + \Delta H_i(s) \tag{3.111}$$

Figure 3.14 General model of the system corresponding to the fault [53].

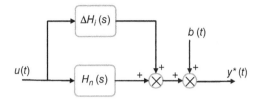

Figure 3.14 shows the general model of the system in a faulty case where $b(t)$ is a random perturbation, $u(t)$ is the input, and $y^*(t)$ is the measured output.

The nominal model $H_n(s)$, or the safe functioning model, summarizes the user expertise on the functioning of the system (i.e. the knowledge on the nominal parameters θ_n and on their variance $\mathrm{Var}\{\theta_n\}$, as well as noises affecting the output; i.e. their variance σ_b^2). In addition, the modeling error $\delta H_i(s)$ must constitute a true signature of the fault, not only by its structure but also its parameters θ_i.

The general model of the system $H(s)$ is then composed of a term of "common mode" (the nominal model $H_n(s)$) and of a term of "differential mode" (the fault model $\delta H_i(s)$) that is only sensitive when a fault d_1 appears. In addition, the nominal model must take into account the foreseeable variations of the parameters, whereas the fault model must, for its part, remain insensitive to the same variations. Finally, the nominal model must include the expertise of the user (i.e. summarized by $\{\theta_n, \mathrm{Var}\{\theta_n\}\}$). Thus, this methodology is naturally linked to identification with a priori information. An extended parameter vector is thus defined as follows:

$$\theta_e = \begin{bmatrix} \theta_n \\ \theta_i \end{bmatrix} \tag{3.112}$$

An extended covariance matrix is defined as

$$\mathrm{Var}\{\theta_e\} = \begin{bmatrix} \mathrm{Var}\{\theta_n\} & 0 \\ 0 & \mathrm{Var}\{\theta_n\} \end{bmatrix} \tag{3.113}$$

The a priori knowledge can be essentially defined on the nominal model. Then, we obtain

$$\theta_{e_0} = \begin{bmatrix} \theta_n \\ 0 \end{bmatrix} \tag{3.114}$$

and

$$\mathrm{Var}\{\theta_{e_0}\} = \begin{bmatrix} \sigma^2_{\theta_{1n}} & & & & \\ & \sigma^2_{\theta_{Nn}} & & & 0 \\ & & \infty & & \\ & 0 & & \infty & \\ & & & & \infty \end{bmatrix} \tag{3.115}$$

Note that $\mathrm{Var}\{\theta_{n0}\}$ takes into account only the diagonal terms resulting from $\mathrm{Var}\{\theta_n\}$. In addition, the terms $\sigma^2_{\theta_{jn}}$, resulting from safe functioning, must be overestimated to tolerate the foreseeable parameter variations (e.g., due to the temperature rise or magnetic saturation). Conversely, as one does not know if the fault will occur, its a priori value θ_i is null, while its initial variance is infinite (or very large). Thus, the optimization algorithm responsible for the minimization of the criterion is as follows [53]:

$$J_c = (\theta_e - \theta_{e_0})^T \mathrm{Var}\{\theta_{e_0}\}^{-1}(\theta_e - \theta_{e_0}) + \frac{1}{\sigma_b^2}\sum_{k=1}^{K}(y_k^* - y_k(\theta_e))^2 \tag{3.116}$$

Characterization of actuator parameter changes is based on a joint estimation algorithm of states and parameters [54]. Fault diagnosis procedures typically include residual generation and residual evaluation. Therefore, the difference between the nominal value of the parameter vector θ and its recursively computed estimate is viewed as a residual vector, and its evaluation is simply based on some thresholds or on more sophisticated decision mechanisms.

3.7 Summary

This chapter was devoted to transformed models and parameter identification of synchronous generators. Simple steady-state mathematical models based on Park's equations of a two-phase, three-phase, multi-phase, and six-phase synchronous generator were presented. In this modeling, some approximations were considered, and some unimportant effects were ignored. Typically, the modeling included symmetrical three-phase sinusoidally distributed stator winding, while neglecting core losses, ignoring winding stray capacitances, representing the distributed windings as concentrated windings, including sinusoidal variations of the stator winding inductances with the rotor angular position (no harmonics), and considering linear magnetization characteristics. Nevertheless, the models are acceptable for systems analysis and parameter identification of synchronous generators.

Due to the requirement for parameter estimation for system analysis and controller design, all existing parameter estimation methods, including offline and online methods, were reviewed. A summary of each one was presented for easier understanding. Two parametric examples were also offered to acquaint the reader with online and offline estimation systems.

Parameter values change in response to varying operating conditions, and magnetic saturation is the main reason for this. Therefore, the consequences of saturation in parameter accuracy and its modeling were studied. The extra advantage of parameter estimation is fault detection, which was carried out using online and offline methods. This fault detection method requires a general model that includes a model of safe functioning and a fault model. Then, by developing the general model using estimations of the new parameters of the generator, faults were identified by comparison with previous information.

References

1 Ghomi, M. and Sarem, Y.N. (2007). *Review of Synchronous Generator Parameters Estimation and Model Identification*, vol. 1, 228–235.

2 Park, R.H. (1929). Two reaction theory of synchronous machines, part 1. *AIEE Transactions* 48: 716–730.

3 Park, R.H. (1933). Two reaction theory of synchronous machines, Part 2. *AIEE Transactions* 52: 352.

4 Levi, E. (2008). Multiphase electric machines for variable-speed applications. *IEEE Transactions on Industrial Electronics* 55 (5): 1893–1909.

5 Bernard, A. and Harley, R.G. (2013). *The General Theory of Alternating Current Machines: Application to Practical Problems*. Springer.

6 Dehkordi, A.B. Development of a symmetrical multi-phase synchronous machine model for real-time digital simulation.

7 Krrause, P.C. (2013). *Analysis of Electric Machinery*. McGraw-Hill.

8 Parsa, L. (2005). On the advantages of multi-phase machines, Industrial Electronics Society. *IECON 2005*, Raleigh.

9 Fuchs, E.F. and Rosenberg, L.T. (1973). Analysis of an alternator with two displaced stator windings. *IEEE Transactions on Power Apparatus and Systems* 93: 1776–1786.

10 Singh, G.K. (2002). Multi-phase induction machine drive research - a survey. *Electric Power Systems Research* 61: 139–147.

11 Schiferl, R.F. and Ong, C.M. (1983). Six-phase synchronous machine with AC and DC stator connections, Part I: Equivalent circuit representation and steady state analysis. *IEEE Transactions on Power Apparatus and Systems* 102: 2685–2693.

12 Schiferl, R.F. and Ong, C.M. (1983). Six-phase synchronous machine with AC and DC stator connections, Part II: Harmonic studies and a proposed uninterruptible power. In: *IEEE Transactions on Power Apparatus and Systems*, vol. 1, 2694–2701.

13 Abuismais, I., Arshad, W.M., and Kanerva, S. (2008). Analysis of VSI-DTC fed six phase synchronous machines. In: *Proceedings of the 13th Power Electronics and Motor Control Conference (EPE-PEMC)*, 883–888.

14 Singh, G.K. (2011). Modeling and analysis of six-phase synchronous generator for stand-alone renewable energy generation. *Energy* 36 (3): 5621–5631.

15 Jahns, T.M. (1980). Improved reliability in solid-state ac drives by means of multiple independent phase drive units. *IEEE Transactions on Industry Applications* 16 (3): 321–331.

16 Robb, D.D. and Krause, P.C. (1975). Dynamic simulation of generator fault using combined abc and 0dq variables. *IEEE Transactions on Power Apparatus and Systems* 94 (6): 2084–2091.

17 Kerkman, R.J., Krause, P.C., and Lipo, T.A. (1977). Simulation of a synchronous machine with an open phase. *Electric Machines and Power Systems* 1 (3): 245–254.

18 Welchko, B.A., Jahns, T.M., and Hiti, S. (2002). IPM synchronous machine drive response to a single-phase open circuit fault. *IEEE Transactions on Power Electronics* 17 (5): 764–771.

19 Zhao, Y. and Lipo, T.A. (1996). Modeling and control of a multi-phase induction machine with structural unbalance. Part I: machine modeling and multi-dimensional current regulation. *IEEE Transactions on Energy Conversion* 11 (3): 570–577.

20 Tleis, N. (2008). *Power System Modelling and Fault Analysis*. Amsterdam: Elsevier Ltd.

21 Lipo, T.A. (2012). *Analysis of Synchronous Machines*, 2e. New York: CRC Press.

22 Iqbal, A., Singh, G.K., and Pant, V. (2014). Steady-state modeling and analysis of six-phase synchronous motor. *Systems Science & Control Engineering* 2 (1): 236–249.

23 Singh, G.K. (2011). A six-phase synchronous generator for standalone renewable energy generation: experimental analysis. *Energy* 36 (3): 1768–1775.

24 Jones, C.V. (1967). *The Unified Theory of Electric Machine*. London: Butterworths.

25 Aghamohammadi, M.R. and Pourgholi, M. (2008). Experience with SSSFR test for synchronous generator model identification using Hook-Jeeves optimization method. *International Journal of Systems Applications, Engineering & Development* 2 (3): 122–127.

26 Monticelli, A. (2000). Electric power system state estimation. *Proceedings of the IEEE* 88, no. 2: 262–282.

27 Abur, A. and Exposito, A. (2004). *Power System State Estimation: Theory and Implementation*. CRC Press.

28 He, G., Dong, S., Qi, J., and Wang, Y. (2011). Robust state estimator based on maximum normal measurement rate. *IEEE Transactions on Power Systems* 26, no. 4: 2058–2065.

29 Qi, J., He, G., Mei, S., and Liu, F. Power system set membership state estimation. In: *IEEE Power and Energy Society General Meeting*, (22–26 July 2012). San Diego, CA, USA.

30 Khalil, I., Doyle, J., and Glover, K. (1996). *Robust and Optimal Control*, 1e. Pearson.

31 Ariff, M., Pal, B., and Singh, A.K. (2015). Estimating dynamic model parameters for adaptive protection and control in power system. *IEEE Transactions on Power Systems* 30, no. 2: 829–839.

32 Du, P. Generator dynamic model validation and parameter calibration using phasor measurements at the point of connection. In: *2014 IEEE PES General Meeting – Conference & Exposition,* (27–31 July 2014). National Harbor, MD, USA.

33 Liu, Y., Ning, P., and Reiter, M.K. (2011). False data injection attacks against state estimation in electric power grids. *ACM Transactions on Information and System Security* 14 (1): https://doi.org/10.1145/1952982.1952995.

34 Kosut, O., Jia, L., Thomas, R.J., and Tong, L. (2011). Malicious data attacks on the smart grid. *IEEE Transactions on the Smart Grid* 2, no. 4: 645–658.

35 Vukovic, O., Sou, K.C., Dan, G., and Sandberg, H. (2012). Network-aware mitigation of data integrity attacks on power system state estimation. *IEEE Journal on Selected Areas in Communications* 30, no. 6: 1108–1118.

36 Yang, Q., Yang, J., Yu, W. et al. (2014). On false data-injection attacks against power system state estimation: Modeling and countermeasures. *IEEE Transactions on Parallel and Distributed Systems* 25, no. 3: 717–729.

37 Arellano-Padilla, J., Sumner, M., and Gerada, C. (2011). Winding condition monitoring scheme for a permanent magnet machine using high-frequency injection. *IET Electric Power Applications* 5 (1): 89–99. https://doi.org/10.1049/iet-epa.2009.0264.

38 Kang, G.-H., Hur, J., Nam, H. et al. (2003). Analysis of irreversible magnet demagnetization in line-start motors based on the finite-element method. *IEEE Transactions on Magnetics* 39, no. 3: 1488–1491.

39 Arellano-Padilla, J., Sumner, M., and Gerada, C. (2011). Winding condition monitoring scheme for a permanent magnet machine using high-frequency injection, C.," *IET Electr. Power Appl.,* 5 (1), 89–99.

40 Rafaq, M.S. and Jung, J. (2019). A comprehensive review of state-of-the-art parameter estimation techniques for permanent magnet synchronous motors in wide speed range. *IEEE Transactions on Industrial Informatics* 16: 1. https://doi.org/10.1109/TII.2019.2944413.

41 Honsinger, V.B. (1982). The fields and parameters of interior type AC permanent magnet machines. In: *IEEE Transactions on Power Apparatus and Systems,* vol. PAS-101, no. 4, 867–876.

42 Arjona, M.A., Hernandez, C., Cisneros-Gonzalez, M., and Escarela-Perez, R. (2012). Estimation of synchronous generator parameters using the standstill step-voltage test and a hybrid Genetic Algorithm. *International Journal of Electrical Power & Energy Systems* 35 (1): 105–111. https://doi.org/10.1016/j.ijepes.2011.10.003.

43 Ghahremani, E. and Kamwa, I. (2011). Online state estimation of a synchronous generator using unscented Kalman filter from phasor. *IEEE Transactions on Energy Conversion* 26 (4): 1099–1108.

44 Ramana, P., Mary, K.A., Kalavathi, M.S., and Swathi, A. (2015). Parameter estimation of permanent magnet synchronous motors – A review, *i-manager's. Journal of Electrical* 9 (2): 49–59.

45 Dang, D.Q., Rafaq, M.S., Choi, H.H., and Jung, J.W. (2016). Online parameter estimation technique for adaptive control applications of interior PM synchronous motor drives. *IEEE Transactions on Industrial Electronics* 63 (3): 1438–1449. https://doi.org/10.1109/TIE.2015.2494534.

46 Hinkkanen, M., Tuovinen, T., Harnefors, L., and Luomi, J. (2012). A combined position and stator-resistance observer for salient PMSM drives. https://aaltodoc.aalto.fi/handle/123456789/30764 (accessed 01 August 2021).

47 Zhang, X., Hou, B., and Mei, Y. (2017). Deadbeat predictive current control of permanent-magnet synchronous motors with stator current and disturbance observer. *IEEE Transactions on Power Electronics* 32, no. 5: 3818–3834.

48 Song, W., Gang, Y., Zhi-Jian, Q. et al. (2009). Identification of PMSM based on EKF and Elman neural network. In: *IEEE International Conference on Automation and Logistics* (5–7 August 2009). Shenyang, China.

49 Ichikawa, S., Tomita, M., Doki, S., and Okuma, S. (2006). Sensorless control of synchronous reluctance motors based on extended EMF models considering magnetic saturation with online parameter identification. *IEEE Transactions on Industry Applications* 42 (5): 1264–1274. https://doi.org/10.1109/TIA.2006. 880848.

50 Kyriakides, E., Heydt, G.T., and Vittal, V. (2005). Online parameter estimation of round rotor synchronous generators including magnetic saturation. *IEEE Transactions on Energy Conversion* 20 (3): 529–537. https://doi.org/10.1109/TEC.2005.847951.

51 Xu, Y., Vollmer, U., Ebrahimi, A., and Parspour, N. (2012). Online estimation of the stator resistances of a PMSM with consideration of magnetic saturation. In: *EPE 2012 – Proceedings of the 2012 International Conference and Expo. on Electrical Power Engineering*, 360–365. https://doi.org/10.1109/ICEPE.2012.6463906.

52 Xu, Y., Parspour, N., and Vollmer, U. (2014). Torque ripple minimization using online estimation of the stator resistances with consideration of magnetic saturation. *IEEE Transactions on Industrial Electronics* 61 (9): 5105–5114. https://doi.org/10.1109/TIE.2013.2279378.

53 Trigeassou, J., Poinot, T., and Bachir, S. (2009). Parameter estimation for knowledge and diagnosis of electrical machines. ISTE Ltd and John Wiley & Sons Inc.

54 Zhang, Q. (2018). Adaptive Kalman filter for actuator fault diagnosis. *Automatica* 93: 333–342. https://doi.org/10.1016/j.automatica.2018.03.075.

4

Introduction to Different Types of Faults in Synchronous Generators

4.1 Reasons for Condition Monitoring of Synchronous Generators

Condition monitoring in electrical machines, particularly in synchronous generators, has increased considerably in recent years. Although much work has focused on induction motor fault diagnosis, synchronous generators have received less attention. The reasons include the complicated structure and faults of synchronous machines and the lesser relevance of the traditional fault detection methods widely used in induction motors.

On-load operation of synchronous generators causes an internal temperature rise that gradually leads to aging of its insulation. This eventual failure of the insulation is the origin of many faults in the generator. For example, the continuation of insulation failure can lead to a turn-fault in the stator winding, which may trigger many additional turn-to-turn faults. Expansion of this type of fault generates asymmetry in the stator magnetic field, leading to unbalanced induced voltage in the generator terminals. This imbalance, in turn, may cause a two-phase fault or an earth-phase fault and damage the stator core. Therefore, having a short circuit fault in a multi-turn synchronous generator can not only destroy the winding system, but it can also destroy the stator core.

The consequence of this fault is severe, since electricity production must be stopped and a huge expenditure must be made on generator repair. Companies generally have no specialist force to repair and change the windings or other damaged parts of a generator when it fails, so the repairs must be transferred to the manufacturer company at a separate expense. The complex structure of the generator itself makes the rewinding difficult, costly, and time consuming. These potentially dire consequences of a simple fault emphasize the importance of condition monitoring in synchronous generators. In addition to insulation failure, faults in generators also create several

Electromagnetic Analysis and Condition Monitoring of Synchronous Generators, First Edition. Hossein Ehya and Jawad Faiz.
© 2023 The Institute of Electrical and Electronics Engineers, Inc. Published 2023 by John Wiley & Sons, Inc.

harmonics or noises that can inject into the system and disrupt the functioning of protective relays or decrease the power quality produced by the system.

In some countries, governments subsidize the electricity produced for industries and home consumers, and the global increase in electricity costs has prompted the shortening of transmission lines to reduce the distance between generation and consumption to reduce losses. These efforts, coupled with the distributed generation (DG) and privatization of electrical power generation and sale, suggest that generator fault diagnosis should have a far more important place in the electrical power generation industry. Condition monitoring methods that do not involve taking a generator offline are needed for early identification of faults and to provide a better understanding of these faults; this monitoring should be an essential part of the generation system [1]. This chapter investigates the different fault types encountered in synchronous generators and the reasons for the creation of these faults.

4.2 Different Faults in Synchronous Generators

The effect of a fault in a generator not only depends on the fault type, but it also depends on the generator type and the environment in which it is operated. However, it is also possible to identify the base mechanism underlying fault development that applies to all generators.

As a first step, the origin of the fault must be quickly recognized, as this is very useful for generator condition monitoring. Any fault with a relevant mechanism begins from an initial fault and expands depending on existing conditions. Some features indicate the presence of a fault and the duty of a condition monitoring system is to track, locate, and announce the fault.

The Electric Power Research Institute (EPRI) and IEEE reports indicate that around 40% of all generator faults are related to the bearings, 35% to the stator, 10% to the rotor, and 15% to other parts [2]. Figure 4.1 shows the different fault types in the stator and rotor of synchronous machines.

The stator faults can be categorized as follows:

1. Short circuit (SC) faults in the stator winding inside the slots.
2. SC fault in the stator end winding.
3. Core-related faults in the stator winding.

The rotor faults can be summarized as follows:

1. SC faults in the rotor field winding.
2. SC fault in the rotor end winding.
3. Broken damper bar.
4. Broken end ring.
5. Rotor eccentricity.

The synchronous generator rotor faults have been widely studied. The SC fault in the rotor winding can be investigated using magnetic flux monitoring [3]. A healthy synchronous generator and the generator excitation winding with an SC fault have been modeled using the Finite Element Method (FEM), and this model has been applied to a salient pole synchronous generator [4]. Most rotor faults in a no-load synchronous generator have been simulated by FEM in reference [5], and several signal frequency tools have been utilized for diagnosing these faults.

The origin of mechanical faults is mainly a bearing fault. Bearing faults originate from inappropriate lubrication, radial and axial stresses, and adjusting with a weak foundation, which cause

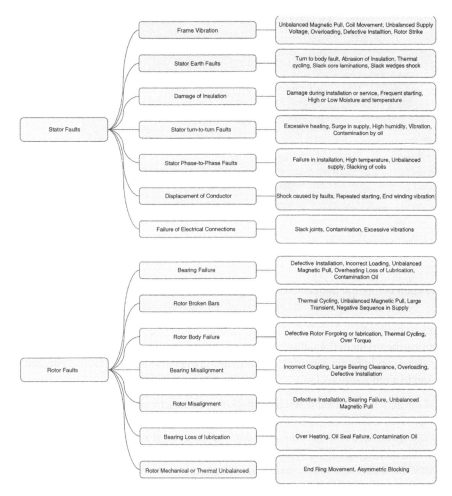

Figure 4.1 Fault types in the stator and rotor of synchronous machines.

wear and fatigue of the bearings. Generally, a bearing fault generates a stator and rotor eccentricity fault, and if this fault is intensified, a collision may occur between the rotor and stator. Several static, adaptive and on-line methods have been introduced to diagnose a bearing fault [6]. Figure 4.2a shows the destruction caused by wearing due to an electric arc in a moving bearing. Figure 4.2b shows the pit-shaped electric destruction on the bearing.

4.3 Main Factors Leading to Electrical Machine Damage

The health of a synchronous generator, its different types of faults, and the origins of the faults depend on the materials used, the types of mechanical and electrical stresses, and the machine's operating conditions [7, 8]. Mechanical and electrical faults are always associated with electrical, mechanical, magnetic, and cooling components. If the damage occurs over a defined period, measuring is possible to determine its origin, and a pattern may emerge that would allow monitoring

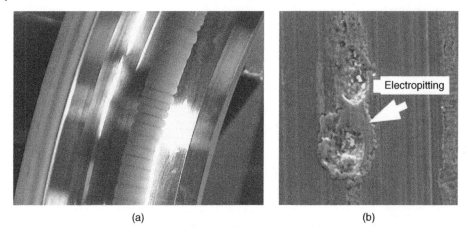

(a)   (b)

Figure 4.2 (a) Damage to a moving bearing due to an electro-erosion track, (b) electro-pitting damage on the bearing.

of the machine condition before the fault occurrence. This condition monitoring can be classified as follows:

1. Direct measurement methods.
2. Indexes determining the origin of the failure.
3. Provision of the alarm condition for repairing or interrupting a generator before fault expansion and/or progression to a more serious fault.

The nameplate of the generator contains the information necessary to identify the generator. It gives some basic information about the operating characteristics of the generator. If the operator exceeds the nameplate operating conditions, this may damage the generator. The need for monitoring and parameters to be monitored can be determined by the operation condition of the generator. Table 4.1 summarizes the different components of an AC generator, its relevant characteristics, inappropriate conditions, and resultant damage [9].

4.4 Major Destruction Factors of Stator Winding

Electrical insulation damage plays a major role in electrical machine breakdown. The published literature shows that more than one-third of rotating electrical machine faults are attributed to stator faults [10]. Various stresses are exerted on the generator stator, and these induce different types of faults. Undesirable changes in the physical and chemical structures of the insulation due to the stresses during servicing are called aging phenomena in the insulation system. Different stresses that act as major factors in the stator winding destruction are categorized as follows [11, 12]:

1. Thermal stresses.
2. Electrical stresses.
3. Mechanical stresses.
4. Ambient stresses.

Stator winding is under the influence of all the above-mentioned stresses. These stresses clearly cause long-term and short-term problems that depend on various factors, such as operating conditions and cooling systems. For example, air-cooled synchronous machines age more quickly from

Table 4.1 General specifications of a generator and relevant characteristics and damage caused by inappropriate conditions.

Operating condition	Normal conditions	Partial condition	Origin and fault types
Mechanical	Characteristics of load and initial starting	Operating period Load variations Repeated drive vibrations Repeated starting	Successive overloading causes extra heat and damage to bearings May lead to damage and fatigue of bearings Repeated starting torque leads to extraordinary heat damage to stator and rotor end windings Leads to fatigue over a long period
Electrical	Characteristics of the generator and its connected system	Low-voltage variations High-voltage variations Unbalanced voltage	Leads to power reduction. Causes locked rotor in motor and pole slip in generator Leads to disruption and interruption of the excitation system and to insulation failure Leads to heat generation in the stator and stator winding
Environmental	The place where the machine is used	Environment temperature Environment humidity Cleanliness of the environment	Accelerates the destruction of insulation and bearings. Low temperature causes frostbite. At high humidity, condensation occurs on the insulation surface, causing a linear breakdown of insulation and corrosion in metal parts. Environmental pollution can penetrate inside the generator and cause insulation pollution and accumulate on it. It can also deposit on the heat exchanger and pollute the brushes.

thermal stress compared with machines directly cooled by oil. Generators cooled by compressed hydrogen have less oxidation risk. In the following sections, each stress and its destructive impacts are separately addressed. The combined mechanism of these stresses is too complicated, and interactions between different stresses are a difficult matter and do not offer the expected results.

4.4.1 Thermal Stress

The insulation system of AC windings of the generator is designed such that it can withstand electrical, mechanical, and thermal tension. It should also endure definite voltage levels and temperature rises, at least over its defined lifespan. International standards provide different classes of electrical insulation systems based on the maximum temperature at which the generator can operate continuously under its nominal voltage and withstand the rated voltage of the system. Table 4.2 summarizes the commonest insulation classes used in generators. As seen in the table, the temperature that is bearable by insulations is very low. The second column of Table 4.2 shows the

Table 4.2 Insulation class, permissible temperature, and rise-able temperature.

Insulation class	Permissible temperature of insulation (°C)	Rise-able temperature of stator winding (°C)
O or Y	90	50
A	105	65
E	120	80
B	130	90
F	155	115
H	180	120
C	220	180

permissible temperature, which is the summation of the winding temperature and the ambient temperature (assumed to be 40 °C). The third column of Table 4.2 indicates the maximum winding temperature rise with the internal operating temperature of the generator. According to the IEC standard, the maximum internal temperature combined with the environmental temperature must be lower than the permissible insulation temperature [3]. However, in practice, normal operating conditions for insulation systems are a lower heat class than the rated heat class.

On-load operation of a synchronous generator causes heat and raises the internal temperature. This gradually weakens the strength of the insulation of the generator and accelerates the aging process of the insulation. Gradually, the insulation system becomes fragile and can break in response to a partial shock. Even in the absence of a shock, continuation of this insulation weakening can lead to an inter-turn fault that, after a long time, can progress to a failure of the entire winding. A standard test shows that for every 10 °C rise in winding temperature above the maximum permissible temperature, the lifespan of its insulation is reduced by half [12].

The temperature variation occurring in a synchronous generator during the on-load operation can also increase mechanical stresses. The connection point between the copper winding and its insulation is one point that can expand or shrink, and even this small displacement can cause mechanical stress [10]. The following factors can impose stress on the synchronous generator:

1. Overvoltage increases the induced voltage of the generator, leading to more core losses. The higher flux also causes higher insulation losses. All these changes increase the temperature within the synchronous generator and impose thermal stress.
2. Unbalanced voltage causes an increase in the winding's current and, consequently, its copper losses. This also leads to a higher temperature rise in the winding. For example, a 2.5% unbalanced voltage causes a 25% rise in the winding temperature [11, 12].
3. Successive starts and stops of the synchronous generator do not permit enough time for cooling the generator and integrating heat due to higher ohmic losses; this causes a temperature rise and thermal stress.
4. Overloading increases the winding temperature in proportion to the load current.
5. Cooling system defects prevent proper cooling of the synchronous generator.
6. Variations in the environmental temperature: The main cooling factor in the synchronous generator is the ambient air. If the air temperature is lower than the designated temperature value, cooling is effective. At an ambient temperature higher than the standard temperature, however, the fans will be useless because of the very small difference between the hot air inside

Main wall insulation puncture

Figure 4.3 A typical stator main wall insulation puncture caused by heat stress in a synchronous generator.

the generator and the substituted air. Heat generated in the windings will remain inside the insulation and may damage the generator. Humidity from the air can also sometimes precipitate on the internal parts, and this decreases the lifespan and strength of the insulation. Thermal aging of the generator has different signs, including: wrinkled insulation, insulation hardness, insulation crispness and fragility, insulation endurance, and insulation color change (in most cases, from the original color to black). Figure 4.3 presents a typical main wall insulation puncture caused by heat stress. The puncture of the insulation indicates its permanent failure [10]. IEEE defines a puncture as a disruptive discharge through the body of a solid dielectric; therefore, a puncture is a somewhat longer process than a flashover. If there is moisture ingress or a small void formation in the insulating medium, a small leakage current will flow; over a longer time, it may result in permanent damage to the insulating material. If a direct short circuit to the ground occurs from a live conductor, this leads to an instant flashover of the insulation system.

The signs of heat stress include weight loss caused by evaporation of volatile parts, oxidation and/or chemical changes due to evaporation of materials, and generation of CO, CO_2, water, and low molecular weight hydrocarbons.

4.4.2 Electrical Stress

Transient and DC voltages can cause insulation aging. However, even in normal cases, AC voltages impose high stresses on the synchronous generator and operation under abnormal and faulty conditions leads to electrical aging. For example, gases generated due to saturation create a partial discharge (PD) in the insulation system. In contrast to thermal stresses originating from internal factors, electrical stresses have an external origin and, in fact, are applied from the system to the generator. Electrical stress may be caused by the fundamental frequency and/or transient impulse voltages. Electrical stresses can be classified into the following three types:

1. **Partial discharge**: Generally, destruction due to PD is generated by holes that appear during manufacturing and/or operation due to mechanical and thermal stresses. The direct impact of PD on insulation sheets is the removal of the solid insulation. Figure 4.4 shows samples of the insulation defects caused by PD.
2. **Sheet and humidity absorbance paths**: Electrical paths are conductive or carbonized paths created on the insulation sheets by AC electrical stress. When an insulation sheet absorbs humidity and dust, it can act as a conducting sheet. Pollution renders membranes non-uniform and conductive, quickly causing dry regions in the high-resistance areas. Many voltages along these dry sheets appear as sparks, and these sparks can lead to partial breakdown in the dry

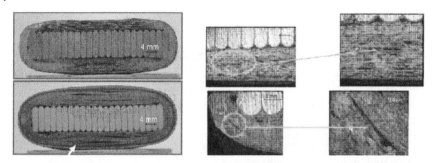

Figure 4.4 Samples of generator insulation defects caused by PD. *Source*: Saba Information Network.

region. The occurrence of an electric arc around the dry region is either quickly turned off or it continues and ignites the insulation. Sometimes, the electric arc may gradually continue and lead to a complete breakdown. Electric arcs in polluted insulation sheets can cause insulation breakdown at stresses of 0.02–0.04 kV/mm [10].

3. **Transient voltages:** Voltage variations in the form of transient impulses caused by lightning, switching, faults, and the adjustable speed drive supplying some machines can impose major electrical stresses. Normally, insulation in a synchronous generator is designed to withstand lightning and switching. However, transient variations caused by drives due to non-uniform voltage distribution and rapid dv/dt are proposed as reasons for the insulation system breakdown. Nevertheless, transient events can be a reason for the breakdown in a generator that has been under stress for a long time. The reasons for electrical stress can be summarized as follows:
   (a) Occurrence of short circuits in the power network.
   (b) Frequent spark return.
   (c) Current-limiter fuse.
   (d) Power system bus maneuver.
   (e) Lightning strike to the system.
   (f) Switching off and on of circuit breakers.
   (g) Power factor compensator capacitors.
   (h) Power system insulation failure.

4.4.3 Mechanical Stresses

The magnetic field of the synchronous generator exerts electromagnetic forces upon the windings that are carrying current. These forces can displace the winding and generate vibration. Mechanical aging makes major direct and indirect contributions to insulation aging. Relative movement of insulation components can be caused by mechanical electromagnetic forces, vibrations, mechanical shocks due to short circuit currents and centrifugal forces in the rotor, resonance, insufficient number of wedges, curvature or bending, wearing, and scratching. These movements produce mechanical stresses in the synchronous generator and on the insulation itself, thereby accelerating insulation breakdown. Mechanical stresses also adversely affect end-winding and winding leads. Connecting the stator end-windings to the supporting loop and injecting resin into this region can prevent the undesirable effects of mechanical stresses. The following are the main factors leading to mechanical stresses [7, 8]:

1. Winding displacement.
2. Eccentricity fault.
3. Rotor stator collision.

Figure 4.5 Typical destruction of the stator winding and tracking of the slot liner caused by moisture contamination.

Moisture contamination and slot liner tracking

4.4.4 Ambient Stress

Synchronous generators have a particular temperature, humidity, and range of pollution tolerance, and changes in any of these can disturb the operation of the generator and reduce its optimal life. The ambient stresses can be summarized as follows:

1. Environmental pollution.
2. Ambient temperature variations and very cold weather.
3. Ambient humidity.
4. Abrasive materials.
5. Acidic vapors and corrosives in the environment.
6. Radiation from nuclear centers.

Pollution of insulation with water, oil, and chemical components leads to mechanical and electrical defects. Some gases, such as hydrogen, can also accelerate thermal aging. The effect of humidity and water absorption is clear and, in practice, depends on the insulation type. The old insulation systems were organic adhesives and bandages, which were very prone to water absorption and subsequent mechanical defects. However, the new generators use inorganic adhesives and synthetic resins, such as epoxy, which reduce the probability of this destruction. Polyester-based materials can reduce mechanical and electrical properties in humid environments. Hydrolysis due to moisture can lead to chemical bond failure in many layered insulations and make them swell, thereby increasing the probability of insulation breakdown in response to other stresses. Oils, acids, alkalis, and solvents can attack the insulating material and its chemical bonds. Pollution caused by dust and its aggregation on the insulation surface can be another threat. Figure 4.5 shows the typical destruction of the stator winding caused by moisture contamination and subsequent tracking of the slot liner.

Insulation materials exposed to nuclear radiation (such as in nuclear power plants) can quickly undergo destruction of their mechanical properties. Nuclear radiation has little impact on some insulating materials, such as ceramics, glass, and mica. However, organic materials are more affected because of ionization. Polymers, such as polyamides, have tolerance and are not destroyed by radiation. X-ray generation sources and neutrons are dangerous for insulation because of electron generation. UV radiation leads to insulation aging. Synchronous generators can be protected against inappropriate ambient conditions by isolation.

4.5 Common Faults in Stator Winding

Faults in stator winding are one of the main reasons for the failure of synchronous generators. Different damage in the stator winding is caused by power lines and failure of related devices, but the internal fault current of the generator is a major contributor to generator damage, whereas factors imposed by the power system are less important. About 30–40% of the main faults in synchronous generators are related to the stator winding. In the case of a fault in the stator winding, the fault current can remain even after circuit-breaker operation and can continue to generate serious damage to the stator winding [13].

The origin of most internal faults is the weakening of the stator winding insulation strength. Different stresses gradually expand weak points in the insulation and in a very short time the insulation breaks down [14, 15]. Depending on the type of winding (random-wound or form-wound) of the stator, the insulation breakdown time ranges from 1/3 seconds up to several minutes [16, 17]. Different faults of the stator winding of a synchronous generator include:

1. The open circuit fault of one of the windings. This creates a two-phase generator and can be easily diagnosed by protective relays.
2. Phase-to-earth SC (earth fault). This is the worst fault in the generator, as the high voltage without resistance connected to the earth leads to a high current in this track. The stator core may melt and this may continue even after the circuit-breaker disconnection. In contrast to the high damage intensity created by this fault, its recognition is easy. The sensitivity of sensors and relays must be high.
3. Phase-to-phase SC (phase fault): This fault occurs due to the connection of one or several points of the stator phase with other phases. This fault is very destructive to the stator winding; however, it occurs less frequently than the earth fault. Recognition of this fault is very easy due to the highly unbalanced and asymmetric condition in the generator. This fault requires rapid operating relays.
4. SC of coil-to-coil (coil fault): This fault is caused by contact of one or several points of a coil with the adjacent coils of the same phase. This creates less imbalance and asymmetry in the generator compared with the previously mentioned faults. Therefore, detection of this fault is difficult.
5. Turn-to-turn fault: The common mechanism of the stator fault in multi-turn synchronous generators is insulation defeat. An SC between the copper winding turns generates a high current, leading to quick destruction and breakdown of the insulation [15]. This fault is very destructive, as it can burn the insulation and melt copper conductors. Insulation breakdown between the internal turns arises as small holes created mostly in the stator core end parts and in the first and second coils of the end line. In the turn-to-turn fault, several turns of a phase coil are short-circuited. In a coil-to-coil fault, two coils of a phase are short-circuited. Since the coils are isolated with respect to each other and also to the stator core by special paper, the possibility of turn-to-turn faults is higher than with other stator faults. In addition, a turn-to-turn fault with a low number of turns has no significant effect on the apparent performance of the machine. By contrast, the other generated faults add extra current to the generator and are easily diagnosable by other protection systems. However, if the turn-to-turn fault is not detected at its initial stages, the high circulating current in the SC turns and the high heat generated may destroy the complete winding and possibly damage the core [15, 18]. Therefore, diagnosis of a turn-to-turn fault is more difficult but highly important. An investigation detecting turn breakdown in equipment showed that the turn SC rise was related to increasing substitution of air circuit breakers with the vacuum circuit breakers [10], as the high amplitude transients and forward slopes during the vacuum circuit-breaker operation decreased insulation strength. The generators with mica paper insulation have a more robust resilience against the forwarding impulses of the vacuum circuit breaker. In addition, these generators are robust against steady-state performance problems (network frequency) [10]. Figure 4.6 shows typically melted stator cores caused by the internal fault.

Figure 4.6 Two typical melted stator cores caused by an internal fault.

4.6 Rotor Field Winding Fault

The rotor of the salient pole synchronous machine consists of two windings: the rotor field winding and the damper winding. The rotor field winding has windings concentrated around each rotor pole. Based on the power rating of the machine, the utilized wire can have a rectangular or circular cross-section. Figure 4.7 shows a rotor from a salient pole synchronous machine. The rectangular wire is commonly used in the field winding of large synchronous machines. An SC fault in the rotor field winding of a salient pole synchronous machine is a common fault. The SC in the rotor field winding can be divided into two categories:

1. Inter-turn short circuit fault.
2. Turn-to-ground fault.

Although the inter-turn short circuit fault in the rotor field winding is a common fault, the fault severity is low due to the small magnitude of the DC current and a safe margin for the tolerated temperature rise of the insulation system [19–21]. The increased inter-turn short circuit fault yields rises in temperature and vibration in the rotor field winding. The consequences of these vibration and temperature rises are a deterioration of the insulation of the neighboring turns and a larger

Figure 4.7 A salient-pole synchronous generator rotor – field winding.

inter-turn short circuit fault. However, this fault can become a rotor body short circuit fault, known as a turn-to-ground short circuit fault. If the inter-turn short circuit fault happens close to the rotor body, the insulation between the body and the rotor winding can be damaged and result, in the long term, in a ground fault in the rotor.

One of the main consequences of having a short circuit fault in the rotor field winding is the asymmetric air gap magnetic field. This asymmetry can create a circulating current in the rotor shaft that closes its path through the bearing. The circulating current in the bearing can damage the insulation system of the bearing and destroy the bearing over the long term.

4.7 Eccentricity Faults

Similar to other rotating electrical machines, the synchronous generator has three longitudinal axes: rotor symmetrical axis O_r, stator symmetrical axis O_s, and rotor rotational axis O_z. In an ideal healthy generator, these three axes coincide with each other; therefore, the air gap distribution function is constant and is not distorted by rotor rotation. However, complete coincidence of these axes is almost impossible. Even in a brand-new synchronous generator, depending on the manufacturing technology and assembly of the parts, a slight but permissible non-coincidence will occur between these axes and is termed the inherent eccentricity fault [22–26]. The reasons why the inherent eccentricity might become amplified are as follows:

- Incorrect locating of the stator and rotor to one another.
- Incorrect locating of the bearings.
- Displaced or bent rotor shaft.
- Abrasion of bearings.
- Loose bolts.
- Internal elliptical cross-section of the stator.
- Misaligned connection of the synchronous machine shaft.

In addition to factors that directly cause eccentricity, some main faults, such as bearing faults, can lead to rotor and stator eccentricity. Therefore, by detecting the root cause of the eccentricity, many faults can be diagnosed in the generator.

The type of eccentricity can be identified by determining the position of the rotor rotation axes. Three eccentricity types are recognized:

1. Static eccentricity (SE).
2. Dynamic eccentricity (DE).
3. Mixed eccentricity (ME).

In the case of an SE fault, the minimal air gap length varies only by position, excluding the natural variations caused by the saliency of the rotor poles. During an SE fault, as shown in Figure 4.8, the center of the rotor is shifted from the stator center, while it rotates around its center axis. The possibility is low of having an SE fault due to high precision in the manufacturing process. However, the SE fault is one of the prevalent fault types in the synchronous generators of hydropower plants located in a mountain, due to the instability of the rock mass.

On the contrary, when a DE fault occurs, the rotor revolves around the stator center axis while having the center point displaced from the stator. As seen in Figure 4.8, the minimum air gap length varies with both position and time. In the case of a DE fault, the air gap length dynamically varies.

Static and DE can exist simultaneously and is termed ME, as shown in Figure 4.8. Therefore, in the case of an ME fault, the rotor is close to the stator on one side, due to the SE fault, while the

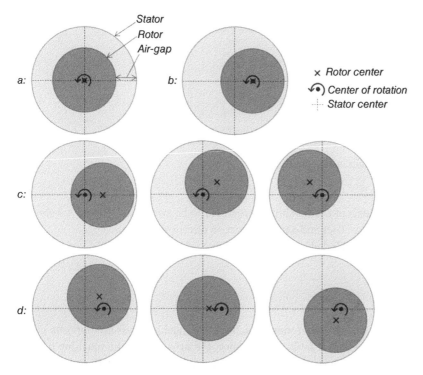

Figure 4.8 Stator and rotor during healthy operation (a), under static eccentricity (b), and under three time instants of dynamic eccentricity (c) and mixed eccentricity (d).

rotor whirling location varies with time and location due to the DE fault. Operating with an SE can cause the development of a DE and thereby cause an ME fault [27, 28].

The asymmetry caused by eccentricity results in Unbalance Magnetic Pull (UMP) with vibration as a possible symptom in the synchronous generators. Eccentricity also wears down the bearings, which can further increase the fault degree and create a vicious loop that eventually ends in rotor-to-stator rubbing and serious damage. The abrasion of the stator and rotor can damage both the core and winding in the stator and rotor, and the machine must be replaced if the damage is extreme. Other consequences are reduced generator efficiency, amplified torque pulsation, and distorted harmonic content in the terminal voltage and currents. The increased vibration level may result in acoustic noise and the development of other faults, such as short circuits due to intensified insulation deterioration and increased temperature rise caused by local saturation. The induced current in the shaft and the vibrations generated by the eccentricity fault can also damage bearings and the insulation system of the rotor [24]. Monitoring the asymmetry of the air gap and bearings can provide important data related to the rotor eccentricity fault. However, the methods used for this purpose are invasive techniques that need extra measurement devices and present challenges in fixing them. The consequences of eccentricity indicate that proper condition monitoring of faults is of great importance in synchronous machines.

4.8 Misalignment Faults

Rotor misalignment faults in an electric machine represent the worst case of eccentricity fault, as the air gap length varies along the longitude axis of the machine. In the case of SE, DE, and ME

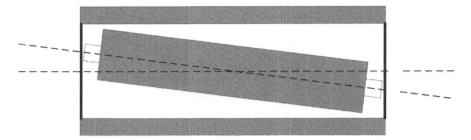

Figure 4.9 Static misalignment fault in electric machines.

faults, the air gap along the z-axis of the machine is assumed to be symmetric. Figure 4.9 shows misalignment based on the terms offset and gap difference in electric machine.

The misalignment fault is also included in three categories:

1. Static misalignment
2. Dynamic misalignment
3. Mixed misalignment

Assuming a horizontally mounted synchronous machine, the machine ends are opposite each other in the case of static misalignment fault. For instance, the drive end of the machine moves up, while the non-drive end of the machine falls down. However, the location of both ends is fixed. The dynamic misalignment fault results in a variation of the rotor both in the radial and axial directions of the synchronous machine, indicating that the rotor location changes with time. The simultaneous existence of both static and dynamic misalignment in the machine leads to an ME fault.

The bearing wear, bent shaft, defects caused during manufacturing, misaligned connection of the prime mover, such as turbine in the generator, and imbalance in the mechanical load in the motor applications are the factors that result in the misalignment fault. The misalignment fault is the most dangerous type of mechanical fault, as it can damage the entire machine, including the bearings, stator, rotor core, and winding. Detecting a misalignment fault in the early stage is essential to prevent a huge economic cost.

4.9 Damper Winding Fault

The rotor of the salient pole synchronous machine consists of two separate electric circuits: the rotor field winding circuit and the damper bar circuit. The damper bar circuit consists of several damper bars in each rotor pole that are short-circuited at both ends by end rings. Damper bars in the salient-pole synchronous machine shown in Figure 4.10 are heavy copper bars, with two copper rings that short circuit the damper bars on two sides of the rotor. There are two types of end ring connections in the salient pole synchronous machine, where the damper bars are short circuits only in each pole, and no connection exists between the adjacent pole using an end connection. The second type is the connected end ring that connects all damper bars in each rotor of a salient pole synchronous generator is shown in Figure 4.11.

The rotor of the synchronous machine rotates at the speed of the stator field to develop electromagnetic torque. Clearly, if these speeds are unequal, the machine is unable to generate a mean

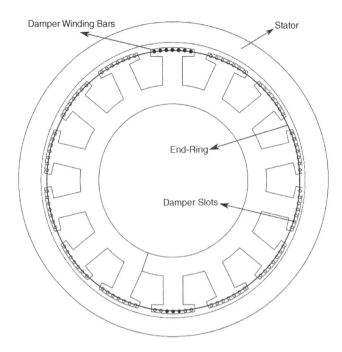

Figure 4.10 A salient pole synchronous generator with damper bars and end ring.

Figure 4.11 The rotor of a salient pole synchronous machine with seven damper bars in each rotor pole.

torque. At any other speed, the rotating field of stator poles will not be synchronized with the rotor poles. However, the concept of using the rotor bars of the induction machine can be similarly applied to the synchronous machine to start the synchronous machines asynchronously. Therefore, the salient pole synchronous motor starts to rotate when the stator magnetic field creates the magnetic field in the air gap, which consequently results in the induced voltage in the damper winding. The induced voltage in the damper bars creates a current since the damper bars are short-circuited using the end rings at both sides. The circulating current in the damper bars creates a magnetic field in the air gap, and its combination with the air gap magnetic field results in the torque that rotates the shaft. The damper bars are used to suppress the transient dynamics in the rotor caused by an external fault in the power system and the sudden load variation. In addition, the damper bars act as a shield to protect the rotor field winding during transients in the stator winding caused by the power network.

Broken damper bars in synchronous machines have not been widely addressed [29–37]. During transient modes, the synchronous machine is similar to asynchronous machines and high current follows in damper bars. The damper bars may be broken due to a large number of starts and stops or to frequent changes in the load or speed. In addition, inadequate connection in the joints between the damper bars and the end ring is the critical section that results in a broken damper bar fault. When a broken damper bar fault occurs, its current flows to the adjacent damper bars. Since the cross-section of the damper bars is designed for a limited current density, the additional current increases the ohmic loss and, consequently, the local temperature rise. The impact of an excessive temperature rise due to the additional current accelerates the process of breaking the bars adjacent to the faulty bar. Therefore, early detection of the broken damper bar is needed to avoid further damage to the neighboring damper bars.

The online diagnosis of broken damper bars is possible using air gap magnetic field monitoring [31]. For the squirrel cage induction machines, several methods of detection of broken rotor bars have been reported in the literature. In squirrel cage induction machines, when a bar breaks, some of the currents that would have flown into that bar will instead flow into the two adjacent bars on either side. This could result in the breakage of several bars [29]. Similar effects have been reported for converter-fed synchronous machines with broken damper bars [38].

4.10 Summary

Regardless of type, a serious generator fault can be critical for a power plant. Repair costs are huge due to the size, weight, and complexity of the machine, and the installation process requires a stoppage of power production, leading to a loss of income. If the defect is not reparable on site, tailored parts must be manufactured and transported, thereby adding further costs and making the repair long lasting and challenging, especially in countries where hydropower stations are often located in remote areas or inside mountains. If a fault leads to a shutdown of the machine, other aggregates may be overloaded in an attempt to compensate for the lost power production, thereby increasing the risk of new incidents. Moreover, a generator in faulty operation can result in a dangerous situation or a harmful work environment for the station's employees. This chapter described the various types of electrical and mechanical faults in synchronous generators. In addition, a detailed descriptions of the faults was investigated which provides a deep understanding of the fault causes and their consequences on the performance of the generators.

References

1 Khan, M., Ozgonenel, O., and Rahman, M.A. (2007). Diagnosis and protection of stator faults in synchronous generators using wavelet transform. In: *IEEE International Electric Machines & Drives Conference*, vol. 1, 184–189. Antalya, Turkey.

2 Cruz, S.M. and Cardoso, A.M. (2001). Stator winding fault diagnosis in three-phase synchronous and asynchronous motors, by the extended Park's vector approach. *IEEE Transactions on Industry Applications* 37 (5): 1227–1233.

3 Stone, G.C., Sasic, M., Stein, J., and Stinson, C. (2012). Using magnetic flux monitoring to detect synchronous machine rotor winding shorts. In: *Annual IEEE Pulp and Paper Industry Technical Conference (PPIC)*, June 17–21, 2012, 1–7. Oregon, USA.

4 Ehya, H., Nysveen, A., Nilssen, R., and Liu, Y. (2021). Static, and dynamic eccentricity fault diagnosis of large salient pole synchronous generators by means of external magnetic field. *IET Electric Power Applications* 15 (4): 890–902.

5 Fiser, R., Lavric, H., Bugeza, M., and Makuc, D. (2011). FEM modeling of inter-turn short-circuits in excitation winding of turbogenerator. *Przegląd Elektrotechniczny* 87 (3): 49–52.

6 Kiani, M., Lee, W., Kenarangi, R., and Fahimi, B. (2007). Frequency domain methods for detection of rotor faults in synchronous machines under no-load condition. In: *39th IEEE North American Power Symposium*, 30 September–2 October 2007, 31–36. Las Cruces, NM, USA.

7 Ehya, H., Sadeghi, I., and Faiz, J. (2017). Online condition monitoring of large synchronous generator under eccentricity fault. In: *12th IEEE Conference on Industrial Electronics and Applications (ICIEA)*, 18–20 June 2017, 19–24. Siem Reap, Cambodia.

8 Sadeghi, I., Ehya, H., Faiz, J., and Akmal, A.A.S. (2017). Online condition monitoring of large synchronous generator under short circuit fault – A review. In: *IEEE International Conference on Industrial Technology (ICIT)*, 18-20 June 2017, 1843–1848. Siem Reap, Cambodia.

9 Bone, J. and Schwarz, K. (1973). Large AC motors. *Proceedings of the Institution of Electrical Engineers (IEE)* 120 (10): 1111–1132.

10 Tavner, P., Ran, L., Penman, J., and Sedding, H. (2008). *Condition Monitoring of Rotating Electrical Machines*. UK: IET.

11 Bonnett, A.H. and Soukup, G.C. (1992). Cause and analysis of stator and rotor failures in three-phase squirrel-cage induction motors. *IEEE Transactions on Industry Applications* 28 (4): 921–937.

12 Bonnett, A. (1991). The cause of winding failures in three phase squirrel cage induction motors. In: *IEEE Annual Pulp and Paper Industry Technical Conference,* 3–7 June 1991. Montreal, QC, Canada.

13 Pillai, P., Bailey, B., Bowen, J. et al. (2004). Grounding and ground fault protection of multiple generator installations on medium-voltage industrial and commercial power systems. Part 1 – The problem defined. In: *57th IEEE Annual Conference for Protective Relay Engineers*, 1 April 2004, 132–138. College Station, TX, USA.

14 Lee, Y. and Habetler, T.G. (2007). An on-line stator turn fault detection method for interior PM synchronous motor drives. In: *22nd IEEE Annual Applied Power Electronics Conference and Exposition*, 5 February 2007–01 March 2007, 825–831. Anaheim, CA, USA.

15 Neti, P. and Nandi, S. (2009). Stator interturn fault detection of synchronous machines using field current and rotor search-coil voltage signature analysis. *IEEE Transactions on Industry Applications* 45 (3): 911–920.

16 Kliman, G., Premerlani, W., Koegl, R., and Hoeweler, D. (1996). A new approach to on-line turn fault detection in AC motors. In: *21st IEEE Industry Applications Conference*, 6–10 October 1996, vol. 1, 687–693. San Diego, CA, USA: IEEE.

17 Kim, K.H., Gu, B.G., and Jung, I.S. (2011). *IET Electric Power Applications* 5 (6): 529–539.

18 Nandi, S. and Toliyat, H.A. (2002). Novel frequency-domain-based technique to detect stator interturn faults in induction machines using stator-induced voltages after switch-off. *IEEE Transactions on Industry Applications* 38 (1): 101–109.

19 Ehya, H. and Nysveen, A. (2021). Pattern recognition of inter-turn short circuit fault in a synchronous generator using magnetic flux. *IEEE Transactions on Industry Applications* 57: 3573–3581.

20 Ehya, H., Nysveen, A., Groth, I., and Mork, B. (2020). Detailed magnetic field monitoring of short circuit defects of excitation winding in hydro-generator. In: *2020 International Conference on Electrical Machines (ICEM)*, 23–26 August 2020, vol. 1, 2603–2609. Gothenburg, Sweden.

21 Ehya, H., Nysveen, A., and Nilssen, R. (2020). Pattern recognition of inter-turn short circuit fault in wound field synchronous generator via stray flux monitoring. In: *International Conference on Electrical Machines (ICEM)*, 23–26 August 2020, vol. 1, 2631–2636. Gothenburg, Sweden.

22 Pennacchi, P. and Frosini, L. (2007). Computational model for calculating the dynamical behaviour of generators caused by unbalanced magnetic pull and experimental validation. In: *International Design Engineering Technical Conferences and Computers and Information in Engineering,* September 4–7, vol. 48027, 1313–1326. Las Vegas, Nevada, USA.

23 Torregrossa, D., Khoobroo, A., and Fahimi, B. (2011). Prediction of acoustic noise and torque pulsation in PM synchronous machines with static eccentricity and partial demagnetization using field reconstruction method. *IEEE Transactions on Industrial Electronics* 59 (2): 934–944.

24 Talas, P. and Toom, P. (1983). Dynamic measurement and analysis of air gap variations in large hydroelectric generators. *IEEE Transactions on Power Apparatus and Systems* 9: 3098–3106.

25 Bruzzese, C., Santini, E., Benucci, V., and Millerani, A. (2009). Model-based eccentricity diagnosis for a ship brushless-generator exploiting the machine voltage signature analysis (MVSA). In: *IEEE International Symposium on Diagnostics for Electric Machines, Power Electronics and Drives,* (31 August–3 September 2009), 1–7. Cargese, France.

26 Nandi, S., Ilamparithi, T.C., Lee, S.B., and Hyun, D. (2010). Detection of eccentricity faults in induction machines based on nameplate parameters. *IEEE Transactions on Industrial Electronics* 58 (5): 1673–1683.

27 Rødal, G.L. (2020). Online condition monitoring of synchronous generators using vibration signal, MSc Thesis, NTNU, Norway.

28 Ehya, H., Rødal, G.L., Nysveen, A., and Nilssen, R. (2020). Condition monitoring of wound field synchronous generator under inter-turn short circuit fault utilizing vibration signal. In: *23rd International Conference on Electrical Machines and Systems (ICEMS)*, (23–26 August 2020), 177–182. Gothenburg, Sweden.

29 Ehya, H. and Nysveen, A. (2022). Comprehensive broken damper bar fault detection of synchronous generators. *IEEE Transactions on Industrial Electronics* 69 (4): 4215–4224.

30 Ehya, H., Nysveen, A., Nilssen, R., and Lundin, U. (2019). Time domain signature analysis of synchronous generator under broken damper bar fault. In: *IEEE 45th Annual Conference* of *Industrial Electronics Society (IECON)*, (14–17 October 2019), vol. 1, 1423–1428. Lisbon, Portugal.

31 Karmaker, H. (2003). Broken damper bar detection studies using flux probe measurements and time-stepping finite element analysis for salient-pole synchronous machines. In: *4th IEEE International Symposium on Diagnostics for Electric Machines, Power Electronics and Drives, (SDEMPED)*, 24–26 August 2003, 193–197. Atlanta, GA, USA.

32 Kliman, G., Koegl, R., Stein, J., and e.a. (1988). Noninvasive detection of broken rotor bars in operating induction motors. *IEEE Transactions on Energy Conversion* 3 (4): 873–879.

33 Jovanovski, S.B. (1969). Calculation and testing of damper-winding current distribution in a synchronous machine with salient poles. *IEEE Transactions on Power Apparatus and Systems* 11: 1611–1619.

34 Rahimian, M.M. and Butler-Purry, K. (2009). Modeling of synchronous machines with damper windings for condition monitoring. In: *IEEE International Electric Machines and Drives Conference* (3–6 May 2009), 577–584. Miami, FL, USA.

35 Bacher, J. (2004). Detection of broken damper bars of a turbo generator by the field winding. *Renewable Energy Power Quality Journal* 1 (2): 199–203.

36 Antonino-Daviu, J.A., Riera-Guasp, M., Pons-Llinares, J. et al. (2012). Toward condition monitoring of damper windings in synchronous motors via EMD analysis. *IEEE Transactions on Energy Conversion* 27 (2): 432–439.

37 Sahoo, S., Holmgren, F., Rodriguez, P. et al. (2018). Damper winding fault analysis in synchronous machines. In: *XIII International Conference on Electrical Machines (ICEM)*, (3–6 September 2018), 1789–1795. Alexandroupoli, Greece.

38 M.M. Rahimian, Broken bar detection in synchronous machines based wind energy conversion system, PhD Thesis, Texas A&M University, 2011.

5

Laboratory Scale Implementation

5.1 Introduction

Although theoretical approaches, such as analytical or numerical methods, provide a deep understanding of machine performance, either in a healthy case or in faulty conditions, experimental verification is always necessary. In addition, the results acquired in a simulation may sometimes lack detailed specifications because each modeling method has its own limitations, leading to results with several assumptions. Moreover, some ideas that can be modeled by simulation are practically impossible. Therefore, validating the simulation results through experiments is essential. A detailed description of the implemented experimental setup is as follows:

1. Laboratory equipment for the test setup includes:
   a. Salient pole synchronous generator.
   b. Induction motor as a prime mover.
   c. Gearbox.
   d. Converter.
   e. Rotor magnetization unit.
   f. DC power supply.
   g. Local passive load.
2. Sensors:
   a. Hall-effect sensors.
   b. Stray magnetic field sensor.

Electromagnetic Analysis and Condition Monitoring of Synchronous Generators, First Edition. Hossein Ehya and Jawad Faiz.
© 2023 The Institute of Electrical and Electronics Engineers, Inc. Published 2023 by John Wiley & Sons, Inc.

    c. Voltage transformer.
    d. Current transformer.
    e. Accelerometer.
3. Data Acquisition:
    a. Oscilloscope.
    b. Lab-view national instrument.
4. Fault implementation:
    a. Stator short circuit fault.
    b. Rotor short circuit fault.
    c. Static and dynamic eccentricity fault.
    d. Misalignment.
    e. Broken damper bar.

5.2 Salient Pole Synchronous Generator

A custom-made 100 kVA salient pole synchronous generator (SPSG), shown in Figure 5.1, is used as a case study in both the experimental and simulation cases in this book. Table 5.1 summarizes the specifications of the SPSG. Accurate fault detection and condition monitoring, in addition to the machine rating power, depend on the SPSG topology. The topology of the 100 kVA SPSG resembles that of an actual hydro-generator in a Norwegian power plant. Its number of poles is 14, which is the second most common number of poles in the Norwegian hydro-power plant after synchronous generators with 16 salient poles. One of the main intentions is to investigate the SPSG performance when the generator is connected to the power grid. Note that the proper synchronous reactance of the machine is achieved by having an air gap length equal to 1.75 mm. Table 5.2 demonstrates the detailed geometry specifications of the SPSG.

Figure 5.1 A 100 kVA salient pole synchronous generator.

Table 5.1 Nameplate of the salient pole synchronous generator.

Parameter	Value
Rating power (kVA)	100
Nominal stator terminal voltage (V)	400
Nominal full load current (A)	144.3
Rated torque (Nm)	2204/2233
Power factor	0.9
Synchronous speed (rpm)	428
Runaway speed (rpm)	685
Frequency (Hz)	50
No-load excitation current (A)	53.2
The full load excitation current (A)	103
Synchronous reactance X_d-unsaturated	≤ 1.3
Transient reactance X'_d-unsaturated	≤ 0.3

Table 5.2 Specifications of the salient pole synchronous generator.

Parameter	Value
Number of poles	14
Number of slots	114
Number of damper bars in each pole	7
Number of turns in rotor pole winding	35
The number of turns per stator slot	8
Winding connection	Wye
Winding layout	Double layer
The outer diameter of the stator	780 mm
The outer diameter of the rotor	646.5 mm
The minimum air gap length	1.75 mm
Length of stack	240 mm

The stator core and the stator slot drawing of the SPSG are depicted in Figure 5.2. The stator of the machine consists of 114 slots filled with copper conductors. The slot opening and the height of the stator slot are 8.5 and 33.5 mm, respectively. The material used in the stator core is electrical steel (M400-50A), which is insulated on both sides. The number of slots (Q_s) per pole (p) and phase (m) for this generator is

$$q_s = \frac{Q_s}{pm} = \frac{114}{14 \times 3} = 2 + \frac{5}{7} \qquad (5.1)$$

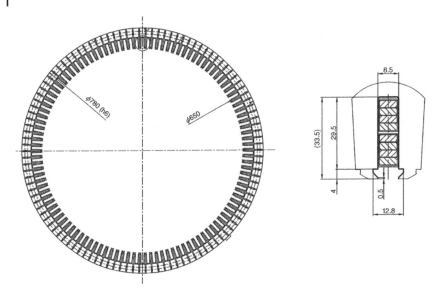

Figure 5.2 Stator and rotor of 100 kVA salient pole synchronous generator.

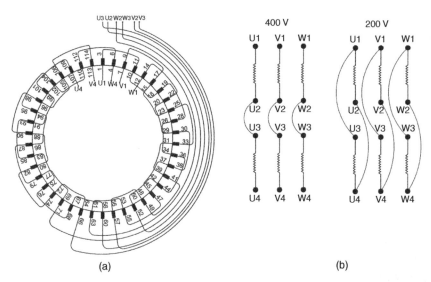

Figure 5.3 Winding layout of the stator winding (left) and type of stator winding connection in (a) series, (b) parallel (right).

The number of slots per pole and phase is equal to 19/7, which leads to a double layer layout. The winding layout of the machine is presented in Figure 5.3 and indicates the possibility of in-series or in-parallel connection. An in-series connection of the stator winding would yield an output RMS (root mean square) value in the stator terminal of 400 V, while a parallel connection would result in an output voltage of 200 V. The parallel connection is desirable, as it will reduce an unbalanced magnetic field arising during manufacturing or due to faults.

The generator rotor consists of the 14 salient poles that are attached to the rotor yoke. Figure 5.4 shows a three-dimensional drawing of the rotor and its photo. The rotor of the SPSG consists of a rotor yoke, shaft, and field windings. Each rotor pole contains seven damper bars with a diameter

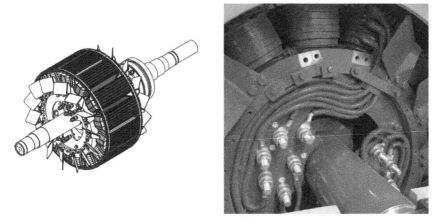

Figure 5.4 Three-dimensional drawing and photo of an SPSG rotor.

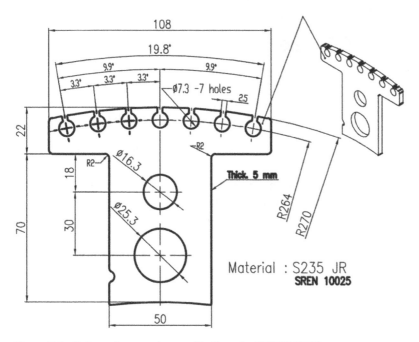

Figure 5.5 Rotor pole geometry specification of a 100 kVA SPSG.

of 7 mm and a length of 310 mm. The rotor field winding has 35 copper turns. The rotor pole shoe is a laminated type, and its geometry, with the detailed specifications, is shown in Figure 5.5.

5.3 Induction Motor

A synchronous generator is normally rotated by a turbine and generates electrical power. In the present experimental setup, a four-pole 90 kW inverter-fed induction motor is used as the prime mover of the synchronous generator. The rated speed of the induction motor is 1482 rpm. The stator winding can be connected in either the delta or the star configuration. The input voltage is 400 V in

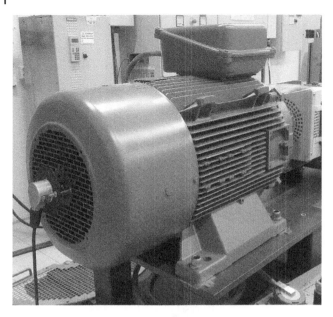

Figure 5.6 A 90 kW induction motor with a rated speed of 1482 rpm.

the delta connection and 690 V in the star connection. The input currents of the induction motor for delta and star connections are 165 and 95.5 A, respectively. Figure 5.6 shows the induction motor in this experimental setup.

The working environment of the induction motor, such as temperature, humidity, and altitude, is dictated by the datasheet. The induction motor is placed in a firm bed to avoid any distortion and unexpected vibration. The foot screws must be tightened according to the recommended torque by the manufacturer. The diameter of the screw must fill the motor feet holes to prevent any movement. The alignment of the motor shaft with the generator or the gearbox must be checked carefully, since any small misalignment may result in serious damage to both the induction motor and the equipment coupled to its shaft.

5.4 Gearbox

The rotating speed of the 14 pole SPSG is 428 rpm and the running speed of the induction motor is 1485 rpm; therefore, the prime mover speed must be reduced. A helical gear unit, which is an industrial gearbox, adjusts the speed. Here, a gearbox guide with a gear ratio of 1485 rpm to 375.95 rpm is used. Since the gearbox output is less than is needed to achieve a 50 Hz frequency in the SPSG, the running speed of the induction motor is increased to achieve a 428 rpm at the gearbox output. The utilized gearbox is a foot-mounted type and is fixed with a screw on to a firm bed. The gearbox is filled with grease by the manufacturer, but the operator must check the oil level frequently and avoid oil leakage, because this results in serious damage to the gearbox. Figure 5.7 shows the industrial gearbox that connects the induction motor shaft to the synchronous generator.

Figure 5.7 Speed adjustment of the industrial gearbox.

5.5 Converter

A three-phase 100 kVA converter made by SINTEF Energi[1] (see Figure 5.8) supplies the induction motor. The converter is also used to adjust the synchronous speed of the generator and the active power. For proper operation of the motor, its parameter motor must be included in the converter software. The speed reference is the only parameter that must be adjusted based on the active power variation in the synchronous generator.

The converter is built based on Semikron SEMIKUBE converter modules. The SEMIKUBE platform is a family of pre-qualified power assemblies. The converter is directly connected to the grid to set up the DC link using the built-in filters and the IGBT rectifier. The input voltage of the converter is 750 V AC and the output is 750 V DC. The converter consists of the following three main parts:

1. Converter core module, including IGBT bridges, drivers, heatsinks, rectifier, and DC link.
2. LCL filters.
3. Switchgear, including contactors, DC link charging, and discharging circuits.

Figure 5.9 shows the schematic diagram of the induction motor connection to the grid. The motor is directly supplied by the converter, and the converter is supplied by the grid via the diodes and filters. The main contractor can be engaged to drive the motor if the DC link is already charged. This is done with a charging circuit consisting of charging resistors, a charging rectifier bridge, and a separate contractor.

5.6 Rotor Magnetization Unit

The magnetization unit shown in Figure 5.10 is used to provide a DC current to feed the rotor field winding. The rating power of the magnetization unit is 20 kW and is connected to the 230 V

---

1 https://www.sintef.no/en/sintef-energy.

Figure 5.8 A three phase 100 kVA converter used to drive an induction motor.

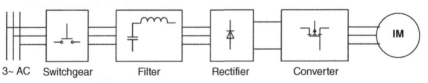

3~ AC   Switchgear   Filter   Rectifier   Converter

Figure 5.9 Schematic diagram of the induction motor connection to the grid.

power network. The magnetization unit consists of a capacitor bank and an IGBT transistor bridge (Semikron SKM400GB125D). Figure 5.11 shows the schematic of the magnetization unit. The unit is directly connected to the grid, and its DC output voltage is connected to the terminal cabinet. This cabinet includes the taps connected to the slip rings on the rotor shaft. The unit can be controlled by the front control panel or it can be connected to the high-speed link and controlled remotely. The magnetization unit can be used to provide no-load to full-load current in two modes:

1. Local passive load.
2. Grid connection.

When the generator supplies the local passive load, the magnetization current behaves like a DC power supply. When it is connected to the grid, synchronization of the generator must be performed

Figure 5.10   Magnetization unit of the rotor field winding.

Figure 5.11   Schematic diagram of the magnetization unit.

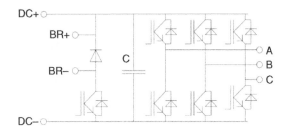

using the following steps:

1. The amplitude of the induced line voltage in the stator terminal must be equal to the grid voltage.
2. The frequency of the generator and the grid must be the same.
3. The phase sequence of both the generator and the grid must be the same.
4. The phase angle must be the same.

All these criteria must be met when the generator is connected to the grid. The criteria are checked with a synchroscope on the front panel of the magnetization unit.

Figure 5.12 DC power source as an exciter.

5.7 DC Power Supply

An exciter supplies a DC current to a rotor field winding. Although programming an exciter is possible to produce a power source with a specific waveform, another possibility is to use a DC power supply as an alternative for a specific application, such as producing a waveform with a specific pattern. The DC power supply used here (i.e., DELTA ELEKTRONIKA SM66-AR-110) has the following features (see Figure 5.12):

1. Output voltage between 0 and 66 V.
2. Output current between 0 and 110 A.
3. Excellent dynamic response to load changes required for a fast rise waveform.
4. Low noise reduces noise impact on sensors and measurements.
5. High-speed programming.
6. It can operate in either constant current mode or constant voltage mode.

The power supply output can be either a constant current source with voltage limits or a constant voltage source with current limits. The output power of the power supply is limited; therefore, it can be connected in series or parallel to have one power source as a master and one or more power supplies as a slave. This feature provides flexible control over the entire DC power supply. The temperature of the operating environment must range from −20 to +50 °C for full power operation, whereas a working environment of 60 °C requires that the power be derated to 75% of the full power. The available DC power supply has numerous approaches for control, including control in a local mode from the front interface or from a remote interface that includes the remote web, ethernet, and sequencer. A local area network (LAN) cable is needed to connect the power supply to the PC to have access to remote control. If a power supply with a specific shape is needed, the sequencer must be used. A sample program for a trapezoidal-shaped current source is used for a broken bar detection. The rise time and fall time of a signal and its amplitude can be adjusted, as shown in Figure 5.13, by changing the parameters:

s = 56 → Amplitude of current
inc sv, 0.1 −→ Increase of voltage with 0.1 step
w = 0.1 −−−−−−−−−−→ Wait for 0.1 seconds
ci1 sv, 14, 2 −−−−−−−→ Compare, if less jump to 2
#j = 50 −−−−−−−−−−→ Count down counter
cjne #j,0,6 −−−−−−−−→ Compare and jump
Label 1:
dec sv,0.1−−−−−−−→ Decrease voltage with 0.1 step
w = 0.1 −−−−−−−−−→ Wait for 0.1 seconds
cjg sv,0,label −−−−→ Compare, if greater jump to label 1
end −−−−−−−−−−→ End

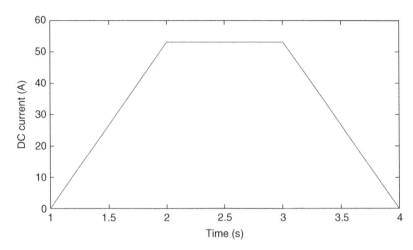

Figure 5.13 Trapezoidal shape DC current generated at the terminal of the DC power source current.

5.8   Local Passive Load

The synchronous generator fault is diagnosed in no-load or full-load operation, and the load effect on the faulty machine is identified. The no-load test is a simple matter. The reference speed of the induction motor gradually increases until the generator reaches the synchronous speed; then, the magnetization current is increased to obtain a no-load phase voltage of 230 V at the stator terminal of the synchronous generator. Based on the network demand, the load is imposed on the generator in the power plants. Having control of the load in the laboratory setup is useful for determining the different load configuration effects on the measured signals.

Since the dissipated power in the resistors is high and can result in a temperature rise in the laboratory, a water-cooled resistor bank can be an appropriate option. A three-phase water-cooled resistor bank is connected to the generator terminals. Figure 5.14 shows the water-cooled resistor bank and its control panel, which is controlled by contactors and relays. The resistor bank consists of two parallel sets of resistors. The total resistance can be adjusted stepwise by contactors from the control panel. The resistor bank is connected to the relays, and in the case where the dissipated loss exceeds the set temperature limit, water cooling is activated. The per-phase resistance can vary from 160 to 2.78 Ω. The maximum resistive loading that can be achieved by a minimum resistance is 57 kW.

The behavior of both healthy and faulty synchronous generators must be investigated under either active power or reactive power or a combination of both loads. The inductors can be used as an inductive load. Two three-phase inductors connected in series (to increase the inductance) are connected to the three-phase transformer. In a case with a lower inductance than the expected value, the use of a transformer is possible to increase the inductance. Figure 5.15 shows the two series-connected three-phase inductors with their outputs connected to the transformer.

5.9   Sensors

Signals are the most important ingredient required for fault detection in electrical machines. Numerous measurable signals are present in electric machines. The measurement techniques are divided into the following two categories:

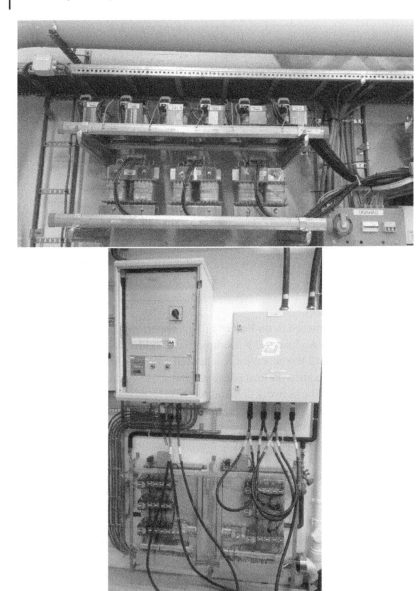

Figure 5.14 A water-cooled resistive bank and its control panel.

1. **Invasive approach**: In this approach, a significant change and modification are required to measure the signal. For instance, to measure the air gap magnetic field, the air gap must be accessible; therefore, the machine must be stopped and dismantled.
2. **Non-invasive approach**: In this method, signals can be measured without any modification to the machine, and the use of an existing sensor is possible. Since most power plants have current transformers and potential transformers, the stator terminal voltage and current are measured non-invasively.

In the following sections, different signals and their required sensors are discussed.

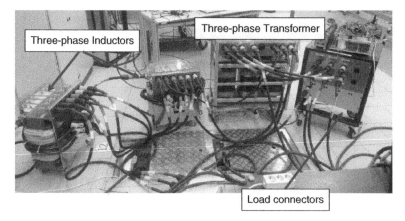

Figure 5.15 Two three-phase inductors are connected in series, and the output is connected to a three-phase transformer.

5.9.1 Hall-Effect Sensors

The air gap magnetic field is a key parameter of electromagnetic field conversion. The air gap magnetic field also contains adequate information regarding the health status of the generators; therefore, monitoring the magnetic field is a useful approach. However, applying this approach is not easy if the manufacturer has not already fixed a Hall-effect sensor inside the machine. Moreover, the length of the air gap and the size of the Hall-effect sensor are crucial because finding a Hall-effect sensor for a machine with a small air gap length is difficult. In addition, adjusting a Hall-effect sensor in a noisy environment by modifying its connections is laborious.

The proper type of Hall-effect sensor must be selected based on the requirement for the generator condition monitoring purpose. The sensor must:

- Be designed to withstand a strong magnetic field.
- Be linear (analog).
- Have high precision and accuracy.
- Have a wide bandwidth.
- Be small enough to fix in the air gap.
- Be capable of withstanding a tangential force.
- Be able to tolerate the operating temperature of the generator.

A suitable Hall-effect sensor is the AST244 sensor, which has a high linear precision. The sensor is able to measure magnetic fields varying from a few μT to more than 10 T. It has low noise and low pick-up Electromagnetic Compatibility (EMC) characteristics. Moreover, it has a low temperature coefficient, which makes it a good choice for operating inside a generator, where the temperature varies from no-load to full-load operation. The upper limit of the sensor temperature is 120 °C, while for applications with a higher temperature, the AST244HT model has a temperature limit of 200 °C. The size of the sensor is $(3.0 \times 5.0 \times 0.8)$ mm. The sensor has high sensitivity and a low offset voltage. According to the data provided by the manufacturer, for a flux density of 1 T with a supply current of 2 mA, the output voltage must be around 200 mV. The sensor has four wires: two for the power supply and two for the sensor output. The length of the wires is 40 cm. Figure 5.16 shows a Hall-effect sensor.

The sensor is supplied by a constant DC current. According to the manufacturer's instructions, the four wires of an HP 34401A multi-meter must be connected to provide a constant current to

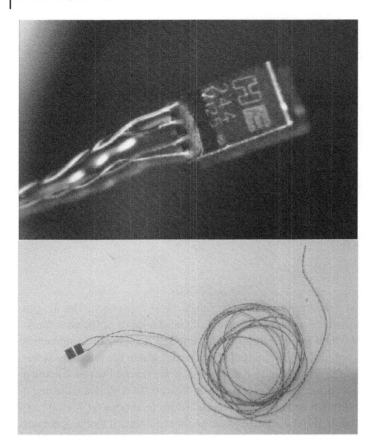

Figure 5.16 Hall-effect sensor AST244. https://www.asensor.eu.

the sensor and measure the Hall-element induced voltage. The multi-meter was able to provide a 1 mA current, but the needed current was at least 2 mA. Moreover, the multi-meter needed to be connected to the grid, which was not appropriate for portable measurement. As an alternative, a portable constant DC current circuit has been proposed in reference [1] that only needs a battery as a power supply.

Figure 5.17 shows the schematic of the used circuit [2]. A simple circuit consists of a 9 V battery, as well as a constant voltage regulator, which is required to avoid an input voltage drop when the battery is discharged. A current regulator, providing a constant current, is connected to the output of the voltage regulator. The circuit consists of multiple diodes, resistors, and potentiometers for adjusting the voltage and current ratio to compensate for the temperature dependency of the current regulator. Figure 5.18 shows the implemented circuit for the final product. Although the recommended constant current of the Hall-effect sensor is 2 mA, increasing the current is possible to reduce the noise impact on the signal. At a 4 mA current, the output voltage is expected to be 400 mV/T.

Although the proposed circuit in reference [2] provides a stable constant current to the sensor, a commercialized current source is required if the purpose of the measurement is a permanent or periodic measurement of the air gap magnetic field in a power plant. The battery voltage drops over the long term and may result in inaccurate magnetic field monitoring.

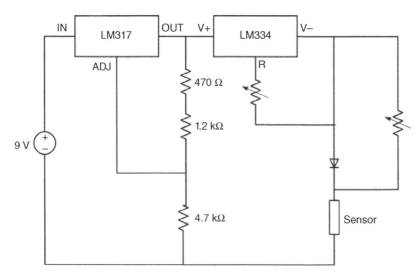

Figure 5.17 Schematic of a constant current circuit. LM334 and LM317 are the current and voltage controllers, respectively. The resistors were chosen to provide a 4 mA constant current. The temperature dependency of the voltage regulator is compensated for by a diode and two potentiometers with a 1/10 ratio.

Figure 5.18 Final prototype of the constant current circuit.

The designed DC current source must be calibrated for accurate measurement of the air gap magnetic field. The circuit is calibrated using a reference magnet and a measuring device (Leybold Mobile-CASSY 2). The reference magnet has an air gap with a uniform magnetic flux density of 0.5 T. A probe from the measuring device is inserted inside the magnet air gap and the measured value is 486 mT. The Hall-effect sensor connected to the DC current power source is placed in the same spot as where the probe of the measuring device had been placed. The measured voltage generated by the reference magnet in both sensors must be the same. Figure 5.19 shows the reference

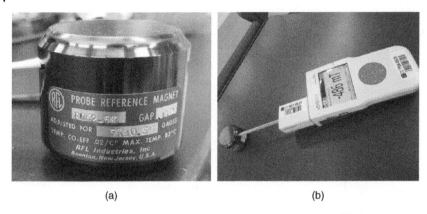

(a)   (b)

Figure 5.19 (a) Reference magnet with a uniform magnetic field of 0.5 T. *Source*: RFL Electronics Inc and (b) magnetic field measuring device. *Source*: Leybold.

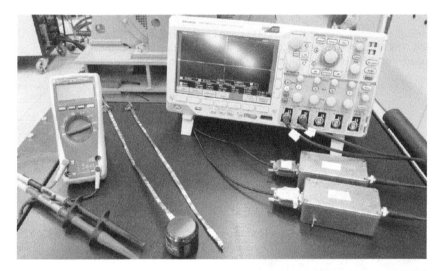

Figure 5.20 Complete setup of the Hall-effect sensor used to measure the air gap magnetic field.

magnet and the measuring device. The possibility remains that a sensor output, even after calibration, will become dissimilar when installed inside the air gap due to changes in the resistivity of the wires. Bending of the sensor wire could result in a variation in the induced voltage. Therefore, making the wire connection of the sensor through the end winding is recommended, rather than bending the wire to pass it through the ventilation ducts. An experiment shows that the magnetic field differs in wires installed in these two different situations. Figure 5.20 shows the complete setup, including the power source and the sensor for the air gap magnetic field measurement [1].

The Hall-effect sensor must be installed close to the end winding and the ventilation ducts because wire bending can markedly alter the magnetic field in the sensor [3]. Assessment of the static eccentricity fault of the generator requires at least two sensors installed exactly opposite each other in a horizontally mounted machine [4]. Figure 5.21 shows the installed sensor in the middle of the stator tooth.

The output voltage, the location of the mounted sensor, and the faces exposed to the magnetic field may result in a voltage offset in the sensor output. Moreover, since the sensors are not identical, the voltage offset is not constant. The voltage offset results in an evident DC component in

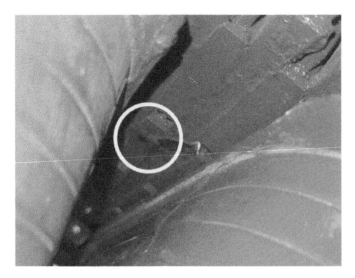

Figure 5.21 Location of the Hall-effect sensor on the stator tooth.

the frequency spectrum that is removable if the average value of each pole is computed. According to the sensor data sheet, it works up to 120 °C without any changes in the sensor output characteristics; however, the temperature rises yield a small increment in the sensor output. Therefore, for an environment with a high temperature variation, a sensor with a higher temperature stability is required.

5.9.2 Search Coil

The search coil is one of the most prevalent types of magnetic field measurement sensors. It can be used either inside the generator to measure the air gap magnetic field or outside the generator to measure the stray magnetic field. The search coil-based measurement has the following advantages over the Hall-effect sensor:

1. The search coil does not need a current power source or a sophisticated calibration.
2. Unlike the case for the Hall-effect sensor, the environment and the temperature do not have any impact on the data measured by the search coil.
3. The output-induced voltage in the search coil is more robust to the small changes on the location of the installation compared with the Hall-effect sensor.

The structure of the search coil varies depending on its location in the generator. The search coil used inside the generator to measure the air gap magnetic field consists of a couple of copper turns around the stator tooth. The induced voltage in the search coil around the stator tooth depends on the number of coil turns, in addition to the tooth area, which acts as a core. Figure 5.22 shows the location of the search coil inside the synchronous generator. If the use of a search coil is preferred over the Hall-effect sensor in the air gap, the sensors must be identical to achieve a uniform output using a printed circuit board technique.

The search coil has several advantages compared with the Hall-effect sensor. Search coils are not sensitive to temperature variations or noise and they are not fragile. However, installation in an operating generator is almost impossible because the rotor must be removed to insert the coils around the stator tooth.

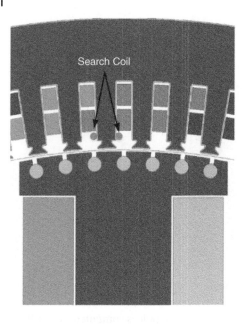

Figure 5.22 Location of a search coil inside a synchronous machine to capture the air gap magnetic field.

Inserting a Hall-effect sensor or a search coil inside the air gap is an invasive method and is not an industry-favored approach because the generator must be stopped and dismantled. Measuring the external magnetic field close to the stator yoke or on the generator frame is an advanced solution for monitoring the magnetic field. The reason is that the external magnetic field is a mirror of the air gap magnetic field.

An external search coil consists of a plastic reel and hundreds of turns of copper wire. The shape of the external search coil must be optimized based on the generator topology, because the order of the magnetic field on the frame or stator back side varies case by case. For instance, a synchronous generator operating in a power plant does not have a frame and the sensor measures the external magnetic field on the stator core [4, 5].

Figure 5.23 shows the external search coil. The output of the sensor is connected to the coaxial cable to avoid any disturbance or noise coming from the working environment. A Bayonet Neil-Concelman or British Naval Connector (BNC) port, or sometimes (British Naval Connector) connector is used to connect a computer to a coaxial cable in a 10BASE-2 Ethernet network. BNC port is used to connect the signal to the data acquisition card or an oscilloscope. The connection point between the search coil wire and the coaxial cable is covered with copper tape to avoid noise leakage into the signal.

The dimension of the sensor depicted in Figure 5.23 is $(100 \times 100 \times 10)$ mm. It contains 300 turns of copper wire with a diameter of 0.12 mm^2. The resistivity and the inductance of the search coil at the terminal are $912\,\Omega$ and 714 mH, respectively. The location of the sensor can differ based on the type of external magnetic field that is required to be measured [6]. Three types of external magnetic fields can be captured:

1. Axial flux, as shown in Figure 5.24 with label A. The sensor only captures the axial flux in this direction.
2. Radial flux, as shown in Figure 5.24 with label C. In this location, only the radial magnetic field is captured.
3. The combination of both axial and radial flux can be achieved by locating a sensor in location B, as shown in Figure 5.24.

Figure 5.23 An external search coil for measuring an external magnetic field in the vicinity of a generator.

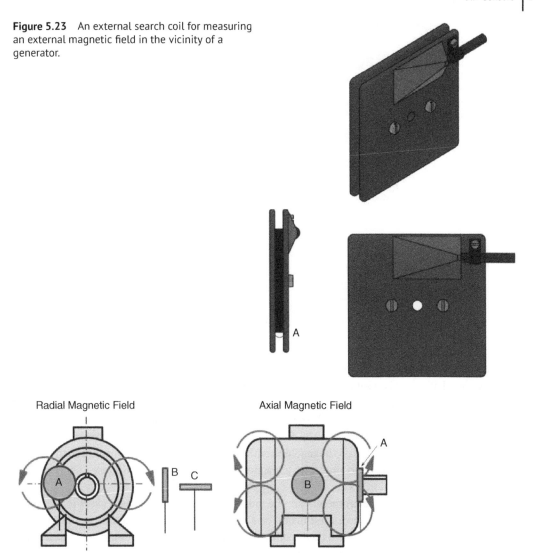

Figure 5.24 Location of external search coils in generators [14] / with permission of IEEE.

The number of sensors needed for each application depends on the type of fault. For instance, only one sensor is required if the generator is exposed to a short circuit or dynamic eccentricity fault. Two or four sensors are required for a static eccentricity fault detection for horizontally and vertically mounted generators, respectively. If the generator is under a misalignment fault, the number of needed sensors is eight because the magnetic field must be assessed through the stator stack length.

5.9.3 Accelerometer

Faults in both the stator and rotor result in an air gap magnetic field distortion. A consequence of the non-uniform air gap magnetic field is a marked variation in the vibration behavior of the generator. Vibration monitoring is therefore a promising approach for fault detection. Vibration

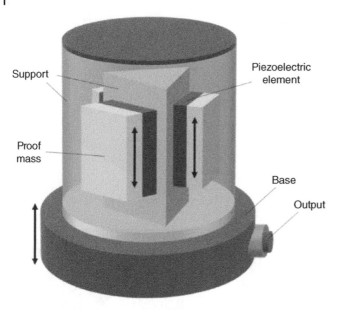

Figure 5.25   Internal components of an accelerometer.

monitoring is a well-established method with the standard covering of various types of generators [7, 8]. An accelerometer is used to measure the vibration. Vibration can be expressed as velocity, displacement, or acceleration without any stoppage or modification of the generator. However, an accelerometer can also be installed inside the generator frame, but this makes it an invasive method. An accelerometer can be mounted on the frame or on the stator backside. Each accelerometer sensor is designed to be mounted on the machine with a screw or a magnet. A sensor to be mounted on the stator backside would be secured with a magnet, as using a screw would cause a short circuit between the stator lamination sheets.

Piezoelectric shear-type accelerometers are the most common types used as sensors to measure vibration [9, 10]. Figure 5.25 shows the inside of a sensor, where proof masses oscillate in the transducer in the vertical direction. The direction of oscillation is perpendicular to the base that is attached to the body of the generator. The piezoelectric element moves in a shear manner and the generated electric charge is transmitted through a cable to the data acquisition system. The natural frequency of the sensor using the piezoelectric element is more than 20 kHz, which is far from the frequency of the vibration of the generator, indicating that the output signal would not be distorted by resonance in the transducer.

Figure 5.26 shows an industrial accelerometer (DYTRAN 3059A), which is mainly used for rotating electrical generators. The sensor is hermetic and has a low noise feature. The sensitivity of the model is 100 mV/g, which gives a full-scale voltage of ± 5 V. This sensor is an integrated piezoelectric with a built-in amplifier that is connected to a constant current power source through the cable. The generated electric charges are fed into an integrated circuit JFET charge amplifier. The output voltage is in mV/g and is compatible with an industrialized or homemade constant power source. Figure 5.27 shows the power source used to feed the constant power into the sensor. It has four channels, where the input channel is connected to the sensor that supplies constant power to the sensor circuit and transfers the signal from the sensor to the power supply, while the output sends a signal to the data acquisition system.

Figure 5.26 Industrialized accelerometer to measure vibration in a generator. https://www.dytran.com/Model-3059A-Industrial-Accelerometer-P1331.

Figure 5.27 Industrial constant current source with four channels, utilized for accelerometers. *Source:* Dytran Instruments, Inc.

Although using an accelerometer is a straightforward approach for measuring the vibration of the generator, the vibration can also be calculated using an air gap magnetic field. Neglecting the tangential magnetic field, the vibration is approximately equal to the square root of the radial magnetic field. Therefore, measuring the air gap magnetic field signal with a Hall-effect sensor or an internal search coil can also capture the vibration signal.

5.9.4 Voltage Transformer

Voltage is one of the commonly measured quantities in the condition assessment of generators. The measured voltage signal on the stator terminal can provide useful information, both in the no-load and on-loaded conditions. A high-voltage differential probe must be used to measure the voltage difference between the stator terminals, as neither of the terminals is at the ground. Several types of high voltage differential probes are available, and a suitable one must be selected based on the terminal voltage level. Figure 5.28 shows a high-voltage differential probe (P5200 Tektronix) that provides a safe measurement of the RMS voltage up to 1000 V for the common mode and 1300 V for the differential probe. The differential probe converts the high-voltage non-grounded floating signal into a low-voltage signal that can be measured by an oscilloscope. This probe has a two-range button. For voltages above 130 V and up to 1300 V for the differential mode, a 1/500 range must be used, while for a measured voltage below 130 V, a 1/50 range must be used for high

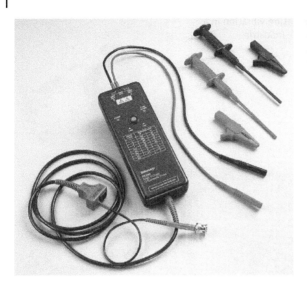

Figure 5.28 High-voltage differential probe. https://www.tek.com/probes-and-accessories/high-voltage-differential-probes. *Source*: Nortelco.

accuracy. The voltage can be calculated based on the multiplication of the oscilloscope V/div to the range set, which is 500 or 50. For example, if a V/div of the oscilloscope is 0.2 and the range set of the differential probe is 50, then the effective volts per division is 10 V. The bandwidth of the differential probe is 100 MHz and provides sufficient bandwidth for fault detection purposes. For safe operation, the high-voltage differential probes with an oscilloscope must not be used with floating inputs (isolated inputs) because the probe needs an oscilloscope or data acquisition device with a grounded input.

5.9.5 Current Transformer

The on-load synchronous generator current is the most prevalent type of signal used for fault detection. The available current transformers (CTs) in power plants are widely used in fault detection of synchronous generators. Although the frequency bands of the utilized CTs are limited to a few kHz, the current signal is adequate for low-frequency analysis of the generator health status. If the available CT is inadequate and a new sensor for a permanent installation is required, a CT, especially one with a Hall-effect sensor, is recommended. A CT has a simple structure: a cable carrying the synchronous generator current forms the primary winding of a CT. The secondary winding of the CT is connected to the OP-AMP to scale-up the current. If the purpose is to investigate high-frequency components, the frequency bandwidth of the CT must be checked. A BNC cable or USB cable must be used to transfer the CT output to the oscilloscope or data acquisition device.

The LEM CT is able to measure both AC and DC currents; therefore, it could measure the DC current that passes through the rotor field winding. Both mechanical faults and electrical faults have a marked impact on the DC current of the magnetization winding. Figure 5.29 shows the LEM CT used to measure the DC current of the rotor winding.

In many cases, disconnecting the synchronous generator connecting cables is not an option. Therefore, AC current probes for oscilloscopes are recommended. Figure 5.30 shows an AC current probe (Fluke i1000s). Some AC current probe specifications are as follows:

1. The current range of 100 mA to 1000 A AC RMS.
2. The output signal is a voltage in a range of mV.

Figure 5.29 An LEM CT with a BNC cable to measure the rotor DC current.

Figure 5.30 An AC current probe. https:// www.fluke.com/en/product/accessories/ probes/fluke-i1000s. *Source*: Fluke Corporation.

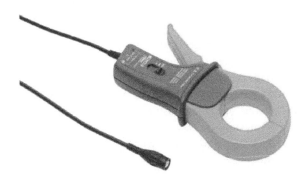

3. The frequency range of the probe is between 5 Hz and 100 kHz.
4. It has three different ranges (for inputs of 10, 100, and 100 A, it gives outputs equal to 100, 10, and 1 mV/A).

5.10 Data Acquisition

Data acquisition is one of the main steps in the fault detection procedure for generators. Although the quality of the measured data mainly depends on the sensor, data acquisition also plays a key role. The data can be measured by:

1. Oscilloscope.
2. Data acquisition device.

Both oscilloscopes and data acquisition devices have their own pros and cons, so the user must decide on the most appropriate one. The number of input channels in oscilloscopes is limited to four, while data acquisition devices can have at least 16 ports. However, in some data acquisition devices, such as AdvanTech PCI-1710 HG, the maximum sampling frequency is divided by the number of channels. For example, in the present case, with a maximum sampling frequency of

Figure 5.31   Oscilloscope with 12-bit resolution. https://www.tek.com/oscilloscope/mso3000-dpo3000. *Source*: TEKTRONIX, INC.

100 kHz, each channel can have a maximum sampling frequency of 6.25 kHz in a single-ended topology. By reducing the number of required channels, the sampling frequency can be increased. The data acquisition device must be chosen based on the signal type that is going to be measured – this could be voltage, current, temperature, sound, vibration, pressure, or force. By contrast, the oscilloscope is a multi-function device that provides online monitoring and in-house fabricated filters can suppress the noise impacts.

A data acquisition device with multiple inputs and outputs may be the only option if a condition assessment based on sensor fusion is the main purpose and requires synchronized data measured at the same time, with the same sampling frequency and resolution. However, an automatic measurement of the same quality could perhaps be obtained using multiple oscilloscopes, with one working as a master and the rest as slaves. However, this is probably not practical, since this approach would require all the oscilloscopes to be of the same brand.

Two different oscilloscopes and one data acquisition device were used to measure the required data in this book. The two oscilloscopes with high resolution were the Tektronix MSO 3054 and the ROHDE & SCHWARZ RTO2044. The key performance specifications of the Tektronix MSO 3054 are the following (see Figure 5.31):

1. Bandwidth up to 500 MHZ.
2. Four analog channels.
3. Sixteen digital channels.
4. Sample rate of 2.5 GS/s in all channels.
5. Record length of 5 mega points in each channel.
6. Maximum waveform capture rate greater than 50.000 wfm/s.
7. Suite of advanced triggers.
8. High-resolution data acquisition (11-bit).

The key features of the ROHDE & SCHWARZ RTO2044 are the following (see Figure 5.32):

1. Bandwidth up to 6 GHz.
2. Fast data acquisition and analysis.

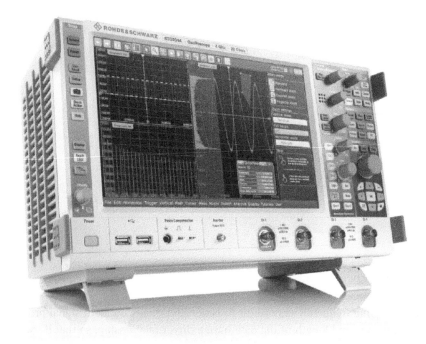

Figure 5.32 Oscilloscope with 16-bit resolution. https://www.rohde-schwarz.com/dk/product/rto-productstartpage_63493-10790.html. *Source*: Rohde & Schwarz.

3. Waveform acquisition rate of 1 million waveforms per second.
4. Sample rate of 20 GS/s in all channels.
5. High sensitivity to noise and a low noise floor.
6. Trigger system with high accuracy.
7. Four analog channels.
8. Highly accurate measurement system because of a single-core analog to digital converter.
9. High acquisition rate and short blind time to capture rare signals.
10. Numerous functions to analyze data (mathematical operations, mask tests, histograms, spectrum analysis, etc.).
11. High trigger sensitivity to capture the closest successive events.
12. Remote network-based operation.
13. High resolution (16-bit).

A data acquisition device, the CompactDAQ from National Instrument (see Figure 5.33), is a compact system that can be used to collect and deliver the data even over a long distance, which is not the case for an oscilloscope. The CompactDAQ can be used when numerous physical and electrical signals are intended to be measured. This data acquisition device has 16 inputs, with a sampling rate of 250 kS/s. The resolution of the CompactDAQ is 16-bit, which is high enough for fault detection purposes. The CompactDAQ can be connected to a PC via a USB port or PCI port. Three software programs are available to configure and automate the data. They are LabVIEW, FlexLogger, and DAQExpress.

Here, the length of the measured data is mostly equal to 100 000 samples. For condition monitoring purposes, the sampling frequency and the length of the data are important. The minimum sampling frequency must be equal to at least twice the maximum frequency, based on Nyquist's

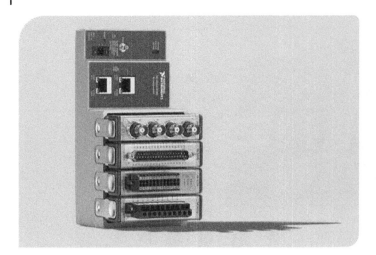

Figure 5.33 Compact DAQ to measure the signal. https://www.ni.com/en-no/shop/compactdaq.html. *Source*: NATIONAL INSTRUMENTS CORP.

law. However, a higher sampling frequency may result in better and more confident signal processing. The sampling frequency and length of the data have a direct relationship, so that if the sampling frequency is high enough, even a few mechanical revolutions of the generator result in an enriched frequency content. However, a trade-off is required to ensure that the length of the data and the number of samples is adequate to achieve an accurate result. At the same time, the amount of sampled data must be small enough to avoid any problems related to data storage and real-time analysis. The number of sampling points also depends on the type of signal processing tool used. For instance, fast Fourier transforms are sensitive to the length of the data.

5.11 Fault Implementation

The main purpose of this section is to provide a practical approach for treatment under different fault scenarios in the laboratory. The different types of faults are:

1. Stator short circuit fault.
2. Rotor inter-turn short circuit fault.
3. Static and dynamic eccentricity fault.
4. Broken damper bar fault.
5. Misalignment.

5.11.1 Stator Short Circuit Fault

The short circuit fault occurs only in the stator winding turns. The majority of large synchronous generators have stator bars; however, to achieve the intended voltage level, the multi-turn winding type of stator is used. A short circuit fault of the stator winding bars inside the stator slot can totally destroy the winding in a few seconds. For fault detection, in this case, PD analysis can be used. Three types of short circuit faults can occur in the stator winding: the inter-turn short circuit fault, the phase-to-phase short circuit fault, and the phase-to-ground short circuit fault.

Figure 5.34 Schematic diagram of different types of short circuit faults in a synchronous generator (top). End winding of a synchronous generator (bottom).

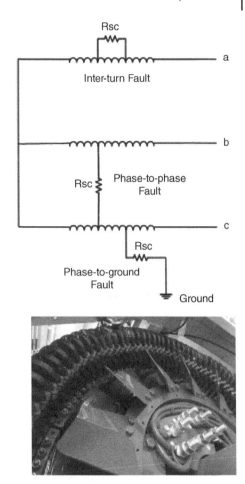

The most prevalent type of fault is an inter-turn short circuit fault. In this case, the insulation of the two turns from the same phase is damaged due to an insulation problem, contamination, aging, or excessive local heating, and a circulating current passes through these turns. In a generator with a fractional slot winding, windings from two phases are in the same slot. If the insulation of the winding from two different phases undergoes a serious problem and a circulating current forms between the turns, the fault is called a phase-to-phase short circuit fault. The last type is a phase-to-ground fault, in which the insulation of the winding and the ground wall has deteriorated and a current flows from the damaged winding to the body of the generator. The phase-to-phase and the phase-to-ground faults are the most serious types of short circuit faults, and they must be detected in the early stage; otherwise, they result in a winding burn out. Figure 5.34 shows the three types of stator short circuit faults.

The only possible access point to these faults is the end winding (see Figure 5.34). The faults all happen inside the stator slots, but getting access to the machine slot to eliminate the short circuit fault is impossible. This indicates that the end-winding must be modified to conduct the short circuit fault. The insulation of the two turns, either from the same phase or two different phases, must be removed. To limit the circulating current, a resistance (Rsc) must be used; otherwise, the current will increase and destroy the winding.

5.11.2   Inter-Turn Short Circuit Fault in a Rotor Field Winding

The inter-turn short circuit fault in the rotor of a synchronous generator is one of the predominant types of fault; however, its severity is low due to the low amplitude of the voltage in the rotor compared with the stator winding. A generator with a short circuit fault in the rotor field winding can operate for a couple of months. The treatment for an inter-turn short circuit fault is similar to that for a stator winding, as the only access point is again the rotor end winding. Therefore, the rotor insulation between two turns must be removed and resistance must be used to limit the circulating current between the two turns. Figure 5.35 shows a schematic diagram of the rotor field winding under an inter-turn short circuit fault.

Figure 5.36 shows a rotor of an SPSG. In this custom-made SPSG, instead of causing any damage to the rotor turns, a number of turns are removed. Different taps connected to the rotor field winding are connected to the bolts on the rotor. There is a common tap and 1, 2, 3, 7, and 10 turns can be removed using a copper plate and connecting the common tap and the desired tap. This is a safe approach that can be taken to treat a short circuit fault on the rotor field winding.

5.11.3   Eccentricity Fault

Static, dynamic, and mixed eccentricity faults in the generator result in a non-uniform air gap. Several methods can be used to implement static and dynamic eccentricity faults. The following can be used to implement the above-mentioned faults:

1. Using a specific flange, as shown in Figure 5.37, two rings are used to regulate the direction and severity of the dynamic eccentricity fault (ring No. 1 and ring No. 2). Two rings are also used to adjust the direction and severity of the static eccentricity fault (ring No. 4 and ring No. 5). The retaining ring, which is fixed to the external ring, and the external ring, which is coupled to the generator, are ring No. 6 and ring No. 7. A ball bearing is also mounted between the static and dynamic eccentricity rings (ring No. 3). The developed advanced flange is able to apply the

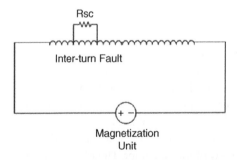

Figure 5.35   A schematic of an inter-turn short circuit fault in a synchronous generator.

Figure 5.36   A rotor of a synchronous generator and a short circuit tap on the rotor.

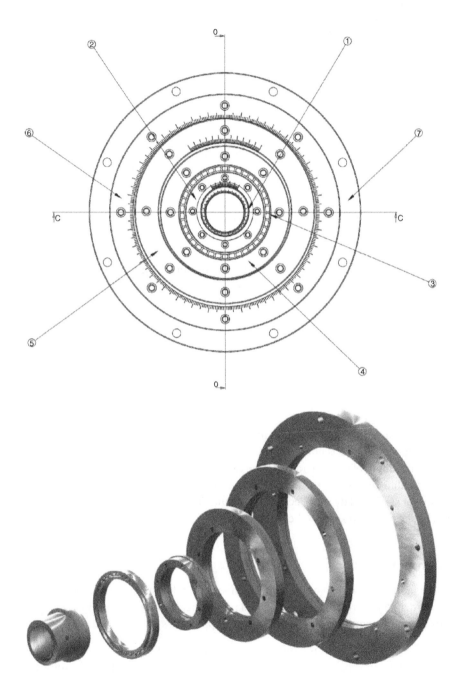

Figure 5.37 Schematic of a flange for applying static and dynamic eccentricity faults. The exploded view of the flange [11].

dynamic and static eccentricity faults with different severity. It is also possible to adjust both static and dynamic eccentricity faults and have a mixed eccentricity fault, which is the most prevalent type of eccentricity fault [11, 12].

2. The second affordable approach, which does not require a specific flange, is the use of a bearing with a bigger inner radius. In this method, a new bearing with a larger inner race diameter and smaller outer race replaces the original bearing. The replaced bearing, having two spaces in the inner and outer races, can be used to emulate both static and dynamic eccentricity faults. A concentric ring is used inside the inner race while an eccentric ring is used on the outer race to fill the empty spaces. In this case, a static eccentricity fault is emulate. For a dynamic eccentricity fault, a concentric ring must be placed on the outer race, where the bearing fits the endplate housing, and an eccentric ring must be used inside the inner race. The degree of both static and dynamic faults depends on the diameter of the ring that is used either inside or outside the bearing.

3. The third approach is implemented in a custom-made synchronous generator to emulate the static eccentricity fault. In this case, a stator core is connected to a movable housing that can move along the horizontal axis (see Figure 5.38). With this method, it is only possible to emulate the static eccentricity fault. Two bolts on both sides of the generator are used to move the stator frame smoothly. To measure the eccentricity fault severity, four measuring clocks are installed on both sides of the generator housing (two on each side). The clocks are set to zero on both sides and by moving the generator the clocks are changed and show the eccentricity fault severity (see Figure 5.38). The unit of the measuring clock is in mm. Therefore, the eccentricity fault severity is easy to measure knowing the air gap length. Figure 5.38 shows the implemented method.

5.11.4 Misalignment Fault

The trickiest type of eccentricity fault is the misalignment fault. In the case of a static or dynamic eccentricity fault, although the air gap is non-uniform in a radial direction, the length of the air gap is constant along the axial length of the generator. In the case of a misalignment fault, the length of the air gap, both in the radial and axial directions, differs. A misalignment fault is treated with the same approaches used for static and dynamic eccentricity faults. The difference is that the bearings or flanges must be changed on both sides of the machine.

5.11.5 Broken Damper Bar Fault

A broken damper bar fault is a serious fault, especially in salient pole synchronous motors. The operation of the direct-drive synchronous motors depends on the health status of the damper bars. To emulate the broken damper bar fault, the following three approaches are available:

1. In an induction motor, the broken bar is emulated by drilling a bar. However, it is not possible to implement this strategy in synchronous generators because the slot opening of the rotor is much smaller than the damper bar width, and drilling the rotor can damage it.
2. Instead, it is possible to break the damper bars from the welded joint on the end ring. This point is the most vulnerable area in which the breakage happens. Later, the damper bar can be welded to the end ring to restore the rotor health. However, that would require the synchronous generator to be dismantled.

Figure 5.38 Measuring clocks installed on both sides. Measuring clock and bolt use for frame movement.

3. In the custom-made synchronous generator used for the experimental purpose of this book, the damper bars can be removed from all poles. The end rings or the connection rings can also be removed to have a separated pole. Figure 5.39 shows the damper bars that were removed from one rotor pole shoe.

5.12   Noise Considerations

Noise is unavoidable in the industrial environment and has a marked impact on the quality of the measured signals. Some signals are very sensitive to noise since the signal-to-noise ratio is low due to the low amplitude of the signal. Therefore, the quality of the signal is very important in condition monitoring of the generator. The majority of sensors have noise rejection features, while some custom-made sensors need some modification to reject or reduce the noise effect. The following are methods that can help [13]:

1. Noise filtering by analog or digital filters.
2. Shielding the noise source from the measurement equipment.

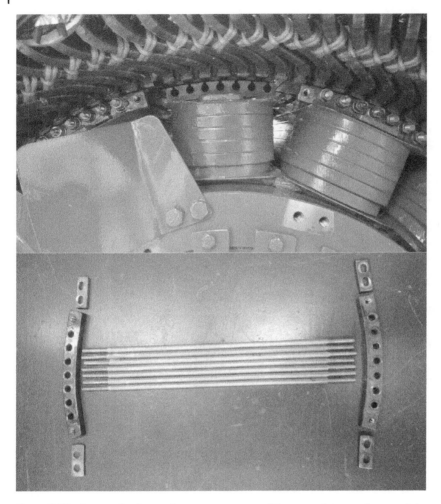

Figure 5.39 Rotor pole of a synchronous generator with removed damper bars, end ring, and connecting ring.

3. Using the low-pass filter of the oscilloscope.
4. Using an isolating transformer between the oscilloscope and the current power source.
5. Using a differential probe.
6. Software noise rejection program such as wavelet transforms.
7. Grounding the body of the equipment.
8. Using a metal box for a constant current source.
9. Using coaxial cable or shielded wire to transfer data.
10. A thin copper tape is the proper choice when there is limited space and coaxial cable is not applicable.

5.13 Summary

This chapter illustrated the equipment required for experiments in the laboratory, since the results obtained by simulations must be verified by experimental results. It also illustrated how to emulate various mechanical and electrical faults in synchronous generators. Different types of signals are utilized in fault detection, indicating that various sensors can be introduced to measure signals, such as magnetic field, voltage, current, and vibration. Some approaches require a pure signal, so some methods were introduced for noise rejection purposes. The data acquisition system and its requirements for signal measurement are explained in detail.

References

1 Groth, I.L. (2019). *On-Line Magnetic Flux Monitoring and Incipient Fault Detection in Hydropower Generators*. NTNU.

2 Solli, J.F. (2014). Measurements of magnetic flux density in rotating machines: Development of a calibrated and portable measurement system. Universitetet i Agder; University of Agder.

3 Kokoko, O., Merkhouf, A., Tounzi, A., et al. Detection of short circuits in the rotor field winding in large hydro generator, in *2018 XIII International Conference on Electrical Machines (ICEM)*. 2018. IEEE.

4 Ehya, H., Nysveen, A., Nilssen, R., & Liu, Y. (2021). *Static and dynamic eccentricity fault diagnosis of large salient pole synchronous generators by means of external magnetic field*. IET Electric Power Applications, 15 (7), 890–902.

5 Ehya, H. and Nysveen, A. (2022). Comprehensive broken damper bar fault detection of synchronous generators. *IEEE Transactions on Industrial Electronics* 69, no.4: 4215–4224.

6 Ehya, H., Nysveen, A., and Nilssen, R. Pattern recognition of inter-turn short circuit fault in wound field synchronous generator via stray flux monitoring. In: *2020 International Conference on Electrical Machines (ICEM)*, 23–26 August 2020. Gothenburg, Sweden.

7 P. Tavner, L. Ran, J. Penman, et al. (2008). *Condition Monitoring of Rotating Electrical Machines*. IET Publisher, UK.

8 Rødal G.L. (2020). *Online Condition Monitoring of Synchronous Generators Using Vibration Signal*. MSc Thesis, Norwegian University of Science and Technology, Faculty of Information Technology and ElectricalEngineering, Department of Electric Power Engineering, Norway.

9 Gieras, J.F., Wang, C., and Lai, J.C. (2018). *Noise of Polyphase Electric Motors*. CRC Press, USA.

10 Ehya, H., Rødal, G.L., Nysveen, A. et al. Condition monitoring of wound field synchronous generator under inter-turn short circuit fault utilizing vibration signal. In: *23rd International Conference on Electrical Machines and Systems (ICEMS)*. 24-27 November 2020. Hamamatsu, Japan.

11 Bruzzese, C. and Joksimovic, G. (2011). Harmonic signatures of static eccentricities in the stator voltages and in the rotor current of no-load salient-pole synchronous generators. *IEEE Transactions on Industrial Electronics* 58, no 5: 1606–1624.

12 Bruzzese, C. (2013, 2014). Diagnosis of eccentric rotor in synchronous machines by analysis of split-phase currents—Part II: Experimental analysis. *IEEE Transactions on Industrial Electronics* 61, no. 8: 4206–4216.

13 Ehya, H., Nysveen, A., and Skreien, T.N. (2021). Performance evaluation of signal processing tools used for fault detection of hydro-generators operating in noisy environments. *IEEE Transactions on Industry Applications* 57, no. 4: 3654–3665.

14 Zamudio-Ramirez, et al., (2019). Detection of winding asymmetries in wound-rotor induction motors via transient analysis of the external magnetic field. *IEEE Transactions on Industrial Electronics* 67, no. 6: pp. 5050–5059.

6

Analytical Modeling Based on Wave and Permeance Method

6.1 Introduction

Although induction motors (IMs) [1] and permanent magnet machines (PMMs) [2–6] are modeled and analyzed analytically under various kinds of failures, no comprehensive analytical method exists for analyzing synchronous machine (SM) faults. Electric machine modeling, whether using the finite element method (FEM) or analytical approaches, may provide deep insight into a faulty machine; however, this modeling process is not a useful tool for fault detection. In other words, fault detection is based on analyzing extracted features from signals, whereas scrutinizing a machine's signal trends cannot yet provide meaningful data.

SM modeling is not as simple as modeling induction machines and PMMs because the SM configuration is sophisticated. The pole saliency is one of the factors that makes SM modeling difficult; by contrast, IMs and PMMs have constant air gap lengths. Therefore, a comprehensive analytical method is required to extract features based on the signal of the machine.

Analytical methods provide approximated results in comparison to FEMs. However, they are not as computationally expensive as FEMs and their solutions can provide a simple representation of the machine signals. In addition, they can provide a physical understanding of the underlying physics, while the commercialized FEM toolbox is covered with a graphical interface that hides the mathematical procedures and physical concepts. Wave and permeance (WP) are used to take out

Electromagnetic Analysis and Condition Monitoring of Synchronous Generators, First Edition. Hossein Ehya and Jawad Faiz.
© 2023 The Institute of Electrical and Electronics Engineers, Inc. Published 2023 by John Wiley & Sons, Inc.

features based on the specifications of SM, with the intention of extracting the frequencies that give rise to the faulty situation in the machine.

The WP method is based on the resultant magneto-motive force (MMF) of stator and rotor windings. These MMFs create the magnetic waves in the machines that interact in the air gap. In the wound field rotor, the rotor MMF and, consequently, its magnetic field, are generated by its windings. By contrast, a magnetic field based on permanent magnets on the rotor core or inside the rotor core is assumed for the PMMs.

An accurate feature based on WP can be obtained by considering detailed machine specifications with few assumptions. Therefore, magnetic saturation, slotting effects, and rotor pole saliency must be taken into account. In addition, in the case of failures that lead to an unbalanced magnetic field, the non-uniform air gap must be considered. The WP method allows the air gap magnetic field, induced voltage, torque, vibration, and unbalanced magnetic pull (UMP) to be obtained.

This chapter is divided into two main sections, with several subsections. In the first section, the saturation and slotting effects are not considered, whereas these effects are included in the second section. Based on the WP method, some features are extracted to demonstrate the vulnerable frequencies that undergo significant changes when the machine is exposed to a fault. These frequencies, as "indexes," have a predictable pattern.

6.2 Eccentricity Fault Definition

The air gap of the eccentric SM varies based on the eccentricity type. The severity and angle of the eccentricity fault are required for modeling the air gap permeance [7]. Figure 6.1 shows an electric machine in a healthy state and under a static eccentricity (SE) and a dynamic eccentricity (DE) fault. Since the saliency of the pole is defined in the air gap function, the definition provided here is valid for the salient pole SM.

According to Figure 6.1, g_0 is the minimum air gap function that is constant for a round rotor SM, whereas it is a function of pole saliency in a salient pole SM. C_s, C_r, and C_ω are the center of the stator core, center of the rotor core, and center of rotation, respectively. The severity of SE and DE is defined as:

$$\delta_{se} = \frac{\left\| \overrightarrow{C_s C_{r,\omega}} \right\|}{g_0} \tag{6.1}$$

According to Equation (6.1), the SE fault is defined as the distance between the stator center to the rotor and the rotational center where they coincide. The air gap length under the SE fault and its corresponding angle are obtained by defining an arbitrary point (P) on the rotor circumference, as shown in Figure 6.2. In this figure, R_r is the radius of the rotor and φ, ψ are the arbitrary angles around the stator frame angle and eccentricity angle, respectively. The distance between the center of the stator to an arbitrary point on the rotor is defined as:

$$\left\| \overrightarrow{C_s P} \right\| = g_0 \delta_{se} \cos(\psi) + \sqrt{R_r^2 - g_0^2 \delta_{se}^2 \sin^2(\psi)} \tag{6.2}$$

In the electric machines like run-off-river synchronous generators that have a small air gap, g_0 is negligible in comparison to the rotor radius. Therefore, Equation (6.2) is simplified as follows:

$$\sqrt{R_r^2 - g_0^2 \delta_{se}^2 \sin^2(\psi)} \approx R_r n \tag{6.3}$$

$$\left\| \overrightarrow{C_s P} \right\| \approx g_0 \delta_{se} \cos(\psi) + R_r \tag{6.4}$$

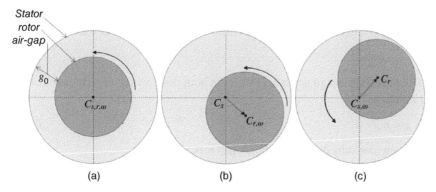

Figure 6.1 (a) Healthy machine, (b) under SE fault, (c) under DE fault.

Figure 6.2 The location of the stator and rotor of an electric machine under a static eccentricity fault.

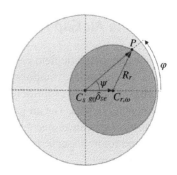

If R_s, the radial air gap length under the SE fault, is written as follows:

$$g_{ec}(\psi) = R_s - \left\|\overrightarrow{C_s P}\right\| = R_s - g_0 \delta_{se} \cos(\psi) - R_r = g_0(1 - \delta_{se} \cos(\psi)) \tag{6.5}$$

where R_s is the inner radius of the stator bore. In the SE fault, the eccentricity severity and an arbitrary angle around the stator are the same; therefore:

$$g_{ec}(\phi) = g_0(1 - \delta_{se} \cos(\phi)) \tag{6.6}$$

The DE is also defined as the distance between the stator center, where it coincides with the rotational center, to the rotor center, according to Figure 6.3:

$$\delta_{de} = \frac{\left\|\overrightarrow{C_{s,\omega} C_r}\right\|}{g_0} \tag{6.7}$$

Figure 6.3 Location of the stator and rotor of an electric machine under a DE fault.

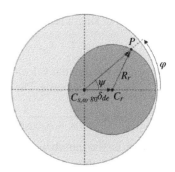

A procedure similar to that was followed for the SE to determine the air gap length and its angle under a DE fault. The air gap length under a DE fault based on SE deduction is as follows:

$$g_{ec}(\psi) = g_0(1 - \delta_{de}\cos(\psi)) \tag{6.8}$$

According to the DE, the minimum air gap length varies in both position and time; therefore, the DE angle is defined as follows:

$$\psi = \psi(\phi, t) = \phi - \omega_r t \tag{6.9}$$

where is ω_r is the rotor angular velocity. Hence, the air gap function is:

$$g_{ec}(\phi, t) = g_0 \left(1 - \delta_{de}\cos(\phi - \omega_r t)\right) = g_0 \left(1 - \delta_{de}\cos\left(\frac{\omega_s t}{p} - \phi\right)\right) \tag{6.10}$$

6.3 The Air Gap Magnetic Field

The air gap magnetic field can provide detailed information about the electric machine's health status. The harmonic components of the air gap magnetic field consist of odd and even harmonics, where odd harmonics represent the main harmonic components of the MMF and even harmonics represent its ripples. The air gap magnetic field includes all harmonics related to the physical and geometrical specifications of the machine. In a healthy machine, the dominant harmonics are the stator and rotor magnetic field, in addition to harmonics related to stator and rotor slotting effects, and saturation.

A fault can change the uniformity of the air gap magnetic field and may cause subharmonics. Therefore, condition monitoring based on the air gap magnetic field can provide more information. By considering the entire harmonic component, the air gap magnetic field, B_{normal}, can be represented as follows:

$$B_{normal} = \sum_{1,3,5,\ldots} B_n\sin(n\omega t) + \sum_{n=2,4,6,\ldots} B_n\cos(n\omega t) \tag{6.11}$$

where B_n is the normal component of the air gap magnetic field; the odd and even harmonics are representative of the main harmonics of the magnetic field and MMF ripples, respectively. Consequently, by having the air gap magnetic field, the required signals, such as induced MMF in the stator terminals, torque, vibrations, and unbalanced magnetic field, can be estimated.

The air gap magnetic field can be measured by a Hall-effect sensor or by placing search coils around the stator slots. By integrating the induced voltage in the search coil, the magnetic field is estimated. However, the noise and disturbance from the environment can significantly change the signal behavior; hence, noise cancelation must be considered during signal measurement.

The air gap length in a case of a DE fault is as follows [8]:

$$g(\varphi, t) = g_0(1 - \delta_{de}\cos(\omega_r t - \varphi)) \tag{6.12}$$

where ω_r is the angular velocity, δ_{de} is the DE severity, t is the time, and φ is the arbitrary angle in the stator reference frame. The air gap length g_0 in SM is a function of rotor saliency, as follows [9]:

$$g_0(\varphi, \theta_r) = \frac{1}{\alpha_1 + \alpha_2\cos2p(\varphi - \theta_r)} \tag{6.13}$$

where p is the number of pole pairs, θ_r is the rotor position, and α_1, α_2 are coefficients depending on the rotor geometry, as described in reference [10]. The air gap permeance (Λ) is proportional to the inverse air gap length:

$$\Lambda(\varphi, \theta_r) = \frac{1}{g} = \frac{\alpha_1 + \alpha_2 \cos 2p(\varphi - \theta_r)}{(1 - \delta_{de}\cos(\omega_r t - \varphi))} \tag{6.14}$$

To approximate the division, the Taylor series expansion is employed as follows:

$$\Lambda(\varphi, \theta_r) = (\alpha_1 + \alpha_2 \cos 2p(\varphi - \theta_r))\left(1 + \delta_{de}\cos(\omega_r t - \varphi) + (\delta_{de}\cos(\omega_r t - \varphi))^2\right.$$
$$\left. + (\delta_{de}\cos(\omega_r t - \varphi))^3 + \ldots\right) \tag{6.15}$$

where ω_r is the angular velocity of the rotor and θ_r is the rotor position. In the air gap permeance, the second and higher-order terms are eliminated because the value of $\delta_{de}\cos(\omega_r t - \varphi)$ is very small. Therefore, the approximated air gap permeance is:

$$\Lambda(\varphi, \theta_r) = (\alpha_1 + \alpha_2 \cos 2p(\varphi - \theta_r))(1 + \delta_{de}\cos(\omega_r t - \varphi)) \tag{6.16}$$

since ω_r is:

$$\omega_r = \frac{\omega_s}{p} \tag{6.17}$$

where ω_s is the angular synchronous frequency. Besides, θ_r is:

$$\theta_r = \omega_r t = \frac{\omega_s t}{p} \tag{6.18}$$

The air gap permeance can be expressed as follows:

$$\Lambda(\varphi, \theta_r) = \left(\alpha_1 + \alpha_2 \cos 2p\left(\varphi - \frac{\omega_s t}{p}\right)\right)\left(1 + \delta_{de}\cos\left(\frac{\omega_s t}{p} - \varphi\right)\right) \tag{6.19}$$

If the MMF of the stator varies sinusoidally, the current density in the stator windings $j_s(\varphi, t)$ is determined as follows:

$$j_s(\varphi, t) = J_s \sin(\omega_s t - p\varphi) \tag{6.20}$$

where J_s is the maximum value of the current density. According to Ampere's law, the magnetic flux density in the air gap $B_s(\varphi, t)$ due to the stator winding is defined as follows:

$$B_s(\varphi, t) = \Lambda(\varphi, t) \int \mu_0 j_s(\varphi, t) \, d\varphi \tag{6.21}$$

where μ_0 is the air permeability. By substituting the air gap permeance and stator current density in Equation (6.21), $B_s(\varphi, t)$ is:

$$B_s(\varphi, t) = \left(\alpha_1 + \alpha_2 \cos 2p\left(\varphi - \frac{\omega_s t}{p}\right)\right)\left(1 + \delta_{de}\cos\left(\frac{\omega_s t}{p} - \varphi\right)\right)\int \mu_0 J_s \sin(\omega_s t - p\varphi) \, d\varphi \tag{6.22}$$

By applying the integral and trigonometric relationships to Equation (6.22), $B_s(\varphi, t)$ yields:

$$B_s(\varphi, t) = \frac{\alpha_1 \mu_0 J_s}{P}\cos(\omega_s t - P\varphi)$$
$$+ \frac{\alpha_2 \mu_0 J_s}{2P}\cos(-\omega_s t + P\varphi)$$
$$+ \frac{\alpha_2 \mu_0 J_s}{2P}\cos(3\omega_s t - 3P\varphi)$$
$$+ \frac{\alpha_1 \mu_0 J_s \delta_{de}}{2P}\cos\left(\left(1 + \frac{1}{P}\right)\omega_s t - (P+1)\varphi\right)$$

$$+ \frac{\alpha_1 \mu_0 J_s \delta_{de}}{2P} \cos \left(\left(1 - \frac{1}{P} \right) \omega_s t - (P-1)\varphi \right)$$

$$+ \frac{\alpha_2 \mu_0 J_s \delta_{de}}{4P} \cos \left(\left(-1 + \frac{1}{P} \right) \omega_s t + (P+1)\varphi \right)$$

$$+ \frac{\alpha_2 \mu_0 J_s \delta_{de}}{4P} \cos \left(\left(3 + \frac{1}{P} \right) \omega_s t - (3P+1)\varphi \right)$$

$$+ \frac{\alpha_2 \mu_0 J_s \delta_{de}}{4P} \cos \left(\left(-1 + \frac{1}{P} \right) \omega_s t + (P+1)\varphi \right)$$

$$+ \frac{\alpha_2 \mu_0 J_s \delta_{de}}{4P} \cos \left(\left(3 - \frac{1}{P} \right) \omega_s t - (3P+1)\varphi \right) \tag{6.23}$$

The same procedure can be applied to the rotor magnetic field, where the main harmonic of the rotor current density $J_s(\varphi', t)$ is assumed to be sinusoidal, as follows:

$$j_r(\varphi', t) = J_r \sin(\omega_s t - p\varphi') \tag{6.24}$$

where J_r is the maximum amplitude of the rotor current density. The rotor magnetic field $B_r(\varphi', t)$ is derived as:

$$B_r(\varphi', t) = \Lambda(\varphi', t) \int \mu_0 j_s(\varphi', t) \, d\varphi' \tag{6.25}$$

where φ' is an arbitrary angle in the rotor reference frame. By substituting the air gap permeance and rotor current density in Equation (6.25), $B_r(\varphi', t)$ is:

$$B_r(\varphi', t) = \left(\alpha_1 + \alpha_2 \cos 2p \left(\varphi' - \frac{\omega_s t}{p} \right) \right) \left(1 + \delta_{de} \cos \left(\frac{\omega_s t}{p} - \varphi' \right) \right) \int \mu_0 J_r \sin(\omega_s t - p\varphi') \, d\varphi' \tag{6.26}$$

Integrating the rotor current density and multiplying it by the air gap permeance gives $B_r(\varphi', t)$ as:

$$B_r(\varphi', t) = \frac{\alpha_1 \mu_0 J_r}{P} \cos(\omega_s t - P\varphi')$$

$$+ \frac{\alpha_2 \mu_0 J_r}{2P} \cos(-\omega_s t + 2P\varphi - P\varphi')$$

$$+ \frac{\alpha_2 \mu_0 J_r}{2P} \cos(3\omega_s t - 2P\varphi - P\varphi')$$

$$+ \frac{\alpha_1 \mu_0 J_r \delta_{de}}{2P} \cos \left(\left(1 + \frac{1}{P} \right) \omega_s t - P\varphi' + \varphi \right)$$

$$+ \frac{\alpha_1 \mu_0 J_r \delta_{de}}{2P} \cos \left(\left(1 - \frac{1}{P} \right) \omega_s t - P\varphi' - \varphi \right)$$

$$+ \frac{\alpha_2 \mu_0 J_r \delta_{de}}{4P} \cos \left(\left(-1 + \frac{1}{P} \right) \omega_s t + P\varphi' + 2P\varphi - \varphi \right)$$

$$+ \frac{\alpha_2 \mu_0 J_r \delta_{de}}{4P} \cos \left(\left(3 + \frac{1}{P} \right) \omega_s t - P\varphi' - 2P\varphi - \varphi \right)$$

$$+ \frac{\alpha_2 \mu_0 J_r \delta_{de}}{4P} \cos \left(\left(-1 + \frac{1}{P} \right) \omega_s t - P\varphi' + 2P\varphi + \varphi \right)$$

$$+ \frac{\alpha_2 \mu_0 J_r \delta_{de}}{4P} \cos \left(\left(3 - \frac{1}{P} \right) \omega_s t - P\varphi' - 2P\varphi + \varphi \right) \tag{6.27}$$

The magnetic field in the air gap $B(\varphi, \varphi', t)$ is a consequence of the stator and the rotor magnetic field. Therefore:

$$B(\varphi, \varphi', t) = \frac{\alpha_1 \mu_0 J}{P} \cos(\omega_s t - P\varphi) + \frac{\alpha_1 \mu_0 J}{P} \cos(\omega_s t - P\varphi')$$

$$+ \frac{\alpha_2 \mu_0 J}{2P} \cos(-\omega_s t + P\varphi) + \frac{\alpha_2 \mu_0 J}{2P} \cos(-\omega_s t + 2P\varphi - P\varphi')$$

$$+ \frac{\alpha_2 \mu_0 J}{2P} \cos(3\omega_s t - 3P\varphi) + \frac{\alpha_2 \mu_0}{2P} \cos(3\omega_s t - 2P\varphi - P\varphi')$$

$$+ \frac{\alpha_1 \mu_0 J \delta_{de}}{2P} \cos\left(\left(1 + \frac{1}{P}\right)\omega_s t - (P+1)\varphi\right) + \frac{\alpha_1 \mu_0 J \delta_{de}}{2P} \cos\left(\left(1 + \frac{1}{P}\right)\omega_s t - P\varphi' + \varphi\right)$$

$$+ \frac{\alpha_1 \mu_0 J \delta_{de}}{2P} \cos\left(\left(1 - \frac{1}{P}\right)\omega_s t - (P-1)\varphi\right) + \frac{\alpha_1 \mu_0 J \delta_{de}}{2P} \cos\left(\left(1 - \frac{1}{P}\right)\omega_s t - P\varphi' - \varphi\right)$$

$$+ \frac{\alpha_2 \mu_0 J \delta_{de}}{4P} \cos\left(\left(-1 + \frac{1}{P}\right)\omega_s t + (P+1)\varphi\right) + \frac{\alpha_2 \mu_0 J \delta_{de}}{4P} \cos\left(\left(-1 + \frac{1}{P}\right)\omega_s t + P\varphi' + 2P\varphi - \varphi\right)$$

$$+ \frac{\alpha_2 \mu_0 J \delta_{de}}{4P} \cos\left(\left(3 + \frac{1}{P}\right)\omega_s t - (3P+1)\varphi\right) + \frac{\alpha_2 \mu_0 J \delta_{de}}{4P} \cos\left(\left(3 + \frac{1}{P}\right)\omega_s t - P\varphi' - 2P\varphi - \varphi\right)$$

$$+ \frac{\alpha_2 \mu_0 J \delta_{de}}{4P} \cos\left(\left(-1 + \frac{1}{P}\right)\omega_s t + (P+1)\varphi\right) + \frac{\alpha_2 \mu_0 J \delta_{de}}{4P} \cos\left(\left(-1 + \frac{1}{P}\right)\omega_s t - P\varphi' + 2P\varphi + \varphi\right)$$

$$+ \frac{\alpha_2 \mu_0 J \delta_{de}}{4P} \cos\left(\left(3 - \frac{1}{P}\right)\omega_s t - (3P+1)\varphi\right) + \frac{\alpha_2 \mu_0 J \delta_{de}}{4P} \cos\left(\left(3 - \frac{1}{P}\right)\omega_s t - P\varphi' - 2P\varphi + \varphi\right)$$

$$\tag{6.28}$$

and $\omega_s = 2\pi f_s$, where f_s is the frequency of the machine in Hz. According to Equation (6.28), the following frequency patterns exist in the air gap magnetic field of the synchronous generator under a DE fault:

$$f_{eccentricity1} = \left(3 \pm \frac{k}{P}\right) f_s \tag{6.29}$$

$$f_{eccentricity2} = \left(1 \pm \frac{k}{P}\right) f_s \tag{6.30}$$

where k is the integer number and $f_{eccentricity1}$ and $f_{eccentricity2}$ are introduced as indexes to detect the DE fault in the salient pole synchronous generator.

6.4 The Electromotive Force in Stator Terminals

The occurrence of an eccentricity fault stimulates harmonics in the air gap magnetic field. The air gap non-uniform magnetic field due to the failure may affect different signals in the machine. As a non-invasive approach to detect the fault, the induced electromotive force (EMF) in the stator terminals is obtained using the air gap magnetic field. The induced EMF in the stator terminals is robust against the disturbance and noise, and can be used to detect the fault. The induced EMF can be used whether the machine is in a no-load or load state, while the current can be used in the on-load machine. The induced EMF in the stator terminal, $V(\varphi, \varphi', t)$, is as follows:

$$V(\varphi, \varphi', t) = A \frac{d}{dt}(B(\varphi, \varphi', t)) \tag{6.31}$$

where the synchronous generator flux passes the effective cross-section A. By substituting Equation (6.28) into Equation (6.31), the EMF is obtained as follows:

$$V(\varphi, \varphi', t) = \frac{-\alpha_1 \mu_0 J_s A \omega_s}{P} \sin(\omega_s t - P\varphi) + \frac{-\alpha_1 \mu_0 J_r A \omega_s}{P} \sin(\omega_s t - P\varphi')$$

$$+ \frac{\alpha_2 \mu_0 J_s A \omega_s}{2P} \sin(-\omega_s t + P\varphi) + \frac{\alpha_2 \mu_0 J_r A \omega_s}{2P} \sin(-\omega_s t + 2P\varphi - P\varphi')$$

$$+ \frac{-\alpha_2 \mu_0 J_s A 3 \omega_s}{2P} \sin(3\omega_s t - 3P\varphi) + \frac{-\alpha_2 \mu_0 J_r A 3 \omega_s}{2P} \sin(3\omega_s t - 2P\varphi - P\varphi')$$

$$+ \frac{-\alpha_1\mu_0 J_s\delta_{de}A\left(1+\frac{1}{P}\right)\omega_s}{2P}\sin\left(\left(1+\frac{1}{P}\right)\omega_s t-(P+1)\varphi\right)$$

$$+ \frac{-\alpha_1\mu_0 J_r\delta_{de}A\left(1+\frac{1}{P}\right)\omega_s}{2P}\sin\left(\left(1+\frac{1}{P}\right)\omega_s t-P\varphi'+\varphi\right)$$

$$+ \frac{-\alpha_1\mu_0 J_s\delta_{de}A\left(1-\frac{1}{P}\right)\omega_s}{2P}\sin\left(\left(1-\frac{1}{P}\right)\omega_s t-(P-1)\varphi\right)$$

$$+ \frac{-\alpha_1\mu_0 J_r\delta_{de}A\left(1-\frac{1}{P}\right)\omega_s}{2P}\sin\left(\left(1-\frac{1}{P}\right)\omega_s t-P\varphi'-\varphi\right)$$

$$+ \frac{-\alpha_2\mu_0 J_s\delta_{de}A\left(-1+\frac{1}{P}\right)\omega_s}{4P}\sin\left(\left(-1+\frac{1}{P}\right)\omega_s t+(P+1)\varphi\right)$$

$$+ \frac{-\alpha_2\mu_0 J_r\delta_{de}A\left(-1+\frac{1}{P}\right)\omega_s}{4P}\sin\left(\left(-1+\frac{1}{P}\right)\omega_s t+P\varphi'+2P\varphi-\varphi\right)$$

$$+ \frac{-\alpha_2\mu_0 J_s\delta_{de}A\left(3+\frac{1}{P}\right)\omega_s}{4P}\sin\left(\left(3+\frac{1}{P}\right)\omega_s t-(3P+1)\varphi\right)$$

$$+ \frac{-\alpha_2\mu_0 J_r\delta_{de}A\left(3+\frac{1}{P}\right)\omega_s}{4P}\sin\left(\left(3+\frac{1}{P}\right)\omega_s t-P\varphi'-2P\varphi-\varphi\right)$$

$$+ \frac{-\alpha_2\mu_0 J_s\delta_{de}A\left(-1+\frac{1}{P}\right)\omega_s}{4P}\sin\left(\left(-1+\frac{1}{P}\right)\omega_s t+(P+1)\varphi\right)$$

$$+ \frac{-\alpha_2\mu_0 J_r\delta_{de}A\left(-1+\frac{1}{P}\right)\omega_s}{4P}\sin\left(\left(-1+\frac{1}{P}\right)\omega_s t-P\varphi'+2P\varphi+\varphi\right)$$

$$+ \frac{-\alpha_2\mu_0 J_s\delta_{de}A\left(3-\frac{1}{P}\right)\omega_s}{4P}\sin\left(\left(3-\frac{1}{P}\right)\omega_s t-(3P+1)\varphi\right)$$

$$+ \frac{-\alpha_2\mu_0 J_r\delta_{de}A\left(3-\frac{1}{P}\right)\omega_s}{4P}\sin\left(\left(3-\frac{1}{P}\right)\omega_s t-P\varphi'-2P\varphi+\varphi\right) \tag{6.32}$$

According to Equation (6.29), the frequency pattern that appeared in the air gap magnetic field is also presented in the induced EMF. The extracted features ($f_{eccentricity1}$, $f_{eccentricity2}$, and $f_{eccentricity3}$) based on the EMF depend only on the number of poles and the frequency of the machine:

$$f_{eccentricity1} = \left(3\pm\frac{k}{P}\right)f_s \tag{6.33}$$

$$f_{eccentricity2} = \left(\pm 1\pm\frac{k}{P}\right)f_s \tag{6.34}$$

$$f_{eccentricity3} = 3kf_s \tag{6.35}$$

The first introduced index contains the other indices as well. However, the machine configuration or load harmonic may affect the third index and fade or reduce the subharmonics level under faulty conditions.

6.5 The Stator Current

The stator current may provide an efficient non-invasive approach to detect the fault. However, it is limited to the on-load machine only. Due to the simplicity of sampling the stator current and the ability to have complete information about the condition of electric machines, the stator current is used as one of the main signals to diagnose a fault in the SM. Therefore, one of the fault detection methods in electric machines is continuous monitoring of the stator phase current. This method involves a continuous comparison of the amplitude of one phase of the stator current of a healthy machine with one phase of the stator current of a faulty machine. The occurrence of a fault is then detected based on the imbalance in the current caused by the fault [11]. This method is not reliable because the current changes arise due to various factors, such as load. For this reason, the methods used in practice are instead based on stator current processing. In this method, the stator current frequency spectrum is used and the amplitudes of the harmonic components arising due to failure are introduced as indicators and frequencies as a frequency pattern [12]. The existing current transformers can be used for measurement purposes if their frequency bands do not limit the frequencies of interest; therefore, no additional sensors are required. The simplicity of the measurement equipment makes this economically justified. The following equation is used to calculate the stator current $I(t)$ [13]:

$$I(t) = \frac{1}{\mu_0 N} B(\varphi, \varphi', t) \frac{(1 - \delta_{de}\cos(\omega_r t - \varphi))}{\alpha_1 + \alpha_2\cos2p(\varphi - \theta_r)} \tag{6.36}$$

Substituting Equation (6.28) in Equation (6.36) gives the stator current as

$$I(\varphi, \varphi', t) = -\frac{\alpha_1{}^2 J}{PN}\cos(\omega_s t - P\varphi) - \frac{\alpha_1{}^2 J}{PN}\cos(\omega_s t - P\varphi') - \frac{\alpha_1\alpha_2 J}{2PN}\cos(-\omega_s t + P\varphi)$$

$$- \frac{\alpha_1\alpha_2 J}{2PN}\cos(-\omega_s t + 2P\varphi - P\varphi') - \frac{\alpha_1\alpha_2 J}{2PN}\cos(3\omega_s t - 3P\varphi) - \frac{\alpha_1\alpha_2 J}{2PN}\cos(3\omega_s t - 2P\varphi - P\varphi')$$

$$- \frac{\alpha_1{}^2 J\delta_{de}}{2PN}\cos\left(\left(1\pm\frac{1}{P}\right)\omega_s t - (P\pm1)\varphi\right) - \frac{\alpha_1{}^2 J\delta_{de}}{2PN}\cos\left(\left(1\pm\frac{1}{P}\right)\omega_s t - P\varphi'\pm\varphi\right)$$

$$- \frac{\alpha_1\alpha_2 J\delta_{de}}{4PN}\cos\left(\left(-1\pm\frac{1}{P}\right)\omega_s t + (P\pm1)\varphi\right)$$

$$- \frac{\alpha_1\alpha_2 J\delta_{de}}{4PN}\cos\left(\left(-1\pm\frac{1}{P}\right)\omega_s t + P\varphi'\pm(2P-1)\varphi\right)$$

$$- \frac{\alpha_1\alpha_2 J\delta_{de}}{4PN}\cos\left(\left(3\pm\frac{1}{P}\right)\omega_s t - (3P\pm1)\varphi\right)$$

$$- \frac{\alpha_1\alpha_2 J\delta_{de}}{4PN}\cos\left(\left(3\pm\frac{1}{P}\right)\omega_s t - P\varphi'\pm(2P-1)\varphi\right)$$

$$+ \frac{\alpha_1{}^2 J\delta_{de}}{2PN}\cos\left(\left(1\pm\frac{1}{P}\right)\omega_s t - (P+1)\varphi\right) + \frac{\alpha_1{}^2 J\delta_{de}}{2PN}\cos\left(\left(1\pm\frac{1}{P}\right)\omega_s t - \varphi - P\varphi'\right)$$

$$+ \frac{\alpha_1\alpha_2 J\delta_{de}}{4PN}\cos\left(\left(-1\pm\frac{1}{P}\right)\omega_s t + (P-1)\varphi\right)$$

$$+ \frac{\alpha_1\alpha_2 J\delta_{de}}{4PN}\cos\left(\left(-1\pm\frac{1}{P}\right)\omega_s t + (2P-1)\varphi - P\varphi'\right)$$

$$+ \frac{\alpha_1\alpha_2 J\delta_{de}}{4PN}\cos\left(\left(3\pm\frac{1}{P}\right)\omega_s t - (3P+1)\varphi\right) + \frac{\alpha_1\alpha_2\delta_{de}}{4PN}\cos\left(\left(3\pm\frac{1}{P}\right)\omega_s t - (2P+1)\varphi - P\varphi'\right)$$

$$+ \frac{\alpha_1{}^2 J\delta_{de}{}^2}{4PN}\cos\left(\left(1\pm\frac{2}{P}\right)\omega_s t - (P-2)\varphi\right) + \frac{\alpha_1{}^2 J\delta_{de}{}^2}{4PN}\cos\left(\left(1\pm\frac{2}{P}\right)\omega_s t - P\varphi' - 2\varphi\right)$$

$$+ \frac{\alpha_1\alpha_2 J\delta_{de}{}^2}{8PN}\cos\left(\left(-1\pm\frac{2}{P}\right)\omega_s t + (P-2)\varphi\right)$$

$$+ \frac{\alpha_1 \alpha_2 J \delta_{de}^{\ 2}}{8PN} \cos\left(\left(-1 \pm \frac{2}{P}\right)\omega_s t + P\varphi' \pm (P-1)2\varphi\right)$$

$$+ \frac{\alpha_1 \alpha_2 J \delta_{de}^{\ 2}}{8PN} \cos\left(\left(3 \pm \frac{2}{P}\right)\omega_s t - (3P-2)\varphi\right)$$

$$+ \frac{\alpha_1 \alpha_2 J \delta_{de}^{\ 2}}{8PN} \cos\left(\left(3 \pm \frac{2}{P}\right)\omega_s t - P\varphi' \pm (P-1)2\varphi\right)$$

$$+ \frac{\alpha_1 \alpha_2 J}{2PN} \cos(-\omega_s t + P\varphi) + \frac{\alpha_1 \alpha_2 J}{2PN} \cos(-\omega_s t + 2P\varphi - P\varphi') + \frac{\alpha_2^{\ 2} J}{4PN} \cos(-3\omega_s t + 2P\varphi)$$

$$+ \frac{\alpha_2^{\ 2} J}{4PN} \cos(-3\omega_s t + 4P\varphi - P\varphi') + \frac{\alpha_2^{\ 2} J}{4PN} \cos(\omega_s t - P\varphi) + \frac{\alpha_2^{\ 2} J}{4PN} \cos(\omega_s t - P\varphi')$$

$$+ \frac{\alpha_1 \alpha_2 J \delta_{de}}{4PN} \cos\left(\left(-1 \pm \frac{1}{P}\right)\omega_s t + (P \pm 1)\varphi\right) + \frac{\alpha_1 \alpha_2 J \delta_{de}}{4PN} \cos\left(\left(-1 \pm \frac{1}{P}\right)\omega_s t - P\varphi' + 3P\varphi\right)$$

$$+ \frac{\alpha_2^{\ 2} J \delta_{de}}{8PN} \cos\left(\left(-3 \pm \frac{1}{P}\right)\omega_s t + (3P \pm 1)\varphi\right) + \frac{\alpha_2^{\ 2} J \delta_{de}}{8PN} \cos\left(\left(-3 \pm \frac{1}{P}\right)\omega_s t + P\varphi' \pm (4P-1)\varphi\right)$$

$$+ \frac{\alpha_2^{\ 2} J \delta_{de}}{8PN} \cos\left(\left(-1 \pm \frac{1}{P}\right)\omega_s t - (5P \pm 1)\varphi\right) + \frac{\alpha_2^{\ 2} J \delta_{de}}{8PN} \cos\left(\left(-1 \pm \frac{1}{P}\right)\omega_s t - P\varphi' \pm (4P-1)\varphi\right)$$

$$+ \frac{\alpha_1 \alpha_2 J}{PN} \cos(3\omega_s t - 3P\varphi) + \frac{\alpha_1 \alpha_2 J}{PN} \cos(3\omega_s t - 2P\varphi - P\varphi') + \frac{\alpha_2^{\ 2} J}{2PN} \cos(\omega_s t - P\varphi)$$

$$+ \frac{\alpha_2^{\ 2} J}{2PN} \cos(\omega_s t - P\varphi') + \frac{\alpha_2^{\ 2} J}{2PN} \cos(5\omega_s t - 5P\varphi) + \frac{\alpha_2^{\ 2} J}{2PN} \cos(5\omega_s t - 4P\varphi - P\varphi')$$

$$+ \frac{\alpha_1 \alpha_2 J \delta_{de}}{2PN} \cos\left(\left(3 \pm \frac{1}{P}\right)\omega_s t - (3P \pm 1)\varphi\right) + \frac{\alpha_1 \alpha_2 J \delta_{de}}{2PN} \cos\left(\left(3 \pm \frac{1}{P}\right)\omega_s t - P\varphi' \pm (2P+1)\varphi\right)$$

$$+ \frac{\alpha_2^{\ 2} J \delta_{de}}{4PN} \cos\left(\left(1 \pm \frac{1}{P}\right)\omega_s t + (-P \pm 1)\varphi\right) + \frac{\alpha_2^{\ 2} J \delta_{de}}{4PN} \cos\left(\left(1 \pm \frac{1}{P}\right)\omega_s t + P\varphi' \pm \varphi\right)$$

$$+ \frac{\alpha_2^{\ 2} J \delta_{de}}{4PN} \cos\left(\left(5 \pm \frac{1}{P}\right)\omega_s t - (5P \pm 1)\varphi\right) + \frac{\alpha_2^{\ 2} J \delta_{de}}{4PN} \cos\left(\left(5 \pm \frac{1}{P}\right)\omega_s t - P\varphi' \pm (4P-1)\varphi\right)$$

$$- \frac{\alpha_1 \alpha_2 J \delta_{de}}{2PN} \cos\left(\left(\pm\frac{1}{P} - 1\right)\omega_s t + (P \pm 1)\varphi\right) - \frac{\alpha_1 \alpha_2 J \delta_{de}}{2PN} \cos\left(\left(\pm\frac{1}{P} - 1\right)\omega_s t + (2P \pm 1)\varphi - P\varphi'\right)$$

$$- \frac{\alpha_2^{\ 2} J \delta_{de}}{4PN} \cos\left(\left(\pm\frac{1}{P} - 3\right)\omega_s t + (3P \pm 1)\varphi\right) - \frac{\alpha_2^{\ 2} J \delta_{de}}{4PN} \cos\left(\left(\pm\frac{1}{P} - 3\right)\omega_s t + (4P \pm 1)\varphi - P\varphi'\right)$$

$$- \frac{\alpha_2^{\ 2} J \delta_{de}}{4PN} \cos\left(\left(\pm\frac{1}{P} + 1\right)\omega_s t + (-P \pm 1)\varphi\right) - \frac{\alpha_2^{\ 2} J \delta_{de}}{4PN} \cos\left(\left(\pm\frac{1}{P} + 1\right)\omega_s t \pm \varphi - P\varphi'\right)$$

$$- \frac{\alpha_1 \alpha_2 J \delta_{de}^{\ 2}}{4PN} \cos\left(\left(-1 \pm \frac{2}{P}\right)\omega_s t - (3P \pm 2)\varphi\right)$$

$$- \frac{\alpha_1 \alpha_2 J \delta_{de}^{\ 2}}{4PN} \cos\left(\left(-1 \pm \frac{2}{P}\right)\omega_s t - P\varphi' + (2P \pm 2)\varphi\right)$$

$$- \frac{\alpha_2^{\ 2} J \delta_{de}^{\ 2}}{8PN} \cos\left(\left(-3 \pm \frac{2}{P}\right)\omega_s t + (3P \pm 2)\varphi\right) - \frac{\alpha_2^{\ 2} J \delta_{de}^{\ 2}}{8PN} \cos\left(\left(-3 \pm \frac{2}{P}\right)\omega_s t + P\varphi' \pm 4P\varphi\right)$$

$$- \frac{\alpha_2^{\ 2} J \delta_{de}^{\ 2}}{8PN} \cos\left(\left(1 \pm \frac{2}{P}\right)\omega_s t - (5P \pm 2)\varphi\right) - \frac{\alpha_2^{\ 2} J \delta_{de}^{\ 2}}{8PN} \cos\left(\left(1 \pm \frac{2}{P}\right)\omega_s t - P\varphi' + (4P \pm 2)\varphi\right)$$

$$- \frac{\alpha_1 \alpha_2 J \delta_{de}}{2PN} \cos\left(\left(\pm\frac{1}{P} + 3\right)\omega_s t + (-3P \pm 1)\varphi\right)$$

$$- \frac{\alpha_1 \alpha_2 J \delta_{de}}{2PN} \cos\left(\left(\pm\frac{1}{P} + 3\right)\omega_s t + (-2P \pm 1) - P\varphi'\right)$$

$$- \frac{\alpha_2^{\ 2} J \delta_{de}}{4PN} \cos\left(\left(\pm\frac{1}{P} + 1\right)\omega_s t + (-P \pm 1)\varphi\right) - \frac{\alpha_2^{\ 2} J \delta_{de}}{4PN} \cos\left(\left(\pm\frac{1}{P} + 1\right)\omega_s t \pm \varphi - P\varphi'\right)$$

$$-\frac{\alpha_2{}^2 J\delta_{de}}{4PN}\cos\left(\left(\pm\frac{1}{P}+5\right)\omega_s t+(-5P\pm1)\varphi\right)-\frac{\alpha_1\alpha_2 J}{2PN}\cos\left(\left(\pm\frac{1}{P}+5\right)\omega_s t+(-5P+1)\varphi-P\varphi'\right)$$

$$-\frac{\alpha_1\alpha_2 J\delta_{de}{}^2}{4PN}\cos\left(\left(3\pm\frac{2}{P}\right)\omega_s t-(3P\pm2)\varphi\right)-\frac{\alpha_1\alpha_2 J\delta_{de}{}^2}{4PN}\cos\left(\left(3\pm\frac{2}{P}\right)\omega_s t-P\varphi'+(-2P\pm2)\varphi\right)$$

$$-\frac{\alpha_2{}^2 J\delta_{de}{}^2}{8PN}\cos\left(\left(1\pm\frac{2}{P}\right)\omega_s t+(-P\pm2)\varphi\right)-\frac{\alpha_2{}^2 J\delta_{de}{}^2}{8PN}\cos\left(\left(1\pm\frac{2}{P}\right)\omega_s t+P\varphi'-\varphi\right)$$

$$-\frac{\alpha_2{}^2 J\delta_{de}{}^2}{8PN}\cos\left(\left(5\pm\frac{2}{P}\right)\omega_s t-(5P\pm2)\varphi\right)-\frac{\alpha_2{}^2 J\delta_{de}{}^2}{8PN}\cos\left(\left(5\pm\frac{2}{P}\right)\omega_s t-P\varphi'-(4P+1)\varphi\right)$$

$$(6.37)$$

According to Equation (6.37), the DE fault creates numerous subharmonics in the on-load condition. In addition to the subharmonics, the harmonics such as 3 and 5, which have been introduced and utilized in many previous studies, also appear [14, 15]. The comprehensive frequency index becomes

$$f_{eccentricity1}=\left(3\pm\frac{k}{P}\right)f_s \tag{6.38}$$

$$f_{eccentricity2}=\left(\pm1\pm\frac{k}{P}\right)f_s \tag{6.39}$$

$$f_{eccentricity3}=3kf_s \tag{6.40}$$

6.6 Force Density and Unbalanced Magnetic Pull

Faults in the machine that cause vibrations are due to the air gap unbalanced magnetic field. The location and the degree of vibration depend on the type and severity of the fault. In addition, the unbalanced magnetic field causes a UMP and its direction relies on the type of fault. In the case of an SE fault, the location of the UMP is perpendicular to the stator movement direction (90° clockwise or counterclockwise).

The operation of the machine under an eccentricity fault is similar to the positive feedback loop. In other words, the eccentricity fault causes a UMP in the machine, which intensifies the non-uniformity of the air gap magnetic field and helps to increase the eccentricity severity. If the action required for maintenance is not applied at an appropriate time, the stator and rotor core collide. The vibration or UMP can be measured on the machine frame or rotor shaft, respectively.

However, other factors, such as a fault in the bearings, load instability, turbines connected to the generator, broken damper bars, stator and rotor SC faults, and faults in the power grid, may cause vibration in the machine. The vibration signal is sensitive to disturbance and noise, and its sensors are expensive.

The radial force $\tau(\varphi,y)$ on a machine can be obtained based on the Maxwell stress tensor as follows [16]:

$$\tau(\varphi,t)=\frac{B(\varphi,t)^2}{2\mu_0} \tag{6.41}$$

According to Equation (6.41), the radial force is proportional to the square root of the air gap magnetic flux density. Therefore, substituting Equations (6.28) in (6.41) and simplifying gives the radial force as:

$$\tau(\varphi,t)=\frac{B(\varphi,t)^2}{2\mu_0}$$

$$\tau(\varphi, t) = \frac{1}{4\mu_0} \begin{cases} = 1 + \dfrac{\alpha_1 \mu_0 J}{P} \cos(2\omega_s t - P\varphi'') \\[2mm] + \dfrac{\alpha_2 \mu_0 J}{2P} \cos(-2\omega_s t + P\varphi'') \\[2mm] + \dfrac{\alpha_2 \mu_0 J}{2P} \cos(6\omega_s t - 3P\varphi'') \\[2mm] + \dfrac{\alpha_1 \mu_0 J \delta_{de}}{2P} \cos\left(2\left(1 + \dfrac{1}{P}\right)\omega_s t - (P+1)\varphi''\right) \\[2mm] + \dfrac{\alpha_1 \mu_0 J \delta_{de}}{2P} \cos\left(2\left(1 - \dfrac{1}{P}\right)\omega_s t - (P-1)\varphi''\right) \\[2mm] + \dfrac{\alpha_2 \mu_0 J \delta_{de}}{4P} \cos\left(2\left(-1 + \dfrac{1}{P}\right)\omega_s t + (P+1)\varphi''\right) \\[2mm] + \dfrac{\alpha_2 \mu_0 J \delta_{de}}{4P} \cos\left(2\left(3 + \dfrac{1}{P}\right)\omega_s t - (3P+1)\varphi''\right) \\[2mm] + \dfrac{\alpha_2 \mu_0 J \delta_{de}}{4P} \cos\left(2\left(-1 + \dfrac{1}{P}\right)\omega_s t + (P+1)\varphi''\right) \\[2mm] + \dfrac{\alpha_2 \mu_0 J \delta_{de}}{4P} \cos\left(2\left(3 - \dfrac{1}{P}\right)\omega_s t - (3P+1)\varphi''\right) \end{cases} \tag{6.42}$$

According to Equation (6.43), the UMP is [17]:

$$\text{UMP} = \int_{\varphi=0}^{\varphi=2\pi} (\tau(\varphi, t) \cos(\varphi)L)d\varphi \tag{6.43}$$

where L is the effective length of the machine. After simplification of Equation (6.43), the UMP under the DE fault is as follows:

$$\begin{aligned}
\text{UMP} = &\frac{L}{4\mu_0} \sin\varphi \\
&+ \frac{\alpha_1 l_m J}{8P(P-1)} \sin(2\omega_s t - (P-1)\varphi'') + \frac{\alpha_1 l_m J}{8P(P+1)} \sin(2\omega_s t - (P+1)\varphi'') \\
&+ \frac{\alpha_2 l_m J}{16P(P+1)} \sin(-2\omega_s t + (P+1)\varphi'') + \frac{-\alpha_2 l_m J}{16P(P-1)} \sin(2\omega_s t + (P-1)\varphi'') \\
&+ \frac{\alpha_2 l_m J}{16P(3P-1)} \sin(6\omega_s t - (3P-1)P\varphi'') + \frac{\alpha_2 l_m J}{16P(3P+1)} \sin(2\omega_s t - (3P+1)\varphi'') \\
&+ \frac{\alpha_1 l_m J \delta_{de}}{16P^2} \sin\left(2\left(1 + \frac{1}{P}\right)\omega_s t - P\varphi''\right) \\
&+ \frac{\alpha_1 l_m J \delta_{de}}{16P(P-2)} \sin\left(2\left(1 + \frac{1}{P}\right)\omega_s t - (P-2)\varphi''\right) \\
&+ \frac{\alpha_1 l_m J \delta_{de}}{16P(P-2)} \sin\left(2\left(-1 + \frac{1}{P}\right)\omega_s t - (P-2)\varphi''\right) \\
&+ \frac{\alpha_1 l_m J \delta_{de}}{16P^2} \sin\left(2\left(-1 + \frac{1}{P}\right)\omega_s t - P\varphi''\right) \\
&+ \frac{-\alpha_2 l_m J \delta_{de}}{32P(P+2)} \sin\left(2\left(-1 + \frac{1}{P}\right)\omega_s t + (P+2)\varphi''\right) \\
&+ \frac{\alpha_2 l_m J \delta_{de}}{32P^2} \sin\left(2\left(-1 + \frac{1}{P}\right)\omega_s t - P\varphi''\right) \\
&+ \frac{-\alpha_2 l_m J \delta_{de}}{32P(P+2)} \sin\left(2\left(-1 - \frac{1}{P}\right)\omega_s t + (P+2)\varphi''\right)
\end{aligned}$$

$$+ \frac{\alpha_2 l_m J \delta_{de}}{32P^2} \sin\left(2\left(-1 - \frac{1}{P}\right)\omega_s t - P\varphi''\right)$$

$$+ \frac{\alpha_2 l_m J \delta_{de}}{32P^2} \sin\left(2\left(3 + \frac{1}{P}\right)\omega_s t - 3P\varphi''\right)$$

$$+ \frac{\alpha_2 l_m J \delta_{de}}{32P(3P + 2)} \sin\left(2\left(3 + \frac{1}{P}\right)\omega_s t - (3P + 2)\varphi''\right)$$

$$+ \frac{\alpha_2 l_m J \delta_{de}}{32P(3P + 2)} \sin\left(2\left(3 - \frac{1}{P}\right)\omega_s t - (3P + 2)\varphi''\right)$$

$$+ \frac{\alpha_2 l_m J \delta_{de}}{32P^2} \sin\left(2\left(3 - \frac{1}{P}\right)\omega_s t - 3P\varphi''\right) \tag{6.44}$$

According to Equation (6.44), DE inserts subharmonics into the vibration and UMP to give a unique frequency pattern that can be used as an index, as follows:

$$f_{eccentricity1} = 2k f_s \tag{6.45}$$

$$f_{eccentricity2} = 6k f_s \tag{6.46}$$

$$f_{eccentricity3} = 2\left(1 \pm \frac{k}{P}\right) f_s \tag{6.47}$$

$$f_{eccentricity4} = 2\left(3 \pm \frac{k}{P}\right) f_s \tag{6.48}$$

As expected, the frequencies extracted from the vibration and UMP are twice those of the extracted magnetic field frequencies.

6.7 Stator Slotting Effects

One of the major factors influencing the air gap length is the stator slot, as it produces a non-uniform air gap. This increases the flux path reluctance and reduces the magnetizing inductance [18], thereby significantly altering the induced EMF, current, and torque. Therefore, the impact of the stator slots on the air gap must be taken into account. Various methods and coefficients have been proposed for inclusion of the slotting effects in an air gap analysis. The best way to consider the effects of slots, including slot shape and depth, is to use the FEM, although some analytical coefficients have provided very good accuracy in considering these effects. A coefficient is introduced in reference [17] to estimate the effects of stator slots [18]. The effective air gap length is as follows:

$$g_e = C_{ca} g_0 \tag{6.49}$$

where the Carter coefficient C_{ca} is

$$C_{ca} = \frac{W_{st}}{W_{st} - \Gamma W_{ss}} \tag{6.50}$$

where W_{st} is the stator slot pitch and W_{ss} is the slot width. The value of Γ depends on the machine specifications and is determined as follows:

$$\Gamma = \frac{\dfrac{W_{ss}}{g_0}}{5 + \dfrac{W_{ss}}{g_0}} \tag{6.51}$$

6.8 Magnetic Saturation Effects

Magnetic saturation has a significant effect on the precision of fault detection. Neglecting magnetic saturation introduces an error of up to 62% in the computationally predicted eccentricity fault compared with the experimental ones [19]. The main reason for this large difference is that the air gap effective length depends on the magnetic saturation level. The effective length of the air gap at different points in the machine depends on the flux density, which explains the non-uniform air gap length. A relatively accurate method in reference [20] is introduced for analytical modeling of magnetic saturation. In this method, the reluctance of the core is assumed to be zero, and the increase in the reluctance of the flux path due to magnetic saturation is equated with the increase in the air gap length. Since the magnetic saturation level at different points is proportional to the magnetic flux density at these points, the increase in the air gap length should be proportional to the distribution of the magnetic flux density around the air gap. The dependence of the air gap function on the magnetic saturation is modeled as follows:

$$g(C_{sa}, \varphi, t) = g_{ee}(1 + \delta_{sa}\cos(2p\omega_r - 2p\varphi)) \tag{6.52}$$

where g_{ee} is the effective air gap length and δ_{sa} is the oscillation amplitude of the air gap length. These two parameters are calculated as follows:

$$g_{ee} = g_e \frac{3C_{sa}}{C_{sa} + 2} \tag{6.53}$$

$$\delta_{sa} = \frac{2(C_{sa} - 1)}{3C_{sa}} \tag{6.54}$$

where g_e is the effective air gap length that takes into account the effects of the stator slots. C_{sa} is the magnetic saturation coefficient value, which is equal to the ratio of the main harmonic amplitude of the air gap magnetic flux density in saturated and unsaturated conditions.

As already discussed, the saturation and rotor slot effects should not be overlooked. In this section, we examine all the equations mentioned in the previous section and consider the saturation and slotting effects for a mixed eccentricity (ME) fault.

6.9 The Mixed Eccentricity Fault

In Figure 6.4, the vector O_sO_r represents the ME fault, where ψ is the angle between the vectors O_sO_r and O_sO_w. The ME fault and its angle are defined as follows:

$$\delta_{me} = \sqrt{\delta_{se}^2 + \delta_{de}^2 + 2\delta_{se}\delta_{de}\cos(\omega_r t - \varphi)} \tag{6.55}$$

$$\psi = \varphi + tg^{-1}\frac{\delta_{de}\sin(\omega_r t)}{\delta_{se} + \delta_{de}\cos(\omega_r t)} \tag{6.56}$$

If only SE exists, then ψ and δ_{me} are converted to φ and δ_{se}. In the case of DE, ψ and δ_{me} are converted to $\varphi + \omega_r t$ and δ_{de}. In the case of an ME fault, SE and DE are considered simultaneously. The representation of the ME includes both SE and DE. In fact, it is most probable that both eccentricity faults present at the same time. Taking into account the ME, slotting and saturation effects give the following for the air gap function:

$$g(\varphi, t) = g(C_{sa}, \varphi, t)(1 - \delta_{me}\cos(\psi)) \tag{6.57}$$

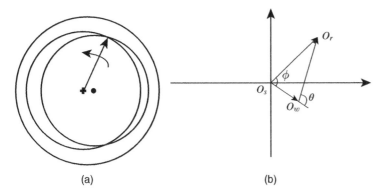

(a)   (b)

Figure 6.4 Mixed eccentricity fault in a synchronous machine: (a) in cross-section, (b) in vector form.

To include the saturation and slotting effects, the function $g(C_{sa}, \varphi, t)$ is replaced by the function $g_{ee}(1 + \delta_{sa} \cos(2p\omega_r t - 2p\varphi))$. Therefore, the air gap function is:

$$g(\varphi, t) = g_{ee}(1 + \delta_{sa}\cos(2p\omega_r t - 2p\varphi))(1 - \delta_{me}\cos(\psi)) \tag{6.58}$$

By replacing ω_r with ω_s/p, Equation (6.58) is converted to:

$$g(\varphi, t) = g_{ee}(1 + \delta_{sa}\cos(2\omega_s t - 2p\varphi) - \delta_{me}\cos(\psi) - \delta_{sa}\delta_{me}\cos(2\omega_s t - 2p\varphi)\cos(\psi)) \tag{6.59}$$

By considering the saturation effect, the effective air gap length is replaced by the air gap function with a saturation coefficient. Therefore:

$$g(\varphi, t) = g_0 \frac{3C_{sa}C_{ca}}{C_{sa} + 2}\left(1 + \delta_{sa}\cos(2\omega_s t - 2p\varphi) - \delta_{me}\cos(\psi) - \frac{1}{2}\delta_{sa}\delta_{me}\cos(2\omega_s t - 2p\varphi \pm \psi)\right) \tag{6.60}$$

For the rotor pole saliency, $\alpha_1 + \alpha_2 \cos 2p(\varphi + \theta_r)$, is included and the saturation coefficient $(3C_{sa}C_{ca})/(C_{sa} + 2)$ is replaced by γ. Therefore:

$$g(\varphi, t) = \gamma(\alpha_1 + \alpha_2\cos2p(\varphi - \theta_r))\left(1 + \delta_{sa}\cos(2\omega_s t - 2p\varphi)\right.$$
$$\left. -\delta_{me}\cos(\psi) - \frac{1}{2}\delta_{sa}\delta_{me}\cos(2\omega_s t - 2p\varphi \pm \psi)\right) \tag{6.61}$$

The air gap permeance is defined as the inverse of the air gap length, as follows:

$$\Lambda(\varphi, t) = \frac{1}{g(\varphi, t)}$$
$$= \frac{(\alpha_1 + \alpha_2\cos2p(\varphi - \theta_r))}{(1 - \delta_{sa}\cos(2\omega_s t - 2p\varphi) - \delta_{me}\cos(\psi) - 0.5\delta_{sa}\delta_{me}\cos(2\omega_s t - 2p\varphi \pm \psi))} \tag{6.62}$$

Using the Taylor expansion series, Equation (6.62) becomes:

$$\Lambda(\varphi, t) = (\alpha_1 + \alpha_2\cos2p(\varphi - \theta_r))\left(1 + \delta_{sa}\cos(2\omega_s t - 2p\varphi)\right.$$
$$\left. +\delta_{me}\cos(\psi) + 0.5\delta_{sa}\delta_{me}\cos(2\omega_s t - 2p\varphi \pm \psi)\right) \tag{6.63}$$

After mathematical manipulation, the simplified form of Equation (6.63) becomes:

$$\Lambda(\varphi, t) = \alpha_1 + \alpha_1\delta_{sa}\cos(2\omega_s t - 2p\varphi) + \alpha_1\delta_{me}\cos(\psi) + 0.5\delta_{sa}\delta_{me}\cos(2\omega_s t - 2p\varphi \pm \psi)$$
$$+ \alpha_2\cos2p(\varphi - \theta_r) + 0.5\alpha_2\delta_{sa}\cos(2p\varphi - 2p\theta_r \pm 2\omega_s t - 2p\varphi)$$
$$+ 0.5\alpha_2\delta_{me}\cos(2p\varphi - 2p\theta_r \pm \psi) \tag{6.64}$$

Equation (6.64) is valid for the ME fault, while Equation (6.65) and Equation (6.66) are valid for the SE and DE faults, respectively:

$$\Lambda_{SE}(\varphi, t) = \alpha_1 + \alpha_1 \delta_{sa} \cos(2\omega_s t - 2p\varphi) + \alpha_1 \delta_{se} \cos(\varphi)$$
$$+ 0.5 \delta_{sa} \delta_{se} \cos(2\omega_s t - 2p\varphi \pm \varphi) + \alpha_2 \cos 2p(\varphi - \theta_r)$$
$$+ 0.5 \alpha_2 \delta_{sa} \cos(2p\varphi - 2p\theta_r \pm 2\omega_s t - 2p\varphi) + 0.5 \alpha_2 \delta_{se} \cos(2p\varphi - 2p\theta_r \pm \varphi) \quad (6.65)$$

$$\Lambda_{DE}(\varphi, t) = \alpha_1 + \alpha_1 \delta_{sa} \cos(2\omega_s t - 2p\varphi) + \alpha_1 \delta_{de} \cos\left(\frac{\omega_s t}{p} + \varphi\right)$$
$$+ 0.5 \delta_{sa} \delta_{de} \cos\left(2\omega_s t - 2p\varphi \pm \frac{\omega_s t}{p} + \varphi\right) + \alpha_2 \cos 2p\left(\varphi - \frac{\omega_s t}{p}\right)$$
$$+ 0.5 \alpha_2 \delta_{sa} \cos\left(2p\varphi - 2p\frac{\omega_s t}{p} \pm 2\omega_s t - 2p\varphi\right) + 0.5 \alpha_2 \delta_{de} \cos\left(2p\varphi - 2p\frac{\omega_s t}{p} \pm \frac{\omega_s t}{p} + \varphi\right)$$
$$(6.66)$$

The simplified form of the air gap permeance in the presence of the DE is as follows:

$$\Lambda_{DE}(\varphi, t) = \alpha_1 + \alpha_1 \delta_{sa} \cos(2\omega_s t - 2p\varphi) + \alpha_1 \delta_{de} \cos\left(\frac{\omega_s t}{p} + \varphi\right)$$
$$+ 0.5 \delta_{sa} \delta_{de} \cos\left(\left(2 \pm \frac{1}{p}\right)\omega_s t - (2p \pm 1)\varphi\right) + \alpha_2 \cos(-2\omega_s + 2p\varphi) + 0.5 \alpha_2 \delta_{sa}$$
$$+ 0.5 \alpha_2 \delta_{sa} \cos(-4\omega_s t + 2p\varphi) + 0.5 \alpha_2 \delta_{de} \cos\left(\left(-2 \pm \frac{1}{p}\right)\omega_s t + (2p \pm 1)\varphi\right) \quad (6.67)$$

6.10 The Air Gap Magnetic Field

The air gap magnetic flux density is first derived, disregarding the saturation and slotting effects. In the case of an eccentricity fault, some subharmonic frequencies have been observed to appear and have been used as an index. A similar procedure (see Section 6.3) is used here, but with consideration of the saturation and slotting effects.

The stator magnetic flux density due to the current density and the permeance of the stator windings is:

$$B_s(\varphi, t) = \Lambda_{DE}(\varphi, t) \int \mu_0 J_s \sin(\omega_s t - p\varphi) d\varphi \quad (6.68)$$

By integrating the current density and substituting Equation (6.67) into Equation (6.68), the stator magnetic field becomes:

$$B_s(\varphi, t) = \left(\alpha_1 + \alpha_1 \delta_{sa} \cos(2\omega_s t - 2p\varphi) + \alpha_1 \delta_{de} \cos\left(\frac{\omega_s t}{p} + \varphi\right)\right.$$
$$+ 0.5 \delta_{sa} \delta_{de} \cos\left(\left(2 \pm \frac{1}{p}\right)\omega_s t - (2p \pm 1)\varphi\right) + \alpha_2 \cos(-2\omega_s + 2p\varphi)$$
$$+ 0.5 \alpha_2 \delta_{sa} + 0.5 \alpha_2 \delta_{sa} \cos(-4\omega_s t + 2p\varphi)$$
$$\left. + 0.5 \alpha_2 \delta_{de} \cos\left(\left(-2 \pm \frac{1}{p}\right)\omega_s t + (2p \pm 1)\varphi\right)\right)\left(\frac{\mu_0 Js}{P} \cos(\omega_s t - p\varphi)\right) \quad (6.69)$$

After some mathematical manipulations, the stator magnetic flux density becomes:

$$B_s(\varphi, t) = \frac{\mu_0 Js \, \alpha_1}{P} \cos(\omega_s t - p\varphi)$$

$$+ \frac{\mu_0 \, Js \, \alpha_1}{2P}(\cos((2\kappa+1)\omega_s t - (2\kappa p+1)\varphi) + \cos((2\kappa-1)\omega_s t - (2\kappa p-1)\varphi))$$

$$+ \frac{\mu_0 \, Js \, \delta_{de}\alpha_1}{2P}\cos\left(\left(1 \pm \frac{1}{p}\right)\omega_s t \pm (p \pm 1)\varphi\right)$$

$$+ \frac{\mu_0 \, Js \, \delta_{de}\delta_{sa}\alpha_1}{4P}\cos\left((1 \pm \left(2 \pm \frac{1}{p}\right)\omega_s t \pm (1 + (2p \pm 1)\varphi)\right)$$

$$+ \frac{\mu_0 \, Js \, \alpha_2}{2P}\cos\left((1 \pm 2\kappa)\omega_s t \pm (1 \pm 2\kappa)\varphi)\right)$$

$$+ \frac{\mu_0 \, Js \, \delta_{sa}\alpha_2}{2P}\cos(\omega_s t - p\varphi)$$

$$+ \frac{\mu_0 \, Js \, \delta_{sa}\alpha_2}{4P}\cos((1 \pm (-4\kappa))\omega_s t \pm ((2\kappa \pm 1)p\varphi))$$

$$+ \frac{\mu_0 \, Js \, \delta_{de}\alpha_2}{4P}\cos\left(\left(1 \pm \left(-2 \pm \frac{1}{p}\right)\right)\omega_s t \pm (-1 + (2p \pm 1))\varphi\right) \tag{6.70}$$

The same procedure is then used to extract the rotor magnetic flux density, as follows:

$$B_r(\varphi, t) = \Lambda_{DE}(\varphi, t)\int \mu_0 \, Jr\sin(\omega_s t - p\varphi)\,\mathrm{d}\varphi \tag{6.71}$$

Integrating the rotor current density and performing some mathematical manipulations gives the final value of the rotor magnetic flux density as:

$$B_r(\varphi, t) = \frac{\mu_0 \, Jr \, \alpha_1}{P}\cos(\omega_s t - p\varphi)$$

$$+ \frac{\mu_0 \, Jr \, \alpha_1}{2P}(\cos((2\kappa+1)\omega_s t - (2\kappa p+1)\varphi) + \cos((2\kappa-1)\omega_s t - (2\kappa p-1)\varphi))$$

$$+ \frac{\mu_0 \, Jr \, \delta_{de}\alpha_1}{2P}\cos\left(\left(1 \pm \frac{1}{p}\right)\omega_s t \pm (p \pm 1)\varphi\right)$$

$$+ \frac{\mu_0 \, Jr \, \delta_{de}\delta_{sa}\alpha_1}{4P}\cos\left((1 \pm \left(2 \pm \frac{1}{p}\right)\omega_s t \pm (1 + (2p \pm 1)\varphi)\right)$$

$$+ \frac{\mu_0 \, Jr \, \alpha_2}{2P}\cos\left((1 \pm 2\kappa)\omega_s t \pm (1 \pm 2\kappa)\varphi)\right)$$

$$+ \frac{\mu_0 \, Jr \, \delta_{sa}\alpha_2}{2P}\cos(\omega_s t - p\varphi)$$

$$+ \frac{\mu_0 \, Jr \, \delta_{sa}\alpha_2}{4P}\cos((1 \pm (-4\kappa))\omega_s t \pm ((2\kappa \pm 1)p\varphi))$$

$$+ \frac{\mu_0 \, Jr \, \delta_{de}\alpha_2}{4P}\cos\left(\left(1 \pm \left(-2 \pm \frac{1}{p}\right)\right)\omega_s t \pm (-1 + (2p \pm 1))\varphi\right) \tag{6.72}$$

The air gap magnetic flux density is the result of the stator and rotor magnetic flux densities, as follows:

$$B(\varphi, t) = \left(\frac{\mu_0 J \alpha_1}{P} + \frac{\mu_0 J \delta_{sa}\alpha_2}{2P}\right)\cos(\omega_s t - p\varphi)$$

$$+ \frac{\mu_0 J(\alpha_1 + \alpha_2)}{2P}(\cos((2\kappa \pm v)\omega_s t \pm (2\kappa \pm v)p\varphi))$$

$$+ \frac{\mu_0 J \delta_{de}\alpha_1}{2P}\cos\left(\left(1 \pm \frac{\kappa}{p}\right)\omega_s t \pm (\kappa \pm 1)p\varphi\right)$$

$$+ \frac{\mu_0 J \delta_{de}\delta_{sa}\alpha_1}{4P}\cos\left(\left(\left(2 \pm \frac{\kappa}{p}\right) \pm v\right)\omega_s t \pm ((2\kappa \pm 1) + v)\,p\varphi\right)$$

$$+ \frac{\mu_0 J \delta_{sa} \alpha_2}{4P} \cos((4\kappa \pm v)\omega_s t \pm ((4\kappa \pm v)p\varphi))$$

$$+ \frac{\mu_0 J \delta_{de} \alpha_2}{4P} \cos\left(\left(\left(-2 \pm \frac{\kappa}{p}\right) \pm v\right)\omega_s t + ((2\kappa \pm 1) \pm v)p\varphi\right) \tag{6.73}$$

According to Equation (6.73), the DE fault causes subharmonics to emerge in the air gap magnetic flux density. The extracted indices are as follows:

$$f_{eccentricity1} = (2\kappa \pm v)f_s \tag{6.74}$$

$$f_{eccentricity2} = \left(1 \pm \frac{\kappa}{p}\right)f_s \tag{6.75}$$

$$f_{eccentricity3} = \left(\left(2 \pm \frac{\kappa}{p}\right) \pm v\right)f_s \tag{6.76}$$

$$f_{eccentricity4} = (4\kappa \pm v)f_s \tag{6.77}$$

Among the introduced indices for the DE fault detection, $f_{eccetricity3}$ can cover all the subharmonic frequencies. This reveals the possibility of using only $f_{eccetricity3}$ for comprehensive diagnosis of the desired frequency components.

6.11 Induced Electromotive Force in Stator Terminals

The induced voltage in the stator is considered a non-invasive approach for fault diagnosis because voltage transformers have already been installed for data acquisition. In this section, saturation and slotting effects are included in the formulas to see their effects on the induced EMF. The induced EMF arising from the magnetic field is as follows [21]:

$$V(\varphi, t) = A\frac{d}{dt}B(\varphi, t) \tag{6.78}$$

Substituting the air gap magnetic field in Equation (6.78) and applying the differential operator to the magnetic field gives the EMF as:

$$V(\varphi, t) = \left(\frac{\mu_0 J \alpha_1 A \omega_s}{P} + \frac{\mu_0 J \delta_{sa} \alpha_2 A \omega_s}{2P}\right)\sin(\omega_s t - p\varphi)$$

$$+ \frac{\mu_0 J(\alpha_1 + \alpha_2)A(2\kappa \pm v)\omega_s}{2P}(\sin((2\kappa \pm v)\omega_s t \pm (2\kappa \pm v)p\varphi))$$

$$+ \frac{\mu_0 J \delta_{de} \alpha_1 A \left(1 \pm \frac{\kappa}{p}\right)\omega_s}{2P}\sin\left(\left(1 \pm \frac{\kappa}{p}\right)\omega_s t \pm (\kappa \pm 1)p\varphi\right)$$

$$+ \frac{\mu_0 J \delta_{de} \delta_{sa}(\alpha_1 + \alpha_2)A\left(v \pm \left(2 \pm \frac{\kappa}{p}\right)\right)\omega_s}{4P}$$

$$\sin\left(\left(\left(2 \pm \frac{\kappa}{p}\right) \pm v\right)\omega_s t \pm ((2\kappa \pm 1) + v)p\varphi\right)$$

$$+ \frac{\mu_0 J \delta_{sa} \alpha_2 A(v \pm (4\kappa))\omega_s}{4P}\sin((4\kappa \pm v)\omega_s t \pm ((4\kappa \pm v)p\varphi)) \tag{6.79}$$

The extracted frequency indices from the spectrum of the induced EMF at the stator terminals are as follows:

$$f_{eccentricity1} = (2\kappa \pm v)f_s \tag{6.80}$$

$$f_{eccentricity2} = \left(1 \pm \frac{\kappa}{p}\right) f_s \tag{6.81}$$

$$f_{eccentricity3} = \left(\left(2 \pm \frac{\kappa}{p}\right) \pm v\right) f_s \tag{6.82}$$

$$f_{eccentricity4} = (4\kappa \pm v) f_s \tag{6.83}$$

A comprehensive investigation of featured indices indicates that $f_{eccentricity3}$ is comprehensive enough to cover the entire indices.

6.12   Force Density and Unbalanced Magnetic Pull

Neglecting the magnetic saturation causes an increase in the magnetic flux density and, therefore, the magnetic potential, especially in the teeth, which leads to local saturation. Because the stress tensor and UMP are directly related to the magnetic potential, the calculated UMP and vibration are much higher than the actual value. Including magnetic saturation in modeling leads to an increase in the magnetic reluctance of the flux path, and a consequent decrease in the magnetic potential and UMP.

Note that the differences in the magnitude of vibration and UMP are seen in the transient state, while their amplitude is in the steady state, even in the case of neglected saturation, and is very close to the actual value. This is because the synchronous generator operates at the knee point of the magnetization characteristic during the steady-state operation. As in the previous section, to obtain the UMP, the magnetic stress tensor must be obtained as follows:

$$\tau(\varphi, t) = \frac{B(\varphi, t)^2}{2\mu_0} \tag{6.84}$$

The stress tensor is proportional to the square root of the magnetic field; therefore, by substituting Equation (6.73) into Equation (6.84), it becomes:

$$\tau(\varphi, t) = \frac{1}{2\mu_0} \begin{pmatrix} B_1 \cos(\omega_s t - p\varphi) \\ + B_2(\cos((2\kappa \pm v)\omega_s t - (2\kappa \pm v)p\varphi)) \\ + B_3 \cos\left(\left(1 \pm \frac{\kappa}{p}\right)\omega_s t \pm (\kappa \pm 1)p\varphi\right) \\ + B_4 \cos\left(\left(\left(2 \pm \frac{\kappa}{p}\right) \pm v\right)\omega_s t + ((2\kappa \pm 1) \pm v)\, p\varphi\right) \\ + B_5 \cos((2\kappa \pm v)\omega_s t + (2\kappa \pm v)p\varphi)) \\ + B_6 \cos((4\kappa \pm v)\omega_s t + ((4\kappa \pm v)p\varphi)) \\ + B_7 \cos\left(\left(\left(-2 \pm \frac{\kappa}{p}\right) \pm v\right)\omega_s t \pm ((2\kappa \pm 1) - v)p\varphi\right) \end{pmatrix}^2 \tag{6.85}$$

The final format of the stress tensor (vibration), after some mathematical manipulations, is as follows:

$$\tau(\varphi, t) = \tau_0 + \tau_1 \cos(2\omega_s t - 2p\varphi)$$
$$+ \tau_2 \cos((4\kappa \pm 2v)\omega_s t - (4\kappa \pm 2v)p\varphi)$$
$$+ \tau_3 \cos\left(\left(2 \pm \frac{2\kappa}{p}\right)\omega_s t + (2\kappa \pm 2)p\varphi\right)$$

$$+ \tau_4 \cos\left(\left(\left(4 \pm \frac{2\kappa}{p}\right) \pm 2v\right)\omega_s t + ((4\kappa \pm 2) \pm 2v)p\varphi\right)$$

$$+ \tau_5 \cos((4\kappa \pm 2v)\omega_s t + (4\kappa \pm 2v)p\varphi)$$

$$+ \tau_6 \cos((8\kappa \pm 2v)\omega_s t + (8\kappa \pm 2v)p\varphi)$$

$$+ \tau_7 \cos\left(\left(\left(-4 \pm \frac{2\kappa}{p}\right) \pm 2v\right)\omega_s t + ((4\kappa \pm 2) \pm 2v)p\varphi\right)$$

$$+ \tau_8 \cos(((2\kappa \pm 1) \pm v)\omega_s t - ((2\kappa \pm 1) \pm v)p\varphi)$$

$$+ \tau_9 \cos\left(2 \pm \frac{\kappa}{p}\right)\omega_s t + ((\kappa \pm 2)p\varphi)$$

$$+ \tau_{10} \cos\left(\pm\frac{\kappa}{p}\right)\omega_s t + ((\kappa \pm 2)p\varphi)$$

$$+ \tau_{11} \cos\left(\left(\left(3 \pm \frac{\kappa}{p}\right) \pm v\right)\omega_s t + ((2\kappa \pm 2) \pm v)p\varphi\right)$$

$$+ \tau_{12} \cos\left(\left(\left(1 \pm \frac{\kappa}{p}\right) \pm v\right)\omega_s t + ((2\kappa \pm v)p\varphi)\right)$$

$$+ \tau_{13} \cos(((2\kappa \pm 1) \pm v)\omega_s t + ((2\kappa \pm 1) \pm v)p\varphi)$$

$$+ \tau_{14} \cos(((4\kappa \pm 1) \pm v)\omega_s t + ((4\kappa \pm 1) \pm v)p\varphi)$$

$$+ \tau_{15} \cos\left(\left(\left(-1 \pm \frac{\kappa}{p}\right) \pm v\right)\omega_s t + (((2\kappa \pm 2) \pm v)p\varphi)\right)$$

$$+ \tau_{16} \cos\left(\left(\left(-3 \pm \frac{\kappa}{p}\right) \pm v\right)\omega_s t + ((2\kappa \pm v)p\varphi)\right)$$

$$+ \tau_{17} \cos\left(\frac{2\kappa p \pm \kappa}{p} \pm (v + 1)\right)\omega_s + ((-\kappa \pm (v + 1))p\varphi)$$

$$+ \tau_{18} \cos\left(\frac{2\kappa p \pm \kappa}{p} \pm (v - 1)\right)\omega_s + ((-3\kappa \pm (v + 1))p\varphi)$$

$$+ \tau_{19} \cos\left(\left(\frac{2\kappa p \pm \kappa}{p} \pm (2v + 2)\right)\omega_s + (2v + 1)p\varphi\right)$$

$$+ \tau_{20} \cos\left(\left(\frac{2\kappa p \pm \kappa}{p} \pm (2v + 2)\right)\omega_s + (-4\kappa \pm (2v + 1))p\varphi\right)$$

$$+ \tau_{21} \cos((4\kappa \pm 2v)\omega_s t + (2v)p\varphi)$$

$$+ \tau_{22} \cos((\pm 2v)\omega_s t + (-4\kappa \pm 2v)p\varphi)$$

$$+ \tau_{23} \cos((6\kappa \pm 2v)\omega_s t + (2\kappa \pm 2v)p\varphi)$$

$$+ \tau_{24} \cos((-2\kappa \pm 2v)\omega_s t + (-6\kappa \pm 2v)p\varphi)$$

$$+ \tau_{25} \cos\left(\left(\frac{2\kappa p \pm \kappa}{p} \pm (2v + 2)\right)\omega_s + (2v + 1)p\varphi\right)$$

$$+ \tau_{26} \cos\left(\left(\frac{2\kappa p \pm \kappa}{p} \pm (2v + 2)\right)\omega_s + (-4\kappa \pm (2v + 1))p\varphi\right)$$

$$+ \tau_{27} \cos\left(\left(3 \pm \frac{2\kappa}{p}\right) \pm v\right)\omega_s + ((3\kappa \pm 1) \pm v)p\varphi)$$

$$+ \tau_{28} \cos\left(\left(\frac{2\kappa}{p} - 1\right) \pm v\right)\omega_s + (-\kappa \pm (v + 2))p\varphi)$$

$$+ \tau_{29}\cos\left(\left(\frac{2\kappa p \pm \kappa}{p} \pm (v+1)\right)\omega_s + (3\kappa \pm (v+1))p\varphi\right)$$

$$+ \tau_{30}\cos\left(\left(\frac{-2\kappa p \pm \kappa}{p} \pm (v+1)\right)\omega_s + (-\kappa \pm (v+1))p\varphi\right)$$

$$+ \tau_{31}\cos\left(\left(\frac{4\kappa p \pm \kappa}{p} \pm (v+1)\right)\omega_s t + (5\kappa \pm (v+1))p\varphi\right)$$

$$+ \tau_{32}\cos\left(\left(\frac{-4\kappa p \pm \kappa}{p} \pm (v+1)\right)\omega_s t + (-3\kappa \pm (v+1))p\varphi\right)$$

$$+ \tau_{33}\cos\left(\left(\frac{2\kappa}{p} \pm (v-1)\right)\omega_s t + (3\kappa \pm (v+1))p\varphi\right)$$

$$+ \tau_{34}\cos\left(\left(\frac{2\kappa}{p} \pm (v+3)\right)\omega_s t + (-\kappa \pm (v+1))p\varphi\right)$$

$$+ \tau_{35}\cos\left(\left(\frac{2\kappa p \pm \kappa}{p} \pm (2v+2)\right)\omega_s t + (4\kappa \pm (v+1))p\varphi\right)$$

$$+ \tau_{36}\cos\left(\left(\frac{-2\kappa p \pm \kappa}{p} \pm (v+2)\right)\omega_s t + (v+1)p\varphi\right)$$

$$+ \tau_{37}\cos\left(\left(\frac{4\kappa p \pm \kappa}{p} \pm (2v+2)\right)\omega_s t + (6\kappa \pm 1)p\varphi\right)$$

$$+ \tau_{38}\cos\left(\left(\frac{-4\kappa p \pm \kappa}{p} \pm (2v+2)\right)\omega_s t + (-2\kappa \pm (2v+1))p\varphi\right)$$

$$+ \tau_{39}\cos\left(\left(\frac{2\kappa}{p} \pm 2v\right)\omega_s t + (4\kappa \pm (2v+1))p\varphi\right)$$

$$+ \tau_{40}\cos\left(\left(\left(4 \pm \frac{4\kappa}{p}\right) \pm 2v\right)\omega_s t + (2v+2)p\varphi\right)$$

$$+ \tau_{41}\cos((6\kappa \pm 2v)\omega_s t + (6\kappa \pm 2v)p\varphi)$$
$$+ \tau_{42}\cos((-2\kappa \pm 2v)\omega_s t + (-2\kappa \pm 2v)p\varphi)$$

$$+ \tau_{43}\cos\left(\left(\frac{2\kappa p \pm \kappa}{p} \pm (v-2)\right)\omega_s t + (4\kappa \pm (2v+1))p\varphi\right)$$

$$+ \tau_{44}\cos\left(\left(\frac{2\kappa p \pm \kappa}{p} \pm (v+2)\right)\omega_s t + (2v+1)p\varphi\right)$$

$$+ \tau_{45}\cos\left(\left(\frac{4\kappa p \pm \kappa}{p} \pm (2v-2)\right)\omega_s t + (6\kappa \pm (2v+1))p\varphi\right)$$

$$+ \tau_{46}\cos\left(\left(\frac{4\kappa p \pm \kappa}{p} \pm (v+2)\right)\omega_s t + (2\kappa \pm (2v+1))p\varphi\right) \tag{6.86}$$

The UMP is the close integration of the stress tensor over the machine surface, as follows:

$$\text{UMP} = \int_{\varphi=0}^{\varphi=2\pi} (\tau(\varphi,t)\cos(\varphi)lm)\,d\varphi$$

$$= ump_1\sin(2\omega_s t + (-2p \pm 1)\varphi)$$

$$+ ump_2\sin((4\kappa \pm 2v)\omega_s t + ((-4\kappa \pm 2v)p \pm 1)\varphi)$$

$$+ ump_3\sin\left(\left(2 \pm \frac{2\kappa}{p}\right)\omega_s t + ((2\kappa \pm 2)p \pm 1)\varphi\right)$$

$$+ ump_4\sin\left(\left(\left(4 \pm \frac{2\kappa}{p}\right) \pm 2v\right)\omega_s t + ((4\kappa \pm 2) \pm 2v)p \pm 1\right)\varphi)$$

$$+ ump_5 \sin((4\kappa \pm 2v)\omega_s t + ((4\kappa \pm 2v)p \pm 1)\varphi)$$

$$+ ump_6 \sin((8\kappa \pm 2v)\omega_s t + ((8\kappa \pm 2v)p \pm 1)\varphi)$$

$$+ ump_7 \sin\left(\left(\left(-4 \pm \frac{2\kappa}{p}\right) \pm 2v\right)\omega_s t + (((4\kappa \pm 2) \pm 2v)p \pm 1)\varphi\right)$$

$$+ ump_8 \sin(((2\kappa \pm 1) \pm v)\omega_s t - ((2\kappa \pm 1) \pm v)p \pm 1)\varphi)$$

$$+ ump_9 \sin\left(\left(2 \pm \frac{\kappa}{p}\right)\omega_s t + ((\kappa \pm 2)p \pm 1)\varphi\right)$$

$$+ ump_{10} \sin\left(\left(\pm\frac{\kappa}{p}\right)\omega_s t + ((\kappa \pm 2)p \pm 1)\varphi\right)$$

$$+ ump_{11} \sin\left(\left(\left(3 \pm \frac{\kappa}{p}\right) \pm v\right)\omega_s t + (((2\kappa \pm 2) \pm v)p \pm 1)\varphi\right)$$

$$+ ump_{12} \sin\left(\left(\left(1 \pm \frac{\kappa}{p}\right) \pm v\right)\omega_s t + ((2\kappa \pm v)p \pm 1)\varphi\right)$$

$$+ ump_{13} \sin(((2\kappa \pm 1) \pm v)\omega_s t + (((2\kappa \pm 1) \pm v)p \pm 1)\varphi)$$

$$+ ump_{14} \sin(((4\kappa \pm 1) \pm v)\omega_s t + (((4\kappa \pm 1) \pm v)p \pm 1)\varphi)$$

$$+ ump_{15} \sin\left(\left(\left(-1 \pm \frac{\kappa}{p}\right) \pm v\right)\omega_s t + (((2\kappa \pm 2) \pm v)p \pm 1)\varphi\right)$$

$$+ ump_{16} \sin\left(\left(\left(-3 \pm \frac{\kappa}{p}\right) \pm v\right)\omega_s t + ((2\kappa \pm v)p \pm 1)\varphi\right)$$

$$+ ump_{17} \sin\left(\left(\frac{2\kappa p \pm \kappa}{p} \pm (v+1)\right)\omega_s + ((-\kappa \pm (v+1))p \pm 1)\varphi\right)$$

$$+ ump_{18} \sin\left(\left(\frac{2\kappa p \pm \kappa}{p} \pm (v-1)\right)\omega_s + ((-3\kappa \pm (v+1))p \pm 1)\varphi\right)$$

$$+ ump_{19} \sin\left(\left(\frac{2\kappa p \pm \kappa}{p} \pm (2v+2)\right)\omega_s + ((2v+1)p \pm 1)\varphi\right)$$

$$+ ump_{20} \sin\left(\left(\frac{2\kappa p \pm \kappa}{p} \pm (2v+2)\right)\omega_s + ((-4\kappa \pm (2v+1))p \pm 1)\varphi\right)$$

$$+ ump_{21} \sin((4\kappa \pm 2v)\omega_s t + ((2v)p \pm 1)\varphi)$$

$$+ ump_{22} \sin((\pm 2v)\omega_s t + ((-4\kappa \pm 2v)p \pm 1)\varphi)$$

$$+ ump_{23} \sin((6\kappa \pm 2v)\omega_s t + ((2\kappa \pm 2v)p \pm 1)\varphi)$$

$$+ ump_{24} \sin((-2\kappa \pm 2v)\omega_s t + ((-6\kappa \pm 2v)p \pm 1)\varphi)$$

$$+ ump_{25} \sin\left(\left(\frac{2\kappa p \pm \kappa}{p} \pm (2v+2)\right)\omega_s + ((2v+1)p \pm 1)\varphi\right)$$

$$+ ump_{26} \sin\left(\left(\frac{2\kappa p \pm \kappa}{p} \pm (2v+2)\right)\omega_s + ((-4\kappa \pm (2v+1))p \pm 1)\varphi\right)$$

$$+ ump_{27} \sin\left(\left(3 \pm \frac{2\kappa}{p}\right) \pm v\right)\omega_s + (((3\kappa \pm 1) \pm v)p \pm 1)\varphi)$$

$$+ ump_{28} \sin\left(\left(\left(\frac{2\kappa}{p} - 1\right) \pm v\right)\omega_s + ((-\kappa \pm (v+2))p \pm 1)\varphi\right)$$

$$+ ump_{29} \sin\left(\left(\frac{2\kappa p \pm \kappa}{p} \pm (v+1)\right)\omega_s + ((3\kappa \pm (v+1))p \pm 1)\varphi\right)$$

$$+ ump_{30}\sin\left(\left(\frac{-2\kappa p \pm \kappa}{p} \pm (v+1)\right)\omega_s + ((-\kappa \pm (v+1))p \pm 1)\varphi\right)$$

$$+ ump_{31}\sin\left(\left(\frac{4\kappa p \pm \kappa}{p} \pm (v+1)\right)\omega_s t + ((5\kappa \pm (v+1))p \pm 1)\varphi\right)$$

$$+ ump_{32}\sin\left(\left(\frac{-4\kappa p \pm \kappa}{p} \pm (v+1)\right)\omega_s t + ((-3\kappa \pm (v+1))p \pm 1)\varphi\right)$$

$$+ ump_{33}\sin\left(\left(\frac{2\kappa}{p} \pm (v-1)\right)\omega_s t + ((3\kappa \pm (v+1))p \pm 1)\varphi\right)$$

$$+ ump_{34}\sin\left(\left(\frac{2\kappa}{p} \pm (v+3)\right)\omega_s t + ((-\kappa \pm (v+1))p \pm 1)\varphi\right)$$

$$+ ump_{35}\sin\left(\left(\frac{2\kappa p \pm \kappa}{p} \pm (2v+2)\right)\omega_s t + ((4\kappa \pm (v+1))p \pm 1)\varphi\right)$$

$$+ ump_{36}\sin\left(\left(\frac{-2\kappa p \pm \kappa}{p} \pm (v+2)\right)\omega_s t + ((v+1)p \pm 1)\varphi\right)$$

$$+ ump_{37}\sin\left(\left(\frac{4\kappa p \pm \kappa}{p} \pm (2v+2)\right)\omega_s t + ((6\kappa \pm 1)p \pm 1)\varphi\right)$$

$$+ ump_{38}\sin\left(\left(\frac{-4\kappa p \pm \kappa}{p} \pm (2v+2)\right)\omega_s t + ((-2\kappa \pm (2v+1))p \pm 1)\varphi\right)$$

$$+ ump_{39}\sin\left(\left(\frac{2\kappa}{p} \pm 2v\right)\omega_s t + ((4\kappa \pm (2v+1))p \pm 1)\varphi\right)$$

$$+ ump_{40}\sin\left(\left(\left(4 \pm \frac{4\kappa}{p}\right) \pm 2v\right)\omega_s t + ((2v+2)p \pm 1)\varphi\right)$$

$$+ ump_{41}\sin((6\kappa \pm 2v)\omega_s t + ((6\kappa \pm 2v)p \pm 1)\varphi)$$

$$+ ump_{42}\sin((-2\kappa \pm 2v)\omega_s t + ((-2\kappa \pm 2v)p \pm 1)\varphi)$$

$$+ ump_{43}\sin\left(\left(\frac{2\kappa p \pm \kappa}{p} \pm (v-2)\right)\omega_s t + ((4\kappa \pm (2v+1))p \pm 1)\varphi\right)$$

$$+ ump_{44}\sin\left(\left(\frac{2\kappa p \pm \kappa}{p} \pm (v+2)\right)\omega_s t + ((2v+1)p \pm 1)\varphi\right)$$

$$+ ump_{45}\sin\left(\left(\frac{4\kappa p \pm \kappa}{p} \pm (2v-2)\right)\omega_s t + ((6\kappa \pm (2v+1))p \pm 1)\varphi\right)$$

$$+ ump_{46}\sin\left(\left(\frac{4\kappa p \pm \kappa}{p} \pm (v+2)\right)\omega_s t + ((2\kappa \pm (2v+1))p \pm 1)\varphi\right) \tag{6.87}$$

According to Equation (6.87), the UMP due to eccentricity faults contains numerous subharmonics. The frequency of these subharmonics can therefore be used for feature extraction. Comparison of Equation (6.44) and Equation (6.87) indicates that the slotting and saturation effects result in more harmonic components in the signals listed below:

$$f_{eccentricity1} = 2\kappa f_s \tag{6.88}$$

$$f_{eccentricity2} = (4\kappa \pm 2v)f_s \tag{6.89}$$

$$f_{eccentricity3} = \left(2 \pm \frac{2\kappa}{p}\right)f_s \tag{6.90}$$

$$f_{eccentricity4} = \left(\left(4 \pm \frac{2\kappa}{p} \right) \pm 2v \right) f_s \tag{6.91}$$

$$f_{eccentricity5} = (8\kappa \pm 2v)f_s \tag{6.92}$$

$$f_{eccentricity6} = ((2\kappa \pm 1) \pm v)f_s \tag{6.93}$$

$$f_{eccentricity7} = \left(2 \pm \frac{\kappa}{p} \right) f_s \tag{6.94}$$

$$f_{eccentricity8} = \left(\pm \frac{\kappa}{p} \right) f_s \tag{6.95}$$

$$f_{eccentricity9} = \left(\left(3 \pm \frac{\kappa}{p} \right) \pm v \right) f_s \tag{6.96}$$

$$f_{eccentricity10} = \left(\left(1 \pm \frac{\kappa}{p} \right) \pm v \right) f_s \tag{6.97}$$

$$f_{eccentricity11} = ((2\kappa \pm 1) \pm v)f_s \tag{6.98}$$

$$f_{eccentricity12} = ((4\kappa \pm 1) \pm v)f_s \tag{6.99}$$

$$f_{eccentricity13} = \left(\frac{2\kappa p \pm \kappa}{p} \pm (v \pm 1) \right) f_s \tag{6.100}$$

$$f_{eccentricity14} = \left(\frac{2\kappa p \pm \kappa}{p} \pm (2v + 2) \right) f_s \tag{6.101}$$

$$f_{eccentricity15} = (\pm 2v)f_s \tag{6.102}$$

$$f_{eccentricity16} = (6\kappa \pm 2v)f_s \tag{6.103}$$

$$f_{eccentricity17} = (-2\kappa \pm 2v)f_s \tag{6.104}$$

$$f_{eccentricity18} = \left(\left(3 \pm \frac{2\kappa}{p} \right) \pm v \right) f_s \tag{6.105}$$

$$f_{eccentricity19} = \left(\left(\frac{2\kappa}{p} - 1 \right) \pm v \right) f_s \tag{6.106}$$

$$f_{eccentricity20} = \left(\frac{(\pm 4\kappa p \pm \kappa)}{p} \pm (v + 1) \right) f_s \tag{6.107}$$

$$f_{eccentricity21} = \left(\frac{2\kappa}{p} \pm (v - 1) \right) f_s \tag{6.108}$$

$$f_{eccentricity22} = \left(\frac{2\kappa}{p} \pm (v + 3) \right) f_s \tag{6.109}$$

$$f_{eccentricity23} = \left(\frac{(\pm 2\kappa p \pm \kappa)}{p} \pm (v \pm 2) \right) f_s \tag{6.110}$$

$$f_{eccentricity24} = \left(\frac{(\pm 4\kappa p \pm \kappa)}{p} \pm (2v \pm 2) \right) f_s \tag{6.111}$$

$$f_{eccentricity25} = \left(\frac{2\kappa}{p} \pm 2v \right) f_s \tag{6.112}$$

$$f_{eccentricity26} = \left(\left(4 \pm \frac{4\kappa}{p} \right) \pm 2v \right) f_s \tag{6.113}$$

$$f_{eccentricity27} = \left(\frac{4\kappa p \pm \kappa}{p} \pm (v+2) \right) f_s \tag{6.114}$$

Among all 27 indices extracted in this section as UMP indicators, $f_{eccentricity27}$ is the most comprehensive one as it can cover the majority of indices.

6.13 Short Circuit Modeling

An inter-turn short circuit fault causes an unbalanced magnetic field to arise from the reduction of effective MMF in one or more poles in the rotor winding. In other words, the reduction in the number of turns in the rotor pole leads to a lower MMF. The effects of shorted turns, by neglecting the saturation effects, can be modeled by superimposing a fictitious coil into the healthy coils of the field winding. Figure 6.5 demonstrates the MMF of the healthy and the faulty machine. The model depicts a machine with six poles that span over 360°. It assumes that each pole produces a constant MMF over the pole span and that the distance between them is due to the distance between the rotor poles in a salient pole machine. The saliency of the rotor pole and damper slot effects are neglected.

As seen in the fictitious MMF model, the net average value of the MMF above and below the *x*-axis must be the same. In other words, the shorted turns affect the neighboring poles of the rotor. The procedure for calculating the amplitude of the fictitious MMF has been addressed in reference [22]. It is proportional to the number of shorted turns and depends on the configuration of the machine. The distribution of the faulty MMF can be written based on the MMF of the healthy coils and the fictitious coils, as follows:

$$F_{sc}^r(\phi, t) = F_h(\phi, t) + F_f(\phi_r) \tag{6.115}$$

$$F_h(\phi, t) = \oint j^r(\phi, t) \, d\phi = \oint J^r \sin(\omega_s t - p\phi) \, d\phi = \frac{J^r}{p} \cos(\omega_s t - p\phi) \tag{6.116}$$

The Fourier series of the fictitious MMF is required to determine the frequency components of the shorted turns. The Fourier series of the fictitious MMF coils are as follows:

$$F_f(\phi_r) = a_0 + \sum_{n=1}^{\infty} a_n \cos(n\phi_r) + b_n \sin(n\phi_r) \tag{6.117}$$

where the Fourier coefficients are calculated as follows:

$$a_0 = \frac{1}{2\pi} \int_0^{2\pi} F_f(\phi_r) \, d\phi_r = \frac{1}{2\pi} \left[\int_0^{\phi_1} F_f \, d\phi_r + \int_{\phi_1}^{2\pi} F_f \, d\phi_r \right] = 0 \tag{6.118}$$

$$
\begin{aligned}
a_n &= \frac{1}{\pi} \int_{-\pi}^{\pi} F_f \cos(n\phi_r) \, d\phi_r \\
&= \frac{1}{\pi} \left[\int_{-\pi}^{-\phi'} F_f \cos(n\phi_r) \, d\phi_r + \int_{-\phi'}^{\phi'} F_f \cos(n\phi_r) \, d\phi_r + \int_{\phi'}^{\pi} F_f \cos(n\phi_r) d\phi_r \right] \\
&\approx \frac{1}{\pi} \left[\int_{-\pi}^{-\phi'} F_P \cos(n\phi_r) \, d\phi_r + \int_{-\phi'}^{\phi'} F_N \cos(n\phi_r) \, d\phi_r + \int_{\phi'}^{\pi} F_P \cos(n\phi_r) \, d\phi_r \right] \\
&= \frac{1}{\pi} \left[\frac{F_P}{n} (\sin(-n\phi') - \sin(-n\pi) + \sin(n\pi) - \sin(n\phi')) + \frac{F_N}{n} (\sin(n\phi') - \sin(n\phi')) \right]
\end{aligned}
$$

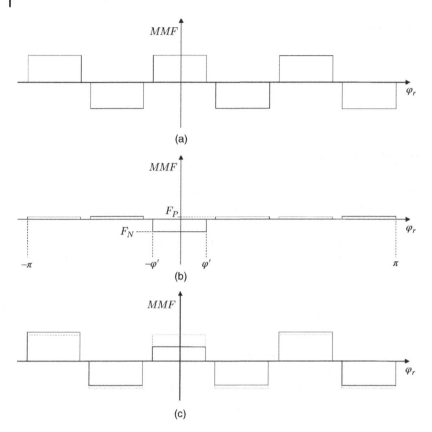

Figure 6.5 (a) The MMF distribution of a rotor pole in a healthy case, (b) fictitious coils for reduced MMF due to shorted turns, (c) net MMF for a faulty case.

$$= \frac{1}{\pi}\left[\frac{F_P}{n}(-\sin(n\phi') - \sin(n\phi')) + \frac{F_N}{n}(\sin(n\phi') + \sin(n\phi'))\right]$$

$$= \frac{1}{\pi}\left[\frac{-2F_P}{n}\sin(n\phi') + \frac{2F_N}{n}\sin(n\phi')\right] = \frac{2(F_N - F_P)}{n}\sin(n\phi') \tag{6.119}$$

$$b_n = \frac{1}{\pi}\int_{-\pi}^{\pi} F_f \sin(n\phi_r)\,d\phi_r$$

$$= \frac{1}{\pi}\left[\int_{-\pi}^{-\phi'} F_f \sin(n\phi_r)\,d\phi_r + \int_{-\phi'}^{\phi'} F_f \sin(n\phi_r)\,d\phi_r + \int_{\phi'}^{\pi} F_f \sin(n\phi_r)\,d\phi_r\right]$$

$$\approx \frac{1}{\pi}\left[\int_{-\pi}^{-\phi'} F_P \sin(n\phi_r)\,d\phi_r + \int_{-\phi'}^{\phi'} F_N \sin(n\phi_r)\,d\phi_r + \int_{\phi'}^{\pi} F_P \sin(n\phi_r)\,d\phi_r\right]$$

$$= \frac{1}{\pi}\left[\frac{F_P}{n}(-\cos(-n\phi') + \cos(n\pi) - \cos(n\pi) + \cos(n\phi') + \frac{F_N}{n}(-\cos(n\phi') + \cos(-n\phi'))\right]$$

$$- \frac{1}{\pi}\left[\frac{F_P}{n}(-\cos(n\phi') + \cos(n\pi) - \cos(n\pi) + \cos(n\phi') + \frac{F_N}{n}(-\cos(n\phi') + \cos(n\phi'))\right] = 0$$

$$\tag{6.120}$$

The Fourier series of the MMF distribution of the faulty coils is:

$$F_f(\phi_r) = \sum_{n=1}^{\infty} \frac{2(F_N - F_P)}{n\pi} \sin(n\phi') \cos(n\phi_r) \tag{6.121}$$

The first harmonic component of the fictitious coils is:

$$F_{f1}(\phi_r) = -\frac{4}{\pi} \sin(4\phi_r) \tag{6.122}$$

Comparison of the first component of the healthy MMF and the MMF of the fictitious coil reveals that the first component of the faulty pole travels four times faster than that of the healthy component. The reference frame for the fictitious coil is the rotor reference frame, which is determined by ϕ_r. However, from the stator reference frame, the fictitious MMF rotation speed is ω_r. Therefore, the net MMF in the air gap of the faulty machine based on a stator reference frame is:

$$F_{sc}^r(\phi, t) = F_h(\phi, t) + F_f(\phi, t) = \frac{J^r}{p} \cos(\omega_s t - p\phi) + \sum_{n=1}^{\infty} F_{f,n} \cos\left(\frac{n}{p}\omega_s t - n\phi\right) \tag{6.123}$$

6.14 Air Gap Permeance Under a Short Circuit Fault

Considering the saturation and slotting effects under an inter-turn short circuit fault in the rotor field winding, the air gap function is as follows [23]:

$$\Lambda_{sc} = \frac{\mu_0}{g_{sc}} = \frac{\mu_0(\alpha_1 + \alpha_2\cos(2\omega_s t - 2p\phi))}{C_c(k_m - k_e\cos(2\omega_s t - 2p\phi))} = \frac{\alpha_1'' + \alpha_2''\cos(2p\phi - 2\omega_s t)}{1 - k_e/k_m\cos(2\omega_s t - 2p\phi)} \tag{6.124}$$

where

$$\alpha_1'' = \frac{\mu_0\alpha_1}{C_c k_m} \tag{6.125}$$

$$\alpha_2'' = \frac{\mu_0\alpha_2}{C_c k_m} \tag{6.126}$$

The Taylor expansion of Equation (6.124), after some mathematical manipulations, is as follows:

$$\Lambda_{sc} = \sum_{\kappa=0}^{\infty} \Lambda_{sc}^\kappa \cos((2\kappa + 2)\omega_s t - 2p(\kappa \pm 1)\phi) \tag{6.127}$$

The stator current and its corresponding MMF are assumed to be sinusoidal. This assumption gives the current density at the inner surface of the stator as follows:

$$j_s(\phi, t) = J_s \sin(\omega_s t - p\phi) \tag{6.128}$$

where J_s is the maximum current density. The MMF of the stator is:

$$F_{sc}^s(\phi, t) = \oint j^s \phi, t)\, d\phi = \oint J^s \sin(\omega_s t - p\phi)\, d\phi = \frac{J^s}{p} \cos(\omega_s t - p\phi) \tag{6.129}$$

The same assumption is made for the rotor field winding because the current density flowing in the rotor field winding causes an MMF wave to travel with synchronous speed when seen from the stator frame. Moreover, the shape of the rotor pole shoe also supports the assumption made above. The MMF of the rotor under a short circuit fault in the rotor pole is as follows:

$$F_{sc}^r(\phi, t) = F_h(\phi, t) + F_f(\phi, t) = \frac{J^r}{p} \cos(\omega_s t - p\phi) + \sum_{n=1}^{\infty} F_{f,n} \cos\left(\frac{n}{p}\omega_s t - n\phi\right) \tag{6.130}$$

The air gap magnetic field is obtained by multiplying the air gap permeance and the MMF. The magnetic field of the air gap, described in Section 6.3, is the combination of the stator and the rotor magnetic fields. The simplified form of the stator magnetic field is calculated as a cosine series, as follows:

$$B_{sc}^s(\phi, t) = \Lambda_{sc}(\phi, t) F_{sc}^s(\phi, t) = \left[\Lambda_{sc} = \sum_{\kappa=0}^{\infty} \Lambda_{sc}^\kappa \cos((2\kappa + 2)\omega_s t - 2p(\kappa \pm 1)\phi) \right] \frac{J^s}{p} \cos(\omega_s t - p\phi)$$

$$= \sum_{\kappa=0}^{\infty} B_{sc}^{\kappa,s} \cos((2\kappa + 1)\omega_s t - ((2\kappa + 1)p \pm 2p)\phi) \tag{6.131}$$

Considering a short circuit fault in the field winding, the rotor magnetic field is as follows:

$$B_{sc}^r(\phi, t) = \Lambda_{sc}(\phi, t) F_{sc}^r(\phi, t) = \left[\sum_{\kappa=0}^{\infty}\sum_{v=0}^{\infty} \Lambda_{sc}^{\kappa,v} \cos((2\kappa + 2)\omega_s t - (2\kappa(p \pm 1))\phi) \right]$$

$$\left[\frac{J^r}{p} \cos(\omega_s t - p\phi) + \sum_{n=1}^{\infty} F_{f,n} \cos\left(\frac{n}{p}\omega_s t - n\phi\right) \right]$$

$$= \sum_{\kappa=0}^{\infty}\sum_{v=0}^{\infty} B_{sc}^{\kappa,v,r} \cos\left(\left(2\kappa \pm \frac{v}{p}\right)\omega_s t - (2p\kappa \pm v)\phi\right) \tag{6.132}$$

The net magnetic field due to the stator and the faulty rotor in the air gap is:

$$B_{sc}(\phi, t) = B_{sc}^s(\phi, t) + B_{sc}^r(\phi, t) = \sum_{\kappa=0}^{\infty}\sum_{v=0}^{\infty} B_{sc}^{\kappa,v} \cos\left(\left(2\kappa \pm \frac{v}{p}\right)\omega_s t - (\pm 2p\kappa \pm v)\phi\right) \tag{6.133}$$

6.15 Force Density and Unbalanced Magnetic Pull under a Rotor Inter-turn Short Circuit Fault

The radial force is proportional to the square of the air gap magnetic field, which describes the total radial force per square meter that is exerted on the stator teeth. Based on the Maxwell stress tensor, only the radial force is obtained, as follows:

$$f(\phi, t) = \frac{B^2(\phi, t)}{2\mu_0} \tag{6.134}$$

By substituting Equation (6.133) in Equation (6.134), the radial force in a radial direction is as follows:

$$f_{sc}(\phi, t) = \frac{B_{sc}^2(\phi, t)}{\mu_0} = \frac{1}{\mu_0}\left[\sum_{\kappa=0}^{\infty}\sum_{v=0}^{\infty} B_{sc}^{\kappa,v} \cos\left(\left(\pm 2\kappa \pm \frac{v}{p}\right)\omega_s t - (\pm 2p\kappa \pm v)\phi\right) \right]^2$$

$$= \sum_{\kappa=0}^{\infty}\sum_{v=0}^{\infty} F_{sc}^{\kappa,v} \cos\left(\left(\frac{2\kappa p \pm \kappa}{p} \pm 2v\right)\omega_s t - (p\kappa \pm v)\phi\right) \tag{6.135}$$

The total force acting on the machine's teeth due to the fault is given by the integral form of the radial force along the machine length, which is the UMP. The UMP under a rotor inter-turn short circuit fault, based on Equation (6.43), is described in the cosine series as follows:

$$\text{UMP}_{sc}(\phi, t) = L_s \int f_{sc}(\phi, t) \cos(\phi) d\phi$$

$$= L_s \int \left[\sum_{\kappa=0}^{\infty} \sum_{\nu=0}^{\infty} F_{sc}^{\kappa,\nu} \cos \left(\left(\frac{2\kappa p \pm \kappa}{p} \pm 2\nu \right) \omega_s t - (p\kappa \pm \nu)\phi \right) \right] d\phi$$

$$= \sum_{\kappa=0}^{\infty} \sum_{\nu=0}^{\infty} U_{sc}^{\kappa,\nu} \sin \left(\left(\frac{2\kappa p \pm \kappa}{p} \pm 2\nu \right) \omega_s t - (p\kappa \pm \nu)\phi \right) \qquad (6.136)$$

According to Equation (6.136), inter-turn short circuit fault give rise to subharmonics that can be characterized in the vibration signal as follows:

$$\left(\frac{2\kappa p \pm \kappa}{p} \pm 2\nu \right) f_s \qquad (6.137)$$

$$2\kappa f_s \qquad (6.138)$$

$$\left(\frac{2\kappa p \pm \kappa}{p} \pm 2\nu \right) f_s \qquad (6.139)$$

6.16 Summary

Analytical modeling is the key approach for a better understanding of the electromagnetic phenomena in electrical machines. WP is one of the simple methods that is utilized frequently in electric machine modeling. The method is based on air gap magnetic field modeling, which consists of a stator and rotor magnetic field. The stator and rotor magnetic field is also a product of the air gap permeance to the MMF. Although the method is simple, it takes into consideration the saturation and slotting effects that have significant impacts on the frequency components. The main use of the WP method in fault detection of electric machines is for frequency signature detection. It is able to detects frequency component variations in various signals based on the physical link that exists between the air gap magnetic field and different signals. In this chapter, several frequency signatures were introduced for the air gap magnetic field, voltage, and vibration, which can be used in a frequency analysis of various types of faults in the salient pole synchronous generator.

References

1 Toliyat, H.A., Nandi, S., Choi, S. et al. (2012). *Electric Machines: Modeling, Condition Monitoring, and Fault Diagnosis*. CRC Press.

2 Ebrahimi, B.M. and Faiz, J. (2010). Diagnosis and performance analysis of three-phase permanent magnet synchronous motors with static, dynamic and mixed eccentricity. *IET Electric Power Applications* 4 (1): 53–66.

3 Ebrahimi, B.M. and Faiz, J. (2012). Magnetic field and vibration monitoring in permanent magnet synchronous motors under eccentricity fault. *IET Electric Power Applications* 6, no. 1: 35–45 1751–8679.

4 Ebrahimi, B.M., Faiz, J., and Araabi, B.N. (2010). Pattern identification for eccentricity fault diagnosis in permanent magnet synchronous motors using stator current monitoring. *IET Electric Power Applications* 4, no. 6: 418–430.

5 Ebrahimi, B.M., Faiz, J., Javan-Roshtkhari, M. et al. (2008). Static eccentricity fault diagnosis in permanent magnet synchronous motor using time stepping finite element method. *IEEE Transactions on Magnetics* 14, no. 11: 4297–4300.

6 Ebrahimi, B.M., Faiz, J., and Javan-Roshtkhari, M. (2008). Static-, dynamic-, and mixed-eccentricity fault diagnoses in permanent-magnet synchronous motors. *IEEE Transactions on Industrial Electronics* 44, no. 11: 4297–4300.

7 Faiz, J., Tabatabaei, I., and Sharifi-Ghazvini, E. (2004). A precise electromagnetic modeling and performance analysis of a three-phase squirrel-cage induction motor under mixed eccentricity condition. *Electromagnetics* 24, no. 6: 471–489.

8 Heller, L. and Hamata, V.C. (1977). *Harmonic Field Effects in Induction Machines*. Elsevier. Science & Technology.

9 Babaei, M., Faiz, J., Ebrahimi, B.M. et al. (2011). A detailed analytical model of a salient-pole synchronous generator under dynamic eccentricity fault. *IEEE Transactions on Magnetics* 17, no. 4: 764–771.

10 Tu, X., Dessaint, L.A., El Kahel, M. et al. A new model of synchronous machine internal faults based on winding distribution. *IEEE Transactions on Industrial Electronics* 53, no. 6: 1818–1828.

11 Al Nuaim, N.A. and Toliyat, H.A. (1998). A novel method for modeling dynamic air-gap eccentricity in synchronous machines based on modified winding function theory. *IEEE Transactions on Energy Conversion* 13, no. 2: 156–162.

12 Strangas, E.G., Aviyente, S., and Zaidi, S.S.H. (2008). Time frequency analysis for efficient fault diagnosis and failure prognosis for interior permanent-magnet AC motors. *IEEE Transactions on Industrial Electronics* 55, no. 12: 4191–4199.

13 Bimbhra, D.P.S. (1996). *Generalized Theory of Electrical Machines*. New Delhi, India: Khanna Publishers.

14 Ehya, H., Sadeghi, I., and Faiz, J. Online condition monitoring of large synchronous generator under eccentricity fault. In: *12th IEEE Conference on Industrial Electronics and Applications (ICIEA)*, 18–20 June 2017. Siem Reap, Cambodia.

15 Sadeghi, I., Ehya, H., Faiz, J. et al. (2018). Condition monitoring of large synchronous generator under short circuit fault – A review. In: *12th IEEE International Conference on Industrial Technology (ICIT)*, 18–20 June 2017. Siem Reap, Cambodia.

16 Salon, S.J. (1995). *Finite Element Analysis of Electrical Machines*. New York, NY, USA: Springer.

17 Smith, A.C. and Dorrell, D.G. (1996). Calculation and measurement of unbalanced magnetic pull in cage induction motors with eccentric rotors. Part 1: Analytical model. *IEE Proceedings-Electric Power Applications* 143, no. 3: 193–201.

18 Faiz, J. and Ebrahimi, B.M. (2007). Influence of magnetic saturation upon performance of induction motor using time stepping finite element method. *Electric Power Components and Systems* 35, no. 5: 505–524.

19 Faiz, J., Ebrahimi, B.M., and Toliyat, H.A. (2009). Effect of magnetic saturation on static and mixed eccentricity fault diagnosis in induction motor. *IEEE Transactions on Magnetics* 45, no. 8: 3137–3144.

20 Moreira, J.C. and Lipo, T.A. (1992). Modeling of saturated AC machines including air gap flux harmonic components. *IEEE Transactions on Industry Applications* 28, no. 2: 343–349.

21 Sadeghi, I., Ehya, H., and Faiz, J. (2017). Analytic method for eccentricity fault diagnosis in salient-pole synchronous generators. In: *International Conference on Optimization of Electrical and Electronic Equipment (OPTIM) & 2017 International Aegean Conference on Electrical Machines and Power Electronics (ACEMP)*, 25–27 May 2017. Brasov, Romania.

22 Yonggang, L., Heming, L., Hua, Z. et al. (2003). Fault identification method of rotor inter turn short-circuit using Sstator winding detection. In: *Sixth International Conference on Electrical Machines and Systems (ICEMS)*, 9–11 November 2003. Beijing, China.

23 Ehya, H., Nysveen, A., and Antonino-Daviu, J.A. (2022). Advanced fault detection of synchronous generators using stray magnetic field. In: *IEEE Transactions on Industrial Electronics*, vol. 69, no. 11, 11675–11685.

7

Analytical Modeling Based on Winding Function Methods

7.1   Introduction

The finite element method (FEM) provides a deep understanding of the magnetic behavior of electrical machines. Nevertheless, its use requires solving the Maxwell and Poisson equations in a combination of difficult boundary conditions related to the complicated geometry of the electrical machines. The computational time needed to solve these complicated equations greatly exceeds the time needed for analytical approaches such as the winding function method (WFM).

Conversely, the WFM analyzes the electrical machine based on coupled electromagnetic circuits rather than the electromagnetic field, which makes it simple. In other words, the WFM pays attention to the terminals of the machines rather than to their internal characteristics. The WFM does not give any vision regarding the spatial distribution of the current or the flux in the machine, so it makes modeling of the machine too simple. To model the magnetic field accurately, the linkage flux in the rotating coupled circuit should be expressed in detail. Since the flux linkage in the machines is proportional to the inductances, close attention is needed to model the winding distribution in the WFM.

Electromagnetic Analysis and Condition Monitoring of Synchronous Generators, First Edition. Hossein Ehya and Jawad Faiz.
© 2023 The Institute of Electrical and Electronics Engineers, Inc. Published 2023 by John Wiley & Sons, Inc.

7.2 History and Usage of the WFM

The winding function approach, or modified winding function method (MWFM), is one of the most accurate and appropriate methods for modeling salient pole synchronous generators (SGs) [1]. The WFM was invented in 1960 and was used to calculate the inductances of induction motors in 1969 and the linear induction motor in 1978. Further development of this method and its combination with mutual circuit theory led to its use in the analysis of induction motors. Most of its use has focused on the analysis of squirrel cage induction motors from 1991 [2].

The computational time for the modeling of electrical machines by WFM is quite short compared to the FEM. For example, calculation of the self-inductance and mutual inductance of a synchronous machine takes one minute using the WFM and eight hours using the FEM.

Many studies conducted over the years on different electrical machines have led to the development of this approach in a way that can produce results with high accuracy. This MWFM is able to calculate the inductances of electrical machines even in a faulty condition [3, 4]. The WFM does not require any hypothesis of symmetric distribution of the armature winding, unlike Park's theory [5]. It can take into account the non-sinusoidal distribution of the stator winding under faulty conditions or spatial harmonics due to the winding distribution in the air gap of the electrical machines. In other words, spatial harmonics, which are created by the winding distribution, are considered by taking into account the net magneto-motive force (MMF). MWFM is extensively used to model induction motors under broken bar faults, broken end rings, stator short circuit faults, and eccentricity. The accuracy of this method is high enough to distinguish defects like stator inter-turn fault and partially broken damper bars of the rotor [5, 6].

The number of researches related to the modeling of synchronous machines using the WFM is not comparable to the number examining induction motors. In the majority of synchronous machines modeled using the WFM, the damper winding in the rotor pole is ignored, which leads to instability of the model [7].

7.3 Winding Function Modeling of a Synchronous Generator

Some hypotheses make WFM modeling straightforward and have no considerable effects on calculated inductances. These assumptions are as follows:

1. The permeability of the iron is assumed to be infinite and the magnetic saturation of the iron is neglected.
2. The air gap length in the case of a uniform air gap is constant.
3. The slotting effect of the stator lamination sheet is neglected.
4. Stator winding, which is distributed in different phase belts, is in a series connection.

Figure 7.1 shows an arbitrary path that includes the stator, rotor, and air gap of the salient pole SG. This arbitrary path (1–2–3–4–1) is located between 0 and 2π ($0 \leq \Phi \leq 2\pi$). By Ampere's law:

$$\oint_{12341} H \, dl = \int_S J \, ds \tag{7.1}$$

where S is the enclosed arbitrary path. Since the current in the enclosed path is assumed to be equal to i, Equation (7.1) can be rewritten as follows:

$$\oint_{12341} H \, dl = n(\varphi, \theta) i \tag{7.2}$$

Figure 7.1 An arbitrary path, including a stator, air gap, and rotor of a salient pole synchronous machine.

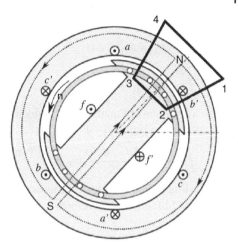

where $n(\varphi, \theta)$ is the turn function, φ is the stator reference angle, and θ is the rotor rotational mechanical angle. For simplicity, the stator reference angle is assumed to be fixed in the x-axis of the Cartesian coordinate system. The extracted MMF of the path 1–2–3–4–1 is as follows:

$$F_{12} + F_{23} + F_{34} + F_{41} = n(\varphi, \theta) \tag{7.3}$$

where the F_{23} and F_{41} are negligible if the permeability of the iron is assumed to be infinite; consequently, F_{12} and F_{34} are:

$$F_{12} = \int_{R_S}^{R_r} H(0, \theta) \, dr \tag{7.4}$$

$$F_{34} = \int_{R_S}^{R_r} H(0, \theta) \, dr \tag{7.5}$$

so Equation (7.3) can be simplified as:

$$F_{12}(0, \theta) + F_{34}(\varphi, \theta) = n(\varphi, \theta) \tag{7.6}$$

The value of H in a radial direction is assumed to be a constant, since the effective air gap length $g(\varphi,\theta)$ of the machine is negligible compared to the rotor diameter. Therefore, Equations (7.4) to (7.6) can be simplified as:

$$F_{12}(0, \theta) = -H(0, \theta)g(0, \theta) \tag{7.7}$$

$$F_{34}(\varphi, \theta) = H(\varphi, \theta)g(\varphi, \theta) \tag{7.8}$$

Equations (7.7) and (7.8) can be solved based on Gauss' law. Equations (7.9) and (7.10) are valid if a closed cylindrical surface is assumed to cover the inner surface of the stator and is located inside the air gap:

$$\oint_S B \, ds = 0 \tag{7.9}$$

$$\mu_0 \int_0^{2\pi} \int_0^L \underset{Rr \leq r \leq Rs}{rH(\varphi, \theta)} \, d\varphi \, dz = 0 \tag{7.10}$$

where L is the effective length of the rotor. The value H is independent of any changes in the coordinate axis. Therefore:

$$\int_0^{2\pi} H(\varphi, \theta) \, d\varphi = 0 \tag{7.11}$$

According to Equation (7.8) and (7.11):

$$\int_0^{2\pi} H_{34}(\varphi, \theta)\, d\varphi = 0 \tag{7.12}$$

$$\int_0^{2\pi} \frac{F_{34}(\varphi, \theta)}{g(\varphi, \theta)}\, d\varphi = 0 \tag{7.13}$$

The integration of Equation (7.6), which is divided into the effective air gap length function $g(\varphi, \theta)$ in integration limits ($0 \leq \varphi \leq 2\pi$) is:

$$\int_0^{2\pi} \frac{F_{12}(0, \theta)F_{34}(\varphi, \theta)}{g(\varphi, \theta)}\, d\varphi = \int_0^{2\pi} \frac{n(\varphi, \theta)}{g(\varphi, \theta)}\, i\, d\varphi \tag{7.14}$$

According to Equation (7.13), the second term of the integral in the left side of the equality in Equation (7.14) is zero. Therefore:

$$F_{12}(0, \theta) \int_0^{2\pi} \frac{d\varphi}{g(\varphi, \theta)} = \int_0^{2\pi} \frac{n(\varphi, \theta)}{g(\varphi, \theta)}\, i\, d\varphi \tag{7.15}$$

$$F_{12}(0, \theta) = i\, \frac{1}{2\pi \langle g^{-1} \rangle} \int_0^{2\pi} n(\varphi, \theta) g^{-1}(\varphi, \theta)\, d\varphi \tag{7.16}$$

$$F_{12}(0, \theta) = i\, \frac{\langle ng^{-1} \rangle}{\langle g^{-1} \rangle} \tag{7.17}$$

where $\langle F \rangle$ is the average value of the function f, defined as:

$$\langle f \rangle = \frac{1}{2\pi} \int_0^{2\pi} f(\varphi, \theta)\, d\varphi \tag{7.18}$$

Consequently, the average of the air gap inverse function is:

$$\langle g^{-1} \rangle = \frac{1}{2\pi} \int_0^{2\pi} g^{-1}(\varphi, \theta)\, d\varphi \tag{7.19}$$

In addition:

$$\langle ng^{-1} \rangle = \frac{1}{2\pi} \int_0^{2\pi} n(\varphi, \theta) g^{-1}(\varphi, \theta)\, d\varphi \tag{7.20}$$

Substituting Equation (7.13) in Equation (7.6) gives:

$$\frac{\langle ng^{-1} \rangle}{\langle g^{-1} \rangle} i + F_{34}(\varphi, \theta) = n(\varphi, \theta)i \tag{7.21}$$

$$F_{34}(\varphi, \theta) = n(\varphi, \theta)i - \frac{\langle ng^{-1} \rangle}{\langle g^{-1} \rangle} i \tag{7.22}$$

Finally, the modified winding function based on Equation (7.14) can be expressed as:

$$M(\varphi, \theta) = n(\varphi, \theta) - \frac{\langle ng^{-1} \rangle}{\langle g^{-1} \rangle} \tag{7.23}$$

$$M(\varphi, \theta) = n(\varphi, \theta) - \langle M \rangle \tag{7.24}$$

$$\langle M \rangle = \frac{\langle ng^{-1} \rangle}{\langle g^{-1} \rangle} \tag{7.25}$$

The self and mutual inductances of the machine can be calculated according to Equations (7.23), (7.24), and (7.26). The distributed MMF created by the current passing through the winding phase A is:

$$F_A(\varphi, \theta) = M_A(\varphi, \theta)i_A \tag{7.26}$$

Differential distribution of the flux that passes through the volume with a cross-section of $g(\varphi,\theta)$, and $r\,dl\,d\varphi$ can be presented as:

$$d\lambda = F_A(\varphi,\theta)\mu_0 rLg^{-1}(\varphi,\theta)\,d\varphi \tag{7.27}$$

The mutual flux of a coil in the winding phase B is:

$$\varphi_{1-2} = \mu_0 rL \int_0^{2\pi} n_{B1}(\varphi,\theta)F_A(\varphi,\theta)g^{-1}(\varphi,\theta)i_A\,d\varphi \tag{7.28}$$

where μ_0 is an air relative magnetic permeability and r is the rotor radius. The mutual linkage flux of phase B due to the current passing through the coil A is as follows:

$$\lambda_{BA} = \mu_0 rL \int_0^{2\pi} n_B(\varphi,\theta)F_A(\varphi,\theta)g^{-1}(\varphi,\theta)i_A\,d\varphi \tag{7.29}$$

The mutual inductance of the phase B (L_{BA}) due to the current of the phase A is:

$$L_{BA} = \frac{\lambda_{BA}}{i_A} = \mu_0 rL \int_0^{2\pi} n_B(\varphi,\theta)M_A(\varphi,\theta)g^{-1}(\varphi,\theta)\,d\varphi \tag{7.30}$$

The self inductance of the phase A is:

$$L_{AA} = \frac{\lambda_{AA}}{i_A} = \mu_0 rL \int_0^{2\pi} n_A(\varphi,\theta)M_A(\varphi,\theta)g^{-1}(\varphi,\theta)\,d\varphi \tag{7.31}$$

According to Equation (7.31), an accurate calculation of the turn function and air gap function is required to determine the inductance of the winding.

7.4 Mutual Inductance Calculation Between the Stator Phases

The self and mutual inductance of a phase winding can be calculated using Equations (7.31) and (7.30), respectively. The step-by-step procedure for determining the self and mutual inductances of a three-phase SG was discussed in depth in the previous section. The mutual inductance between phases A and B were driven according to Equation (7.30).

The easiest way to calculate the integral term of Equation (7.30) is the term-wise integration according to the integral contour, in a way that the functions in the integration limits become a constant value. Therefore, the principle is to calculate the inductances at any time step, which makes the calculation faster, since the analytical calculation turns into the numerical calculation. In Equation (7.30), three functions should be considered:

7.4.1 Turn Function of Winding Phase B

The magnitude of the turn function $n_B(\varphi,\theta)$ in the stator slot is assumed to be constant and its value in the middle of the slot is assumed to vary as a step function. For instance, the variation in the stator turn function in a stator with 36 slots occurs each $(2\pi/36)$ in radians. No changes occur in the value of the turn or its function between these points, and its magnitude is constant.

7.4.2 The Modified Winding Function of Phase A

The value of the $\langle M_A(\varphi,\theta)\rangle$ in Equation (7.24) is zero since the average value of the turn function in a healthy SG is zero. Therefore, the MWF is equal to the turn function and Equation (7.24) can be modified as:

$$M_A(\varphi,\theta) = n_A(\varphi,\theta) \tag{7.32}$$

7.4.3 The Inverse Air Gap Function

The turn and MWF is a function of the stator angle, since the inverse air gap function has a direct relationship with the rotor rotating mechanical angle. It has two constant values.

For example, the stator mutual inductance of winding phases A and B in a stator with 36 slots, where the rotor position in the direction of the x-axis is assumed to be fixed ($\theta = 0$), is:

$$
\begin{aligned}
L_{BA} &= \mu_0 rL \int_{0+\theta}^{360+\theta} n_B(\varphi, \theta) n_A(\varphi, \theta) g^{-1}(\varphi, \theta) d\varphi \\
&= \mu_0 rL \int_0^{360} n_B(\varphi) n_A(\varphi) g^{-1}(\varphi) \, d\varphi
\end{aligned}
\tag{7.33}
$$

$$\text{if } \theta = 0$$

Figure 7.2 shows the stator lamination sheet, with the location of the rotor assumed to be at $\theta = 0$. The constant value of the inverse air gap function for contours between $[0\text{–}60]$, $[90\text{–}150]$, $[180\text{–}240]$, and $[270\text{–}330]$ is equal to 500 ($k_1 = 500$), and the second constant for contour $[60\text{–}90]$, $[150\text{–}180]$, $[240\text{–}270]$, and $[330\text{–}360]$ is 16.95 ($k_2 = 16.95$). Consequently, the integral term in Equation (7.33) can be divided into the following eight parts:

$$
\begin{aligned}
L_{BA} &= \mu_0 rL \int_0^{360} n_B(\varphi) n_A(\varphi) g^{-1}(\varphi) \, d\varphi \\
&= \mu_0 rL k_1 \left\{ \begin{aligned} &\int_0^{60} n_B(\varphi) n_A(\varphi) \, d\varphi + \int_{90}^{150} n_B(\varphi) n_A(\varphi) \, d\varphi \\ &+ \int_{180}^{240} n_B(\varphi) n_A(\varphi) \, d\varphi + \int_{270}^{330} n_B(\varphi) n_A(\varphi) \, d\varphi \end{aligned} \right\} \\
&+ \mu_0 rL k_2 \left\{ \begin{aligned} &\int_{60}^{90} n_B(\varphi) n_A(\varphi) \, d\varphi + \int_{150}^{180} n_B(\varphi) n_A(\varphi) \, d\varphi \\ &+ \int_{240}^{270} n_B(\varphi) n_A(\varphi) \, d\varphi + \int_{330}^{360} n_B(\varphi) n_A(\varphi) \, d\varphi \end{aligned} \right\}
\end{aligned}
\tag{7.34}
$$

The constant value of the turn function for phases A and B, corresponding to each slot and tooth in an integral limit, should be calculated. In this case, the constant values $n_A(\varphi)$ and $n_B(\varphi)$ can be multiplied in the integral term. The slot pitch and tooth pitch are called X and Y, respectively. The summation of slot and tooth pitch is equal to Z. In addition, the turn function $n_B(1)$ is the turn function of the phase B winding in the first slot and tooth. In the same way, $n_A(2)$ is a turn function of the phase A winding in the second slot and tooth. The simplified form of the first term of Equation (7.34) is:

$$
\begin{aligned}
\int_0^{60} n_B(\varphi) n_A(\varphi) \, d\varphi &= \left\{ \begin{aligned} &n_B(1) n_A(1) \int_0^{z_1} d\varphi + n_B(2) n_A(2) \int_{z_2}^{z_1} d\varphi \\ &+ n_B(3) n_A(3) \int_{z_2}^{z_3} d\varphi + n_B(4) n_A(4) \int_{z_3}^{z_4} d\varphi \\ &+ n_B(5) n_A(5) \int_{z_4}^{z_5} d\varphi + n_B(6) n_A(6) \int_{z_5}^{z_6} d\varphi \end{aligned} \right\} \\
&= \sum_{i=1}^{6} n_B(i) n_A(i) Z
\end{aligned}
\tag{7.35}
$$

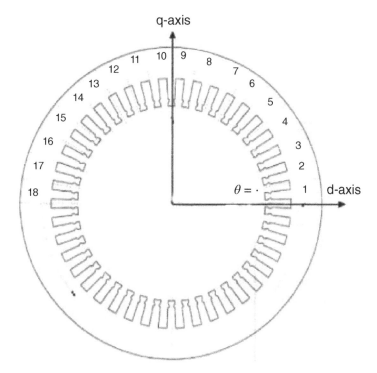

Figure 7.2 Stator lamination sheet of the synchronous machine and the assumed location of the rotor at $\theta = 0$.

The same procedure can be applied to the other equations, which leads to:

$$\int_{90}^{150} n_B(\varphi)n_A(\varphi)\, d\varphi = \sum_{i=10}^{15} n_B(i)n_A(i)Z \tag{7.36}$$

$$\int_{180}^{240} n_B(\varphi)n_A(\varphi)\, d\varphi = \sum_{i=19}^{24} n_B(i)n_A(i)Z \tag{7.37}$$

$$\int_{270}^{330} n_B(\varphi)n_A(\varphi)\, d\varphi = \sum_{i=28}^{33} n_B(i)n_A(i)Z \tag{7.38}$$

The second term of Equation (7.34), after the same assumption and simplification as its first term, can be derived as:

$$\int_{60}^{90} n_B(\varphi)n_A(\varphi)\, d\varphi = \sum_{i=7}^{9} n_B(i)n_A(i)Z \tag{7.39}$$

$$\int_{150}^{180} n_B(\varphi)n_A(\varphi)\, d\varphi = \sum_{i=16}^{18} n_B(i)n_A(i)Z \tag{7.40}$$

$$\int_{240}^{270} n_B(\varphi)n_A(\varphi)\, d\varphi = \sum_{i=25}^{27} n_B(i)n_A(i)Z \tag{7.41}$$

$$\int_{330}^{360} n_B(\varphi)n_A(\varphi)\, d\varphi = \sum_{i=34}^{36} n_B(i)n_A(i)Z \tag{7.42}$$

Finally, the mutual inductance of the phase A and B at $\theta = 0$ can be established as:

$$
\begin{aligned}
L_{BA} &= \mu_0 rL \int_0^{360} n_B(\varphi)n_A(\varphi)g^{-1}(\varphi)\,d\varphi \\
&= \mu_0 rL k_1 Z \left\{ \begin{array}{l} \displaystyle\sum_{i=1}^{6} n_B(i)n_A(i) + \sum_{i=10}^{15} n_B(i)n_A(i) \\[2mm] \displaystyle + \sum_{i=19}^{24} n_B(i)n_A(i) + \sum_{i=28}^{33} n_B(i)n_A(i) \end{array} \right\} \\
&\quad + \mu_0 rL k_2 Z \left\{ \begin{array}{l} \displaystyle\sum_{i=7}^{9} n_B(i)n_A(i) + \sum_{i=16}^{18} n_B(i)n_A(i) \\[2mm] \displaystyle + \sum_{i=25}^{27} n_B(i)n_A(i) + \sum_{i=34}^{36} n_B(i)n_A(i) \end{array} \right\}
\end{aligned}
\tag{7.43}
$$

If the rotor rotates for θ degree, as shown in Figure 7.3, then Equations (7.34) to (7.43) are no longer valid. In this condition, two terms must be added to the previous equation due to rotor movement. These terms are due to the air gap variation in front of one slot and tooth pitch. Consequently, the limits of the integral should be changed to two new limits:

$$
\begin{aligned}
L_{BA} &= \mu_0 rL \int_0^{360} n_B(\varphi)n_A(\varphi)g^{-1}(\varphi)\,d\varphi \\
&= \mu_0 rL k_1 \left\{ \begin{array}{l} \displaystyle\int_{0+\theta}^{60+\theta} n_B(\varphi)n_A(\varphi)\,d\varphi + \int_{90+\theta}^{150+\theta} n_B(\varphi)n_A(\varphi)\,d\varphi \\[2mm] \displaystyle + \int_{180+\theta}^{240+\theta} n_B(\varphi)n_A(\varphi)\,d\varphi + \int_{270+\theta}^{330+\theta} n_B(\varphi)n_A(\varphi)\,d\varphi \end{array} \right\} \\
&\quad + \mu_0 rL k_2 \left\{ \begin{array}{l} \displaystyle\int_{60+\theta}^{90+\theta} n_B(\varphi)n_A(\varphi)\,d\varphi + \int_{150+\theta}^{180+\theta} n_B(\varphi)n_A(\varphi)\,d\varphi \\[2mm] \displaystyle + \int_{240+\theta}^{270+\theta} n_B(\varphi)n_A(\varphi)\,d\varphi + \int_{330+\theta}^{360+\theta} n_B(\varphi)n_A(\varphi)\,d\varphi \end{array} \right\}
\end{aligned}
\tag{7.44}
$$

The same procedure as in Equation (7.36) can be applied to the turn function in Equation (7.44). Therefore, Equation (7.44) can be divided, based on new limits, as:

$$
\int_{0+\theta}^{60+\theta} n_B(\varphi)n_A(\varphi)\,d\varphi = \left\{ \begin{array}{l} n_B(1)n_A(1)\displaystyle\int_{0+\theta}^{z_1} d\varphi + n_B(2)n_A(2)\int_{z_1}^{z_2} d\varphi \\[2mm] + n_B(3)n_A(3)\displaystyle\int_{z_2}^{z_3} d\varphi + n_B(4)n_A(4)\int_{z_3}^{z_4} d\varphi \\[2mm] + n_B(5)n_A(5)\displaystyle\int_{z_4}^{z_5} d\varphi + n_B(6)n_A(6)\int_{z_5}^{z_6} d\varphi \\[2mm] + n_B(7)n_A(7)\displaystyle\int_{z_6}^{z_6+\theta} d\varphi \end{array} \right.
\tag{7.45}
$$

$$
\int_{0+\theta}^{60+\theta} n_B(\varphi)n_A(\varphi)\,d\varphi = n_B(1)n_A(1)(Z-\theta) + \sum_{i=2}^{6} n_B(i)n_A(i)Z + n_B(7)n_A(7)(\theta)
\tag{7.46}
$$

where $z_1, z_2, \ldots, z_6$ is the angle between the slots. In Equation (7.46), two terms contain θ, which is due to rotor movement. The same procedure can be used for the rest of the terms, as in Equation

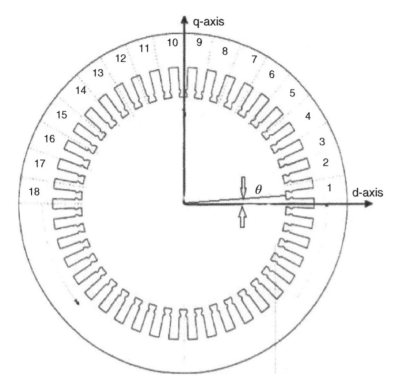

Figure 7.3 The rotor is rotated θ degrees with respect to the d-axis.

(7.44). The simplified equations are:

$$\int_{90+\theta}^{150+\theta} n_B(\varphi)n_A(\varphi)\,d\varphi = n_B(10)n_A(10)(Z-\theta) + \sum_{i=11}^{15} n_B(i)n_A(i)Z + n_B(16)n_A(16)(\theta) \qquad (7.47)$$

$$\int_{180+\theta}^{240+\theta} n_B(\varphi)n_A(\varphi)\,d\varphi = n_B(19)n_A(19)(Z-\theta) + \sum_{i=20}^{24} n_B(i)n_A(i)Z + n_B(25)n_A(25)(\theta) \qquad (7.48)$$

$$\int_{270+\theta}^{330+\theta} n_B(\varphi)n_A(\varphi)\,d\varphi = n_B(28)n_A(28)(Z-\theta) + \sum_{i=29}^{33} n_B(i)n_A(i)Z + n_B(34)n_A(34)(\theta) \qquad (7.49)$$

The same assumption is used to derive the equations in Equation (7.44), as below:

$$\int_{60+\theta}^{90+\theta} n_B(\varphi)n_A(\varphi)\,d\varphi = n_B(7)n_A(7)(Z-\theta) + \sum_{i=8}^{9} n_B(i)n_A(i)Z + n_B(10)n_A(10)(\theta) \qquad (7.50)$$

$$\int_{150+\theta}^{180+\theta} n_B(\varphi)n_A(\varphi)\,d\varphi = n_B(16)n_A(16)(Z-\theta) + \sum_{i=17}^{18} n_B(i)n_A(i)Z + n_B(19)n_A(19)(\theta) \qquad (7.51)$$

$$\int_{240+\theta}^{270+\theta} n_B(\varphi)n_A(\varphi)\,d\varphi = n_B(25)n_A(25)(Z-\theta) + \sum_{i=26}^{27} n_B(i)n_A(i)Z + n_B(28)n_A(28)(\theta) \qquad (7.52)$$

$$\int_{330+\theta}^{360+\theta} n_B(\varphi)n_A(\varphi)\,d\varphi = n_B(34)n_A(34)(Z-\theta) + \sum_{i=35}^{36} n_B(i)n_A(i)Z + n_B(1)n_A(1)(\theta) \qquad (7.53)$$

Comparison of Equations (7.43) and (7.44) proves that they have a systematic pattern. According to this principle, a unified equation for the mutual inductance of the stator can be defined. To that end, a new variable θ is defined as below:

$$a = \left[\frac{\theta}{Z}\right] \tag{7.54}$$

$$sum_1 = \sum_{i=j}^{j+4} n_B(i)n_A(i)Z + \sum_l n_B(l)n_A(l)((a+1)Z - \theta) + \sum_m n_B(m)n_A(m)(\theta - aZ)$$

$$j = a + 2, a + 11, a + 20, a + 29 \tag{7.55}$$

$$l = a + 1, a + 10, a + 19, a + 28$$

$$m = a + 7, a + 16, a + 25, a + 34$$

$$sum_2 = \sum_{i=jj}^{jj+1} n_B(i)n_A(i)Z + \sum_{ll} n_B(ll)n_A(ll)((a+1)Z - \theta) + \sum_{mm} n_B(mm)n_A(mm)(\theta - aZ)$$

$$jj = a + 8, a + 17, a + 26, a + 35 \tag{7.56}$$

$$ll = a + 10, a + 19, a + 28, a + 1$$

$$m = a + 7, a + 16, a + 25, a + 34$$

where [] is a bracket function. The mutual inductance of a phase A and B is:

$$L_{AB} = \mu_0 rL(sum_1 k_1 + sum_2 k_2) \tag{7.57}$$

Equations (7.31) to (7.57) are valid for the calculation of the mutual inductance of the other stator winding phases, since the required turn function of the considered phase must be substituted. For example, the turn function of phase C must be substituted in Equations (7.31) to (7.57) instead of phase B for calculation of the mutual inductance between phase winding A and C, L_{AC}. These equations can be used to calculate the stator self inductance, since both turn functions must be the same for that phase winding.

7.5 The Mutual Inductance Between the Stator and Rotor

7.5.1 The Mutual Inductance Between the Stator Phase Winding and Rotor Field Winding

To minimize the computation complexity, the integral limits must be determined in a way that converts the turn function to a constant value. In the first step, the equation must be derived for the rotor in the stationary position ($\theta = 0$). Figure 7.4 shows the turn function of the rotor field winding.

$$
\begin{aligned}
L_{FA} &= \mu_0 rL \int_{0+\theta}^{360+\theta} n_F(\varphi, \theta)n_A(\varphi, \theta)g^{-1}(\varphi, \theta)\,d\varphi \\
&= \mu_0 rLk_1 \int_{0+\theta}^{360+\theta} n_F(\varphi)n_A(\varphi)g^{-1}(\varphi)\,d\varphi \\
&= \mu_0 rLk_1 \left\{ \begin{array}{l} \int_{0+\theta}^{60+\theta} n_F(\varphi)n_A(\varphi)\,d\varphi - \int_{90+\theta}^{150+\theta} n_F(\varphi)n_A(\varphi)\,d\varphi + \\[2ex] + \int_{180+\theta}^{240+\theta} n_F(\varphi)n_A(\varphi)\,d\varphi + \int_{270+\theta}^{330+\theta} n_F(\varphi)n_A(\varphi)\,d\varphi \end{array} \right\}
\end{aligned} \tag{7.58}
$$

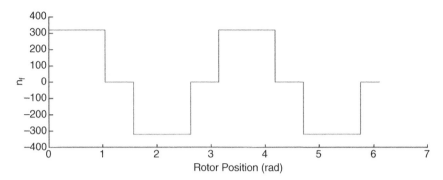

Figure 7.4 The turn function of the field winding.

$$+ \mu_0 rLk_2 \left\{ \begin{array}{l} \displaystyle\int_{60+\theta}^{90+\theta} n_F(\varphi)n_A(\varphi)d\varphi - \int_{150+\theta}^{180+\theta} n_F(\varphi)n_A(\varphi)\,d\varphi \\[3mm] + \displaystyle\int_{240+\theta}^{270+\theta} n_F(\varphi)n_A(\varphi)d\varphi + \int_{330+\theta}^{360+\theta} n_F(\varphi)n_A(\varphi)\,d\varphi \end{array} \right\}$$

The value of the rotor turn function in a period between [60–90], [150–180], [240–270], and [330–360] degrees is zero; therefore, the second term of Equation (7.58) is zero. In addition, in the first term of Equation (7.58), the value of the turn function of rotor field winding is equal to n_F. The positive and negative signs of the field winding turn function can be extracted according to Figure 7.4. The simplified form of Equation (7.58) is therefore:

$$L_{FA} = n_F\mu_0 rLk_1 \left\{ \begin{array}{l} \displaystyle\int_{0}^{60} n_A(\varphi)\,d\varphi - \int_{90}^{150} n_A(\varphi)\,d\varphi \\[3mm] + \displaystyle\int_{180}^{240} n_A(\varphi)\,d\varphi + \int_{270}^{330} n_A(\varphi)\,d\varphi \end{array} \right\} \tag{7.59}$$

The value of the turn function for different slots and tooth pitches must be a constant value to change the format of the analytical equation into a numerical one. Therefore, Equation (7.59) becomes:

$$\int_{0}^{60} n_A(\varphi)\,d\varphi = n_F\mu_0 rLk_1 \sum_{i=j}^{j+5} n_A(i)Z \tag{7.60}$$

Equation (7.60) is similar to Equation (7.35). Therefore, the mutual inductance of the rotor field winding and the stator phase winding can be derived as in Equation (7.45):

$$\begin{aligned} sum = & \sum_{i=a+2}^{a+6} n_A(i)Z + n_A(a+1)((a+1)Z-\theta) + n_A(a+7)(\theta-aZ) \\[2mm] & + \sum_{i=a+11}^{a+15} n_A(i)Z + n_A(a+10)((a+1)Z-\theta) + n_A(a+16)(\theta-aZ) \\[2mm] & + \sum_{i=a+20}^{a+24} n_A(i)Z + n_A(a+19)((a+1)Z-\theta) + n_A(a+25)(\theta-aZ) \\[2mm] & + \sum_{i=a+29}^{a+33} n_A(i)Z + n_A(a+28)((a+1)Z-\theta) + n_A(a+34)(\theta-aZ) \end{aligned} \tag{7.61}$$

The final format of the mutual inductance of the rotor field winding and stator phase winding is:

$$L_{FA} = n_F \mu_0 rLk_1 \, sum \tag{7.62}$$

7.5.2 The Mutual Inductance of the Stator Phase Winding and Rotor Damper Winding

Unlike the four primary windings (stator three-phase winding and rotor field winding) that work during the transient and steady-state operation of the SG, the damper windings that are located in the rotor pole shoes are active during transient operation. Damper windings consist of damper bars that are distributed in rotor pole shoes and are short-circuited in both ways with end rings. For example, a short circuit fault in the network or stator winding leads to an inrush of current into the stator winding and could damage the field windings. The induced voltage in the damper circuit creates a large amount of current that, in turn, produces a strong magnetic field precisely opposite to the stator magnetic field that could limit the stator current. Therefore, modeling the damper winding is essential, especially for fault detection purposes. For the sake of simplification, two damper circuits that are perpendicular to each other are assumed for this modeling and are located in the direct and quadrature axis [8]. The damper windings are defined as:

$$N_{kd}(\varphi, \theta) = -\omega_{kd} \sin\left(p\left(\varphi - \theta + \frac{\pi}{6}\right)\right) \tag{7.63}$$

$$N_{kq}(\varphi, \theta) = -\omega_{kq} \cos\left(p\left(\varphi - \theta + \frac{\pi}{6}\right)\right) \tag{7.64}$$

$$(\varphi - \theta) \in [0, 2\pi]$$

where ω_{kd} and ω_{kq} are the number of turns for damper windings located in the d-axis and q-axis and p is the number of pole pairs. The mutual inductance between the stator phase A winding and damper winding located in the d-axis (kd) when $\theta = 0$ is as follows:

$$
\begin{aligned}
L_{kdA} &= \mu_0 rL \int_0^{360} n_{kd}(\varphi) n_A(\varphi) g^{-1} \, d\varphi \\
&= \mu_0 rLk_1 \left\{ \begin{array}{l} \int_0^{60} -\omega_{kd} n_A(\varphi) \sin p\left(\varphi + \frac{\pi}{6}\right) d\varphi + \int_{90}^{150} -\omega_{kd} n_A(\varphi) \sin p\left(\varphi + \frac{\pi}{6}\right) d\varphi + \\ + \int_{180}^{240} -\omega_{kd} n_A(\varphi) \sin p\left(\varphi + \frac{\pi}{6}\right) d\varphi + \int_{270}^{330} -\omega_{kd} n_A(\varphi) \sin p\left(\varphi + \frac{\pi}{6}\right) d\varphi \end{array} \right\} \\
&\quad + \mu_0 rLk_2 \left\{ \begin{array}{l} \int_{60}^{90} -\omega_{kd} n_A(\varphi) \sin p\left(\varphi + \frac{\pi}{6}\right) d\varphi + \int_{150}^{180} -\omega_{kd} n_A(\varphi) \sin p\left(\varphi + \frac{\pi}{6}\right) d\varphi \\ + \int_{240}^{270} -\omega_{kd} n_A(\varphi) \sin p\left(\varphi + \frac{\pi}{6}\right) d\varphi + \int_{330}^{360} -\omega_{kd} n_A(\varphi) \sin p\left(\varphi + \frac{\pi}{6}\right) d\varphi \end{array} \right\}
\end{aligned}
\tag{7.65}
$$

The turn function of the damper winding in the d-axis is close to zero for a period of [60–90], [150–180], [240–270], and [330–360] degrees. Therefore, the magnitude of k_2 is so small that we can neglect the second term of Equation (7.52). The expansion of the first term of Equation (7.65) is:

$$
\begin{aligned}
\int_0^{60} -\omega_{kd} n_A(\varphi) \sin p\left(\varphi + \frac{\pi}{6}\right) d\varphi &= -\omega_{kd} \int_0^{60} n_A(\varphi) \sin p\left(\varphi + \frac{\pi}{6}\right) d\varphi \\
&= -\omega_{kd} n_A(1) \int_0^{z_1} \sin p\left(\varphi + \frac{\pi}{6}\right) d\varphi - \omega_{kd} n_A(2) \int_{z_1}^{z_2} \sin p\left(\varphi + \frac{\pi}{6}\right) d\varphi \\
&\quad - \omega_{kd} n_A(3) \int_{z_2}^{z_3} \sin p\left(\varphi + \frac{\pi}{6}\right) d\varphi - \omega_{kd} n_A(4) \int_{z_3}^{z_4} \sin p\left(\varphi + \frac{\pi}{6}\right) d\varphi
\end{aligned}
$$

$$
- \omega_{kd} n_A(5) \int_{z_4}^{z_5} \sin p \left(\varphi + \frac{\pi}{6} \right) d\varphi - \omega_{kd} n_A(6) \int_{z_5}^{z_6} \sin p \left(\varphi + \frac{\pi}{6} \right) d\varphi
$$

$$
= \frac{\omega_{kd}}{p} n_A(1) \left[\cos p \left(z_1 + \frac{\pi}{6} \right) - \cos p \left(0 + \frac{\pi}{6} \right) \right]
$$

$$
+ \frac{\omega_{kd}}{p} n_A(2) \left[\cos p \left(z_2 + \frac{\pi}{6} \right) - \cos p \left(z_1 + \frac{\pi}{6} \right) \right]
$$

$$
+ \frac{\omega_{kd}}{p} n_A(3) \left[\cos p \left(z_3 + \frac{\pi}{6} \right) - \cos p \left(z_2 + \frac{\pi}{6} \right) \right]
$$

$$
+ \frac{\omega_{kd}}{p} n_A(4) \left[\cos p \left(z_4 + \frac{\pi}{6} \right) - \cos p \left(z_3 + \frac{\pi}{6} \right) \right]
$$

$$
+ \frac{\omega_{kd}}{p} n_A(5) \left[\cos p \left(z_5 + \frac{\pi}{6} \right) - \cos p \left(z_4 + \frac{\pi}{6} \right) \right]
$$

$$
+ \frac{\omega_{kd}}{p} n_A(6) \left[\cos p \left(z_6 + \frac{\pi}{6} \right) - \cos p \left(z_5 + \frac{\pi}{6} \right) \right]
$$

$$
= \sum_{i=1}^{6} \frac{\omega_{kd}}{p} n_A(i) \left[\cos p \left(z_i + \frac{\pi}{6} \right) - \cos p \left(z_{i-1} + \frac{\pi}{6} \right) \right] \tag{7.66}
$$

The same procedure is applicable to the remaining terms in Equation (7.65), as:

$$
\int_0^{60} -\omega_{kd} n_A(\varphi) \sin p \left(\varphi + \frac{\pi}{6} \right) d\varphi = \sum_{i=1}^{6} \frac{\omega_{kd}}{p} n_A(i) \left[\cos p \left(z_i + \frac{\pi}{6} \right) - \cos p \left(z_{i-1} + \frac{\pi}{6} \right) \right] \tag{7.67}
$$

$$
\int_{90}^{150} -\omega_{kd} n_A(\varphi) \sin p \left(\varphi + \frac{\pi}{6} \right) d\varphi = \sum_{i=10}^{15} \frac{\omega_{kd}}{p} n_A(i) \left[\cos p \left(z_i + \frac{\pi}{6} \right) - \cos p \left(z_{i-1} + \frac{\pi}{6} \right) \right] \tag{7.68}
$$

$$
\int_{180}^{240} -\omega_{kd} n_A(\varphi) \sin p \left(\varphi + \frac{\pi}{6} \right) d\varphi = \sum_{i=19}^{24} \frac{\omega_{kd}}{p} n_A(i) \left[\cos p \left(z_i + \frac{\pi}{6} \right) - \cos p \left(z_{i-1} + \frac{\pi}{6} \right) \right] \tag{7.69}
$$

$$
\int_{270}^{330} -\omega_{kd} n_A(\varphi) \sin p \left(\varphi + \frac{\pi}{6} \right) d\varphi = \sum_{i=28}^{33} \frac{\omega_{kd}}{p} n_A(i) \left[\cos p \left(z_i + \frac{\pi}{6} \right) - \cos p \left(z_{i-1} + \frac{\pi}{6} \right) \right] \tag{7.70}
$$

Finally, the mutual inductance of the stator phase A winding and direct axis damper winding in $\theta = 0$ is:

$$
L_{dkA} = \mu_0 r L k_1 \left\{
\begin{aligned}
& \sum_{i=1}^{6} \frac{\omega_{kd}}{p} n_A(i) \left[\cos p \left(z_i + \frac{\pi}{6} \right) - \cos p \left(z_{i-1} + \frac{\pi}{6} \right) \right] \\
& + \sum_{i=10}^{15} \frac{\omega_{kd}}{p} n_A(i) \left[\cos p \left(z_i + \frac{\pi}{6} \right) - \cos p \left(z_{i-1} + \frac{\pi}{6} \right) \right] \\
& + \sum_{i=19}^{24} \frac{\omega_{kd}}{p} n_A(i) \left[\cos p \left(z_i + \frac{\pi}{6} \right) - \cos p \left(z_{i-1} + \frac{\pi}{6} \right) \right] \\
& + \sum_{i=28}^{33} \frac{\omega_{kd}}{p} n_A(i) \left[\cos p \left(z_i + \frac{\pi}{6} \right) - \cos p \left(z_{i-1} + \frac{\pi}{6} \right) \right]
\end{aligned}
\right\} \tag{7.71}
$$

In a case where θ varies between zero and $z_1 (0 \leq \theta \leq z_1)$, the value of L_{kdA} can be derived as:

$$
L_{kdA} = \mu_0 rL \int_{0+\theta}^{360+\theta} n_{kd}(\varphi, \theta) n_A(\varphi, \theta) g^{-1}(\varphi, \theta) \, d\varphi
$$

$$
= \mu_0 rL \int_{0+\theta}^{360+\theta} n_{kd}(\varphi, \theta) n_A(\varphi) g^{-1}(\varphi) \, d\varphi
$$

$$
= \mu_0 rL k_1 \left\{
\begin{array}{l}
\displaystyle \int_{0+\theta}^{60+\theta} -\omega_{kd} n_A(\varphi) \sin p \left(\varphi - \theta + \frac{\pi}{6}\right) d\varphi \\[2mm]
\displaystyle + \int_{90+\theta}^{150+\theta} -\omega_{kd} n_A(\varphi) \sin p \left(\varphi - \theta + \frac{\pi}{6}\right) d\varphi \\[2mm]
\displaystyle + \int_{180+\theta}^{240+\theta} -\omega_{kd} n_A(\varphi) \sin p \left(\varphi - \theta + \frac{\pi}{6}\right) d\varphi \\[2mm]
\displaystyle + \int_{270+\theta}^{330+\theta} -\omega_{kd} n_A(\varphi) \sin p \left(\varphi - \theta + \frac{\pi}{6}\right) d\varphi
\end{array}
\right.
$$

$$
+ \mu_0 rL k_2 \left\{
\begin{array}{l}
\displaystyle \int_{60+\theta}^{90+\theta} -\omega_{kd} n_A(\varphi) \sin p \left(\varphi - \theta + \frac{\pi}{6}\right) d\varphi \\[2mm]
\displaystyle + \int_{150+\theta}^{180+\theta} -\omega_{kd} n_A(\varphi) \sin p \left(\varphi - \theta + \frac{\pi}{6}\right) d\varphi \\[2mm]
\displaystyle + \int_{240+\theta}^{270+\theta} -\omega_{kd} n_A(\varphi) \sin p \left(\varphi - \theta + \frac{\pi}{6}\right) d\varphi \\[2mm]
\displaystyle + \int_{330+\theta}^{360+\theta} -\omega_{kd} n_A(\varphi) \sin p \left(\varphi - \theta + \frac{\pi}{6}\right) d\varphi
\end{array}
\right.
\tag{7.72}
$$

The first term in Equation (7.72) can be expanded as:

$$
\int_{0+\theta}^{60+\theta} -\omega_{kd} n_A(\varphi) \sin p \left(\varphi - \theta + \frac{\pi}{6}\right) d\varphi = -\omega_{kd} \int_{0+\theta}^{60+\theta} n_A(\varphi) \sin p \left(\varphi - \theta + \frac{\pi}{6}\right) d\varphi
$$

$$
= -\omega_{kd} n_A(1) \int_{0+\theta}^{z_1} \sin p \left(\varphi - \theta + \frac{\pi}{6}\right) d\varphi - \omega_{kd} n_A(2) \int_{z_1}^{z_2} \sin p \left(\varphi - \theta + \frac{\pi}{6}\right) d\varphi
$$

$$
- \omega_{kd} n_A(3) \int_{z_2}^{z_3} \sin p \left(\varphi - \theta + \frac{\pi}{6}\right) d\varphi - \omega_{kd} n_A(4) \int_{z_3}^{z_4} \sin p \left(\varphi - \theta + \frac{\pi}{6}\right) d\varphi
$$

$$
- \omega_{kd} n_A(5) \int_{z_4}^{z_5} \sin p \left(\varphi - \theta + \frac{\pi}{6}\right) d\varphi - \omega_{kd} n_A(6) \int_{z_5}^{z_6} \sin p \left(\varphi - \theta + \frac{\pi}{6}\right) d\varphi
$$

$$
- \omega_{kd} n_A(7) \int_{z_6}^{z_6+\theta} \sin p \left(\varphi - \theta + \frac{\pi}{6}\right) d\varphi
$$

$$
= \frac{\omega_{kd}}{p} n_A(1) \left[\cos p \left(z_1 - \theta + \frac{\pi}{6}\right) - \cos p \left(0 + \theta - \theta + \frac{\pi}{6}\right)\right]
$$

$$
+ \frac{\omega_{kd}}{p} n_A(2) \left[\cos p \left(z_2 - \theta + \frac{\pi}{6}\right) - \cos p \left(z_1 - \theta + \frac{\pi}{6}\right)\right]
$$

$$
+ \frac{\omega_{kd}}{p} n_A(3) \left[\cos p \left(z_3 - \theta + \frac{\pi}{6}\right) - \cos p \left(z_2 - \theta + \frac{\pi}{6}\right)\right]
$$

$$
+ \frac{\omega_{kd}}{p} n_A(4) \left[\cos p \left(z_4 - \theta + \frac{\pi}{6}\right) - \cos p \left(z_3 - \theta + \frac{\pi}{6}\right)\right]
$$

$$
+ \frac{\omega_{kd}}{p} n_A(5) \left[\cos p \left(z_5 - \theta + \frac{\pi}{6}\right) - \cos p \left(z_4 - \theta + \frac{\pi}{6}\right)\right]
$$

$$+ \frac{\omega_{kd}}{p} n_A(6) \left[\cos p \left(z_6 - \theta + \frac{\pi}{6} \right) - \cos p \left(z_5 - \theta + \frac{\pi}{6} \right) \right]$$

$$+ \frac{\omega_{kd}}{p} n_A(7) \left[\cos p \left(z_6 + \theta - \theta + \frac{\pi}{6} \right) - \cos p \left(z_6 - \theta + \frac{\pi}{6} \right) \right] \tag{7.73}$$

The simplified form of Equation (7.73) is:

$$\int_{0+\theta}^{60+\theta} -\omega_{kd} n_A(\varphi) \sin p \left(\varphi - \theta + \frac{\pi}{6} \right) \, d\varphi$$

$$= \frac{\omega_{kd}}{p} n_A(1) \left[\cos p \left(z_1 - \theta + \frac{\pi}{6} \right) - \cos p \left(0 + \frac{\pi}{6} \right) \right]$$

$$+ \sum_{i=2}^{6} \frac{\omega_{kd}}{p} n_A(i) \left[\cos p \left(z_i - \theta + \frac{\pi}{6} \right) - \cos p \left(z_{i-1} - \theta + \frac{\pi}{6} \right) \right] \tag{7.74}$$

$$+ \frac{\omega_{kd}}{p} n_A(7) \left[\cos p \left(z_6 + \frac{\pi}{6} \right) - \cos p \left(z_6 - \theta + \frac{\pi}{6} \right) \right]$$

The same procedure used for Equation (7.73) should be applied to the remaining terms in Equation (7.72), leading to:

$$\int_{0+\theta}^{60+\theta} -\omega_{kd} n_A(\varphi) \sin p \left(\varphi - \theta + \frac{\pi}{6} \right) \, d\varphi$$

$$= \frac{\omega_{kd}}{p} n_A(1) \left[\cos p \left(z_1 - \theta + \frac{\pi}{6} \right) - \cos p \left(0 + \frac{\pi}{6} \right) \right]$$

$$+ \sum_{i=2}^{6} \frac{\omega_{kd}}{p} n_A(i) \left[\cos p \left(z_i - \theta + \frac{\pi}{6} \right) - \cos p \left(z_{i-1} - \theta + \frac{\pi}{6} \right) \right] \tag{7.75}$$

$$+ \frac{\omega_{kd}}{p} n_A(7) \left[\cos p \left(z_6 + \frac{\pi}{6} \right) - \cos p \left(z_6 - \theta + \frac{\pi}{6} \right) \right]$$

$$\int_{90+\theta}^{150+\theta} -\omega_{kd} n_A(\varphi) \sin p \left(\varphi - \theta + \frac{\pi}{6} \right) \, d\varphi$$

$$= \frac{\omega_{kd}}{p} n_A(10) \left[\cos p \left(z_{10} - \theta + \frac{\pi}{6} \right) - \cos p \left(z_9 + \frac{\pi}{6} \right) \right]$$

$$+ \sum_{i=11}^{15} \frac{\omega_{kd}}{p} n_A(i) \left[\cos p \left(z_i - \theta + \frac{\pi}{6} \right) - \cos p \left(z_{i-1} - \theta + \frac{\pi}{6} \right) \right] \tag{7.76}$$

$$+ \frac{\omega_{kd}}{p} n_A(16) \left[\cos p \left(z_{15} + \frac{\pi}{6} \right) - \cos p \left(z_{15} - \theta + \frac{\pi}{6} \right) \right]$$

$$\int_{180+\theta}^{240+\theta} -\omega_{kd} n_A(\varphi) \sin p \left(\varphi - \theta + \frac{\pi}{6} \right) \, d\varphi$$

$$= \frac{\omega_{kd}}{p} n_A(19) \left[\cos p \left(z_{19} - \theta + \frac{\pi}{6} \right) - \cos p \left(z_{18} + \frac{\pi}{6} \right) \right]$$

$$+ \sum_{i=20}^{24} \frac{\omega_{kd}}{p} n_A(i) \left[\cos p \left(z_i - \theta + \frac{\pi}{6} \right) - \cos p \left(z_{i-1} - \theta + \frac{\pi}{6} \right) \right] \tag{7.77}$$

$$+ \frac{\omega_{kd}}{p} n_A(25) \left[\cos p \left(z_{24} + \frac{\pi}{6} \right) - \cos p \left(z_{24} - \theta + \frac{\pi}{6} \right) \right]$$

$$\int_{270+\theta}^{330+\theta} -\omega_{kd} n_A(\varphi) \sin p \left(\varphi - \theta + \frac{\pi}{6} \right) d\varphi$$

$$= \frac{\omega_{kd}}{p} n_A(28) \left[\cos p \left(z_{28} - \theta + \frac{\pi}{6} \right) - \cos p \left(z_{27} + \frac{\pi}{6} \right) \right]$$

$$+ \sum_{i=29}^{33} \frac{\omega_{kd}}{p} n_A(i) \left[\cos p \left(z_i - \theta + \frac{\pi}{6} \right) - \cos p \left(z_{i-1} - \theta + \frac{\pi}{6} \right) \right] \tag{7.78}$$

$$+ \frac{\omega_{kd}}{p} n_A(34) \left[\cos p \left(z_{33} + \frac{\pi}{6} \right) - \cos p \left(z_{33} - \theta + \frac{\pi}{6} \right) \right]$$

Finally, the mutual inductance L_{kdA} is:

$$L_{kdA} = \mu_0 r L k_1 \, sum \tag{7.79}$$

where *sum* is:

$$sum = \sum_l \frac{\omega_{kd}}{p} n_A(l) \left[\cos p \left(lz - \theta + \frac{\pi}{6} \right) - \cos p \left((l-1-a)z + \frac{\pi}{6} \right) \right]$$

$$+ \sum_{i=j}^{j+4} \frac{\omega_{kd}}{p} n_A(i) \left[\cos p \left(z_i - \theta + \frac{\pi}{6} \right) - \cos p \left(z_{i-1} - \theta + \frac{\pi}{6} \right) \right]$$

$$+ \sum_m \frac{\omega_{kd}}{p} n_A(m) \left[\cos p \left((m-1-a)z + \frac{\pi}{6} \right) - \cos p \left((m-1)z - \theta + \frac{\pi}{6} \right) \right] \tag{7.80}$$

$$j = a+2, a+20, a+29$$

$$l = a+1, a+10, a+19, a+28$$

$$m = a+7, a+16, a+25, a+34$$

The mutual inductance of the stator phase winding and damper winding in the quadrature axis can be derived as for L_{kdA}, since the definition of the winding function for the kq axis is:

$$N_{kq}(\varphi, \theta) = \omega_{kq} \cos \left(p \left(\varphi - \theta + \frac{\pi}{6} \right) \right) \tag{7.81}$$

L_{kdA} can be extracted as:

$$L_{kqA} = \mu_0 r L k_1 \, sum \tag{7.82}$$

$$sum = \sum_l \frac{\omega_{kq}}{p} n_A(l) \left[\sin p \left(lz - \theta + \frac{\pi}{6} \right) - \sin p \left((l-1-a)z + \frac{\pi}{6} \right) \right]$$

$$+ \sum_{i=j}^{j+4} \frac{\omega_{kq}}{p} n_A(i) \left[\cos p \left(z_i - \theta + \frac{\pi}{6} \right) - \cos p \left(z_{i-1} - \theta + \frac{\pi}{6} \right) \right] \tag{7.83}$$

$$+ \sum_m \frac{\omega_{kq}}{p} n_A(m) \left[\cos p \left((m-1-a)z + \frac{\pi}{6} \right) - \cos p \left((m-1)z - \theta + \frac{\pi}{6} \right) \right]$$

$$l = a+1, a+10, a+19, a+28$$

$$j = a+2, a+11, a+20, a+29$$

$$m = a+7, a+16, a+25, a+34$$

7.6   The Self Inductance of the Rotor

7.6.1   The Self Inductance of the Rotor Field Winding

The rotor field winding is a single phase located inside a stator cylindrical core. The value of the self inductance of the rotor field winding should be a constant value. The rotor self inductance L_{FF} can be calculated according to Equation (7.31) as:

$$
\begin{aligned}
L_{FF} &= \mu_0 rLk_1 \int_{0+\theta}^{360+\theta} n_F(\varphi)n_F(\varphi)g^{-1}(\varphi)\,d\varphi \\
&= \mu_0 rLk_1 \left\{ \begin{aligned} & \int_0^{60} n_F(\varphi)^2\,d\varphi + \int_{90}^{150} n_F(\varphi)^2\,d\varphi \\ & + \int_{180}^{240} n_F(\varphi)^2\,d\varphi + \int_{270}^{330} n_F(\varphi)^2\,d\varphi \end{aligned} \right\}
\end{aligned}
\tag{7.84}
$$

where n_F is the rotor turn function, which is a constant value. The limits of the integral must be changed from degrees to radians.

7.6.2   The Self Inductance of the Rotor Damper Winding

The self inductance of the damper winding is a constant value, as is the inductance of the rotor field winding. The self inductance of the d-axis can be derived as:

$$
\begin{aligned}
L_{kd} &= \mu_0 rL \int_0^{360} n_{kd}(\varphi)n_{kd}(\varphi)g^{-1}(\varphi)\,d\varphi \\
&= \mu_0 rLk_1 \left\{ \begin{aligned} & \int_0^{60} \omega_{kd}{}^2 \sin^2 p\left(\varphi+\frac{\pi}{6}\right)d\varphi + \int_{90}^{150} \omega_{kd}{}^2 \sin^2 p\left(\varphi+\frac{\pi}{6}\right)d\varphi \\ & + \int_{180}^{240} \omega_{kd}{}^2 \sin^2 p\left(\varphi+\frac{\pi}{6}\right)d\varphi + \int_{270}^{330} \omega_{kd}{}^2 \sin^2 p\left(\varphi+\frac{\pi}{6}\right)d\varphi \end{aligned} \right\} \\
&+ \mu_0 rLk_2 \left\{ \begin{aligned} & \int_{60}^{90} \omega_{kd}{}^2 \sin^2 p\left(\varphi+\frac{\pi}{6}\right)d\varphi + \int_{150}^{180} \omega_{kd}{}^2 \sin^2 p\left(\varphi+\frac{\pi}{6}\right)d\varphi \\ & + \int_{240}^{270} \omega_{kd}{}^2 \sin^2 p\left(\varphi+\frac{\pi}{6}\right)d\varphi + \int_{330}^{360} \omega_{kd}{}^2 \sin^2 p\left(\varphi+\frac{\pi}{6}\right)d\varphi \end{aligned} \right\}
\end{aligned}
\tag{7.85}
$$

If the second term of Equation (7.85) is neglected, the final format of the self inductance can be written as:

$$
\begin{aligned}
L_{kd} &= \mu_0 rL \int_0^{360} n_{kd}(\varphi)n_{kd}(\varphi)g^{-1}(\varphi)\,d\varphi \\
&= \mu_0 rLk_1 \left\{ \begin{aligned} & \int_0^{60} \omega_{kd}{}^2 \frac{1-\cos 2p\left(\varphi+\frac{\pi}{6}\right)}{2}\,d\varphi + \int_{90}^{150} \omega_{kd}{}^2 \frac{1-\cos 2p\left(\varphi+\frac{\pi}{6}\right)}{2}\,d\varphi \\ & + \int_{180}^{240} \omega_{kd}{}^2 \frac{1-\cos 2p\left(\varphi+\frac{\pi}{6}\right)}{2}\,d\varphi + \int_{270}^{330} \omega_{kd}{}^2 \frac{1-\cos 2p\left(\varphi+\frac{\pi}{6}\right)}{2}\,d\varphi \end{aligned} \right\}
\end{aligned}
\tag{7.86}
$$

The integrals in Equation (7.86) can be simplified as:

$$\omega_{kd}^2 \int_0^{60} \frac{1 - \cos 2p\left(\varphi + \frac{\pi}{6}\right)}{2} \, d\varphi = \omega_{kd}^2 \left(\frac{1}{2}\int_0^{60} d\varphi - \frac{1}{2}\int_0^{60} \cos 2p\left(\varphi + \frac{\pi}{6}\right) d\varphi\right)$$

$$= \omega_{kd}^2 \left(\frac{1}{2}\left(\frac{\pi}{3}\right) - \frac{1}{4p}\left(\sin 2p\left(\frac{\pi}{3} + \frac{\pi}{6}\right) - \sin 2p\left(0 + \frac{\pi}{6}\right)\right)\right) \tag{7.87}$$

$$\int_{90}^{150} \omega_{kd}^2 \frac{1 - \cos 2p\left(\varphi + \frac{\pi}{6}\right)}{2} \, d\varphi$$

$$= \omega_{kd}^2 \left(\frac{1}{2}\left(\frac{\pi}{3}\right) - \frac{1}{4p}\left(\sin 2p\left(\frac{5\pi}{6} + \frac{\pi}{6}\right) - \sin 2p\left(\frac{\pi}{2} + \frac{\pi}{6}\right)\right)\right) \tag{7.88}$$

$$\int_{180}^{240} \omega_{kd}^2 \frac{1 - \cos 2p\left(\varphi + \frac{\pi}{6}\right)}{2} \, d\varphi$$

$$= \omega_{kd}^2 \left(\frac{1}{2}\left(\frac{\pi}{3}\right) - \frac{1}{4p}\left(\sin 2p\left(\frac{8\pi}{6} + \frac{\pi}{6}\right) - \sin 2p\left(\pi + \frac{\pi}{6}\right)\right)\right) \tag{7.89}$$

$$\int_{270}^{330} \omega_{kd}^2 \frac{1 - \cos 2p\left(\varphi + \frac{\pi}{6}\right)}{2} \, d\varphi$$

$$= \omega_{kd}^2 \left(\frac{1}{2}\left(\frac{\pi}{3}\right) - \frac{1}{4p}\left(\sin 2p\left(\frac{11\pi}{6} + \frac{\pi}{6}\right) - \sin 2p\left(\frac{3\pi}{2} + \frac{\pi}{6}\right)\right)\right) \tag{7.90}$$

The final value of the *d*-axis self inductance of the damper winding can be written as

$$L_{kd} = \mu_0 rLk_1 \, sum \tag{7.91}$$

where the *sum* is:

$$sum = \int_0^{60} \omega_{kd}^2 \frac{1 - \cos 2p\left(\varphi + \frac{\pi}{6}\right)}{2} \, d\varphi + \int_{90}^{150} \omega_{kd}^2 \frac{1 - \cos 2p\left(\varphi + \frac{\pi}{6}\right)}{2} \, d\varphi$$

$$+ \int_{180}^{240} \omega_{kd}^2 \frac{1 - \cos 2p\left(\varphi + \frac{\pi}{6}\right)}{2} \, d\varphi + \int_{270}^{330} \omega_{kd}^2 \frac{1 - \cos 2p\left(\varphi + \frac{\pi}{6}\right)}{2} \, d\varphi \tag{7.92}$$

The same procedure can be applied to the self inductance of the damper bar in the *q*-axis, whose winding function is:

$$N_{kq}(\varphi, \theta) = \omega_{kq} \cos p\left(\varphi - \theta + \frac{\pi}{6}\right)$$

$$L_{kq} = \mu_0 rL \int_0^{360} n_{kq}(\varphi)n_{kq}(\varphi)g^{-1}(\varphi) \, d\varphi$$

$$= \mu_0 rLk_1 \left\{ \begin{aligned} &\int_0^{60} \omega_{kq}^2 \cos^2 p\left(\varphi + \frac{\pi}{6}\right) d\varphi + \int_{90}^{150} \omega_{kq}^2 \cos^2 p\left(\varphi + \frac{\pi}{6}\right) d\varphi \\ &+ \int_{180}^{240} \omega_{kq}^2 \cos^2 p\left(\varphi + \frac{\pi}{6}\right) d\varphi + \int_{270}^{330} \omega_{kq}^2 \cos^2 p\left(\varphi + \frac{\pi}{6}\right) d\varphi \end{aligned} \right\}$$

$$+ \mu_0 rLk_2 \left\{ \begin{aligned} &\int_{60}^{90} \omega_{kq}^2 \cos^2 p\left(\varphi + \frac{\pi}{6}\right) d\varphi + \int_{150}^{180} \omega_{kq}^2 \cos^2 p\left(\varphi + \frac{\pi}{6}\right) d\varphi \\ &+ \int_{240}^{270} \omega_{kq}^2 \cos^2 p\left(\varphi + \frac{\pi}{6}\right) d\varphi + \int_{330}^{360} \omega_{kq}^2 \cos^2 p\left(\varphi + \frac{\pi}{6}\right) d\varphi \end{aligned} \right\}$$

$$\tag{7.93}$$

The second term of Equation (7.93) can be neglected, since k_2 is so small. Therefore:

$$L_{kq} = \mu_0 r L \int_0^{360} n_{kq}(\varphi) n_{kq}(\varphi) g^{-1}(\varphi) \, d\varphi$$

$$+ \mu_0 r L k_1 \left\{ \begin{array}{l} \displaystyle\int_0^{60} \omega_{kq}{}^2 \frac{1 - \cos 2p\left(\varphi + \frac{\pi}{6}\right)}{2} \, d\varphi + \int_{90}^{150} \omega_{kq}{}^2 \frac{1 - \cos 2p\left(\varphi + \frac{\pi}{6}\right)}{2} \, d\varphi \\[4mm] \displaystyle+ \int_{180}^{240} \omega_{kq}{}^2 \frac{1 - \cos 2p\left(\varphi + \frac{\pi}{6}\right)}{2} \, d\varphi + \int_{270}^{330} \omega_{kq}{}^2 \frac{1 - \cos 2p\left(\varphi + \frac{\pi}{6}\right)}{2} \, d\varphi \end{array} \right\}$$

$$(7.94)$$

Each integral inside the limits can be calculated as:

$$\int_0^{60} \omega_{kd}{}^2 \frac{1 + \cos 2p\left(\varphi + \frac{\pi}{6}\right)}{2} \, d\varphi = \omega_{kd}{}^2 \int_0^{60} \frac{1 + \cos 2p\left(\varphi + \frac{\pi}{6}\right)}{2} \, d\varphi$$

$$= \omega_{kd}{}^2 \left(\frac{1}{2} \int_0^{60} d\varphi + \frac{1}{2} \int_0^{60} \cos 2p\left(\varphi + \frac{\pi}{6}\right) d\varphi \right) \qquad (7.95)$$

$$= \omega_{kd}{}^2 \left(\frac{1}{2}\left(\frac{\pi}{3}\right) + \frac{1}{4p}\left(\sin 2p\left(\frac{\pi}{3} + \frac{\pi}{6}\right) - \sin 2p\left(0 + \frac{\pi}{6}\right) \right) \right)$$

$$\int_{90}^{150} \omega_{kq}{}^2 \frac{1 + \cos 2p\left(\varphi + \frac{\pi}{6}\right)}{2} \, d\varphi$$

$$= \omega_{kq}{}^2 \left(\frac{1}{2}\left(\frac{\pi}{3}\right) + \frac{1}{4p}\left(\sin 2p\left(\frac{5\pi}{6} + \frac{\pi}{6}\right) - \sin 2p\left(\frac{\pi}{2} + \frac{\pi}{6}\right) \right) \right) \qquad (7.96)$$

$$\int_{180}^{240} \omega_{kq}{}^2 \frac{1 + \cos 2p\left(\varphi + \frac{\pi}{6}\right)}{2} \, d\varphi$$

$$= \omega_{kq}{}^2 \left(\frac{1}{2}\left(\frac{\pi}{3}\right) + \frac{1}{4p}\left(\sin 2p\left(\frac{8\pi}{6} + \frac{\pi}{6}\right) - \sin 2p\left(\pi + \frac{\pi}{6}\right) \right) \right) \qquad (7.97)$$

$$\int_{270}^{330} \omega_{kd}{}^2 \frac{1 + \cos 2p\left(\varphi + \frac{\pi}{6}\right)}{2} \, d\varphi$$

$$= \omega_{kd}{}^2 \left(\frac{1}{2}\left(\frac{\pi}{3}\right) + \frac{1}{4p}\left(\sin 2p\left(\frac{11\pi}{6} + \frac{\pi}{6}\right) - \sin 2p\left(\frac{3\pi}{2} + \frac{\pi}{6}\right) \right) \right) \qquad (7.98)$$

The self inductance damper bar winding in the quadrature axis is:

$$L_{kq} = \mu_0 r L k_1 \, sum \qquad (7.99)$$

where the *sum* is:

$$sum = \int_0^{60} \omega_{kq}{}^2 \frac{1 + \cos 2p\left(\varphi + \frac{\pi}{6}\right)}{2} \, d\varphi + \int_{90}^{150} \omega_{kq}{}^2 \frac{1 + \cos 2p\left(\varphi + \frac{\pi}{6}\right)}{2} \, d\varphi$$

$$+ \int_{180}^{240} \omega_{kq}{}^2 \frac{1 + \cos 2p\left(\varphi + \frac{\pi}{6}\right)}{2} \, d\varphi + \int_{270}^{330} \omega_{kq}{}^2 \frac{1 + \cos 2p\left(\varphi + \frac{\pi}{6}\right)}{2} \, d\varphi \qquad (7.100)$$

7.6.3 The Mutual Inductance Between the Rotor Field Winding and Damper Winding in the *d*-Axis

The mutual inductance of the rotor field winding and damper winding in the *d*-axis is defined as:

$$
\begin{aligned}
L_{kdF} &= \mu_0 rL \int_0^{360} n_{kd}(\varphi) n_F(\varphi) g^{-1}(\varphi) \, d\varphi \\
&= \mu_0 rLk_1 \left\{ \begin{array}{l}
n_F \int_0^{60} -\omega_{kd} \sin p \left(\varphi + \dfrac{\pi}{6} \right) d\varphi - n_F \int_{90}^{150} -\omega_{kd} \sin p \left(\varphi + \dfrac{\pi}{6} \right) d\varphi \\[2ex]
+ n_F \int_{180}^{240} -\omega_{kd} \sin p \left(\varphi + \dfrac{\pi}{6} \right) d\varphi - n_F \int_{270}^{330} -\omega_{kd} \sin p \left(\varphi + \dfrac{\pi}{6} \right) d\varphi
\end{array} \right\}
\end{aligned}
\tag{7.101}
$$

where the integrals in Equation (7.101), according to their limits, are:

$$
\begin{aligned}
\int_0^{60} -\omega_{kd} \sin p \left(\varphi + \frac{\pi}{6} \right) d\varphi &= -\omega_{kd} \int_0^{60} \sin p \left(\varphi + \frac{\pi}{6} \right) d\varphi \\
&= \frac{\omega_{kd}}{p} \left(\cos p \left(\frac{\pi}{3} + \frac{\pi}{6} \right) - \cos p \left(0 + \frac{\pi}{6} \right) \right)
\end{aligned}
\tag{7.102}
$$

$$
\begin{aligned}
\int_{90}^{150} -\omega_{kd} \sin p \left(\varphi + \frac{\pi}{6} \right) d\varphi &= -\omega_{kd} \int_{90}^{150} \sin p \left(\varphi + \frac{\pi}{6} \right) d\varphi \\
&= \frac{\omega_{kd}}{p} \left(\cos p \left(\frac{5\pi}{6} + \frac{\pi}{6} \right) - \cos p \left(\frac{\pi}{2} + \frac{\pi}{6} \right) \right)
\end{aligned}
\tag{7.103}
$$

$$
\begin{aligned}
\int_{180}^{240} -\omega_{kd} \sin p \left(\varphi + \frac{\pi}{6} \right) d\varphi &= -\omega_{kd} \int_{180}^{240} \sin p \left(\varphi + \frac{\pi}{6} \right) d\varphi \\
&= \frac{\omega_{kd}}{p} \left(\cos p \left(\frac{8\pi}{6} + \frac{\pi}{6} \right) - \cos p \left(\pi + \frac{\pi}{6} \right) \right)
\end{aligned}
\tag{7.104}
$$

$$
\begin{aligned}
\int_{270}^{330} -\omega_{kd} \sin p \left(\varphi + \frac{\pi}{6} \right) d\varphi &= -\omega_{kd} \int_{270}^{330} \sin p \left(\varphi + \frac{\pi}{6} \right) d\varphi \\
&= \frac{\omega_{kd}}{p} \left(\cos p \left(\frac{11\pi}{6} + \frac{\pi}{6} \right) - \cos p \left(\frac{9\pi}{6} + \frac{\pi}{6} \right) \right)
\end{aligned}
\tag{7.105}
$$

The simplified term for L_{kdf} is:

$$
L_{kdF} = \mu_0 rLk_1 \, sum
\tag{7.106}
$$

where *sum* is:

$$
\begin{aligned}
sum = &\int_0^{60} -\omega_{kd} \sin p \left(\varphi + \frac{\pi}{6} \right) d\varphi + \int_{90}^{150} -\omega_{kd} \sin p \left(\varphi + \frac{\pi}{6} \right) d\varphi \\
&+ \int_{180}^{240} -\omega_{kd} \sin p \left(\varphi + \frac{\pi}{6} \right) d\varphi + \int_{270}^{330} -\omega_{kd} \sin p \left(\varphi + \frac{\pi}{6} \right) d\varphi
\end{aligned}
\tag{7.107}
$$

7.6.4 The Mutual Inductance Between the Rotor Field Winding and Damper Winding in the *q*-Axis

The mutual inductance between the rotor field winding and the damper bar winding in the quadrature axis is zero, as they are perpendicular.

7.7 Derivative Forms of Synchronous Generator Inductances

7.7.1 Derivative Form of Stator Mutual Inductance

The derivative format of the inductances is required to calculate the electromagnetic torque of the machine. The inductances must be differentiated based on θ. Calculation of the derivative forms of the stator inductances requires that Equations (7.35) to (7.37) be differentiated with regards to θ. For example, the derivative form of mutual inductance between the stator phase A winding and the phase B winding is:

$$\frac{dL_{AB}}{d\theta} = \mu_0 rL(sum_1\,k_1 + sum_2\,k_2) \tag{7.108}$$

where

$$sum_1 = -\sum_l n_B(l)\,n_A(l) + \sum_m n_B(m)\,n_A(m)$$
$$l = a+1, a+10, a+19, a+28 \tag{7.109}$$
$$m = a+7, a+16, a+25, a+34$$

$$sum_2 = -\sum_{ll} n_B(ll)\,n_A(ll) + \sum_{mm} n_B(mm)\,n_A(mm)$$
$$ll = a+7, a+16, a+25, a+34 \tag{7.110}$$
$$mm = a+10, a+19, a+28, a+1$$

7.7.2 Derivative Form of Stator and Rotor Mutual Inductance

The derivative form of mutual inductance between the rotor and the stator winding can be derived according to Equations (7.58) to (7.62), as follows:

$$\frac{dL_{FA}}{d\theta} = \mu_0 rL n_F k_1\,sum \tag{7.111}$$

where

$$sum = -n_A(a+1) + n_A(a+7) + n_A(a+10) - n_A(a+16)$$
$$- n_A(a+19) + n_A(a+25) + n_A(a+28) - n_A(a+34) \tag{7.112}$$

The derivative form of the mutual inductance between the stator and rotor damper winding in the d-axis is:

$$\frac{dL_{kdA}}{d\theta} = \mu_0 rL k_1\,sum \tag{7.113}$$

where:

$$sum = \sum_l \omega_{kd} n_A(l)\left[\sin p\left(lz - \theta + \frac{\pi}{6}\right)\right]$$
$$+ \sum_{i=j}^{j+4} \omega_{kd} n_A(i)\left[\sin p(z_i - \theta) - \sin p\left(z_{i-1} - \theta + \frac{\pi}{6}\right)\right]$$
$$+ \sum_m \omega_{kd} n_A(m)\left[\sin p\left((m-1)z - \theta + \frac{\pi}{6}\right)\right] \tag{7.114}$$
$$l = a+1, a+10, a+19, a+28$$

$$m = a + 7, a + 16 + a + 25 + a + 34$$

$$j = a + 2, a + 11, a + 20, a + 29$$

The same procedure used for the d-axis can be applied to the q-axis as follows:

$$\frac{dL_{kqA}}{d\theta} = \mu_0 rLk_1 \ sum \tag{7.115}$$

where

$$
\begin{aligned}
sum = &\sum_l - \omega_{kq} n_A(l) \left[\cos p \left(lz - \theta + \frac{\pi}{6} \right) \right] \\
&- \sum_{i=j}^{j+4} \omega_{kq} n_A(i) \left[\cos p \left(z_i - \theta + \frac{\pi}{6} \right) - \cos p \left(z_{i-1} - \theta + \frac{\pi}{6} \right) \right] \\
&+ \sum_m \omega_{kd} n_A(m) \left[\cos p \left((m-1)z - \theta + \frac{\pi}{6} \right) \right]
\end{aligned}
$$

$$l = a + 1, a + 10, a + 19, a + 28$$

$$m = a + 7, a + 16 + a + 25 + a + 34$$

$$j = a + 2, a + 11, a + 20, a + 29 \tag{7.116}$$

Since the self and mutual inductances of the rotor field winding and damper winding (L_{FF}, L_{kd}, L_{kq}, L_{kdfd}) are constant values, their derivative with respect to θ is zero.

7.7.3 Dynamic Equations Governing the Synchronous Machines

The dynamic equation governing the MWF of the synchronous machine under healthy conditions is the same as the equations in references [7] and [8]. To satisfy the constraint $I_a + I_b + I_c = 0$ in a case of a star connection, the machine's main voltages must be written into line-to-line voltages. In a faulty machine under an inter-turn short circuit fault, the reduced turns from the phase winding are represented as a separate winding with an applied voltage equal to zero. The detailed schematic of the self inductance and mutual inductances of a synchronous machine considering a field winding and damper winding in both the direct and quadrature axis of the rotor in addition to the stator phase winding under an inter-turn short circuit fault in phase A is shown in Figure 7.5, and the governing dynamic equations are:

$$[V_{sl}] = [R_{sl}][I_{sl}] + \frac{d}{dt}[\lambda_{sl}] \tag{7.117}$$

$$[V_r] = [R_r][I_r] + \frac{d}{dt}[\lambda_r] \tag{7.118}$$

$$T_e - T_L = J \frac{d\omega}{dt} \tag{7.119}$$

where $[V_{sl}]$, $[I_{sl}]$, and $[\lambda_{sl}]$ are the vectors of the line-to-line voltage, current, and flux-linkages of the stator; $[V_r]$, $[I_r]$, and $[\lambda_r]$ are the vectors of the rotor voltage, current, and flux-linkages; $[R_{sl}]$ and $[R_r]$ are the modified matrices of the stator and rotor resistance; T_i and T_e are the applied and electromagnetic torque; J is the moment inertia; and ω is the rotor speed. The vectors and matrices are defined as below:

$$[V_{sl}] = [V_{ab} \ V_{bc} \ 0]^T \tag{7.120}$$

$$[I_{sl}] = [i_a \ i_b \ i_d]^T \tag{7.121}$$

Figure 7.5 Detailed schematic of a self inductance and mutual inductance of the synchronous machine under an inter-turn short circuit fault.

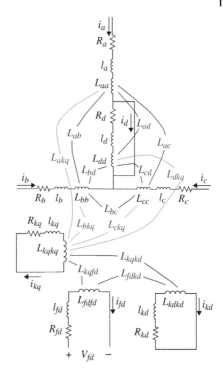

$$[I_r] = [i_{kq}\ i_{fd}\ i_{kd}]^{\mathrm{T}} \tag{7.122}$$

$$[\lambda_{sl}] = [\lambda_a - \lambda_b\ \lambda_b - \lambda_c\ \lambda_d]^{\mathrm{T}} \tag{7.123}$$

$$[\lambda_r] = [\lambda_{kq}\ \lambda_{fq}\ \lambda_{kd}]^{\mathrm{T}} \tag{7.124}$$

$$[R_{sl}] = \begin{bmatrix} R_{sa} & -R_{sb} & 0 \\ R_{sc} & R_{sb} + R_{sc} & 0 \\ 0 & 0 & R_{sd} \end{bmatrix} \tag{7.125}$$

$$[R_r] = \begin{bmatrix} R_{kq} & 0 & 0 \\ 0 & R_{fd} & 0 \\ 0 & 0 & R_{kd} \end{bmatrix} \tag{7.126}$$

The stator and rotor linkage flux are defined as:

$$[\lambda_{sl}] = [L_{ssl}][I_{ssl}] + [L_{srl}][I_r] \tag{7.127}$$

$$[\lambda_r] = [L_{rsl}][I_{sll}] + [L_{rr}][I_r] \tag{7.128}$$

where $[L_{ssl}]$ is the modified matrix of mutual and self inductances of the stator, $[L_{srl}]$ and $[L_{rsl}]$ are the modified matrices of mutual inductances between the stator and rotor winding, and $[L_{rr}]$ is the self inductance of the rotor. The mentioned matrices are defined as:

$$[L_{ssl}] = \begin{bmatrix} L_{aa} + (1-\mu)l_{ls} - L_{ca} - L_{ab} - L_{bc} & L_{ab} - L_{ca} - L_{bb} - l_{ls} + L_{bc} & L_{ad} - L_{bd} \\ L_{ab} - L_{bc} - L_{ca} + L_{cc} + l_{ls} & L_{bb} + l_{ls} - L_{bc} - l_{bc} + L_{cc} + l_{ls} & L_{bd} - L_{cd} \\ L_{ad} - L_{cd} & L_{bd} - L_{cd} & L_{dd} - \mu l_{ls} \end{bmatrix} \tag{7.129}$$

$$[L_{srl}] = \begin{bmatrix} L_{kqa} - L_{kqb} & L_{fda} - L_{fdb} & L_{kda} - L_{kdb} \\ L_{kqb} - L_{kqc} & L_{fdb} - L_{fdc} & L_{kda} - L_{kdb} \\ L_{kqd} & L_{fdd} & L_{kdd} \end{bmatrix} \tag{7.130}$$

$$[L_{rsl}] = \begin{bmatrix} L_{kqa} - L_{kqc} & L_{kqb} - L_{kqc} & L_{kqd} \\ L_{fda} - L_{fdc} & L_{fdb} - L_{fdc} & L_{fdd} \\ L_{kqd} - L_{kdc} & L_{kdb} - L_{kdc} & L_{kdd} \end{bmatrix} \tag{7.131}$$

$$[L_{rr}] = \begin{bmatrix} L_{kq} & 0 & 0 \\ 0 & L_{fd} & L_{fdkd} \\ 0 & L_{kdfd} & L_{kd} \end{bmatrix} \tag{7.132}$$

where L_{ij} is the inductance between two arbitrary coils, called i and j. The voltage of each phase in a star connection is:

$$V_a = R_a I_a$$
$$V_b = R_b I_b$$
$$V_c = R_c I_c \tag{7.133}$$

The phase current of each phase can be derived as:

$$I_b = \frac{V_{ab} - V_{bc}}{3R_l} \tag{7.134}$$

$$I_a = I_b + \frac{V_{ab}}{R_l} \tag{7.135}$$

The electromagnetic torque of the machine (T_e) is defined as:

$$T_e = \frac{1}{2}\left([I_s]\frac{d[L_s]}{d\theta}[I_s]^{\mathrm{T}}\right) + [I_s]\frac{d[L_{sr}]}{d\theta}[I_r]^{\mathrm{T}} \tag{7.136}$$

where $[I_s]$ is the stator phase current, $[L_s]$ is the self inductances of the stator windings, and $[L_{sr}]$ is the mutual inductances between the rotor and the stator winding, as defined below:

$$[I_s] = [i_a\ i_b\ i_c\ i_d] \tag{7.137}$$

$$i_c = i_a + i_b \tag{7.138}$$

$$L_s = \begin{bmatrix} L_{aa} + (1+\mu)l_{ls} & L_{ab} & L_{ac} & L_{ad} \\ L_{ba} & L_{bb} + l_{ls} & L_{bc} & L_{bd} \\ L_{ca} & L_{cd} & L_{cc} + l_{ls} & L_{cd} \\ L_{da} & L_{db} & L_{dc} & L_{dd} + \mu l_{ls} \end{bmatrix} \tag{7.139}$$

$$L_{sr} = \begin{bmatrix} L_{akq} & L_{afd} & L_{akd} \\ L_{bkq} & L_{bfd} & L_{bkd} \\ L_{ckq} & L_{cfd} & L_{ckd} \\ L_{dkq} & L_{dfd} & L_{dkd} \end{bmatrix} \tag{7.140}$$

7.8 A Practical Case Study

In this section, a practical case study of a 3.7 kVA synchronous machine is studied to demonstrate the application of the modeling method. The dynamic modeling of the synchronous machine is studied for both a healthy case and under a short circuit fault. The first step in this study is parameter identification of the synchronous machine. A detailed explanation of parameter identification is also discussed in Chapter 3.

7.8.1 Parameter Identification

Parameter identification is a key point in the modeling of electrical machines. Many scientific articles and standards, such as the IEC, IEEE, and NEMA, describe the parameter determination procedure in detail [9, 10]. The following are required parameters for synchronous machine modeling:

r_s: Resistance of stator phase winding
r_{fd}: Resistance of rotor field winding
r_{kd}: Resistance of damper winding in the direct axis (7.d)
r_{kq}: Resistance of damper winding in the quadrature axis (7.q)
x_{ls}: Leakage inductance of stator phase winding
x_{lkd}: d-axis leakage inductance of damper winding
x_{lkq}: q-axis leakage inductance of damper winding
ω_{fd}: Number of rotor field winding turns
ω_{kd}: Number of damper windings in the d-axis
ω_{kq}: Number of damper winding turns in the q-axis
x_{lfd}: Rotor field winding leakage inductance
x_d: d-axis reactance
x_d': d-axis transient reactance
x_d'': q-axis transient reactance
L: Rotor length
R: Radius of rotor
K: Air gap length

Nameplate data are also required for modeling. Table 7.1 shows the nameplate and dimensions of the four-pole salient pole SG examined in this section.

7.8.1.1 Resistance of the Stator Phase Winding
A direct current (DC) voltage is applied to one of the stator phase windings, the current passing through the winding is measured, and the DC resistance of the stator phase winding is calculated in the cold condition. Ideally, the resistance of the winding should be measured immediately after the machine is switched off, as a significant difference exists between the cold and warm measured resistance. The measured stator phase winding resistance in the proposed SG is 4.15 Ω.

7.8.1.2 Rotor Field Winding Resistance
The procedure described in Section 2.1 for the stator phase winding can be applied to measure the rotor field winding resistance. Its measured value is 23.3 Ω.

7.8.1.3 The Direct Axis (d) Reactance
Two tests are required to obtain the d-axis reactance:

1. Short circuit test.
2. Open circuit test.

These tests are used to determine the characteristics of the synchronous generator.

Table 7.1 Nameplate and the dimensions of a three-phase four-pole salient pole synchronous generator.

Parameter	Value
Power (kW)	3.7
Phase voltage (V)	380
Frequency (Hz)	50
Rotor field voltage (V)	70
Nominal stator current (A)	5.7
Rotor length (mm)	98
Rotor radius (mm)	72
Air gap length (mm)	2
Distance between adjacent pole (mm)	60
Number of turns of the stator winding	20
Number of slots	36

The open-circuit characteristic of the SG depicts a variation in the induced voltage in the armature winding versus the field winding current in a no-load condition and fixed rotation speed. In this case, the induced voltage in the armature winding is proportional to the magnetic flux φ:

$$E = k\varphi\omega \tag{7.141}$$

where ω is the angular velocity and k is a constant that depends on the configuration of the SG winding. Increasing the field current increases the magnetic flux of the machine core and therefore increases the induced voltage. The core magnetic residual causes a non-zero induced voltage, even when the field current is zero. Note that the initial electrical power supplied to the field winding must be zero. When the SG approaches its rated speed, the field current must be increased stepwise until it reaches the rating for the field winding.

The short circuit characteristic of the SG shows the variation in the armature winding current versus the field winding current, while the armature windings are short-circuited and the generator rotates at a steady speed. In this condition, an ammeter must be inserted in series with the shorted terminals. The prime mover should reach the SG speed while the field current is zero. The field winding current must be increased stepwise. At the same time, the armature current is also increased. The field winding current should be increased up to the point that the nominal current passes through the armature winding while the armature terminals are short circuited. The short circuit and open circuit characteristics of the proposed SG in pu (per unit) are depicted in Figure 7.6.

The d-reactance is calculated based on the open circuit and short circuit characteristics [9]:

$$x_{dpu} = \frac{AC}{BC} \tag{7.142}$$

$$x_B = \frac{V_n}{\sqrt{3}I_n} \tag{7.143}$$

$$x_d = x_B x_{dpu} \tag{7.144}$$

where AC is equal to 1 and BC is equal to 0.6569. Therefore, the calculated d-reactance is 57.73 Ω.

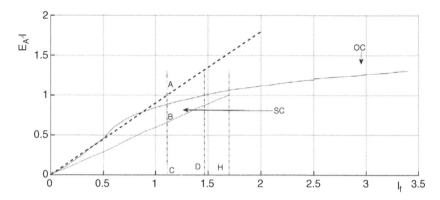

Figure 7.6 Open circuit and short circuit characteristics of a synchronous generator.

7.8.1.4 Sub-transient Reactance of the Direct Axis

The applied test voltage is used to calculate the sub-transient d-reactance in the SG [9]. The procedure for applying the test voltage is as follows:

1. The field winding terminals are short-circuited.
2. The AC voltage with the rated generator frequency is applied to the armature winding.
3. The rotor shaft must be rotated to find the direct axis (maximum field winding current) and the quadrature axis (minimum field winding current) of the generator.
4. In the position determined for the rotor, the applied voltage to the armature winding, the field winding current, and input power must be measured.
5. Based on the recorded data, the subtransient d-reactance is estimated as follows:

$$z_d'' = \frac{V}{2I} \tag{7.145}$$

$$R_d'' = \frac{P}{2I^2} \tag{7.146}$$

$$x_d'' = \sqrt{\left(z_d''\right)^2 + \left(R_d''\right)^2} \tag{7.147}$$

The subtransient d-reactance for the proposed generator is 12.3 Ω.

7.8.1.5 Number of Turns of Rotor Field Windings

The number of turns of a rotor field winding must be adjusted so that the induced voltage in the armature winding becomes equal to the nominal value after applying a nominal rotor field winding voltage. This can be done by trial and error. The number of turns for a rated field voltage of 70 V and an armature RMS voltage of 380 V is 320 turns.

7.8.1.6 Transient Direct Axis Reactance

The transient d-axis reactance x_d' is as follows [11]:

$$x_d' = x_{ls} + \frac{x_{md}x_{lfd}}{x_{md} + x_{lfd}} \tag{7.148}$$

where

$$x_d = x_{ls} + x_{md} \tag{7.149}$$

Generally, the leakage flux reactance x_{ls} is approximately 10% of the d-axis reactance x_d. For our case, $x_{md} = 51.955\,\Omega$. The rotor self inductance L_{FF} is calculated as follows:

$$L_{FF} = \mu_0 r L k_1 \int_{0+\theta}^{360+\theta} n_F(\varphi)\, n_F(\varphi)\, g^{-1}(\varphi)\, \mathrm{d}\varphi \tag{7.150}$$

Considering the known values of the rotor radius r, machine length L, and the rotor number of turns n_F, and based on Equation (7.150), the L_{FF} is equal to 1.189 H. The rotor field winding leakage inductance is 0.119 H. The field winding reactance x_{lfd} is calculated as follows:

$$xl_{fd} = 2\pi f l_{fd} \tag{7.151}$$

The calculated $x_d{}'$ is 27.5 Ω.

7.8.1.7   Number of d-Axis Damper Winding Turns

The following procedure has been proposed in reference [11] for calculating the number of damper winding turns in the d-axis:

1. The damper winding leakage inductance in the d-axis is calculated.
2. The d-axis reactance is approximately $x_{kd} = 0.1 x_{lkd}$.
3. The d-axis damper winding inductance is obtained from its reactance.
4. Based on the following equations, all parameters on both sides of the equation are known except the number of d-axis damper windings.

$$L_{kd} = \mu_0 r L k_1 \int_0^{60} \omega_{kd}{}^2 \frac{1 - \cos 2p\left(\varphi + \dfrac{\pi}{6}\right)}{2}\, \mathrm{d}\varphi + \int_{90}^{150} \omega_{kd}{}^2 \frac{1 - \cos 2p\left(\varphi + \dfrac{\pi}{6}\right)}{2}\, \mathrm{d}\varphi$$
$$+ \int_{180}^{240} \omega_{kd}{}^2 \frac{1 - \cos 2p\left(\varphi + \dfrac{\pi}{6}\right)}{2}\, \mathrm{d}\varphi + \int_{270}^{330} \omega_{kd}{}^2 \frac{1 - \cos 2p\left(\varphi + \dfrac{\pi}{6}\right)}{2}\, \mathrm{d}\varphi \tag{7.152}$$

According to the following equation, the d-axis leakage reactance of the damper winding is 8.77 Ω and the d-axis self inductance of the damper winding is 0.28 H. Based on the proposed procedure, the number of d-axis damper winding turns is 217:

$$x_d{}'' = x_{ls} + \frac{x_{md} x_{lfd} x_{lkd}}{x_{md} x_{lfd} + x_{md} x_{lkd} + x_{lfd} x_{lkd}} \tag{7.153}$$

The DC current decay in the armature winding at standstill is used to calculate the number of turns of the damper winding in the q-axis (ω_{kq}). The q-axis and d-axis damper winding resistances are r_{kq} and r_{kd}, respectively [9]. In the DC decay test procedure, the two phases of the stator windings are fed with a DC power supply while the field winding is short-circuited and the machine is at a standstill. The two phases of the stator windings are short-circuited at a given time, which causes the decay of the stator DC. This induces a transient voltage in field winding 4 [12]. The d–q model of the machine consequently gives $\omega_{kd} = 30$, $r_{kd} = 11.91\,\Omega$, and $r_{kq} = 5.6\,\Omega$.

7.9   Healthy Case Simulation

7.9.1   Stator and Rotor Winding Function

A salient pole SG with the specifications shown in Table 7.1 is utilized to determine the accuracy of the method proposed in the previous sections. An understudy SPSG has 36 slots with double layer windings. Each winding layer contains 20 turn wires. Figure 7.7 depicts the winding layout

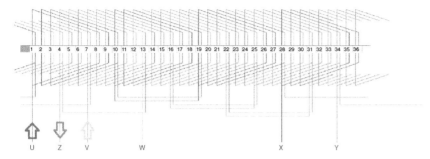

Figure 7.7 Winding layout of an SPSG with 36 slots, double layer.

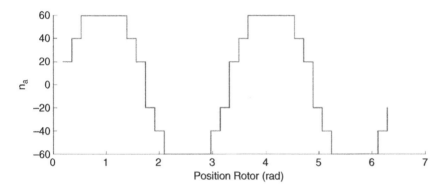

Figure 7.8 The turn function of stator phase A winding.

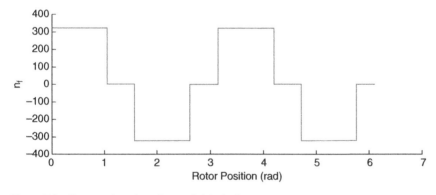

Figure 7.9 The turn function of rotor field winding.

of the SPSG as having a d-axis. Figure 7.8 demonstrates the winding function of the stator phase A winding. Based on the direction of the current in each slot, the net number of turns is 20 or 40 turns. The winding functions of the rotor field winding and damper windings are depicted in Figures 7.9 to 7.11.

7.9.2 Stator Phase Windings Mutual Inductances

Figure 7.12 demonstrates the mutual inductance between phase A and phase B of the SPSG and its corresponding inductance derivative. Figure 7.13 shows the self inductance and its derivative for phase A of the SPSG.

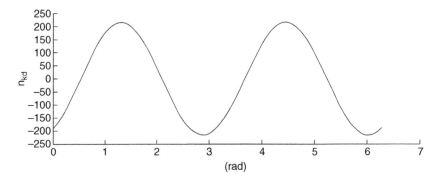

Figure 7.10 The turn function of damper winding in the *d*-axis.

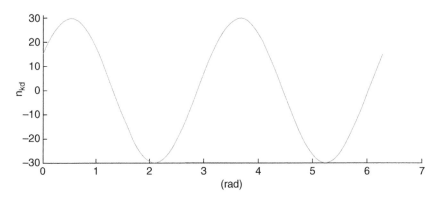

Figure 7.11 The turn function of damper winding in the *q*-axis.

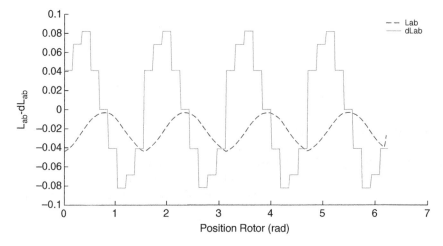

Figure 7.12 The mutual inductance between phase A and phase B and their derivative.

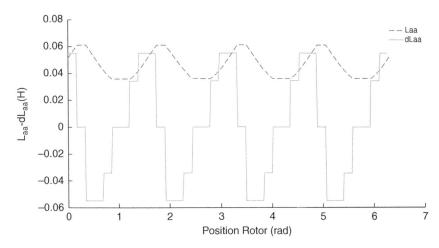

Figure 7.13 The self inductance of phase A and its derivative.

7.9.3 Mutual Inductance Between Stator and Rotor Windings

7.9.3.1 Mutual Inductance Between Stator Phase Windings and Rotor Field Windings

Figure 7.14 shows the mutual inductance between stator phase A and the rotor field winding and its derivative. Figure 7.15 presents the mutual inductance between the stator phase windings and the rotor field winding of the SPSG.

7.9.3.2 Mutual Inductance Between Stator Phase Windings and Damper Windings

Figures 7.16 and 7.17 show the mutual inductance between the stator phase winding (phase A) and the damper winding in the d- and q-axes, and its mutual inductance derivatives, respectively.

7.9.4 Dynamic Model Simulation in the Healthy Case

The dynamic simulation of the machine is achievable based on Equations (7.117) to (7.140). The simulation includes the following three stages:

1. The no-load machine rotates at a synchronous speed in which the stator terminal currents are zero and only the rotor current passes through the field winding. In a simulation, the stator

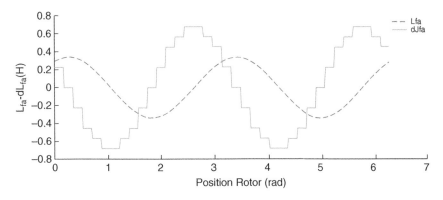

Figure 7.14 The mutual inductance between phase A and the rotor field winding and its derivative.

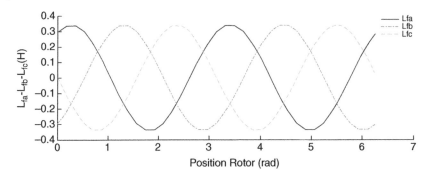

Figure 7.15 The mutual inductance between the three phases and the rotor field winding in the SPSG.

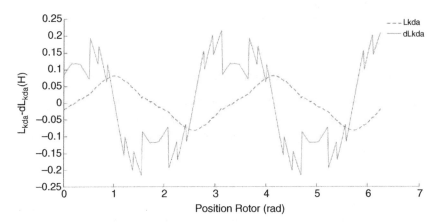

Figure 7.16 The mutual inductance between the stator phase winding (phase A) and the damper winding located in the *d*-axis and its derivative.

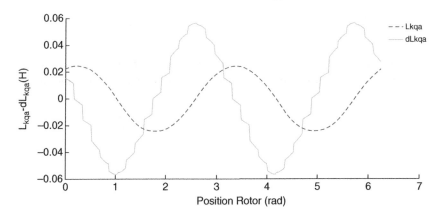

Figure 7.17 The mutual inductance between the stator phase winding (phase A) and the damper winding located in the *q*-axis and its derivative.

terminals are open and the input torque into the shaft is used to change the stator voltage to its nominal value. This stage occurs between 0 and 1 second.

2. In the second stage, the SPSG is loaded; therefore, the current that passes through the stator winding creases the magnetic field in the air gap. In turn, this results in a reduction in the stator phase voltage and synchronous speed. This stage occurs between 1 and 2 seconds.

3. In the third stage, the field winding voltage and input torque are increased to compensate for the speed and terminal voltage. This stage occurs from 2 seconds to the end of the simulation.

Figures 7.18 to 7.21 demonstrate the current in the rotor field winding, damper winding in the d-axis and q-axis, and the stator current of the SPSG. In Figure 7.22, the line-to-line voltage of phase A to B is depicted. The dynamic response of the SPSG speed to changes in the load condition is shown in Figure 7.23.

The damper winding has a crucial impact on converging the winding function modeling. By excluding the damper winding in the simulation, the modeled machine does not converge into the desired value and becomes unstable. Figures 7.23 to 7.25 demonstrate the changes in the speed and field winding current of the SPSG caused by removing the damper winding from the model.

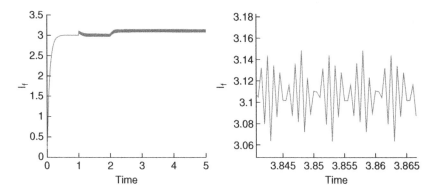

Figure 7.18 Rotor field winding current.

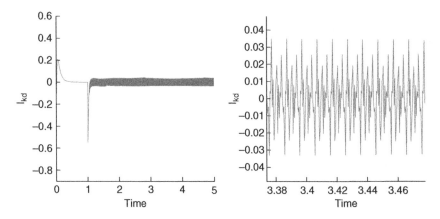

Figure 7.19 The damper winding current in the d-axis.

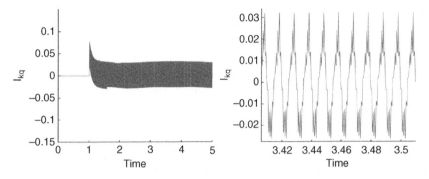

Figure 7.20 The damper winding current in the *q*-axis.

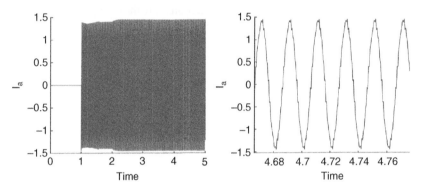

Figure 7.21 Current in phase A of the stator winding.

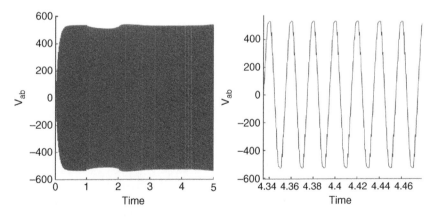

Figure 7.22 Induced line voltage in the stator terminal.

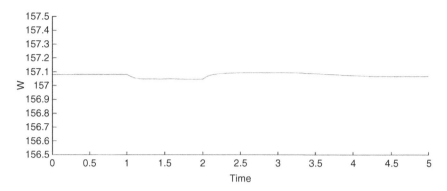

Figure 7.23 The dynamic response of the SPSG speed from no-load to full-load operation.

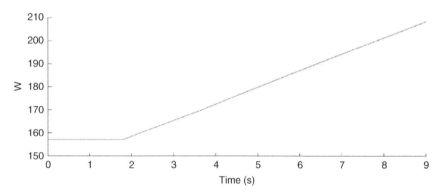

Figure 7.24 The dynamic response of speed by excluding the damper winding from modeling.

7.10 Faulty Case Simulation

7.10.1 Turn Functions and Inductances

Modeling of the SPSG under a stator inter-turn short circuit fault using WFM is similar to healthy case modeling. The worst-case scenario, which is an inter-turn short circuit fault in one of the stator windings, is evaluated with the assumption that nine turns of phase A are short-circuited. The short circuit fault can be considered as a separate phase, so-called D, and including it in the dynamic formulation of the machine. Note that a separate turn function must be considered for phase D, and the self inductance and mutual inductance between this phase and other stator phase windings, damper windings, and rotor field windings must be calculated.

The turn function of the faulty phase must be modified. Figure 7.26 shows the turn function of the faulty phase A. The reduction of nine turns in phase A results in a reduction of the turn function from slot 1 to 8. Figure 7.27 also depicts the turn function of phase D (faulty phase). The amplitude of phase D is equal to the number of shorted turns and is between slots 1 and 8. The superposition of Figures 7.26 and 7.27 leads to a turn function of the healthy phase.

Due to the fault, the SPSG is no longer electrically symmetric; therefore, the average turn function for the faulty phase is not zero. Hence, the average value of phase D must be subtracted from the faulty phase. The procedure for the turn function and inductance calculation for the two other phases is the same as for the healthy case. The self inductance of phase A is shown in Figure 7.28 in

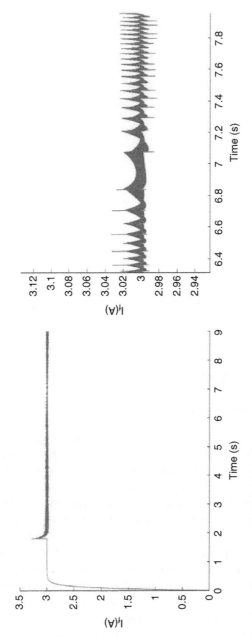

Figure 7.25 The dynamic response of the field winding current by excluding the damper winding from the modeling.

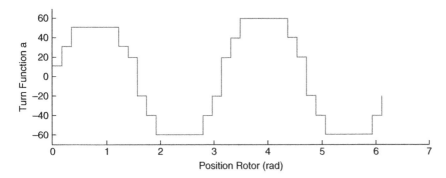

Figure 7.26 The turn function of faulty phase A (nine turns are reduced between slots 1 to 8).

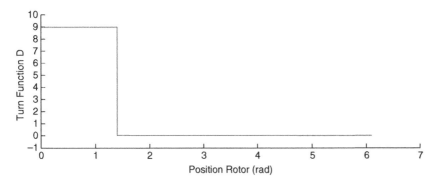

Figure 7.27 The turn function of the faulty phase (phase D).

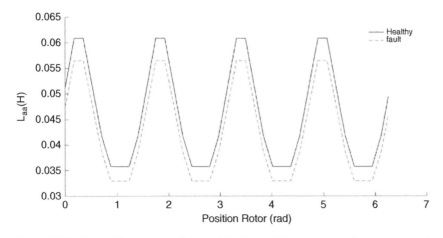

Figure 7.28 The self inductance of phase A in the healthy case and under a nine-turn short circuit fault.

comparison to the healthy case. Figure 7.29 presents the self inductance of phase D. The asymmetry of the turn function and, consequently, the inductance is increased by increasing the number of short circuited turns.

Figure 7.30 shows the mutual inductance between the windings of phase A and phase B in a healthy state and under a nine-turn short circuited fault. Figure 7.31 presents the mutual inductance between phase D and other stator phase windings.

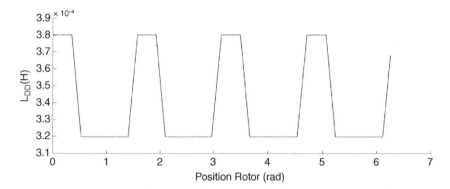

Figure 7.29 The self inductance of phase D.

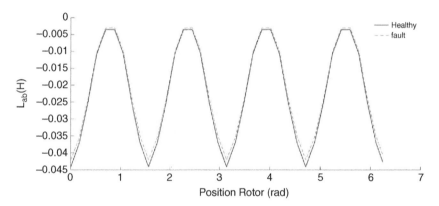

Figure 7.30 The mutual inductance between phase A and phase B in a healthy and a faulty case.

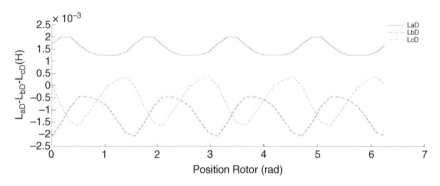

Figure 7.31 The mutual inductance between phase D and three other phases.

7.10.2 Dynamic Model Simulation in the Faulty Case

The dynamic simulation of SPSG under a stator inter-turn short circuit fault includes the following three stages:

1. The no-load machine rotates at the synchronous speed. For the no-load machine, the stator terminal currents are zero, and only the rotor current passes through the field winding. In the

simulation, the stator terminals are open, and the input torque into the shaft is used to increase the stator voltage to its nominal value. This stage occurs between 0 and 1 second.

2. In the second stage, the SPSG is loaded; therefore, the current that passes through the stator winding increases the magnetic field in the air gap, thereby resulting in a reduction in the stator phase voltage and synchronous speed. This stage occurs between 1 and 2 seconds.

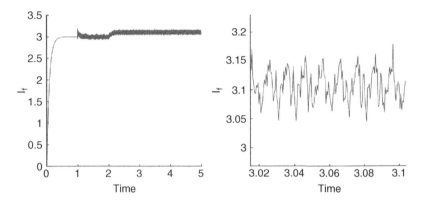

Figure 7.32 The rotor field winding current in a faulty SPSG under a short circuit fault.

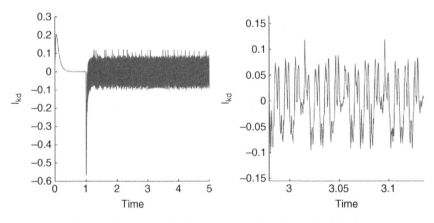

Figure 7.33 The *d*-axis damper winding current under a short circuit fault.

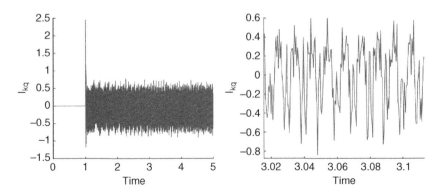

Figure 7.34 The *q*-axis damper winding current under a short circuit fault.

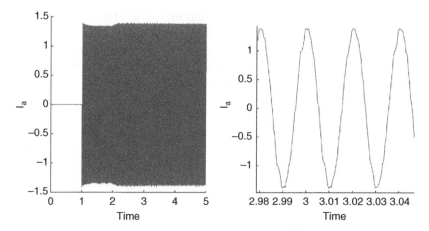

Figure 7.35 The stator winding current (phase A) under a short circuit fault.

3. In the third stage, the field winding voltage and input torque are increased to compensate for the speed and terminal voltage. This stage occurs from two seconds until the end of the simulation.

The rotor field winding current, damper winding current in the *d*-axis and *q*-axis, and the stator phase A current under the stator short circuit fault are shown in Figures 7.32 to 7.35. Figure 7.36 shows the line voltage between phase A and phase B under an inter-turn short circuit fault. Figure 7.37 and Figure 7.38 depict the electromagnetic torque and speed variations of the SPSG under the faulty situation.

7.11 Algorithm for Determination of the Magnetic Saturation Factor

The optimized operating point of the salient pole synchronous machine is at the knee point of the magnetization curve during a steady-state operation. Therefore, the electromechanical energy conversion in the synchronous machine is conducted under saturated conditions. The fault detection is performed in the steady-state operation of the synchronous machine; therefore, modeling of the healthy and faulty synchronous machine must be performed by including the saturation effect. The first step in the analytical method for saturation modeling is determination of the saturation factor. An analytical-numerical algorithm is proposed to calculate the saturation factor of the synchronous machine. The machine geometry and specifications and the magnetic properties of the lamination sheets are included in the process of a saturation factor calculation to increase the accuracy of the model.

7.11.1 Algorithm

An algorithm is proposed based on a MEC that uses the magnetization characteristic of the magnetic material to calculate the saturation factor in the synchronous machine. Figure 7.39 shows the proposed algorithm.

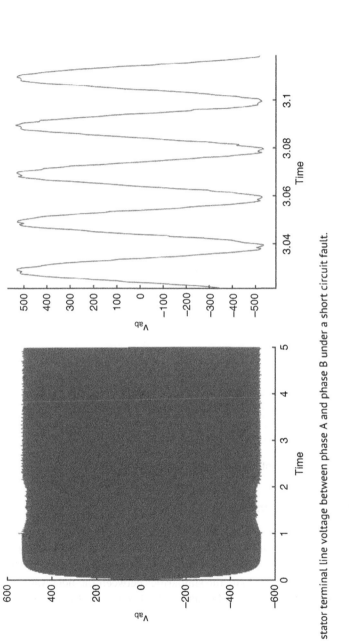

Figure 7.36 The stator terminal line voltage between phase A and phase B under a short circuit fault.

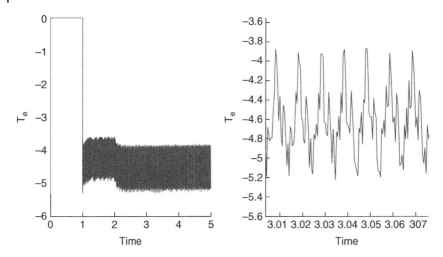

Figure 7.37 Electromagnetic torque in a faulty SPSG under a short circuit fault.

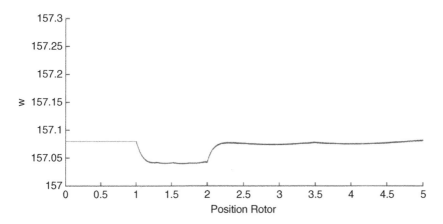

Figure 7.38 The speed variation of the SPSG under a short circuit fault.

7.11.2 Excitation Field Factors Plot

An excitation field factor is one of the required parameters in the process of calculating the magnetic saturation algorithm. These factors are required to solve the MEC for the synchronous machine. Excitation field factors are extracted based on the no-load air gap magnetic field. The main intention in obtaining these factors is to evaluate the similarities between the field waveform and the ideal sinusoidal waveform. The excitation field factors plot is obtained by first defining three factors related to the magnetic field distribution in a rotor pole shoe, as shown in Figure 7.40. The three factors are:

1. Field form factor.
2. Excitation flux form factor.
3. EMF form factor.

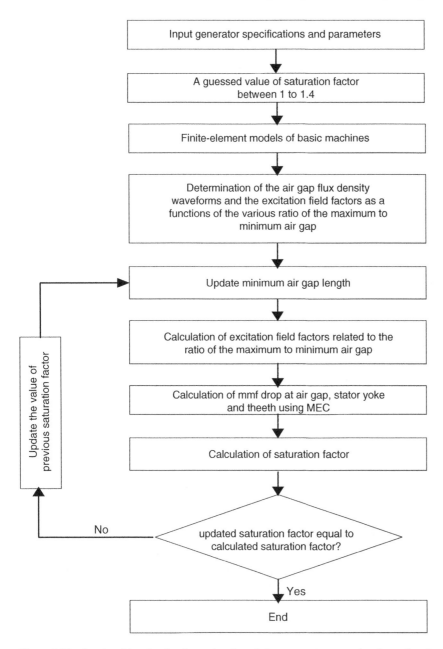

Figure 7.39 An algorithm for the determination of the magnetic saturation factor in a healthy no-load SPSG.

The field form factor is defined as:

$$K_f = \frac{B_{\delta 1,m}}{B_\delta} \tag{7.154}$$

where B_δ and $B_{\delta 1,m}$ are the maximum value of the main component of the flux density and the maximum value of the radial component of the magnetic flux density in the air gap, respectively.

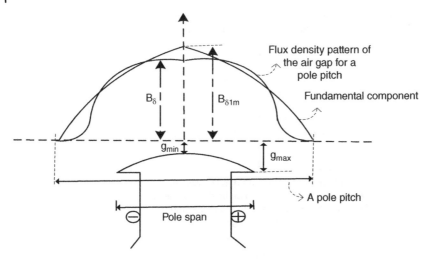

Figure 7.40 The magnetic field distribution in the air gap under one rotor pole shoe to define magnetic field factors (only one rotor field winding is excited).

The location of the maximum value of the radial component of the air gap flux density in the rotor pole is where the air gap length is the minimum. The excitation field factor is:

$$K_\Phi = \frac{\Phi_{fm}}{\Phi_{f1,m}} = \frac{\pi\alpha_\delta}{2K_f} = \frac{\pi B_{\delta,mean}}{2K_f B_\delta} \tag{7.155}$$

where α_δ and $B_{\delta,mean}$ are the pole span factor and the average value of the air gap flux density, respectively. The electromotive force (EMF) form factor is defined as:

$$K_B = \frac{B_{\delta,rms}}{B_{\delta,mean}} = \frac{\pi}{2\sqrt{2}K_\Phi} \tag{7.156}$$

where $B_{\delta,rms}$ is the effective value of the main harmonic component of the magnetic flux density in the air gap. In an ideal case, the field form factor, excitation flux form factor, pole span factor, and EMF form factor are equal to 1, 1, $2/\pi$, and 1.11, respectively.

The optimal ratio between the maximum and minimum air gap lengths in the salient pole SGs could improve the magnetic field distribution in the air gap. The magnetic saturation factor is modeled in the air gap of the machines, and it could vary due to the excitation and load variations. Therefore, the excitation field factor plot that could determine the saturation factor is required, as it could show the variation in the excitation field factors as a function of the variation of the ratio between the maximum air gap length to its minimum value. This could demonstrate the air gap length variation due to the saturation.

The excitation field distribution waveform in the air gap with a different ratio of the maximum air gap length to minimum air gap length must be available to obtain the excitation field factor plot. One accurate way to calculate these waveforms is to use the FEM. Consequently, it is not a purely analytical approach, since it requires numerical inputs. A basic SG is modeled in an FE environment to obtain the excitation field distribution waveform. Since FEM is time consuming, only four options of the maximum to minimum air gap ratio are considered to cover the saturation effect from 1 to 2.2.

The saturation effect in the magnetic material, the saliency of the tooth, and the slotting effect in the rotor and stator of the salient pole SG are neglected to ensure a magnetic field distribution that

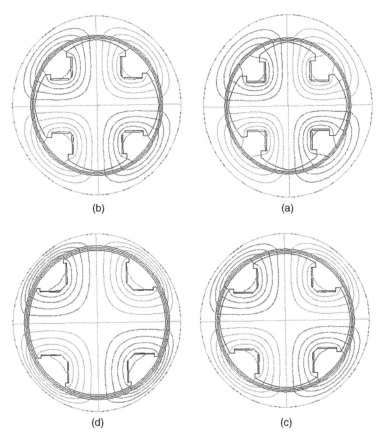

(b)   (a)

(d)   (c)

Figure 7.41 Magnetic flux distribution of a basic SPSG by changing the maximum air gap length to a minimum air gap length: (a) 2.2, (b) 1.8, (c) 1.4, and (d) 1.

wholly depends on rotor pole saliency in the no-load condition. Therefore, the magneto-motive force (MMF) drops in the iron parts are disregarded (where the relative permeability $\mu_r = \infty$), and the rotor and the stator are slotless in the FE modeling of the SG. Figure 7.41 shows the magnetic field distribution of four basic salient pole SGs with a different ratio of a maximum to minimum air gap. These different ratios lead to the various shapes of rotor pole shoes, which could significantly change the magnetic field distribution. The air gap length must be divided into three sections to increase the accuracy of magnetic field modeling. The modeling of one pole is adequate in the case of a healthy SG that has a uniform magnetic field in the air gap. Figure 7.42 shows the variation in the air gap magnetic flux density in one pole pitch under a different maximum to minimum air gap length ratio. The harmonic analysis of each waveform is extracted to obtain the harmonic contents of various ratios. According to Figure 7.42, the air gap magnetic field distribution and its harmonic content depend on the rotor pole saliency.

In a healthy SG, the maximum and minimum air gap length in each pole pitch is constant; however, in the case of an eccentricity fault, the maximum and minimum air gap length varies in at least two poles in the case of a four-pole machine. Therefore, three different values are obtained for the maximum to minimum air gap length ratio, which leads to different magnetic field distribution waveforms. Consequently, the harmonic analysis of each waveform must indicate different harmonic contents.

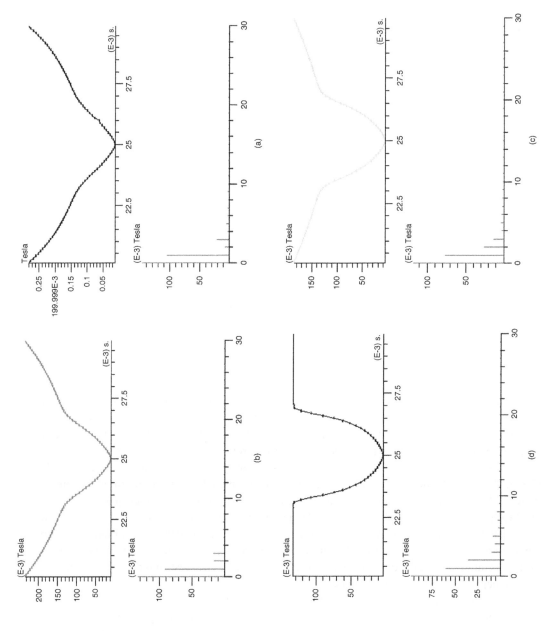

Figure 7.42 The magnetic flux density of one rotor pole for various ratios of the maximum to minimum air gap length (β): (a) 2.2, (b) 1.8, (c) 1.4, and (d) 1.

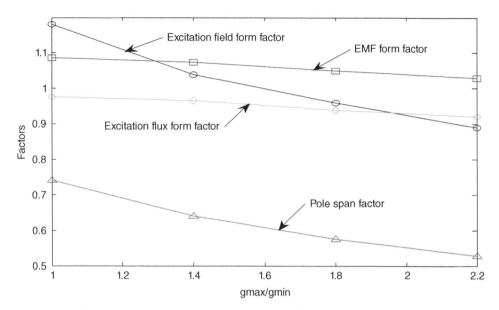

Figure 7.43 The air gap magnetic field graph factors in an SPSG.

According to the underlying assumption in the FE modeling of the salient pole SG, the super-position principle is applicable in the case of an eccentricity fault affecting the magnetic field distribution. Therefore, the harmonic contents of each magnetic waveform in the case of an eccentricity fault must be considered. Figure 7.43 shows the field graph factors for a salient pole synchronous generator.

7.11.3 Magnetic Equivalent Circuit Modeling Under the No-Load Condition

The MEC is one of the fundamental and frequently used approaches in modeling of electrical machines. This approach involves the calculation of the approximate ampere-turn drops of machine parts, including the stator or rotor yoke, stator or rotor slots, and air gap according to the flux path. Determining the flux path and its distribution in the machine geometry is the first step in MEC modeling and an idea can be obtained using FEM.

The total ampere-turns drop in the stator yoke F_{sy}, stator tooth F_m, air gap F_{gap}, rotor core F_{RC}, and the rotor yoke F_{Ry} are equal to the total ampere-turns generated by the rotor excitation field F_{field} in the no-load condition. The equation is as follows [13]:

$$F_{field} = F_{sy} + F_{th} + F_{gap} + F_{Rc} + F_{Ry} \qquad (7.157)$$

The total flux density ϕ_f generated by a rotor field winding to induce the no-load voltage E_f in the armature winding is [13]:

$$\phi_f = K_\phi \phi_{f1,m} = \frac{E_f}{2\lambda K_B f_1 N_{ph,1} K_{w1}} \qquad (7.158)$$

where f_1 is the frequency of the main component, N_{ph1} is the number of armature winding turns for the main component, and K_{w1} is the phase winding distribution coefficient. The magnetic flux

Table 7.2 Value of coefficient ζ for different values of flux density in the stator yoke.

B_{sy} (T)	1	1.1	1.2	1.3	1.4	1.5
ζ	0.57	0.54	0.50	0.46	0.40	0.33

density in the stator yoke for a constant value of flux ϕ_f is:

$$B_{sy} = \frac{\phi_f}{2h_{sy}L_{c1}K_f} \tag{7.159}$$

where h_{sy} is the length of the radial path that flux traverse in the stator yoke, L_{c1} is the axial length of the stator core, and k_f is the stator filling factor that for lamination sheets with the width of 0.5 mm is equal to 0.93 [13]. The magnetic field intensity in the stator yoke (H_{sy}) can be determined based on the value of B_{sy}, according to the stator B–H lamination sheet. The ampere-turn drops in the stator yoke can be derived as:

$$F_{sy} = H_{sy}L_{sy}\zeta \tag{7.160}$$

where L_{sy} is the length of the peripheral path of the flux that traverses the stator yoke and ζ is a coefficient that can be changed according to the magnetic field intensity variation in the peripheral path of the stator yoke and can be extracted from Table 7.2.

The air gap flux is assumed to enter the stator teeth without any changes; therefore [13]:

$$\phi_{th} = \phi_{gap} \tag{7.161}$$

The magnetic flux density B_{th} in the stator tooth is [13]:

$$B_{th} = B_f \frac{L_\delta t_{th}}{K_f L_{c1} b_{th}} \tag{7.162}$$

where L_δ, t_{th}, and b_{th} are the radial length of the air gap, the stator slot pitch, and the width of the stator tooth, respectively. The air gap magnetic flux ϕ_{gap} is [13]:

$$\phi_{gap} = t_{th}L_\delta B_f \tag{7.163}$$

The magnetic flux intensity in the stator tooth can be determined by the B–H characteristic based on the tooth magnetic flux density. Therefore, the MMF of the stator slot can be determined as:

$$F_{th} = H_{th}h_{th} \tag{7.164}$$

where h_{th} is the height of the stator tooth.

The magnetic field analysis of the salient pole SGs proves that the saliency of the rotor pole shoe and the magnetic saturation (especially in the stator yoke and the tooth) can significantly reduce the magnetic field in the minimum air gap area. Since the rotor pole saliency and the magnetic saturation have a considerable impact on the amplitude of the air gap magnetic field, their effects can be considered by increasing the minimum air gap length. However, the effects of saliency and the magnetic saturation in the corners of adjacent rotor poles and between the two poles are neglected. The maximum air gap length is assumed to be a constant value and the effects of saturation and saliency are assumed to lead to an increase in the minimum air gap length. The air gap increment

δ' due to these effects can be written as [13]:

$$\delta' = \delta K_\delta K_{za} \tag{7.165}$$

where K_δ and K_{za} are the saliency and the saturation coefficient. The saliency coefficient is defined as [13]:

$$K_\delta = \frac{t_{z1}}{t_{z1} - \gamma_1 \delta} \tag{7.166}$$

where

$$\gamma_1 = \frac{bs/\delta}{5 + bs/\delta} \tag{7.167}$$

where b_s and t_{z1} are the stator tooth pitch and the width of the stator slot, respectively. The ampere-turn drops in the air gap must be calculated to obtain the magnetic saturation factor k_{za}. The ampere-turn drop in the air gap is determined as [13]:

$$F_{gap} = \frac{1}{\mu_0} B_{gap} K_\delta \delta \tag{7.168}$$

where B_{gap} is the magnetic field density of the minimum air gap. It can be derived as [13]:

$$B_{gap} = \frac{\phi_{gap}}{\alpha_\delta \tau L_\delta} \tag{7.169}$$

The ampere-turn drops of the stator tooth, stator yoke, and air gap must be calculated to obtain the magnetic saturation factor, as follows [13]:

$$K_{za} = \frac{F_{th} + F_{gap} + F_{sy}}{F_{gap}} \tag{7.170}$$

The study shows that the range of the magnetic saturation factor, the salient pole SG, is between 1 and 1.4. In the analytical modeling, the initial value is determined for the magnetic saturation factor and its value must be updated after each iteration.

7.12 Eccentricity Fault Modeling Considering Magnetic Saturation Under Load Variations

A model is necessary to analyze more accurately and practically the eccentricity fault. Since the saturation effect is important in operation of SG, taking into account this effect in the modeling can help to obtain more efficient and real results. Besides, one of the more common methods in eccentricity fault analysis is the use of different harmonic spectra of the faulty machine, as considering the saturation effect in the faulty SG helps to recognize and analyze the possible harmonics generated by interaction between these phenomena with existing faults in the machine.

In this chapter an attempt is made to include the magnetic saturation effect in the faulty salient-pole SG. In this method variations of the air gap length instead of the variations of the permeability of the iron parts are considered. This can simply and precisely take into account the saturation effect in the air gap distribution function of the generator. This function depends on the size and position of the saturating magnetic flux of the machine. Also, variations of the minimum air gap length are estimated for different eccentricity severities and are then added to the resultant air gap length while considering the saturation effect.

7.12.1 Calculation of Inverse Air Gap Length by Considering the Saturation Effect

The air gap distribution function of the SG could be altered in a way that considers the magnetic field saturation effect. Figure 7.44 shows the magnetic saturation modeling in a cross-section of the salient pole SG. The air gap distribution function should be able to indicate the variation in the minimum and maximum air gaps and the gaps between the rotor poles. The air gap distribution function of the salient pole SG can be defined based on a binomial function as [14, 15]:

$$g(\varphi, \theta_r) = \frac{1}{\alpha_1 + \alpha_2 \cos 2p(\varphi - \theta_r)} \tag{7.171}$$

where α_1 and α_2 are the geometrical parameters of the salient pole SG, and p, φ, and θ_r are the number of pole pairs, the angle along the inner surface of the stator, and the rotor position, respectively. Equation (7.171) does not have sufficient accuracy in modeling the air gap distribution function, especially under the rotor pole shoe that has a minimum and maximum air gap length. According to Equation (7.171), the inverse air gap function is assumed to be an ideal sinusoidal waveform; however, this is not the case in reality. Therefore, to have an accurate inverse air gap function

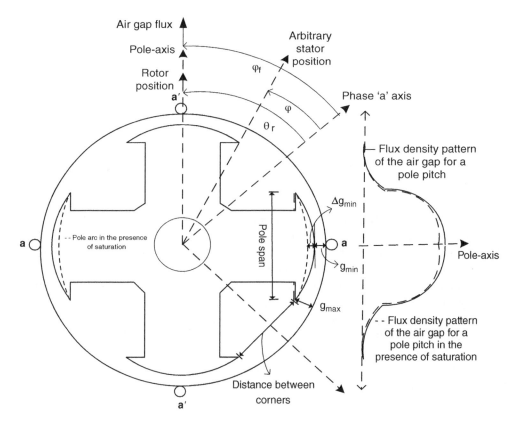

Figure 7.44 The cross section of an SPSG showing modeling of the saturation effect in the rotor pole.

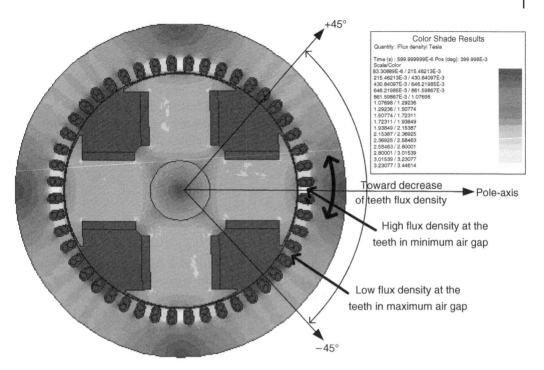

Figure 7.45 The magnetic flux density distribution of an SPSG in a no-load operation.

distribution, the following equation is used:

$$g(\varphi, \theta_r) = \frac{1}{\alpha_1 + \alpha_2 \cos 2p(\varphi - \theta_r) + \alpha_3 \cos 4p(\varphi - \theta_r)} \qquad (7.172)$$

The shape of the air gap magnetic field distribution in the salient pole SG at the no-load case that is created by the rotor field winding current is a function of rotor pole shoe shapes. The maximum amplitude of the flux density is at the minimum air gap length. The air gap reluctance is increased by increasing the air gap length, which leads to a reduction in the air gap flux density. The flux density between the rotor poles is negligible, since the reluctance between the poles is too large.

The first part of an SG that is vulnerable to magnetic saturation is the stator slot, since its cross-section is small and it is located close to the air gap magnetic field. However, the stator yoke is also susceptible to magnetic saturation in the case of deep saturation. Figure 7.45 shows the magnetic flux density distribution of the salient pole SG in a no-load case. The highest amplitude of the magnetic saturation is in the stator tooth in the case of a minimum air gap length.

The magnetic field of the on-load SG is created by the current that passes through the rotor field winding (I_f) and stator windings (I). In turn, the magnetic field phasor is the result of the joint action of the excitation MMF phasor (F_{fm}) and the armature MMF phasor (F_{am}). The amplitude and direction of the magnetic field phasor depend on the SG load, whereas the MMF phasors depend on the loading condition. Figure 7.46 shows a phasor analysis of the magnetic field and the magneto-motive force in one of the on-load salient pole SGs for an arbitrary operation mode. The phase angle (β) is the ratio of the reactive impedance to the resistance as follows:

$$\beta = \tan^{-1} \frac{X_1 + X_L}{R_1 + R_L} \qquad (7.173)$$

where X_1, X_L, R_1 and R_L are the reactance and resistance of the armature winding and load, respectively. According to Equation (7.173), the phase angle could vary between 0 and 2π, based on the load variation.

The armature MMF (F_{am}) can be resolved into a direct (F_{dm}) and quadrature (F_{qm}) axis for an arbitrary value of the phase angle, as shown in Figure 7.46. The comparison between the resultant MMF phasor (F_T) in the on-load condition versus the no-load case (F_{fm}) reveals that load can lead to MMF reduction. Therefore, the pattern of the air gap flux density at full load due to F_T must be decreased and shifted in comparison to the no-load case. The shifted angle ζ can be calculated as:

$$\xi = \tan^{-1}\frac{F_{qm}}{F_{dT}} = \frac{F_{am} \times \cos\beta}{F_{fm} - F_{am} \times \sin\beta} \tag{7.174}$$

The resultant MMF can be determined as follows:

$$F_T = \sqrt{(F_{fm} - F_{am} \times \sin\beta)^2 + (F_{am} \times \cos\beta)^2} \tag{7.175}$$

The flux density of the one-pole pitch due to the rotor excitation field winding and armature winding are shown as full and dashed lines, respectively, in Figure 7.46. The net MMF, shift angle, and phase angle in the on-load SG can be calculated according to Equations (7.174) and (7.175). The FE modeling of the on-load SG also confirms that the load shifts the net MMF phasor, since magnetic saturation occurs in the no-load case in the slots that face the minimum air gap

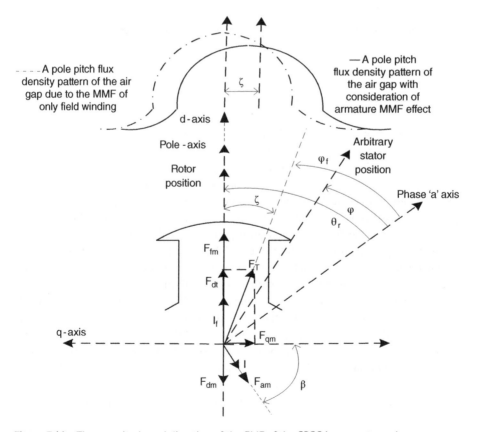

Figure 7.46 The magnitude and direction of the EMF of the SPSG in one rotor pole.

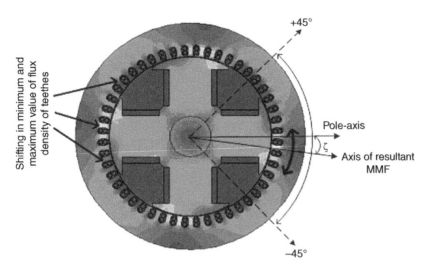

Figure 7.47 The magnetic field distribution of an on-load SPSG.

length (Figures 7.45 and 7.47). However, no magnetic saturation occurs in the same slots with the minimum air gap in the on-load condition, and magnetic flux is shifted based on the load angle.

The magnetic flux density in the stator tooth in the no-load case is only the function of the air gap reluctance; however, in the on-load case, it is also a function of the SG loading condition. Consequently, the magnetic saturation factor for the no-load case is no longer valid for the on-load condition.

Therefore, due to the net MMF, the on-load induced voltage in the stator terminals must be replaced by the no-load voltage in the algorithm of the magnetic saturation factor. The flux paths in the no-load and on-load conditions in the salient pole SG are the same; therefore, the MEC used for the no-load case is also valid for the on-load case. The induced voltage in the stator terminals is:

$$E_s = E_f - R_1 I_s - jX_1 I_s \tag{7.176}$$

The synchronous reactance in the linear and non-linear parts of the saturation characteristics varies, since the saturation effect can reduce the synchronous reactance of the SG. The synchronous reactance can be recalculated based on the short circuit and the no-load characteristics of the SG, as follows:

$$X_1 = \sqrt{\frac{E_f^2}{I_s^2} - R_s^2} \tag{7.177}$$

where E_f and I_s can be derived based on the no-load condition and the short circuit characteristic of the SG at different excitation currents.

The saturation effect can be replaced by a proportional increase in the air gap length; therefore, in the case of saturation in the SG, the rotor pole shoe, as shown in Figure 7.46 with a full line, could be replaced by the dashed line by taking into account the increased air gap. The diversity of the air gap due to the saturation effects relies on the direction of the flux in the magnetic materials. It is therefore predicted that the fictitious air gap length fluctuates a complete cycle every half cycle of the air gap flux density. Therefore, the air gap length is assumed to be a function of the position and the level of the flux density in the air gap. By taking into account the saturation, the air gap distribution function is:

$$g_{sat}(\varphi, \theta_r, \varphi_f, K_{za}) = g(\varphi, \theta_r) + g_{\min} (K_{za} - 1) \cos 2p(\varphi - \varphi_f) \tag{7.178}$$

where φ_f is the position of the air gap flux with respect to the reference phase ζ. According to Figure 7.46, φ_f can be written as:

$$\varphi_f = \theta_r - \xi \tag{7.179}$$

According to Equation (7.174), the load angle ζ is a function of load and the electrical parameters of the SG; consequently, φ_f depends on the rotor angular position. By this assumption, the air gap distribution function is:

$$g_{sat}(\varphi, \theta_r, \xi, K_{za}) = g(\varphi, \theta_r) + g_{\min} (K_{za} - 1) \cos 2p(\varphi - \theta_r - \xi) \tag{7.180}$$

Equation (7.181) shows that the air gap distribution function is wholly based on the angular position and the magnitude of the air gap flux density from the no-load to the full-load conditions in the SG.

The inverse air gap function is required to calculate the air reluctance and the inductances of the windings. Equation (7.180) is a non-linear and periodic function and can be estimated using the Fourier series. The Fourier representation of $g^{-1}{}_{sat}(\varphi, \theta_r, \xi, K_{za})$ on the $1/g_{sat}(\varphi, \theta_r, \xi, K_{za})$ distribution could be achieved by taking into account even multiple harmonics of the generator's pole numbers. Figure 7.48 shows the air gap and inverse air gap function of the saturated and unsaturated conditions in the SG for a saturation factor of 1.2. It shows that the proposed function is able to predict the variation in the air gap length, especially under the rotor pole shoes. In addition, it confirms the excellent agreement between the approximated $g^{-1}{}_{sat} (\varphi, \theta_r, \xi, K_{za})$ and $1/g_{sat}(\varphi, \theta_r,$

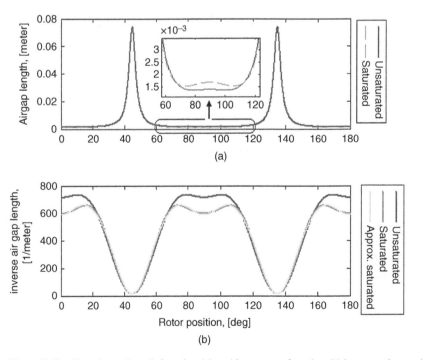

Figure 7.48 The air gap length function (a) and its reverse function (b) in saturation and unsaturated conditions.

ξ, K_{za}). For the Fourier series, only the second, fourth, sixth, and eighth harmonics of the function $g^{-1}{}_{sat}(\varphi, \theta_r, \xi, K_{za})$ are considered.

7.12.2   The Air Gap Length Calculation in the Presence of the Eccentricity Fault

Modeling the eccentricity fault using the winding function approach in SG requires that its effects on the air gap distribution function be considered. The modeling procedure of the dynamic eccentricity fault and its effect on the air gap distribution function is studied, and its result can be generalized for static eccentricity. According to Figure 7.49, the dynamic eccentricity degree (DED) is defined as:

$$\text{DED} = \frac{|\overrightarrow{O_S O_R}|}{g_{\min}} \tag{7.181}$$

where O_S and O_R are the center of the stator and the rotor, respectively, and $O_S O_R$ is the dynamic eccentricity vector that rotates around the stator centerline with a velocity equal to the rotor angular velocity. The top position of the rotor pole shoe in the case of dynamic eccentricity inside the air gap is shown by point P in Figure 7.49. The projection of the dynamic eccentricity fault vector on $O_S P$ is:

$$O_S O_R = \text{DED} \times g_{\min} \ \cos p(\varphi - \alpha) \tag{7.182}$$

where α is the angular position of the dynamic eccentricity vector with respect to the stator α reference phase, which is equal to $\theta_R - \theta$, where θ is the initial angle of the dynamic eccentricity vector. The value of θ is constant, since the rotational velocity of the vectors of $O_S P$ and $O_S O_R$ are equal. Therefore, α follows the variation in the rotor position angle. The magnitude of the vector $O_R P$ can be determined as follows:

$$O_R P = \sqrt{R_r{}^2 - \text{DED}^2 \times g_{\min}{}^2 \sin^2(\varphi - \alpha)} \tag{7.183}$$

Since the value of the second term under the square root in the equation ($\text{DED}^2 \times g_{\min}{}^2 \sin^2(\varphi - \alpha)$) is much smaller than the radius of the rotor (R_r), the magnitude of the vector $O_R P$ is considered

Figure 7.49   The location of the center stator, rotor, and rotating center in a case of a dynamic eccentricity fault.

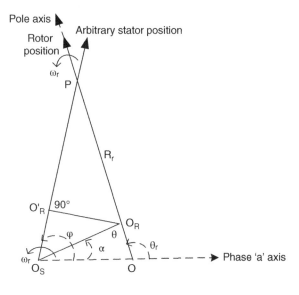

equal to R_r. Therefore, the distance of point P from the stator center is as follows:

$$O_S P = \text{DED} \times g_{\min} \cos(\varphi - \alpha) + R_r \tag{7.184}$$

Therefore, the length of the minimum air gap in the case of dynamic eccentricity is:

$$g_{min,DE} = R_s - \text{DED} \times g_{\min} \cos(\varphi - \alpha) + R_r \tag{7.185}$$

where R_s is the inner radius of the stator. Equation (7.186) can be rewritten as:

$$g_{min,DE} = g_{\min} - \Delta g_{\min,DE} \tag{7.186}$$

$$\Delta g_{min,DE} = \text{DED} \times g_{\min} \cos(\varphi - \alpha) \tag{7.187}$$

Equation (7.186) shows that the minimum air gap length and, therefore, the air gap distribution function in the presence of the dynamic eccentricity fault varies by Δg_{min}. This variation depends on the fault severity and the angular position of the rotor. Since θ_r depends on the rotor mechanical angular velocity, Δg_{min} is also time-dependent. The air gap distribution function in the salient pole SG determined by taking into account the saturation in the presence of dynamic eccentricity is as follows:

$$g_{sat}(\varphi, \theta_r, \xi, K_{za}, \text{DED}) = g_{sat}(\varphi, \theta_r, \xi, K_{za}) - \Delta g_{min,DE} \tag{7.188}$$

The variation in the inverse air gap distribution function for different severities of dynamic eccentricity and saturation factor is shown in Figure 7.50 for the period of $[0 - 2\pi]$. To increase the accuracy of the pattern, the first 10 terms of the Fourier transform are used to approximate the inverse air gap distribution function.

In the case of a static eccentricity fault, $\alpha = 0$, since the center of rotation and the rotor center are the same. Figures 7.51 and 7.52 show the air gap distribution function and its inverse function in the presence of the static eccentricity fault.

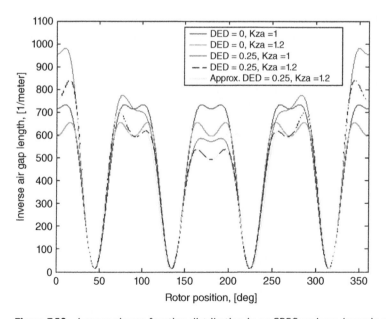

Figure 7.50 Inverse air gap function distribution in an SPSG under a dynamic eccentricity fault.

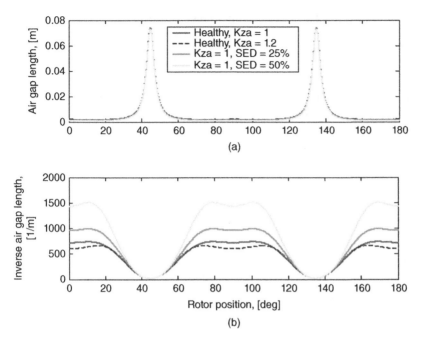

Figure 7.51 The air gap length distribution (a) and the inverse air gap length distribution (b) under a static eccentricity fault without a saturation effect.

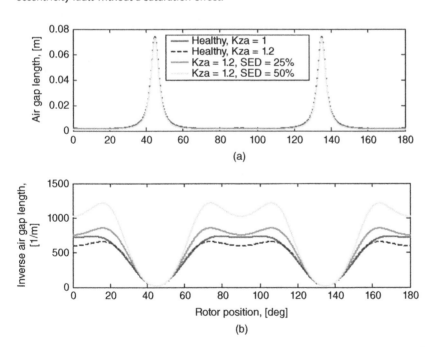

Figure 7.52 The air gap length distribution (a) and the inverse air gap length distribution (b) under a static eccentricity fault with a saturation effect.

7.12.3 Mutual and Self Inductance Calculations under an Eccentricity Fault

The prerequisite step to obtain the inductances of the machine in WFM is to calculate the turn functions. Figure 7.53 shows the stator winding layout. Figure 7.54 depicts the turn function of the stator phase winding of the salient pole machine with four poles. According to Figure 7.53, the turn function is calculated based on the machine's winding layout. The second and the sixth order harmonics, which are odd multiples of the machine pole pair number, are considered in turn function modeling. The turn functions of all three phases have an AC component, but the phase shifts between phases B and C with respect to phase A are 60 and 120 mechanical degrees since the machine has four poles. Figure 7.55 demonstrates the Fourier series approximation of the rotor turn function. In this turn function, the second and the sixth order harmonics are also used. The rotor turn function is not a stationary function and varies with respect to the rotor angular position, since the stator turn function is stationary.

The procedures of the mutual and self inductance calculation under an eccentricity fault are based on the equations in Sections 7.5 and 7.6. The only difference is the air gap distribution function, which should take into account the eccentricity term based on Equation (7.188). The mutual inductance of the stator phase winding is calculated as:

$$L_{AB}(\theta_r, \xi, K_{za}, \text{DED}) = \mu_0 rl \int_0^{2\pi} n_A(\varphi, \theta_r) N_B(\varphi, \theta_r, \xi, K_{za}, \text{DED}) g_{sat,DE}^{-1}(\varphi, \theta r, \xi, K_{za}, \text{DED}) \, d\varphi$$

(7.189)

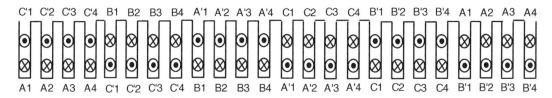

Figure 7.53 Stator phase winding distribution layout (half of the total slots are demonstrated).

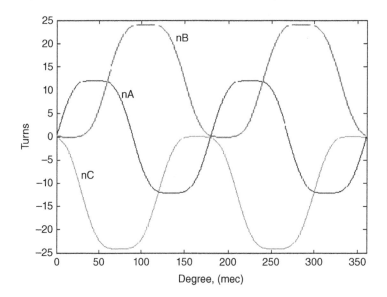

Figure 7.54 The Fourier series approximation of the turn function of the stator phase windings.

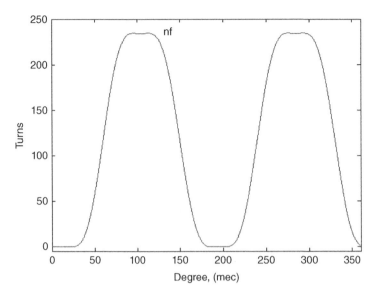

Figure 7.55 The Fourier series approximation of the rotor field winding.

where A and B represent the stator phase windings of the machine, μ_0 is the free space permeability, l is the stack length of the iron, and r is the average radius of the air gap. In addition, N_B and n_A are the modified winding function and the turn function of phases A and B, respectively. The MWF based on the proposed turn function in a case of symmetrical and asymmetrical air gap length is as follows:

$$N(\varphi, \theta_r, \xi, K_{za}, \text{DED}) = n_A(\varphi, \theta_r) - \langle H(\theta_r, \xi, K_{za}, \text{DED}) \rangle \tag{7.190}$$

where

$$H(\theta_r, \xi, K_{za}, \text{DED}) = \frac{1}{2\pi \left\langle g_{sat}^{-1}(\varphi, \theta_r, \xi, K_{za}, \text{DED}) \right\rangle}$$

$$\times \int_0^{2\pi} n_A(\varphi, \theta_r) g_{sat,\text{DE}}^{-1}(\varphi, \theta_r, \xi, K_{za}, \text{DED}) \, d\varphi \tag{7.191}$$

where $\left\langle g_{sat,\text{DE}}^{-1}(\varphi, \theta r, \xi, K_{za}, \text{DED}) \right\rangle$ represents the inverse average air gap distribution function, which can be calculated as follows:

$$\left\langle g_{sat}^{-1}(\varphi, \theta_r, \xi, K_{za}, \text{DED}) \right\rangle = \frac{1}{2\pi} \int_0^{2\pi} g_{sat,\text{DE}}^{-1}(\varphi, \theta_r, \xi, K_{za}, \text{DED}) \, d\varphi \tag{7.192}$$

The self inductance of the stator winding is calculated as:

$$L_{AA} = \mu_0 r L \int_0^{2\pi} n_A(\phi, \theta_r) N_A(\phi, \theta_r, \xi, K_{za}, \text{DED}) g_{sat,\text{DE}}^{-1}(\phi, \theta_r, \xi, K_{za}, \text{DED}) \, d\phi \tag{7.193}$$

Figures 7.56 to 7.58 show the variation in the phase A inductance for different eccentricity severity and saturation factors. The findings prove:

1. The saturation effect can reduce the average value of self inductance in healthy and faulty cases. The reluctance of the iron part of the machine due to saturation increases and leads to an increase in the net value of the reluctance in the equivalent circuit, which means a reduction in the inductance.

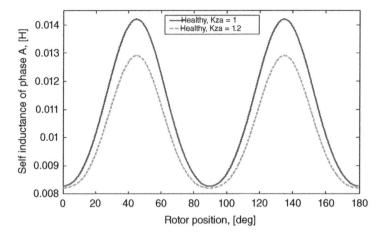

Figure 7.56 Self inductance of the stator phase winding with and without a saturation effect in a healthy operation.

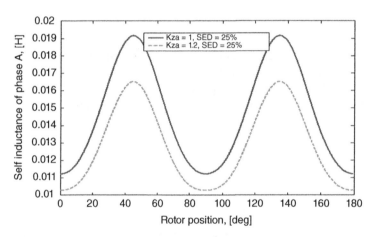

Figure 7.57 Self inductance of the stator phase winding with and without a saturation effect under a 25% static eccentricity fault.

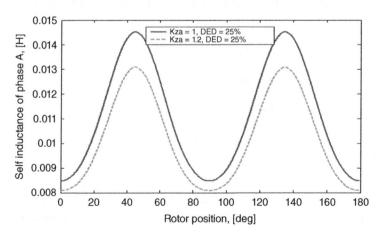

Figure 7.58 The self inductance of the stator phase winding with and without a saturation effect under a 25% dynamic eccentricity fault.

2. Although the eccentricity fault can increase the minimum, maximum, and average values of the inductance, the increment in the minimum and the maximum values in the inductance waveform is not the same, since the phase windings are distributed inside the machine slots. In the case of an eccentric air gap, the value of the reluctance in front of each part of the phase winding varies, leading to a variation in the peak point of the inductance waveform.

The mutual inductance between the stator phase *A* winding and rotor field winding is shown in Figures 7.59 to 7.61 with and without saturation effect modeling. The eccentricity fault results in a symmetric increment in the maximum and minimum magnitudes of the inductances. The variation is increased by increasing the severity of the fault. The average mutual inductances under the faulty condition remains zero. The self inductance under the eccentricity fault is increased by increasing the severity of the fault. The amplitude of the self inductance is decreased, as expected, by including the saturation effect.

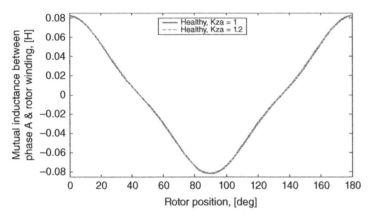

Figure 7.59 The mutual inductance of the stator winding between phase A and the rotor field winding in a healthy operation with and without a saturation effect.

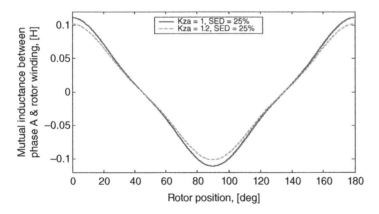

Figure 7.60 The mutual inductance of the stator winding between phase A and the rotor field winding under a 25% static eccentricity fault with and without a saturation effect.

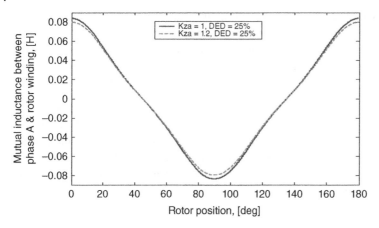

Figure 7.61 The mutual inductance of the stator winding between phase A and the rotor field winding under a 25% dynamic eccentricity fault with and without a saturation effect.

7.13 Dynamic Modeling under an Eccentricity Fault

The air gap function in the no-load and on-load conditions must be modified to consider the saturation factor in WF modeling. Since the magnitude and angle of the electromotive force on-load condition is different from the no-load case, the saturation factor must be changed. The saturation factor is constant because it must be calculated for a specific operating condition.

An algorithm is suggested for modeling the dynamic behavior of the machine by including the saturation effect operating under an eccentricity fault. According to Figure 7.62, the first step in the proposed algorithm is the determination of the loading condition and the no-load and short circuit curves. An initial value is assumed for the armature and field winding currents, and then the synchronous reactance of the machine is calculated based on an assumed value of the current. The resultant MMF and its corresponding phase shift angle are determined. The obtained terminal voltage is used as an input to determine the saturation factor. The air gap and inverse air gap functions are calculated and their values, based on the eccentricity fault severity, are updated and their Fourier series are extracted. The self inductance and mutual inductance of the machine are determined based on the Fourier series in a previous step. The inductance matrix is used to achieve dynamic modeling of the machine based on the equations provided in Section 7.10. The output of this section is fed into the algorithm again to update the current value and the saturation factor. Hence, the saturation factor value is updated based on variations in the machine operating condition.

The variation in the electromagnetic torque, excitation current, and angular velocity of the SPSG in a healthy state and under an eccentricity fault and either excluding or including the saturation effect are shown in Figures 7.63 and 7.64, respectively. The variation in the transient signal increases under the faulty condition compared to the healthy case. However, the variation of the signal in steady-state operation under the faulty condition is not noticeable compared to the healthy case. In Figure 7.64, where the saturation factor is 1.2, the variation in the signal under the eccentricity fault is more than that observed for a case where the saturation factor is considered to equal 1 (Figure 7.63). The change of the three-phase current during the transient condition, where the saturation effect is neglected under the eccentricity fault, is significant. However, the variation in the stator current under the eccentricity fault and including the saturation effect is greater than

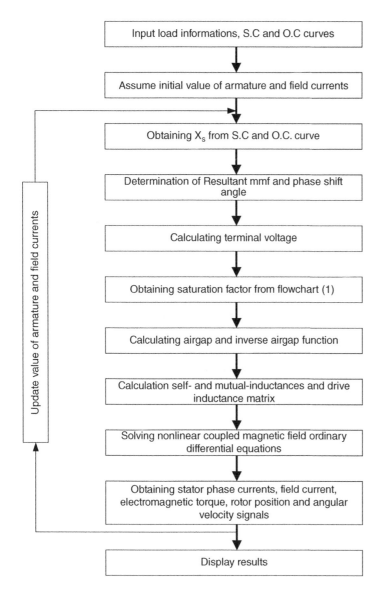

Figure 7.62 An algorithm for modeling an SPSG including the saturation effect and operation under an eccentricity fault.

in the case where the saturation was neglected. The reason is that the harmonics due to saturation contribute to and create a large variation in the signal.

7.14 Summary

In this chapter, a detailed description and formulation of the synchronous generator using the winding function approach were explained. The high accuracy and low computational complexity of the winding function approach make it a popular method for modeling electrical machines, especially for fault detection purposes. In this chapter the procedure of modeling both short circuit

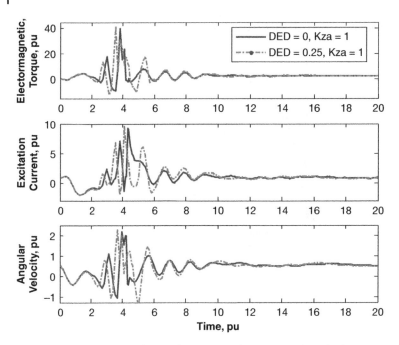

Figure 7.63 The variation in the electromagnetic torque (top), excitation current (middle), and the angular velocity of an SPSG in a healthy state and under a 25% eccentricity fault without a saturation effect.

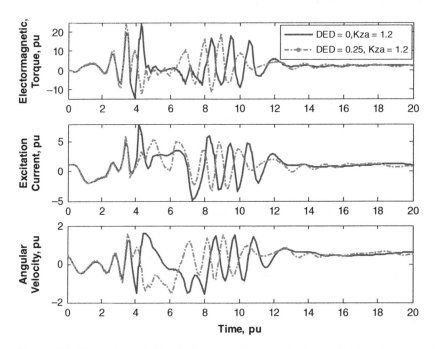

Figure 7.64 The variation in the electromagnetic torque (top), rotor field winding current (middle), and the angular velocity of an SPSG in a healthy state and under a 25% eccentricity fault with a saturation effect.

and eccentricity faults was described. Since saturation has a significant impact on the generator performance and the signals, an algorithm was described that has high accuracy when compared with FEM results.

References

1 Babaei, M., Faiz, J., Ebrahimi, H. et al. (2011). A detailed analytical model of a salient-pole synchronous generator under dynamic eccentricity fault. *IEEE Transactions on Magnetics* 47 (4): 764–771.

2 Obe, E.S. (2009). Direct computation of ac machine inductances based on winding function theory. *Energy Conversion and Management* 50 (3): 539–542.

3 Toliyat, H.A. (2002). Condition monitoring and fault diagnosis of electric machinery. Advanced Electric Machines & Power Electronics Laboratory Department of Electrical Engineering, Texas A & M University (USA).

4 Al-Nuaim, N.A. and Toliyat, H. (1998). A novel method for modeling dynamic air-gap eccentricity in synchronous machines based on modified winding function theory. *IEEE Transactions on Energy Conversion* 13 (2): 156–162.

5 Tu, X., Dessaint, L.A., El Kahel, M., and Barry, A. (2006). Modeling and experimental validation of internal faults in salient pole synchronous machines including space harmonics. *Mathematics and Computers in Simulation* 71 (4–6): 425–439.

6 Faiz, J. and Tabatabaei, I. (2002). Extension of winding function theory for nonuniform air gap in electric machinery. *IEEE Transactions on Magnetics* 38 (6): 3654–3657.

7 Ojaghi, M. and Bahari (2017). V.,Rotor damping effects in dynamic modeling of three-phase synchronous machines under the stator interturn faults – winding function approach. *IEEE Transactions on Industry Applications* 53 (3): 3020–3028.

8 Tu, X., Dessaint, L.A., El Kahel, M., and Barry, A.O. (2006). A new model of synchronous machine internal faults based on winding distribution. *IEEE Transactions on Industrial Electronics* 53 (6): 1818–1828.

9 IEC 60034, Rotating Electrical Machines – Part 4-1: Methods for determining electrically excited synchronous machine quantities from tests. IEC Std, 2018.

10 IEEE Guide for Test Procedures for Synchronous Machines (2010). Part I – Acceptance and performance testing, Part II – Test procedures and parameter determination for dynamic analysis. In: *IEEE Std 115-2009 (Revision of IEEE Std 115-1995)*, 1–219.

11 Krause, P.C., Wasynczuk, O., Sudhoff, S.D., and Pekarek, S.D. (2002). *Analysis of Electric Machinery and Drive Systems*, vol. 2. Wiley Online Library.

12 Cisneros-Gonzalez, M., Hernandez, C., Escarela-Perez, R., and Arjona, M.A. (2011). Determination of equivalent-circuit parameters of a synchronous generator based on the standstill DC decay test and a hybrid optimization method. *Electric Power Components & Systems* 39 (7): 645–659.

13 Smolensky, A.I. (1983). *Electrical Machines*. Moscow: Mir.

14 Tu, X., Dessaint, L.A., Fallati, N., and De Kelper, B. (2007). Modeling and real-time simulation of internal faults in synchronous generators with parallel-connected windings. *IEEE Transactions on Industrial Electronics* 54 (3): 1400–1409.

15 Lin, X., Tian, Q., Gao, Y., and Liu, P. (2007). Studies on the internal fault simulations of a high-voltage cable-wound generator. *IEEE Transactions on Energy Conversion* 22 (2): 240–249.

8

Finite Element Modeling of a Synchronous Generator

8.1 Introduction

As far as engineers are concerned, no essential difference exists between analytical and numerical approaches [1]. Numerical methods are generally simple, but they may, of course, be time-consuming. However, the use of a computer allows the development of accurate numerical methods for magnetic field computation.

The modeling of electrical machines typically employs an analytical method based on the winding function or its magnetic equivalent circuit (MEC). Other numerical techniques, such as finite difference, boundary element, and finite element (FEM) methods, have also been adopted. Of these methods, FEM has the highest accuracy and is used in problems dealing with complicated structures, different material properties, and nonlinearity. Generally, FEM is used where analytical methods are unable to provide solutions. FEM allows a general computer program arrangement [2] and is developed to give a numerical solution with high accuracy.

In FEM, the first geometry is decomposed to "small elements," which have two or three common points with their neighbors. Equations are written for the small elements and, finally, by aggregation of these equations, a whole system of equations is formed. Besides these equations, FEM

Electromagnetic Analysis and Condition Monitoring of Synchronous Generators, First Edition. Hossein Ehya and Jawad Faiz.
© 2023 The Institute of Electrical and Electronics Engineers, Inc. Published 2023 by John Wiley & Sons, Inc.

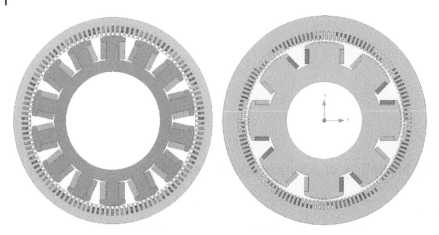

Figure 8.1 Cross-section of a typical 14-pole and 8-pole salient-pole synchronous generator.

considers the boundary elements and boundary conditions of the geometry and the equations are solved based on these conditions.

FEM was introduced for electrical machines in 1970 [3–6], when most work was related to synchronous machines. Later, the method was adapted for use with round rotors and salient pole generators. FEM is used for the analysis of electrical machines [7], for the use of external circuits for utilizing voltage sources instead of current sources [8, 9], in combination with mechanical and thermal parts for their combined analysis [10], in hysteresis phenomena [11], and to study nonlinearity properties of materials and saturation effects [12–14].

Figure 8.1 shows the cross-section of a typical salient pole synchronous generator. In many FEMs, an initial current is considered to be the input parameter. The output results are then given to state-space: if the computation error is smaller than a specific value, this current is used again as the initial value of the next stage. The major problem with this method is that the current in FEM is one of the unknown parameters of the problem, whereas it is considered one of the inputs in the time-stepping FEM of a synchronous machine. Therefore, in synchronous machine modeling, only the voltage is given directly as an excitation input on the rotor, while the load current, induced voltage, and torque are considered unknown parameters. However, using DC voltage instead of current as a rotor field winding input results in a significant increase in computation time.

In the FEM of a synchronous machine, the external circuit showing the supply and electrical load is combined with the field equations; movement equations are also taken into account. Finally, all equations are combined in the FEM. Electrical circuit and load equations, field equations, and movement equations are combined in each step and to finally determine the unknown parameters.

Despite the requirements of large computer storage, long computation time, and tedious data preparation stages, the power of FEM for the study of a synchronous generator, either as a partial model or for the complete machine, has been demonstrated by many applications involving machine design parameters [15, 16]. FEM can be used to analyze a synchronous generator as a whole, with a prediction view of the performance through magnetic field analysis within the generator. Mapping the magnetic field is a basic task for the synchronous generator designer. The unique ability of FEM is that it takes into account the complicated boundary geometries and material nonlinear characteristics that exist in synchronous generators.

Although FEM is time-consuming, its application in fault detection of electric machines is useful and economical, so it is widely used. Different fault types, including electrical and mechanical faults, can be simulated in FEM. The fault impact can then be studied on various machine parameters, such as inductances, and/or on signals, such as voltage, current, and magnetic field. Therefore, the potential impact of a fault on various machine functions can be anticipated before performing destructive tests on the actual electric machines. The purpose of this chapter is to provide a better understanding of FEM, synchronous machine modeling in FEM, and fault modeling in FEM.

8.2 Electromagnetic Field Computation

The magnetic field problem in the synchronous generator is described in terms of a vector potential A, whose Curl, $\nabla \times A$, gives the magnetic flux density B. From Ampere's law, $\nabla \times H = J$, where J is the current density vector. By substituting B in these equations, the vector potential is governed by

$$\nabla \times (\gamma \times A) = J \tag{8.1}$$

where the magnetic reluctivity γ is a nonlinear function of the flux density in the iron part and is taken to be constant and equal to that of air and to occur through the non-ferrous parts of the magnetic circuit.

Suppose a two-dimensional (2D) magnetic field exists in the xy plane and the current density vector J has a component only in the z-direction of magnitude J. Equation (8.1) can be written as follows:

$$\frac{\partial}{\partial x} \left(\gamma \frac{\partial A}{\partial x} \right) + \frac{\partial}{\partial x} \left(\gamma \frac{\partial A}{\partial y} \right) = -J \tag{8.2}$$

subject to the appropriate boundary conditions. The γ is field-dependent; hence, Equation (8.2) is a nonlinear partial differential equation or Poisson equation in the vector potential formulation. The FE approach tries to solve Equation (8.2) by minimizing the energy function (F). The function is expressed by the difference between the stored energy and the input energy in the system volume:

$$F = \int v \left[\int_0^B H \, dB - \int_0^A J \, dA \right] dv \tag{8.3}$$

F is minimized when $\partial F / \partial A = 0$, which leads to the following nonlinear matrix equation:

$$[S][A] = [T] \tag{8.4}$$

in which the coefficient matrix $[S]$ represents the shape and physical properties of the materials and $[T]$ is the forcing function (current/voltage source) matrix containing terms in A. A contour of equal values of A is a flux line, and the line integral of the vector potential A around any given contour $\oint A \, dl$ gives the flux-linkage of that contour. When the field solution given by the static analysis program has been obtained, the post-processor can perform the line integration.

8.3 Eddy Current and Core Loss Considerations

Generally, iron losses consist of hysteresis and eddy current losses. The electricity-conducting materials of the generator are exposed to an alternating magnetic flux, which induces eddy currents. The eddy current losses are generated by the induced current in the lamination sheets. Note that these lamination sheets have thin insulation to reduce losses. An attempt has been made to minimize these losses by locating the lamination sheet normal to the induced current. During the manufacturing and assembly process of the synchronous generator, cutting and making holes in the machine can increase these losses, and some repairs are necessary to decrease or remove these additional losses.

Eddy current losses can be estimated using analytical methods; however, these methods can only tackle simple configurations and linear problems. Synchronous generators have no simple structure and they also operate in the nonlinear region of the magnetization characteristics. However, FEM can precisely handle these problems. Nevertheless, since the FE solution is an iterative procedure, it requires a long computation time.

To design and predict the performance of high-power synchronous generators, eddy current losses must be accurately determined. These losses must be minimized to improve the performance and efficiency of the generator. In a wound rotor synchronous generator, the damper bars located on the rotor are also subject to iron losses caused by the slots of the rotor and stator [17]. Precise estimation of these losses is an important factor in the design process of the generator. A three-dimensional (3D) analytical network method with reasonable accuracy may also be applied to calculate the eddy current losses [18].

Rotor losses consist of eddy current losses and hysteresis losses. Eddy current losses tend to increase more quickly with frequency than hysteresis losses, and hence are typically the most critical at very high frequencies. To analyze eddy current losses, the distribution of eddy currents in the rotor must be carefully considered and modeled. A steady-state two-dimensional (2D) FE program has been developed [14, 19] to calculate steady-state eddy current rotor losses for solid-rotor machines.

FEM can estimate the iron loss in the synchronous generator and obtain its pattern. The application of 3D FEM also allows accurate consideration of the hysteresis effects [20]. Hysteresis modeling takes into account the impacts of the minor loops caused by the high-frequency components of the magnetic field. Analytical techniques can be used to estimate the losses in a post-processing stage of FEM. For example, iron losses account for a considerable portion of the losses that depend on the material properties and configuration of a synchronous generator. These losses occur from the iron parts, including the stator yoke, stator teeth, and rotor yoke. The use of power electronics components for adjusting the excitation current of the synchronous generator may also increase the iron losses because of the high harmonics content of the magnetic field of the core. These features can be modeled by FEM.

The iron losses P_i can be decomposed into the quasi-static hysteresis P_{qsh}, classic eddy current P_{cec}, and excess losses P_{el} [21]. Suppose that the time derivative of the time-dependent flux density $B(t)$, $dB(t)/dt$ is denoted by B_d; the iron losses can be evaluated as follows [22]:

$$P_i = k_{qsh} f B_m + k_{cec} \frac{1}{T} \int_0^T B_d^2 dt + k_{el} \frac{1}{T} \int_0^T B_d^{1.5} dt \qquad (8.5)$$

where k_{qsh}, k_{cec}, k_{el}, f, B_m, and T are the loss coefficients corresponding to P_{qsh}, P_{cec}, and P_{el}, the frequency, peak flux density, and the waveform period, respectively. These loss coefficients are determined by curve fitting to the measured data. Note that Equation (8.5) estimates P_{cec} and P_{el} using B_d, and P_{qsh} evaluate the hysteresis losses only for symmetrical waveforms. However, for symmetrical waveforms without minor loops, the approximate hysteresis losses can be estimated as follows [23]:

$$P_{qsh} = \frac{1}{2} k_{qsh} (B_{max} - B_{min})^\alpha \qquad (8.6)$$

Note that Equation (8.6) cannot be applied to the non-sinusoidal case. To overcome this shortage, the single equivalent frequency f_{eq} is considered as follows:

$$f_{eq} = \frac{1}{2} \frac{1}{(B_{max} - B_{min})} \left(\frac{1}{T} \int_0^T B_d dt \right) \qquad (8.7)$$

The losses caused by the minor loops also cannot be estimated by this technique. However, Fourier transform can be applied to the magnetic field and the hysteresis losses can then be estimated by adding the losses of different frequencies (i = 1, 2, 3, …) [24]. In this case, only now with minor loops, the iron losses are determined as follows:

$$P_i = k_{qsh} \sum_{i=1}^{n} f_i B_{mj}^{\alpha} + k_{cec} \frac{1}{T} \int_0^T B_d^2 dt + k_{el} \frac{1}{T} \int_0^T B_d^{1.5} dt \qquad (8.8)$$

Equations (8.5) and (8.8) are used to estimate the iron losses of the synchronous generators in the post-processing stage. By storing $B(t)$ and approaching the steady state, each element of the mesh is evaluated. In addition to the alternating magnetic field inducing the core losses, a large portion of the ferromagnetic core of the stator may be subject to the local rotational magnetic field, which generates extra hysteresis losses. The rotational flux patterns exist in a large portion of the stator of a synchronous generator. It is an elliptically polarized flux density distribution at the back of the stator slots and along the inner periphery of the stator core, and an almost circular flux density polarization at the roots of the stator teeth [25].

The major and minor axes corresponding to the maximum and minimum flux density loci must be defined for every mesh. FEM evaluates the field in the xy system, which must be changed to a local coordinate system for estimation of iron losses of every element. At every time step, the flux density vector is decomposed along the major and minor axes. The iron losses are then estimated as described above.

8.4 Material Modeling

Analysis of synchronous generators generally requires knowledge of the physical properties of different materials used in the structure of the generator. They must be prepared in a form that is usable in computer programming. The properties of the materials are provided by manufacturers and published in their catalogs. They should be made ready in convenient forms for computation purposes.

Many modeling techniques have been introduced for elucidating the magnetization characteristics of materials, particularly ferromagnetic materials. Although the shape of the magnetization characteristic seems simple, its modeling is not easy.

The stator and rotor of the generator are formed from assemblies of thin laminations that are punched from a sheet of silicon steel with known magnetization characteristics that give the material properties of the stator and rotor. This characteristic is defined for the FE program by pairs of numbers specifying B and H, entered separately, and related to the regions by a material code number. The effect of lamination is simply modeled by a packing factor scaling of the material characteristic. Generally, the intermediate values of B and H are calculated by cubic-spline interpolation to ensure slop continuity. In the solution process, each iteration updates the permeability concerning the magnetization characteristic of the iron. Figure 8.2 shows the B-H curve of a steel M400 material utilized in the stator core of the synchronous machine.

8.5 Band Object, Motion Setup, and Boundary Conditions

The band object must be defined in order to separate the moving objects from the stationary objects. The moving band encompasses the rotor geometry and must be located in the middle of the air

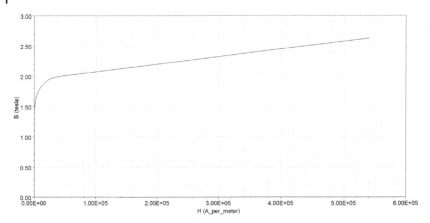

Figure 8.2 The *B-H* curve of the M400 lamination is utilized in the stator core of the synchronous machine.

gap. However, it can be brought closer to the stator or rotor while applying static and dynamic eccentricity with a severity above 50%. The objects inside the moving band must be rotated based on the number of poles of synchronous machine to provide the required frequency in the induced voltage of the stator terminal. The mechanical angular velocity of the objects inside the moving band can be obtained as follows:

$$N = \frac{120f}{p} \tag{8.9}$$

where f is the frequency of the machine in Hz and p is the number of poles pairs. For instance, the mechanical angular velocity for a synchronous machine with 14 poles is 3000/7 in order to achieve 50 Hz voltage in the generator terminals. In the radial flux synchronous machine, the direction of the rotation can be clockwise or counterclockwise, but is around the z-axis.

A circle that encompasses the entire geometry must be defined with a property of air to define the spaces where materials are not defined, such as spaces between the conductor and stator core in the stator slots. This implies that the air gap, as well as the wedges and the insulation between the armature coils, is modeled as air in the FE analysis. This simplification was made because the analysis does not include an electrostatic solver and because the wedges and the winding insulation are made from non-magnetic materials with a relative permeability in the same order as air.

A boundary condition must be set to define the magnetic field situation on the stator core backside. A zero-flux boundary condition can be assigned to the outer rim of the stator core, which indicates that no leakage flux exists on the stator back side. This assumption is valid as long as the stator core is not saturated.

8.6 Mesh Consideration

The shape of the synchronous generator is implied by the FE mesh, which is used in the mathematical analysis. Therefore, every geometric object is considered a composite structure consisting of elements. The magnetic field inside any element is represented by a simple equation with a few unknown coefficients. These coefficients are obtained considering:

1. The magnetic field satisfies the boundary conditions.
2. The Maxwell field equations are satisfied over the whole region.

Generally, triangular elements are used, with quadrilateral elements employed in some cases. In 3D models, tetrahedral and hexahedra elements are utilized. For FE-based packages, any problem is defined in terms of a mesh. A triangular element is specified by the coordinates of the vertices that identify the location and shape of the element. An attribute label provides the nature of the element.

The synchronous generator has a complicated structure; therefore, the majority of time spent in solving the problem by FEM must be expended in the creation and modification of the model. The generator has a repetitive structure, so modeling only a part of the machine is possible in a healthy generator. The magnetic flux also has a repetitive pattern; for example, in the synchronous machine with four poles, modeling one-quarter of the generator can give an accurate solution. Of course, a faulty synchronous generator has no such magnetic field pattern and the whole structure of the generator must be modeled.

To discretize the geometrical structure, the regions are defined using both Cartesian and polar coordinates. For some parts, such as the stator slots, the local coordinates are convenient. The problem space must be divided into suitable regions. The regions have curvilinear quadrilaterals or sectors of annuli. They have constant permeability in the air and copper winding area and one magnetization characteristic in the iron parts. Each region is specified by the coordinates of its vertices, property of the material, and number of subdivisions on the region's faces. Within each phase, discretization into triangles is automatic. First-order triangle elements are used, giving constant flux density and permeability in each triangle. The neighboring regions must match and the FE program generally displays an error for any that do not. Figure 8.3 shows a typical meshing model of a complete salient-pole synchronous generator. The number of meshes can vary based on the machine size; for instance, an adequate number of meshes in a synchronous machine with a diameter of 780 mm is between 200 000 and 250 000 elements. Performing a mesh sensitivity analysis of the synchronous machine is recommended before utilizing it for fault detection purposes, since the number of meshes can significantly influence the simulation time.

The air gap between the stator pole and rotor core is finally discretized in two layers: one with the stator and one with the rotor. The discretization of the air gap, in comparison with other regions, is very fine. If a rotor with its own discretized air gap rotates in a clockwise/counterclockwise direction, a new discretized model of the generator is obtained. However, the air gap region, due to the stator air gap, must be modified for the new position. Note also that convenient boundary conditions must be given for the new rotor position.

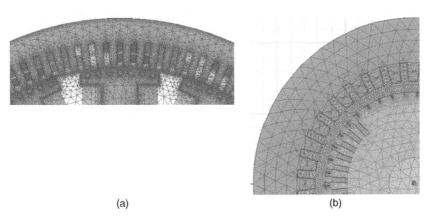

(a) (b)

Figure 8.3 The meshing of a synchronous generator: (a) salient-pole and (b) round rotor.

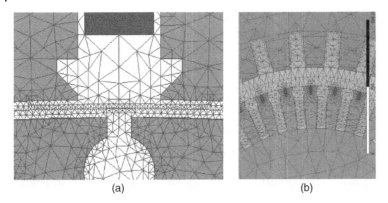

Figure 8.4 The air gap mesh of a synchronous generator: (a) salient pole and (b) round rotor.

A combination of two different mesh configuration techniques is recommended when creating the FEM mesh of the synchronous machine. First, determine the total number of elements and the maximum size of all the elements within a selected object. Afterward, the size of the elements only needs to be restricted along the edges of the object. This approach allows for the generation of a high-resolution mesh in the air gap, while the remaining part of the stator and the rotor objects are assigned a coarser mesh to decrease the computational time. Figure 8.4 shows the mesh in the air gap of the synchronous machine.

8.7 Time Steps and Simulation Run Time

The computation time of the FEM simulation depends on several factors, including:

1. Length of the time step.
2. Required mechanical revolutions to achieve an adequate resolution for the post-processing.
3. Resolution of the finite element mesh.
4. Healthy or faulty operation of the synchronous machine.
5. Computational resources.

At least one mechanical revolution is required for signal processing, but more mechanical revolutions result in a higher resolution in the measured signal and can demonstrate the subharmonics. The minimum number of revolutions that can provide a high-resolution signal is 8. Therefore, a machine with 8 poles requires at least 640 ms, while a machine with 14 poles requires 1120 ms.

A fixed time step must be used for fault detection purposes in FEM analysis. Furthermore, based on the Nyquist sampling theorem, an inspection of the frequency spectrum reveals that up to half the sampling frequency can be achieved. Therefore, the frequency of interest for the selection of the sampling frequency is important. In the fault detection of the electric machines, the maximum frequency utilized around the slot harmonic frequency is approximately up to 800–900 Hz. Therefore, the use of a sampling frequency of 2000 Hz (time step of 0.5 ms) can provide a good resolution to identify the harmonics based on the Nyquist sampling theorem. However, the sampling frequency must be increased to observe the impact of damper bars in the rotor pole.

8.8 Transient and Steady-State Modeling

Traditional methods for modeling a synchronous generator can provide the approximate magnetic field within the generator. These techniques generally offer reasonable results for steady-state operation, but they are unreliable for transient operation. The magnetic field distribution in the synchronous generator is determined by the system of Maxwell's equations. Since the core materials are nonlinear, the system of equations is iteratively solved [26]. To achieve a fast convergence over a short computation time, the equations should be solved simultaneously in the same iteration loop. In the 2D case, the axial length of the generator is taken to be infinite and its geometry is invariant along the z-axis. The synchronous generator is treated as a quasi-static magnetic system. The following three steps are used for computation of the magnetic field by FEM:

1. Pre-processing – the derivation of the FE model of the synchronous generator, defining material properties, boundary conditions, and mesh generation.
2. Processing – solving the problem by the relevant Maxwell equations and obtaining the field distribution in the analyzed domain of the synchronous generator, at arbitrary chosen excitations and loading conditions.
3. Post-processing – calculation and presentation of characteristics, as well as parameters, of the synchronous generator. The open-circuit characteristic of the synchronous generator is determined for different field currents. For each field current, the flux-linkage ψ is calculated from the FE solution. The phase voltage E is then calculated as follows:

$$E = 4.44\psi N k_w f \tag{8.10}$$

The synchronous generator air gap flux characteristics depend on the flux in the middle of the air gap against the rotor angular position for different armature currents. In the post-processing step, the flux density distribution is obtained for an arbitrarily selected region in the synchronous generator.

Since the synchronous generator operates under steady-state conditions most of the time, its detailed steady-state model is essential in the design and testing stages. The steady-state model predicts the behaviors of the generator under various operating conditions and determines their influence on the power system. Furthermore, the model can estimate the on-load excitation requirements, reactance, and different losses.

The steady-state FE analysis of synchronous generators is mostly carried out in the time domain using a time-stepping method. The frequency-domain technique, using the harmonic balance FEM (HBFEM) [27], is not widely adopted, as it requires the cumbersome assembly and resolution of a very large system of nonlinear algebraic equations. Furthermore, HBFEM requires the use of a Eulerian representation of the continuum variables, where the material is moving with respect to a fixed coordinate system. This introduces a term of the velocity of material through the coordinate of interest, which would create numerical difficulties in FEM due to the Peclet effect. An alternative approach to obtain steady-state performance is to perform transient analysis until convergence, but this can take a long time.

The d-q axis system is usually used for dynamic and transient modeling of the synchronous generator. However, the synchronous generator can be modeled by FE, as FE analysis can be incorporated into the dynamic model of synchronous generators [28–30]. For accurate modeling of a synchronous generator, an efficient FE solver can be directly coupled with the dynamic model and correlation of optimized parameters to synchronous generator traits can be easily generated. A numerically efficient static solver can be developed. The solver is implemented as a block in a large-scale dynamic simulation. Steady-state FE analysis can provide even more information about synchronous generator operation, but it is so computationally expensive that it is often only used in the design verification stage.

In the above-mentioned case, the magneto-static FEM is employed to provide performance/field data for nonlinear circuit-based simulation. However, instead of directly coupling the solver within the model, the FE-generated generator characteristics are used as performance lookup tables. With the available FE formulation, the computation time will be long. Another disadvantage of these models is that optimized parameters from the dynamic simulation are difficult to correlate with specific generator traits because of the irreversibility of the FE analysis that created the equivalent circuit.

Different techniques, including approximate electric circuit models [31], FEMs [32, 33], and dynamic magnetic models, are available for modeling synchronous generators. FEM needs a long computation time; therefore, it is more suited for static studies than for transient studies. FE analysis requires physical geometric data, which is available for a synchronous generator for its routine check. Internal faults, such as shorted turns or either the stator or rotor faults, can be efficiently and satisfactorily modeled by FEM, but the computation time is extremely long. For instance, a 100 kVA synchronous generator with a diameter of 780 mm and an approximate number of 250 000 elements in a server computer with 128 cores and 900 GB memory takes approximately 24 hours to simulate a faulty operation with a sampling frequency of 10 kHz and time duration of 1140 ms.

8.9 No-Load and On-Load Modeling

In the no-load operation of the synchronous generator, a nominal no-load current must be set to the rotor field winding in FEM. By applying a no-load magnetizing current, the measured induced voltage in the stator terminals must be equal to the nominal no-load voltage in the nameplate. If the machine has a distributed three-phase winding in the stator, the three-phase induced voltage must have a 120° electrical phase shift.

Although the majority of reported studies regarding various fault impacts on the performance of the synchronous machine are demonstrated in the no-load operation, the synchronous machine operates in an on-load case most of the time. Therefore, investigating the load impact on healthy and faulty machine operation in FEM is crucial. The external circuit is a valuable toolbox in the FEM simulation packages of the commercial software. Various configurations of the stator winding layout, parallel winding connection, and assorted load connections can be performed using the external circuit feature in FEM. Figure 8.5 shows that the external circuit of a synchronous generator consists of the stator winding, the resistive-inductive load, and the grounding point. The value of the resistance and inductance must be set in such a way to achieve a nominal voltage in the stator terminals. Since the end winding inductance is ignored in the 2D modeling, the end winding inductance is added in the external circuit model.

8.10 2D and 3D FEM

Since the synchronous generator has a 3D topology and since 3D flux paths of the complete transient characteristics and parameters cannot be obtained without an extensive computation effort,

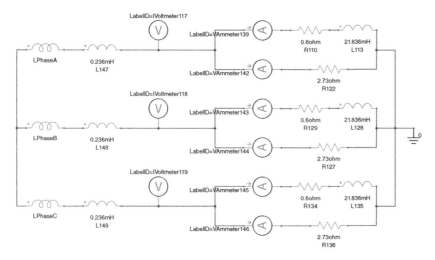

Figure 8.5 The external load circuit of the 100 kVA synchronous generator model.

the FE analysis of the synchronous generator can be done using a commercial software platform. When applied to a healthy synchronous generator, the platform is usually reduced to cover only one pole or one pole pair with the help of boundary and symmetry conditions to shorten the computation time. A 3D FE is preferred when only some key values or verification values are required that cannot be obtained using 2D FEM. Since the synchronous machine topology has a 3D field, the model requires a 3D magneto-static model solver. Using the 3D FEM tool is still time-consuming, which limits its use in the fault detection process.

The solution of the FE model using a 3D FE solver lasts several hours, even per step. Therefore, only magneto-static parameters, such as airgap flux density (the absolute value of B) and inductance, can be determined [34]. In addition to a good accuracy of the 2D-FE model, even for the transient analysis, some cases still have some limits in 2D modeling, so 3D modeling must inevitably be applied. The following are some of these limits [35]:

1. In 2D FE modeling it is not possible to model the magnetic flux fringing effect, which plays a major role in the reduction of the flux density.
2. Although analysis of the leakage fluxes and mutual fluxes in the radial direction of the machine is possible, modeling the leakage flux and mutual flux of the end-winding is impossible.
3. If the material used in the axial direction varies, the use of 2D-FE modeling is not recommended.

Figure 8.6 shows a 3D model of a salient-pole synchronous generator. The stator, rotor, shaft, stator windings, and excitation winding models are shown in Figure 8.7 and Figure 8.8. As indicated, the stator windings have been modeled as number bars beside each other, while in the 2D modeling, the stator slot is filled by conductors.

8.11 3D-FE Equations of the Synchronous Generator

Unlike the 2D case, modeling the stator end-winding and end-winding of the rotor excitation winding is possible by 3D-FEM (Figure 8.7). The stator windings provide the output voltages, while the path of the input and output currents within the conductors are specified by defining a plane as input or output in each conductor.

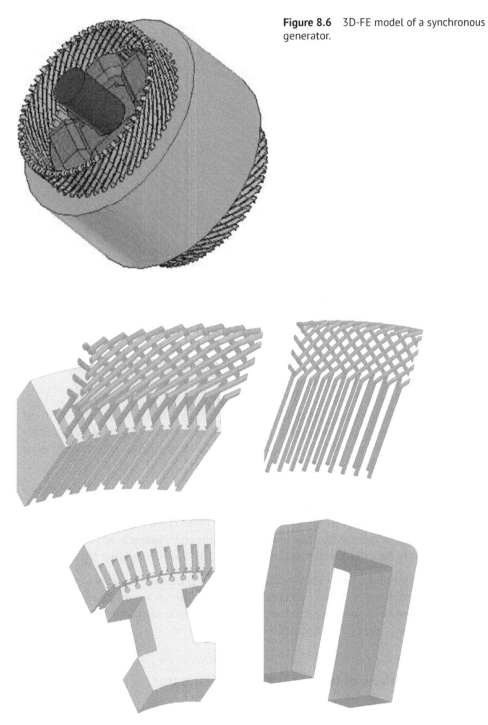

Figure 8.6 3D-FE model of a synchronous generator.

Figure 8.7 Different parts of the synchronous generator, including the stator core, rotor core, stator winding, and rotor field winding.

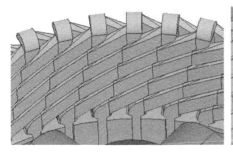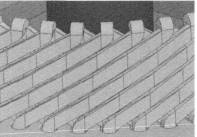

Figure 8.8 Stator end-winding of a salient-pole synchronous generator.

FE meshing is very important, especially in 3D modeling. In many cases, the main reason for the lack of convergence of equation systems is improper meshing. The air gap meshing is the most important because all equations are solved based on the magnetic field of the air gap. Therefore, very fine meshes are considered in this region. Our experience shows that a typical number of meshes using a server computer with 64-core and 128 GB of memory is around 1 385 479 for a 50 kVA synchronous machine. For calculation of a 3D magnetic field by A-V formulation, we have [36, 37]:

$$\nabla[v]\nabla\vec{A} - \nabla v_e\nabla\vec{A} + [\sigma]\frac{\partial\vec{A}}{\partial t} + [\sigma]\nabla V = 0 \tag{8.11}$$

$$-\nabla[\sigma]\left(\frac{\partial\vec{A}}{\partial t} + \nabla V\right) = 0$$

where $\vec{A}$ is the magnetic potential, V is the electric scalar potential, $[v]$ is the reluctance tensor, $[\sigma]$ is the electric conductivity tensor, and v_e is one-third of the reluctance tensor. In a large synchronous generator, where the stator slots contain stator winding bars, the current density J is as follows:

$$\vec{J} = \sigma\frac{\partial\vec{A}}{\partial t} - \sigma\nabla V \tag{8.12}$$

If the conductors are in strand form, J will be as follows:

$$\vec{J} = \hat{t}\frac{n_c}{S_c}i(t) \tag{8.13}$$

where S_c is the conductor cross-section, n_c is the number of strands, $i(t)$ is the current of every strand, and $\hat{t}$ is the vector normal to the conductor surface, which indicates the current direction. Therefore, by a proper choice of the current vector, Equation (8.10) in the strand is as follows:

$$\nabla[v]\nabla\vec{A} - \nabla v_e\nabla\vec{A} - \vec{J} = 0 \tag{8.14}$$

8.12 Modeling of the Stator and Rotor Windings of the Generator and Its Load

Modeling different conditions of the generator operation, including no-load and on-load conditions, and providing the correct connection of the FE region to the external load requires definition of the stator phase winding and rotor winding distribution types, as well as their corresponding electrical circuit. Figure 8.9 presents the generator phase winding distribution inside half of the stator slots. According to Figure 8.9, the distribution is a full-pitch one and the distribution of the

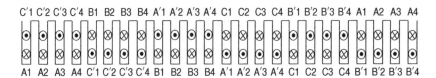

C′1 C′2 C′3 C′4 B1 B2 B3 B4 A′1 A′2 A′3 A′4 C1 C2 C3 C4 B′1 B′2 B′3 B′4 A1 A2 A3 A4

A1 A2 A3 A4 C′1 C′2 C′3 C′4 B1 B2 B3 B4 A′1 A′2 A′3 A′4 C1 C2 C3 C4 B′1 B′2 B′3 B′4

Figure 8.9 Phase winding distribution inside half of the stator slots.

conductors generates four poles in the stator. The electrical circuit components for FE modeling consist of 96 coils for three-phase stators and their connection procedure corresponds to the winding distribution diagram in Figure 8.9.

Connection of these coils is shown in Figure 8.10, which depicts a rotor consisting of two coils connected and supplied by a DC voltage. A three-phase star-connected RL load is connected to the stator coils. All modeled coils are coupled to the proposed regions inside the slots and regions related to the rotor winding in the FE region to establish the real link between the electrical and magnetic equations. Having the vector magnetic potential, the current density at every point is then estimated. In this case, the 2D equation of the transient magnetic field in the generator is expressed as follows:

$$\nabla \upsilon \nabla \vec{A} = \vec{J} \tag{8.15}$$

where J is the current density vector along the z-axis. To relate the external circuit and the field equations, the passing current from each conductor is calculated. The current density is determined, knowing the vector A, as follows:

$$\vec{J} = -\sigma \frac{\sigma \vec{A}}{\partial t} - \sigma \frac{\delta V}{L} \tag{8.16}$$

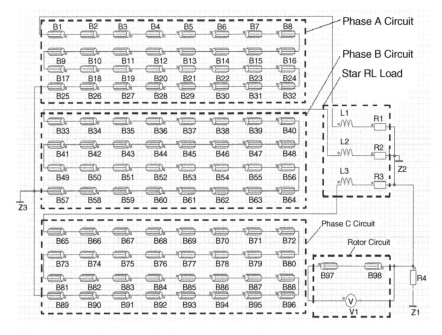

Figure 8.10 Electric circuit modeling of the stator phases, rotor winding, and three-phase Resistance-Inductance (RL) load.

where δV is the potential difference along the conductor, σ is the conductor electric conductivity, and L is the conductor length. Establishing an appropriate relation between the circuit and field equations requires estimation of the total current of each conductor. This is done by integrating Equation (8.17) over the conductor cross-section:

$$i = \iint \sigma \left(\frac{\partial \vec{A}}{\partial t} + \frac{\Delta V}{L} \right) ds \tag{8.17}$$

Substituting Equation (8.15) into Equation (8.16), we have

$$\nabla V \nabla \vec{A} + \sigma \frac{\partial \vec{A}}{\partial t} + \sigma \frac{\Delta V}{L} = 0 \tag{8.18}$$

The mathematical model of the external electrical circuit in matrix form is as follows:

$$[E] = [R][i] + [L]\frac{d[i]}{dt} \tag{8.19}$$

where $E, R,$ and L are the induced voltage, resistance, and inductance of the rotor and stator winding matrices, respectively. Applying the standard Galerkin method [38], the general matrix equations will be as follows:

$$\begin{bmatrix} G & H & 0 \\ 0 & W & D \\ 0 & D & -Z \end{bmatrix} \begin{bmatrix} A \\ \Delta V \\ i \end{bmatrix} + \begin{bmatrix} Q & 0 & 0 \\ H^{\mathrm{T}} & 0 & 0 \\ 0 & 0 & -L_0 \end{bmatrix} \frac{\partial}{\partial t} \begin{bmatrix} A \\ \Delta V \\ i \end{bmatrix} = \begin{bmatrix} J \\ 0 \\ -E \end{bmatrix} \tag{8.20}$$

where

$$G_{ij} = \int \nabla N_i \nabla N_j \, ds \tag{8.21}$$

$$H_{ij} = \int_k \sigma N_i \, ds \tag{8.22}$$

$$W_{kk} = \int_k \sigma \, ds \tag{8.23}$$

$$Q_{ij} = \int \sigma N_i N_j \, ds \tag{8.24}$$

$$L_0 = \int J N_i \, ds \tag{8.25}$$

Integrations of $J, Q,$ and G over the whole area of the model are performed, while integration of W and H are performed over the conductor k. Matrix D is also a sparse matrix in which $+1$ and -1 are shown for passing current through the conductor [14]. Equation (8.20) has the following differential form:

$$MX + K\frac{dx}{dt} = V \tag{8.26}$$

which can be solved using conventional numerical methods. In this equation:

$$M = \begin{bmatrix} G & H & 0 \\ 0 & W & D \\ 0 & D^{\mathrm{T}} & -Z \end{bmatrix}, K = \begin{bmatrix} Q & 0 & 0 \\ H^{\mathrm{T}} & 0 & 0 \\ 0 & 0 & -L_0 \end{bmatrix} \tag{8.27}$$

Also:

$$V = \begin{bmatrix} J \\ 0 \\ -E \end{bmatrix}, X = \begin{bmatrix} A \\ \Delta V \\ i \end{bmatrix} \tag{8.28}$$

Solving Equation (8.12), the stator current and magnetic potential vector of the healthy and faulty synchronous generators are determined.

8.12.1 Modeling Movement of Movable Parts and Electromechanical Connections

The movable parts in the generator somehow determine the modeling and magnetic forces and influence the magnetic field inside the generator. Therefore, modeling movement in the movable parts of the generator is necessary. The following equation expresses the relation between the electromagnetic torque and speed in the generator:

$$T_l - T_e = J \frac{d\omega_r}{dt} + B\omega_r \tag{8.29}$$

$$\omega_r = \frac{d\theta_r}{dt} \tag{8.30}$$

where T_e is the electromagnetic torque, T_l is the load torque, ω_r is the generator speed, J is the moment inertia, B is the friction coefficient, and θ_r is the rotor angular position.

8.13 Air Gap Magnetic Field Measurements

In the commercialized software, the air gap magnetic field can be saved for all the time steps. This is beneficial since the magnetic field can be assessed for any arbitrary point; however, it is time consuming and can extensively increase the computation time. Therefore, a better recommendation is to save the magnetic field of the air gap at the desired point. That point in the air gap is configured in a way to indicate the radial flux density. Figure 8.11 shows the dedicated measuring point on the stator tooth that is assigned to record the radial flux density. The measuring point should not be included inside the moving band. In addition, the measuring point should not be very close to the stator core. The materials of the measuring point and the stator core differ and putting the measuring point very close to the stator core could result in an inaccurate air gap magnetic field due to the sudden change in the material properties in the FEM. This measuring point resembles the Hall-effect sensor. Mounting two measuring points on two sides of the stator slot is also possible for simulating the passive search coil and determining the air gap magnetic field.

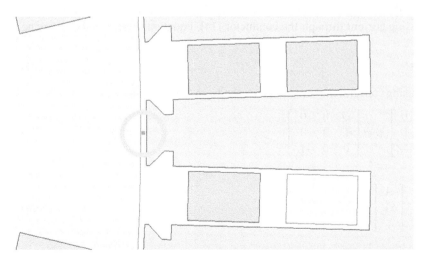

Figure 8.11 Defined measuring point on the stator tooth surface of a synchronous generator.

8.14 Stray Flux Measurements

Measuring stray magnetic field analysis for fault detection purposes has become popular recently because it provides valuable information that mirrors the air gap magnetic field. The stray magnetic field in the synchronous machine is generated due to the leakage flux that closes its path on the stator backside. The stray magnetic field on the stator backside of the synchronous generator can be picked up using the following steps in the FEM:

1. The magnetic field on the stator backside is assumed to be zero; therefore, a zero-flux boundary condition is set on the stator core. Setting the stray magnetic field requires that the boundary be set at least a few centimeters away from the stator core. A sensitivity analysis is required to analyze the impact of the distance of the boundary from the stator core on the pattern and amplitude of the stray magnetic field.
2. A coil, as shown in Figure 8.12, needs to be mounted on the stator backside. The winding size has an impact on the induced voltage due to the stray magnetic field. In addition, the winding should not touch the stator core as the different material properties can cause inaccurate results in the FEM simulation.

Two different types of stray magnetic field can be measured in the electric machine:

1. Radial stray flux
2. Axial stray flux

In the 2D FEM, only the radial flux can be modeled, while axial flux modeling can be achieved only in 3D FEM.

8.15 Eccentricity Fault Modeling

Eccentricity faults influence the synchronous generator geometry by disturbing the symmetry of the healthy generator. This asymmetrical magnetic field pattern must be determined first. An appropriate geometrical computer-aided design model and a mathematical model are required for prediction of the performance of a faulty synchronous generator. In the case of dynamic eccentricity, the rotational axis coincides with the origin of the stator; however, the rotational axis is displaced with respect to the geometric origin of the rotor.

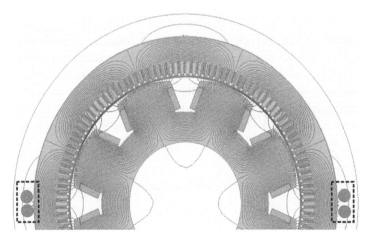

Figure 8.12 Location of the search coil on the stator backside to measure the stray magnetic field in the synchronous generator.

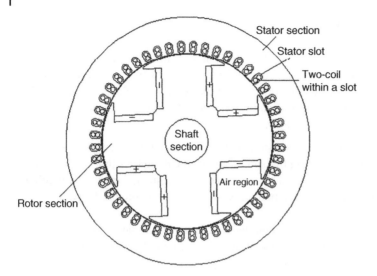

Figure 8.13 Cross-section of a salient pole synchronous generator model for FE analysis.

The dynamic eccentricity fault can be modeled in FEM by moving the rotor parts, including the shaft, rotor core, field windings, and damper bars, in a preferred direction. The distance between the global origin of the model and the origin of the displaced rotor corresponds to the level of dynamic eccentricity. The following steps must be checked in the case of a dynamic eccentricity fault:

1. The band that encompasses the rotating objects must be checked to avoid any moving object placed outside the moving band.
2. The quality of the mesh in the air gap must be checked to avoid incorrect results.

A typical three-phase, four-pole synchronous generator for FE modeling and analysis is shown in Figure 8.13. Figure 8.14 presents the magnetic flux pattern in the healthy state and under a static eccentricity fault in the synchronous generator. The dynamic eccentricity degree is defined as follows [39]:

$$\text{Dynamic Eccentricity Degree} = \frac{r}{g} \times 100 \tag{8.31}$$

where g is the radial air gap length in the case of no eccentricity and r is the displacement of the rotor in the horizontal direction.

In a machine suffering from a static eccentricity fault, the rotational axis coincides with the geometric origin of the rotor, but the rotational axis is permanently displaced with respect to the origin of the stator. Implementation of the static eccentricity fault in the synchronous machine requires that the stator core and stator winding must be moved based on the degree of eccentricity along the x- or y-axis. Figure 8.15 shows the rotor and stator cross-section in a static eccentricity fault. The degree of the static eccentricity fault δ_{se} can be calculated as follows:

$$\delta_{se} = \frac{|O_w O_s|}{g} \tag{8.32}$$

where O_s is the stator symmetrical center, δ_{se} is the static eccentricity degree, and $O_w O_s$ is the static eccentricity transfer vector.

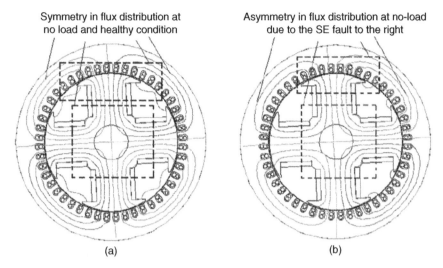

Figure 8.14 Magnetic flux pattern in a synchronous generator: (a) healthy and (b) with 40% SE to the right at no-load and arbitrary rotor position with 2 A excitation current.

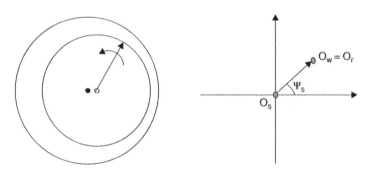

Figure 8.15 Stator and rotor cross-section in a static eccentricity fault.

The stator core and its winding become closer to the rotor pole shoe and the moving band in the case of a static eccentricity fault, indicating that, in the case of a severe static eccentricity fault above 50%, the radius of the moving band must be reduced to avoid contact with the stator and the moving band. In addition, the air gap mesh must also be checked.

The computation of the field in the generator operating under an eccentricity fault is started from the system of Maxwell's equations, which describes the magnetic fields as closed and bounded systems. The magnetic field equations are solved in the given structure via the FE approach. Flux distribution within a generator with a 40% static eccentricity fault to the right at no-load and an arbitrary rotor position with a 2A excitation current is shown in Figure 8.14, while Figure 8.16 presents the corresponding flux distribution at a rated excitation. The lower excitation current at no-load results in a larger sensitivity of the flux distribution due to the eccentricity compared to the rated excitation current. The reasons are:

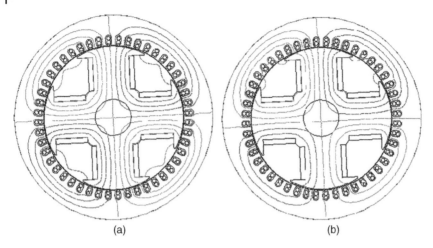

(a)   (b)

Figure 8.16 Magnetic flux distribution in the generator: (a) in a healthy state and (b) with 40% SE to the right and an arbitrary rotor position with a 12A excitation current.

1. The air gap of the salient pole synchronous generator is inherently large, due to the large gap between the two adjacent poles. Therefore, the eccentricity will not affect the total magnetic reluctance of the generator magnetic circuit or the flux distribution within the generator.
2. The eccentricity is capable of a greater effect on the total reluctance of the generators in a low excitation current in which the generator's magnetic circuit is linear.

Therefore, the full reluctance paths of the generators are predominantly determined by the air gap reluctance. In a rated excitation, the generator saturates intentionally (with a 12 A DC excitation current in the 50 kVA synchronous machine). Therefore, the reluctance of the iron core increases significantly and the variation in the air gap reluctance does not have a noticeable effect on the total reluctance of the flux line paths. As seen in Figure 8.16, eccentricity cannot greatly affect the generator flux pattern when operating at a high excitation current.

To predict the performance of the synchronous generator under an eccentricity fault, the inductances of the rotor and stator circuits are evaluated in references [40] and [41]. A symmetrical distribution of the windings in the generator's slots results in an identical variation in the stator windings inductance profile, with a phase shift in the presence of the eccentricity. Consequently, one phase of the stator can be considered.

Transient analysis can be used to model and analyze a round-rotor synchronous generator that is mechanically coupled. The electrical equations of the external circuits exhibiting electrical circuits and supply are combined with a magnetic field equation in FEM and the mechanical coupling equations of motion. Figure 8.17a presents the flux pattern in the round-rotor synchronous generator and shows that an eccentricity fault clearly influences the flux distribution within the round-rotor generator. In the faulty generator, the magnetic flux distributions are not identical on both sides of the generator because the flux path reluctance is mainly determined by the air gap length. Since the rotor is displaced with respect to the stator, the gap length between the rotor poles and stator core decreases in one half of the rotor surrounding air, while rising in the other half. This generates unbalanced magnetic reluctance paths for the flux and leads to asymmetric flux distribution. The eccentricity fault affects the generator's total reluctance in a low excitation current because the generator's magnetic circuit is linear. Therefore, the air gap reluctance determines the full reluctance path of the generator.

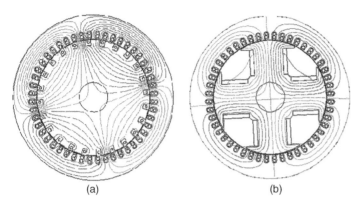

(a) (b)

Figure 8.17 Magnetic flux pattern in a synchronous generator cross-section at no-load and rated excitation current, arbitrary rotor position, and with 40% DE: (a) round-rotor type and (b) a salient-pole type.

The phase A self-inductance (L_{aa}) can be estimated using the flux-linkage λ_{aa} shown by phase A, where a DC current passes through it, and windings of the other phases are open-circuited:

$$L_{aa} = \frac{\lambda_{aa}}{I_a} \bigg|_{i_b, i_c, i_f = 0} \tag{8.33}$$

Figure 8.18a presents L_{aa} in the healthy state and in a case of different dynamic eccentricity (DE) in a round-rotor synchronous generator. These values have been determined for various rotor angular positions between 0° and 360° with a step-angle of 3°. Referring to Figure 8.18, the minimum and maximum L_{aa} clearly rise in response to increasing DE severity. Faults with 10% DE and 40% DE increase L_{aa} by 4% and 12%, respectively.

Figure 8.18b shows the self-inductance of phase A in the salient-pole and round-rotor synchronous generator with different dynamic eccentricity severities for different rotor positions from 0° to 360°, taking rotational steps of 3°. The minimum and maximum amplitudes of the self-inductance of phase A rise with the increase in dynamic eccentricity severity. As shown, unequal minima of the inductance profiles occur in the profile due to the increase in dynamic eccentricity severity. The reason for this phenomenon is that phase A shows the minimum self-flux twice instead of four times with a DE fault.

Figure 8.19 shows the mutual-inductance profile between phase A and B in healthy and different dynamic eccentricity fault severities for salient-pole and round-rotor synchronous generators:

1. The peak amplitudes of the mutual inductance in different eccentricity severities are approximately identical and sinusoidal variants in salient-pole and round-rotor generators.
2. The minimum amplitude of the mutual inductance decreases with increasing DE degree in both generators.
3. The double periodicity that occurred in the self-inductance profile also exists in the mutual-inductance profile, especially in the salient-pole type, because the flux produced by phase B is seen by phase A two times per cycle instead of four times.
4. The percent of the variation in the minimum amplitude of the mutual-inductance in different dynamic eccentricity fault degrees is larger in the salient-pole type than in the round-rotor type.

The 2-D FE analysis reveals that the eccentricity fault has greater effects on the flux pattern within the round-rotor synchronous generator cross-section than within the salient-pole type. This

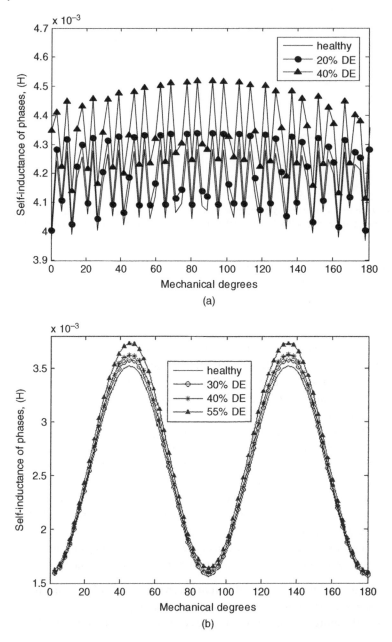

Figure 8.18 Self-inductance profiles of phase A of (a) a round-rotor and (b) a salient-pole synchronous generator.

is due to the mismatch between the two structures and the inherently large air gap between the poles in the salient-pole generator. The self- and mutual-inductance profiles of the stator phases are also changed, which can result in a double periodic phenomenon, especially in the salient-pole type. This can lead to errors in the prediction of generator characteristics. The dynamic eccentricity severity also increases the average value of the inductances profile in the round rotor synchronous machine.

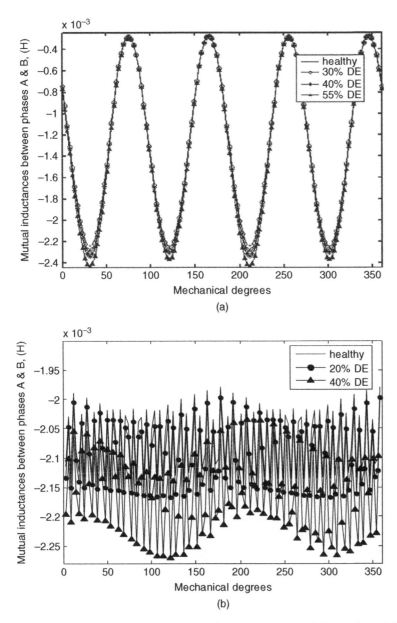

Figure 8.19 Mutual-inductance profiles of stator phase A and B in healthy and different DE severities for synchronous generators: (a) salient-pole type and (b) round-rotor type.

8.16 Stator and Rotor Short Circuit Fault

The FEM can be employed to model short circuit (SC) faults in the stator windings of a synchronous generator, and it can evaluate the equivalent circuit of the synchronous generator parameters in a faulty case. Generally, stator winding faults consist of: (a) a phase-to-earth fault, (b) a phase-to-phase fault, and (c) an inter-turn fault [42]. Asymmetrical phase-to-earth and phase-to-phase faults in synchronous generators have been simulated by FEM in references [42] and [43]. To perform this, FEM with moving air-band and electric circuit coupling has been applied. This coupling, in combination with FEM, provides a powerful tool for time-dependent problems. The electric circuit coupling method injects the voltage sources into FE equations using Kirchhoff's equations, thereby allowing the use of additional elements, such as inductance and resistance, in this analysis. The moving air band method is used to apply the rotor motion in the FE equations, with the assumption that the existing layer in the air gap is a single-layer mesh. The main advantage of this method is that it fixes the FE meshing in the stator and rotor after rotor rotation. Mesh regeneration is done only in the air gap region.

The transient performance of a salient pole synchronous generator with an internal fault and a different number of short-circuited turns has been addressed in references [43] and [44], where the short circuit of a one-phase turn and the short circuit of two-phase turns of stator windings have been examined. The minimum number of SC turns in a single phase and two phases were considered in references [44–46]. In these cases, the number of SC turns is taken to be the same where this number is equal in both cases.

Transient phenomena of a salient-pole synchronous generator with field winding SC can be addressed using FEM [47, 48]. The performance of a large hydro-generator under a phase-to-phase fault has been predicted by FEM in references [48] and [49]. A method for diagnosis and classification of the inter-turn fault in the stator winding of a synchronous generator based on the existing harmonics in the terminal voltage waveform has been introduced in references [50] and [51]. The results were obtained using a decision tree algorithm, the fundamental harmonic, and subharmonics. Simulations were performed for a power plant unit using FE software to achieve a desirable precision, with all relevant quantities, such as a magnetization characteristic, air gap asymmetry poles saliency, teeth, and slots, taken into account. Different inter-turn faults have been diagnosed based on the voltage distortion in the terminal of the phase voltage. The results in reference [52] indicate that the fourth and third harmonics of the voltage waveform can diagnose the number of SC turns, while also detecting the faulty phase.

The implementation of the stator short circuit fault depends on the short circuit fault type in the stator winding. In the case of an inter-turn short circuit fault inside the phase winding, the number of turns can be reduced based on the number of shorted turns in FEM. However, if the short circuit fault happens between two phases or as a phase-to-ground fault, an external circuit that exhibits the connection of the different coils is required. A procedure for the short circuit faults for a synchronous machine with 36 slots is presented in Figure 8.20. A typical two-layer winding with 21-full-pitch winding is considered. Each phase has 25 coils. Figure 8.20 shows the coil locations in the slots and Figure 8.21 presents the coil connections. The vertical lines in Figure 8.21 indicate a coil, where T and B represent the upper coil and lower coil, respectively, in the slots. The numbers represent the slot numbers.

8.16.1 Phase-to-Earth Fault

In the phase-to-earth fault, the fault current enters the earth through the neutral of the synchronous generator. According to standard C37-101, the neutral is connected to the earth in different forms. Here, the synchronous generator has been connected to the earth by a high resistance, which can limit the earth current. In the external circuit, a short circuit connection can be conducted between one of the stator phase windings and the grounding point to simulate the phase-to-earth fault in FEM. In this fault, all waveforms are similar to those of the healthy generator due to the limitation of the fault current. The only difference between the faulty case and the healthy case is the neutral current, as seen in Figure 8.22.

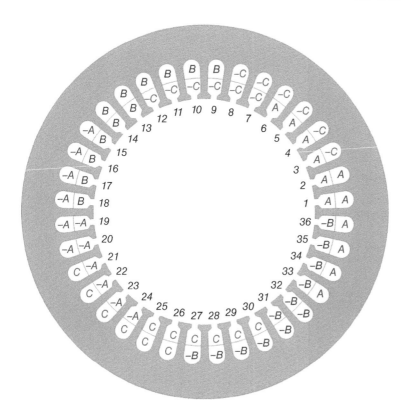

Figure 8.20 Arrangement of the coils in the stator core slots.

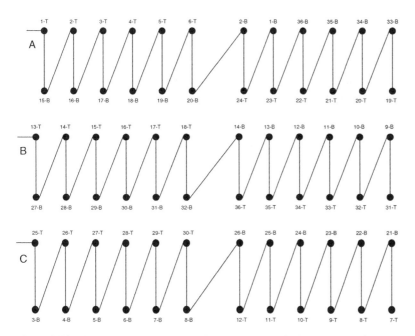

Figure 8.21 Connection of phase coils for three phases, including phase A (top), phase B (middle), and phase C (bottom).

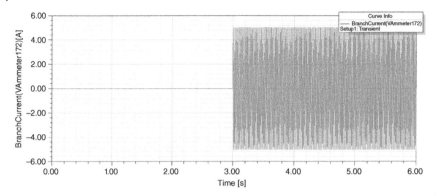

Figure 8.22 Neutral current in a phase-to-earth fault simulated in FEM.

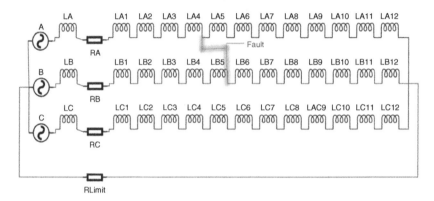

Figure 8.23 The phase-to-phase fault implementation in the stator winding of the synchronous machine.

8.16.2 Phase-to-Phase Fault

The phase-to-phase short circuit fault happens when windings of two phases are in the same slot of the stator core, indicating that no possibility exists for having a phase-to-phase short circuit fault in all stator slots. Figure 8.23 shows the implementation of the phase-to-phase short circuit fault in the external circuit of the FEM. The impact of the phase-to-phase short circuit fault is severe and can result in serious damage to both the stator core and stator winding, since a large current flows in the stator winding, as shown in Figure 8.24.

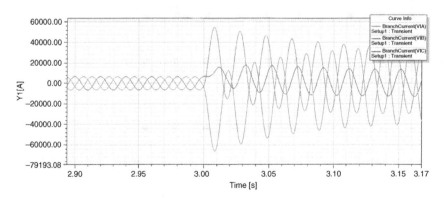

Figure 8.24 The impact of a phase-to-phase fault on the stator current of the synchronous machine.

8.16.3 Inter-turn Fault

Simulation of the inter-turn fault in a synchronous generator stator winding assumes that this fault occurs only in some slots of a phase, where winding of the same phase is in the same slot. Two coils of a phase are present in these slots. Figure 8.25 shows the implementation of the inter-turn short circuit fault in the external circuit of the FEM software. Suppose the inter-turn fault occurred in slot number 20 of the synchronous generator with a severity of 58.34% in phase A. Figure 8.26 shows the current waveforms in this fault. The faulty phase has a fault current higher than that of the other two phases and the faulty current is distributed almost equally between the two healthy phases. The fault phase current has a 180 phase angle difference from the healthy phases. The amplitudes of the induced voltages of the coils of the faulty phase drop due to a high fault current and saturation. The induced voltages of the healthy phase coils have no large variations and only their amplitude slightly decreases. Therefore, the faulty phase voltage variations are correct.

8.16.4 Inter-turn Fault in Field Windings of the Synchronous Generator

The field winding inter-turn short circuit is a common fault in synchronous generators and may cause field current increments, reactive power output decreases, and generator-set vibration aggravation. The SC of the rotor winding turns in a turbo-generator is analyzed by FEM in references [46] and [47]. The inter-turn fault in the rotor winding of the salient pole synchronous generator has been proposed in reference [50] based on FEM.

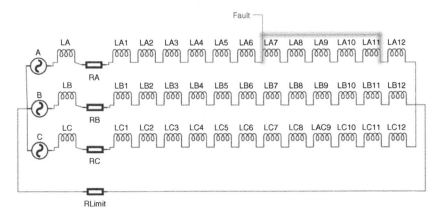

Figure 8.25 Inter-turn short circuit fault in the stator winding of the synchronous machine.

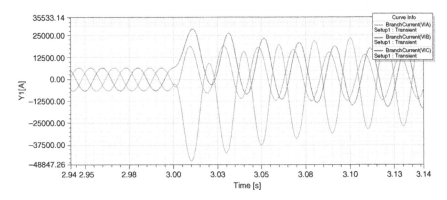

Figure 8.26 Waveforms of the phase currents in an inter-turn fault in phase A with 56.34% severity.

Generally, both multi-layer wire-wound and strip-on-edge field windings are employed in salient-pole synchronous generators [53]. In the wound-field winding design, a rectangular cross-section wire is usually wrapped around the pole, which has many layers of wire and turns. The wire insulation insulates the turns from each other. The strip-on-edge field winding design is usually used in moderate- and high-power hydro-generators and is capable of withstanding extensive rotational forces [54]. The strip-on-edge design involves shaping thin copper strips into rectangular "frames," which can then be slid on to the pole body and connected in series. Epoxy glass laminates and insulating tape separators are mounted between the copper "frames," forming the turn insulation. In these winding designs, insulating washers and strips insulate the turns of the field winding from the pole body and the pole shoe.

The field winding of synchronous generators is normally very reliable. Note that the thermal, mechanical, electrical, and chemical stresses during its operation can deteriorate the turn-to-turn and ground insulation of the field winding. Overheating and thermal aging, conductive or abrasive contamination, load cycling, centrifugal forces, and/or vibration may decrease the dielectric strength and eventually cause puncturing of the insulation [55].

In the case of an inter-turn short circuit fault in the field winding, the magnetomotive force of the relevant pole is reduced, resulting in a weaker magnetic field. This decreases the air gap magnetic flux density between the pole shoe of the faulty pole and the stator core, and leads to an asymmetric magnetic field. This results in unbalanced magnetic forces, leading to higher bearing vibrations and noise [56]. In addition, the local overheating is observed in the field winding, and the output of the synchronous generator is reduced due to the lower average magnetic field in the air gap [57]. Notably, the SC turns have a significant impact on the synchronous generator with a low number of poles.

Generally, the insulation in a turn-to-turn fault is damaged faster than the ground insulation [55]. The single-turn short circuit fault or a few turns can only marginally affect the synchronous generator operation. However, the fault may gradually progress and damage more turns. An inter-turn short circuit fault can also eventually develop into an earth fault [57].

The modeling methods for the short circuit fault in the field winding are similar to those used for the stator winding of the synchronous generator. Excitation of proper polarity is assigned to each conductor object, generating alternating positive and negative polarity poles. All coil excitations are linked to the field winding group. The number of conductors of each coil excitation is set equal to the number of turns wound on each pole. An inter-turn fault in the field winding can be simulated by reducing the total number of conductors in the excitation coil. In this case, the specified excitation currents are applied at no-load or on-load conditions [58–61].

8.17 Broken Damper Bar Modeling

Broken damper bar diagnosis methods have been introduced in [62–67]. A broken damper bar or a fractured end ring can be diagnosed by measuring and analyzing the air gap magnetic field of a generator at synchronous and balanced operating conditions. FEM analysis of the air gap magnetic field, both during steady-state and transient operation of the machine, can also help to understand the fault.

Simulation of a broken damper bar fault in a synchronous generator requires the FE model of the complete generator. The reason is that the broken damper bar fault leads to an asymmetrical field distribution in arbitrary locations in one or more poles. However, the applied method can be verified by introducing a broken damper bar in one of the two poles in the FE model. Figure 8.27 shows the external circuit of the synchronous machine with a detailed resistance and inductance

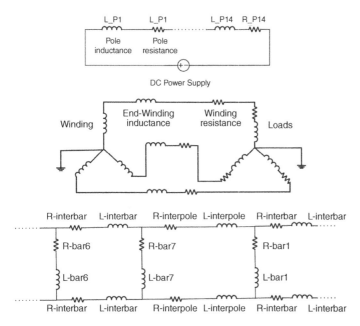

Figure 8.27 The external circuit of the salient pole synchronous generator is linked to a finite element, including the rotor magnetization circuit, stator windings with connected loads, and damper bar circuit connections.

of the damper bars and inter-connections. To simulate the broken damper bar fault in the FEM, the resistivity of the faulty bar must be increased to kΩ, while its healthy resistivity is in the order of $\mu\Omega$. Therefore, a low current passes through the damper bar, even during a broken damper bar fault. In Figure 8.27, damper bar No. 1 (counting from the rightmost bar in the model) is simulated as broken by increasing the external resistance of the damper loop circuit containing the broken bar to a very large value. Figure 8.28 shows the calculated flux distribution with the broken bar and healthy operation of the synchronous generator [59]. The simulated flux signals by FE analysis, with and without a broken damper bar, indicate the concentration of the flux lines in the case of a broken damper bar fault.

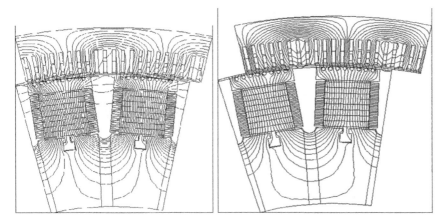

Figure 8.28 Flux distribution in the salient pole synchronous machine operating in the healthy case (left) and under one broken damper bar fault (right).

8.18 Summary

The finite element method can be applied for very precise modeling of synchronous generators, particularly because of the availability of some commercial FE software. This chapter introduced the concept of modeling by FE, particularly in faulty synchronous generators. Electromagnetic field computations with their related mathematical equations were addressed. Methods for iron losses as major losses in the generators were discussed. In FE modeling, different materials, including iron, copper, insulation, and air, must be properly modeled for the computation process in which the ferromagnetic material nonlinearity is the more important part. In the meshing process, more attention must be paid to parts in critical regions, such as air gaps and poles. Both transient and steady-state operations of the synchronous generators are considered in the condition monitoring of the generator and their modeling procedures were studied. Three-dimensional modeling is essential, particularly in a faulty synchronous generator. Although applying the 3D FE is a very time-consuming process, the need for accurate fault diagnosis means that 3D FEM must eventually be applied. Some faults, including the eccentricity fault, the inter-turn fault in the stator winding and excitation winding, and the broken damper bar fault, were evaluated by FEM in this chapter.

References

1 Wexler, A. (1969). Computation of electromagnetic fields. *IEEE Transactions on Microwave Theory and Techniques* MTT-17 (8): 416–439.

2 Rafinejas, P. and Sabonnadiere, J.C. (1976). Finite element computer programs in design of electromagnetic devices. *IEEE Transactions on Magnetics* MAG-12 (5): 575–578.

3 Clough, R.W. (1990). Original formulation of the finite element method. *Finite Elements in Analysis and Design* 7 (2): 89–101.

4 Chari, M.V.K. and Silvester, P. (1971). Finite-element analysis of magnetically saturated DC machines. *IEEE Transactions on Power Apparatus and Systems* 90 (5): 2362–2372.

5 Silvester, P., Cabayan, H.S., and Browne, B.T. (1973). Efficient techniques for finite element analysis of electric machines. *IEEE Transactions on Power Apparatus and Systems* 92, 4: 1274–1281.

6 Fahmy, S.M., Browne, B.T., Silvester, P. et al. (1975). The open-circuit magnetic field of a saturated synchronous machine, an experimental and computational study. *IEEE Transactions on Power Apparatus and Systems* 94 (5): 1584–1588.

7 Bianchi, N. (2005). *Electrical Machine Analysis Using Finite Elements*. CRC Press.

8 Nakata, T., Takahashi, N., Fujiwara, K. et al. (1988). 3-D finite element method for analyzing magnetic fields in electrical machines excited from voltage sources. *IEEE Transactions on Magnetics* 24 (4): 2582–2584.

9 Righi, L.A., Sadowski, N., Carlson, R. et al. (2001). A new approach for iron losses calculation in voltage fed time stepping finite elements. *IEEE Transactions on Magnetics* 37 (5) Part 1: 3353–3356.

10 Salon, S., Chari, M., Sivasubramaniam, K. et al. (2009). Computational electromagnetics and the search for quiet motors. *IEEE Transactions on Magnetics* 45 (3): 1694–1699.

11 Ha, K.H. and Hong, J.P. (2001). Dynamic rotor eccentricity analysis by coupling electromagnetic and structural time-stepping FEM. *IEEE Transactions on Magnetics* 37 (5): 3452–3455.

12 Saitz, J. (2001). Magnetic field analysis of induction motors combining Preisach hysteresis modeling and finite element techniques. *IEEE Transactions on Magnetics* 37 (5) Part 1: 3693–3697.

13 Isaac, F.N., Arkadan, A.A., and El-Antably, A. (1998). Magnetic field and core loss evaluation of ALA-motor synchronous reluctance machines taking into account material anisotropy. *IEEE Transactions on Magnetics* 34 (5) Part 1: 3507–3510.

14 Li, S. and Hofmann, H. (2003). Numerically efficient steady-state finite element analysis of magnetically saturated electromechanical devices using a shooting newton/gmres approach. *IEEE Transactions on Magnetics* 39 (6): 3481–3485.

15 Tandon, S., Richter, E., and Chari, M. (1980). Finite elements and electrical machine design. *IEEE Transactions on Power Apparatus and Systems* MAG-16, 5: 1020–1022.

16 Preston, T.W. and Reece, A.B.J. (1983). The contribution of the finite-element method to the design of electrical machines: an industrial viewpoint. *IEEE Transactions on Magnetics* MAG-19, 6: 2375–2380.

17 Hiramatsu, D., Tokumasu, T., Fujita, M. et al. (2012). A study of rotor surface losses in small to medium cylindrical synchronous machines. *IEEE Transactions on Energy Conversion* 27 (4): 813–821.

18 Davidson, J. and Balchin, M. (1983). Three-dimensional eddy-current calculation by network methods. *IEEE Transactions on Magnetics* 19 (6): 2325–2328.

19 Hofmann, H. and Sanders, S.R. (2000). High-speed synchronous reluctance machine with minimized rotor losses. *IEEE Transactions on Industry Applications* 36 (2): 531–539.

20 Dlala, E. (2011). Efficient algorithm for the inclusion of the Preisach hysteresis model in nonlinear finite element methods. *IEEE Transactions on Magnetics* 47 (2): 395–408.

21 Bertotti, G. (1988). General properties of power losses in soft ferromagnetic materials. *IEEE Transactions on Magnetics* 24 (1): 621–630.

22 Fratila, M., Benabou, M.A., Tounzi, A. et al. (2017). Iron loss calculation in a synchronous generator using finite element analysis. *IEEE Transactions on Energy Conversion* 32 (1): 640–648.

23 Tacca, H.E. (2002). Extended Steinmetz equation. Thayer School of Engineering, India, Postdoctoral Research Report.

24 Hargreaves, P.A., Mecrow, B.C., and Hall, R. (2011). Calculation of iron loss in electrical generators using finite element analysis. In: *International Electric Machines Drives Conference (IEMDC)*, 1368–1373.

25 Findlay, R.D., Stranges, N., and MacKay, D.K. *Losses Due to Rotational Flux in Three-Phase Induction Motors*. Boston, MA, USA: Springer.

26 Salon, S.J. (1995). *Finite Element Analysis of Electrical Machines*. Norwell, MA, USA: Kluwer Academic Publishers.

27 Yamada, S. and Bessho, K. (1988). Harmonic field calculation by the combination of finite element analysis and harmonic balance method. *IEEE Transactions on Magnetics* 24 (6): 2588–2590.

28 Lee, W. and Hong, J.P. (2006). Object-oriented modeling of an interior permanent magnet synchronous motor drives for dynamic simulation of vehicular propulsion. In: *IEEE Vehicle Power and Propulsion Conference, VPPC'06*, 6–8 September 2006, 1–6. Windsor, United Kingdom.

29 Michon, M., Calverley, S.D., Powell, D.J. et al. (2005). Dynamic model of a switched reluctance machine for use in a SABER based vehicular system simulation. In: *Industry Applications Conference, 4th IAS Annual Meeting, Conference Record*, vol. 4, 2280–2287. Hong Kong, China.

30 Mohammed, O.A., Liu, S., and Liu, Z. (2006). FE-based physical phase variable models of electric machines and transformers for dynamic simulations. In: *IEEE Power Engineering Society General Meeting*, 18–22 June 2006. Montreal, QC, Canada.

31 Greenwood, A. (1991). *Electrical Transient in Power System*, 2e. New York: Wiley-Interscience.

32 Demerdash, N., Hamiltom, H., and Brown, G. (1972). Simulation for design purposes of magnetic fields in turbogenerators with symmetrical and asymmetrical rotors, Part I: Model development and solution techniques. *IEEE Transactions on Power Apparatus and Systems* PAS-91 (5): 1985–1992.

33 Nehl, T.W., Fouad, F.A., Demerdash, N. et al. (1982). Dynamic simulation of radially oriented permanent magnet type electrically operated synchronous machines with parameters obtained from finite element field solution. *IEEE Transactions on Industry Applications* IA-18 (2): 172–182.

34 Deaconu, S.I., Topor, M., Tutelea, L.N. et al. (2018). Wind or hydro homo-heteropolar synchronous generators: Equivalent magnetic circuit and FEM analysis. In: *MATEC Web of Conferences 210*, 02008, 1–7. CSCC.

35 Chan, T.F. and Lai, L.L. (2010). Performance of an axial flux permanent magnet synchronous generator from 3-D finite element analysis. *IEEE Transaction on Energy Conversion* 25 (3): 669–676.

36 Wang, J.S. (1996). A nodal analysis approach for 2D and 3D magnetic circuit coupled problems. *IEEE Transactions on Magnetics* 32 (3): 1074–1077.

37 Zhong, D. (2010). Finite element analysis of synchronous machines, PhD Thesis, The Pennsylvania State University, USA.

38 Tu, X., Tian, Q., Gao, Y. et al. (2007). Studies of the internal fault simulation of high voltage cable-wound generator. *IEEE Transactions on Energy Conversion* 22 (2): 240–249.

39 Faiz, J. and Ebrahimi, B.M. (2006). Mixed fault diagnosis in three-phase squirrel-cage induction motor using analysis of the air-gap magnetic field. In: *Progress In Electromagnetics Research*, vol. 64, 239–255. PIER.

40 Tu, X., Dessaint, L.A., Kahel, M.E. et al. (2006). A new model of synchronous machine internal faults based on winding distribution. *IEEE Transactions on Industrial Electronics* 53 (6): 1818–1828.

41 Reichmeider, P.P., Gross, C.A., Querrey, D. et al. (2000). Internal faults in synchronous machines. Part I: The machine model. *IEEE Transactions on Energy Conversion* 15 (4): 376–379.

42 Kanjun, H.Z. and Yong, Z. Detection of turbine generator field winding serious inter-turn short circuit based on the rotor vibration feature. In: *44th International Universities Power Engineering Conference (UPEC)*, 1–4 September 2009, 1–5. Glasgow, United Kingdom.

43 Nabeta, S.L., Foggia, A., Coulomb, J.-L. et al. (1996). Finite element simulations of unbalanced faults in a synchronous machine. *IEEE Transactions on Magnetics* 32 (3): 1561–1564.

44 Dallas, S.E., Safacas, A.N., and Kappatou, J.C. (2012). A study on the transient operation of a salient pole synchronous generator during an inter-turn stator fault with the different number of short-circuited turns using FEM. In: *International Symposium on Power Electronics, Electrical Drives, Automation and Motion (SPEEDAM)*, June 19–22, 2012, 1372–1377. Italy: Sorrento.

45 Dallas, S.E., Safacas, A.N., and Kappatou, J.C. (2011). Inter-turn stator faults analysis of a 200-MVA hydro-generator during transient operation using FEM. *IEEE Transactions on Energy Conversion* 26 (4): 1151–1160.

46 Dallas, S.E., Kappatou, J.C., and Safacas, A.N. (2010). Analysis of transient phenomena of a salient pole synchronous machine during inter-turn stator short-circuits using FEM. In: *XIX International Conference on Electrical Machines (ICEM)*, September 6–8, 2010, 1–6. Rome, Italy.

47 Šašić, M., Lloyd, B., and Elez, A. (2012). Finite element analysis of turbine generator rotor winding shorted turns. *IEEE Transactions on Energy Conversion* 27 (4): 930–937.

48 Dallas, S.E., Kappatou, J.C., and Safacas, A.N. (2010). Investigation of transient phenomena of a salient pole synchronous machine during field short-circuit using FEM. In: *International*

Symposium on Power Electronics Electrical Drives Automation and Motion (SPEEDAM), June 14–16, 2010. Pisa, Italy.

49 Wamkeue, R., Kamwa, I., and Chacha, M. (2002). Line-to-line short-circuit-based finite-element performance and parameter predictions of large hydrogenerators. *IEEE Transactions on Energy Conversion* 22 (11): 370–378.

50 Iamamura, B.A.T., Le Menach, Y., Tounzi, A. et al. (2010). Study of inter-turn short circuit in rotor windings of a synchronous generator using FEM. In: *14th Biennial IEEE Conference on Electromagnetic Field Computation (CEFC)*, 9–12 May 2010. Chicago, Illinois, USA.

51 Fayazi, M. and Haghjoo, F. Turn to turn fault detection and classification in stator winding of synchronous generators based on terminal voltage waveform components. In: *Power Systems Protection and Control Conference (PSPC)*, January 14–15, 2015. Tehran, Iran.

52 Barendse, P.S. and Pillay, P. A new algorithm for the detection of faults in permanent magnet machines. In: *32nd IEEE Annual Conference on Industrial Electronics*, IECON7, 7–10 November 2006. Paris, France.

53 Faiz, J., Mahmoodi, A., Keravand, M. et al. (2019). Diagnosis of inter-turn fault in stator winding of turbo-generator. *International Transactions on Electrical Energy Systems* 29 (12) https://doi.org/10.1002/2050-7038.12132.

54 Guo, W., Ge, B.-J., Gao, Y.-P. et al. (2015). Negative-sequence component analysis of an AP1000 nuclear turbo-generator in an internal short-circuit condition. *Electric Power Components & Systems* 43: 633–643.

55 Stone, G.C., Sasic, M., Stein, J. et al. (2011). Using magnetic flux monitoring to detect synchronous machine rotor winding shorts. In: *58th Annual IEEE Petroleum and Chemical Industry Conference (PCIC)*, Record of Conference Papers Industry Applications Society, Toronto, ON, Canada, 9–11 September 2011, 1–7.

56 Kokoko, O., Merkhouf, A., Tounzi, A. et al. (2018). Detection of short circuits in the rotor field winding in large hydro generator. In: *XIII International Conference on Electrical Machines*. Alexandroupoli, Greece.

57 Tavner, P., Ran, L., Penman, J. et al. (2008). *Condition Monitoring of Rotating Electrical Machines*. United Kingdom, London: The Institution of Engineering and Technology.

58 H. Ehya, A. Nysveen and J. A. Antonino-Daviu, "Advanced Fault Detection of Synchronous Generators Using Stray Magnetic Field," in *IEEE Transactions on Industrial Electronics*, vol. 69, no. 11, pp. 11675–11685, Nov. 2022, doi: 10.1109/TIE.2021.3118363.

59 H. Ehya, A. Nysveen, J. A. Antonino-Daviu and B. Akin, "Inter-turn Short Circuit Fault Identification of Salient Pole Synchronous Generators by Descriptive Paradigm," 2021 *IEEE Energy Conversion Congress and Exposition (ECCE)*, 2021, pp. 4246–4253, doi: 10.1109/ECCE47101.2021.9595993.

60 H. Ehya and A. Nysveen, "Pattern Recognition of Interturn Short Circuit Fault in a Synchronous Generator Using Magnetic Flux," in *IEEE Transactions on Industry Applications*, vol. 57, no. 4, pp. 3573–3581, July-Aug. 2021, doi: 10.1109/TIA.2021.3072881.

61 H. Ehya, A. Nysveen and R. Nilssen, "Pattern Recognition of Inter-Turn Short Circuit Fault in Wound Field Synchronous Generator via Stray Flux Monitoring," 2020 *International Conference on Electrical Machines (ICEM)*, 2020, pp. 2631–2636, doi: 10.1109/ICEM49940.2020.9270986.

62 Yun, J., Park, S., Yang, C. et al. (2019). Airgap search coil-based detection of damper bar failures in salient pole synchronous motors. *IEEE Transactions on Industry Applications* 55 (4): 3640–3648.

63 Karmaker, H.C. Broken damper bar detection studies using flux probe measurements and time-stepping finite element analysis for salient-pole synchronous machines. In: *4th IEEE*

International Symposium on Diagnostics for Electric Machines, Power Electronics and Drives (SDEMPED), 193–197. Stone Mountain, Georgia, USA, 24–26 August 2003.

64 Elkasabgy, N., Eastham, A., and Dawson, G. (1992). Detection of broken bars in the cage rotor on an induction machine. *IEEE Transactions on Industry Applications* 28 (1): 165–171.

65 Vetter, W. and Reichert, K. (1994). Determination of damper winding and rotor iron currents in converter- and line- fed synchronous machine. *IEEE Transactions on Energy Conversions* 9 (4): 709–716.

66 H. Ehya and A. Nysveen, "Comprehensive Broken Damper Bar Fault Detection of Synchronous Generators," in *IEEE Transactions on Industrial Electronics*, vol. 69, no. 4, pp. 4215–4224, April 2022, doi: 10.1109/TIE.2021.3071678.

67 Ehya, H., Nysveen, A., Nilssen, R. et al. Time domain signature analysis of synchronous generator under broken damper bar fault. In: *IECON 2019 – 45th Annual Conference of the IEEE Industrial Electronics Society*, 1423–1428. Lisbon, Portugal, 14–17 October.

9

Thermal Analysis of Synchronous Generators

9.1 Introduction

Design and analysis of electrical generators is a multi-disciplinary process, generally comprising electromagnetic, thermal, and mechanical analysis in a highly repetitive process [1]. However, less attention has been paid to its thermal analysis. A convincing cause is that electrical generator designers usually have electrical engineering knowledge, whereas thermal analysis is a mechanical engineering science [2]. Furthermore, this subject has high intricacy and non-linearity because of its fluid flow nature [3]. However, due to the powerful capability of modern computers, the complicated and long computations can be performed in a short time. The general aim of thermal analysis of electrical generators is enhancement of their efficiency. The impact of faults on the thermal behavior and thermal aging of the synchronous machine is also an important issue that must be considered for maintenance and replacement planning.

Since synchronous generators are widely used in electrical power plants, their optimal thermal analysis is very important. Analysis of the thermal behavior of the generator is necessary for the energy-efficiency enhancement, energy generation cost reductions, power density increments, and reductions in size and weight of the equipment in some applications, such as in aircraft and ships.

The load of the synchronous generator may vary during its operation. It may operate under unbalanced, overload, over-excited (leading power factor), or high reactive power absorption conditions. The distorted magnetic flux distribution generates higher core losses in the stator. These mentioned conditions cause thermal stress in different parts of the generator, leading to over-heating and heat concentration, especially in the windings. This can lead to insulation system failure, demagnetization of the permanent magnets (PMs) in synchronous generators, short circuit turns, coil displacement, and rotor thermal instability [4]. These conditions ultimately lead to a reduction in the generator lifespan. Thermal analysis and precise prediction of the temperature distribution in the generators are therefore essential for the optimal design and good operation of the generator. The proper thermal design of generators requires precise calculation of power losses, reasonable modeling of the cooling system, and material property checking due to temperature rises.

Electromagnetic Analysis and Condition Monitoring of Synchronous Generators, First Edition. Hossein Ehya and Jawad Faiz.
© 2023 The Institute of Electrical and Electronics Engineers, Inc. Published 2023 by John Wiley & Sons, Inc.

Figure 9.1 Concept of coupled electromagnetic and thermal analysis [6].

A combination of thermal and electromagnetic analysis is essential to achieve the best results. The reason is that the losses in electromagnetic analysis and thermal analysis results are interdependent. Electromagnetic and thermal models can be coupled by calculating the losses in the various parts of the generator and then feeding them into the thermal model; this process reiterates to converging solutions [5]. Figure 9.1 presents the concept of coupled electromagnetic and thermal analysis.

Traditionally, the thermal behavior of a synchronous generator is predicted based on a few simple parameters, such as the housing heat transfer coefficient, the winding current density, the generator thermal resistance, or the use of simple and elegant empirical equations based on data fitting like Brostrom's formula [7, 8]. Many experienced synchronous generator manufacturing companies use spreadsheet calculations based on equivalent circuits to predict their thermal behavior [9]. Therefore, designers need a faster modeling technique with relatively high accuracy and cost-effectiveness.

Analysis of the thermal aspects of synchronous generators using existing methods can generally be divided into analytical and numerical methods. The analytical method includes the lumped parameter thermal network (LPTN) method, also called the nodal or thermal network method [10]. This fast calculation method has reasonable accuracy. However, in this method, thermal resistances are estimated based on semi-empirical equations prepared by many experiments. The reason is that many complex thermal phenomena in electric generators cannot be solved by purely mathematical approaches. In most cases, empirical data must be used to calibrate analytical models and achieve accurate results [10].

Unlike the analytical method, the numerical method can be further divided into finite element analysis (FEA) and computational fluid dynamics (CFD). Extensive studies using CFD and FEA on numerous electrical machines and operating conditions have generated a high level of insight regarding their thermal behavior, which is now taken into account in the creation of generator thermal models [11]. Both numerical methods have high accuracy, but they are time-consuming techniques and are usually carried out by different conventional software tools in which useful equations and processes are already combined. One of the main advantages of numerical analysis is that any electrical generator geometries can be modeled, although this is limited by model setup and computational time [2]. Analytical and numerical modeling techniques must not compete against each other, but they can be combined in the same design process. The calculation speed advantages that characterize analytical tools should be integrated with the high accuracy provided by numerical methods [9].

9.2 Overview of Thermal Modeling and Analysis

Thermal modeling of a 22.5 kVA synchronous generator has been proposed by coupling FEA analysis and the lumped parameters network. Indeed, the use of FEA was used to calculate iron-loss distribution in the generator and lumped circuit coefficients (LCCs) were used in the lumped parameters model, which calculates both the air flow and heat transfer in the generator. The maximum differences between the stator and rotor temperature rise were determined as 10% using the model nodes with the experimental results [12]. A thermal analysis of a 5 kW generator with salient pole and claw pole rotors has been carried out using the analytical lumped model and FEA software, but no comparison has been provided [13].

To speed up the CFD computations in large hydro-generators, some techniques have been applied, and the results are compared in reference [14]. Some impressive points are evident in the CFD computations. They include the meshing order, mesh distribution through the geometrical and flow features, and cell size adjustment near solid wall boundaries to ensure that the selected turbulence modeling options are within acceptable limits.

The thermal design of superconducting generators has been discussed in reference [15]. The FEA has been applied to analyze coupled electromagnetic and thermal analysis, but CFD analysis did not apply to this model despite its complex cooling system. Critical parameters for the calculation of conduction paths and for modeling the convection paths of electrical open self-ventilated (OSV) cooling systems have been investigated in detail in reference [3] and the OSV system has been subjected to analytical thermal analysis. This cooling method is used widely in large salient-pole synchronous generators for marine and industrial applications.

Many papers have been published in the field of thermal analysis and modeling of PM and synchronous reluctance machines (SYNRMs) for both generators and motors. Due to the similarity of their structures, the published information could be useful for enhancement of the accuracy of thermal modeling of synchronous generators. For instance, the lumped parameter thermal model of a typical permanent magnet synchronous generator (PMSG) for gas turbine applications has been proposed in reference [16], but friction and windage losses have been passed up and the final results did not compare with others for validation. The thermal model of a dual stator five-phase PMSG in a lumped circuit and FEA mode have also been investigated in reference [17]. The lumped analysis has been done just in the radial direction, but the FEA was in a 2D model. Also, the results did not match each other, especially in transient modes.

An LPTN has been developed in reference [10] for PMSM with 12 nodes and 17 thermal resistances between them with some simplifying assumptions such as uniformity of losses distribution in the winding and core and ignoring thermal conductivity in the axial direction toward the radial direction. The results have been validated with temperature measurements. Another LPTN of a typical PMSG has also been used in two simple modes with seven nodes and a complicated mode with 15 nodes [7]. The obtained results have been compared with FEA. A simplified LPTN has been introduced for a double excitation synchronous generator (DESG) in reference [18], where 9 nodes in steady-state and transient modes have been considered. Radiative heat transfer has also been taken into account in this model. Finally, some experimental calibrations have been applied for some coefficients. The thermal behavior of two SynRMs and two induction motors with the same output powers and same stators has been compared in reference [19] using the lumped parameters thermal model.

Two types of analytical analysis containing a lumped parameter thermal model and an analytical model, derived from Brostrom's formula, have been reported in reference [20] for a typical SynRM machine. The methods were then compared and the pros and cons of each method were investigated. The apparent consensus reached in references [10], [21], and [22] was that accurate determination of the parameters in lumped parameter thermal modeling requires a calibration using optimization techniques.

9.3   Thermal Modeling and Analyzing Synchronous Generators

Several analytical and numerical methods with varying complexities are available for thermal modeling and analyzing synchronous generators.

9.3.1   Analytical Method

The analytical LPTN is based on building up a thermal flow network using the lumped model. This method could be complicated and large design and geometrical dimensions data of the generator are required. This is similar to the permeance network of electromagnetic design in the simulation of electrical circuits [6]. The foundation of this method is based on the theory of heat transfer. The heat-transfer analysis is similar to an electrical network analysis. Table 9.1 shows the analogs of thermal and electrical quantities [7].

Application of the method requires determining several temperature nodes inside the generator components, such as the stator yoke, stator teeth, rotor, and winding. The thermal resistances and capacitances, which accurately model the nature and path of the heat transfer, are then defined between these nodes. A conflict always arises between the number of nodes and the accuracy of the network; therefore, a node increase in all parts of the generator can improve the model accuracy.

One point to note is that increasing the number of nodes from one place to another does not significantly improve the accuracy; it only increases the complexity of the network and leads to a long computation time. However, for a simpler thermal model, more empirical data are required to solve it. The defined thermal resistances are calculated based on conduction, convection, and radiation. The heat sources, depending on their physical location in the generator, are then connected to the identified nodes. The heat sources correspond to losses near the different generator parts and can include copper loss in the windings, core loss in the laminations, windage loss in the air gap, or mechanical loss in the shaft (bearings) [5]. Figure 9.2a shows a simple lumped parameters circuit network and Figure 9.2b displays the output thermal network of a typical synchronous generator obtained by running Ansys Motor-CAD software [23]. The nodes placed in each generator component represent a lumped area temperature. These nodes are connected with thermal resistances due to conduction, convection, or radiation heat transfer. Thermal capacitance is connected within nodes and ambient heat for transient thermal simulations of synchronous generators under faulty or duty cycle load conditions. The heat generation losses are then input to these nodal locations.

The Ansys Motor-CAD software is a novel lumped-circuit-based computer-aided design software package dedicated to thermal analysis of many types of rotating machines, including induction machines (IMs), switched reluctance machines (SRMs), synchronous machines (SYNCs), etc. The software was developed in 1999 and has since been tested in a wide range of scenarios in

Table 9.1   Thermal and electrical analysis equalities.

Electrical quantities	Symbol	Unit	Thermal quantities	Symbol	Unit
Potential	V	V	Temperature	T	°K
Current	I	A	Heat	Q	W
Current density	J	A/m^2	Heat flux	φ	W/m^2
Capacitance	C_{el}	F	Thermal capacitance	C_{th}	J/K
Conductivity	σ	S/m	Thermal conductivity	k	W/m.K
Resistance	R_{el}	Ω	Thermal resistance	R_{th}	K/W

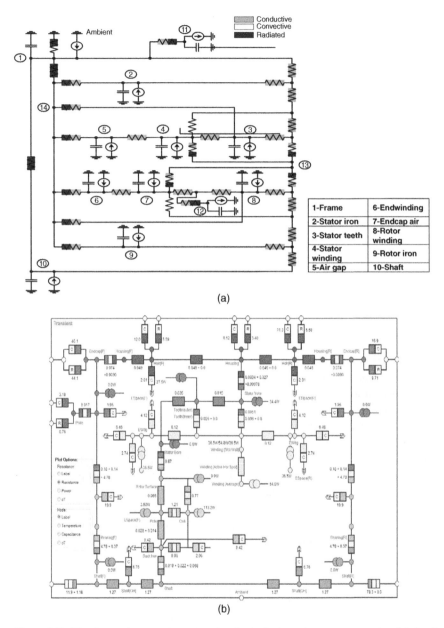

Figure 9.2 Two lumped parameter networks of a typical synchronous generator: (a) simple lumped parameter circuit network [5]/SAE International and (b) output thermal network.

the automotive, aerospace, and industrial areas [24]. It offers a quick computation that considers heat transfer paths through conduction, convection, and radiation mechanisms. All types of cooling methods, such as fan ventilation, spray cooling, and water jackets, have been included in the package. For instance, by setting up a synchronous generator geometric dimensions and material composition, a thermal network containing thermal resistances, power sources, thermal capacitances, and nodal temperatures is created. At their end, the designers can receive numerical (FEA) and analytical (lumped parameters) outputs that facilitate the design procedure and help to provide a broad analysis of the results [9]. A package that couples electromagnetic and thermal

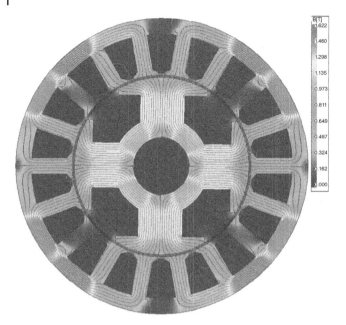

Figure 9.3 A detailed magnetic flux distribution across the synchronous generator.

analysis is also available in this software. In fact, the losses and temperatures are passed between the electromagnetic and thermal models until the program converges on the solution. Figure 9.3 shows a magnetic flux distribution across the synchronous generator that is closely related to iron magnitude and distribution of iron loss.

The users can modify the generator geometric features and all types of material properties defined in the database, such as density, thermal conductivity, specific heat capacity, etc. Therefore, this software can be a very useful tool in synchronous generator thermal design and analysis, especially for non-specialists in thermal analysis of electrical machines [24]. Another attractive option offered by this tool is that it can be coupled to ANSYS Fluent for CFD thermal analysis.

As previously reported, the implementation of the lumped thermal parameter model first requires the determination of the thermal resistances, which are based on heat transfer theory and involve the three types of heat: conduction, convection, and radiation.

9.3.1.1 Heat Conduction

Heat is transferred in solids by the conduction mechanism because of molecular vibration. Typically, materials with a superior crystalline structure have high thermal conductivity and noticeably less thermal resistivity [3]. The fundamental law for conduction is

$$\varphi = -k\nabla T \tag{9.1}$$

where φ is the heat flux, k is the thermal conductivity, and T is the temperature. The heat flux in a complex body is expressed by a partial differential equation (PDE); however, in the case of a simple and homogeneous body, the equation can be a simple algebraic equation. The conduction heat transfer occurs in both the solid and motionless gaseous parts and can be calculated as follows:

$$R_{conduction} = \frac{L}{kA} \tag{9.2}$$

where $R_{conduction}$ is the thermal resistance, L is the length, and A is the area. Table 9.2 shows the range of thermal conductivities of common materials at 20 °C [25].

The challenging issue is assigning an admissible value to L for thermal resistances because of the interface gap between generator components [2]. This parameter is determined based on

Table 9.2 Thermal conductivities of common materials at 20 °C [25]/EOMYS Engineering.

Material	k (W/m/K)
Air	0.026
PVC	0.15
Epoxy	0.25
Water	0.6
Stainless steel	30
Cast iron	50
Aluminum	230
Copper	390

the material type and hardness of the interconnected components of synchronous generators. For instance, the interface gap for aluminum–aluminum ranges from 0.0005 to 0.0025 mm, aluminum–stainless steel from 0.006 to 0.009 mm, and stainless steel–stainless steel from 0.007 to 0.015 mm. These values can be used as a first approximation for the interface gaps in synchronous generator analysis. Detailed investigation into the calculation of interface gaps has been done in reference [26].

Another challenge regarding conduction heat transfer is calculation of the thermal conductivity of the slots in the synchronous generator. The slots consist of materials with a high thermal sensitivity and the thermal conductivity of the slot is lower than other parts, so this has a significant effect on the lifespan of the synchronous generator. Thermal modeling of the slot does not require modeling of the position of every single conductor, and nor is this even possible. Accordingly, calculating the thermal conductivity of the slot has been addressed by different approaches, such as composite thermal conductivity, direct equations based on the conductor geometries, and T-equivalent circuits for the thermal resistance [3, 26].

9.3.1.2 Heat Convection

A significant part of the heat generation in synchronous generators that is dissipated to the environment is due to the heat convection phenomenon. Modeling and analyzing convection heat transfer is challenging due to the involvement of fluid flow movement; thus, for the calculation of convection heat transfer, empirical dimensionless correlations must be used [27].

The convection resistance $R_{convection}$ is calculated as follows:

$$R_{convection} = \frac{1}{hA} \tag{9.3}$$

where h is the convection coefficient, which is the challenging parameter in the calculation of the convection resistance. The accurate prediction of h leads to a higher accuracy of the analytical LPTN. Generator manufacturers usually have their own empirically determined numbers [28]. These empirical correlations have been obtained by the dimensionless number, e.g., Prandtl (P_r) and Reynolds (R_e) numbers [29].

The correlation between h and the Nusselt number (N_u) (the ratio of convective to conductive heat transfer across a boundary) is defined as follows:

$$h = \frac{N_u K}{L} \tag{9.4}$$

where K is the fluid flow conductivity and L is the characteristic length of the surface. Various empirical correlations suitable for the calculation of the convection coefficient of different parts of

Table 9.3 Range of convective heat transfer coefficient [25]/EOMYS Engineering.

Material	h (W/m^2/K)
Air (natural convection)	5–10
Air (forced convection)	10–300
Water (forced convection)	500–10 000

the generator have been discussed in reference [3]. Table 9.3 shows the range of convective heat transfer coefficients for air and water [25].

9.3.1.3 Heat Radiation

Thermal radiation is heat dissipation into the environment through electromagnetic waves. Radiation can occur regardless of the material type, including solids or fluids [20]. The radiative thermal resistance $R_{radiation}$ is as follows:

$$R_{radiation} = (T_1 - T_0)/\left(\sigma \varepsilon F\left(T_1^4 - T_0^4\right) A\right) \tag{9.5}$$

where σ is the Stefan–Boltzmann constant ($\sigma = 5.67; 10^{-8}$ W/m^2); and ε is the surface emissivity ($0 < \varepsilon < 1$) relative to a black body. The data for calculating emissivity is given in references [7] to [14]. In radiative heat transfer, a view factor F is the proportion of the radiation that leaves the surface, which is equated to 1 for the dissipating surface and 2 for the absorbing surface, and T_1 and T_2 are, respectively, the temperatures in Kelvin of surfaces 1 and 2, and A is the surface area. The F can be estimated as reported in references [27] and [28]. Table 9.4 shows the range of emissivity values of common materials at 20 °C [25].

Heat radiation strongly depends on the surface and ambient temperatures and on the surface properties, which are characterized by the emissivity coefficient. This coefficient is not constant; instead, it varies with the temperature, color, and surface of the radiating material, which leads to a very complicated analysis. For a quick calculation, heat radiation is taken to be constant in all simulations or is ignored because of the poor effect of this phenomenon compared to others due to the relatively low temperature levels of generator surfaces [25].

Table 9.4 Emissivity for common materials at 20 °C [25]/EOMYS Engineering.

Material	ε
Aluminum (polished)	0.05
Aluminum (strongly oxidized)	0.25
Black electrical tape	0.95
Cast iron (polished)	0.21
Copper (polished)	0.01
Copper (oxidized)	0.65
Galvanized steel	0.28
Ideal Black Body	1
Matt paint (oil)	0.9–0.95
Water	0.98

For transient simulation of thermal analysis, thermal capacitances C should be estimated; they must be equal to the considered nodes. Each capacitor in the thermal network represents the heat capacity of the respective part of the machine, C_{th}, as follows [8]:

$$C_{th} = V\rho c \tag{9.6}$$

where V is the volume, ρ is the density, and c is the heat capacity of the material.

The temperature deviation in the axial direction has been ignored in reference [17] and the generator is modeled only in the radial (2D) direction. This thermal analysis model is better suited to approximate design applications and does not have the appropriate level of detail. However, some have proposed a three-dimensional (3D) thermal model for heat transfer in a generator in both the radial and axial directions to improve the accuracy of the model and assess the temperature variations of the coolant [30].

Designing and manufacturing synchronous generators, especially the high-power types, requires the investigation of the design variables that could have a significant impact on the thermal performance of the whole generator. This is where sensitivity analysis becomes crucial, as it allows the recognition of crucial design areas [9]. The lumped parameters model is perfect for sensitivity analysis on manufacturing data, such as interface gaps between components or regions like the stator lamination and external frame. This is very important because most generator losses cross these surfaces [2]. These areas have a significant impact on the analysis, particularly in highly rated generators or generators with complex duty cycle loads.

9.3.2 Synchronous Generator Loss Calculation

The first step in temperature prediction in synchronous generators is to obtain an accurate calculation of the losses [31]. Different operating losses include electrical, magnetic, and mechanical losses generated in various parts that are heat sources in the thermal analysis. These consist of DC losses and AC losses.

The first loss is the stator winding loss. The stator winding AC resistance loss, $P_{copper-s}$, is [32]

$$P_{copper-s} = 3K_R (R_s)_{DC} I_s^2 \tag{9.7}$$

where R_s and I_s are the armature resistance and current, respectively, and the coefficient $K_R > 1$ accounts for the frequency (skin) effect on the stator resistance.

The second loss is the field winding DC loss, $P_{copper-f}$ [32]:

$$P_{copper-f} = R_f I_f^2 \tag{9.8}$$

where R_f and I_f are the field resistance and field current, respectively. For electrical losses, copper losses are temperature dependent because the winding resistance depends on the temperature. To calculate the change in the copper resistivity with temperature, the following equation has been used [31]:

$$R_e = R_0(1 + \alpha(T - T_0)) \tag{9.9}$$

where R_e is the effective electrical resistance at T °C, R_0 is the resistivity of copper at $T_0 = 20$ °C, and α is the temperature coefficient, whose value is 3.93×10^{-3} K^{-1}.

The magnetic losses consist of iron losses in the stator yoke, teeth, and rotor back that can be calculated analytically [32–34] or numerically by FEA. Of course, the iron losses from the rotor and stator lamination are significantly more difficult to estimate than the copper losses, due to their dependency on the frequency and magnetic flux densities. In comparison with the behavior of iron losses in IMs, where these vary significantly with machine loading, iron losses in synchronous generators remain constant with changing load and fluctuate with alterations in the armature voltage [12]. The stator core consists of soft magnetic material, so rotor rotation causes eddy current loss, hysteresis loss, and additional losses from the stator core because of the alternating magnetic field. The core loss in the stator core, P_{core-s}, is calculated by the Steinmetz equation [32]:

$$P_{core-s} = P_h + P_e + P_a = k_h f B_m^n + k_e f^2 B_m^2 + k_a f^{1.5} B_m^{1.5} \tag{9.10}$$

where k_h, k_e, and k_a are the coefficients of hysteresis loss, eddy current loss, and additional loss, respectively, f is the power supply frequency, B_m is the maximum flux density, and n is Steinmetz constant. The unknown values for Equation (9.10) are calculated using the experimental information of the lamination material by the vector fields technique utilized in loss calculations, as it has given good results in the past for a wide range of machines, including synchronous generators [12].

Another additional core loss is the rotor surface magnetic losses produced by the field current due to air gap flux density variation on the rotor pole surface due to the stator slot openings. The following general analytical equation has been proposed [32]:

$$P_{rs} = 0.232 \times 10^6 \Delta \left[(K_{C1} - 1) B_{g0} \tau_s \right]^2 \times 2 p_1 A_{pole} (N_s n)^{1.5} \tag{9.11}$$

where:

Δ is the lamination thickness in the pole shoe (m)
K_{C1} is the stator slotting Carter coefficient
B_{g0} is the no-load air gap flux density (T)
τ_s is the stator slot pitch (m)
A_{pole} is the rotor pole shoe area (m^2)
N_s is the number of stator slots
n is the rotor speed in rps

Additional losses are approximately 0.05–0.2% of the input power in synchronous generators [33]. A more exact analytical approach, now verified by FEA and by some experiments, is given for the rotor surface losses [32].

Mechanical losses are a consequence of bearing friction and windage. Bearing losses depend on the shaft speed, bearing type, properties of the lubricant, and the load on the bearing. According to reference [35], the bearing friction loss $P_{bearing}$ are as follows:

$$P_{bearing} = 0.5 \Omega \mu F D_{bearing} \tag{9.12}$$

where Ω is the angular frequency of the shaft supported by a bearing, μ is the friction coefficient (typically 0.0010–0050), F is the bearing load, and $D_{bearing}$ is the inner diameter of the bearing. Windage losses become more considerable with increasing generator speed. These losses arise due to the friction between the rotating surfaces and the surrounding gas. The equation for calculating windage losses $P_{windage}$ is obtained by modeling the synchronous generator rotor as a rotating cylinder in an enclosure [36]. This equation is

$$P_{windage} = \frac{1}{32} k C_M \pi \rho \Omega^3 D_r^4 l_r \tag{9.13}$$

where k is a roughness coefficient (for a smooth surface $k = 1$, usually $k = 1$–1.4), C_M is the torque coefficient, ρ is the density of the coolant, Ω is the angular velocity, D_r is the rotor diameter, and l_r is the rotor length. The C_M is determined by measurements [12]. More comprehensive explanations and various experimental analytical equations for loss calculations in synchronous generators have been reported in references [32] and [33]. In parallel, FEA could be applied to calculate various component losses in synchronous generators more precisely, but this would require a prohibitive computation time.

Note that, in addition to the losses, the distribution of losses is another important factor in determining the temperature of various parts of the generator, especially the generator hot spot temperature [37]. In fact, a better distribution of losses in the generator leads to a lower temperature of the hot spots and a longer lifespan for the generator. For instance, high conductivity ceramic potting material can be used in the end windings regions to reduce heat concentration in this region (hot spots usually occur in the end winding regions of synchronous generators) [38]. Even the winding pattern has a considerable effect on the temperature distribution around the stator slots. For instance, the influence of the concentrated and distributed windings in the heat distribution of stator slots has been reported in reference [39], which includes details of thermal management considerations.

9.3.3 Numerical Methods

The FEA and CFD are the main numerical methods applicable to synchronous generator heat analysis. The FEA precisely models the heat distribution in solid materials. It is the best technique for conducting electromagnetic and thermal analysis and provides a convenient solution for very complex geometry, which is not possible using lumped parameters [2]. It can model solid component conduction more accurately than the thermal network.

The simulation procedures based on the FEA consist of geometric modeling, material defining, boundary condition settings, meshing or discretizing, and finally solving the problem. This method is especially useful for analyzing new topologies and large power generators [27]. It can also be used along with the lumped parameters method to increase the thermal network accuracy by meticulous accounting of the loss distribution.

Figure 9.4 displays the example of thermal analysis by FEA of two synchronous generators in 2D in Figure 9.4a and 3D in Figure 9.4b. The 3D model shows temperature changes in the axial direction in addition to the radial direction, but it has higher computational processing and is much more time consuming to solve compared to the 2D one.

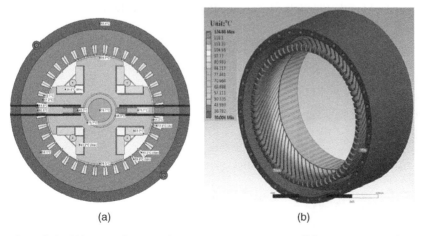

(a)   (b)

Figure 9.4 FEA thermal model of synchronous generators: (a) 2D model and (b) 3D model [40].

The FEA suffers from uncertainty in the computation of thermal resistances due to interface gaps and calculating convection at open surfaces. It is also relatively difficult and slow when inputting a new geometry and has long computation times. Some commercialized FEA software that has thermal packages are ANSYS MECHANICAL [23], ANSYS MOTORCAD [23], JMAG Designer [41], and Siemens Simcenter MotorSolve [42].

CFD is the most precise technique for modeling the heat transfer between two different media. It is employed to model the fluid flow distribution and to determine the coolant flow rate and the convection coefficient. Indeed, when the cooling system involves complex fluid flow systems, FEA is no better than the Lumped Parameter Thermal Network (LPTN). For these cooling systems, CFD can be used successfully [1]. It is a computer-based simulation system involving fluid flow, heat transfer, and associated phenomena, such as chemically reacting flows. It is also a powerful analysis tool for generator design and has the ability to generate vast amounts of air flow and thermal data in a conjugate environment with a higher resolution than alternative solution methodologies [43]. However, this approach has a high setup time and computation costs compared with the FEA or thermal network analysis. Its use is increasing in the modeling of conjugate heat transfer.

Conjugate heat transfer modeling may be used to simultaneously solve conductive and convective heat transfer through neighboring solid and fluid regions, respectively [44]. This method could be 2D or 3D, like FEA, but the 3D analysis is more widely used in thermal modeling and analyzing synchronous generators. Figure 9.5 shows an example of CFD thermal analysis of a synchronous generator and displays the air flow velocity variations in the generator outlet ventilation [45].

The advantage of CFD is that it can be used to predict air flow in complex regions around the generator end windings. Other benefits of using CFD include the ability to obtain the local heat transfer coefficient and local mass flow rate, which can be used in the lumped parameters network [46]. An important point to emphasize is that the data obtained using CFD are useful for improving the analytical algorithms used in the FE model or the thermal networks [2].

From an industrial viewpoint, CFD can also be used to identify possibilities of segmentation of product ranges (e.g. for high efficiency). The importance of mesh independency as part of the CFD process indicates that it should be routinely carried out [43]. For instance, thermal analysis of the convective heat transfer at the end winding bars, stator ducts, pole coils, all sharp and narrow parts, and boundary regions needs a fine mesh toward other surfaces. Changes in meshing alone for a vented, complex synchronous generator topology can affect the heat transfer coefficient by 128%, as shown in reference [43]. Consequently, only after the correct execution of this process can the CFD model be compared to other models or experimental data for validation.

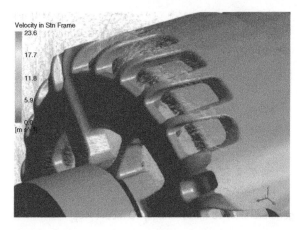

Figure 9.5 CFD thermal analysis of a synchronous generator [46]/HAL Open Science.

Modeling the whole generator geometry with its large dimensions is not possible using fine meshes because the number of elements increases enormously and the computation time increases significantly. Some state-of-the-art methods can be proposed to simplify these numerical models [14].

Some conventional standard software programs for thermal CFD analysis of synchronous methods are ANSYS FLUENT [23], COMSOL METAPHYSICS [47], POWER FLOW [48], SIMSCALE [49], and AUTODESK CFD [50]. These use the finite volume method instead of the finite difference method to numerically solve discretized forms of the Reynolds Averaged Navier–Stokes (RANS) equations at each cell in the solution domain [51]. Recently, this method has been made more attractive by the enhancement of PC computational capability and the availability of more user-friendly pre- and post-processing software; some of which are mentioned above.

Note that all thermal modeling and analyzing methods proposed above are effective with high precision and depend on the project time and costs, the result accuracy required, available resources, and generator type. Therefore, the designer must decide which methods can be used.

Table 9.5 Summary of thermal modeling and analysis results for synchronous generators.

References	Synchronous generator power	Simulation type	Thermal modeling or analysis methods	Software used	Maximum error (%)	Experimental
[27]	400 MW	2D	FEA	FEMM 4.2	3.33	No
[52]	250 MW	2D/3D	FEA	MAGNET 7.5	2	Yes
[53]	40 MW	3D	FEA-CFD	ANSYS MAXWELL, ANSYS MECHANICAL, ANSYS FLUENT	13.3	Yes
[31]	12 MW (superconducting)	2D/3D	FEA	–		No
[28]	6 MW	3D	LPTN-FEA	MODELICA	20	No
[5]	1 MW	2D	LPTN-FEA	MATLAB	–	No
[44]	1 MVA	2D/3D	CFD	ANSYS FLUENT 6	–	
[37]	1 MVA	2D/3D	FEA	–	–	Yes
[54]	250 kVA	3D	CFD	ANSYS FLUENT 13	4	Yes
[55]	30 kW	2D/3D	LPTN	MATLAB	4	Yes
[51]	27.5 kVA	3D	CFD	STAR-CD VERSION 3.05A	6	Yes
[56]	25 kVA	2D	FEA	–	19	No
[12]	22.5 kVA	2D/3D	LPTN-FEA-CFD	MATLAB-Opera-ANSYS FLUENT	9	Yes
[9]	18 kVA	3D	LPTN-CFD	Motor-CAD	19	Yes
[13]	5 kW	2D	FEA	THERMNET	–	No
[18]	3 kW	3D	LPTN		3	Yes

For instance, for thermal modeling or analysis of large power generators or superconductive synchronous generators, the CFD is more practical and accurate for turbulence modeling of cooling fluids and air flow. The lumped parameters method is widely used in sensitivity analysis due to its quick calculation, but it sometimes needs to apply an optimized algorithm like PSO for strategic fitting of uncertain parameters to improve estimation accuracy.

The FEA is also placed between the lumped parameters and CFD in terms of simulation time and accuracy. It could be used to modify lumped parameter thermal networks for accurate determination of loss distributions in the generator or coupled to CFD analysis to incorporate convection heat transfer coefficients determined by CFD into FEA. This method needs a shorter time compared to CFD analysis. Table 9.5 shows a synopsis of the results of thermal modeling and analysis for some synchronous generators investigated, ranging from large high-power models to small low-power models.

Finally, all these methods could be well suited for thermal analysis of all types of synchronous generators, but the difficulty arises when years of operation and periodic maintenance and repair change the generator component characteristics from their primary design specifications. Hence, direct experimental and testing methods or practical methods are applied to determine the thermal performance of the generator.

9.4 Modeling and Analysis of Faulty Synchronous Generators

Normal operation of synchronous generators is very important. To keep the natural conditions of the system, prediction and diagnosis of the faults is an essential part of operation of the system. In particular, for a generator operating continuously for a long duration, a fault must be diagnosed as soon as possible, because operating the generator under a fault leads to breakdown and damage. The impact of any faults must also be fully studied to improve the thermal design of the generator based on the fault tolerance up to the cut-off time of the relays. This is especially important in sensitive applications in which immediate disconnection of the generator is impossible, as this requires particular attention to thermal modeling and analysis of the synchronous generator in the faulty case.

9.4.1 Reasons for Faults in Synchronous Generators

Excessive wear and tear, poor and improper designs, improper assembly, improper use, many fluctuations, overload, over-speed, inadequate ambient temperature, and/or any combination of these can cause a fault in a generator. The main faults in the generator can be classified as follows:

1. Turn-to-turn and grounding of stator windings.
2. Stator faults due to open circuits or different phase types of short circuit (single-phase, two-phase, and three-phase).
3. Abnormal connection of stator windings, such as conductor transposition.
4. Static and dynamic eccentricities.
5. Rotor winding fault consisting of short circuit, turn-to-turn fault, and rotor winding disconnection.
6. Bearing faults.

7. Spark generation and partial discharge.
8. Core faults that destroy the core layer structure, and burning and melting of part of the core due to eddy current losses.
9. External faults, such as a line short circuit near the generator, serious overload, unbalanced load, over-voltage due to disconnection for an unexpected and unpredicted large load on the generator.

These faults in the stator windings and their high impact on the stator winding temperature, and even on the overall generator temperature over a short time, are more predictable than other faults and can be detected by thermal detection methods. The overwhelming percentage of generator insulation faults occur due to thermal problems.

9.4.1.1 Single-Phase Open-Circuit Fault

In this type of fault, one phase of the stator windings at its start or end is open-circuited. Thus, a fault exists at star-connected windings, in which an open circuit of the coils at the center of the star is more likely. In this case, a three-phase synchronous generator is converted to a two-phase generator.

9.4.1.2 Conversion of Three-Phase to Two-Phase

Three-phase symmetrical currents in the stator N-turn windings with $120°$ phase difference generates a rotating MMF with a constant amplitude of $1.5\,I_m N$. The MMF of three-phase and two-phase systems can be made equal by the following changes:

1. Changing the amplitude of the two-phase currents.
2. Changing the number of two-phase winding turns.
3. Changing both the current amplitude and the turn number.

In case one, the current in the two phases must be 1.5 times that of the three-phase current. This can be shown by resolving the three-phase instantaneous MMF along the α-β axes [12].

Considering an equal MMF in two-phase and three-phase windings, and the same number of turns in the two-phase model, the electrical parameters (such as equivalent resistance, etc.) of the two-phase synchronous generator is equal to 2–3 times that of the corresponding parameters of the three-phase synchronous generator. Therefore, the stator winding losses in the two-phase generator is 1.5 times that of the three-phase generator:

$$\frac{P_{loss-3phase}}{3_{phase}} = RI^2 \tag{9.14}$$

$$Z_{thermal} = \frac{T^2}{P_{loss}} \tag{9.15}$$

By applying Equations (9.14) and (9.15) to the lumped equivalent circuit parameter, the synchronous generator in two-phase operation is analyzed thermally. Care must be taken that the designed lumped equivalent circuit parameter is a thermal equivalent circuit of a three-phase symmetrical generator. This equivalent circuit must be converted to a single-phase equivalent circuit and then the thermal analysis is performed for one of the healthy phases. To convert the three-phase lumped equivalent circuit parameter into the single-phase circuit, consider $Z = V^2/P$ in electric circuits, where its equivalent in the thermal equivalent circuits is $Z_{thermal} = T^2/P_{loss}$, and the conversion is done.

Table 9.6 Comparison of errors of different methods results in thermal analysis of a healthy synchronous generator.

Error (%)	FEM	Lumped parameter	Test	Proposed part
		Maximum temperature (°C)		
8.19	38.3	35.4	35.2	Stator winding
2.5	28.3	27.9	28.6	Frame
18.0	31.6	37.1	30.9	Ventilated air
2.0	32.8	32.0	32.3	Stator body
3.5	34.2	35.4	DNA	Tooth
7.2	49.8	53.6	DNA	Shaft
8.0	60.1	56.6	DNA	Rotor winding
7.5	50.6	53.4	DNA	Rotor body

DNA = data not available

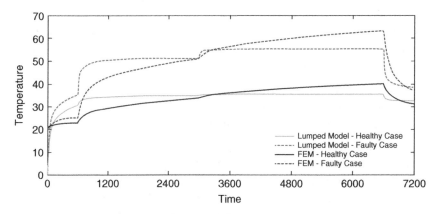

Figure 9.6 Stator winding temperature variations in healthy and faulty (stator phase OC) cases using the lumped parameters method and FEM.

The duty cycle of the generator is chosen according to the test results, as summarized in Table 9.6. The fault was on the stator windings, which have major impacts on the winding; therefore, the two lumped parameter methods and FEM are compared in the healthy and faulty cases. Note that only temperature variations in the stator windings have been proposed.

Figure 9.6 presents the stator winding temperature variations in healthy and faulty (stator phase OC) cases using the lumped parameters method and FEM. It shows the reasonable agreement between the results of applying the two methods.

A 30 V DC supply is used to supply the excitation windings of the synchronous generator. Heat sensors (PT100) from a 12-channel LUTRON BTM-4208SD were attached to different parts of the generator to measure the temperatures of these parts, which include stator end-windings, generator frame, stator yoke, and the generator internal air outlet location (output ventilation). Table 9.6 includes the test results. The impossibility of access to the internal parts and rotor of the generator is due to the impossibility of dismantling the generator and connecting the physical sensors to the rotational parts. Therefore, measuring the temperature of these parts is not possible. According to Table 9.6, the maximum error for the stator winding temperature obtained by the two methods in the proposed healthy synchronous generator is 8.19%.

Table 9.7 Data of thermal test of the proposed generator with single-phase open-circuit fault.

Duty cycle	Period of time (min)	Generator output power (kW)	Generator input torque (Nm)
First	10	3	19
Second	40	12	76
Third	60	14	88
Fourth	10	0	0

Table 9.8 Comparison of the error of the lumped parameters method and FEM in two faulty cases of the synchronous generator.

Error (%)	Maximum temperature in different methods (°C)		Type of fault	Proposed part of synchronous generator
	FEM	LPM		
14.2	63.2	55.3	Open circuit of a phase	Stator windings
9.8	170.6	155.2	Three-phase SC fault	

The above-mentioned test was performed for two hours with four duty cycles on the proposed synchronous generator. The speed of the generator in this period was adjusted at about 1500 rpm. Table 9.7 summarizes the data of the thermal test of the proposed generator with a single-phase open-circuit fault.

Table 9.8 shows that in the generator single-phase OC fault, the maximum temperature in FEM is 63.2 °C and the lumped parameter method is 55.2 °C or 14.2% error.

The temperature difference between the maximum temperature of the stator winding in the healthy and faulty cases using the lumped parameter method and FEM is 23.2 and 19.8 °C, respectively.

Considering the maximum electrical load (14 kW) on the synchronous generator during duty cycles, this fault does not appear to cause high stress on the generator because insulation class B has a maximum temperature tolerance of up to 120 °C. Note that operating the generator as a two-phase machine has a very destructive effect on the output torque ripple and accordingly generates mechanical stress on the rotor windings, shaft, and bearings. This shows the importance of a mechanical analysis in addition to a thermal analysis.

9.4.1.3 Three-Phase Short Circuit Fault

The three-phase SC has the least probability of occurrence (5%) in a synchronous generator compared to other external faults, but this is the most serious type of fault. Generally, the time from the beginning of the SC up to the steady-state mode in a synchronous generator can be divided into three time intervals, as follows (Figure 9.7):

1. Subtransient period: The reason for this period is the reaction of damper bars against the induced current in the initial time of the SC, which mainly lasts for 2–3 time cycles, depending on the generator type.
2. Transient period: The reason for this period is the reaction of the excitation windings against the induced current in them over the initial time of the SC (after the subtransient period). This mainly lasts for 15–20 time cycles, depending on the generator type.
3. Steady-state period: This occurs after the subtransient and transient periods.

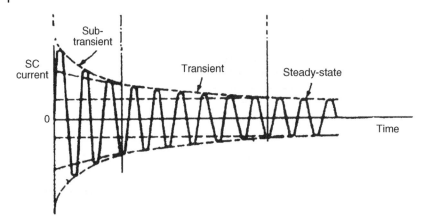

Figure 9.7 Stator phase current waveform of a synchronous generator during the SC period.

For thermal analysis of the proposed 22 kW synchronous generator during the three-phase SC period, two transient and subtransient periods are neglected in the thermal analysis due to the very short time of the occurrence, considering the large time constant (inertia) of the heat phenomenon in the mass. Normally, the over-current relays and/or differential relays are adjusted such that they provide the relay disconnection comment before the end of the transient period. In the case of a weak performance of the relays, the generator enters into the SC steady-state region, which needs thermal modeling and analysis to determine the tolerance level of the stator windings.

To determine the SC current in the steady state, with ignoring the armature winding resistance, we have [21]

$$r_a = 0 \tag{9.16}$$

Under the SC condition, the terminal voltage is also zero, so

$$I_q = 0 \tag{9.17}$$

Therefore,

$$I_{SC} = I_q \tag{9.18}$$

$$I_{SC} = \frac{E_f}{X_d} \tag{9.19}$$

The excitation voltage can be assumed to be almost constant under the steady-state SC condition; therefore, its value is taken to be 1 pu. The estimated X_d in the proposed synchronous generator is 0.47 pu, so the SC current in the steady state is $I_{sc} = 1/0.47 = 2.12$ pu.

The duty cycle of the synchronous generator in the test is chosen from Table 9.5. According to Table 9.6, the only difference is the change in the fourth duty cycle from no-load to a three-phase SC case. For this faulty case, the time period, power, and torque of the generator are not predictable in advance.

Figure 9.8 shows the experimental results. Similar to the previous test in the faulty case, in this diagram, only temperature variations in the stator windings are investigated.

According to Figure 9.8, the results agreed well for the lumped parameter analysis and the FE analysis in the healthy and faulty cases. Due to the same duty cycles of the generator up to 6600 seconds in the healthy and faulty cases, the results of both methods coincide. In the healthy case, after 6600 seconds, the generator is unloaded.

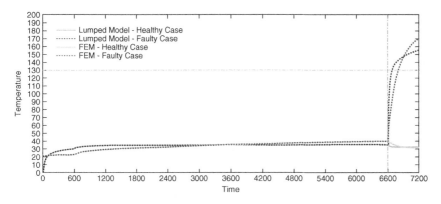

Figure 9.8 Stator winding temperature variations in healthy and faulty (stator three-phase SC fault) cases using analytical lumped thermal parameters and FEM.

The maximum temperature is 179.61 °C in FEM and 155.26 °C in the lumped parameters method, which means a 13% error. Considering the class B insulation, a temperature tolerance higher than 130 °C for stator windings is not possible. At that temperature, the insulation is rapidly destroyed, raising the possibility of SCs between the turns and coils of the stator windings. Therefore, the difference between the time each model takes to approach the temperature of 130 °C can be another criterion for investigating the error between these two models. The lumped parameter model reaches this temperature after 85 seconds and the FE model after 205 seconds. This shows the higher reaction speed of the lumped parameter model to the thermal variations, which agrees with the results reported in reference [9]. Generally, in a lumped parameter model, the convergence speed is higher due to fewer computations and its simplicity. Table 9.8 compares the error of the lumped parameters method and FEM in two faulty cases of the synchronous generator.

The reason for the higher error in the predicted temperature in both methods is that in the phase open circuit compared to the three-phase SC fault in the synchronous generator is the asymmetrical distribution of the heat between the stator three-phase windings in the phase open circuit compared with the three-phase SC.

9.5 Summary

This chapter serves as a reference guide for those who are interested in thermal modeling, analyzing, and managing synchronous generators. Methodologies and evolution of thermal design and analysis of synchronous generators were investigated briefly. Some advantages and disadvantages of the proposed thermal analysis methods were discussed. Some recommendations were provided for those who deal with the thermal modeling or analysis of synchronous generators and even other types of electrical machines [57].

References

1 Mendes, G., Ferreira, A., and Miotto, E. (2020). Coupled electromagnetic and thermal analysis of electric machines. In: *MATBUD'2020 – Scientific-Technical Conference: E-mobility, Sustainable Materials and Technologies*. Cracow, Poland, 19–21 October 2020, Article Number 01052.

2 Boglietti, A., Cavagnino, A., Staton, D. et al. (2009). Evolution and modern approaches for thermal analysis of electrical machines. *IEEE Transactions on Industrial Electronics* 56 (3): 871–882.

3 Ghahfarokhi, P.S., Kallaste, A., Belahcen, A. et al. (2020). Analytical thermal model and flow network analysis suitable for open self-ventilated machines. *IET Electric Power Applications* 14 (6): 929–936.

4 Narain Singh, A. and Cronje, W.A. (2018). Thermal instability analysis of a synchronous generator rotor using direct mapping. *SAIEE Africa Research Journal* 109 (1): 4–14.

5 Loop, B.P., Amrhein, M., Pekarek, S.D. et al. (2009). Modeling, analysis, and control design for an intermittent Megawatt generator. *SAE International Journal of Aerospace* 1 (1): 843–851.

6 Andersson, B. (2013). Analysis of an interior permanent magnet synchronous machine for vehicle applications, MSc Thesis, Department of Energy, Environment Division of Electrical Power Engineering, Chalmers University of Technology, Goteborg, Sweden.

7 Chin, Y.K., Nordlund, E., and Staton, D.A. (2003). Thermal analysis – Lumped-circuit model and finite element analysis. In: *Proceedings of Power Engineering Conference*, 435–440. The Pan Pacific, Singapore, 27–29 November 2003.

8 Zhang, Z., Matveev, A., Øvrebø, S. et al. Review of modeling methods in electromagnetic and thermal design of permanent magnet generators for wind turbines. In: *3rd International Conference on Clean Electrical Power Renewal, Energy Resource Impact, (ICCEP)*. Ischia, Italy, 14–16 June 2011, pp. 377 for author –382.

9 Mejuto, C., Mueller, M., Staton, D. et al. (2006). Thermal modelling of TEFC alternators. In: *IECON Proceedings of the Industrial Electronnics Conference*, 4813–4818. France, 7–10 November 2006.

10 Kačenka, A., Pop, A.C., Vintiloiu, I. et al. (2019). Lumped parameter thermal modeling of permanent magnet synchronous motor. In: *IEEE Electric Vehicles International Conference (EV)*, 1–6. Bucharest, Romania, 3–4 October, 2019.

11 Mejuto, C., Mueller, M., Shanel, M. et al. (2008). Improved synchronous machine thermal modelling. In: *Proceedings of the International Conference on Electric Machines, ICEM*, 1–6. Vilamoura, Portugal, 6–9 September, 2008.

12 Mejuto, C. (2010)., Improved lumped parameter thermal modelling of synchronous generators, PhD Thesis, Department of Engineering, Edinburgh, Scotland.

13 Arumugam, D., Logamani, P., and Karuppiah, S. (2018). Electromagnetic and thermal analysis of synchronous generator with different rotor structures for aircraft application. *Alexandria Engineering Journal* 57 (3): 1447–1457.

14 Klomberg, S., Farnleitner, E., Kastner, G. et al. (2014). Comparison of CFD analyzing strategies for hydro generators. In: *Proceedings of International Conference on Electric Machines, ICEM*, 1990–1995. Berlin, Germany, 2–5 September, 2014.

15 Komiya, M., Sugouchi, R., Sasa, H. et al. (2020). Conceptual design and numerical analysis of 10 MW fully superconducting synchronous generators installed with a novel casing structure. *IEEE Transactions on Applied Superconductivity* 30 (4): 1–7.

16 Anita, S. and Chellamuthu, C. (2015). Thermal modeling of PMSG generator for gas turbine applications Indian. *Journal of Science and Technology* 8 (12): 1–5.

17 Kumar, R.R., Singh, S.K., and Srivastava, R.K. (2018). Thermal modelling of dual-stator five-phase permanent magnet synchronous generator. In: *IEEE Transport Electrification Conference ITEC – India*, 1–6. Pune, India, 13–16 December 2018.

18 T.K. Hoang, L. Vido, F. Gillon, et al. (2017). Thermal Aanalysis of a Ddouble Eexcitation Ssynchronous Mmachine, Electrimacs, Toulouse, France.

19 Boglietti, A., Cavagnino, A., Pastorelli, M. et al. (2006). Thermal analysis of induction and synchronous reluctance motors. *IEEE Transactions on Industry Applications* 42 (3): 675–680.

20 Mahmoudi, M. (2012). Thermal modelling of the synchronous reluctance machine. MSc Thesis, School of Electrical and Computer Engineering, Uiversity of Tehran.

21 Yang, Y., Bilgin, B., Kasprzak, M.M. et al. (2017). Thermal management of electric machines. *IET Electrical Systems in Transportation* 7 (2): 104–116.

22 Boglietti, A., Cossale, M., Popescu, M. et al. (2019). Electrical machines thermal model: advanced calibration techniques. *IEEE Transactions on Industry Applications, May–June* 55 (3): 2620–2628.

23 Ansys Motor–CAD Software, EMag, Therm and Lab. https://www.ansys.com/products/electronics/ansys-motor-cad.

24 Staton, D., Hawkins, D., and Popescu, M. (May, 2009). Thermal behaviour of electrical motors – An analytical approach program. *CWIEME, Berlin, Germany* 5–7: 1–8.

25 Jandaud, P.O. and Le Bensneraisl, J. (2017). Heat transfer in electric machines, *EOMYS Engineering*. In: , 1–41. www.eomys.com.

26 Staton, D., Boglietti, A., and Cavagnino, A. (2003). Solving the more difficult aspects of electric motor thermal analysis. In: *IEMDC 2003 – IEEE International Electric Machine Drives Conference*, vol. 2, no. 3, 747–755. Wisconsin, USA, June 1–4, 2003.

27 Alam, H.S., Djunaedi, I., and Soetraprawata, D. (2017). Thermal study for power uprating of 400 MW generator using numerical simulation. *International Journal of Materials, Mechanics and Manufacturing* 5 (3): 191–195.

28 Centner, M. and Sabelfeld, I. (2012). Coupled fluid-thermal network modeling approach for electrical machines. In: *Proceedngs 20th International Conference on Electric Machines, ICEM*, 1238–1241. Marseille, France, 2–5 September 2012.

29 Gai, Y., Kimiabeigi, M., Chuan Chong, Y. et al. (2019). Cooling of automotive traction motors: schemes, examples, and computation methods. *IEEE Transactions on Industrial Electronics* 66 (3): 1681–1692.

30 Traxler-Samek, G., Zickermann, R., and Schwery, A. (2010). Cooling airflow, losses, and temperatures in large air-cooled synchronous machines. *IEEE Transactions on Industrial Electronics, Jan.* 57: 172–180.

31 Zhang, B., Qu, R., Wang, J. et al. (2013). Stator thermal design of a 12 MW superconducting generator for direct-drive wind turbine applications. In: *International Conference on Electrical Machine Systems, ICEMS*, 1473–1477. Busan, South Korea, 26–29 October 2013.

32 Boldea, I. (2016). *The Electric Generators Handbook, Synchronous Generators*, 2e. Boca Raton, FL, United States: CRC Press.

33 Pyrhönen, J., Jokinen, T., and Hrabovcová, V. (2008). *Design of Rotating Electrical Machines*. United Kingdom: Wiley.

34 Xiong, K., Li, Y., Li, Y.Z. et al. (2020). Power loss and efficiency analysis of an onboard three-level brushless synchronous generator. *International Journal of Electronics* 108 (1): 1–20.

35 Ghods, M., Faiz, J., Pourmoosa, A.A. et al. (2021). Analytical evaluation of core losses, thermal modelling and insulation lifespan prediction for induction motor in presence of harmonic and voltage unbalance. *International Journal of Engineering Transactions B Applications* 34 (5): 1213–1224.

36 SKF (1994). *General Catalogue from SKF*, Catalogue 4000/IV E.

37 Saari, J. (1998). Thermal modelling of high-speed induction machines. PhD Thesis, Laboratory of Electromechanics, Helsinki University of Technology, Espoo, Finland.

38 Matosevic, V. and Štih, Ž. (2014). 2D Magneto-thermal analysis of synchronous generator. *Przeglad Elektrotechniczny* 90 (12): 157–160.

39 Pyrhönen, J., Lindh, P., Polikarpova, M. et al. (2015). Heat-transfer improvements in an axial-flux permanent-magnet synchronous machine. *Applied Thermal Engineering* 76: 245–251.

40 Zhao, J., Liu, Y., Xu, X. et al. (2018). Comparisons of concentrated and distributed winding PMSM in MV power generation. In: *Proceedings – 23rd International Conference on Electric Machines, ICEM*, 2437–2443. Alexandroupoli, Greece, 3–6 September 2018.

41 Kim, J.H., Hyeon, C.J., Chae, S.H. et al. (2017). Design and analysis of cooling structure on advanced air-core stator for megawatt-class HTS synchronous motor. *IEEE Transactions on Applied Superconductivity* 27 (4): Article Sequence Number: 5202507.

42 Jmag Designer Software, Multiphysics https://www.jmag-international.com/products/jmagdesigner.

43 Simcenter Motorsolve Software, FEA – Thermal and performave analysis of electric machines. https://www.plm. automation.siemens.com/global/en/ products/simcenter/motorsolve.

44 Connor, P.H., Eastwick, C.N., Pickering, S.J. et al. (2016). Stator and rotor vent modelling in a MVA rated synchronous machine. In: *Proceedings of the 22nd International Conference on Electric Machines, ICEM*, 571–577. Lausanne, Switzerland, 4–7 September 2016.

45 Shanel, M., Pickering, S.J., and Lampard, D. (2003). Conjugate heat transfer analysis of a salient pole rotor in an air cooled synchronous generator, in *IEMDC–IEEE International Electric Machine Drives Conference*. In: vol. 2, 737–741. Madison, WI, USA, 1–4 June, 2003.

46 Jandaud, P. (2016). Étude et optimisation aérothermique d'un alterno-démarreur. To cite this version: HAL Id: tel-01332851,Université Val. du Hainaut-Cambresis, France.

47 Lancial, N., Torriano, F., Beaubert, F. et al. (2017). Taylor-Couette-Poiseuille flow and heat transfer in an annular channel with a slotted rotor. *International Journal of Thermal Sciences* 112: 92–103.

48 Comsol Multiphysics software, Understand, Predict, and Optimize Physics-Based Designs and Processes. https://www.comsol.com/comsolmulti-physics.

49 3ds software, Fluid & Computational Fluid Dynamics Simulation. https://www.3ds.com/productsservices/simulia/products/fluid-cfd-simulation/.

50 Simscale software, CFD, FEA, and Thermal Simulation. https://www.simscale. com.

51 Autodesk CFD software, Computational Fluid Dynamics Simulation. https://www.autodesk .com/products/cfd.

52 Maynes, B.D.J., Kee, R.J., Tindall, C.E. et al. (2003). Simulation of airflow and heat transfer in small alternators using CFD. *IEE Proceedings – Electric Power Applications* 150 (2): 146–152.

53 Carounagarane, C., Chelliah, T.R., and Khare, D. (2020). Analysis on thermal behavior of large hydrogenerators operating with continuous overloads. *IEEE Transactions on Industry Applications* 56 (2): 1293–1305.

54 Franc, J. and Pechanek, R. (2018). Ventilation – Thermal calculation of 40 MW synchronous machine. In: *Proceedings – 23rd International Conference on Electric Machines, ICEM*, Alexandroupoli, Greece, 3–6 September 2018, 1344–1349.

55 Connor, P.H., Pickering, S.J., Gerada, C. et al. (2013). Computational fluid dynamics modelling of an entire synchronous generator for improved thermal management. *IET Electric Power Applications* 7 (3): 231–236.

56 Madonna, V., Walker, A., Giangrande, P. et al. (2019). Improved thermal management and analysis for stator end-windings of electrical machines. *IEEE Transactions on Industrial Electronics* 66 (7): 5057–5069.

57 Fujinami, Y., Noguchi, S., Yamashita, H. et al. Thermal analysis of a synchronous generator taking into account the rotating high-frequency magnetic field harmonics. In: *International Conference on Electrical Machines*, 28–30 August 2000, 353–357. Espoo, Finland.

10

Signal Processing

10.1 Introduction

With the advent of powerful computational resources, the field of advanced digital signal processing has become an attractive research area that continues to broaden and mature. However, the fast growth of powerful computational resources also simplifies the use of real-time computing and makes it more accessible to researchers in the signal processing field. Signal processing is the act of applying mathematical tools to raw data to obtain deep insights into the data components. Various signal processing tools have been developed over the years based on the needs in certain fields, and these new tools, with their capacity to handle certain signal properties, have been adopted for use in other fields. In this chapter, prior to an exploration of the use of signal processing in fault detection, a concise description of "signal" is provided. This is followed by a brief introduction to the advanced signal processing tools that have already been used in fault diagnosis.

10.2 Signal

In signal processing, a "signal" $S(t)$ is defined as a function that characterizes some attributes about a process or phenomenon [1]. In the context of the condition monitoring field, the signals are those sampled from electric machines that demonstrate a machine's magnetic field, its phase, line–line voltage or current, its vibration, its external magnetic field, or its thermal or other physical characteristics. The signals are acquired using discrete sampling methods, and their digital characterization is stored as time series data. Therefore, the term "signal" in this book is referred to

Electromagnetic Analysis and Condition Monitoring of Synchronous Generators, First Edition. Hossein Ehya and Jawad Faiz.
© 2023 The Institute of Electrical and Electronics Engineers, Inc. Published 2023 by John Wiley & Sons, Inc.

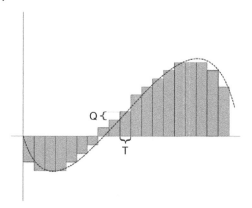

Figure 10.1 A discretized signal characteristics acquired with sampling frequency and resolution of T and Q, respectively.

as the digital characterization of a discretely sampled signal. A continuous signal sampled with a periodicity of (T) is represented as follows:

$$S(kT) = S(k) \tag{10.1}$$

The digital characterization of an analog signal has limitations in the following:

1. Quantization.
2. Sampling rate.

Figure 10.1 demonstrates the two principal limitations of digital signal acquisitions. The signal is divided into a discrete number representing the so-called "quantization." The precision of the quantization is determined by gauging the difference between the main form of the signal and its reproduced version [2]. The accuracy of the discrete number with respect to its true value is proportional to its memory size. The high sampling frequency provides a sufficient resolution that may minimize the adverse effects of the quantization error.

The use of a non-uniform reproduction codebook in the procedure of signal processing or for increasing its size may reduce the quantization error [2]. In a modern analog-to-digital (AD) conversion, the effects are mainly dominated by analog noise [1]. The frequency at which the signal is sampled is the so-called "sampling rate." The higher sampling frequency rate principally leads to a more precise representation of the main signal. The Nyquist–Shannon theorem determines the minimum sampling rate of the signal. The minimum sampling rate is also called the Nyquist rate. The theorem states that in order to thoroughly represent the main signal, a sampling rate of twice the highest frequency in the signal is imperative [3]. In other words, the sampling frequency of the signal must be at least twice the rate of the highest frequency component of the signal. In fault detection, the sampling frequency of the signal is determined based on the highest sampling frequency of interest [4].

The linearity and stationarity are two important process characteristics that have significant effects on signal processing techniques. An ability of a process that can be described as a superposition of its components is called "linearity." The linearity property allows us to use linear transforms to demonstrate the resulting process. Equation (10.1) describes the linearity of the signal "$S(t)$."

Having a perfect linear combination of the signal components is rarely possible when a signal is sampled because data leakage and the noise from the environment and sensors always reduce the linear superposition of the signal components. However, many signals can be approximated as linear without losing its precision [1]:

$$S(t) = k_1 s_1(t) + k_2 s_2(t) + \cdots + k_n s_n(t) \tag{10.2}$$

A time invariance in a probability distribution of a process is called "stationarity." If the mean value of the stochastic component and autocovariance of a signal is constant and its variance is finite, then the signal is stationary [5]. If the signal, in practice, can be represented by the same stochastic components throughout its defined period, it is stationary. For example, white Gaussian noise is a stationary process, since its distribution is time-invariant.

If a signal repeats itself within a predefined period, it is a periodic signal and shows the property called "periodicity." According to Equation (10.2), the signal is periodic if it is valid for any choice of t and k [1]. For a deterministic signal like a sine wave, the periodicity is identical to the stationarity. The characteristics of an entire signal can be defined if one period of a periodic signal becomes available:

$$S(t) = S(p + t) \tag{10.3}$$

$$S(k) = S(N + k) \tag{10.4}$$

In the field of signal processing, the terms stationary and periodic are able to be substituted to some extent. In a case where the deterministic component of the signal demonstrates a lack of periodicity, the signal is non-stationary. Nevertheless, no signal is rigorously periodic if it entails a stochastic component. However, if the signal's stochastic elements are stationary, it is referred to as periodic.

10.3 Fast Fourier Transform

The Fourier transform is a mathematical transform that decomposes a function that could be a signal or a function of time into its constituent frequencies. The Fourier transform also represents the signal in a frequency domain. According to Equation (10.5), the Fourier transform is a convolutional operation. Since the Fourier series is a periodic function, the targeted signal or time function should also be periodic. However, many non-periodic functions are periodic within a defined period. Consequently, for a non-periodic function, the Fourier series is applied with an assumption that the entire function is one period.

$$\hat{s}(\omega) = \frac{1}{\sqrt{2\pi}} s(t) e^{-j\omega t} dt = \frac{1}{2} a_0 + \sum_{k=1}^{\infty} (a_k \cos(2\pi kt) + b_k \sin(2\pi kt)) \tag{10.5}$$

The following are some conditions for the existence of the Fourier transform for a definite function [1]:

1. The function must have a finite number of minima and maxima.
2. The function must have a certain finite number of discontinuities.
3. The function must be absolutely integrable within any time interval.

A real signal obtained in measurement fulfills these criteria.

The discrete Fourier transform (DFT) is formulated to apply to discrete signals. The DFT converts a discrete signal sampled at a fixed sampling frequency into a frequency domain. The fast Fourier transform (FFT) is also generally applied to attain the DFT [6]. The use of the FFT reduces the computational complexity of the DFT from $O(N^2)$ to $O(N \log(N))$, where N is the number of data samples [6].

The fast Fourier transformed signal is represented in a series of spectra, which is the collection of periodic elements in a frequency domain. The resulting spectra has a specific frequency, amplitude, and phase angle. Equation (10.6) represents the DFT of a sampled signal [1]:

$$S(s) = \sum_{n=0}^{N-1} s(k) \, e^{-i \, 2\pi \frac{xk}{N}} \tag{10.6}$$

The FFT is a suitable signal processing tool that is always utilized in fault detection to obtain an overview of the frequency components of the measured signal. Loss of time information of the processed signal by the DFT is the main disadvantage of this approach [1]. Therefore, the DFT is not an appropriate approach for processing non-stationary signals if the temporal location of the frequencies is required.

10.4 Fast Fourier Transform with an Adjusted Sampling Frequency

The DFT is a suitable approach for representing discrete signals in a frequency domain as a frequency spectrum. The FFT is a convenient way to achieve the DFT. The quality of the spectrum density acquired based on the FFT algorithm can be improved by applying such techniques as zero paddings and windowing functions. However, this creates side-lobe distortion in the spectrum shape and error in the amplitude of the spectrum as unavoidable parts of the FFT algorithm called artifacts. The "spreading" phenomenon happens if the time step of the signal is disproportional to the period of the signal and leads to an error in the magnitude of the frequency spectrum [7]. Resampling the data can therefore eliminate this artifact and its spread, and resampling works well when the expected frequency components in the frequency spectrum are known. The length of the input data based on the time step of the signal and expected frequency components must also be adapted. The proposed approach in reference [7] eliminates the need for zero-padding, windowing functions, and other approaches typically applied to improve FFT results.

The sampling frequency of the signal must be chosen based on the following equation [7]:

$$\Delta f = \frac{1}{NSPT\Delta t} \tag{10.7}$$

where Δf is the frequency spacing, NSPT is the number of sampled points to be transferred to FFT, and Δt is the inverse of sampling frequency. Based on the "radix 2" FFT algorithm, the number of points must be a power of two. To avoid errors in the FFT spectrum, the value of Δf and its integer multiples coincide exactly with the fundamental frequency. Consequently, the sampling frequency of the signal and the time step must be adjusted according to Equation (10.7).

Figure 10.2 presents the result obtained using different approaches, such as FFT, FFT with a Hanning window, and FFT with adjusted sampled data, where the sampling frequency of the data is 50 kHz. The spectrum density based on the first two methods has a significant error in magnitude and frequency location, as the frequency components are scattered in the spectra. The reason for having an artifact in the spectrum is that the length of the data set or the signal does not correspond well with the integer number of periods. The side lobe and spreading effects are resolved when the length of the sampled, time step, and frequency spacing are chosen according to Equation (10.7), as seen in Figure 10.2.

The measurement instruments provide a few ranges of standard sampling frequencies (e.g., 10, 25, 50 kHz). Hence, to meet the required conditions governed by Equation (10.7), resampling of the measured data is required. By contrast, no data set resampling is required if the original sampling frequency and the time step of the sampled data are specified manually or if an external trigger is used in the oscilloscope. Two functions provided in MATLAB, called Cubic Spline interpolation

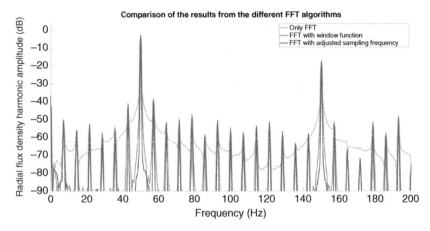

Figure 10.2 The results obtained using different approaches when performing the FFT.

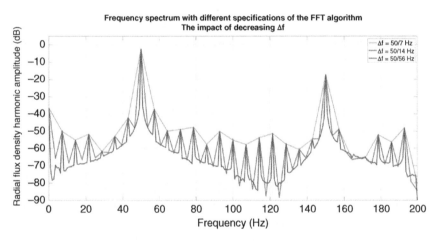

Figure 10.3 The impact of decreasing Δf on the FFT spectrum.

and resampled function, give the same results based on both functions. The sampling frequency, the number of points, and the frequency spacing have a considerable effect on the spectral density [8].

The variation of the frequency spacing has a marked effect on the spectral of the data set. Figure 10.3 depicts the response of the signal spectrum to variation of the Δf, while the number of points is constant and the value of the time step varies by changing the frequency spacing. The denominator of Equation (10.7) must be an integer number to obtain the expected frequency components. For an electric machine with seven pole pairs, the smallest mechanical frequency observed for the main frequency of 50 Hz is 50/7. As a consequence, the spacing frequency (Δf) is set at 50/7.

Figure 10.3 demonstrate the spectral density of a data set with several spacing frequencies. The amplitude of the frequency components is the same, but the spectrum for the highest spacing frequency is floating. Recognition of the frequency components in a floating spectrum is arduous, so the shape and spectrum and the distinguished harmonic components are ameliorated by decreasing the spacing frequency to half (50/14) and one-eighth (50/56).

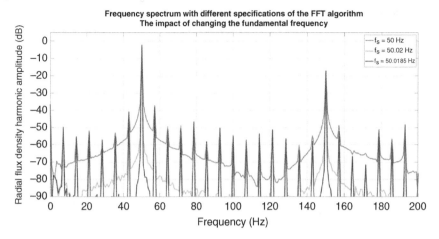

Figure 10.4 The impact of adjusting the fundamental frequency.

The lowest value of the frequency spacing is determined by the length of the data set. The number of data points for a signal sampled at a rate of 50 Hz for a duration of two seconds is equal to 100 000. The spline function used for resampling purposes becomes unstable if the inverse value of the frequency spacing becomes greater than the length of the data. Thus, in order to decrease the value of Δf, the length of the data set must be increased. The required time interval for a spacing frequency equal to 50/56 Hz is 1.12 seconds, while for the Δf equal to 50/112 Hz, the duration of the signal must be 2.24 seconds.

In addition to these factors related to Equation (10.7), the main frequency of the data set also plays a pivotal role in achieving an accurate spectrum. In the simulation, the fundamental frequency of the machine is set by the user, whereas, in reality, many factors can result in the generator's prime over-speed deviation, which causes the fundamental frequency to differ from the set point. Hence, the fundamental frequency of the data set must be checked before feeding the data into the FFT algorithm. Figure 10.4 depicts the consequence of adjusting the fundamental frequency of the spectral density of the signal where it is seen that adjusting the fundamental frequency to 50.02 Hz and 50.0185 Hz leads to more precise and defined peaks at the expected frequencies.

The magnitude of some frequency components varies in response to changes in the fundamental frequency. The difference between the spectral density with a fundamental frequency equal to 50 Hz and 50.0185 Hz is approximately 5 dB. In fact, the value of 5 dB in a fault diagnosis is large and simply leads to a false alarm. Hence, adjusting the fundamental frequency to reduce the error and having detectable frequency components in the spectrum is indispensable. The actual fundamental frequency varies from one sampled data set to another. An algorithm is required to fix the fundamental frequency of all data sets and then their spectrums are valid for comparison.

According to Equation (10.7), the number of sampled points also affects the output of the FFT algorithm. As shown in Figure 10.5, changing the number of sampled points while the spacing frequency is assumed to be constant results in variation in the magnitude of the frequency components. A comparison between the amplitude of the frequency component in the spectrums with NSPT equal to 2^{10}, 2^{12}, and 2^{16} suggests that increasing the NSPT decreases the magnitude of the components. The sampling frequency for a spacing frequency equal to 50.0185/56 Hz and NSPT equal to 2^{10} is 914 Hz, whereas it is 58.5 kHz when NSPT is 2^{16}.

Therefore, the NSPT can be increased until a new sampling frequency is reached that is equivalent to the original sampling frequency. The amplitude of the frequency components then increases,

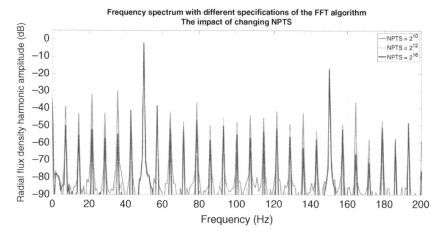

Figure 10.5 The impact of adjusting NPTS.

reaches, and surpasses the original sampling frequency, and the magnitude of the frequency component becomes stabilized. A comparison between NSPT at 2^{16} and 2^{22} proves that the new sampling frequency is higher than $50\,\text{kHz}$, the original sampling frequency of the data set, and the frequency components do not show significant changes. Note that if a variation occurs in the amplitude of the frequency component, this may be due to the low value of the NSPT, the spline interpolation, or the FFT function.

The rate of sampling frequency depends on the frequency range of interest in the spectrum density. However, the sampling frequency must be high enough to have a good resolution to distinguish the fault components, but it should not be too high, as this can create difficulties in data storage and its analysis, especially for online condition monitoring. Therefore, a trade-off based on end-user needs must be defined. Nonetheless, having a long data set does not always lead to a rich and satisfactory spectrum, since the long data set is vulnerable to fluctuating frequencies that may adversely affect the frequency components.

Figure 10.6 demonstrates the spectral density of the sampled data at the frequency rate of 50, 25, and $10\,\text{kHz}$. All steps, such as choosing the spacing frequency, adjusting the fundamental

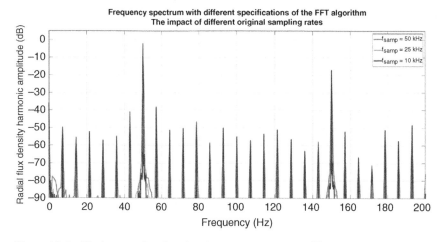

Figure 10.6 The impact of performing the measurements at different sampling rates.

frequency, and choosing the highest value of NSPT, were performed before applying the FFT algorithm to apprehend the effect of sampling frequency on the spectrum density. The difference between the various sampling frequencies was marginal and the amplitude of the side-lobes was almost the same, with a negligible difference of less than 0.5 dB in some cases. The only difference was a broadening and scattering of the harmonic components with increasing sampling frequency, since the resolution of the lowest frequency components is increased. Overall, the sampling frequency had no appreciable effect on the frequency spectral, so it should be adjusted based on demand.

10.5 Short-Time Fourier Transform

The short-time Fourier transform (STFT) is introduced to compensate for the FFT's lack of temporal resolution. The STFT provides a time-frequency resolution for scrutinizing the non-stationary signals, but it only analyzes a small portion of the signal by utilizing the FFT [9]. The STFT applies the FFT to a section of the signal using a window function. The window function is then swiped across the entire data set to find the magnitude for given frequencies and time instants [1]. Therefore, in a case where the signal is non-stationary and FFT is not an applicable processing tool, STFT is a suitable choice [4]. Equation (10.8) shows the mathematical formulation of the STFT for a signal $S(t)$:

$$\text{STFT}(f, t) = \frac{1}{2\pi} S(t) h(t - \tau) e^{-i\,2\pi f\tau} d\tau \tag{10.8}$$

Obtaining an optimal result based on signal characteristics requires that the shape of the window function $h(t)$ and the length of the data set be adjusted. The frequency resolution of the STFT depends on the length of the window, whereas its time resolution is inversely proportional to the length of the time window [1]. The result of the STFT is depicted in a spectrogram, an image where the horizontal axis is time, the vertical axis is frequency, and the intensity of the spectral is designated by color. According to the uncertainty principle, which states that one cannot know with a high degree of certainty both a property and its integration in the time interval at the same instant [1], attaining a high resolution in both time and frequency domain is not possible.

The resolution grid is uniform across all time and frequency values, and is called a rectangular time-frequency resolution. Figure 10.7 shows the rectangular time-frequency resolution. The uniform frequency resolution is practical when the frequencies of interest are immeasurably scattered at high frequencies and are hardly separated at the lower frequency level. The mentioned limitation is formulated concisely by the use of the uncertainty principle according to:

$$\Delta f \Delta t \geq \frac{1}{4\pi} \tag{10.9}$$

Figure 10.8 shows an STFT applied to the air gap magnetic field time series using a default setting. No specific information is extractable from Figure 10.8; therefore, its parameters must be adjusted. The best window length must be selected based on the frequency of interest. The lower frequencies are detectable if the window length increases. The window length must span at least half or one period to obtain a certain frequency, such as $f_{lowerbound}$. Therefore, the relationship between the window length, sampling frequency, and lower frequency of interest is as follows:

$$\Delta n = \frac{f_{samp}}{2f_{lowerbound}} \tag{10.10}$$

Figure 10.7 The resolution grid of a short-time Fourier transform of a signal.

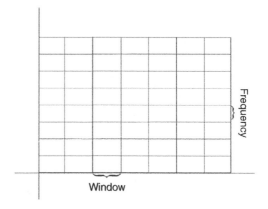

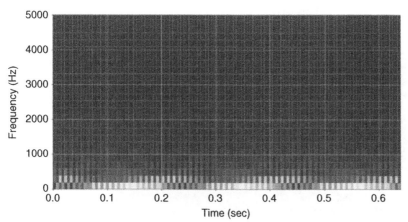

Figure 10.8 The spectrogram of SFTF for the air gap sample series with the default setting.

Figure 10.9 shows a sample series of healthy and faulty data where the difference is discernible by considering the window length of 1000 samples.

Numerous windowing functions, including Hamming, Gaussian, Hann, triangular, Dirichlet, Blackman, Blackman–Harris, Bartlett, Parzen, Bohman, Kaiser, Exponential, Chebwin, Slepian, Bartlett–Hann, flat top, Tukey, and Nuttall, are available for mitigating the end effect phenomena. The windowing function can also be used as a filter, since the multiplication in the time domain is equal to convolution in the frequency domain. This feature leads the windowing function to selectively intensify the largest frequency components, which is undesirable. Figure 10.10 shows the result of using Hann and Blackman windowing functions over the air gap magnetic field time series data [4].

10.6 Continuous Wavelet Transform

To resolve the lower time-frequency resolution acquired by STFT at higher frequencies, the Wavelet transform is introduced. As shown in Figure 10.11, the temporal resolution is enhanced at smaller scales. The signal is convoluted with a "wavelet" function, unlike STFT, which uses the Fourier transform [10]. The wavelet has a compact support, which means that the signal is zero outside the compact set and its integral along its axis is zero [10]. Hence, the wavelet is a windowing function.

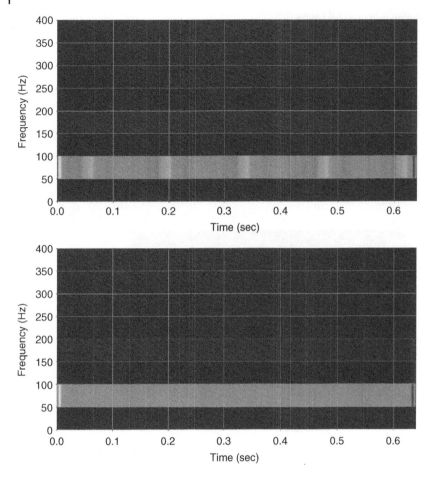

Figure 10.9 The spectrogram of SFTF for a healthy and faulty sample series with a window length of 1000 samples.

By modifying the scaling factor a, the wavelet can be compressed or stretched. This feature enables the convolutional integral to select various frequencies. Equation (10.10) demonstrates the convolution computation of the wavelet transform [10]:

$$S(a, b) = \int_{-\infty}^{\infty} s(t) \Psi_{a,b}^* \, dt \qquad (10.11)$$

where $s(t)$ is the targeted analyzed signal and $\Psi_{a,b}^*$ is the wavelet function, which is associated with the coefficients a and b. The wavelet coefficients a and b are used to adjust the scale and the temporal center of the wavelet, respectively. The notation used in Equation (10.11) is for the continuous wavelet transform, which is also valid for the discrete wavelet transform. For several values of a, the convolution is applied to the signal and the result is merged into a scalogram that demonstrates the signal components. The wavelet function is based on the following equation:

$$\Psi_{a,b}(t) = \frac{1}{\sqrt{a}} \Psi \left(\frac{t - b}{a} \right) \qquad (10.12)$$

where Ψ is the mother wavelet. Numerous types of mother wavelets, such as Haar, Daubechies, Biorthogonal, Coiflets, Symlets, Morlet, Mexican Hat, Meyer, Gaussian derivatives, frequency

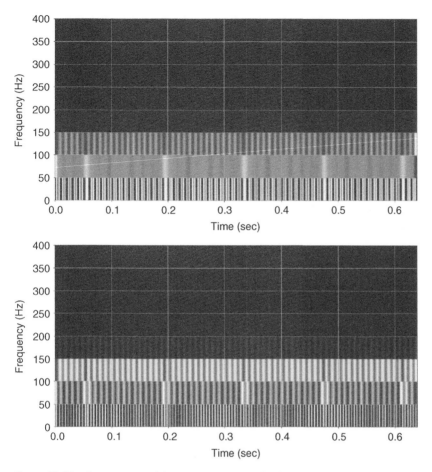

Figure 10.10 Spectrogram of the air gap magnetic field time series using the Hann and the Blackman windows.

B-spline, and Shannon, are available. The choice of mother wavelet depends on the characteristics of the signal and the properties of interest. For instance, the Morlet wavelet is utilized to determine a smooth variation, while the Haar wavelet is a suitable choice for selecting a sudden transition [11]. Figure 10.12 depicts the Morlet and Haar wavelets [4].

Four parameters, namely the continuous wavelet transform (CWT) scales, the normalization function, the mother wavelet, and the color map, must be selected to achieve an optimal result. A set of scales for CWT is computed to define which frequencies are to be evaluated. The relationship between the scale's magnitude and the frequencies is not linear, as shown in Figure 10.13. A solution to this challenge is to define a numerical function that contains an array of frequencies of interest and to compute the corresponding scales for the related continuous mother wavelet. Therefore, an upper and lower limit is defined for scales for each frequency that increase and decrease until either of the frequencies are within a threshold of the desired frequency. Figure 10.13 shows the relationship of scale and the frequency for the Shannon, Gaussian, and Mexican hat mother wavelets.

An illustratable visualization is a key point for interpreting the results obtained with CWT. A series of various color maps is generated, as shown in Figure 10.14. A procedure for an appropriate

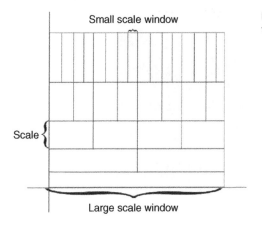

Figure 10.11 The resolution grid of a wavelet transform of a signal.

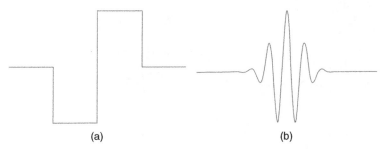

Figure 10.12 The shape of the Morlet mother wavelet (a) and the Haar mother wavelet (b).

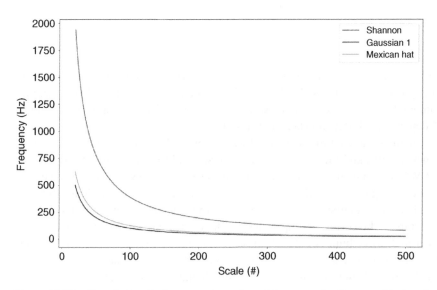

Figure 10.13 The relationship between the scale and frequency for Shannon, Mexican hat, and Gaussian mother wavelets.

Figure 10.14 A selection of color maps utilized for CWT.

selection of CWT color maps is presented in reference [12]. According to reference [4], a proper color map for a meaningful illustration of CWT is achieved using a color map of uniform luminance and an opposing pair of hues. However, no certain color map exists for each data set, so various color maps must be tried to obtain a promising result.

As already mentioned, numerous mother wavelets exist that control the shape of the scalogram. In addition to the mother wavelets listed here, a new mother wavelet can also be defined based on a particular need. Achieving a promising result then depends on testing variations of mother wavelets. Figure 10.15 shows a CWT applied to an air gap magnetic field using the Complex Gaussian and B-spline mother wavelets.

Mapping a CWT coefficient on to a range between zero and one uses a normalization function. Various normalization functions, such as linear normalization, logarithmic normalization, exponential normalization, and symmetric logarithmic normalization, are available. Exponential normalization depends on the degree of the exponent: an exponent equal to one is identical to a linear normalization. Figure 10.16 shows the CWT of the air gap magnetic field time series for the B-spline mother wavelet for the two normalization functions [4].

10.7 Discrete Wavelet Transform

The theory behind the discrete wavelet transform (DWT) is based on the same principle as the continuous wavelet. The signal is convoluted with a selected mother wavelet that can draw out unique features from the signal. The implementation of the DWT differs from the CWT. The filter bank implementation is the most widespread approach to perform the DWT. A cascade filter is the algorithm function in which each filter conforms to a scale or a level. High-pass and low-pass filters are available in parallel at each level, followed by downsampling by 2. Figure 10.17 presents the filter bank of the DWT. The signal is passed through both the low-pass and high-pass filters at the same time, and both filtered signals are downsampled and then stored as detailed and approximate coefficients of that level, respectively. The output signals of each level feed into the next level. Figure 10.18 demonstrates the sequence of several filters that form the DWT.

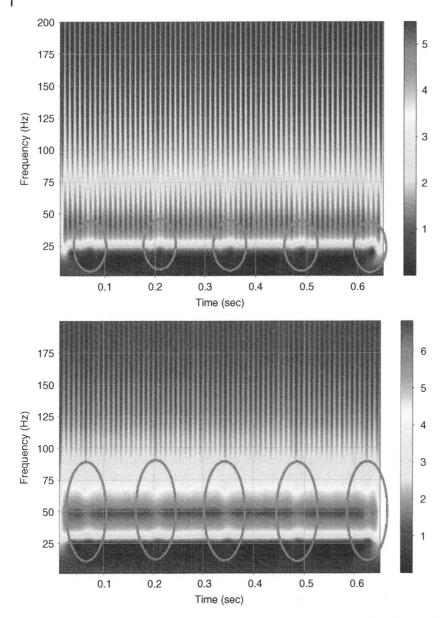

Figure 10.15 The scalogram of the air gap magnetic field using Complex Gaussian and B-spline mother wavelets.

The process of filtering and downsampling is continued until the output signal reaches the targeted decomposition level. The last level returns the low-pass filter with the rest of the decomposition. The selection of the filters is based on the type of mother wavelet. Since the signal in each level is downsampled, the length of the filter can be kept the same length. To avoid redundancy or overlapping, the filter is shifted by an entire filter length.

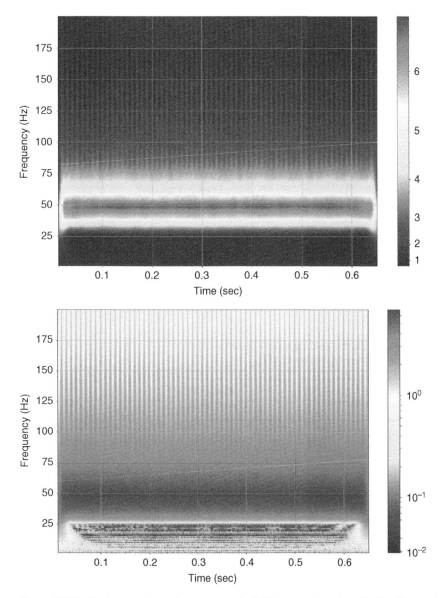

Figure 10.16 Scalogram of an air gap magnetic field time series using the B-spline mother wavelet and a second-degree exponential normalization function and symmetric logarithmic normalization.

The frequency band for each decomposition level with a sampling frequency of the f_{samp} for three-level DWT is given in Table 10.1. The computational complexity of the DWT is less than CWT because the signal is downsampled at each level. The computational complexity of the DWT is $O(n)$, while it is $O(n \log(n))$ for FFT. The temporal resolution is better for CWT than for the DWT, since it can shift the filter by only one sample. However, less storage is required for DWT than for CWT. The smaller requirement for data storage is a high-value benefit of DWT, since it can generate numerous features with low data storage needs for online condition monitoring. Some features, such as

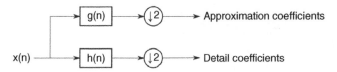

Figure 10.17 One level of the DWT.

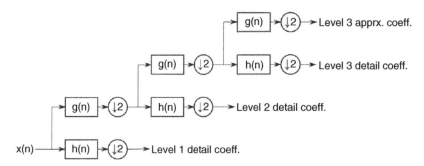

Figure 10.18 A filter bank of cascading filters, equivalent to a three-level DWT.

Table 10.1 The frequencies contained within each DWT decomposition level of a signal of length N and sampling frequency of f_{samp}.

Level	Frequencies	Number of coefficients
4	$0-f_{samp}/8$	$N/16$
3	$f_{samp}/8-f_{samp}/4$	$N/8$
2	$f_{samp}/4-f_{samp}/2$	$N/4$
1	$f_{samp}/2-f_{samp}$	$N/2$

entropy, kurtosis, variance, mean, standard deviation, median, skewness, and various energies, can be utilized to extract unique features based on each level of DWT decomposition. The following are various energy content extractors based on the DWT sub-band that can be used to discriminate a faulty machine from the healthy one [4, 13]:

1. Teager Wavelet Energy (TWE).
2. Relative Wavelet Energy (RWE).
3. Hierarchical Wavelet Energy (HWE).
4. Instantaneous Wavelet Energy (IWE).

Although the method is called the "discrete" wavelet transform, both CWT and DWT are implemented discretely in the field of signal processing. However, the difference is that CWT is defined continuously and performs an infinite number of swipes on an infinitesimal length, whereas DWT, as the discrete algorithm, shifts the length of the wavelet. Figure 10.19 shows the application of DWT to an air gap time series.

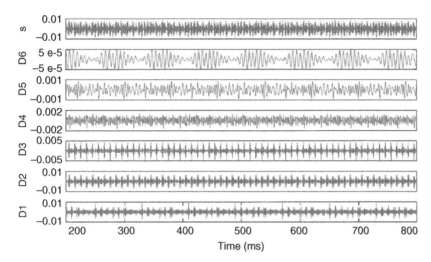

Figure 10.19 Application of the DWT to a magnetic field time series of an electric machine.

10.7.1 Wavelet Energies

Several energies can be extracted to demonstrate some of the frequency band properties based on each of the discrete wavelet decomposition levels with N_j coefficients $w\_j(r)$, $r = 1,...N\_j$. To indicate the amplitude of each decomposition level, IWE is introduced as:

$$\text{IWE}_j = \log_{10}\left(\frac{1}{N_j}\sum_{r=1}^{N_j}(w_j(r))^2\right) \tag{10.13}$$

If the measured data set is distorted by noise, the TWE is noise-robust and can be formulated as:

$$\text{TWE}_j = \log_{10}\left(\frac{1}{N_j}\sum_{r=1}^{N_{j-1}}|(w_j(r))^2 - w_j(r-1)\,w_j(r+1)|\right) \tag{10.14}$$

The downsampled signal may require padding to become an integer number of filter applications. This process may cause end effects that may alter the energies acquired by IWE and TWE. The HWE analyzes the center of each sub-band and ignores the first and last portion of that level's coefficient. The HWE is represented by the formulation:

$$\text{HWE}_j = \log_{10}\left(\frac{1}{N_j}\sum_{r=\frac{N_j-N_J}{2}}^{\frac{N_j+N_J}{2}}(w_j(r))^2\right) \tag{10.15}$$

where N_J is the number of coefficients in the level over the current level.

The RWE is used to compare the energy distribution among the frequency bands and is represented by [14]:

$$\text{RWE}_j = \frac{E_j}{E_{total}} \tag{10.16}$$

where E_j is the energy of each decomposition sub-band and E_{total} is the sum of energies of all sub-bands:

$$E_j = \sum_{r=1}^{N_j} (w_j(r))^2 \tag{10.17}$$

$$E_{total} = \sum_{j=1}^{K} E_j \tag{10.18}$$

10.7.2 Wavelet Entropy

The transient behavior of the non-stationary signal is identified using wavelet entropy, which is now applied in various fields, ranging from psychology to power system analysis [15, 16]. The wavelet entropy of a measured data set demonstrates the degree of disorder in each of the discrete wavelet decomposition levels. The magnitude of the entropy is measured between zero and one, where zero conveys the perfect order, while one represents some degree of disorder. However, the upper limit of the wavelet entropy is not restricted to one and, if the value of the upper limit exceeds one, this means that the data have a higher degree of disorder. The Shannon entropy provides a formulation to evaluate the rate of the disorder in the probability distribution of each sub-band level, defined as follows [17]:

$$\text{Entropy}_{sh}(n) = -\sum_{i=1}^{j} P_i \log P_i \tag{10.19}$$

where P_i is a relative normalized value of each wavelet sub-band energy to the total energy of the signal.

10.8 Hilbert–Huang Transform

The Hilbert–Huang transform (HHT) is a useful time-frequency processor that is a combination of the empirical mode decomposition (EMD) and Hilbert transforms (HTs). The HHT works by applying the HT to the generated intrinsic mode decomposition by EMD. The HHT approach is a completely adaptive method that provides a greater temporal resolution than is obtained with STFT, MUSIC, CWT, or DWT. These properties make HHT a suitable choice for decomposing a nonlinear and non-stationary signal [18]. The HHT approach is based on EMD and HT, which are described in the next two sections.

10.8.1 Hilbert Transform

The Hilbert transform (HT) is a convolutional-based method and is a linear transform. The HT generates an imaginary 90° phase-shifted signal based on the real input signal as is represented as:

$$\hat{x}(t) = \frac{1}{\pi} \int_{-\infty}^{\infty} \frac{x(\tau)}{t - \tau} \, d\tau \tag{10.20}$$

Despite its name, the HT is not a transform since it does not convert the signal to another domain. This feature is desirable, as the combination of the original signal and its HT provides an analytic

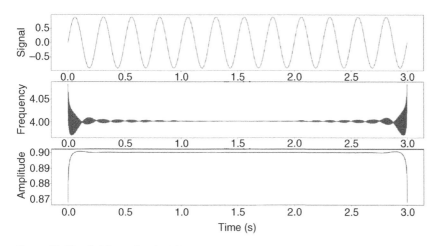

Figure 10.20 A 0.9 amplitude, 4 Hz sine wave and its Hilbert transform.

signal. An analytic signal is a complex-valued function that has no negative frequency components. The analytic signals can be described by phasors, as follows:

$$h(t) = x(t) + j\hat{x}(t) = A(t)\,e^{j\phi(t)} \tag{10.21}$$

The signal is described by its amplitude $A(t)$ and phasor $\phi(t)$ using the phasor representation of the signal and its HT. Equations (10.22) and (10.23) demonstrate the amplitude and the phasor of the signal and its HT:

$$A(t) = \sqrt{x^2(t) + \hat{x}^2(t)} \tag{10.22}$$

$$\phi(t) = \arctan\left(\frac{\hat{x}(t)}{x(t)}\right) \tag{10.23}$$

The instantaneous frequency, $\omega(t)$, based on the phasor angle shows the rate of change of the signal's phasor angle, as shown by

$$\omega(t) = \frac{d\phi(t)}{dt} \tag{10.24}$$

The use of instantaneous frequency provides perfect frequency resolution when it is computed along with the signal in each step [18].

The end effect is the main weakness of the HT, as the signal contains more than one frequency component, making data interpretation challenging [18]. Figures 10.20 to 10.22 depict this challenge. The HT discriminates the correct frequency and the amplitude of the 4 Hz with the wave in Figure 10.20, with some boundary effects. According to Figure 10.21, adding white Gaussian noise to the same 4 Hz sine wave causes the HT to fail to detect the main frequency and introduces uncertainty into the estimated amplitude.

The HT starts to oscillate when the signal contains more than one main frequency component, as seen in Figure 10.22. The addition of more components further worsens the HT performance. The HT provides a perfect time-frequency with a high resolution, even if the signal has some features like non-linearity or non-stationarity. However, it is exceedingly sensitive to noise and requires a monocomponent signal to present precise results [18]. The process of noise rejection is achieved using the EMD algorithm [18].

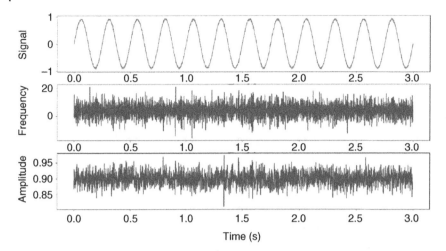

Figure 10.21 A 0.9 amplitude, 4 Hz sine wave with white noise added, and its Hilbert transform.

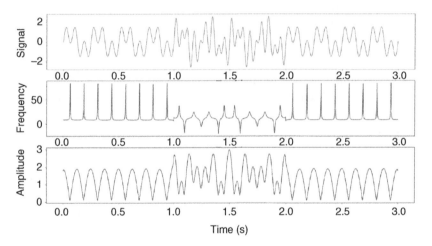

Figure 10.22 A signal and its Hilbert transforms. The signal is a linear combination of sine waves at 4 and 8 Hz, with a 12 Hz sine wave added in from time one to two seconds.

10.8.2 Empirical Mode Decomposition

Separating a signal into its harmonic components, intrinsic mode functions (IMFs), and a residual is achieved using the EMD algorithm [18]. The sifting and IMF extraction are the two primary stages in the EMD algorithm [19], which are performed until the algorithm reaches a monotonic function [19].

The sifting process is an iterative procedure, where the signal is enclosed within an envelope by splines. The mean value of the envelope, m_1, is subtracted from the signal $x(t)$. The new envelope is traced around the remainder h_1, and another average value of the envelope m_2 is calculated. The procedure is halted when the stopping criteria are met. For instance, the stopping factor can be the point at which the standard deviation of the mean value becomes less than a defined threshold θ [18]. The remaining part is an IMF, c:

$$x(t) - m_1 = h_1 \tag{10.25}$$

$$h_1 - m_2 = h_2$$

$$...$$

$$h_{k-1} - m_k = h_k \tag{10.26}$$

$$SD\,(m_{k+1}) < \theta \rightarrow h_k = c_1 \tag{10.27}$$

The IMF is derived from the data set and its residual r is fed into the next sifting operation. The procedure of sifting and IMF extraction lasts until the residual signal becomes monotonic by nature [19]:

$$r_1 = x(t) - c_1$$

$$r_2 = r_1 - c_2$$

$$...$$

$$r_n = r_{n-1} - c_n \tag{10.28}$$

The signal processed by the EMD algorithm includes a set of IMFs and residuals. This set should contain the harmonic components and their trends. A summation of the IMFs and the residual must reconstruct the main signal as follows:

$$x(t) = \sum_{j=1}^{n} c_j + r_n \tag{10.29}$$

The set of generated IMFs are orthogonal and complete [18, 19]. The noise is mostly distributed in the first IMFs and may be discarded. The remaining IMFs can be analyzed further with a signal processing tool or used as a noise rejection approach.

The EMD approach has numerous technical challenges:

1. The IMFs are vulnerable to containing more than one frequency component if their frequencies are the same [20].
2. The boundary effects are due to the behavior of the splines at the ends and produce a particular decomposition due to many tunable parameters [19].
3. Several criteria, such as the choice of a spline function, stopping criteria, and the number of iterations, affect the results of IMFs.

10.9 Time Series Data Mining

Time series data mining (TSDM) is a nonlinear signal processing approach that is found on discrete stochastic models of reconstructed phase space based on the dynamical system theory [21]. The TSDM approach does not have limitations, such as stationarity and linearity, that traditional time series data analysis methods deal with. The TSDM methods may be able to discover hidden patterns in the time series based on dynamical system theory [22]. A metrically equivalent state space can be regenerated by a single sampled state variable. In other words, dynamical invariants are also preserved in the reconstructed state space. The acquired signal (time series) is considered to be a state variable in order to recover the state space. Alternatively, the acquired signal could reproduce a topologically equivalent state space similar to the original systems [23].

Time-delay embedding and derivative embedding are two approaches that can be used to reconstruct the state space. Derivative embedding is an impractical method for results obtained in laboratory or field tests, since the obtained results might contain noise, and the algorithm of the mentioned method includes higher-order derivatives that are sensitive to noise. Time delay embedding is used to overcome these problems. Time delay embedding performs transformations to find the invariant of the dynamical system by transforming scalar points into a vector form [22, 24]. For a given time series, the transformation is as follows:

$$I = \{i(k), k = 1, \dots, N\} \tag{10.30}$$

where k is the time index and N is the number of observations. A two-dimensional representative of the time series in state space is constructed by plotting $i(k - 10)$ and $i(k)$ on the x–y plane abscissa and ordinate, respectively. However, the reconstructed state space does not reveal any criteria for signal processing specifically for feature extraction purposes. Therefore, a feature for quantification must be introduced. The radius from the center of mass is introduced as a radius of gyration for quantification purposes. The radius of gyration is calculated as:

$$r = \sqrt{\frac{\sum\limits_{k=1+l}^{N} d(k)^2}{N - l}} \tag{10.31}$$

where the distance between the center of mass and the kth point in the state space is defined as:

$$d(k)^2 = (x(k) - \mu_0)^2 + (x(k - l) - \mu_l)^2 \tag{10.32}$$

where μ_m is the center of mass:

$$\mu_m = \frac{\sum\limits_{k=l+m}^{N-l+m} x(k)}{N - l} \tag{10.33}$$

where N is the number of observations and l is the time lag of the state space.

10.10 Spectral Kurtosis and Kurtogram

Spectral kurtosis (SK) is a statistical tool introduced by Dwyer [25] that may determine the presence of non-Gaussian components and their location in a signal. SK was introduced to overcome the inefficiency of power spectral density that eradicates the non-stationary–non-Gaussian data from the signal [25]. The first definition of SK was based on the normalized fourth-order moment of the real part of the STFT [25]. A more formal definition of the SK, based on higher-order statistics, is presented in reference [26], where a normalized fourth-order cumulant of Fourier transform was utilized to measure the distance of the process for Gaussianity. Although this approach shows satisfactory results on stationary signals, it encounters some difficulties with non-stationary signals. To resolve this problem, a new definition of SK, based on the Wold-Cramer decomposition of conditionally non-stationary processes, is introduced in reference [27]. The new definition not only covers the superiority of the definitions, it also provides a means to measure the energy density of the temporal dispersion of a process.

10.10.1 Kurtosis

The centered reduced variable for a random variable like x with a mean value and standard deviation equal to μ and σ is defined as follows [28]:

$$X_{CR} = \frac{x - \mu}{\sigma} \tag{10.34}$$

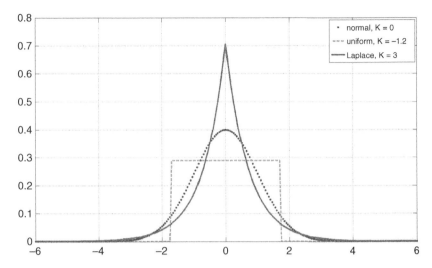

Figure 10.23 The probability distribution function and its corresponding value of kurtosis for three distribution functions: normal, uniform, and Laplace [28]/with permission of IEEE.

The kurtosis of a random variable x in higher-order statistics is defined based on the fourth-order moment of the centered variable, as follows [28]:

$$\kappa_x = E\left[X_{CR}^{\,4}\right] = E\left[\left(\frac{x - \mu}{\sigma}\right)^4\right] = \frac{\kappa_4}{\kappa_2^{\,2}} \tag{10.35}$$

where κ_i is an ith cumulant of the random variable x. A standardized kurtosis must be defined based on cumulants instead of moments to have kurtosis equal to zero in the case of a Gaussian distribution function. In statistics, the relationship between the cumulants and the moments is described as follows [28]:

$$\kappa_i = m_i - \sum_{k=0}^{i-1} \binom{k-1}{i-1} \kappa_k m_{i-k} \tag{10.36}$$

where m_i is an ith moment of x.

Kurtosis is a second shape factor after skewness in statistics and is used to define the sharpness of the distribution function. In the case where the distribution function shape is sharp, the value of kurtosis is high, whereas its value is lower if the distribution function is flatter. Figure 10.23 shows the kurtosis values for three different distribution functions. The value of kurtosis for a uniform distribution function is equal to −1.2, while its value for the Laplace distribution is 3. The value of kurtosis for a uniform Gaussian distribution is zero.

However, kurtosis is also used to find the outlier element within a distribution function. If all components of a distribution function are within a sample's mean value, as in Figure 10.24, its kurtosis value is zero if the distribution function is assumed to be Gaussian. Conversely, the presence of an outlier in a probability distribution function increases the kurtosis value. As shown in Figure 10.24, the amplitude of kurtosis by having approximately 10 outliers is increased to 15. This approach is used to determine a non-stationary component within a signal using kurtosis.

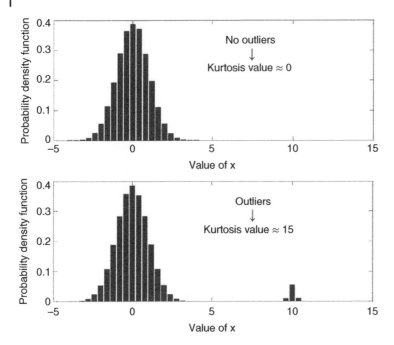

Figure 10.24 The probability density function for a normal distribution with kurtosis equal to zero (a), and the probability distribution function containing outliers; the kurtosis is 15 [28]/with permission of IEEE.

10.10.2 Spectral Kurtosis

A new definition of SK based on Wold-Cramer decomposition is proposed in reference [27] for non-stationary signals. The Wold-Cramer decomposition of a random variable signal $x(n)$ in a frequency domain is as follows [27]:

$$x(n) = \int_{-1/2}^{+1/2} H(n,f)e^{j\,2\pi\,fn}\,dZ_x(f) \tag{10.37}$$

where $H(n, f)$ is the Fourier transform of the time-varying signal and $dZ_x(f)$ is an orthogonal x spectral increment at a frequency of f. The SK based on the fourth-order normalized cumulant is defined as [27]

$$K_x(f) = \frac{\langle |H(n,f)|^4 \rangle}{\langle |H(n,f)|^2 \rangle^2} - 2 \tag{10.38}$$

In the usual definition of cumulants, the factor in Equation (10.38) is 3. Using the present definition, it is equal to 2 since $dZ_x(f)$ is a circular random variable. SK has some interesting properties that make it valuable for signal processing of non-stationary signals [29].

1. The SK value of a stationary Gaussian signal is zero:

$$K_x(f) = 0 \tag{10.39}$$

2. The SK value of the stationary Gaussian signal is a function of frequency:

$$K_x(f) = \text{constant} \tag{10.40}$$

3. The SK value in a case of stationary additive noise $\omega(n)$ is

$$K_{(x+\omega)}(f) = \frac{K_x(f)}{[1 + \rho(f)]^2} \tag{10.41}$$

where $\rho(f)$ is a noise-to-signal ratio as a function of frequency.

The properties of SK allow the detection, characterization, and localization of non-stationary characteristics within a signal.

10.10.3 Kurtogram

Spectrum kurtosis is a stable tool for signal processing purposes, since it is invariant to small parameter changes. However, the value of frequency resolution and the frequency estimator have a considerable impact on the value of SK, especially for transient components of the signal. The selection of frequency resolution on an arbitrary signal at a given frequency is a challenging task. The amplitude of the SK for highly non-stationary signals depends on the choice of frequency resolution. For instance, the value of SK becomes zero in cases where the frequency resolution is infinite. Alternatively, the narrow band transient components buried in the signal are undetectable if the frequency resolution is too coarse. However, a specific optimal frequency f and frequency resolution Δf may exist that maximize the value of SK. This results in the concept of a frequency–frequency resolution dyad that is defined as a kurtogram [30].

The principle of the kurtogram algorithm is based on an arborescent multi-rate filter bank structure [31] with quasi-analytic filters. The filter bank consists of two low-pass ($h_0(n)$) and high-pass ($h_1(n)$) analytic filters:

$$h_0(n) = h(n)\, e^{j\,\pi n/4} \tag{10.42}$$

$$h_1(n) = h(n)\, \pi^{j3\pi n/4} \tag{10.43}$$

with frequency bands limited to [0; 1/4] and [1/4; 1/2]. Figure 10.25 shows the low-pass and high-pass filters used to decompose the signal components at each level. The filters are implemented in a pyramidal shape to construct the arborescent filter bank, as shown in Figure 10.26. The output coefficient of each level of filters, such as $c_k^{\ i}(n)$, is fed into low-pass and high-pass filters and then downsampled by a factor of two to result in two new coefficients, $c_{k+1}^{\ 2i}(n)$ and $c_{k+1}^{\ 2i+1}(n)$, in level $K-1$. The high-pass filter is converted into a low-pass filter by multiplying the output of $h1$ by $(-j)^n$ before downsampling of the coefficient.

The coefficient $c_k^{\ i}(n)$ can be interpreted as a complex envelope of the input signal that is located on the central frequency of $f_i = (i + 2^{-1})2^{-k-1}$ with a frequency resolution of $(\Delta f)_k = 2^{-k-1}$. The kurtosis of all sequences must be calculated to estimate the kurtogram, as follows [30]:

$$K_k^{\ i} = \frac{\left\langle \left| c_k^{\ i}(n) \,|\, 4 \right\rangle \right.}{\left\langle \left| c_k^{\ i}(n) \,|\, 2 \right\rangle 2 \right.} - 2 \tag{10.44}$$

The kurtogram representation of the $(f, \Delta f)$ plane is demonstrated in Figure 10.27. The complexity of the method is on the order of FFT and wavelet transforms. The explained algorithm for kurtogram estimation is based on a dyadic grid, which is too coarse for some transient components with a narrow band. Therefore, to achieve a finer sampling over the $(f, \Delta f)$ plane, the algorithm must be refined based on a 1/3 binary tree of filter banks. However, the computation time for the

Figure 10.25 The high-pass and low-pass filters used for signal decomposition in addition to downsampling.

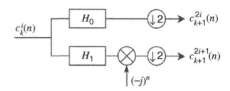

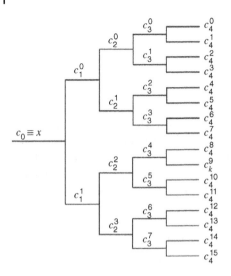

Figure 10.26 An arborescent filter bank decomposition used for fast computation of the kurtogram [27].

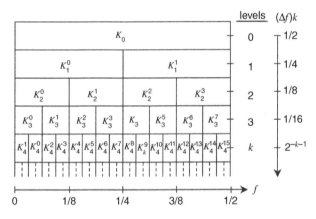

Figure 10.27 Fast kurtogram plane for frequency–frequency resolution [27].

algorithm is higher than that for the binary tree filter bank. The $(f, \Delta f)$ plane of a 1/3 binary tree is shown in Figure 10.28.

10.11 Noise

In the field of signal processing, the technical term "signal" is referred to as only the demanded data that are measured. However, the data acquired during the process of acquisition, storage, and conversion are susceptible to noise. Noise is unwanted data that leak into the desired data and may disturb the data components. Noise in the working environment of electric machines is unavoidable, especially in power plants and production lines. The noise generated by electric equipment has its specific characteristic that falls into a specific category of noise. The quality of the signal is measured based on the signal-to-noise ratio, which represents the ratio of the signal amplitude to the standard deviation of the noise [32].

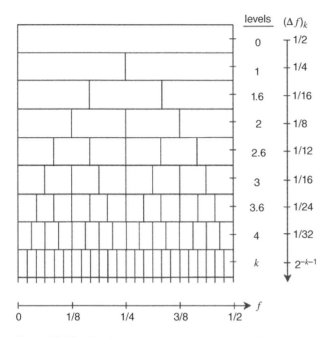

Figure 10.28 Fast kurtogram plane for frequency–frequency resolution in a case of a 1/3 binary tree [27].

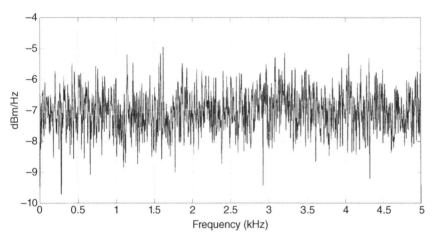

Figure 10.29 The frequency spectrum of white noise.

10.11.1 Various Types of Noise

Various types of noise have their own specific characteristics that may affect a specific part of a signal. Discriminating the noise from the signal is possible based on their frequency properties. The majority of acquired signals in the power industries may contain low-frequency components, whereas the noise may contain high-frequency components or may spread over a wide frequency range. The type of noise is characterized based on its frequency spectrum and is designated as white noise, pink noise, Brownian noise, blue noise, or violet noise, according to its frequency content. The definition of each type of noise based on the frequency distribution of its power spectral density is as follows [33, 34]:

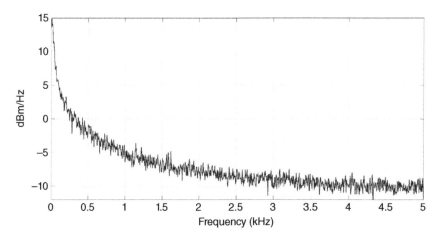

Figure 10.30 The frequency spectrum of pink noise.

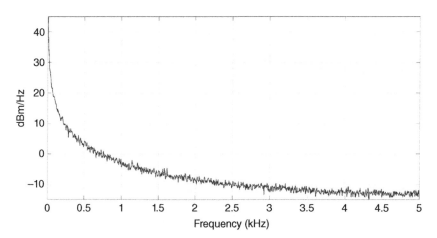

Figure 10.31 The frequency spectrum of Brownian noise.

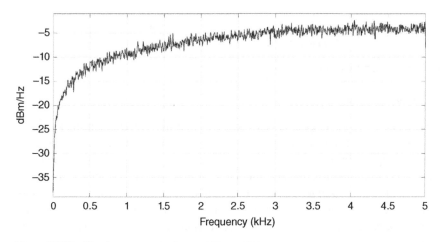

Figure 10.32 The frequency spectrum of blue noise.

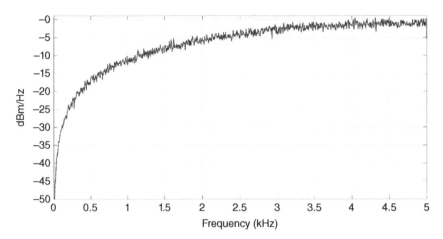

Figure 10.33 The frequency spectrum of violet noise.

1. White noise is a random noise that has an identical power density over the whole frequency span, as shown in Figure 10.29.
2. Pink noise has a high-power density over low frequencies and its amplitude is decreased by increasing the frequency, as shown in Figure 10.30.
3. The amplitude of Brownian noise is proportional to the square root of a frequency. Although its shape is similar to that of pink noise, it has a higher amplitude than pink noise at low frequencies, as shown in Figure 10.31.
4. Blue noise has a strong power density over a high frequency range and its amplitude is increased by increasing the frequency, as shown in Figure 10.32.
5. Violet or purple noise is a derivative form of white noise and its power spectral density is proportional to the square root of frequency, as shown in Figure 10.33.

10.11.2 Sources of Noise in Industry

The noise of electric equipment is generally restricted to low-frequency components. However, stationary or rotating electric machines, like transformers, electric motors, or generators, can produce broadband frequency components due to their cooling systems. Additional high-frequency noise due to power electronic devices in converters can be created that may interfere with the acquired signal. The noise in electric machines is categorized as follows [32, 35]:

1. The noise created is due to the electromagnetic source and to the radial force created by the interaction between the stator and the rotor magnetic field. The structure of the rotor and the stator or to slots creates high-frequency noise components inside the machine. Notably, the frequency range of this noise is almost always below 1 kHz.
2. The cooling system of an electric machine creates aerodynamic noise with a broad frequency range due to air flow through the inlet and outlet.
3. Improper installation of the machine, natural frequency of the stator, and bearing vibration cause mechanical noise. Noise with a high-power density is created in cases where the exciting frequency of the machine coincides with the natural frequency of the stator.

Many electrical motors and variable speed pumps in storage power plants utilize electric converters that contain power electronic devices. In addition to the amount of noise created due to

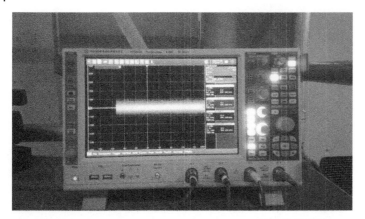

Figure 10.34 The measured noise in a hydropower plant.

switching, the power source supplied by power electronics to machine windings contains a various range of harmonics. This results in significant noise if the supplied harmonics coincide with the natural frequency of the machine [12]. The amount of noise created by power transformers in comparison to the rotating electric machine is remarkable. The noise sources in power transformers are divided into two categories [36]:

1. Magnetic noise due to the magnetic field of the core.
2. Load noise caused by the interaction of leakage flux and the current passing through the windings.

The metallic body of the power transformer suppresses noise emission into the working environment. However, the power transformers are located in separate rooms in the power plants, which reduces their effect on data from the rotating machine.

10.11.3 Noise Recognition

Figure 10.34 shows the noise measured by an installed sensor in a hydropower plant. The frequency analysis of the measured signal, as shown in Figure 10.35, confirms the existence of white noise in the operating environment of the power plant. Figure 10.35 shows that the amplitude of the noise is the same and that its power density throughout the frequency span is the same, which is a white noise property. Figure 10.36 shows the air gap magnetic field of the synchronous generator in another operating environment that indicates noise interference.

The air gap magnetic field is vulnerable to internal and external noise. The internal noise is mostly due to the slotting effect and rotor configuration. External noise leaks into the measured data from environmental noise sources like power converters. Noise effects can be avoided by the use of copper foil or coaxial cables to reject the noise. The analysis of a noisy air gap magnetic field signal confirms the existence of white noise. Therefore, the existence of noise in the operating environment of an electric machine is unavoidable. The following sections will examine noise effects on signal processing tools by combining white noise with the air gap magnetic field in order to investigate its effect on signal processing tools.

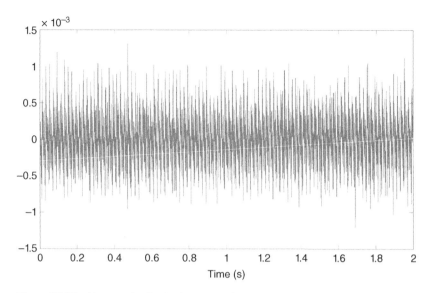

Figure 10.35 Measured noise in the power plant.

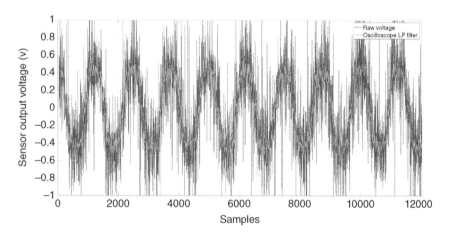

Figure 10.36 The measured air gap magnetic field in a noisy environment and reduction of noise with an oscilloscope LP filter.

10.11.4 Noise Effect on FFT

Fast Fourier transform (FFT) was applied to the air gap magnetic field in a healthy and faulty salient pole synchronous generator (SPSG). The amplitude of the sidebands is increased under faulty conditions. Increasing the severity of the fault also increased the amplitude of the sidebands (a detailed explanation of the detection procedure and index for fault diagnosis is covered in Chapter 6 and Chapter 11). Figure 10.37a shows the FFT applied to the air gap magnet without noise interference in the healthy and faulty cases. The amplitude of the sidebands was discernible. White noise, with a rating of 20 dB, was added to the measured air gap magnetic field signal; its FFT results are shown in Figure 10.37b. The noise level is increased significantly compared with the no-noise data set, and the sidebands are masked. Table 10.2 shows the results of some of the sideband frequencies used for fault detection purposes under the various levels of noise. Table 10.2

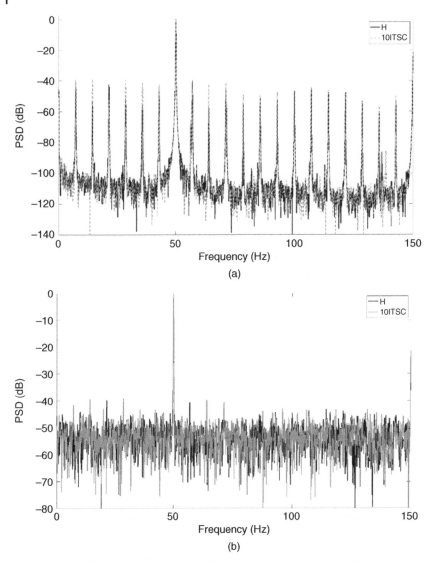

Figure 10.37 The spectral density of an air gap magnetic field in a healthy (solid line) state and under an inter-turn short circuit fault (dashed line): (a) without noise interference, (b) with 20 dB noise.

shows that increasing the noise level from 60 to 30 dB does not have an impact on the fault index. Alternatively, the FFT processor is robust to noisy data.

10.11.5 Noise Effect on the STFT

Figure 10.38 demonstrates the result of applying STFT to air gap magnetic field signals in a healthy and faulty SPSG condition. A simple comparison between a healthy and faulty case in Figure 10.38 reveals that the fault changes the profile of the STFT and the intensity of some of the time-frequency components confined up to 100 Hz. Both healthy and faulty cases are investigated under 40 and 20 dB noise levels, and a qualitative comparison of the STFT results of the air gap magnetic field

Table 10.2 The effect of various signal-to-noise ratios on the nominated index under an ITSC fault.

Cases	Noise level	7.15 Hz	14.3 Hz	28.6 Hz	35.7 Hz	64.3 Hz	78.6 Hz
H	NN	−44.6	−53.1	−51.3	−54.9	−57.4	−54.5
	60 dB	−44.6	−52.9	−51.2	−55.0	−57.6	−54.4
	50 dB	−44.8	−53.1	−51.4	−54.1	−57.8	−54.4
	40 dB	−45.1	−53.0	−51.1	−55.3	−58.2	−54.4
	30 dB	−45.3	−55.1	−51.5	−54.8	−57.3	−54.6
F	NN	−39.6	−38.9	−39.4	−40.8	−41.8	−49.3
	60 dB	−39.6	−38.9	−39.4	−40.8	−41.8	−49.3
	50 dB	−39.5	−38.9	−39.4	−40.9	−41.9	−49.5
	40 dB	−39.8	−39.1	−39.2	−40.6	−42.1	−49.5
	30 dB	−40.3	−39.1	−39.9	−40.2	−41.8	−49.1

under different noisy conditions implies that the noise has no marked impact on it. However, a qualitative comparison of the results indicates that they are not reliable for signal processing. Therefore, the obtained STFT results for the healthy case with 40 dB noise and for the faulty case without noise were processed with an image processing tool, as shown in Figure 10.39. The noise has the same pattern and impact on the healthy case and on the fault components. Therefore, noise with a high ratio may lead to a fault indication if the signal is processed by STFT. Although a practical solution to reject the noise is to increase the window length of the STFT, this may lead to a reduction in temporal resolution, thereby limiting the usefulness of STFT.

10.11.6 Noise Effect on CWT

CWT is a qualitative processing tool used for fault diagnosis of electric machines and is performed using a B-spline wavelet. The interpretation of the results obtained with CWT is arduous, since an image processing tool or convolutional neural network is required. Figure 10.40 shows the application of CWT to a healthy and faulty air gap magnetic field signal in no-noise, 40 dB, and 20 dB white noise. A periodic notch appears upon introducing the fault to the signal and comparing the healthy and faulty cases in a frequency band between 25 and 50 Hz.

The noise impact on the CWT is discernible since the patterns become blurry, and the intensity of the colors decreases as the noise level changes from 40 to 20 dB. The notching of the 50 Hz band is severely weakened by the introduction of noise. The CWT, unlike STFT, is uniformly affected by noise since it has a greater time-frequency resolution. However, the type of mother wavelet has a significant impact on the result, and the analysis must be repeated with different signal processing tools, like Shannon's mother wavelet [34].

10.11.7 Noise Effect on DWT

DWT is a useful tool that is frequently used for feature extraction of electrical machines under faulty operation. Various types of mother wavelets can be used for fault detection purposes, as described in Section 10.6. In the present section, Daubechies 8 is used to study the noise effect on the air gap magnetic field signal. The sampling frequency of the signal is 10 kHz; consequently, the wavelet frequency band is limited to 5 kHz. The first and second sub-bands are between 5 and 2.5 kHz and 2.5 and 1.25 kHz, respectively.

However, understanding the impact of noise on DWT sub-bands is not possible from a qualitative perspective. Therefore, an index based on the energy of each sub-band is used to evaluate the noise impact. The energy of each sub-band is calculated as follows [34]:

$$E = \int_{-\infty}^{+\infty} |D_n|^2 \, dt \tag{10.45}$$

where D_n is the magnitude of each of wavelet sub-band. The occurrence of a fault has a significant impact on D1 and D6 sub-bands compared with a healthy case, as shown in Table 10.3. Figure 10.41 shows the application of DWT to the air gap magnetic field signal in a healthy case without noise interference and with a 20 dB noise effect. The shape of the sub-bands changed dramatically due to the noise effect. According to Table 10.3, the energy of wavelet sub-bands for the healthy state that

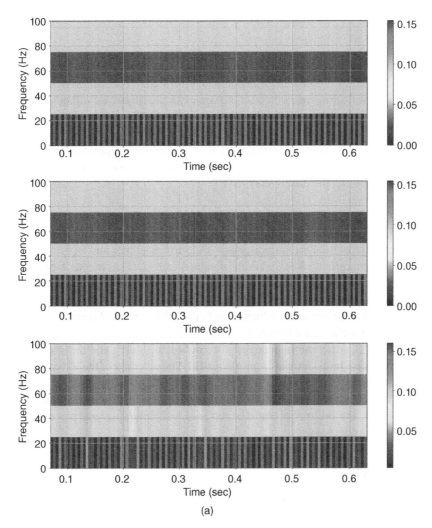

(a)

Figure 10.38 Application of STFT to an air gap magnetic field in healthy SPSG (a) and under a 10 ITSC fault (b) in no-noise, 40 dB, and 20 dB SNR conditions.

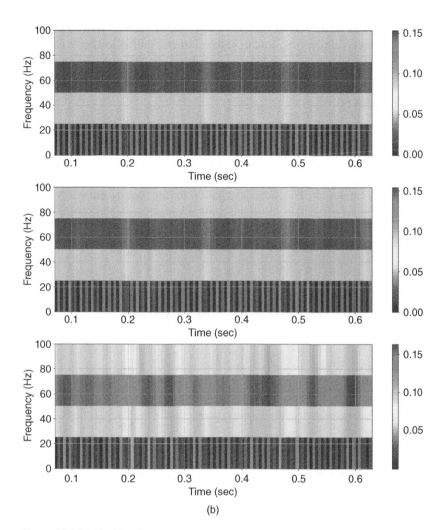

(b)

Figure 10.38 (*Continued*)

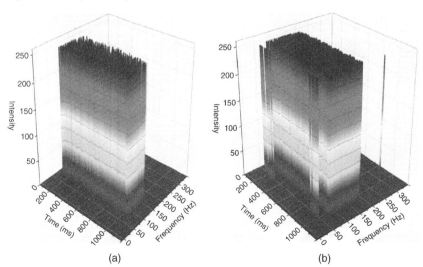

(a)   (b)

Figure 10.39 The impact of 40 dB noise on data processed by STFT (a) and the impact of a 10 ITSC fault on a signal processed by STFT (b) using the image processing tool.

is influenced by 20 dB noise is equal to that of a faulty case. Therefore, the use of signal processing based on DWT in noisy operating conditions may lead to a false alarm for fault detection in electric machines.

10.11.8   Noise Effect on TSDM

The TSDM method is applied to the air gap magnetic field signal to evaluate the changes due to faults that cause field asymmetry. The radius of gyration created by TSDM may change due to variations in the time series data of the air gap magnetic field [34]. Figure 10.42 shows the application of the TSDM method to the air gap magnetic field time series in a healthy and faulty case for the no-noise and 20 dB white noise conditions. The radius of gyration for the healthy and faulty cases is 61 and 78, respectively. The radius of gyration was significantly increased by adding 20 dB white

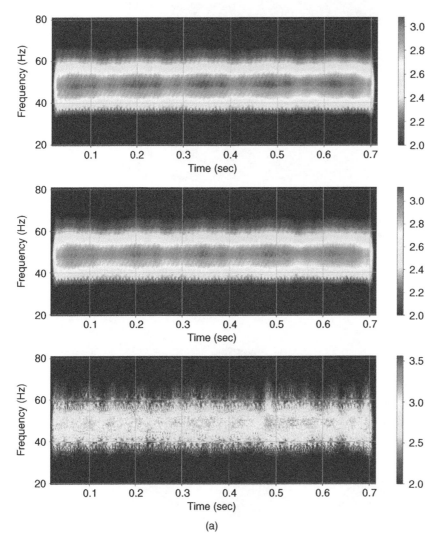

(a)

Figure 10.40   The applied CWT to an air gap magnetic field in healthy (a) and under a 10 ITSC fault (b) in no-noise, 40 dB, and 20 dB SNR.

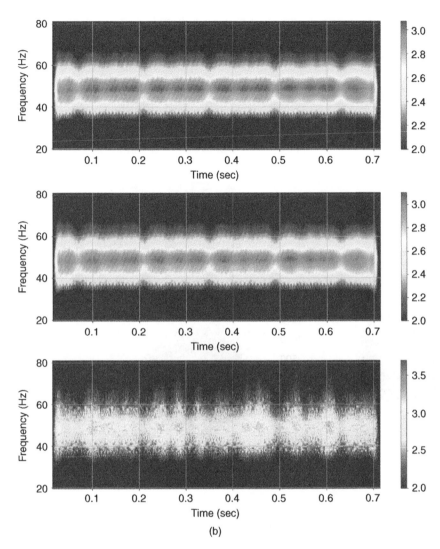

Figure 10.40 (*Continued*)

Table 10.3 The energy of various wavelet sub-bands energy in healthy and faulty cases with noise and without noise effect.

Energy	D8	D7	D6	D5	D4	D3	D2	D1
Healthy no-noise	0.70	0.38	0.83	0.94	0.97	1	1.06	1.35
Healthy 20 dB	0.70	0.38	0.81	0.88	0.83	0.79	0.72	0.65
Faulty no-noise	0.70	0.38	0.81	0.88	0.83	0.79	0.72	0.65

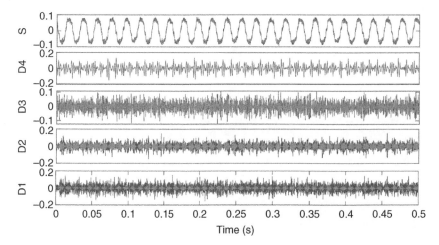

Figure 10.41 The discrete wavelet transforms of an air gap magnetic field in a healthy case (black) and with 20 dB white Gaussian noise (gray).

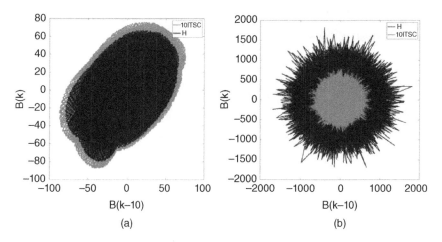

(a)   (b)

Figure 10.42 The gyration radius of the air gap magnetic field in a healthy and a 10 ITSC fault state without noise (a) and with 20 dB white Gaussian noise (b).

noise to the time series. Moreover, the area of the mass and its radius in a healthy case were larger than in the faulty case, indicating a false discrimination of the fault. Therefore, the data analyzed by the TSDM method must be noiseless, otherwise the method results in a false-positive alarm [33].

10.12 Summary

Signals are the main utilized tool in fault detection of electric machines. In this chapter, a basic definition of the signal was described and various signal processing methods to extract useful information from the raw signals were explained. Various signal processing methods were introduced and their formulation methods were described. The signal processing tools were categorized as time domain, frequency domain, and time-frequency domain. Noise is an unavoidable part of experimental and field test data and can mask the extracted features in signal processing. Therefore, various noises that exist in the industry were introduced and their impact on signal processing tools were examined. Indications were that noise has a devastating impact on the measured data and their results. Several practical approaches were suggested to reduce the noise level of the measured signal in a real-world application.

References

1 Priemer, R. (1991). Introductory Signal Processing. In: *World Scientific*. Singapore: Publishing Co Pte Ltd.

2 Gray, R.M. and Neuhoff, D.L. (1998). Quantization. *IEEE Transactions on Information Theory* 44 (6): 2325–2383.

3 Shannon, C.E. (1949). Communication in the presence of noise. *Proceedings of the IRE* 37 (1): 10–21, 0096–8390.

4 Skreien, T.N. (2020). Application of signal processing and machine learning tools in fault detection of synchronous generators. MSc Thesis, Norwegian University of Science and Technology (NTNU), Norway.

5 Yates, R.D. and Goodman, D.J. (1999). *Probability and Stochastic Processes*. USA: Wiley.

6 Bergland, G.D. (1969). A guided tour of the fast Fourier transform. *IEEE Spectrum* 6 (7): 41–52.

7 Mork, B.A. (1997). Comparison measures for benchmarking time domain simulations. In: *International Conference on Power Systems Transient (IPST)*, 22–26 June 1997. Seattle, USA.

8 Groth, I.L. (2019). On-line magnetic flux monitoring and incipient fault detection in hydropower generators. MSc Thesis, Norwegian University of Science and Technology (NTNU), Norway.

9 Yun, J., Park, S., Chanseung Yang, C. et al. (2019). Comprehensive monitoring of field winding short circuits for salient pole synchronous motors. *IEEE Transactions on Energy Conversion* 34 (3): 1686–1694.

10 Torrence, C. and Compo, G.P. (1998). A practical guide to wavelet analysis. *Bulletin of the American Meteorological Society* 79 (1): 61–78.

11 Lee, B.Y. (1999). Application of the discrete wavelet transform to the monitoring of tool failure in end milling using the spindle motor current. *The International Journal of Advanced Manufacturing Technology* 15 (4): 238–243.

12 Yang, S.J. (1988). Effects of voltage/current harmonics on noise emission from induction motors. In: *Vibrations and Audible Noise in Alternating Current Machines*, 457–468. Germany: Springer.

13 Kia, S.H., Mpanda Mabwe, A.M., Henao, H. et al. (2006). Wavelet based instantaneous power analysis for induction machine fault diagnosis. In: *32nd IEEE Industrial Electronics Annual Conference*, 6–10 November 2006. Paris, France.

14 Guo, L., Rivero, D., Seoane, J.A. et al. (2009). Classification of EEG signals using relative wavelet energy and artificial neural networks. In: *Proceedings of the First ACM/SIGEVO Summit on Genetic and Evolutionary Computation*, 177–184.

15 Rossoa, O.A., Blancoa, S., Yordanova, J. et al. (2001). Wavelet entropy: A new tool for analysis of short duration brain electrical signals. *Journal of Neuroscience Methods* 105 (1): 65–75.

16 Zheng-You, H.E., Xiaoqing, C., and Guoming, L. (2006). Wavelet entropy measure definition and its application for transmission line Fault detection and identification (Part I: Definition and methodology). In: *International Conference on Power System Technology*, 22–26 October 2006. Chongqing, China.

17 Shannon, C.E. (1948). A mathematical theory of communication. *The Bell System Technical Journal* 27 (3): 379–423.

18 Huang, N.E. (2014). *Hilbert-Huang Transform and its Applications*, vol. 16. World Scientific.

19 Rilling, G., Flandrin, P., and Goncalves, P. (2003). On empirical mode decomposition and its algorithms. In: *IEEE-EURASIP Workshop on Nonlinear Signal and Image Processing*. Baltimore, Maryland, USA.

20 Fosso, O.B. and Molinas, M. (2019). Method for mode mixing separation in empirical mode decomposition. arXiv:1709.05547v2 [stat.ME].

21 Rand, D.A. and Young, L.S. Dynamical systems and turbulence. In: *Warwick 1980: Proceedings of a Symposium Held at the University of Warwick 1979/80*, vol. 898. Berlin, Heidelberg: Springer.

22 Takens, F. (1981). Detecting strange attractors in turbulence. In: *Dynamical Systems and Turbulence, Warwick 1980*, 366–381. Berlin, Heidelberg: Springer.

23 Ehya, H., Nysveen, A., Nilssen, R. et al. (2019). Time domain signature analysis of synchronous Ggnerator under broken damper bar fault. In: *45th Annual Conference of the IEEE Industrial Electronics Society*, 1423–1428. Lisbon, Portugal, October 14–17, 2019, IEEE.

24 Webb, A.R. and Copsey, K.D. (2011). *Statistical Pattern Recognition*. Wiley.

25 Dwyer, R.F. (1983). A technique for improving detection and estimation of signals contaminated by under ice noise. *The Journal of the Acoustical Society of America* 74 (1): 124–130.

26 Vrabie, V., Granjon, P., and Serviere, C. (2003). Spectral kurtosis: From definition to application. In: *6th IEEE International Workshop on Nonlinear Signal and Image Processing (NSIP)*. Grado-Trieste, Italy.

27 Antoni, J.R.M. (2006). The spectral kurtosis: A useful tool for characterising non-stationary signals. *Mechanical Systems and Signal Processing* 20 (2): 282–307.

28 Fournier, E., Picot, A., Régnierl, J. et al. (2013). On the use of spectral kurtosis for diagnosis of electrical machines. In: *9th IEEE International Symposium on Diagnostics for Electric Machines, Power Electronics and Drives (SDEMPED)*, 27–30 August 2013. Valencia, Spain.

29 Ehya, H., Nysveen, A., and Nilssen, R. (2020). A practical approach for static eccentricity fault diagnosis of hydro-generators. In: *2020 International Conference on Electrical Machines (ICEM)*, 23–26 August 2020. Gothenburg, Sweden.

30 Antoni, J. (2007). Fast computation of the kurtogram for the detection of transient faults. *Mechanical Systems and Signal Processing* 21 (1): 108–124.

31 Peng, Z.K. and Chu, F.L. (2004). Application of the wavelet transform in machine condition monitoring and fault diagnostics: a review with bibliography. *Mechanical Systems and Signal Processing* 18 (2): 199–221.

32 Vijayraghavan, P. and Krishnan, R. (1999). Noise in electric machines: A review. *IEEE Transactions on Industry Applications* 35 (5): 1007–1013.

33 Ehya, H., Nysveen, A., Skreien, T.N. et al. (2020). The noise effects on signal processors ssed for fault detection purpose. In: *23rd International Conference on Electrical Machines and Systems (ICEMS)*, 24–27 November 2020. Hamamatsu, Japan.

34 Ehya, H., Nysveen, A., and Skreien, T.N. (2021). Performance evaluation of signal processing tools used for fault detection of hydro-generators operating in Noisy environments. *IEEE Transactions on Industry Applications* 57 (4): 3654–3665.

35 Tischmacher, H. et al. (2011). Case studies of acoustic noise emission from inverter-fed asynchronous machines. *IEEE Transactions on Industry Applications* 47 (5): 2013–2022.

36 Girgis, R.S., Bernesjo, M., and Anger, J. (2009). *Comprehensive Analysis of Load Noise of Power Transformers*, 26–30 July 2009. Calgary, AB, Canada.

11

Electromagnetic Signature Analysis of Electrical Faults

11.1 Introduction

One of the most important concerns in the manufacturing industry and in power plants is the prevention of unplanned stoppage of electric machines. The role of high-voltage rotating electric machines, such as synchronous generators and synchronous motors, with various rated power in industries and power plants is very important. This is because power generation in power plants and production lines in industries depend on the reliable operation of these machines. Moreover, synchronous machines are among the pieces of expensive equipment in power plants and industries and any damage to them may incur significant costs. Synchronous machines can experience several faults during operation and need to be protected. Faults in synchronous machines are divided into electrical faults and mechanical faults. In general, the electrical faults of synchronous machines can be divided into the following categories:

1. Insulation and electrical defects in the stator winding and the end winding.
2. Defects in the rotor field winding.

Initially, most industries use reactive maintenance, which is repair after equipment failure. The high cost of repairs (or the cost of replacing a failed machine with a new synchronous one) and, more importantly, the cost of reducing output or not producing due to the stoppage of electric machines generally means that reactive maintenance should be replaced with periodic maintenance. Various industries and power plant operators consider that the use of maintenance systems increases reliability and reduces maintenance costs. Most power plants and industries use periodic maintenance programs; however, another method is also available: predictive maintenance.

Electromagnetic Analysis and Condition Monitoring of Synchronous Generators, First Edition. Hossein Ehya and Jawad Faiz.
© 2023 The Institute of Electrical and Electronics Engineers, Inc. Published 2023 by John Wiley & Sons, Inc.

In predictive maintenance, the condition of the machine is monitored online and a notification is sent to the operator at the first signs of a fault to allow the operator to take appropriate preventive action. This proper action depends on the type of fault and its severity. Upon considering this notification, an operator may decide to stop the machine or to allow the rotation to continue until further fault notification.

In general, faults can occur due to external factors, such as short circuit faults in the power grid or internal faults due to gradual defects in the insulation or mechanical components. An external fault that has a high severity must be identified and quick action taken to prevent any damage to the machine; this is the responsibility of the protection systems. Synchronous machines can be protected from rapid and destructive faults, such as single-phase or three-phase short circuit faults in a power grid, using a current- or voltage-based protection system that takes the machine out of service. In addition to sudden failure, minor defects in synchronous generators can result in severe damage over time. Since the protection system cannot detect the occurrence and growth of internal faults in a synchronous machine, another supplementary system must be used to perform a condition monitoring task.

A low-severity fault profile can be detected in its early stage if a condition monitoring system is installed on the synchronous machine; this reduces the rapidly occurring fault and, in turn, increases the reliability of the synchronous machine and its related system. Accordingly, extensive studies have been conducted by industry and scientific research centers around the world on online condition monitoring of synchronous machines. Various methods have been used for condition assessment of synchronous machines, such as:

1. An electromagnetic-based approach based on various signal measurements.
2. Partial discharge measurement.
3. Vibration measurement.
4. Chemical particle measurement, including analysis of cooling gas.
5. Temperature analysis.

This chapter focuses on condition monitoring methods based on electromagnetic signals, while also providing a brief introduction to some methods where detection is not based on electromagnetic analysis. The methods discussed in detail, based on electromagnetic analysis to detect different types of electrical faults, are:

1. Induced voltage in the stator terminal.
2. Current in the stator winding.
3. Current in the rotor field winding.
4. Magnetic field in the air gap.
5. External magnetic field.

11.2 General Introduction to Short Circuit Fault Detection Methods in Synchronous Machines

Several diagnostic techniques are available for the short circuit fault of the stator winding. Various parameters of synchronous generators, such as magnetic flux, vibration, noise, instantaneous power, temperature, air gap torque, induced voltage, speed, partial discharge, gas analysis, surge testing, and motor circuit analysis, can be used.

The detection coil is fixed on a synchronous generator stator [1]. When a fault occurs, the online fault is diagnosed by detecting the air gap flux change. The detection coil method has high diagnostic accuracy, but adding additional detection equipment inside large synchronous generators is complicated and invasive, and therefore difficult to apply in practice. The artificial neural network (ANN) has been applied as an efficient diagnostic tool for estimating a short circuit fault in a three-phase stator winding [2]. However, using this method needs substantial operation data from the faulty machine to support and train the ANN; therefore, this method is not generally applicable to large synchronous generators. Some researchers have proposed a multi-loop analysis method for fault detection [3, 4]. This method deduces the resistance and inductance matrix according to the connection of each loop of the generator and calculates the parameters required to obtain the related waveform. Large hydro-generators contain many poles and damping rings, which complicate solving the related equations, affect the identification accuracy, and limit the scope of application. Therefore, an accurate diagnosis may not be possible using the multi-loop detection method alone.

A zero-sequence voltage component-based method has been introduced in reference [5] to monitor the stator winding short circuit faults in a synchronous machine. The impacts of different stator winding configurations on the stator currents and the zero-sequence voltage component spectra of healthy and faulty machines have been addressed in reference [6]. When an inter-turn short circuit occurs in a synchronous generator, its parameters will change; therefore, a parameter identification technique has been considered to minimize the power consumption of a city bus equipped with a synchronous machine [7]. A two-stage Kalman filter, combined with a robust and optimal algorithm, has been proposed in reference [8] for state and disturbance estimation. The non-parametric Volterra model has been established in references [9–11] to identify its features. However, this model is more suitable for nonlinear systems and does not deal with the inter-turn short circuit fault of a rotor winding. Any continuous time-invariant nonlinear dynamic system can be described completely by a Volterra series as long as the input and output of the system are analytic functions. Most nonlinear systems can be described by three-order transitive relations. The inter-turn short circuit fault is a nonlinear fault and leads to a nonlinear change in many electric quantities of synchronous generators. Analysis of some work [11, 12] reveals that the Volterra series model can be introduced to describe the system characteristics of the fault. Following the inter-turn short circuit fault of the rotor winding, a stator unbalance branch current is generated in the synchronous generator.

The inter-turn short circuit fault of rotor windings is also one of the common faults in synchronous generators with salient rotor poles. Many factors result in rotor field winding failure, with rotational vibration over a long time as one of the main factors of the rotor short circuit fault. The rotor generally carries an excitation winding, and an inter-turn short circuit fault can occur in it [13, 14]. In a large synchronous generator, this fault leads to a higher excitation current, intensifying rotor vibration [15, 16] and having a demagnetization effect. If a few shorted turns in the rotor field winding are not detected in the early stage, this can result in maloperation of the generator.

Offline fault diagnosis techniques are frequently used in synchronous generators. In the AC impedance method [17], an AC voltage is applied to the rotor winding, making the current flowing through the SC winding much higher than the rated current. This short circuit current has a strong demagnetization effect. If the AC impedance value of one magnetic pole in the rotor winding is significantly lower than that of the other magnetic poles, and the loss significantly increases, this can be judged as an inter-turn SC in the rotor winding.

The original offline detection method has difficulty meeting the requirements of modern real-time monitoring. Therefore, many online detection methods so far have addressed the rotor inter-turn short circuit fault of synchronous generators. For example, the detection coil method is used to diagnose inter-turn SC faults. The impact of the fault on the unbalanced magnetic pull (UMP) force and the air gap magnetic density has been calculated in reference [18], and is the basis of the detection coil method.

As already mentioned, using the invasive fault diagnosis method in a synchronous generator is difficult in practice. The Volterra series model has been established and the stator unbalance current has been analyzed. The existing difference in the Volterra kernel function between the nonlinear system of a stator branch voltage and the branch current can be an appropriate index for the detection of a rotor inter-turn short circuit fault. This method combines the advantages of the Volterra method and the multi-loop method, thereby reducing the analysis of synchronous generators and the computational burden [19].

11.3 Stator Short Circuit Fault Types

Short circuit fault detection of synchronous machines depends on the type of stator winding. For example, the fault detection of an inter-turn short circuit fault is only applicable to a multi-turn coil winding type, whereas the same detection approach as the multi-turn coil is impractical for a synchronous generator with a copper bar in the winding. In that winding, detection is based on partial discharge monitoring. In a generator with copper bars, the short circuit fault is more severe and mostly happens between the phases or the winding and the ground. A short circuit fault in a generator with copper bars results in severe damage and may burn the winding within a few seconds [20]. The following are the reasons for insulation damage that causes a short circuit in the stator winding [21]:

1. Operation of synchronous machines at high temperatures.
2. Insulation problems due to the vibration of coils.
3. Insulation problems due to mechanical–thermal stresses due to a continuous load variation.
4. Partial discharge in the coils due to insulation problems or an insufficient distance between the coils or the body.

The insulation degradation of the stator winding in the synchronous machine is the main reason for winding defects and usually begins with the loss of insulation between two turns or between the winding and the ground. Electrical, mechanical, and thermal stresses are the main factors that introduce insulation degradation. The consequence of a turn-to-turn short circuit fault is the loss of symmetry in the magnetic field, which affects the output torque and the reduction of the main harmonic [22].

A short circuit fault in the stator is relatively dangerous, since it progresses in a few seconds and causes substantial damage to the machine in a short time [23]. Stator faults in synchronous machines can be divided as follows:

1. Stator unbalanced phases.
2. Single-phase to ground fault.
3. Phase-to-phase fault.
4. Multi-phase to ground fault.
5. Turn-to-turn fault.

These faults can lead to extensive damage with large short circuit currents [24]. Although the ground faults do not cause significant damage to the stator winding, they can result in an internal short circuit fault if not fixed in a short time. Serious damage can be avoided with proper short circuit ground protection.

11.3.1 Stator Unbalanced Phases

This fault occurs when the winding parameters of the generator phases differ. Various factors, such as inter-turn short circuit faults, may cause this type of fault.

11.3.2 Single-Phase Fault to Ground

The insulation degradation over time in a high-voltage synchronous machine results in a single-phase short circuit to ground fault. A capacitive current passes through the machine to the ground when a single phase to ground short circuit fault happens. The current amplitude is very low and usually cannot exceed a few amps. No adverse effects will be seen on the machine core and it cannot damage the stator iron if the current exists only for a short time. However, currents above 20 amps will have an adverse effect on the stator iron, even if the effective time is short.

This type of fault is unavoidable for many reasons; however, although low current on the iron lamination sheet of the stator cannot have serious effects on the stator core, this current can destroy and burn the insulation of the wire at the connection points. The circulating current, even with low amplitude, must be avoided since the local temperature rise may destroy the insulation of the other conductors in the slot. If the neighboring conductor of the damaged conductor belongs to the same phase, an inter-turn short circuit fault happens, whereas if the neighboring conductor belongs to the other phase, a phase-to-phase short circuit fault happens. Increased phase voltage with respect to the ground is another consequence of the phase-to-ground fault. The protection system should take acute action to prevent further damage [25].

11.3.3 Phase-to-Phase Fault

The connection of two phases inside the generator is called a phase-to-phase short circuit fault. The location of the phase-to-phase short circuit fault can be inside the stator slot, where two phases are located, or in the overhang, where the coils are placed next to each other. Insulation degradation due to several factors, such as vibration, insufficient quality of insulation, and environmental factors, is the main reason for a short circuit fault. Another cause is that the winding of two different phases that are not in the same stator slot may be connected separately to the metal body of the generator, thereby causing contact between two phases through the metal body of the generator. The connection of a single phase to the machine body and the connection of another phase to the ground can also occur and is known as a two-phase short circuit fault.

11.3.4 Turn-to-Turn Short Circuit Fault

Most short circuit faults start with an inter-turn short circuit fault. Figure 11.1 shows the inter-turn short circuit fault in the synchronous machine. This fault is due to the insulation degradation of the stator windings and can quickly become a serious fault. Insulation failure has several reasons [26]:
1. Thermal stresses.
2. Voltage stresses.
3. Aging.

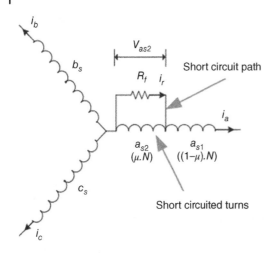

Figure 11.1 The inter-turn short circuit fault in the stator winding of the synchronous machine.

4. Vibration and displacement.
5. Mechanical defects imposed on insulation during manufacturing.
6. Working environment.

11.4 Synchronous Generator Stator Fault Effects

A fault in a synchronous machine has the following consequences:
1. Non-uniform air gap magnetic field.
2. Unbalanced voltages and line current.
3. Distortion in the waveform of voltage, current, and air gap flux.
4. Increase losses and efficiency reduction.
5. Vibration.
6. Excessive heating.

Accurate detection of an inter-turn short circuit fault is challenging in the electrical machine diagnostics field. The main intention of the fault diagnostics algorithm is to provide a reliable and efficient solution with low-priced equipment. Moreover, the algorithm should also be able to detect the location and the severity of the short circuit fault in the stator winding. Precise detection of a stator short circuit fault by the algorithm prevents serious damage and reduces machine downtime. Fault location detection is also an important issue that both speeds up repairs and optimizes the repair planning schemes. However, proposing an algorithm for early-stage detection of an inter-turn short circuit fault detection is challenging.

11.5 Fault Diagnosis Methods in the Stator Winding

Although extensive research has been conducted on short circuit fault diagnosis of induction motor windings [22, 27, 28], few papers have so far considered the stator winding of synchronous generators for the detection of short circuit faults [29–33]. Condition monitoring of the generators indicates that diagnosis of this fault remains the issue of the day and needs more research [34, 35]. The effective methods for using the stator winding for short circuit fault diagnosis can be classified into two general groups: invasive and non-invasive [36]. This classification has been derived based on the use of relevant tools in fault detection.

The invasive method diagnoses faults using sensors that need some modification or stoppage of the machine for the sensor installations. By contrast, the non-invasive group detects the fault based on different data analysis, such as the current or voltage of a generator in which sensors are already installed. These groups and various fault diagnosis techniques are addressed in the following sections.

11.5.1 Invasive Methods

A short circuit fault in the stator winding turns causes variations in the generator parameters. These variations occur in physical quantities, such as temperature, sound, and vibration, and in magnetic quantities, such as the air gap magnetic field of the generators. These variations can be a basis for detecting different faults, such as short circuit faults. In an invasive technique, diverse sensors detect these variations, which are introduced as the fault diagnosis index. Detection sensors were initially used and are still utilized for several faults. Fault sensors have been widely used due to their simple operation; however, because of their low sensitivity, they are unable to detect faults at their initial stages. In addition, because of the extra expense incurred for fixing sensors, they cannot be used in all machines. They tend to be employed in expensive and sensitive machines, and are generally put in place from a customer's order. These methods are also prone to problems arising from faults in the sensor itself, and a sensor fault can seriously damage a generator. Some of these methods are discussed below.

11.5.1.1 Thermal Analysis

Condition monitoring based on temperature measurement has been widely used in drives and generators. It is used for monitoring the temperature condition of some parts of the stator core and windings and the cooling fluids in large generators, such as turbo-generators. Temperature variations can be an index for fault diagnosis in the internal parts of the generator; however, these sensors are influenced by their fixing locations. Temperature variations caused by loading and by generator temperature limitations represent information that must be taken into account. One of the main signs of a fault in the generator is overheating [37–39]. Generally, three temperature measurement methods are utilized [40]:

- Resistance temperature detection.
- Thermistors.
- Thermocouples.

Figure 11.2 presents typical thermal measurement tools. To detect a temperature rise, the required sensor is fixed on the winding and/or insulation and is isolated electrically [27]. A higher than permissible temperature weakens the stator, rotor conductors, core, insulation, and end winding; therefore, thermal protection is one of the most important features of the protection and monitoring system of the generator [41-43]. In generators, the stator insulation is the component most vulnerable to temperature rise. Therefore, monitoring the stator temperature is effective in prolonging the generator's lifetime.

Fixing temperature sensors for diagnosing a stator winding fault, particularly in the initial stages of the short circuit fault, does not provide sufficient accuracy. The fault is affected by other factors, such as the environment temperature, maloperation of generator ventilation, unbalanced supply

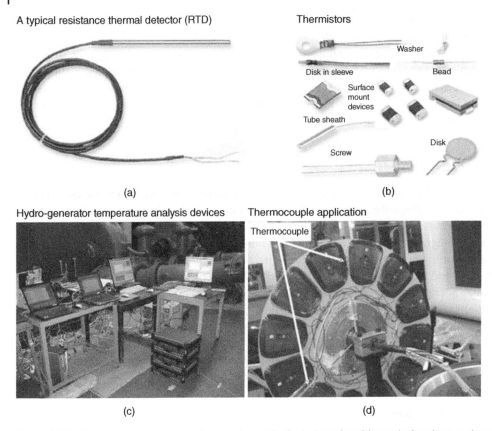

Figure 11.2 Temperature measurement sensors used for fault detection: (a) a typical resistance thermal detector (RTD), (b) thermistors, (c) hydro-generator temperature analysis devices, and (d) thermocouple application.

voltage (in the motor mode operation), overloading, and repeated starting. In addition, thermocouple sensors with detectors are expensive, so they are used only in large generators due to fixing problems and relevant cable disturbances.

11.5.1.2 Vibration Analysis

Vibration sensors are widely used and have now achieved a good level of technology. These sensors can be employed for condition monitoring of the generator and of the associated moving parts related to synchronous machines, such as gearboxes, couplings, and shafts. Rotation can be determined by separately measuring the three components of displacement, speed, and acceleration. Of course, the accuracy of the measured quantities depends on the monitoring equipment and its range of frequency. Identical generators of the same size have a constant vibration speed. By increasing the frequency of the vibrations, the possibility arises of a decrease in the displacement levels while the slope level increases. Therefore, in some cases, the measurement of displacement has been recommended. To measure the vibration level, piezoelectric sensors can measure the vibration acceleration. This is then converted into speed vibrations by integrator modules. Displacement is in the range of 0–8 kHz, velocity between 8 Hz and 8 kHz, and acceleration between 30 Hz and 500 kHz [40].

Figure 11.3 A typical vibration sensor fixed on the stator winding of a hydro-generator.

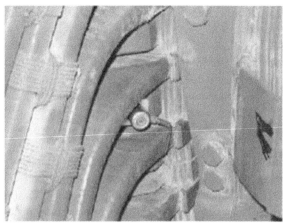

Conversely, the vibration of the stator frame is a function of winding, single-phase, and unbalanced supply voltage faults [37]. Resonance of the magnetomotive force and the stator is one of the reasons for vibration [44–46]. Based on the vibration index, synchronous generator fault detection has been proposed in reference [47] and a stator winding fault has been considered in reference [48]. This index has also been applied to both synchronous and asynchronous machines in reference [43]. Analysis and measurement of vibrations in rotating electrical machines with power higher than 1 MW, such as turbo-generators and gas-generators, are very important tools for the evaluation of operating conditions and the quality of electric machines and their parts. Figure 11.3 presents a typical vibration sensor fixed on the stator winding of a 75 MVA hydro-generator. A detailed explanation and discussion of the detection methods based on vibration analysis are provided in Chapter 13.

11.5.1.3 Acoustic Noise Analysis

Acoustic noise can be used as a stator winding fault index [43–45]. The noise spectrum of generators covers ventilation noise and magnetic noise. Air ventilation noise is related to air disturbance, which arises from periodic distortion of the air pressure in rotating parts, such as the rotor or ventilation blades. Magnetic noise is related to the Maxwell stresses caused by a magnetic field on the iron surfaces. These stresses induce vibrations in the generator structure, leading to noise emission. The acoustic level due to the increase in mechanical and aerodynamic noise is in the range of 12 dB per steer (dabbling). An increase in machine speed causes an increase in magnetic noise [46]. Fixing the sensor on the body of the generator allows the operation disturbance to be picked up. In addition, the sensor may be affected by the operation of adjacent equipment, so the fault detection is distorted. Therefore, many factors can interfere with magnetic noise and reduce the reliability of short circuit fault detection based on noise analysis.

11.5.1.4 Partial Discharge Analysis

Partial discharge analysis (PDA) is one of the primary online methods for health monitoring of the winding. This method was used for the first time in 1976 in hydro-generators. Since then, this online method has been extended using special sensors. The partial discharge (PD) occurs because of insulation degradation and existing holes inside the insulation [48–50]. Experimental evidence indicates that insulation weakness begins as an electric PD and terminates as full discharge (electric arc). By measuring the PD level and its trend, the remaining life of the generator insulation can be predicted. This is a special method and process using advanced computer software introduced by well-known generator manufacturers [3]. In addition, the stator winding fault can be identified by an online PD test, and the health level is determined by measuring the PD of a winding [47, 51–55]. A quick diagnosis of the onset of insulation breakdown between internal turns is possible by analyzing the PD [54, 55]. Although this method has high reliability and accuracy in early-stage detection of a short circuit fault in the stator winding, it is a costly method. A detailed explanation and discussion of the detection methods based on PD analysis are provided in Chapter 15.

11.5.1.5 Output Gas Analysis

Fault diagnosis of a synchronous generator can also be determined by monitoring output gases, such as hydrogen, carbon monoxide, and carbon dioxide, caused by chemical reactions. The destruction of electrical insulation within the generator causes these generated gases to be sucked toward the air-cooling system. These gases attack insulation, leading to its early aging [40, 56, 57]. Therefore, detection of these gases can confirm the start of an insulation defect and a fault in the stator winding. This determination can be performed either online or offline. Infrared absorption is one of the techniques used for detection of these gases [37].

11.5.1.6 Impulse Test

This is a predictive method for the detection of insulation weaknesses before a short circuit occurrence. A stator winding fault in a synchronous generator is diagnosed by applying a voltage impulse to the winding. In this test, voltage impulses with precisely the same frequency spectrum are simultaneously applied to the two windings of the stator. The reflected pulses from the two windings are then detected with an oscilloscope and compared. The results are then used to detect insulation faults between the two windings, coils, and earth [40]. Performing this test involves fixing extra equipment and is too costly. Performing this test is also not easy, and pulse signals cannot be applied during machine operation.

11.5.1.7 Air Gap Magnetic Field Monitoring

A search coil around the stator tooth is used in reference [58] to capture the induced voltage in both healthy and short circuit conditions in the stator winding. An internal short circuit fault changes the induced voltage waveform in the search coils installed around the stator teeth. Therefore, a comparison between the waveform of the induced voltage in each search coil can give a clue to a potential fault occurrence and its location. The use of search coils can reveal a wide variety of faults, including turn-to-turn, phase-to-ground, and phase-to-phase faults in the stator winding of the synchronous machine. The study shows that load variation does not change the fault impact in the induced voltage waveform. Although the method provides detailed data about the status of the machine winding, it is impractical for machines operating in power plants or industries, since the rotor must be removed to gain access to the stator.

11.5.2 Non-invasive Methods

Stator faults originate from turn-to-turn faults or are due to different stresses, and they prevail in other parts. The main aim of condition monitoring is to identify a fault at its initial stages and prevent its failure [22]. In addition to introducing the complexities of applying an invasive method, these methods are unable to detect the fault at the initial stages and cannot diagnose a fault with low severity. Equipment adjacent to the sensor can also influence the measured signal, thereby disturbing the reliability of the method. In addition, the sensor location is very important and must be taken into account in invasive methods [40]. Improper location of the sensor or its displacement during the fault diagnosis period generates problems in fault detection. Since an additional piece is added inside the system, a possibility arises for disturbance in the proper operation of the generator. The sensor impact must be considered as a separate parameter in the precise simulation of the generator. This can specify the mutual impact of the sensor and generator.

Sometimes, the use of invasive methods and fixed sensors is simply not possible. For example, some methods require that the sensor be fixed in particular parts of the air gap or insulation; this must be done during manufacturing, as fixing it in the device after manufacturing may be difficult. Buying and fixing the sensors used in invasive methods may not be justifiable economically. All these factors force the user to apply non-invasive methods for condition monitoring and fault detection of the generators. However, the combination of invasive and non-invasive methods is sometimes considered for condition monitoring or to verify the correctness of the non-invasive method using invasive methods [59].

With non-invasive methods, normal output signals, such as the current, the voltage of the stator or rotor terminals, and the stray magnetic field, can be used to follow the fault impacts. By studying these signals, the generator can be monitored without the need for sensor installation inside the synchronous machine. Therefore, there is no need to buy and fix sensors in non-invasive methods because these methods do not disturb the normal conditions of the generator [29, 31]. Moreover, the signals do not influence the performance of other equipment because they are variables of the generator itself. These signals depend only on the parameters of the generator, and the effect of the fault on these signals and related frequency spectra are clear. With these methods, measurement and analysis can be done remotely and in the control room. The simplicity of the analyses using non-invasive methods makes investigation of the fault impacts easier. A fault in a synchronous generator will have at least one of the following signs [60]:

- Unbalanced air gap magnetic field.
- Unbalanced line currents.
- Distortion of flux, voltage, and current waveforms.
- Variation in the stray magnetic field.
- Rising losses and falling efficiency.
- Generating excessive heat.

Each of these cases can introduce a suitable index for fault diagnosis. The next section presents the indexes used in non-invasive methods and their relevant analyses.

11.5.2.1 Field Current Signature Analysis

An inter-turn short circuit fault, even with a low severity in the stator winding, causes magnetic flux distortion [60]. As a result, this distorted flux will induce a voltage in the rotor field winding. Based on this reasoning, some indexes for stator winding fault diagnosis determine the voltage and current of the rotor winding. Testing the field current harmonics is an appropriate and applicable pattern for stator inter-turn fault diagnosis of synchronous generators [54, 61] because of the transfer of the distorted flux to the rotor winding and the continuity of the effect of this flux in the field current. Increasing some of the even-numbered harmonics of the field current for inter-turn fault diagnosis in a synchronous generator has been reported in references [62] and [63]. In particular, increasing the 8th harmonic in the field current of a four-pole synchronous generator has been introduced in references [63] as an index for the stator inter-turn fault.

One of the attributes of a safe index is that the existing side disorders in a healthy machine do not interfere with the indicator signs [64]. Unfortunately, this is not true for this index. Although an inter-turn fault in stator winding increases the even harmonics in the rotor current, the unbalanced voltage also increases the even harmonics in the rotor current of the healthy generator. This causes ambiguity in the validity of the index operation. Conversely, these harmonics and time harmonics have an identical frequency, making separation of the origin of the even harmonics difficult. In addition, the unbalanced voltage in the rotor winding is also an aggravating factor for these harmonics [59].

One solution to this problem is to use a voltage unbalanced factor in the governing equations and simulations before applying the index. Modification of this unbalanced supply in induction machines has been applied to investigate the impact of the unbalanced voltage [65–72], but this modification has not been used in a synchronous generator.

A number of definitions are proposed for an unbalanced voltage, and well-known definitions have been given by the National Electrical Manufacturers Association (NEMA) and the International Electromechanical Commission (IEC) standards. According to NEMA, the general definition of unbalanced voltage is as follows:

$$\text{Unbalanced voltage } (\%) = (\Delta V_{max}/V_{avg}) \times 100 \tag{11.1}$$

where ΔV_{max} is the greatest difference from the average voltage V_{avg} of the three-phase voltage. Considering these ambiguities, eight types of simulations have been carried out, based on a modified winding function [59]. Note that the fundamental frequency in the simulations and tests is 60 Hz. The tests and simulations consist of the following:

- Healthy machine with balanced voltage and symmetrical and asymmetrical winding in rotors.
- Healthy machine with unbalanced current and symmetrical and asymmetrical winding in rotors.
- Machine with a stator inter-turn fault, balanced voltage, and symmetrical and asymmetrical winding in rotors.
- Machine with a stator inter-turn fault, unbalanced voltage, and symmetrical and asymmetrical winding in rotors.

In a faulty stator winding, a significant increase in the field current is observed at 90 and 150 Hz [59]. By contrast, no traces of these frequencies are observed in the healthy machine with either balanced or unbalanced voltages. Therefore, a field current with a 150 Hz frequency is a proper index that is independent of the unbalanced voltage, power factor, and load variations. The advantage of this index is that it can diagnose a three-turn fault among the 160 turns of the stator winding. The only drawback of this index is its high sensitivity compared with other methods, such as the search coil method, which can detect a one-turn fault without ambiguity [59].

11.5.2.2 Stator Winding Currents

One of the most applicable signals in fault diagnosis of the stator winding of synchronous generators is the stator winding currents. The inter-turn short circuit fault leads to an asymmetrical magnetic field in the generator, which appears in the current waveforms [22, 33]. Consequently, the direct impact of a short circuit fault is reflected in the stator currents. In many models, the stator current is used for condition monitoring [31], which is why the stator current can be a unique signal for the diagnosis of a stator inter-turn fault [70]. This diagnostic method involves no manipulation, so no additional equipment is needed. The reason is that the stator winding acts like a search coil [31].

The prevalent short circuit fault types in a stator winding can occur in three ways, as shown in Figure 11.4. In the first case, a short circuit fault occurs among the turns of a branch of one of the stator phase windings. In this case, the amplitude of the short circuit current depends on the

Figure 11.4 Schematic of three prevalent types of short circuit faults in stator windings.

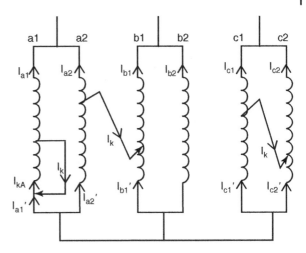

number of shorted turns in the winding. In the second case, for a machine with parallel winding, a short circuit occurs between two branches of the parallel winding of one phase. In the third case, a short circuit occurs between the windings of the two phases.

The amount of transient current compared to the steady-state current will be much higher compared with the steady-state current due to the increase in the magnetic reluctance of the armature reaction in the transient state. The amount of the increase in transient current depends on several factors, including the spatial distribution of the winding, winding layout, number of shorted turns, and changes in the magnetic field due to the sudden changes in the effect of the short circuit, as well as the short circuit location and short circuit time [73]. In large synchronous machines, the occurrence of the third type of short circuit fault in the stator winding causes the operation of the protection system. However, the occurrence of the first and second cases can be detected by monitoring systems if the severity of the fault is low and does not require tripping of the synchronous machine. Therefore, timely diagnosis and taking the required actions can prevent insulation failure that would otherwise result in a severe short circuit fault.

The performance of a 200 MVA synchronous generator is examined under a short circuit fault in reference [74]. The current inside the winding is sinusoidal during the healthy operation of the machine, whereas the waveform shape becomes distorted due to a short circuit and the amplitude of some harmonics increases compared to the healthy case.

A main consequence of the short circuit fault in a synchronous machine is the non-uniform magnetic field in the air gap. Figure 11.5 shows the magnetic field in a healthy machine and under a short circuit fault in one phase and two phases. Figure 11.6 shows the magnetic field of the air gap in the healthy state under the short circuit fault of one phase and two phases in a 50 kW machine. Short circuit faults in the windings not only distort the form of a magnetic field, but they also reduce the magnitude of the average magnetic field due to the opposite magnetic field created by the short circuit conductors. The amplitude of the magnetic field in the healthy state is 0.863 T, which decreases to 0.832 and 0.833 T with the occurrence of a short circuit fault in one phase and two phases, respectively, while the average amplitude of the magnetic field decreases by 3.592 and 3.476%. Examination of the magnetic flux density of a synchronous generator modeled in a finite element environment shows that a part of the stator core, where the short-circuited turns are located, becomes saturated in the case of a short circuit fault in one phase of the stator winding. The reason for the local saturation is the high amount of current that passes through the faulty conductor, in addition to the circulating current in the parallel path [58, 75]. The occurrence of a

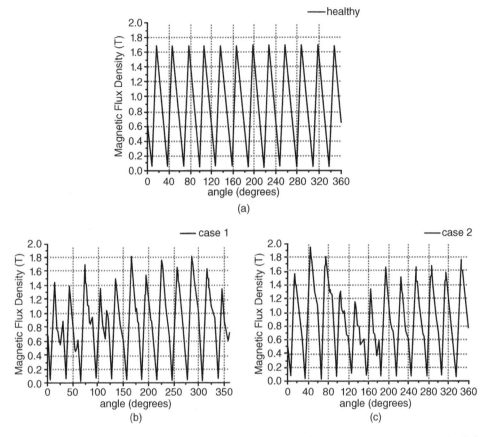

Figure 11.5 Magnetic flux density of a synchronous machine: (a) healthy, (b) under a short circuit fault in one stator phase winding, and (c) under a short circuit fault between two stator phase windings.

short circuit fault in one phase also causes a current with a phase 180° opposite to the main current in the stator winding, as shown in Figure 11.7.

The air gap magnetic field of the synchronous machine under an internal short circuit fault in the stator winding includes both subharmonics and either odd or even multiples of the main frequency harmonic. A harmonic component examination of the stator current and rotor field winding current due to an internal short circuit fault of the stator winding indicates the existence of new harmonic components due to the fault. The rotor field winding current contains the second and fourth harmonic components due to the occurrence of the internal short circuit fault in the stator winding, in addition to its direct current component. The stator current also has third and fifth harmonic components that are more noticeable than other harmonic components. The short circuit fault of the stator, the main component of the negative sequence, induces the second-order harmonic component in the rotor field winding current, and this harmonic causes the generation of the third harmonic component in the stator current, which then generates several harmonic components in the stator and rotor currents [76, 77].

11.5.2.3 Current Park Vector

The Park vector fault diagnosis method was introduced in reference [6], which uses machine current Park vector components as a fault index. The stator current Park vector components are as

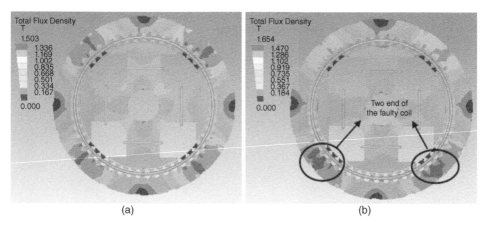

Figure 11.6 Magnetic flux density distribution in a salient pole synchronous generator: (a) healthy state and (b) under a short circuit fault in the stator winding.

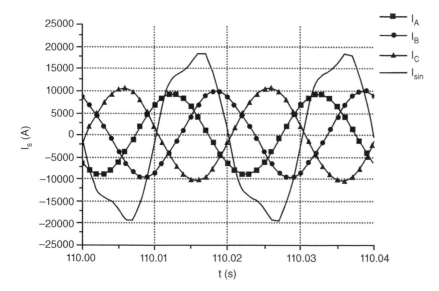

Figure 11.7 Stator current and short circuit current in the stator winding of a synchronous machine.

follows:

$$I_d = \sqrt{2/3}\, I_a - \sqrt{1/6}\, I_b - \sqrt{1/6}\, I_c \tag{11.2}$$

$$I_q = \sqrt{1/2}\, I_b - \sqrt{1/2}\, I_c \tag{11.3}$$

where d and q are the direct and quadratic components of the Park vector and a, b, and c are the phases of the generator. In the ideal case, the components of the Park vector are as follows [6]:

$$I_d = \sqrt{1/2}\, I_{+m}\ \sin \omega t \tag{11.4}$$

$$I_q = \sqrt{1/2}\, I_{+m}\ \sin(\omega t - \pi/2) \tag{11.5}$$

where I_{+m} is the maximum value of the positive sequence of the supply phase current in ampere, ω is the supply angular frequency in rad/s, and t is the time in seconds. The polar presentation corresponding to these two components is a circle with the origin of coordinates centered as shown in

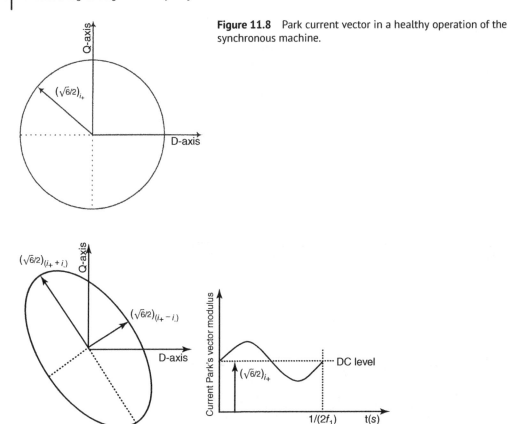

Figure 11.8 Park current vector in a healthy operation of the synchronous machine.

Figure 11.9 Park current vector in a faulty synchronous machine (left) and park current vector modulus (right).

Figure 11.8. In non-ideal conditions, the drawn figure differs from this reference pattern. Of course, in practice, the polar curve related to the healthy case slightly differs from a complete circle, because the voltage has no full sinusoidal waveform. For a short circuit fault, the pattern is elliptical and is directly related to increases in the fault severity, whereas the main axis corresponds to the faulty phase. Figure 11.9 shows the results for the Park vector under a faulty operation of a synchronous generator. Although the method based on the dq-axis representation of the stator current is informative and easy to interpret, any kind of fault or unbalanced voltage in the stator winding and in the power network harmonics that result in an unbalanced current lead to an elliptical representation of the dq current. Therefore, the method is unsuitable for fault detection of a stator short circuit fault.

However, another method based on the current Park vector uses a component of twice the fundamental frequency in the spectrum of the Park vector current [70], which is called the extended Park vector approach (EPVA). Based on Equations (11.4) and (11.5), in a healthy machine, the Park vector size is constant. In non-ideal conditions, Equations (11.4) and (11.5) are not valid because they contain negative and positive components. Under these conditions, in addition to a constant value, the Park vector contains an oscillatory component with twice the frequency of the fundamental frequency in the rotor field winding current. Figure 11.10 shows the frequency spectrum and the EPVA for a healthy and short circuit fault in the stator winding of the synchronous machine.

Figure 11.10 EPVA index for the case of the short circuit fault in the stator winding of a synchronous machine: (a) healthy, (b) 12 turns, and (c) 36 turns.

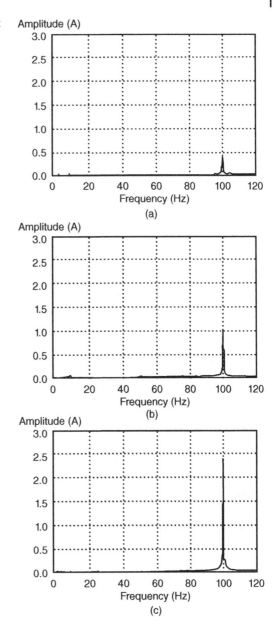

This component is directly related to the asymmetry of the machine or unbalanced supply voltage. Table 11.1 summarizes the oscillatory component of the Park vector current under different short circuit faults and an unbalanced supply voltage. The evidence indicates that the index is sensitive to the unbalanced supply voltage and that interference will occur with an unbalanced supply voltage for a low-severity fault. Meanwhile, under these conditions, the generator has no inherent symmetry and the existing inherent asymmetry can amplify this problem.

Attempts have been made to diagnose a stator winding fault using stator current harmonics, but a specific model has not been introduced to detect this fault [31]. This index has been used to diagnose stator winding faults in synchronous, asynchronous, and large industrial machines up

Table 11.1 Oscillatory components of the Park vector current.

Short circuit fault					
Voltage unbalance (%)	0.2	0.4	0.6	0.8	1
Current amplitude (A)	0.02	0.04	0.06	0.08	0.10

Unbalanced supply voltage			
Fault severity (%)	1	2	3
Current amplitude (A)	0.26	0.47	0.69

to 5 MW, tested with the EPVA method [70]. The proposed harmonics of the stator current have been addressed, and the amplitude of twice the fundamental frequency was found to increase significantly due to the stator inter-turn fault, and this can be considered a proper index. The higher number of short-circuited turn cases leads to a larger index value. The fault normalized severity factor (NSF) for this index is defined as follows:

$$NSF = \frac{\text{Amplitude of twice the frequency component voltage}}{\text{DC value in EPVA}} \tag{11.6}$$

In condition monitoring of a synchronous generator, a rise in the NSF indicates an extension of the inter-turn fault in the stator winding. For a high NSF, a solution must be given to prevent serious damage to the generator. On-site evidence has shown that this is a simple and cheap method. Notably, the current harmonics in a synchronous machine are influenced by different factors, including rotor symmetry, supply voltage, dynamic and static conditions of the load, noise, and structure of the generator [67]. These factors can ambiguously affect the performance of the index. Therefore, a more expert vision is required to address the impacts of load variations, voltage variations, power factor changes, and other factors in the index. Unfortunately, these factors have not been clearly considered in reference [70] so they can be future research questions for researchers in this field.

11.5.2.4 Rotor Current
The frequency spectrum of the rotor field winding current compared with a frequency spectrum of the stator phase current indicates that the first component of the field current compared with the third harmonic of the stator phase current shows a higher sensitivity to an internal short circuit fault in the stator winding. Tables 11.2 and 11.3 show the results of this study. The first column (α) indicates the position of the short circuit in relation to the terminal of the synchronous generator. The amplitude of the first component of the rotor field winding current under an internal short circuit fault is decreased from 1.19 to 0.003 by increasing the fault severity from 10 to 90%. The amplitude of the third component of the stator phase current is decreased from 1.56 to 0.001 by increasing the fault severity from 10 to 90%, while the amplitude of the first component of the stator phase current for a 10% internal short circuit fault is 351, indicating that the rotor components show a higher sensitivity compared with the stator current [78].

11.5.2.5 Using the Negative Sequence Current of the Stator
The methods based on the current sequence components are the initial methods for fault diagnosis of the stator winding of generators. First, a method for an analysis of the induction motor performance under a stator winding fault has been introduced in reference [22]. Figure 11.11 presents the

Table 11.2 The frequency spectrum of the rotor field current per unit.

α (%)	Single phase internal fault		Double phase internal fault	
	Zero component	First component	Zero component	First component
10	0.75	1.19	0.748	1.18
20	0.75	1.6	0.748	1.57
30	0.75	1.37	0.748	1.23
40	0.75	0.86	0.748	0.76
50	0.75	0.53	0.748	0.48
60	0.75	0.345	0.748	0.32
70	0.75	0.21	0.748	0.202
80	0.75	0.07	0.748	0.073
90	0.749	0.003	0.748	0.0032

Table 11.3 The frequency spectrum of the stator phase current per unit.

α (%)	Single phase internal fault		Double phase internal fault	
	First component	Third component	First component	Third component
10	351	1.56	353	1.58
20	160	2.99	162	2.9
30	72	1.69	72	1.47
40	31	0.7	31	0.572
50	14	0.31	15.26	0.26
60	7.8	018	8.69	0.137
70	4.56	0.089	5.16	0.075
80	2.43	0.02	2.63	0.027
90	1.7	0.001	1.73	0.001

results of this analysis as negative and positive sequences of current and fault current. Therefore, the amplitude of the component of the negative sequence current can be used for stator winding fault diagnosis. However, this fault diagnosis method has some basic limitations. The components of the three-phase current sequence can be obtained as follows:

$$\begin{bmatrix} X_+ \\ X_- \\ X_0 \end{bmatrix} = \begin{bmatrix} 1 & a & a^2 \\ 1 & a^2 & a \\ 1 & 1 & 1 \end{bmatrix} \begin{bmatrix} X_a \\ X_b \\ X_c \end{bmatrix} \tag{11.7}$$

where $a = \exp(j\,2\pi/3)$. For any three-phase phasor quantity, X components of positive, negative, and zero sequences can be considered a matrix (Equation (11.7)). For a symmetrical generator supplied by a three-phase voltage supply, the negative sequence current (I_-) will be zero. An inter-turn fault changes the symmetry of the generator and imposes an increased negative current sequence.

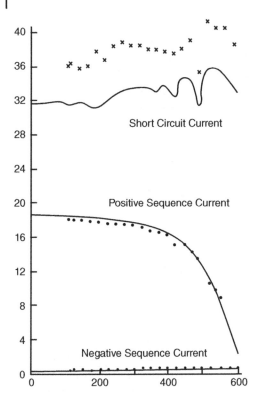

Figure 11.11 Simulation and measurement results for the faulty case using variations of positive and negative sequence currents. Adapted from [12, 22].

The higher severity fault leads to a higher current, while the rise of the positive sequence current is very much smaller than that of the negative sequence current [22]. This can be considered a fault diagnosis pattern or an innovative protective method for disconnecting command to the circuit breaker [22]. The negative sequence current index has been used to diagnose the stator winding of induction machines and synchronous machines [68].

Raising the negative sequence current is proposed as a stator winding fault index in AC machines [79], and other factors, such as supply balanced voltage, load variations, voltage magnitude variations, and equipment measurement errors, have also been taken into account. Testing has confirmed that this is applicable in large machines [79].

The limitation of this index is its dependency on a balanced voltage. The reason is that the index increases for an unbalanced voltage and in fault-less machines. The amplitude of the fundamental component of the current, the negative sequence current under short circuit fault, and the unbalanced voltage are summarized in Table 11.4. The fault severity varies from 0 to 3%, which is equivalent to 10 shorted turns, and exhibits an unbalanced voltage between 0 and 1%. Table 11.4 also indicates that the negative sequence current is sensitive to the unbalanced voltage. In fact, determining the base level of fault diagnosis is necessary so that it does not interfere with the unbalanced supply voltage. The base level must be displaced upward, but the possibility of an initial stage diagnosis for a low number of the inter-turn faults of the stator winding cannot then be monitored. In practice, the inherent asymmetry of the machine causes a negative sequence current, which intensifies the difficulty of fault detection.

The negative sequence current is unable to diagnose a low-severity fault, and increasing the unbalanced voltage level causes variation in the current amplitude, leaving the correctness of the index in doubt and making fault detection difficult [59]. Although this index is not influenced by

Table 11.4 The current amplitude of the fundamental component and negative sequence component under short circuit and unbalanced voltage.

Severity (%)	I_(A)	Unbalanced voltage (%)	I_(A)
0.0	0.00	0.0	0.0000
0.5	0.10	0.2	0.0180
1.0	0.20	0.4	0.0300
2.0	0.38	0.6	0.0500
2.5	0.46	0.8	0.0700
3.0	0.55	1.0	0.0820

load variations [79], it is sensitive to the inherent asymmetry of the machine. In reference [68], an attempt was made to remove this drawback of the index. This index guarantees the identification of up to a three-turn fault as well as an unbalanced voltage [68].

11.5.2.6 The Injected Negative Sequence Current

The injected negative sequence current method has been introduced to diagnose a stator winding fault in which unbalanced voltage and some other problems have been resolved [79]. The major impact of a mild short circuit fault is a small increase in the phase current and displacement of the negative and positive sequence current levels. In fact, the injected negative sequence current does not depend on the supply voltage, and it is estimated using the method depicted in Figure 11.12 [79]. In this circuit, Z is the negative sequence impedance of a healthy machine, which must be already determined. I_i is the injected current and I_r is the residual current, which is extracted from the previously stored data tables [79]. The injected negative sequence current is estimated by measuring the negative components of the current and voltage. This method consists of an initial learning stage in which the residual current under different load and supply voltages are measured and stored in data tables. The points between the measured points or out of the measured limit must be determined by interpolation and extrapolation, respectively.

The measurable negative sequence current in the output of the generator arises from several factors, such as the fault, unbalanced voltage, inherent asymmetry of the generator, and measured faults, and is called the residual current. The residual current depends on the load and voltage and must be measured as a function of these two in the healthy machine and is stored. The injected negative sequence current is zero for different unbalanced voltages between 0 and 1%. The residual current component varies linearly between 0 and 0.17 for a fault severity between 0 and 3%. Therefore, this fault index has no sensitivity to the unbalanced voltage. In addition, by increasing the number of short-circuited turns, its value increases. Although the method has a difficult initial stage for operation, its performance is desirable.

Figure 11.12 Equivalent circuit for estimation of injected current.

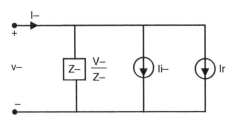

11.5.2.7 Second Component of Current in the *q*-Axis

This index has been proposed in PM synchronous machines [69]. In the faulted cases, the resistance, inductance, and stator EMF of the machine change compared with the healthy case. Conversely, the resistance and EMF depend on the SC-turn fault, and the inductance is proportional to the square of the number of faulty turns. A new feature, the fault turn ratio (FTR), is defined as follows:

$$\text{FTR} = (N_{\text{fault}}/N_{\text{total}}) \tag{11.8}$$

where N_{fault} is the number of faulty turns and N_{total} is the total number of turns. In the case of an inter-turn fault in a phase, the three-phase system becomes unbalanced, and the third harmonic component is generated. This effect is transferred to the synchronous reference frame by the fundamental components and amplifies the second harmonic component of the current in the dq axis. An increase in this component in the q-axis indicates a fault in the stator winding. Increasing the FTR and back-EMF increases the amplitude of the second harmonic.

11.5.2.8 Stator Terminal Voltage

The air gap magnetic field and the magnetic electromotive force are used to diagnose the internal short circuit fault in the stator winding [62]. The internal short circuit fault in one phase of the stator winding creates a circulating current that generates a magnetic field opposite to the main magnetic field. The magnetic field due to the magnetomotive force rotates at a synchronous speed and in the rotor direction, while the magnetomotive force due to the short-circuited turns is stationary in space and the amplitude of this field oscillates sinusoidally with the main frequency of the machine.

The third harmonic order of current is used to detect an internal short circuit fault in the stator winding [78]. Similar to the current, the third harmonic order of the terminal voltage can be used to detect the internal fault in the stator winding. Figure 11.13 shows the variation of the third harmonic in the positive sequence voltage by increasing the short circuit fault percentage. In the

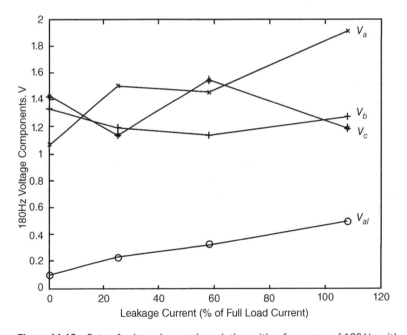

Figure 11.13 Rate of voltage harmonic variation with a frequency of 180 Hz with respect to load changes.

full-load current, the third harmonic of the voltage positive sequence due to the short circuit fault is markedly increased. Therefore, the comparison between the third-order harmonic of the positive sequence voltage and the phase voltage indicates a high sensitivity of the positive sequence voltage to the occurrence of a short circuit fault in the stator winding. The impact of load variation on the nominated feature also shows no sensitivity.

A fault in the excitation system of the rotating diode of brushless synchronous generators excites the third-order harmonic in the voltage of the positive sequence component, which is the same indicator provided for detecting an internal short stator fault in the stator winding. Therefore, diagnosing faults using this feature is not easy when the machine has excitation with rotating diodes. In reference [62], a harmonic component twice the size of the main harmonic in the rotor field winding current was used as an indicator to detect the fault of the diode. The authors concluded that, if the harmonic components are increased, both indicators should be examined simultaneously to make a correct diagnostic.

11.5.2.9 Voltage Sequences
Faults cause distortion in voltage waveforms [60]. The existing second and third harmonics in the stator voltage and current of the generator indicate its unbalanced operation [70]. Therefore, the voltage criteria in [22, 59] have been used as stator fault indexes. In reference [22], the voltage sequence components of particular machines have been considered. Variations of zero, negative, and positive sequence components can individually make an index, but they have some weaknesses and strengths. The zero-sequence voltage has been used as a stator inter-turn fault index in reference [22]. In the analysis of the zero-sequence component, some conditions must be taken into account: stator windings must be star-connected, the neutral point accessible, and line voltages measured. The first two conditions reflect that the zero-sequence voltage mostly applies to star-connected machines, which is considered a restriction for this index. However, most synchronous generators have star-connected stator windings.

Negative sequence voltage can also be used as an index for stator inter-turn fault, but apart from estimation of the negative sequence voltage, the balanced voltage must be insured and estimates of other sequence components must not be neglected. Unfortunately, this increases the cost of applying this index and reduces its economic justification [13, 59].

Reduction of the positive voltage sequence is one of the citable indexes for the existence of a stator inter-turn fault [22]. This index does not need a measurement of the terminal voltages because this component appears as a DC voltage in the rotor reference frame voltage and is accessible with a low-pass filter. This index is safe, inexpensive, and online, and it does not need to estimate other voltage sequences. In reference [63], the existing third harmonic positive sequence voltage of the stator of a brushless synchronous generator has been considered to address a stator fault.

11.5.2.10 Impedance Sequence
Using a negative impedance sequence is one of the first methods for stator winding fault detection [71]. This index has mostly been used in induction machines [22, 68] and only in some cases has it been applied to other AC machines for the detection of an inter-turn short circuit fault [65, 80]. The voltage equations versus current are as follows:

$$\begin{bmatrix} V_z \\ V_p \\ V_n \end{bmatrix} = \begin{bmatrix} Z_z & Z_{zp} & Z_{zn} \\ Z_{pz} & Z_p & Z_{pn} \\ Z_{nz} & Z_{np} & Z_n \end{bmatrix} \begin{bmatrix} I_z \\ I_p \\ I_c \end{bmatrix} \tag{11.9}$$

where V, Z, and I are the voltage, impedance, and current, respectively. Indexes z, p, and n are the zero, positive, and negative sequences, respectively. In a healthy machine, the impedance matrix is diagonal, and the non-diagonal components are zero. The current sequences are independent of each other, as shown below:

$$V_z = Z_z I_z, \; V_p = Z_p I_p, \; V_n = Z_n I_n \tag{11.10}$$

The disarray of this impedance sequence matrix in reference [22] has been proposed as a stator inter-turn fault. In low-voltage machines, stator inter-turn faults with almost zero resistance of the stator winding turns are connected. In this case, the faulty winding has a smaller number of turns compared with the real number of turns. Decreasing the number of turns reduces the positive sequence impedance and increases the negative sequence and coupling impedances. Variations in the machine parameters due to the stator turn fault are identical. Therefore, a simplified model is obtained that ignores the magnetic coupling between phases, leakage flux, and mutual magnetic coupling. The simplified model, assuming an inter-turn fault of phase a, can be written as follows:

$$\begin{bmatrix} V_{a0} \\ V_{b0} \\ V_{c0} \end{bmatrix} = \begin{bmatrix} \mu R & 0 & 0 \\ 0 & R & 0 \\ 0 & 0 & R \end{bmatrix} \begin{bmatrix} I_a \\ I_b \\ I_c \end{bmatrix} + \begin{bmatrix} \mu^2 L & 0 & 0 \\ 0 & L & 0 \\ 0 & 0 & L \end{bmatrix} \frac{\mathrm{d}}{\mathrm{d}t} \begin{bmatrix} I_a \\ I_b \\ I_c \end{bmatrix} + \begin{bmatrix} \mu E_{a0} \\ E_{b0} \\ E_{c0} \end{bmatrix} \tag{11.11}$$

The subscripts a, b, and c indicate the phases and 0 shows the neutral point; μ is the ratio of the healthy turn number of each phase winding and E is the rotor-induced voltage (EMF). Since the zero-sequence current in the star-connection in a three-phase system is zero, applying sequence component factors yields the following equation:

$$\begin{bmatrix} V_{a0} \\ V_{b0} \\ V_{c0} \end{bmatrix} = \begin{bmatrix} \mu R + 2R & (\mu - 1)R \\ (\mu - 1)R & \mu R + 2R \\ (\mu - 1)R & (\mu - 1)R \end{bmatrix} \begin{bmatrix} I_p \\ I_n \end{bmatrix} + 0.5j\omega \begin{bmatrix} \mu^2 L + 2L & (\mu^2 - 1)L \\ (\mu^2 - 1)L & \mu^2 L + 2L \\ (\mu^2 - 1)L & (\mu^2 - 1)L \end{bmatrix} \begin{bmatrix} I_p \\ I_n \end{bmatrix} + \frac{1}{3} \begin{bmatrix} (\mu + 2)E_{a0} \\ (\mu - 1)E_{b0} \\ (\mu - 1)E_{c0} \end{bmatrix} \tag{11.12}$$

where μ is zero in the healthy machine and is positive and less than 1 in the faulty generator; therefore, if the fault severity rises, the μ value decreases. Considering this and Equation (11.13) below, a fault occurrence decreases the positive sequence impedance and EMF, while the negative sequence impedance increases [22].

At synchronous speed, the negative sequence impedance is very much smaller than the positive sequence impedance. Thus, the unbalanced voltage leads to a high negative sequence current in the winding, which indicates that measuring the negative sequence current is unsuitable. In addition, the negative sequence impedance of the generator is almost constant during its operation. Thus, the effective or apparent negative sequence impedance obtained by dividing the negative sequence voltage by the negative sequence current indicates variations in the negative sequence impedance. In this case, variations in the index during continuous monitoring must be determined. If fault diagnosis is carried out based on the comparison with a base level, this base may be changed due to the unbalanced voltage, and fault detection will not be possible.

Table 11.5 presents different tests of unbalanced voltage and short circuit faults. In the case of unbalanced voltage, the index level increases. By increasing the unbalanced level, the index variations also decrease and lose their initial sensitivity. With a fault occurrence, the index decreases significantly. In addition, by increasing the level of the unbalanced voltage, the index variations decrease and, in fact, the index sensitivity declines. The change in the base level value of the index must be adjusted for maximum unbalanced supply voltage conditions. Of course, the effective negative sequence impedance variations do not express fault severity. Therefore, the index is slightly sensitive to the unbalanced supply voltage and, by its increase, the index sensitivity to the fault rises. The index also does not reveal the fault severity; it only indicates a fault occurrence. This

Table 11.5 Impact of unbalanced voltage on the negative sequence impedance index.

| No. | Unbalanced voltage (%) | Healthy machine | Machine with stator winding inter-turn | |
		One-turn fault	Two-turn fault	Three-turn fault
1	0.1	14.22	1.59	0.77	0.48
2	0.2	15.12	3.00	1.61	1.07
3	0.3	15.21	4.18	2.35	1.63

impedance is independent of the load variations. Thus, this index is more secure than the voltage and current sequence for short circuit fault detection. However, increasing the unbalanced voltage causes variation in the sequence impedance index, so the index correctness is doubtful.

A stator winding fault can be diagnosed based on monitoring a non-diagonal component sequence impedance matrix [22]. The sequence impedance matrix components are zero for the healthy machine. Of course, in practice, non-diagonal components are non-zero small values due to the inherent asymmetry of the machine. For a fixed slip synchronous generator, Z_{np} is independent of the level of unbalanced voltage, and can be estimated for a healthy synchronous machine and used as a reference for fault diagnosis. For Z_{np} estimation, the following equation is used:

$$Z_{np} = (I_{n2}V_{n1} - I_{n1}V_{n2})/(I_{p1}V_{n2} - I_{p2}V_{n1}) \tag{11.13}$$

The unbalanced voltage, current, and voltage sequence components are measured at two levels and then Z_{np} is then estimated. In this case, the inherent unbalanced voltages are collected as initial data, while the current and voltage sequence values are also stored to estimate the sequence impedance during monitoring. Therefore, this method requires an initial training stage. Figure 11.14 presents the simulation results under different short circuit faults and an unbalanced supply voltage. Referring to Figure 11.14, this index is quite robust against fault severity. The index under different unbalanced voltages differs considerably. However, its value does not determine the fault severity, since increasing the short-circuited turns still leaves its value almost constant. Therefore, stator winding fault diagnosis based on the monitoring non-diagonal component of the

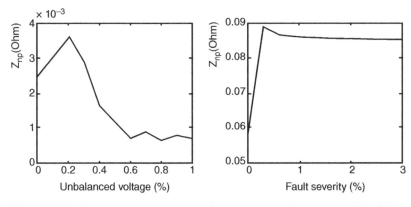

Figure 11.14 Non-diagonal component of sequence impedance versus: (a) unbalanced voltage and (b) fault severity.

sequence impedance matrix is able to detect a fault under an unbalanced supply voltage, but it does not reveal the fault severity and needs an initial stage of data collection.

Hybrid impedance, as the ratio of the positive sequence voltage phasor and negative sequence voltage phasor, can also be used as a stator inter-turn fault index [72]. Since the unbalanced supply and inherent asymmetry of the generator also generate a negative sequence current, this index does not seem to have enough reliability for fault diagnosis. This index requires two sensors to measure the voltage and current and is more expensive than the current index.

11.5.2.11 Instantaneous Power Index

Instantaneous power can be used for fault diagnosis [70]. It is the product of the voltage and current of the phase and consists of two oscillatory and constant components. The constant component is the real power. The oscillatory component is the apparent part and its frequency is twice the supply frequency. The fault index in this method is the oscillatory component of the instantaneous power in every three phases. Filtering the oscillatory components in the three phases and adding their constant components generates a constant summation of the instantaneous torque. The summation is zero in a healthy machine and will even be constant with load variations. However, in faulty cases, and particularly for a short circuit fault of the stator winding, the estimated instantaneous power will be non-zero. Applying the instantaneous power increases the reliability of the fault diagnosis [81, 82]. Similar to the previous case, this method is also sensitive to unbalanced voltage. In this case, an empirical equation for unbalanced voltage reduction has been recommended [24]. The following equations give this phase modification factor:

$$V = (V_a + V_b + V_c)/3 \tag{11.14}$$

$$K_a = 1 - (V_a - V)/V \tag{11.15}$$

For phase a, K_a is calculated, and before estimation of the sum of the three components of constant power, K_a is multiplied by their values. Doing this largely compensates for the unbalanced voltage. Table 11.6 shows the component of the oscillatory current of three phases for a different severity of the short circuit fault and unbalanced voltage. As shown, despite the compensation, the problem of index sensitivity to unbalanced voltage has not been fully resolved. The index sensitivity to the supply voltage with an existing inherent asymmetry of the machine is amplified and fault separation is impossible. Thus, the method is not fully practical.

Some components of the electric power frequency spectrum can also be used as a stator winding fault index because the frequencies of these components depend on the fault and are independent

Table 11.6 Current amplitude of fundamental component and negative sequence component under short circuit and unbalanced voltage.

Severity(%)	I_(A)	Unbalanced voltage (%)	I_(A)
0.0	0	0.0	1.0
0.5	25	0.2	2.5
1.0	50	0.4	7.0
2.0	100	0.6	10.0
2.5	125	0.8	12.5
3.0	150	1.0	15.0

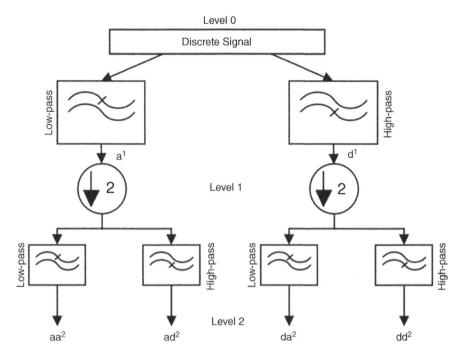

Figure 11.15 The second stage for resolving the discrete signal of instantaneous power by the wavelet function.

of the synchronous speed [80]. Considering that the power index requires a simultaneous measurement of the current and voltage, the cost of this index will be higher than the current indexes. The wavelet coefficients of the instantaneous power have been used for the stator winding fault of a synchronous generator [83]. The wavelet coefficient second stage ($dd'x$) is obtained by two-stage filtering of the instantaneous power, as shown in Figure 11.15 and the following equations:

$$dd'[n] = \Sigma_{k=0}^{N/2-1} d'[k]h[n-k] \tag{11.16}$$

$$P_{j,k} = \Pi_{m=1}^{N} V(m)\, I(m) \tag{11.17}$$

Calculating the power wavelet coefficients from the output terminal and comparing them with the normal values make an online diagnosis of the fault possible. This index has been investigated experimentally in a 1.6 kW synchronous generator. The index can also facilitate online monitoring of the unbalanced stator phases, earth fault, phase-to-phase fault, and stator winding inter-turn fault. This index is applied easily and quickly and uses little computational memory and can also be used in protection methods.

11.5.2.12 Analysis of Transient Operation of the Salient Pole Synchronous Generator

Since the starting stage of the synchronous generator contains informative data, these data can be analyzed for fault diagnosis. This analysis has been carried out in references [32] and [59].

In this criterion, a multi-loop method is utilized [31]. Each stator winding can be divided into two strains. During fault simulation, one of the strains (50% of the winding) is short-circuited. The advantage of this simulation is that the rotor rotation equation is also included in the model. According to the obtained results, in the inter-turn fault, the current, voltage, and torque distort. The starting current is then higher than the normal condition, and after reaching a steady-state level, it becomes oscillatory. In the case of an internal fault, the spatial harmonics are noticeable and, consequently, the electromagnetic torque has large variations and oscillates even in the steady-state condition. By increasing the number of faulty turns, the amplitude of oscillations will be larger and damping the acceleration shortens the transient period. Although this method demonstrates promising results, it has not been validated experimentally.

Air gap torque

Air gap torque monitoring is another stator fault diagnosis method [29]. The air gap torque contains a combination of slot, stator current, and rotor current effects. This torque is sensitive to any asymmetry in the generator and unbalanced voltage and is estimated as follows:

$$T_a = 3p \left\{ (2I_a + I_c) \int [V_{ca} - R(I_c - I_a)]dt + (I_c - I_a) \int [V_{ba} - R(2I_a + I_c)]dt \right\} \tag{11.18}$$

where R is the measured line-to-line resistance in star/delta connection and P is the number of pole pairs of the generator. In a healthy generator, the estimated air gap torque is constant, but in the stator winding fault, the oscillatory component with twice the supply frequency appears in the torque. Table 11.7 shows that the oscillatory torque is increased by raising the number of shorted turns. It also increases by increasing the unbalanced voltage level (i.e., it is sensitive to the unbalanced voltage).

Investigation of the stator current harmonic index

To examine this harmonic index, a synchronous generator has been simulated by FEM in reference [84]. In the theoretical part, the general equation has been considered and, simultaneously, the following have been taken into account:

- External circuit equation: supply.
- Internal circuit equation related to the magnetic field.
- Magnetic coupling leading to rotational equation.

The general air gap field equation is as follows:

$$B(\emptyset, t) = \sum_{K_{cu}=1}^{\infty} \sum_{K_{css}=0}^{\infty} \sum_{K_{sa}=0}^{\infty} B_{K_{cu}, K_{ss}, K_{cu,}} COS \left[\left(K_{cu} \pm \frac{2K_{sa}}{p} \right) \omega_s t - \lambda(\emptyset) \right] \tag{11.19}$$

Considering the supply frequency coefficient, the $(K_{cu} \pm 2K_{sa}/p)f_s$ index is extracted [84], where K_{sa} is the saturation profile, K_{cu} is the harmonic coefficient, p is the pole pair, and f_s is the supply frequency. In this method, increasing the number of faulted turns causes an increase in the faulty

Table 11.7 Oscillatory torque amplitude.

Unbalanced level	Torque amplitude change (Nm)	No. of SC turns	Torque amplitude change (Nm)
0.2	0.045	1	0.15
0.4	0.085	2	0.30
0.6	0.140	3	0.43
0.8	0.180		
1.0	0.220		

harmonic amplitude. This method can detect even one shorted turn and load variations have no impact on the index. In addition, saturation, stator slots, and unbalanced voltage effects have been considered.

11.5.2.13 Stray Magnetic Field

Manufacturing defects, iron core non-homogeneity, stator and rotor asymmetry, and asymmetrical magnetic paths can lead to the unbalanced performance of a generator [59]. The inter-turn short circuit fault of the stator winding also imposes asymmetry on the magnetic field of the synchronous generator. This distortion can be detected by a search coil fixed on to the generator shaft [37, 85, 86]. Many of the online indices in the generators consist of the fixing coil in the air gap [87, 88]. However, enhancement of the sensitivity of the search against faults is recommended. Existing 30 and 60 Hz frequencies in the open-circuit voltage of the search coil can detect a one-turn fault [89]. However, the main weakness of this index is that connecting the search coils for measuring the axial flux is not possible in every generator, as their fixation is too complicated [90]. The search coil can safely detect the turn fault of the stator winding [59] using the existing 90 Hz frequency in the induced voltage. This index even enables the unambiguous diagnosis of a one-turn fault and is less costly than other methods because it uses only one sensor. This index can be employed in salient pole and round rotor synchronous generators.

11.5.2.14 Axial Leakage Flux

The specified frequency components in the axial leakage flux can be used to diagnose a stator winding fault. This method requires an internal sensor and an external sensor for measurement [91]. In addition, this method can locate a short circuit fault. This can be done by fixing several coils inside the generator. In an ideal stator and rotor, the currents are balanced; therefore, no axial leakage flux occurs. However, the circuits can be unbalanced due to an existing slight asymmetry in the winding geometry and in the core due to non-uniform core materials. The axial leakage flux generates significant asymmetry in the stator winding and this fault can lead to a change in the distribution of the air gap spatial harmonics. Analysis of the spatial harmonics and their relationship with the time harmonics is as follows:

$$F = [k \pm n/p]f_1 \tag{11.20}$$

where $k = 1, 3, n = 1, 2, 3, \ldots, 2p - 1, p$ is the number of pole pairs, and f_1 is the supply frequency. The fundamental frequency and third harmonic of the voltage are considered. Equation (11.20) determines the axial leakage flux harmonics frequency and stator winding fault severities. Table 11.8 summarizes the severities of harmonics 49, 72, and 84 Hz for the healthy and short circuit conditions and the no-load and full-load cases [61]. The harmonic severity in the healthy case shows an unbalanced level of generator operation. The fault severity in the no-load and full-load conditions increases in the faulty case. Therefore, the fault can be diagnosed under different loads. The main problem is that this is an invasive method. In addition, the sensitivity of the method against the unbalanced voltage has not been investigated.

11.6 Stator Short Circuit Fault Detection of Brushless Synchronous Machines

Auxiliary coils are distributed inside the stator winding of the brushless synchronous generator to provide voltage for rotor field winding. The coils are widely distributed in all stator slots;

Table 11.8 Flux harmonics severities in healthy and short circuit cases.

Frequency(Hz)	No. of SC turns	Different load	Full SC	Difference load
48	2.7	1.05	55	7
72	10	0.80	170	29
84	7.35	1.45	84	12

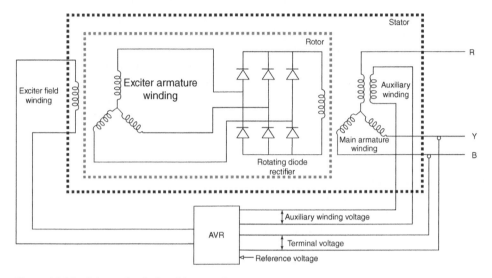

Figure 11.16 Schematic of a brushless synchronous generator.

therefore, the windings can be used to detect an internal short circuit fault in the stator winding. Figure 11.16 shows the schematic of the brushless synchronous generator. Investigation of the frequency spectrum of the rotor field winding coil voltage in the healthy state and under a short circuit fault indicates an increase in the amplitude of the third and seventh harmonic components due to the short circuit fault. Here, the amplitude of the third and seventh harmonic components have changed from 64.34 and 14.68 to 81.3 and 10.36, respectively, under a stator short circuit fault (see Figure 11.17).

Figure 11.18 shows the sensitivity of the eighth and ninth harmonic components to increasing the short circuit fault of the stator winding. As the fault occurs and increases, the third harmonic component increases, whereas the amplitude of the seventh harmonic component decreases. These components can be used as indicators for fault detection [92, 93].

11.7 Stator Short Circuit Fault Detection of Powerformers

The structure of the powerformer differs slightly from conventional generators as a cable is used instead of copper bars inside the stator slots. Stator internal fault detection in a powerformer is also different due to the capacitive nature of the cables. The capacitance is created by the current that passes through the cables, as the inner and outer layers of the cable behave like a capacitor. The electric charges on the conductors of the powerformers are 30 times higher than those of a

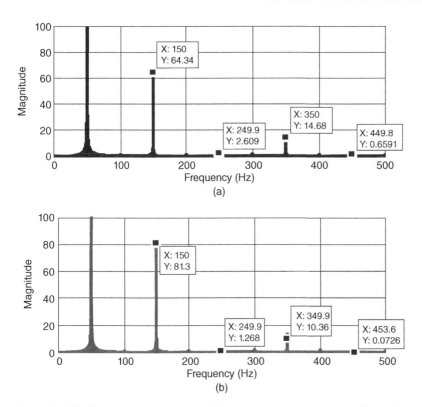

Figure 11.17 The frequency spectrum of the induced voltage in the auxiliary winding in a healthy case (a) and under a short circuit fault (b).

conventional synchronous generator. Therefore, the study of the performance and effect of electric charges and capacitive current in the performance of powerformers has received considerable attention [94, 95]. Figure 11.19 shows an internal short circuit fault in the stator of a powerformer. Figure 11.20 shows the phase currents due to the phase-to-ground fault in phase a of the stator winding. Figures 11.21 and 11.22 show rotor field winding currents, as well as capacitive currents, expressed in per-units (pu). The current i_m and i_n are the same before the short circuit fault, but due to the short circuit fault, the phase of the current i_m is suddenly reversed and the amount of current passing through increases sharply.

The occurrence of a short circuit fault not only affects the phase in which the fault occurs, but it also affects the other two phases of the powerformer and increases the current through them. As with conventional synchronous generators, due to the internal short circuit fault of the stator winding, the harmonic is induced to a level twice the main frequency component in the rotor field winding current, causing the excitation current to deviate from the direct current (see Figure 11.21). Figure 11.22 shows the capacitive current of the cables before and after the occurrence of the fault.

11.8 Stator Short Circuit Fault Detection of Turbo-generators

In this section, a quick, simple, and efficient method is introduced for the short circuit fault detection of a turbo-generator. This method consists of three parts:

1. Using a rising slope of phase current for fault diagnosis.

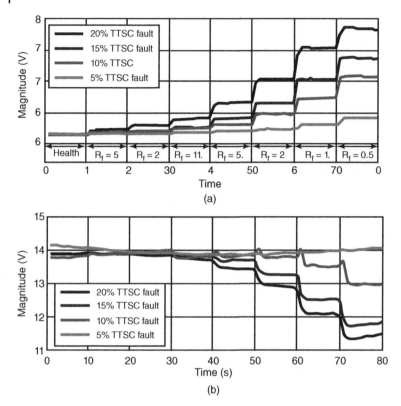

Figure 11.18 The variation rate of the third harmonic (a) and seventh harmonic (b) in the induced voltage of the auxiliary winding due to a short circuit fault in the stator winding.

2. Obtaining Lissajous curves to design the current set.
3. Determining whether the differences in different current phases can distinguish between an inter-turn fault and other short circuit faults of the stator winding.

In this method, the measured current waveform of the faulty generator is considered to be the initial signal and the current waveform in one preceding cycle in a healthy machine to be the second signal. The difference between two successive data values from each signal is estimated and are called ΔX and ΔY. Obtaining the ratio of ΔX and ΔY and its sign allows the diagnosis of the internal fault very quickly in less than a half cycle. Then, by comparing the difference in the generator phase currents, the inter-turn fault is distinguished from other faults.

11.8.1 The Inter-turn Fault Detection Algorithm of the Stator Winding

The winding of the generators consists of several coils in series that form the phase winding of the synchronous generator. By rotating the generator rotor and creating a rotating magnetic field within the generator air gap, voltage is induced inside the coils. As a result, each coil can be considered a voltage source and each phase of the generator can be considered a series of multiple voltage sources. The windings of high-power synchronous generators are usually one-turn, and in the case of two-layer winding, two windings occur in each slot; these can be from two different phases or from one phase.

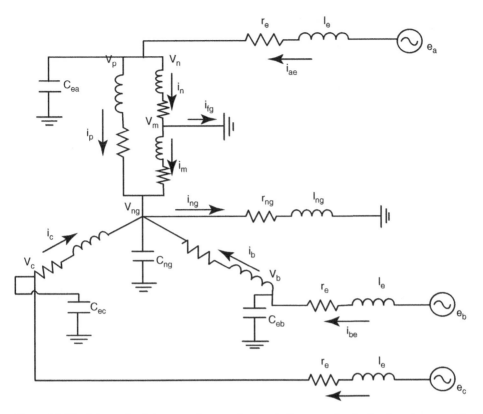

Figure 11.19 Schematic of the powerformer windings in the case of a single-phase to ground fault.

11.8.1.1 Circuit Analysis

In a synchronous generator, the induced voltage angle in the coils depends on the physical location of the coil, so a slight change in the induced voltage angle can occur when a fault occurs. However, due to the fault current and the saturation of the machine, the induced voltage range in the coils can be reduced. Therefore, the coils can be considered as a voltage source with variable amplitude and constant angle connected in series. This indicates that the phase-to-phase and inter-turn short circuit fault in the stator winding can be investigated based on circuit analysis.

11.8.1.2 Turn-to-Turn Fault

Figure 11.23 shows the equivalent electrical circuit of a synchronous generator under an inter-turn short circuit fault in the stator winding. The windings are considered to be a voltage source with a constant phase angle and a variable amplitude and equivalent impedance. The other healthy phase windings are also considered to be a constant voltage source with variable amplitude and equivalent impedance. In Figure 11.23, m is the number of short-circuited coils and n is the number of healthy coils. The transformer in Figure 11.23 has a delta connection, indicating that the current passing through the neutral is negligible. The fault current is almost evenly distributed in the two healthy phases, as follows:

$$IF_B = IF_C = -\frac{IF_A}{2} \tag{11.21}$$

$$IF_{AF} = \gamma IF_A \tag{11.22}$$

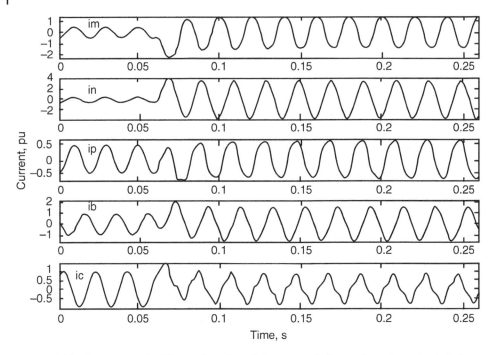

Figure 11.20 The currents in different branches of the stator winding under a short circuit fault.

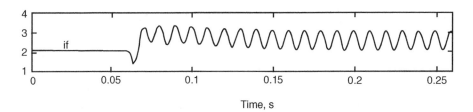

Figure 11.21 The rotor field winding current in the case of a stator short circuit fault.

$$-VA_{grid} + IF_A(Z_{grid} + Z_{gen}(n)) - IF_B X_{AB}(n)$$
$$-IF_C X_{AC}(n) + VA_{gen}(n) + X_{AF\_A}\, \gamma\, IF_A = 0 \qquad (11.23)$$

In Equations (11.21) and (11.22), IF_A, IF_B, and IF_C are the fault currents of phases A, B, and C, respectively. IF_{AF} is the short circuit fault current of the faulty coils and γ is the ratio of the short circuit coil fault current to the phase A current. The number of shorted turns varies depending on the slot in which the inter-turn fault occurs. Due to the reduction in the faulty phase voltage, a current is injected from the power network into the faulty phase. Therefore, the current passing through the faulty phase can be obtained by using Equations (11.21), (11.22), and (11.23).

In Equation (11.23), VA_{grid} is the voltage of the phase A of the grid, Z_{grid} is the power grid impedance, $Z_{gen}(n)$ is the generator healthy coils impedance, $X_{AB}(n)$ is the mutual reactance between phase A and phase B, $X_{AC}(n)$ is the mutual reactance between phase A and phase C, $VA_{gen}(n)$ is the sum of the induced voltages in the healthy coils, and X_{AF_A} is the mutual reactance

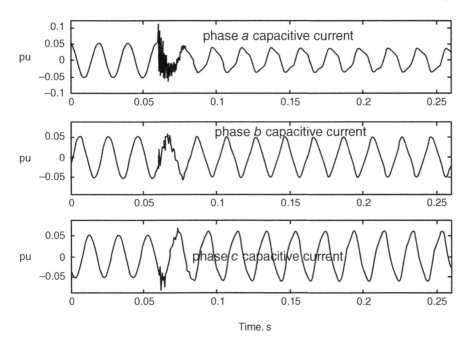

Figure 11.22 Three-phase capacitive currents under an internal short circuit fault in the stator winding.

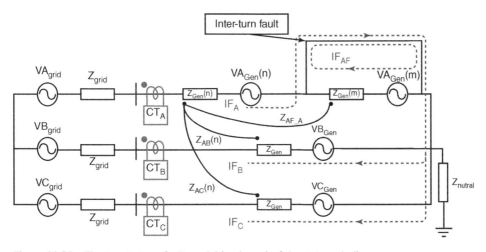

Figure 11.23 The turn-to-turn fault model in phase A of the stator winding.

between the short circuited coils and the healthy coils of phase A. IF_A can be achieved as follows:

$$IF_A = \frac{VA_{grid} - VA_{gen}(\text{n})}{X_{grid} + X_{gen}(n) + \frac{X_{AB}(n)}{2} + \frac{X_{AC}(n)}{2} + X_{AF\_A}\,\gamma IF_A} \tag{11.24}$$

Figure 11.24a shows how the faulty current of phase A is calculated. First, the power grid voltage vector and the generator terminal voltage vector are drawn. The difference between the voltage vectors of the power grid and the generator terminal is then calculated. The faulty current of phase A is calculated based on the net voltage vector and the generator impedance. The same procedure is applied to find the faulty current in all phases, as shown in Figure 11.24b. The faulty current

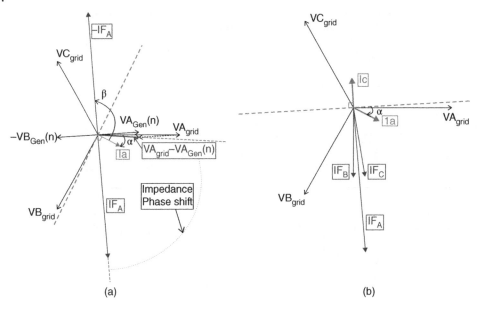

(a)

(b)

Figure 11.24 Phasor diagram of the synchronous generator under an inter-turn short circuit fault in phase A of the winding. The phasor diagram of the faulty phase A (a) and three-phase current phasor diagram of a synchronous generator (b).

vector of phases B and C is almost half the amplitude of phase A. In Figure 11.24, the angle α is the phase difference between the generator terminal voltage and the load current and the angle β is the angle difference between the fault current and the load current. The faulty current vector of each phase can be divided into two groups relative to the healthy state current vector of the same phase, as follows:

Category I: In this category, the faulty current vectors have a phase difference of fewer than 90° compared to the healthy current vector ($|\beta| < 90$).

Category II: In this category, the faulty current vectors have a phase difference of more than 90° compared to the healthy current vector ($|\beta| > 90$).

Since the phase voltage of the generator and power grid have a small phase difference, the net voltage difference of the phase voltage is a vector with a phase angle similar to the faulty phase voltage. Moreover, depending on the network impedance and the impedance of the healthy windings of the generator, the fault current vector has a phase shift of less than 90° relative to the output voltage difference vector. According to the equivalent circuit, as shown in Figure 11.23, the calculated faulty current, IF_A, relative to the healthy case current of phase A, is categorized in category *I* according to its direction, as shown in Figure 11.24. The faulty current of phase B relative to its healthy state current is in category *I* and the faulty current of phase C relative to its healthy state current is in category II.

If the direction of the assumed current in the circuit is in line with the direction of the current in the current transformer (CT), the current measured by the CT is the same as the calculated current. However, if the direction of the assumed current in the circuit is opposite to the direction of the CT, the current measured by the CT is 180° different from the calculated current. Considering that the direction of the assumed faulty current of phase A is opposite to the current in the CT, the measured faulty current of phase A with respect to the phase A healthy current is categorized as category II. The direction of the assumed faulty current is the same as the current of phase B and phase C measured in the CT; therefore, the measured fault current is the same as the calculated fault current. Conclusively, the current of the faulty phase and the current of the phase that has undergone a voltage phase shift of $+120$ with respect to the faulty phase voltage are categorized in category II, while the other phase current that has undergone a voltage phase shift of -120 with respect to the faulty phase voltage is categorized in category I.

11.8.1.3   Factors Affecting the Proposed Index

The difference in electrical angle between the induced voltage of the first coil and the last coil of one phase covering one pole is calculated as follows:

$$\delta = \frac{360}{R_s} \times \left(\frac{R_s}{3P} - 1 \right) \times \frac{P}{2} = \frac{60(R_s - 3P)}{R_s}, \qquad R_s \geq 3P \qquad (11.25)$$

where R_s is the number of stator slots of a synchronous generator and P is the number of poles. The electric angle difference of a healthy generator is equal to $\pm \delta/2$, since the machine has symmetry. In theory, the difference in the electric angle of the induced voltage of healthy coils of a faulty phase and the electric angle of the induced voltage of the healthy case is $\pm \delta/2$. If $R_s \to +\infty$, the electric angle becomes $\pm 30°$.

For instance, for a synchronous generator with 36 slots and one pole pair, $\pm \delta/2$ is equal to 25°. The phase shift becomes equal to 25° if only one of the coils in the faulty phase remains healthy. In practice, the possibility of having an inter-turn short circuit fault in the stator winding that leads to failure of the majority of the coils is impossible due to the winding distribution in the slots. The phase shift under an inter-turn short circuit fault in the stator winding is almost less than $\pm 10°$ since the number of healthy coils is greater than the number of faulty coils. According to Figure 11.24, by changing the angle of I_F to such an extent, the category of the faulty currents is not changed compared to the healthy state.

Figure 11.25 shows the induced voltage in the stator terminal of the synchronous machine. It consists of voltage induced in several coils that have a phase shift of 10° with respect to each other. The induced voltage in phase A of the machine is shown as VA_{GEN}, both in a healthy case and under an inter-turn short circuit fault in Figure 11.25. Although the amplitude of the induced voltage in the faulty case is reduced compared with the healthy case, the phase angle is not changed under faulty conditions.

Another discussion in determining the category of currents is the generator output power factor in the healthy state. In the normal operation of generators, the power factor is between 0.95 and 0.8, indicating that the induced voltage is between 18.2 and 36.9° ahead of its phase current. Figure 11.24 is drawn for a phase angle of 30° and a power factor of 0.866. The power factor impact must be considered in the calculation, otherwise the category of faulty currents can change compared to the healthy state.

11.8.1.4   External Phase-to-Phase Fault

The external phase-to-phase fault can be modeled as shown in Figure 11.26. All the coils in one phase are considered a single voltage source with variable amplitude and constant phase angle in series with an equivalent impedance. The power grid is modeled as a voltage source in series with an equivalent power grid impedance. The external phase-to-phase fault in this context is a fault that occurs in the distance between the CT of the generator terminal and the power grid.

The KVL and KCL are applied to the circuit, as shown in Figure 11.26. The direction of the currents that circulate in the generator side and grid side in the circuit is shown in Figure 11.26. The

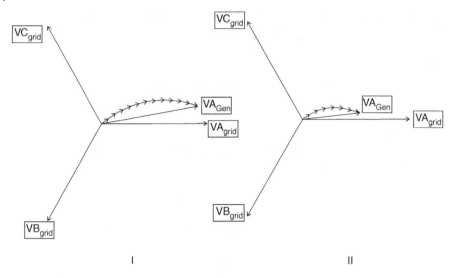

Figure 11.25 The sum of the induced voltage of the coils below a single-pole related to phase A of the synchronous generator: (I) in the healthy state and (II) in the faulty state.

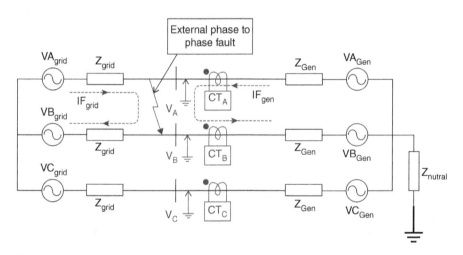

Figure 11.26 Schematic circuit of the external phase-to-phase fault between phase A and phase B.

governing equation in the circuit of Figure 11.26 is as follows:

$$V_A = V_B = V_F \tag{11.26}$$

$$V_A + V_B + V_C = 0 \rightarrow V_F = \frac{-V_C}{2} \tag{11.27}$$

$$-VA_{gen} + (IF_{gen} \times Z_{gen}) + V_F = 0 \tag{11.28}$$

$$IF_{gen} = \frac{VA_{gen} - V_F}{Z_{gen}} \tag{11.29}$$

where V_A, V_B, and V_C are the terminal voltages of the generator of phases *A*, *B*, and *C*, respectively. V_F is the voltage at the fault point, VA_{gen} is the induced voltage in phase *A*, and IF_{gen} and IF_{grid}

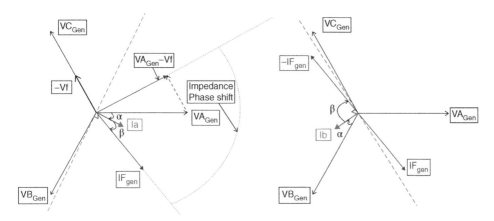

Figure 11.27 The phasor diagram of the external phase-to-phase fault between phase A and phase B.

are the faulty currents of the generator and the power grid, respectively. The stator phase winding impedance is Z_{Gen}. The IF_{gen} current can be presented as a vector diagram, as shown later in Figure 11.30.

According to the circuit of Figure 11.26, the IF_{gen} current passes through the CT; therefore, the vector diagram of Figure 11.27 can be used to calculate the faulty current on the generator side. In Figure 11.27, the phase angle between the load voltage and current is shown as α and the phase angle between the faulty current and the load current is shown as β.

According to the governing equations in the case of phase-to-phase fault, the output net vector of the grid voltage and the voltage at the fault location are plotted first. In order to obtain the IF_{gen} current vector, a phase angle shift equivalent to the impedance of the generator is required, and since the impedance of the generator is inductive, the output voltage vector is shifted maximum to 90° and forms the IF_{gen} current vector.

The fault current calculated in this case is in category I, relative to the healthy phase current of phase A. If the direction of the assumed current in the circuit is in line with the CT current, the current measured by CT is the same as the calculated current. By contrast, if the direction of the assumed current in the circuit is opposite to the direction of the CT current, the current measured by CT has a 180° phase shift. Considering that the direction of the assumed faulty current is in line with the CT direction of phase A, the faulty current of phase A, which is measured by CT, is in category I compared to the measured healthy current of phase A.

In phase B, the calculated faulty current is in category II relative to the healthy state current of phase B. However, because the assumed faulty current is opposite to the CT current of phase B, the measured current of phase B in CT is classified in category I compared to the measured healthy current of phase B. Therefore, in the phase-to-phase fault, the measured current of the faulty phases is in category I compared to their healthy state.

The saturation in the synchronous generator occurs simultaneously with the phase-to-phase short circuit fault due to the high amplitude current that circulates inside the windings. Saturation in the synchronous machine reduces the induced voltage amplitude in the coils. However, the induced voltage angle in the coils is constant, since it depends only on the location of the coils in the machine.

According to Equation (11.25), the maximum phase angle variation is 25°. Since no changes occur in the amplitude of the induced voltage due to the phase-to-phase fault in the machine, the magnetic distribution is symmetrical and is expected to have no changes in the phase angle. In the

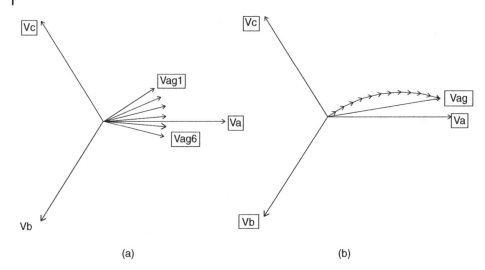

(a)   (b)

Figure 11.28   The total induced voltage of coils in phase A of a synchronous generator: (a) voltage of coils and (b) sum of induced voltages.

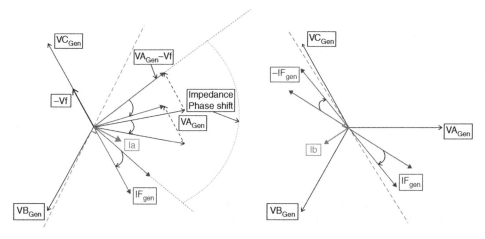

Figure 11.29   Fault current vector diagram considering the interval of the voltage angle change due to saturation.

worst case, the voltage changes are 25° and the sum of the induced voltages of coil *A* is plotted in Figure 11.28.

According to Figure 11.28, the variation in the terminal voltage angle in the faulty operation of the synchronous generator is proportional to which of the induced voltage amplitudes decreases in which of the coils. Since the phase-to-phase fault in the machine cannot disturb the magnetic distribution of the machine, the amplitude of the induced voltage in the coils is decreased simultaneously, indicating that the phase angle of the induced voltage cannot change dramatically. The faulty current diagram based on this analysis is shown in Figure 11.29, which indicates that the current category does not change due to the induced voltage phase angle variation.

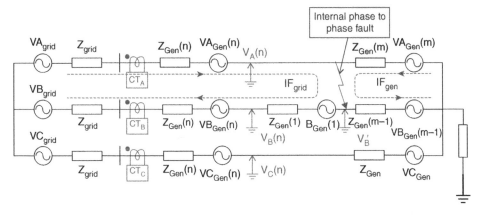

Figure 11.30 Schematic of the synchronous generator under a phase-to-phase fault.

11.8.1.5 Internal Phase-to-Phase Fault

The phase-to-phase fault may occur inside the generator. In this case, the number of coils from the end of the phase to the faulty location is modeled as a voltage source with variable amplitude and constant angle and with equivalent impedance. The rest of the healthy coils are also considered as a separate voltage source with variable amplitude and a fixed angle and equivalent impedance, as shown in Figure 11.30.

In Figure 11.30, n represents the number of coils between the fault location and the machine terminal and m is the number of coils between the neutral and the fault location, where $m + n$ is the total number of coils in the machine. In phase B, the number of coils is equal to $(m + 1)$. The number 1 can be changed to $(m - 1)$ depending on the winding arrangement or fault location of the generator.

Internal phase-to-phase fault behavior can be predicted based on external phase-to-phase faults. The difference between these two types of faults is the direction taken by the faulty current to pass through the CT. In an external phase-to-phase fault, the faulty current of phase A exits from the dotted point of the CT while the faulty current of phase B enters from the dotted point of CT. Therefore, the fault currents are classified in category I compared to the healthy state. In the internal phase-to-phase fault, the faulty current of phase A enters from the dotted point of the CT, while the faulty current of phase B exits from the dotted point of the CT. This change in direction of the current causes the faulty current to classify as a category II relative to the healthy state. Therefore, by indicating the direction of the current and the category of the faulty current, the type of internal or external phase-to-phase fault can be determined.

11.8.1.6 Turn-to-Turn Fault Detection Algorithm

In the previous section, for an internal phase-to-phase short circuit fault and inter-turn short circuit fault, the measured faulty current was classified as category II relative to the healthy current. The current of two phases in these cases are influenced as follows: in the internal phase-to-phase fault, the current of both faulty phases and in the case of an inter-turn short circuit fault, the faulty phase and a phase with a + 120 phase angle difference. Conclusively, the type of fault can be identified by determining the category of the faulty current compared with the healthy case.

According to the stated characteristics, identifying the inter-turn fault is possible by recognizing the category of the healthy state current compared to the faulty state current. The algorithm shown in Figure 11.31 is proposed for fault detection and consists of three parts. In the first part, the currents of all three phases are measured and the gradient of the measured current is calculated to

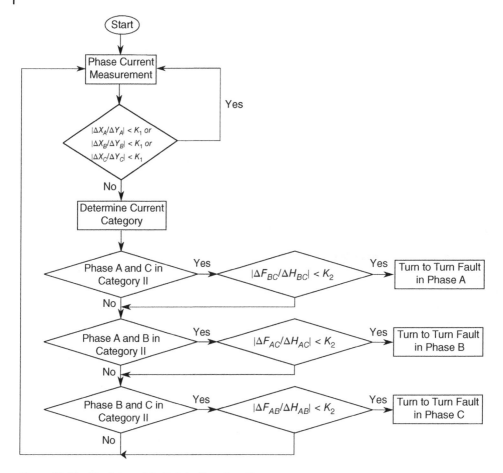

Figure 11.31 The internal fault detection algorithm.

determine whether the machine is healthy or under a short circuit fault. In the second part, the current categories are determined. The difference in each phase current is calculated separately in the third part to determine whether the variation is due to an inter-turn fault or a transient in the power grid.

11.8.1.7 Increasing the Gradient of the Current

The first step in the fault detection algorithm is to determine the increase in the terminal current gradient. In two sinusoidal signals with the same frequency, the gradient of the signal is larger, with a higher amplitude. The higher gradient is because it has to reach a higher peak value at the same time. Therefore, the current peak value can be predicted by measuring the gradient of the current. According to Figure 11.32, the signal that has a higher peak also shows an increase in the difference between its consecutive data.

The short circuit fault in the stator can be determined by measuring the current gradient. To determine the increase in the current gradient, obtaining the difference between two consecutive sampled data values is sufficient. Due to the constant sampling frequency, the difference between two consecutive data values indicates the gradient variation of the signal. Therefore, by increasing

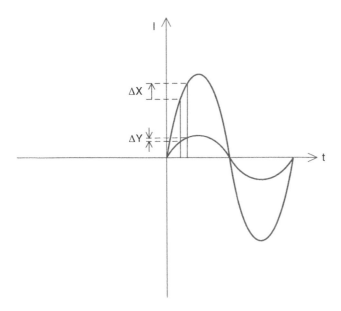

Figure 11.32 The difference between two consecutive data in healthy and faulty modes.

the $\Delta X/\Delta Y$ ratio, the current gradient is also increased. The value of the $\Delta X/\Delta Y$ ratio in this algorithm is compared with the parameter called K_1, which has a value greater than 1. The $\Delta X/\Delta Y$ ratio is 1 if the generator is healthy. If the $\Delta X/\Delta Y$ ratio is greater than K_1, this indicates the presence of perturbation in the current.

11.8.1.8 Current Category Determination

Determining the current category requires obtaining the phase shift of the healthy state and the faulty state relative to each other. If the phase difference of both currents is less than 90° relative to each other, then both current signals classify in category I, whereas if the phase difference of both currents is more than 90°, both are in category II. As a result, in order to determine the current category, the phase shift between the healthy state and the faulty state must be obtained. Several methods are available for detecting phase differences between two signals. One of these methods is the use of Lissajous curves.

If two different sinusoidal signals are plotted on the coordinate plane with respect to each other, the Lissajous curve is obtained. The parametric representation of the Lissajous curve can be presented as follows:

$$
\begin{cases}
x(t) = A_x \sin(\omega_x t + \phi) \\
y(t) = A_y \sin(\omega_y t + \phi + \delta)
\end{cases}
\tag{11.30}
$$

In Equations (11.3) to (11.10), $x(t)$ and $y(t)$ are two sinusoidal signals that have amplitude A, frequency ω, phase angle ϕ, and phase difference δ. In Equation (11.30), by removing the time variable t, the following Lissajous curve equation can be obtained:

$$
y(t) = A_y \sin\left(\frac{\omega_y}{\omega_x} \left(\arcsin\left(\frac{x}{A_x} \right) - \phi \right) + \phi + \delta \right)
\tag{11.31}
$$

Since the two signals used are the current signal before and after the fault, and because the fault will be detected in less than half a cycle, the frequency of the two signals can be equalized. As a

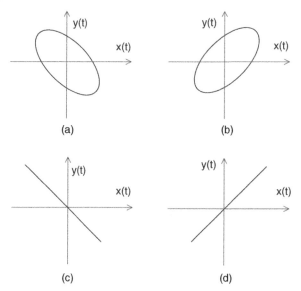

Figure 11.33 Lissajous curves for different phase angles: (a) ellipse with negative gradient for $90° < \delta < 270°$, (b) an ellipse with positive gradient for $-90° < \delta < 9°$, (c) straight line with negative gradient for $\delta = 180°$, and (d) straight line with positive gradient for $\delta = 0$.

result, the following equation can be obtained from Equation (11.31):

$$y(t) = A_y \sin\left(\arcsin\left(\frac{x}{A_x}\right) + \delta\right) \tag{11.32}$$

Equation (11.32) only depends on the amplitude and the phase angle difference of the signals. If Equation (11.30) is drawn for a complete cycle of sine signals, different curves are obtained according to the phase difference. These curves are shown in Figure 11.33. If the phase difference between the two signals is in category I, the obtained diameter of the ellipse has a positive gradient, and if it is in category II, the obtained diameter of the ellipse has a negative gradient.

Determining the gradient of the large diameter of the ellipse first requires obtaining the rate of variation of X relative to Y. The difference between two consecutive data values from the signal is called ΔX and the difference between two consecutive data values from the other signal is called ΔY. The ratio between ΔX and ΔY determines the diameter of the ellipse.

According to the algorithm, the first signal is the current measured by CT and the second signal is the current measured by CT in a previous cycle. If n samples are taken in each cycle and the mth sample of the first signal is considered to be a first signal, then the $m - n$ sample is the second signal. It can be concluded that

$$X = I(m); Y = I(m - n) \tag{11.33}$$

$$\Delta X = I(m) - I(m - 1) \tag{11.34}$$

$$\Delta Y = I(m - n) - I(m - n - 1) \tag{11.35}$$

The $\Delta X/\Delta Y$ defines the sign of the ellipse diameter, where ΔX and ΔY are shown in Figure 11.34.

11.8.1.9 Calculating the Difference Between Two Currents

To distinguish the inter-turn fault from other disturbances and faults, such as the internal phase-to-phase or the external phase-to-phase fault, or the transient in the power grid, the difference between the phase currents of the synchronous generator must be assessed. In the case of an inter-turn fault, almost no difference exists between the currents of the healthy phases, whereas the difference between the current of the faulty phase and healthy phases is noticeable.

Figure 11.34 The difference between two consecutive data values is measured and the difference between two consecutive data values from a previous cycle.

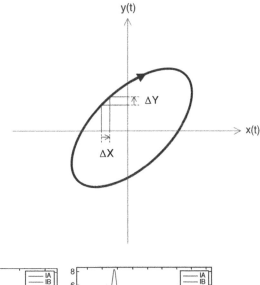

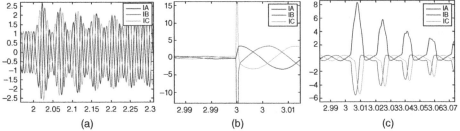

Figure 11.35 The transient disturbances in the power system, the power fluctuation in the synchronous generator (a), the switching in the transmission line (b), and the transformer inrush current (c).

However, in the case of a phase-to-phase fault (both internal and external) and the transient in the power grid, the difference between the current of the two phases is significant. Therefore, the scale of current variation can be used as an indicator for an inter-turn short circuit fault detection in the stator winding. Conclusively, if the current difference between phases A and B in the faulty state is ΔF_{AB} and the current difference between phases A and B in the previous cycle is ΔH_{AB}, then the ratio $\Delta F_{AB}/\Delta H_{AB}$ can be used to distinguish an inter-turn fault from other faults and disturbances in the power grid.

Several disturbances might interfere with the proposed algorithm, since they impose the same variation in the current signal. Among the most important of these disturbances are the oscillation of the generator power (Figure 11.35a), the switching of the transmission line (Figure 11.35b), and the inrush current of the transformers in the power network (Figure 11.35c). The inrush current, with a range of about 8–10 times the rated current, passes through the winding when the power transformer is electrified. The amplitude of the inrush current depends on the moment it is connected to the power grid and the amount of residual magnetic field in the core [96, 97].

In addition, in the case of switching in the power lines and the generator power fluctuations, a large current may pass through the generator. Note that sometimes protection relays in the generator are affected by these disturbances and operate incorrectly [98, 99]. Studies in references [99] show that two features can be used to distinguish these disturbances from an inter-turn fault:

- Two-phase currents are in category II.

- The difference between the instantaneous current during the fault must be less than the value of K2 compared to the current difference of the previous cycle in two healthy phases. The value of K2 in the case of a healthy generator is equal to 1, and by setting this parameter, the inter-turn fault can be distinguished from other faults and disturbances.

11.8.2 Algorithm Applications

11.8.2.1 Single-Phase to Ground Fault

In a phase-to-ground fault, the fault current enters the ground from the generator neutral. According to standard IEEE C37–101, several ways are possible to connect the generator neutral to the ground. The generator can be connected to the ground with high resistance to limit the ground-fault current. In a single-phase-to-ground type of fault, all the waveforms are like the waveforms of the healthy state of the machine. This is due to the resistance that limits the current. The only difference between a faulty state and a healthy state is the current at the neutral point of the machine while it operates under faulty conditions. This indicates that the phase-to-ground fault can be determined by monitoring the neutral current.

11.8.2.2 Inter-Turn Fault

The inter-turn short circuit fault can occur in four slots of one phase, according to Figure 11.36. For example, in phase A, an inter-turn short circuit fault can occur in slots 1, 2, 19, and 20. In these slots are two coils with the same phase. The various slots of the generator under consideration where an inter-turn fault can occur are listed in Table 11.9. In the first scenario, five coils are shorted, and in the other case, seven coils are shorted, which is equivalent to 41.67 and 58.34% of one phase, respectively. A fault with an intensity of 58.34% will occur in phase A if an inter-turn short circuit fault occurs in the 20th slot of the generator (Figure 11.37).

Figure 11.38 shows the inter-turn short circuit fault in slot number 20 and subsequently the fault in phase A. According to Figure 11.38, the faulty phase has a fault current greater than the other two phases and the fault current is distributed almost equally between the two healthy phases. The faulty phase current has a 180° phase difference compared to the healthy phases. The amplitude of the induced voltage in the faulty coils decreases due to the high fault current and saturation. The induced voltages of healthy phase coils do not change considerably, and their amplitude decreases only slightly.

The proposed algorithm is applied to the measured results to show the ability of the algorithm in inter-turn fault detection. In order to implement the stated algorithm, the $\Delta X / \Delta Y$ ratio must be obtained. The sampling frequency of 4 kHz is selected (80 samples per cycle) to have adequate data, but in order to see noticeable changes in the shapes, the samples are removed one by one, and only 40 samples per cycle are selected. The proposed algorithm is tested in three steps, as follows:

Table 11.9 The slot number and the coil numbers in the slots of the synchronous generator are susceptible to the inter-turn short circuit fault.

		1	2	3	4	5	6	7	8	9	10	11	12
Cases		1	2	3	4	5	6	7	8	9	10	11	12
Phases		A	A	A	A	B	B	B	B	C	C	C	C
Coil number	Bottom coil	8	7	5	6	8	7	5	6	8	7	5	6
	Top coil	1	2	12	11	1	2	12	11	1	2	12	11
Slot number		1	2	19	20	13	14	31	32	25	26	7	8

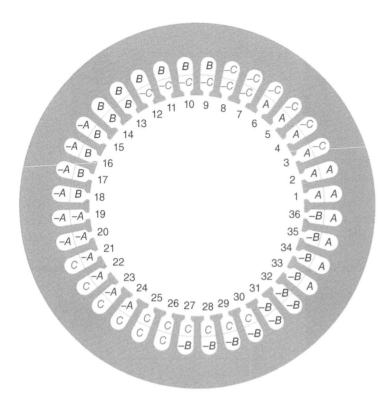

Figure 11.36 The winding layout inside the stator slots of the synchronous machine.

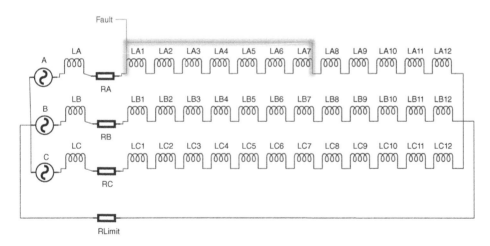

Figure 11.37 Schematic of the inter-turn short circuit fault of a synchronous generator with a severity of 58.34%.

1. The algorithm detects the existence of the fault or disturbance in the generator or power system according to the variation of $\Delta X/\Delta Y$. The variation of $\Delta X/\Delta Y$ values for the generator different phases are given in Figure 11.39., where a fault applied in the third second is also shown. The value of K1 is assumed to be equal to 2 in the algorithm (dashed line). When the generator is healthy, the ratio $\Delta X/\Delta Y$ is 1 and when the value of the $\Delta X/\Delta Y$ ratio in each of the phases is

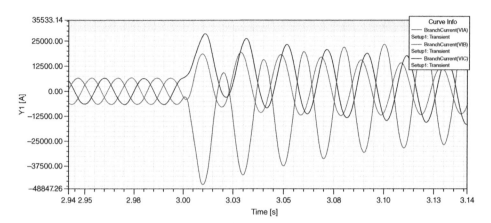

Figure 11.38 The phase current of the synchronous generator under an inter-turn short circuit fault with a severity of 58.34%.

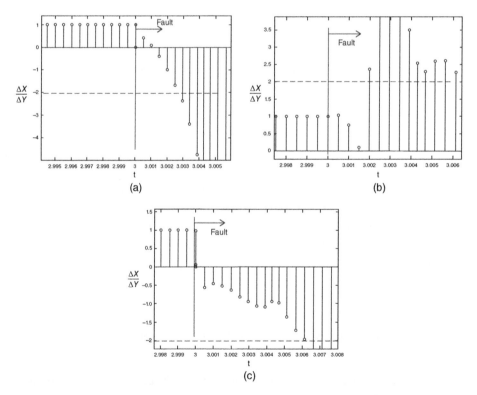

Figure 11.39 The ratio of the sampled current to the current of a previous cycle in the inter-turn short circuit fault in phase A.

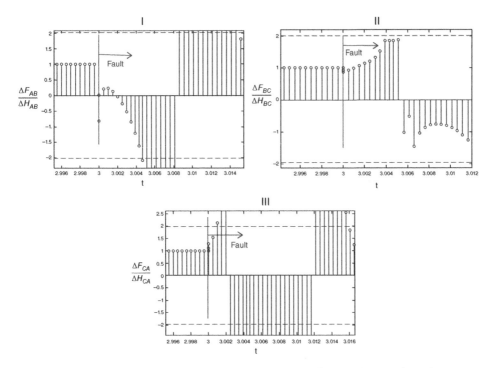

Figure 11.40 The ratio of the sampled current to the current of a previous cycle in the inter-turn short circuit fault: (I) current difference between phase A and B; (II) current difference between phase B and C; and (III) current difference between phase A and C.

more than 2, a fault or disturbance has occurred in the system or the generator. In the case of an inter-turn fault, 2 milliseconds after the fault occurs in phase B, the value of the ratio is greater than 2 and the occurrence of the fault or disturbance is detected. The algorithm moves to the next step to detect the faulty current category relative to the healthy state.

2. The current category is determined based on the sign of $\Delta X/\Delta Y$. In phase A, after 1 millisecond of fault occurrence, the $\Delta X/\Delta Y$ ratio changes and the sign of $\Delta X/\Delta Y$ for phase C changes in less than 1 ms after the fault occurs, while the sign of the $\Delta X/\Delta Y$ for phase B has not shown any changes. As a result, the phase A and C currents are in category II.

3. In order to distinguish the type of fault or disturbance, the third part of the algorithm is examined. In this third part, the ratio $\Delta F/\Delta H$ must be assessed and $\Delta F/\Delta H$ is compared to the K2 factor. The value of K2 is considered to be 2. Figure 11.40 shows the $\Delta F/\Delta H$ ratios for all phases of the generator. According to Figure 11.40, the value of $\Delta F/\Delta H$ for phases B and C is less than 2, while the amplitude of the ratio $\Delta F/\Delta H$ for both phases A and B and A and C is greater than 2, which indicates the occurrence of the fault in phase A. The inter-turn short circuit fault is identified in less than 5 ms, as shown in Figures 11.39 and 11.40.

11.8.2.3 Internal Phase-to-Phase Fault

The internal phase-to-phase fault is another type of short circuit fault that occurs in synchronous generators. The internal phase-to-phase fault occurs when two coils from different phases are present in the slot. The winding layout diagram in Figure 11.36 shows the possibility of identifying the slots in which a phase-to-phase fault can occur. The slots where the internal phase-to-phase fault can occur are listed in Table 11.10. Each slot has two coils, one on top and the other at the

Table 11.10 The slot number and the coil numbers in the slots of the synchronous generator are susceptible to the internal phase-to-phase short circuit fault.

	Coil number						Coil number			
Slot number	Upper coil in the slot	Lower coil in the slot	Phase	State	Slot number	Upper coil in the slot	Lower coil in the slot	Phase	State
27	C-3	B-1	BC	13	15	B-3	A-1	AB	1
28	C-4	B-2	BC	14	16	B-4	A-2	AB	2
29	C-5	B-3	BC	15	17	B-5	A-3	AB	3
30	C-6	B-4	BC	16	18	B-6	A-4	AB	4
3	A-3	C-1	CA	17	33	B-10	A-12	AB	5
4	A-4	C-2	CA	18	34	B-9	A-11	AB	6
5	A-5	C-3	CA	19	35	B-8	A-10	AB	7
6	A-6	C-4	CA	20	36	B-7	A-9	AB	8
21	A-10	C-12	CA	21	9	C-10	B-12	BC	9
22	A-9	C-11	CA	22	10	C-9	B-11	BC	10
23	A-8	C-10	CA	23	11	C-8	B-10	BC	11
24	A-7	C-9	CA	24	12	C-7	B-9	BC	12

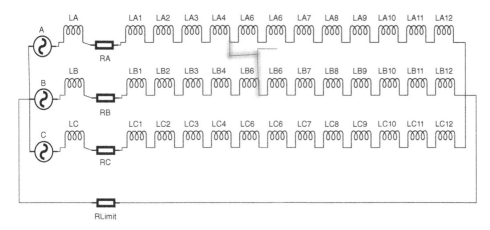

Figure 11.41 The internal phase-to-phase fault circuit in slot 18.

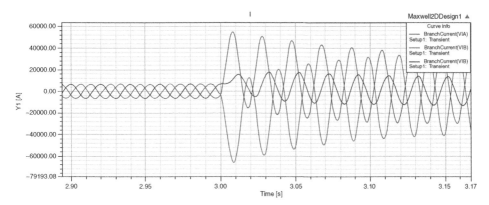

Figure 11.42 Three-phase currents of the synchronous generator under an internal phase-to-phase current in slot 18.

bottom of the slot. The faulty coils are indicated based on the phase and the slots where they are located. The amplitude and intensity of the fault are higher when the location of the fault is closer to the machine terminal and the CT. The circuit in Figure 11.41 shows the phase-to-phase fault in slot 18 and the waveform of the current is shown in Figure 11.42. As shown in Figure 11.42, the two faulty phase currents have large amplitudes and 180° phase difference relative to each other. The phase difference is due to the direction of the current in the CT.

The proposed algorithm is applied to the internal phase-to-phase short circuit fault in the stator winding. In the first step, the $\Delta X/\Delta Y$ ratio is calculated as shown in Figure 11.43. The amplitude of the $\Delta X/\Delta Y$ ratio at 0.5 milliseconds after the fault has occurred for phase A and phase B has reached -2.4 and -3.1, which is more than the predefined factor of 2. This indicates a disturbance in the system. The current phase category of the faulty case must be compared with a healthy state to discriminate between the inter-turn fault and the disturbances. Therefore, according to the $\Delta X/\Delta Y$ ratio sign, it is clear that in the phase-to-phase short circuit fault, the faulty phase currents are in category II. However, in the case of an inter-turn short circuit fault, the faulty current is in category II. In the third part of the algorithm, the difference between each pair of phase currents is calculated. The diagram of the difference between each pair of the faulty currents relative to the

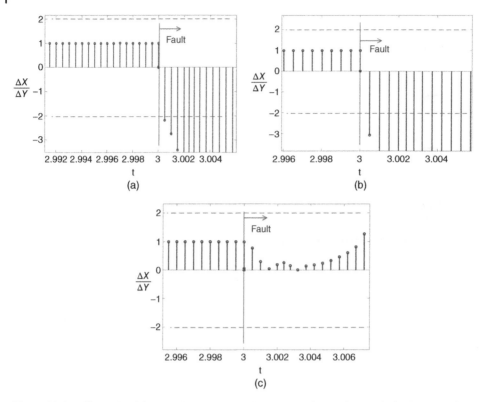

Figure 11.43 The ratio of the sampled current to the current of a previous cycle in the case of a phase-to-phase fault between phase A and phase B.

healthy state is shown in Figure 11.44. According to Figure 11.44, the amplitude of all $\Delta F/\Delta H$ ratios is greater than 2, indicating no inter-turn short circuit fault in the winding.

11.8.2.4 External Phase-to-Phase Fault

An external phase-to-phase short circuit fault is possible at a location between the CT terminal of the generator and the power grid. This fault causes a large current to flow in the two faulty phases, as shown in Figure 11.45. In this section, a phase-to-phase fault is applied between phase A and phase B. The amplitude of the faulty phases is markedly increased, while the amplitude of the current in the healthy case remains almost unchanged. The fault currents of the faulty phases differ by 180°, since the faulty current enters the CT dotted point and exits from the dotted point of the other CT.

The proposed algorithm is applied to the currents of the synchronous generator under an external phase-to-phase short fault (Figure 11.46). In the first step, the $\Delta X/\Delta Y$ ratio is calculated and the amplitude of the ratio becomes 6.5 and 11 in 0.5 milliseconds after the fault occurs in phase A and phase B. The amplitude of the index, since it is more than 2, indicates the presence of disturbances in the system. In the second step, the category of phase currents is determined. According to the sign of the current $\Delta X/\Delta Y$ ratio, all phases are in category I, so the algorithm detects the absence of an inter-turn short circuit fault and the algorithm does not enter the next steps.

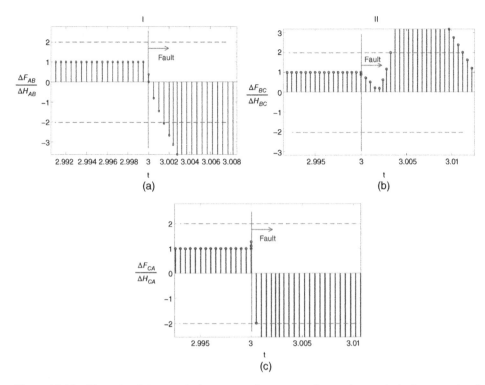

Figure 11.44 The ratio of the sampled current to the current of a previous cycle in the case of an internal phase-to-phase short circuit fault between phase A and phase B: (I) current difference between phase A and B; (II) current difference between phase B and C; and (III) current difference between phase A and C.

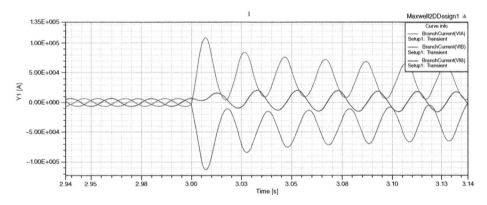

Figure 11.45 The synchronous generator current in the case of an external phase-to-phase fault.

11.8.2.5 Transformer Inrush Current

The inrush current of the power transformer creates a situation similar to that of a fault in the power grid. The proposed algorithm is applied to investigate the behavior during the inrush current of the power transformer. The current $\Delta X / \Delta Y$ ratio is calculated as shown in Figure 11.47.

According to Figure 11.47, the disturbance in phase B is detected 2 milliseconds after electrification of the power transformer. In order to detect the current category of the algorithm, the sign of $\Delta X / \Delta Y$ is determined and the currents of phase A and phase B are classified in category II.

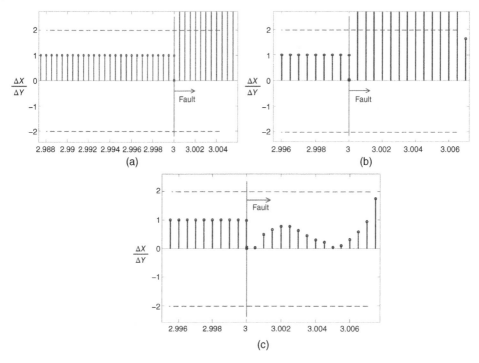

Figure 11.46 The ratio of the sampled current to the current of a previous cycle in the case of an external phase-to-external fault.

In the third part of the algorithm, according to Figure 11.48, the magnitude of the $\Delta F/\Delta H$ ratio of the difference between the faulty phase current and the healthy current is greater than 2. Therefore, the algorithm detects the absence of an inter-turn short circuit fault in the synchronous machine. As a result, the proposed algorithm performs correctly, despite the transformer inrush current.

11.8.2.6 Performance of the Proposed Algorithm in the Face of Various Types of Faults

Table 11.11 shows the effects of different types of faults, including an inter-turn short circuit fault in the different slots of the synchronous generator, on power fluctuation, and on switching in the transmission lines. According to Table 11.11, the algorithm is stopped while classifying the current category in the case of an external phase-to-phase short circuit fault. When an internal phase-to-phase fault occurs, the two-phase current falls into category II, so the algorithm stops at the phase current difference calculation stage. Finally, for power fluctuation and electrification of the transmission line, the algorithm stops at the steps of determining the current category and calculating the phase difference, respectively. The numbers given in Table 11.11 are the first instance in which the $\Delta X/\Delta Y$ ratio is greater than 2. According to the results in Table 11.11, in both the inter-turn short circuit fault and internal phase-to-phase fault, two phase currents from three phases are in category II compared to the healthy state. The third part of the algorithm distinguishes the inter-turn short circuit fault from the internal phase-to-phase short circuit fault. The inter-turn short circuit fault has been detected in less than half a cycle.

Table 11.11 The performance of the algorithm for different types of faults and disturbances.

Fault types	Faulty phases	Slot	$\frac{\Delta X_A}{\Delta Y_A}$		$\frac{\Delta X_B}{\Delta Y_B}$		$\frac{\Delta X_C}{\Delta Y_C}$		$\frac{\Delta F_{AB}}{\Delta H_{AB}}$		$\frac{\Delta F_{BC}}{\Delta H_{BC}}$		$\frac{\Delta F_{CA}}{\Delta H_{CA}}$		Fault detection time t(ms)
			t(ms)	Amplitude	t(ms)	Amplitude	t(ms)	Amplitude	t(ms)	amplitude	t(ms)	amplitude	amplitude	t(ms)	
External phase-to-phase	AB	–	5.0	6.5	0.5	11	7	<2	0.5	2.6	–	<2	5/1	8/5	–
Internal phase-to-phase	AB	18	0.5	−2.4	0.5	−3	8	1.5	1.5	−2.3	3.7	−3.4	1	33.35	–
	AB	34	9/3	−2.35	1	−2.34	51.5	1.8	3.9	−2.1	3.4	−2.8	1.5	2.83	–
	BC	9	1	1.5	3	−2.3	6	−3	6	2.3	5.5	8	2.5	4.3	–
	BC	12	2	1.6	1	−3.3	3	−4.2	3.5	2.8	3.5	−2.45	2	−7	–
Inter-turn	A	20	3	−2.35	2	2.4	6	−1.8	4.5	−2.1	–	<2	1	2.1	5
	B	14	4	−2.9	0.5	−3.8	0.5	3	3.5	−2.5	3	−2.7	–	<2	2.5
	C	7	3	1.4	5	−2.4	3	−3.1	–	<2	1.5	2.8	4.5	−2.5	3
Power variation	–	–	8	1.5	7	3	4	3.5	4	5	9.3	2.6	6.8	2.5	–
Transmission line electrification	–	–	2	5	0.2	−134	/20	323	0.5	15	1	−2.2	0.5	−15	–

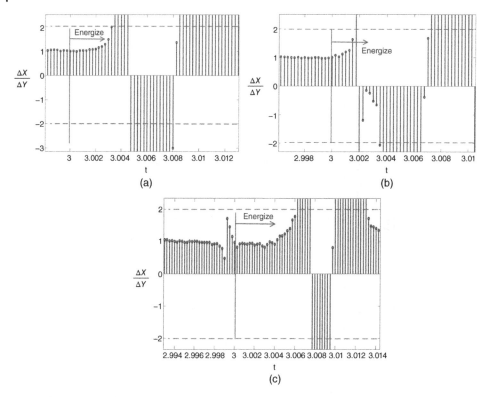

Figure 11.47 The ratio of the sampled current to the current of a previous cycle in the case of a transformer inrush current mode.

11.9 Inter-Turn Short Circuit Fault in Rotor Field Winding

11.9.1 Introduction

The main principle of condition monitoring is fault detection at an early stage. A comprehensive diagnostic approach must include fault type and location detection and severity estimation. One of the most prevalent fault types in a synchronous machine is an inter-turn short circuit fault in the rotor field winding, which accounts for more than 40% of the faults occurring in electric machines [20]. However, the voltage level in the rotor field winding is not high as in the stator winding, and the possibility of the inter-turn short circuit fault is low compared with the stator short circuit fault. Moreover, a synchronous generator with a few inter-turn short circuit faults might be able to operate for a long time. Nevertheless, detection of the inter-turn short circuit fault at its early stage is recommended to avoid vibration increases and consequent forced reduction of the rated power of the synchronous generator.

Two types of rotor windings are available in salient pole synchronous machines: the multi-layer wire wound design and the strip-on-edge design. The first type is the most common rotor winding type, in which the wire is wound around the rotor pole body, forming several layers of wires on top of each layer. The shape of the wire is rectangular and the wire insulation also acts as insulation between the wires. The second rotor winding type is commonly used in recently manufactured large salient pole synchronous machines that can withstand extensive rotational forces. The strip-on-edge design involves shaping thin copper strips into rectangular frames that slide on

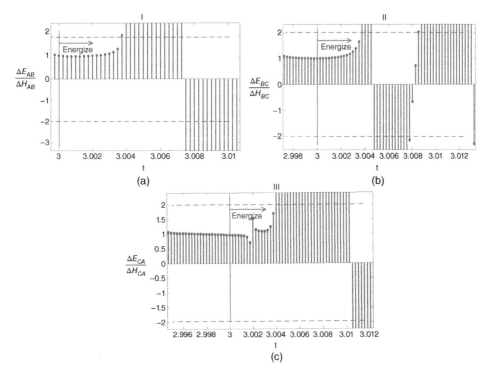

Figure 11.48 The ratio of the sampled current to the current of a previous cycle in the case of a transformer inrush current: (I) current difference between phase A and B; (II) current difference between phase B and C; and (III) current difference between phase A and C.

to the pole body and can be connected in series. The turn insulation in the strip-on-edge design consists of separators made from epoxy glass laminates and insulating tape, which are mounted among the copper frames. In both designs, the insulating washers and strips are used to isolate the winding from the body and the pole shoe.

The main reason for the short circuit fault in electrical machines is a defect in the insulation system of the machine. The reasons for the fault in the winding system can include insulation degradation due to excessive heating, vibration, stresses due to the unbalanced voltage distribution in the winding, over-voltage, humidity, load cycling, centrifugal force, and contamination. The insulation class of the rotor field winding is most likely B and the insulation quality must be high to withstand these tensions.

The magnetomotive force produced by a faulty pole with shorted turns is weaker than that of a healthy pole due to the reduced ampere-turn caused by the inter-turn fault. The reduced magnetomotive force results in a flux density reduction between the rotor pole and the stator core, while the reduced magnetic field in the air gap causes magnetic field distortion. The distorted air gap magnetic field due to the rotor field winding fault results in some of the following symptoms:

1. Unbalanced magnetic force.
2. Bearing vibration.
3. Local temperature rise.
4. Enhanced noise level.
5. Efficiency reduction.
6. Increased harmonics in the voltage and current.

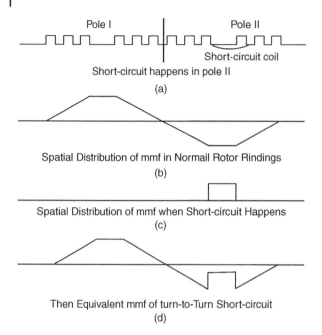

Figure 11.49 Spatial distribution of a magnetomotive force in the rotor field winding of a non-salient pole synchronous generator.

All of these symptoms result in more insulation degradation in the faulty area of the rotor field winding and further increases in the number of damaged winding turns. Therefore, fault detection in an early stage of the short circuit fault may avoid further fault extensions. The increase in rotor shorted turns ultimately results in a rotor ground fault that causes the forced stoppage of machine operation. The methods for inter-turn short circuit fault detection based on electromagnetic signals are divided into two categories: invasive and non-invasive methods. In the invasive method, air gap magnetic field monitoring is used to diagnose the inter-turn short circuit fault.

11.9.2 Invasive Method

11.9.2.1 Airgap Magnetic Field

The spatial distribution of the magnetomotive force in a non-salient pole synchronous generator has a trapezoidal shape without considering the slot effect, as shown in Figure 11.49. A current with a direction opposite to the main current creates a magnetomotive force with opposite polarity when an inter-turn short circuit fault happens in the rotor field winding. The combination of the faulty magnetomotive force with the healthy magnetomotive force reduces the main magnetomotive force, indicating that the net magnetomotive force will be reduced. Therefore, the effect of an inter-turn short circuit fault can be considered as a demagnetization factor [100].

A magnetomotive force of a synchronous generator with four salient poles is investigated under an inter-turn short circuit fault in one of the rotor poles [101]. An equivalent magnetomotive force of a shorted turn is modeled with an opposite polarity to the main magnetomotive force, resulting in an increment or decrement of the net magnetomotive force of the other poles. This indicates that the neighboring poles are also affected by the faulty pole. However, the amplitude of the magnetomotive force for neighboring poles, based on their polarity, is either increased or decreased. According to Figure 11.50, the amplitudes of the magnetomotive force are decreased in the first

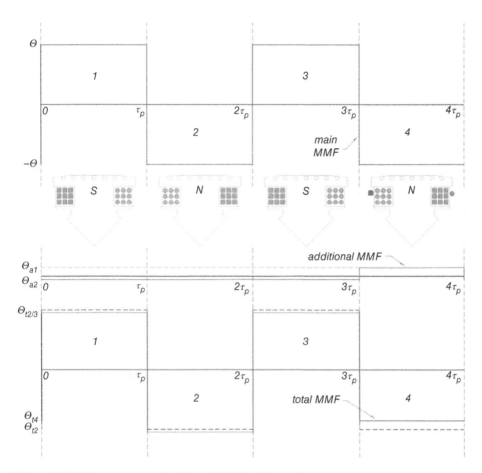

Figure 11.50 The magnetomotive force of a four-pole synchronous generator with the reduced number of turns in the rotor field winding. A healthy magnetomotive force (top), the added magnetomotive force to represent the faulty pole (middle), and the net magnetomotive force (bottom).

and third poles and increased in the second and fourth poles. However, the impact of an inter-turn short circuit fault on the magnetomotive force can be traced by examining the air gap magnetic field.

Measuring the magnetic flux of the air gap is one of the most promising methods for detecting the inter-turn short circuit fault in the rotor [20]. Air gap magnetic field monitoring involves one or more magnetic field sensors inside the electric machine. In this method, a coil called the search coil is installed in the air gap between the rotor and the stator. The search coils are mainly installed around the stator tooth. In addition, a Hall-effect sensor can be used to monitor the airgap magnetic field. The Hall-effect sensor must be installed on the stator tooth to capture most of the magnetic field. Figure 11.51 shows the installed Hall-effect sensor on the stator tooth of a synchronous generator. The induced voltage in the sensor is integrated to provide a measured signal in the form of a magnetic field [102]. The principles of both methods are based on the integration of the induced voltage in the search coil to achieve the magnetic flux density waveform [103].

Faults such as the inter-turn short circuit fault distort the magnetic field in the air gap and fault-related signatures and patterns can be identified by analyzing the air gap magnetic field.

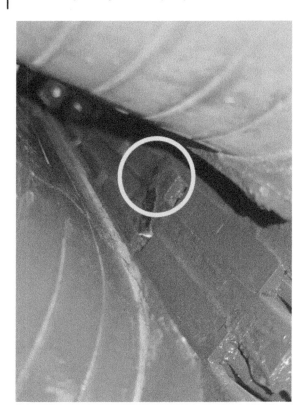

Figure 11.51 A flux sensor placed on the tooth of the stator core to measure the magnetic flux.

However, the detection method based on air gap magnetic field monitoring is highly invasive, especially if it requires removing the rotor to gain access to the generator inside.

Figure 11.52 shows the radial flux in the air gap of a synchronous generator with 14 poles. Two sensors are installed across from each other on opposite stator teeth. The impact of the rotor slot is also visible in Figure 11.52, where the rotor slots cause a small reduction in the air gap magnetic field waveform. The magnitude of the magnetic field component of the faulty pole is less than that of the healthy poles, which indicates the occurrence of a short circuit fault in the rotor field winding. The reduced magnetic field of the faulty pole is due to the reduced number of turns that impose the reduction of the ampere turns. The 20% inter-turn short circuit fault in the rotor field winding results in a 6% reduction in the average magnetic field of the faulty pole compared with the healthy pole. Since two sensors are installed opposite each other, the measured magnetic field polarity of the two poles must be opposite, since the number of pole pairs is odd in the generator under study ($2P = 14$). This indicates why the two poles have opposite polarity, as shown in Figure 11.52 [104].

The neighboring poles of the faulty rotor pole under an inter-turn short circuit fault encounter a variation in the magnitude of the magnetic field. Figure 11.53 shows the flux density and flux line distribution of a synchronous generator with eight poles, where the top rotor pole is an under 20% inter-turn short circuit fault. The generator is simulated in the no-load operation to avoid the armature reaction. The flux that passes through the faulty pole is decreased due to the reduction of the magnetomotive force. The same reduced flux returns through the two adjacent healthy poles of opposite polarity, resulting in a decreased flux density in the two neighboring poles. The next two poles of the same polarity as the faulty pole experience an increase in flux density to compensate for the reduced flux density of the neighboring poles of the faulty pole. Figure 11.54 shows the impacts

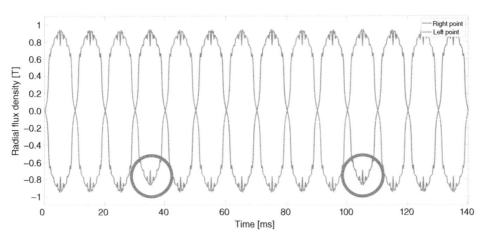

Figure 11.52 The air gap magnetic field of the synchronous generator with 14 poles operating at no-load under 20% inter-turn short circuit fault in one of rotor field windings. Modified from [104].

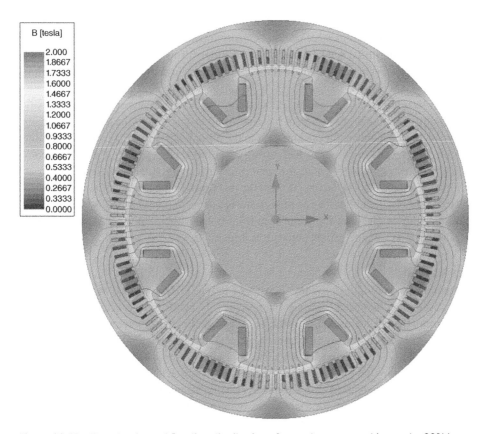

Figure 11.53 Flux density and flux line distribution of a synchronous machine under 20% inter-turn short circuit fault in the top pole.

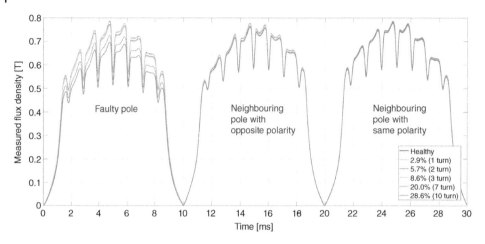

Figure 11.54 The radial flux density of a faulty pole and its two neighboring poles with the same and opposite polarities in a synchronous machine operating at no-load.

of the inter-turn short circuit fault on the faulty pole and two poles with the same and opposite polarities. In Figure 11.54, the number of shorted turns is increased from 2.9 to 28.6%, where the magnitude of the flux density in the faulty pole and the neighboring pole with the opposite polarity is decreased. By contrast, the amplitude of the flux density for the pole with the same polarity is increased, although the amplitude is negligible. The rotor inter-turn short circuit fault ultimately results in a reduction in the average flux density of the synchronous machine, as well as a reduction in the machine output power [104].

One solution for the inter-turn short circuit fault detection is to compare the average flux density of all the poles in one mechanical revolution of the synchronous machine. If the average flux density in one rotor pole is noticeably reduced compared to the other poles, this can be a sign of a rotor short circuit fault. The test must be performed during machine operation since the off-line conventional "pole-drop" test is unable to detect the rotor inter-turn short circuit fault, indicating that the short circuits that appear during the operation of the machine are not evident during standstill. Furthermore, the short circuit fault with one or two shorted turns in the rotor field winding is difficult to detect since the average flux density of each rotor pole is dissimilar, since manufacturing similar poles with a similar magnetic field behavior is impossible.

11.9.2.2 Polar Diagram

Observing the amount of variation in the flux density or induced sensor voltage installed on the stator tooth and comparing them with each other cannot provide an informative visualization of the inter-turn short circuit fault. The flux density of each pole can be shown in polar coordinates, which is a useful method for observing the simultaneous rate of flux density variation. If the amplitude of the flux density decreases for a designated pole, this is an indication of the inter-turn short circuit fault. In this method, a machine must be rotated for at least one mechanical revolution, and the measured magnetic field of each pole is then averaged and plotted in a polar diagram.

Figure 11.55 shows a polar diagram for a simulated synchronous machine with 14 salient poles operating at no-load for a healthy state and under several degrees of inter-turn short circuit faults. The inter-turn short circuit fault is applied to the first rotor pole and, as shown in Figure 11.55, the flux density of the pole is decreased by increasing the fault severity. In an ideal electric machine, as simulated in the finite element, the averaged flux density of each pole is equal. However, this is

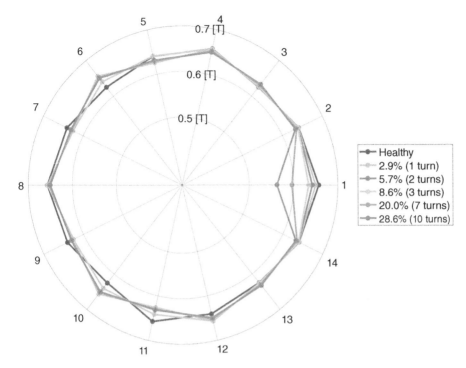

Figure 11.55 The averaged flux density of a simulated synchronous machine with 14 poles operating in the no-load condition in a healthy case and under several degrees of an inter-turn short circuit fault.

not the case in reality. Table 11.12 indicates the average flux density of the faulty pole with respect to the average flux density of all the poles. Therefore, the average flux density of a healthy pole is 100%, while the average flux density is decreased to 99% by having one inter-turn short circuit fault. The average flux density drops to 88.8% by having 10 inter-turn short circuit faults on the rotor pole [104].

Figure 11.56 shows the measured flux density of the 14-pole synchronous machine in the laboratory in the format of a polar diagram. The inter-turn short circuit fault is applied to pole number 6 with the severity of 1 to10 shorted turns in one rotor pole. The polar diagram of Figure 11.56, compared with Figure 11.55, shows a significant difference, where the polar diagram is moved toward pole number 13. The reason for the rotor incline toward pole number 13 is the inherent eccentricity fault.

The average flux density of the entire poles is about 0.508 T, as shown in Figure 11.56. The average flux density for the poles between pole number 3 and pole number 10 varies from 0.493 to 0.507 T, while the average flux density for poles 11, 12, 13, 14, 1, and 2 ranges between 0.515 and 0.525 T. The deviation of the measured average flux density of the poles is approximately ±3.5% with respect to the average flux density of all the poles. The degree of the eccentricity fault is approximately 3%, assuming that the variation in flux density is proportional to the inverse air gap length. The degree of eccentricity is only an estimate, since the saturation, stator, and rotor slotting effects, as well as the rotor pole saliency, are ignored. These factors may cause a fringing effect of the flux in the air gap due to the decreased permeability of the stator iron, thereby causing a nonlinear relation between the air gap length and the air gap flux density [104].

The impact of no-load and various types of on-load operations of the synchronous generator on the average flux density of the faulty rotor pole under a short circuit fault is presented in Table 11.13.

Table 11.12 The average flux density of the faulty pole with respect to the average flux density of all the poles together.

Number of turns	Average faulty pole (%)
0	100
1	99.0
2	97.9
3	96.8
7	93.7
10	88.8

Note: The averaged value of some poles differs, such as pole number 6 in the polar diagram of Figure 11.55. This is due to the different time steps in the FEM simulation for different fault severities.

Note: Plotting the averaged flux density of several mechanical revolutions of the machine is recommended to avoid unexpected fluctuations in the polar diagram that may result in false identification.

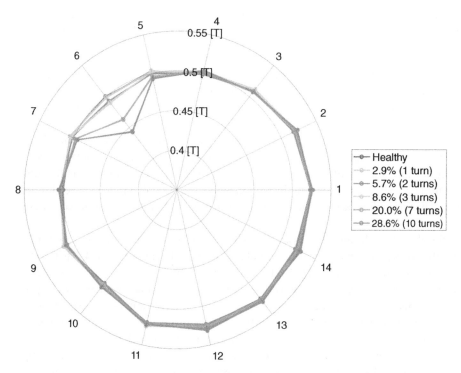

Figure 11.56 The averaged measured flux density of a synchronous machine with 14 poles operating in the no-load condition in a healthy case and under several degrees of an inter-turn short circuit fault. The inter-turn short circuit fault is applied to pole number 6.

Table 11.13 Percentage reduction of pole 6 with respect to the average value of all poles.

Number of turns	Load-case 1 (%)	Load-case 2 (%)	Load-case 3 (%)	Load-case 4 (%)
0	97.6	98.0	97.8	97.7
1 (2.86%)	97.3	97.5	97.6	97.4
2 (5.71%)	96.4	96.7	96.6	96.5
3 (8.57%)	95.6	95.9	95.5	95.5
7 (20.0%)	90.9	91.2	90.2	89.6
10 (28.57%)	87.1	87.4	85.5	84.5

Four operating conditions are investigated: the first case is the no-load operation, the second case is the resistive load, the third case is the low resistive-inductive load, and the fourth is the high resistive-inductive load. Compared with Table 11.12, the average flux density of pole number 6 in a healthy case is not 100%, due to the eccentricity fault in the machine. In addition, the average value of the neighboring poles of pole number 6 is not 100%, due to the eccentricity fault. Table 11.13 shows that in the no-load and in all three on-load cases, a single-turn short circuit fault causes a reduction of only 0.2–0.5% of the faulty pole with respect to the average flux density of all the poles. This is expected to be far less than the inherent and natural variations between the poles. For example, the average flux density of pole number 13 is 1.1% larger than the average value of pole number 14 at a high resistive-inductive load operation with no applied short circuit fault in the field winding. However, determining exactly how much the average flux density of each pole naturally differs from another pole is difficult, due to the additional impact of the eccentricity fault. The detection of one and two inter-turn short circuit faults is not straightforward, since the rate of average flux reduction is between 1 and 2% and can be due to load variations and the inherent eccentricity fault. Therefore, detection of the severity of low inter-turn short circuit faults is difficult using polar diagrams [105].

The impact of the different on-load operations of the synchronous machine on the average flux density of the faulty rotor pole with one, two, and three inter-turn short circuit faults indicates that the average flux density is almost the same regardless of the loading condition. However, the average flux density of the faulty rotor pole under 7 and 10 shorted turns exhibits a higher reduction when the machine supplies a high resistive-inductive load. However, eccentricity also plays a critical role during an on-load operation and results in deviation of the average flux density. At full-load operation, the radial force between the stator and rotor is increased and results in an eccentricity degree increment. Therefore, the magnetic pull on the rotor may occur due to the increase in the air gap length at pole number 6 due to high loading, and while the rotor pole number 6 is under a short circuit fault, it shows a higher reduction of the average flux density. However, the impact of different loading conditions, shown in Table 11.13, indicates that the magnitude of the average flux density does not change significantly, even for severe faults.

Figure 11.57 shows the polar diagram of the synchronous generator in a healthy state and under 1, 2, 3, 7, and 10 turns of a short circuit fault in the rotor winding. Figure 11.57 also shows the impact of a high resistive-inductive load on the average flux density in the polar diagram. This density is almost constant, although the measurement results in Table 11.13 show some slight variations. The polar diagram in Figure 11.57 shows the applied short circuit fault to pole number

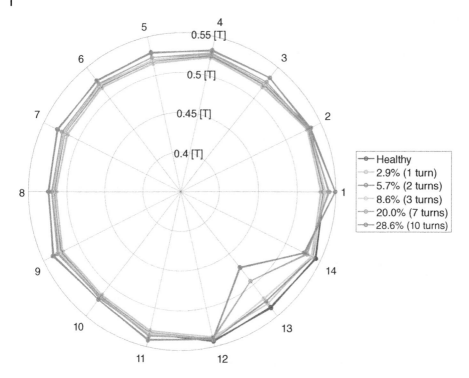

Figure 11.57 The measured average flux density of the synchronous generator operating at a high resistive-inductive load in a healthy state and under various degrees of an inter-turn short circuit fault. The inter-turn short circuit fault is applied to pole number 13.

13, which is the pole located opposite to pole number 6. Having a short circuit on pole number 13 results in a reduction in the air gap magnetic field asymmetry and, consequently, in the severity of the eccentricity fault in the on-load condition. This indicates that the coexistence of the short circuit fault and eccentricity fault in the same direction can excessively increase the air gap magnetic field asymmetry in the on-load condition and increase the severity of the eccentricity fault.

One impact of having an inter-turn short circuit fault on two rotor poles located opposite each other is shown in the polar diagram of Figure 11.58. There, the polar diagram indicates that the average flux density value of pole number 13 is approximately the same as that of pole number 6, regardless of whether pole number 6 has a fault.

The application of the polar diagram method is simple since it only requires one sensor installed on the stator tooth and can be used for condition assessment of the synchronous machines regardless of their power rating or the number of poles. Figure 11.59a shows the polar diagram of a synchronous machine with 64 salient poles. Although the polar diagram depicted in Figure 11.59a is for a healthy operation of the machine, the amplitude of the flux density of each pole is not the same and shows some fluctuations in the magnitude of the magnetic field. The reason is the manufacturing process and the material properties utilized in the machine. The occurrence of the short circuit and its visualization using the polar diagram of the synchronous machine with 64 poles indicate a reduction in the average magnetic field in pole numbers 8 and 48 (see Figure 11.59b) [102].

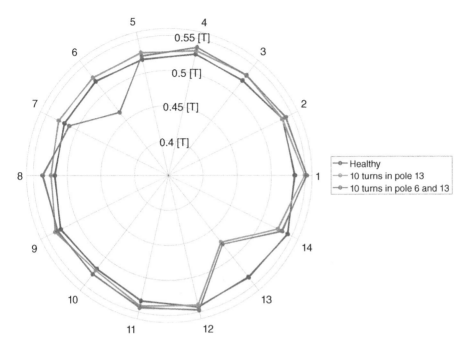

Figure 11.58 The polar diagram of the measured average radial flux of a synchronous generator operating at a high resistive-inductive load in a healthy case and under short circuit faults in pole number 6 and number 13 [102]/with permission of IEEE.

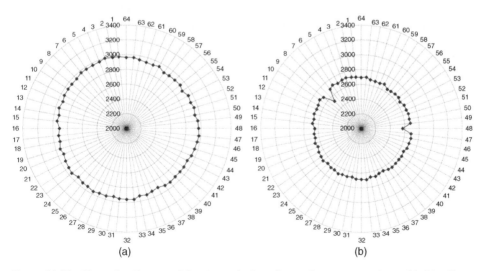

Figure 11.59 The polar diagram of the magnetic flux of a synchronous generator with 64 salient poles operating under the rated load in a healthy condition (a) and a faulty condition (poles 8 and 48 are under an inter-turn short circuit fault). The radial scale is in terms of the average magnetic flux and the circumferential scale is in terms of the number of machine poles [102]/with permission of IEEE.

11.9.2.3 Application of the Frequency Spectrum in the Inter-turns Short Circuit Fault Using the Air Gap Magnetic Field

One straightforward approach for fault detection is measuring the air gap magnetic field using installed sensors in the air gap and detecting the inter-turns in a short circuit fault in the rotor field winding by comparing the average flux density of each pole. However, the occurrence of short circuit faults in the rotor field winding causes an asymmetry in the air gap magnetic field and gives rise to some harmonics in the air gap magnetic field. Figure 11.60 shows the frequency spectrum of the air gap magnetic field of a synchronous generator with 14 salient poles. Some frequency components in Figure 11.60 have higher amplitudes than the other frequency components. The harmonic components are the main frequency and their odd multiples exist even in the frequency spectrum of the healthy machine. The odd multiples of the main frequency in the healthy machine are due to the machine topology, the flux path, and the magnetomotive force [106]. However, other harmonics exist in the frequency spectrum of the air gap magnetic field of the healthy synchronous machine with an amplitude of −65 dB at frequencies of 64.29 and 164.29 Hz. These harmonic components are most likely to appear due to the design parameters of the synchronous machine, such as the winding layout, slot shape, damper bar distribution in the rotor pole shoe, and stator slots.

The frequency spectrum of the air gap magnetic field for a synchronous machine with 14 poles indicates that the occurrence of the inter-turn short circuit fault results in the appearance of new harmonic components. The amplitude of the fault-related harmonic components is increased by increasing the number of short-circuited turns in the rotor field winding. The frequency index that can be used to extract the fault-related harmonics is as follows [105]:

$$f_{fault} = \left(1 \pm \frac{k}{p}\right) f_s \tag{11.36}$$

where f_s is the main frequency in Hz, k is an integer, and p is the number of poles. The f_{fault} is also called a mechanical frequency of the machine, where the harmonic components appear as a multiple of the mechanical frequency. The lowest mechanical frequency for a synchronous machine with 14 poles is equal to 7.14 Hz. The fault-related harmonics are distributed in the entire frequency domain, where the frequencies with the highest sensitivity to even the smallest degree of fault are below the main frequency component. A wave-shaped pattern exists where the amplitude of the fault harmonics is decreased by increasing the frequency. Moreover, the magnitude of the components, even at multiples of the main frequency component, is remarkably smaller than the remaining components.

A comparison between the frequency spectrum of the air gap magnetic field for a machine with 8 poles and a machine with 14 poles indicates that the amplitude of the frequency components is higher for the machine with 8 poles than for the machine with 14 poles for the same degree of inter-turn short circuit faults. For example, the amplitude of the main mechanical frequency component for 20 inter-turn short circuit faults is −32.2 dB and −41.4 dB for the machine with 8 poles and 14 poles, respectively. The reason is that the ratio of the shorted turns to the total number of turns in all the poles is higher in the machine with 8 poles than with 14 poles.

The amplitude of the main mechanical frequency component (7.14 Hz) in a machine with 14 poles under one inter-turn short circuit fault is −60.22 dB, or about 1 mT. The amplitude of that frequency is increased to 2 mT by having two inter-turn short circuit faults, and it increases to 10 mT by having 10 inter-turn short circuit faults. Although the pattern of the frequency component indicates a linear increment of the component amplitude by increasing the fault degree, this is not always the case.

The load impact on the air gap magnetic field indicates that the high resistive-inductive load can slightly increase the amplitude of the air gap magnetic field. The frequency spectra of the air gap

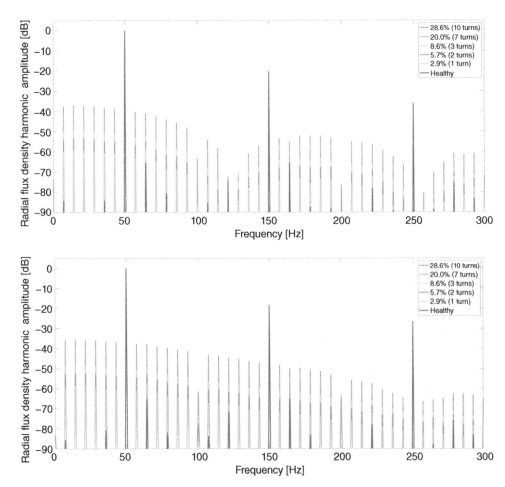

Figure 11.60 The frequency spectrum of the air gap magnetic field of a synchronous machine with 14 salient poles in the no-load (top) and full-load (bottom) operation in a healthy state and under various degrees of an inter-turn short circuit fault.

magnetic fields of a 14-pole machine are shown in Figure 11.60 for various degrees of inter-turn short circuit faults for the full-load operation of the machine. The amplitude of the frequency components below the main frequency of the machine follows the same pattern as the no-load operation, while a slight increment in the amplitude occurs due to the load. A significant difference is observed for the frequency components between the main frequency and the first odd multiple of the main frequency, where the amplitude of the fault components is markedly increased due to the full-load operation.

The frequency spectrum of the measured flux density is shown in Figure 11.61. The following differences are evident between the simulation and experimental results [104]:

1. The frequency spectrum of a healthy machine has mechanical frequency components and its multiples in the measured air gap magnetic field. The reason is the inherent eccentricity fault and imperfections in the manufacturing of the electric machine.
2. Although all the mechanical frequency components and their multiples show a linear increase with the severity of the fault in the simulation result, this is not the case for the measured results.

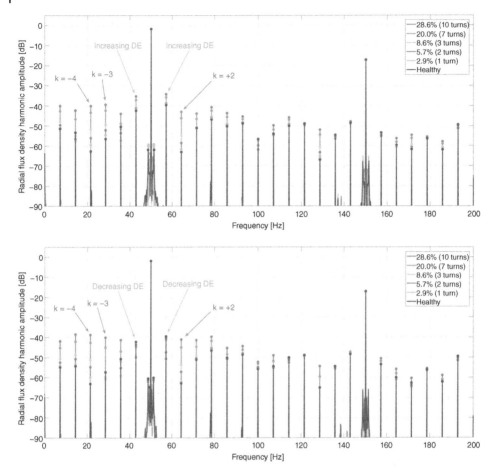

Figure 11.61 The frequency spectrum of the radial flux density of a synchronous machine with 14 poles at a maximum load operation with different degrees of short circuit fault in the rotor field winding for pole number 6 (top) and pole number 13 (bottom).

As shown in Figure 11.61, some of the frequency components are increased and some of them do not increase or are decreased by increasing the fault severity.

3. Few frequency components respond as expected to the increase in the inter-turn short circuit fault. For example, the frequency components with $k = -4$ and $k = 2$ in Equation (11.36) respond similarly to the simulation results. These components show a consistent increase in the amplitude by increasing the number of shorted turns in the rotor field winding.

According to Figure 11.61, the increased inter-turn short circuit fault results in the increment of the eccentricity fault-related components; this is also discussed in the polar diagram. In Figure 11.61, the fault is applied to the opposite rotor pole and the eccentricity-related components are decreased by increasing the number of shorted turns. The decrement in the eccentricity fault-related components by increasing the inter-turn short circuit fault severity in the opposite direction of the eccentricity fault is in agreement with the arguments discussed in the polar diagram.

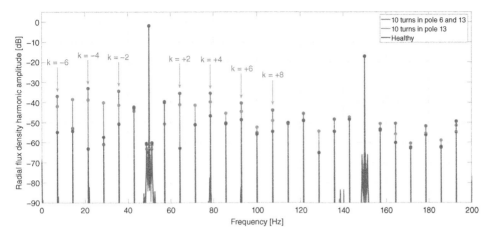

Figure 11.62 The frequency spectrum of the radial flux density of a synchronous machine operating at a maximum load with a short circuit fault in two poles located opposite each other (pole number 6 and pole number 13).

The frequency component that is twice the main frequency of the machine ($k = 7$) shows an inconsistent response to the occurrence of the inter-turn short circuit fault in the rotor field winding. In Figure 11.61, the amplitude of the 100 Hz component is decreased by increasing the short circuit fault severity in the rotor field winding, while its amplitude shows a consistent increase when the opposite pole is under a short circuit fault. Moreover, the frequency component for $k = 10$, which is a 121.4 Hz component, seems to be immune to the inter-turn short circuit fault.

Figure 11.62 shows the frequency spectrum of a magnetic field where an inter-turn short circuit fault is applied to two opposite rotor poles. The occurrence of a short circuit fault in two rotor poles located opposite each other causes a completely different frequency component response when compared with a case where only one pole has a fault. When a short circuit with 10 turns occurs in one of the rotor poles, the mechanical frequency component and both its odd and even multiples are increased. Having 10 short circuit faults in two rotor poles opposite each other causes the frequency components at the even multiples of k to increase even more, while the components at the odd multiples of k remain the same or even decrease with respect to the healthy case.

Conclusively, the application of a frequency analysis to the air gap magnetic field demonstrates that the frequency components that are multiples of the mechanical frequency can be used to detect the inter-turn short circuit fault regardless of the machine topology and the number of poles. In addition, frequency analysis of the air gap magnetic field in the no-load and on-load operation of the synchronous machine shows that the amplitude of the fault components below the main frequency of the machine is almost the same. Therefore, comparison of the fault-related frequency spectrum and the healthy frequency spectrum does not require the same load operation of the machine if a mechanical frequency below the main frequency is used. In addition, the fault-related harmonics below the main frequency show the highest sensitivity to the inter-turn short circuit fault occurrence. Moreover, the detection of the fault for the low-severity inter-turn short circuit fault, such as one or two shorted turns, is very difficult since the magnitude variation is small and the same variation can occur due to the load variation in the machine. In addition, having a symmetrical short circuit fault in the rotor pole makes the detection more difficult since the patterns do not follow the previously observed patterns.

11.9.3 Non-invasive Methods

Fault detection based on analyzing the air gap magnetic field is an invasive method, but it shows promising results for the detection of a short circuit fault in the rotor winding. Nevertheless, the need for a sensor installed inside the machine makes this method difficult to use in reality. Therefore, several methods using non-invasive sensors, based on the application of current, voltage, and stray magnetic fields, have been developed to fulfill the non-invasive need for the detection of inter-turn short circuit faults in the rotor field winding. The following sections are based on analysis of the current or voltage of the stator and rotor windings, power, and stray magnetic fields.

11.9.3.1 The Stator and Rotor Current

The magnetic field of the synchronous machine becomes non-uniform due to inter-turn short circuit faults, which results in the generation of some harmonics, both in the stator winding current and rotor field winding current. A detection method based on analysis of the stator current must be performed while the machine is connected to the power grid or the local load, otherwise no current is present in the stator winding of the synchronous generator.

Having a short circuit fault in the rotor field winding causes an asymmetry in the air gap magnetic field and frequency spectrum analysis indicates the existence of new harmonics due to the fault. An induced voltage in the stator winding contains the harmonics due to the asymmetry in the air gap magnetic field. Therefore, when the terminals of the synchronous generator are connected to the load, a current passes through the stator winding that also contains the voltage harmonics. This indicates that the analysis of the current spectrum of the synchronous generator can provide insight into the health of the rotor field winding. During the healthy operation of a synchronous machine, the current in the stator winding creates a magnetic field. Since the stator magnetic field rotates at synchronous speed and has symmetry, no electromotive force is induced in the rotor field winding. The occurrence of a short circuit fault in the rotor field winding causes a symmetric magnetic field in the air gap and, consequently, a current in the stator winding that contains the fault harmonics. A faulty current in the stator winding generates a magnetic field that rotates with different relative speeds with respect to the synchronous speed, thereby inducing an electromotive force in the rotor field winding. Therefore, a direct current of the rotor contains harmonics due to the short circuit fault in the field winding. Table 11.14 shows the harmonics induced in the rotor field winding [54].

The first, third, and fifth harmonics of current induce a voltage in the parallel winding due to the occurrence of a short circuit fault in the synchronous generator with a parallel winding connection. The induced voltage creates harmonics in the current with a frequency of 30, 90, and 150 Hz. These harmonics can then be used to detect the short circuit fault in the synchronous generators [62]. The sensitivity of the frequency components concerning the severity of the fault is shown in Figure 11.63. A harmonic with a frequency of 150 Hz indicates a lower sensitivity and a harmonic with a frequency of 30 Hz indicates a higher sensitivity to a fault. The impact of loading on the frequency harmonics is depicted in Figure 11.64. The amount of the load increase does not have

Table 11.14 Induced harmonics in the rotor field winding (Hz).

	Harmonic Component (Hz) at synchronous speed of 50								
Stator	25	50	75	100	125	150	175	200	225
Rotor	−25	0	25	50	75	100	125	150	175

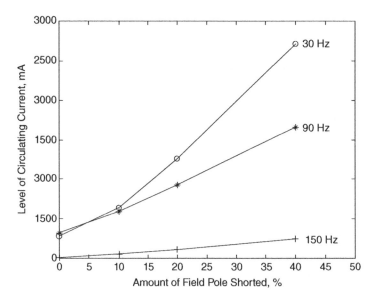

Figure 11.63 The variation of the frequency harmonics of a no-load generator by increasing the fault severity.

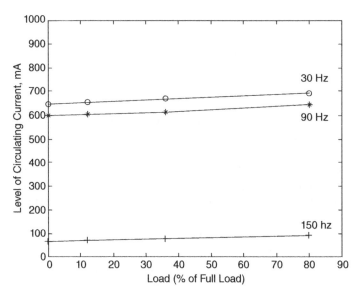

Figure 11.64 The effects of load on the frequency harmonics in the case of a 10% short circuit fault in the field winding.

a significant effect on the harmonics changes, indicating that the index can be used both in the no-load operation and under loading conditions [62].

In the healthy operation of the generator, the number of pole pairs of the stator and rotor is equal and the magnetic field of the air gap rotates at a speed of $120f/p$, where p is the number of pole pairs and f is the frequency of the stator terminals. In this case, the rotor field does not inject any harmonics into the air gap magnetic field. The harmonic components (nf_r) are injected into the stator winding when a short circuit fault happens, where n is equal to 1, 2, 3, … and f_r is the

Table 11.15 The amplitude of current harmonics of the rotor field winding under short circuit fault.

Generator condition (Hz)	1	2	7.5	10
f_r (16.67)	0.0124	0.0210	0.0360	0.0380
$2f_r$ (33.33)	0.0020	0.0046	0.0138	0.0175
$5f_r$ (83.33)	0.0010	0.0010	0.0016	0.0016
$7f_r$ (116.7)	0.0024	0.0032	0.0040	0.0040
$8f_r$ (133.2)	0.0010	0.0022	0.0110	0.0170

rotor frequency. If n is not equal to mP ($m = 1, 2, 3, \ldots$), an asynchronous rotating magnetic field is created. This field is divided into two components: a positive sequence component and a negative sequence component.

The velocity of the positive sequence component of the magnetic field is equal to $60nf_r/p$ and the rotational velocity of the excitation field of the negative sequence component is equal to $-60nf_r/p$. These rotational components cause harmonics in the rotor field winding current equal to $(n + p)f_r$ and $(n - p)f_r$. Moreover, if n is equal to mp, the induced harmonic component in the rotor field winding is equal to $(n - p)f_r$. Table 11.15 shows the results of a laboratory test setup of a 30-kW synchronous generator with six poles. In the present study, the first, second, fifth, and seventh harmonics of the rotor frequency are investigated in a healthy case and in a case with a 10% short circuit fault in the rotor winding. The results indicate that f_r is more sensitive to the occurrence of the short circuit fault and that the amplitude of the index is increased sharply by increasing the fault intensity [107].

One method, introduced in reference [108], is to study the frequency spectrum of the stator current for inter-turn short circuit fault detection. This method is a general method for fault detection but cannot detect the location of the fault or the number of short circuits. The method is also applicable during the loaded operation of the generator, where the armature reaction may affect the harmonic contents.

Another way to diagnose an inter-turn short circuit fault in the rotor field winding is to analyze the circulating current of the stator windings. The presence of a short circuit in the rotor winding generates a circulating current in the stator winding. The reason for this circulating current is that the fault in the rotor results in an unbalanced magnetic field and the reluctance in the machine becomes variable, resulting in a circulating current in the parallel winding to compensate for the unbalance magnetic field in the air gap. The subharmonics in the stator current circulating in a parallel winding of the 350 MW synchronous generator can be used to detect the short circuit fault in the rotor winding [109]. The main harmonics are removed in the frequency spectrum and the remaining harmonics appeared due to a fault. Subharmonics that originate due to the external source can be circulated in the stator winding and may result in false detection of the fault. The problem can be solved if a filter is placed in the stator terminal to identify the external harmonics.

Detection of the inter-turn short circuit fault of the rotor winding is not possible by measuring the rotor current in a synchronous generator with a damper winding in the rotor pole shoes since the damper bars reduce and vanquish the second-order harmonics that generate the oscillation in the rotor current. Figure 11.65 shows the rotor field current under an inter-run short circuit fault in the rotor winding. A second-order harmonic is evident in the rotor current where the damper bars are removed, indicating that the impact of the damper bars reduce the transient or dynamic oscillation in a synchronous machine. The second harmonic component is more visible

Figure 11.65 The rotor field winding current of a synchronous machine without damper bars under a short circuit fault in the rotor winding.

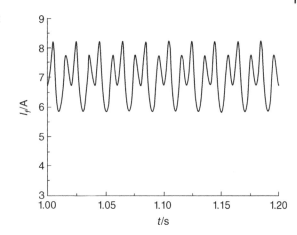

in the case of a rotor without a damper than in the rotor with damper bars. Therefore, in synchronous machines that do not use damper bars, this harmonic component can be used to detect a short circuit fault of the rotor winding.

11.9.3.2 Stator Voltage

The stator phase voltage of a synchronous generator is a suitable non-invasive signal that can be used for the inter-turn short circuit fault in the rotor winding both during no-load and on-load operation. Figures 11.66 and 11.67 show the frequency spectrum of the phase voltage in the no-load and on-load operation of the synchronous machine with eight salient poles in a healthy machine and under one inter-turn short circuit fault. The impact of the one inter-turn short circuit fault results in a remarkable increase in the 25 and 75 Hz frequency components. This indicates that these harmonic components can be used for the early-stage detection of the short circuit fault in the rotor field winding. The harmonic components in the phase voltage are due to the asymmetric air gap magnetic field components; therefore, the same harmonic must be present in the phase voltage. Equation (11.36) can be used to extract the harmonic components of the stator phase voltage [110]. Using Equation (11.36) shows that several other frequency components, in addition to the 25 and 75 Hz frequency components, exist due to the short circuit fault in the stator voltage. These include the frequency harmonics of 100, 125, 225, 325, 375, 1325, and 1725 Hz. Tables 11.16 and 11.17 show the impact of a rotor short circuit fault on these harmonic components. Although the amplitude of these harmonic components is increased due to the fault, in both the no-load and on-load operations of the synchronous generator, the 25 and 75 Hz frequency components show the highest sensitivity to the fault occurrence.

The impact of the three-phase synchronous machine connection on the fault-related harmonic component is shown in Figure 11.68, where the synchronous machine feeds a three-phase balanced resistive load. The 150 Hz harmonic component is suppressed due to the star-connection of the stator three-phase winding without a grounded neutral. When the star-connection point is not grounded, a 360° phase shift occurs between the third harmonic currents of the different phases. In addition, the harmonic component of 75 Hz also disappears for the same reason [110].

The design of synchronous machines, both as motors and generators, is based on having two circuits in the rotor, one as a rotor field winding and the other as a damper circuit. However, for some applications, such as variable speed operation, damper-less designs are offered. The impact of the damper bar circuit on the fault-related harmonic components under an inter-turn short circuit fault is shown in Figure 11.69. The main introduced fault-related harmonics under a short circuit

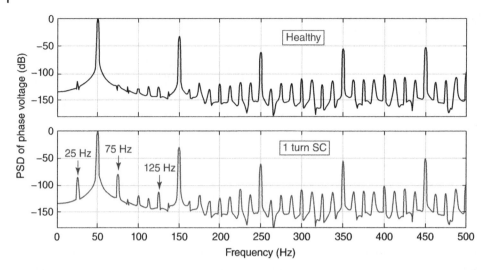

Figure 11.66   The frequency spectrum of the stator phase voltage of a synchronous generator with eight poles operating at no-load in a healthy case (top) and under one inter-turn short circuit fault in the rotor field winding.

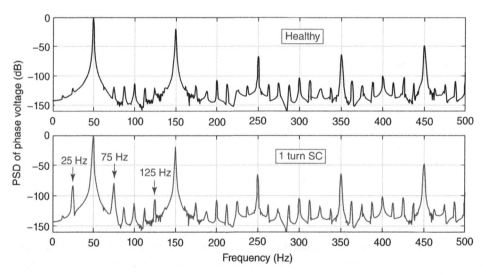

Figure 11.67   The frequency spectrum of the stator phase voltage of a synchronous generator with eight poles operating at full load in a healthy case (top) and under one inter-turn short circuit fault in the rotor field winding.

fault in the stator phase voltage (25 and 75 Hz) are unchanged in a damper-less design of the rotor circuit. However, two harmonic components of 125 and 1625 Hz are changed, as the rate of increase in the 125 Hz harmonic components with increasing fault severity is increased in the absence of a damper bar circuit in the rotor. By contrast, the 1625 Hz harmonic component is unchanged, indicating that the damper does not have any impact on its performance.

The stator winding connection, both in series and parallel, depends on the voltage level required by the generator design. The study shown for the short circuit fault in the rotor winding based on a stator phase voltage analysis is for a synchronous machine with windings connected in series. The

Table 11.16 Amplitude of the fault related harmonics at the no-load operation (dB).

Hz	25	75	125	225	1625	1725
Healthy	−112.3	−127.6	−123.1	−117.3	−103.8	−103.9
1 turn SC	−85.25	−80.1	−112.9	−107.3	−94.2	−107.8
2 turn SC	−79.1	−74	−107.9	−102.1	−88.9	−107.8
5 turn SC	−71	−66	−100.4	−94.7	−81.4	−96.8
10 turn SC	−64.9	−59.8	−94.6	−88.7	−75.1	−90.6
20 turn SC	−58.6	−53.6	−88.6	−82.7	−68.8	−83.6

Table 11.17 The amplitude of the fault related harmonics at the full load operation (dB).

Hz	25	75	125	225	1625	1725
Healthy	−115.9	−121.8	−121.1	−122.0	−107.0	−99.6
1 turn SC	−82.6	−76.6	−102.7	−107.5	−101.4	−95.3
2 turn SC	−76.3	−70.5	−97.3	−103.0	−97.6	−80.9
5 turn SC	−68.2	−62.5	−87.8	−93.6	−91.6	−80.6
10 turn SC	−62.0	−56.3	−83.3	−86.9	−86.1	−81.4
20 turn SC	−55.6	−50.0	−74.2	−79.6	−80.1	−76.0

impact of a rotor short circuit fault for both winding connection types is shown in Figure 11.70, where 20 turns of the rotor field windings are short circuited. Although the two main 25 and 75 Hz frequency components for the parallel winding connection are unchanged and are similar to a series connection, new harmonic components are generated at 12.5, 375.5, 62.5, and 87.5 Hz. The amplitude of these harmonics is changed when at least five inter-turn short circuit faults occur in one of the rotor field windings, as shown in Figure 11.71 [110].

11.9.3.3 Rotor Coil Impedance Index

Two current flows exist inside the winding of the faulty rotor pole and the direction of the current in the short-circuited turns is exactly opposite the main current. This current creates a magnetic field opposite the rotor main magnetic field, and consequently reduces the impedance of the rotor winding. Therefore, detection of the inter-turn short circuit fault in the rotor field winding is possible by comparing the impedance variation. The amplitude of the impedance depends on the machine speed, indicating that the impedance comparison must be either at a standstill or at synchronous speed. The impedance is decreased by increasing the generator speed. Normally, the amplitude of the impedance and the reactance of the synchronous generator running at nominal speed is decreased by 10 and 20%, respectively, compared with a standstill case [20, 40].

The method applied to the synchronous machine in a standstill condition and the amplitude of the impedance of the machine in the healthy and faulty conditions is measured and is equal to 5.55 and 4.03 Ω, respectively. The reduction in amplitude of the impedance indicates the occurrence of a fault. The amplitude of the resistance and the reactance of the synchronous machine is decreased from 3.19 to 2.92 Ω and from 4.55 to 2.78 Ω, respectively. The reduction in the machine reactance

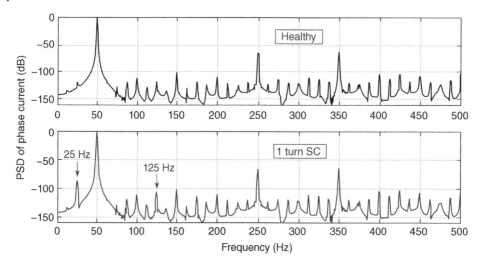

Figure 11.68 The frequency spectrum of the stator phase voltage of a synchronous generator with eight poles operating at full load in a healthy case (top) and under one inter-turn short circuit fault in the rotor field winding (the star point of the stator winding is not grounded).

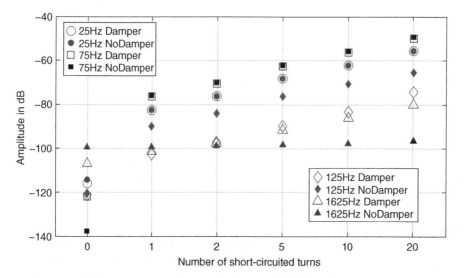

Figure 11.69 The amplitude of inter-turn short circuit fault-related harmonics at the full-load operation of a synchronous machine in two cases: with and without damper bars.

under faulty conditions is reduced by 39%, while the amplitude of the resistance is only reduced by 8.5%, indicating sensitivity of the reactance to the short circuit fault [40].

The sensitivity of impedance variation to a short circuit fault is high at standstill, whereas the sensitivity due to acceleration is reduced. Although the impedance variation provides information regarding the short circuit fault, some consideration must be given to the following when this method is applied:

1. The number of shorted turns must be high for determination of the impedance variation, indicating that the method is not applicable for early-stage fault detection.

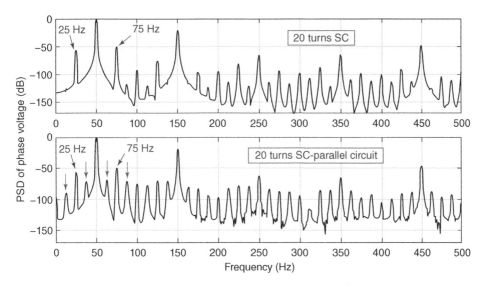

Figure 11.70 The frequency spectrum of the phase voltage at full-load operation of a synchronous machine in the case of 20 short-circuited turns in two cases: with and without parallel circuits.

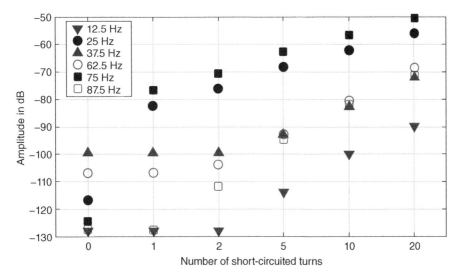

Figure 11.71 The amplitude of fault-related harmonics at full load in the presence of parallel circuits in the stator windings [110]/with permission of IEEE.

2. Different types of fault may also result in impedance variations, such as eccentricity faults.
3. The method cannot estimate the number of shorted turns.
4. Determining the faulty pole is impossible.
5. The impedance of the machine during commissioning must be available for comparison.

11.9.3.4 Electromagnetic Power Index

The occurrence of a short circuit fault in the rotor of a synchronous generator causes a magneto-motive force reduction that leads to a reduction in the electromagnetic power of the machine. This method, based on measured signals such as voltage and current of the generator, calculates the

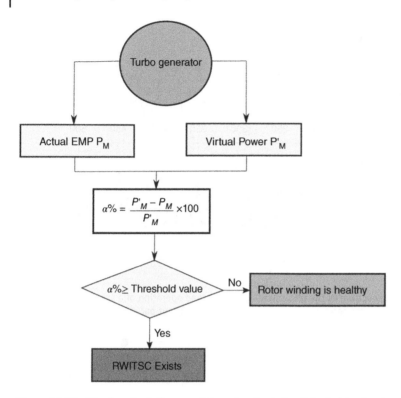

Figure 11.72 The flowchart diagram of the rotor short circuit fault detection based on electromagnetic power.

electromagnetic power of the generator, also called the virtual power. The electromagnetic power of the generator is also measured and the difference between the measured and the calculated electromagnetic power is used as a fault indicator [15, 16]. Figure 11.72 shows the flowchart of the rotor short circuit fault detection based on electromagnetic power. A threshold is defined per unit, based on the virtual electromagnetic power and measured electromagnetic power, as follows:

$$\alpha\% = \frac{P'_M - P_M}{P'_M} \times 100 \tag{11.37}$$

Rotor short circuit fault detection based on electromagnetic power was performed in three power plants with a power rating of 300, 330, and 600 MW [15]. The results of the 600 MW generator test are shown in Table 11.18. According to the measured data before and after the fault, the value of the index in 2010 was equal to 0.29% and increased to 13.23% before maintenance, indicating the occurrence of a severe short circuit fault in the rotor field winding. The results obtained after the maintenance show that the value of the index was reduced by 0.37%, indicating the elimination of the fault. The maintenance report showed the occurrence of the short circuit fault between the eighth slot and the rotor retaining ring. The method based on electromagnetic power is able to diagnose the fault occurrence; however, it has the following limitations:

1. The measurement and the calculated power must be in the same loading condition.
2. The occurrence of the mechanical fault also results in power reduction of the machine, indicating that the method cannot determine the fault type.
3. Fault detection in the early stage is not possible since the low fault severity does not change the power rating of the machine.

Table 11.18 The measured and virtual electromagnetic power of a synchronous generator and the corresponding index.

Time	N	Reactive power (MVar)	Active power (MW)	Exciting current (A)	End voltage (V)	P'_M(W)	P_M(W)	α(%)
Before tripping	1	−19.19	356.9	2410	21.88	358 387 992	357 342 422	0.29
	2	−23.56	357.27	2400	21.86	359 259 914	357 714 794	0.43
	3	−17.74	357.70	2400	21.93	356 567 654	358 142 193	−0.44
	4	−19.63	356.69	2400	21.92	357 033 969	357 130 351	−003
Before overhaul	5	−24.56	361.54	2770	21.95	417 163 621	361 991 885	13.23
	6	−24.70	364.88	2779	21.87	418 958 771	365 343 631	12.80
	7	−20.35	365.67	2820	21.05	424 414 060	366 131 569	13.73
	8	−23.98	364.95	2779	21.89	418 594 970	365 412 840	12.70
After overhaul	9	−16.44	356.54	2420	21.93	358 311 765	356 979 184	0.37

11.9.3.5 Generator Capability Curve

One of the methods for diagnosing short circuit faults in the rotor winding is to check the power capability curve of the synchronous generator, as shown in Figure 11.73. The active and reactive power of the generator when it is connected to the power grid must remain constant if the generator terminal voltage and synchronous reactance of the machine are constant. The only available variable that the power of the machine follows is the terminal voltage of the machine. The magnetomotive force is the same for a stator and rotor in the healthy operation of the machine. If a short circuit occurs in the rotor winding, the amount of current passing through the rotor field winding must be increased to compensate for the number of shorted turns that do not contribute to the active magnetomotive force of the rotor [111].

According to Figure 11.74, if the voltage of the machine terminals is assumed to be constant, the operating range of the machine is limited by the following constraints:

1. The heat generated by the armature winding.
2. The heat generated by the rotor field winding.
3. Limitations of static stability.

The rotor current is increased to maintain a constant terminal voltage of a generator that has a short circuit fault in the rotor field winding. The rotor field winding current increment results ultimately in a temperature rise, and the temperature rise in the rotor limits the capability of the generator. The power delivered to the power grid is reduced by increasing the number of shorted turns in the rotor field winding, as shown in Table 11.19. The main problem of the method is that other types of faults, such as static or dynamic eccentricity, misalignment, and short circuit faults in the stator winding, also result in the same behavior seen in a generator with a rotor short circuit fault [112]. Furthermore, a low-severity fault cannot be detected using this method.

11.9.3.6 Shaft Flux

In the healthy operation of a synchronous machine, the air gap magnetic field is symmetrical and the net flux density in the rotor shaft is zero. When a short circuit fault happens in the rotor winding, the air gap magnetic field in the air gap is no longer uniform; consequently, the net flux density

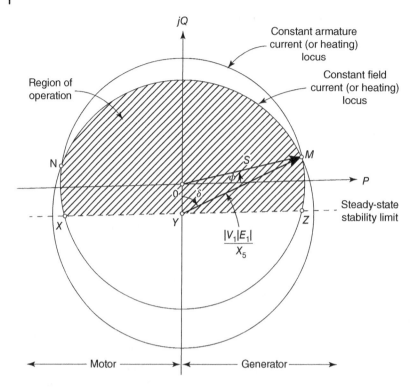

Figure 11.73 Synchronous machine power capability curve.

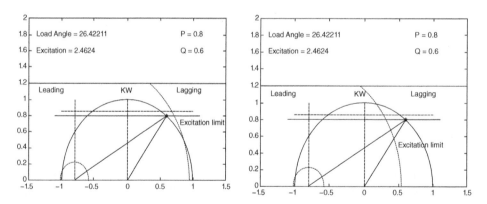

Figure 11.74 Operating conditions of a synchronous machine in a healthy case (left) and under a 30-turn short circuit in the rotor field winding (right).

in the rotor is not zero. The rotor of the synchronous generator has a closed circuit, and when the net flux on the shaft is not zero, a voltage is induced and a current, due to the closed path, is circulated in the rotor shaft. The amplitude of the shaft flux linkage in the healthy operation of the machine is not zero due to the inherent eccentricity caused by the manufacturing tolerance of the synchronous generator. The occurrence of a short circuit fault in the rotor results in an increase in the magnitude of the shaft flux linkage waveform. The study of the frequency spectrum of the shaft flux linkage indicates an increase in the amplitudes of the component of 20 and 160 Hz harmonics [61].

Table 11.19 The value of the active and reactive power of a synchronous generator in a healthy case and under different severities of a short circuit fault in a rotor field winding.

Active power (pu)	Reactive power (pu)	SC turns
0.8	0.6	0
0.8	0.6	5
0.8	0.54	10
0.8	0.48	15
0.8	0.39	20
0.8	0.29	25
0.8	0.20	30
0.7	0.31	35
0.7	0.24	40
0.6	0.32	45
0.6	0.26	50
0.6	0.22	55
0.5	0.30	60

11.9.3.7 Stray Magnetic Field

The stray magnetic field is a non-invasive method that can be used in the fault detection of synchronous generators. Although, in theory, the magnetic field outside the electric machine is assumed to be zero, a weak magnetic field exists in the vicinity of the electric machine due to the leakage flux. In fact, the path by which the flux closes its path is the shaft body, rotor yoke, rotor pole body, air gap, stator tooth, stator yoke, frame, and finally the air at the vicinity of the frame. This indicates that the magnetic field of the machine can be monitored without any need for the installation of intrusive sensors inside the electric machine. However, the amplitude of the stray magnetic field is very weak – on the order of microTeslas [113, 114]. Therefore, careful sensor parameter selection must be considered during the design procedure to ensure detection of most of the stray magnetic field. The sensor location is also a crucial factor that must be considered, based on the type of stray magnetic field to be measured. Figure 11.75 shows a synchronous generator setup with two locations of stray magnetic field sensors. In the case of location A, only the axial stray flux is detected, while mounting a sensor on location B allows the detection of both axial and radial stray flux. In this chapter, location B is used for all measurements, indicating that both axial and radial flux are used, since location B can give the highest quality signal with the lowest amount of noise. A detailed explanation of the sensor design and the impact of location on the signal is explained in Chapter 5.

In all electrical machines, a frame encompasses the stator core; therefore, the sensor must be mounted on the frame. However, the frame can shield the stray magnetic field and further complicate picking up a high-quality stray flux. The designs of the synchronous generator used in hydropower plants are different, since the housing of the synchronous generator is the wall of the generator pit and is usually made of concrete. This specific design provides direct access to the stator backside; consequently, the stray magnetic field sensor can be mounted directly on

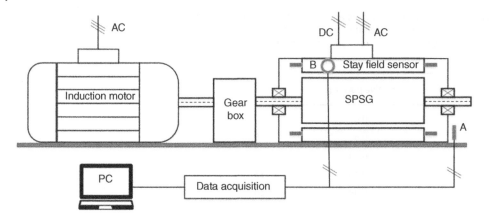

Figure 11.75 A synchronous generator setup with the location of a stray magnetic field sensor. Location A is used when only the axial stray flux is required, while location B is used when both axial and radial stray flux is required.

Figure 11.76 The installed stray magnetic field sensor on the backside of a synchronous machine.

to the stator core [115]. Figure 11.76 shows the stray magnetic field sensor installed on the backside of the synchronous generator. In some cases, the generator has a frame, but a gap still exists between the stator core and the frame, as shown in Figure 11.77.

A series of measurements using stray magnetic field sensors was carried out to determine the shape of the stray magnetic field of the synchronous generator. Figure 11.78 shows the induced sensor voltage on the sensor installed on the backside of a 100 kVA synchronous with 14 salient poles. This voltage depends on the following factors:

1. The topology of the machine.
2. The winding layout.
3. The closeness of the sensor to the vertical pillar that supports the stator core, as shown in Figure 11.77.
4. The ventilation duct.
5. The area that the sensor covers on the stator backside.

Figure 11.77 The installed stray magnetic field sensor on the backside of a synchronous machine located in a hydropower plant.

These factors mean that the shape of the stray magnetic field in synchronous machines is not the same. However, the different shapes of the induced sensor voltage can be measured in a synchronous generator based on the location of the installed sensor. Therefore, a similar pattern should appear for each measurement of the synchronous machine. Figure 11.79 presents induced sensor voltage from a sensor installed on the backside of a 100 kVA synchronous generator with 14 poles operating at no-load under 10 shorted turns in the rotor field winding.

Figure 11.80 shows the impact of 1 turn and 10 turn short circuit faults on the frequency spectrum of the stray magnetic field. The subharmonics and inter-harmonics are intriguing due to the

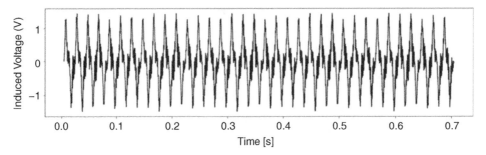

Figure 11.78 The induced sensor voltage on the sensor is installed on the backside of a 100 kVA synchronous.

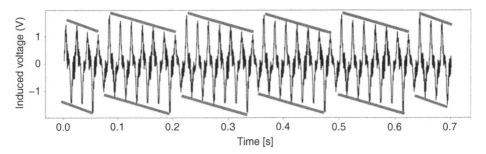

Figure 11.79 The induced sensor voltage from a sensor installed on the backside of a 100 kVA synchronous generator with 14 poles operating at no-load under 10 shorted turns in the rotor field winding.

inter-turn short circuit fault compared with a healthy case. The frequency of the fault-related harmonics can be extracted using Equation (11.36), where the number of poles in this case study is seven. The variation in the fault-related harmonics does not follow the same trend of increment and some of them show high sensitivity, even for a low fault degree. Some frequency components, such as 85.7, 92.9, 107.2, and 114.3 Hz, show a great degree of sensitivity to the short circuit fault. The amplitudes of these harmonic components in a healthy case are −41.5, −34.8, −37.2, and −45.1 dB. The occurrence of a one inter-turn short circuit fault in a one rotor field winding, which accounts for 2.86% of a one rotor winding, increases the amplitude of these harmonics to −39.4, −33.1, −35.1, and −42.8 dB, respectively. The amplitude of the harmonics is increased when a 10-turn short circuit fault happens in the winding, as shown in Figure 11.80. The amplitudes of the side-bands for the harmonic components of 85.7, 92.9, 107.2, and 114.3 Hz with a 10 inter-turn short circuit fault are increased from −41.7, −32.6, −33.9, and −45.5 to −22.4, −19.2, −20.4, and −27.9 dB, respectively. The results indicate that the introduced harmonics have a high sensitivity to short circuit faults. Table 11.20 shows the impact of different fault severities on the side-band components in a no-load operation of the synchronous generator [113, 114]. In reference [115], high-frequency components around the slotting harmonics were introduced for the inter-turn short circuit fault, while the mentioned harmonics were sensitive to mechanical faults as well.

The same trend as the no-load operation of the synchronous generator for the occurrence of the short circuit fault is seen in the full-load operation in the frequency spectrum of the stray magnetic field. The selected frequency components at the frequencies of 85.7, 92.9, 107.2, and 114.3 Hz in a case of one inter-turn short circuit fault increase from −41.7, −32.6, −33.9, and −45.5 to −36.8, −32.1, −32.6, and −43.7 dB, respectively. The occurrence of a 10 inter-turn short circuit fault increases the amplitude of the aforementioned frequency components to −22.4, −19.2, −20.4,

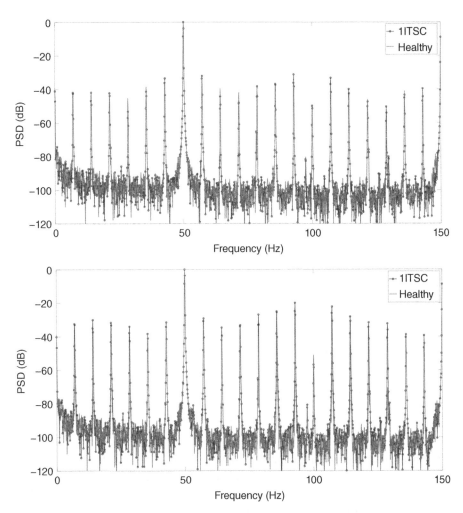

Figure 11.80 The frequency spectrum of the induced sensor voltage in a 14-pole synchronous generator operating at no-load under one inter-turn short circuit fault (top) and 10 inter-turn short circuit fault (bottom).

and −27.9 dB, respectively. These results indicate that the on-load operation of the synchronous machine does not change the selected features for the inter-turn short circuit fault detection. The variation in the selected feature for the various severities of the inter-turn short circuit fault in the on-load condition is shown in Table 11.21 [112, 113].

Fault type detection is one of the features lacking in fault detection based on the application of frequency spectrum analysis, since the fault-related harmonics of different electrical and mechanical faults give rise to the same side-band components. A time-frequency signal processing tool can be used to solve this problem [116]. The short-time Fourier transform (STFT) is one of the time-frequency tools used for the inter-turn short circuit fault [113]. Here, the STFT has been applied to the induced sensor voltage of the installed stray magnetic field sensor on the stator back-side of a 100 kVA synchronous generator, as shown in Figure 11.81. The synchronous generator has 14 poles and one mechanical revolution of the machine takes 140 ms, indicating that the 14 poles need 140 ms to pass once over the sensor. A window in the time-frequency plot is introduced

Table 11.20 The variation of the side-bands harmonics in (Hz) to number of inter-turn short circuit during no-load operation in (dB). Adapted from [114, 115].

F (Hz)	Healthy	1 ITSC	2 ITSC	3 ITSC	7 ITSC	10 ITSC
7.10	−44.9	−43.6	−41.9	−40.3	−34.8	−32.4
14.3	−45.4	−43.9	−41.8	−39.7	−33.1	−30.1
85.7	−41.5	−39.4	−36.8	−34.5	−27.9	−25.0
92.9	−34.8	−33.1	−31.2	−29.9	−22.9	−19.9
107.2	−37.2	−35.1	−33.2	−31.4	−25.0	−22.1
114.3	−45.1	−42.8	−39.9	−37.5	−31.0	−28.1

Table 11.21 The variation of the side-bands harmonics in (Hz) to number of inter-turn short circuit during full load operation in (dB).

F (Hz)	Healthy	1 ITSC	2 ITSC	3 ITSC	7 ITSC	10 ITSC
7.10	−44.3	−43.1	−41.2	−39.5	−32.5	−31.2
14.3	−44.2	−42.8	−40.6	−38.5	−31.5	−28.3
85.7	−41.7	−36.8	−33.4	−31.1	−24.7	−22.4
92.9	−32.6	−32.1	−30.7	−28.9	−23.2	−19.2
107.2	−33.9	−32.6	−30.9	−29.2	−24.4	−20.4
114.3	−45.5	−43.7	−41.5	−39.3	−32.1	−27.9

that shows one mechanical revolution of the machine. The time-frequency plot has three bands: two are depicted in green and one in red. The green bands are expanded in the frequency range of 35–50 Hz and 65–85 Hz, while the red band is expanded between 50 and 65 Hz. During a healthy operation of the machine, no variation will be seen in the time-frequency bands of the induced sensor voltage. However, introducing an inter-turn short circuit fault results in a repetitive pattern in the time-frequency plot. The inter-turn short circuit fault results in an intensity reduction in the frequency band of 50–65 Hz. The time-width of the fault-related pattern is 10 ms, indicating one faulty rotor pole. Figure 11.81 shows the time-frequency of the induced sensor voltage for 10 inter-turn short circuit faults. A severe short circuit fault results in the same pattern seen for a low-degree fault, but in all three frequency bands.

The load impact must also be examined, since the pattern observed in the no-load operation can be changed due to the magnetic field contribution of the stator winding and the armature reaction. Figure 11.82 shows the time-frequency plot of an induced sensor voltage of a synchronous generator operating at full-load in the healthy case and under 1 and 10 inter-turn short circuit faults. A comparison between the time-frequency plots of the no-load and full-load operation of the synchronous machine in Figures 11.81 and 11.82 shows the same pattern, indicating that the on-load condition of the machine does not change the short circuit fault patter.

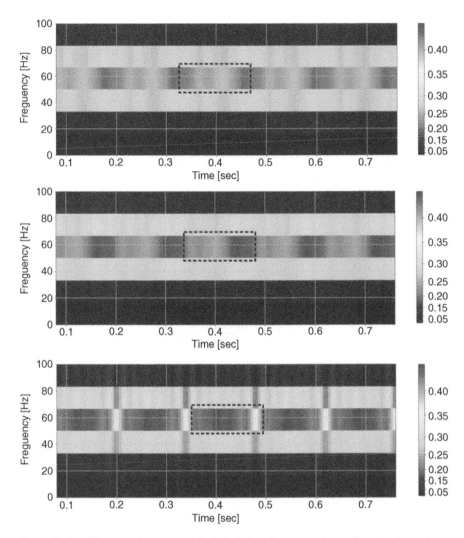

Figure 11.81 The time-frequency plot of the induced sensor voltage of a 14-pole synchronous generator operating at no-load in healthy, one inter-turn short circuit, and 10 inter-turn short circuit fault conditions – Experimental results.

Continuous wavelet transform (CWT) is also in the category of time-frequency signal processing tools that are widely used in fault detection of electrical machines [117, 118]. The shape of the pattern distinguished for each type of electric machine and the fault can be adjusted based on the CWT parameters [119, 120]. CWT is applied to the induced sensor voltage of the 100 kVA synchronous machine, where one stray magnetic field sensor is installed on the stator backside. Figure 11.83 shows the time-frequency plot of the induced sensor voltage of the synchronous machine operating at no-load in a healthy case and under a 10 inter-turn short circuit fault. The dashed window in Figure 11.83 indicates one mechanical revolution of the machine; in this case, this is equal to 140 ms for a 14-pole synchronous machine. Each window has 14 stalks, where each stalk represents

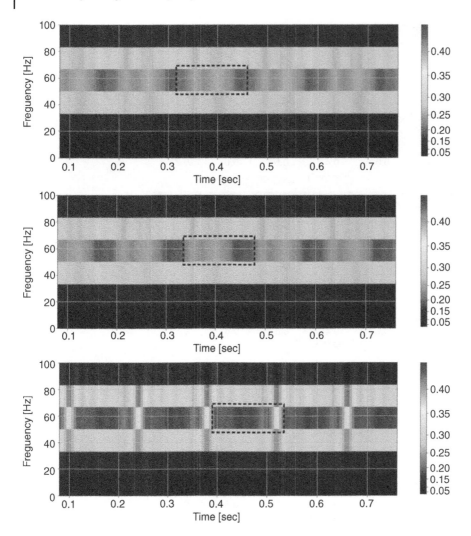

Figure 11.82 The time-frequency plot of the induced sensor voltage of a 14-pole synchronous generator operating at full-load in healthy, one inter-turn short circuit, and 10 inter-turn short circuit fault conditions – Experimental results.

one rotor pole. In a healthy operation of the synchronous machine, the length and intensity of the stalks are the same. However, the lengths and intensities of the stalks are reduced by introducing a 10 inter-turn short circuit fault in the rotor winding. The pattern distinguished due to the short circuit fault is a repetitive pattern that appears every time the faulty pole passes over the installed sensor [121].

The synchronous machine always operates in the on-load condition. The load type is always a resistive-inductive load, whereas a fully resistive loading is also fed by the synchronous generator in some local loads. Figure 11.84 shows the time-frequency plot of the induced sensor voltage of the synchronous machine operating in a healthy case and under a 10 inter-turn short circuit fault with a

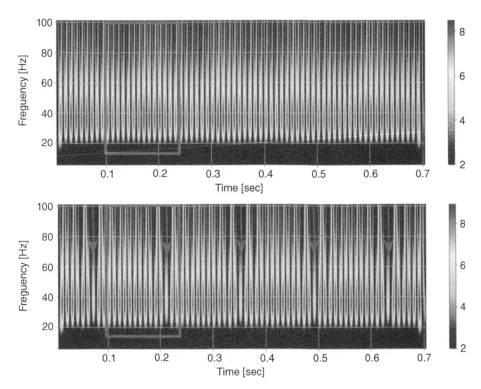

Figure 11.83 Time-frequency plot of the induced sensor voltage in a sensor installed on the backside of a 100 kVA synchronous machine operating in no-load in a healthy case (top) and under a 10 inter-turn short circuit fault (bottom) – Experimental results.

resistive load. The time-frequency pattern of the resistive load is similar to the no-load operation of the synchronous machine, indicating that the resistive load does not change the introduced pattern for short circuit fault detection. The resistive-inductive load impact on the time-frequency plot of the induced sensor voltage is also examined. The results indicate that the resistive-inductive load also does not change the fault pattern, although a magnetic field contribution occurs from the stator winding and the armature reaction during the on-load operation of the synchronous machine.

11.10 Summary

The majority of the approaches used in the inter-turn short circuit fault monitoring of synchronous generators use electromagnetic methods (the measurement of current, voltage, and magnetic flux). Some of the indicators examined in this chapter are the result of research conducted on small synchronous generators. Given that the structure of large synchronous generators operating in an industry and in power plants are similar to a small-scaled synchronous generator, the methods can be applied to a large synchronous generator.

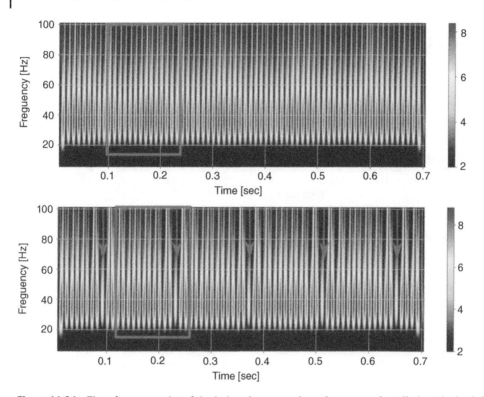

Figure 11.84 Time-frequency plot of the induced sensor voltage in a sensor installed on the backside of a 100 kVA synchronous machine operating in a resistive load in a healthy case (top) and under a 10 inter-turn short circuit fault (bottom) – Experimental results.

References

1 Chen, Y., Chen, X., and Shen, Y. (2018). On-line detection of coil inter-turn short circuit faults in dual-redundancy permanent magnet synchronous motors. *Energies* 11 (3): 662.

2 Yucai, W., Minghan, M., and Yonggang, L. (2017). A new detection coil capable of performing online diagnosis of excitation winding short-circuits in steam-turbine generators. *IEEE Transactions on Energy Conversion* 33 (1): 106–115.

3 Maraaba, L., Al-Hamouz, Z., and Abido, M. (2018). An efficient stator inter-turn fault diagnosis tool for induction motors. *Energies* 11 (3): 653.

4 Wang, X.-H., Sun, Y.G., Gui, L. et al. (2007). Reasonable simplification of the multi-loop model of the large hydro generators. In: *Zhongguo Dianji Gongcheng Xuebao, Proceedings of the Chinese Society of Electrical Engineering*.

5 Zhao, H., Du, Z., Liu, X. et al. (2014). An on-line identification method for rotor resistance of squirrel cage induction motors based on recursive least square method and model reference adaptive system. *Proceedings of the CSEE* 34 (30): 5386–5394.

6 Urresty, J.-C., Riba, J.-R., and Romeral, L. (2012). Application of the zero-sequence voltage component to detect stator winding inter-turn faults in PMSMs. *Electric Power Systems Research* 89: 38–44.

7 Saavedra, H., Urrestyc, J., Riba, J. et al. (2014). Detection of interturn faults in PMSMs with different winding configurations. *Energy Conversion and Management* 79: 534–542.

8 Mercorelli, P. (2014). Parameters identification in a permanent magnet three-phase synchronous motor of a city-bus for an intelligent drive assistant. *International Journal of Modelling, Identification and Control 5* 21 (4): 352–361.

9 Chen, L., Mercorelli, P., and Liu, S. (2005). A Kalman estimator for detecting repetitive disturbances. In: *Proceedings of the 2005 American Control Conference,* 8–10 June 2005. Portland, OR, USA.

10 Wang, Z. et al. (2018). Hourly solar radiation forecasting using a volterra-least squares support vector machine model combined with signal decomposition. *Energies* 11 (1): 68.

11 Stepniak, G., Kowalczyk, M., and Siuzdak, J. (2018). Volterra kernel estimation of white light LEDs in the time domain. *Sensors* 18 (4): 1024.

12 Xu, S., Li, Y., Huang, T. et al. (2017). A sparse multiwavelet-based generalized laguerre–volterra model for identifying time-varying neural dynamics from spiking activities. *Entropy* 19 (8): 425.

13 Rahimian, M.M. (2011). Broken bar detection in synchronous machines based wind energy conversion system. PhD Thesis, Texas A&M University, USA.

14 Zhang, G., Wu, J., and Hao, L. (2018). Analysis on the amplitude and frequency characteristics of the rotor unbalanced magnetic pull of a multi-pole synchronous generator with inter-turn short circuit of field windings. *Energies* 11 (1): 60.

15 Yucai, W. and Yonggang, L. (2014). Diagnosis of rotor winding interturn short-circuit in turbine generators using virtual power. *IEEE Transactions on Energy Conversion* 30 (1): 183–188.

16 Hao, L., Wu, J., and Zhou, Y. (2014). Theoretical analysis and calculation model of the electromagnetic torque of nonsalient-pole synchronous machines with interturn short circuit in field windings. *IEEE Transactions on Energy Conversion* 30 (1): 110–121.

17 Zhang, G., Wu, J., and Hao, L. (2017). Fast calculation model and theoretical analysis of rotor unbalanced magnetic pull for inter-turn short circuit of field windings of non-salient pole generators. *Energies* 10 (5): 732.

18 Li, Y. et al. (2017). Field winding short circuit fault diagnosis on turbine generators based on the ac impedance test. *Large Electric Machine and Hydraulic Turbine* 4: 10–14.

19 Cheng, C. (2015). *Volterra Series Based Nonlinear System Identification and its Application.* Shanghai, China: Shanghai Jiao Tong University.

20 Sadeghi, I., Ehya, H., Faiz, J.I. et al. (2018). Online condition monitoring of large synchronous generator under short circuit fault – A review. In: *IEEE International Conference on Industrial Technology (ICIT),* 20–22 February 2018. Lyon, France.

21 Stone, G. (2013). Condition monitoring and diagnostics of motor and stator windings – A review. *IEEE Transactions on Dielectrics and Electrical Insulation* 20 (6): 2073–2080.

22 Neti, P. and Nandi, S. (2009). Stator interturn fault detection of synchronous machines using field current and rotor search-coil voltage signature analysis. *IEEE Transactions on Industry Applications* 45 (3): 911–920.

23 Gandhi, A., Corrigan, T., and Parsa, L. (2010). Recent advances in modeling and online detection of stator interturn faults in electrical motors. *IEEE Transactions on Industrial Electronics* 58 (5): 1564–1575.

24 Siddique, A., Yadava, G., and Singh, B. (2005). A review of stator fault monitoring techniques of induction motors. *IEEE Transactions on Energy Conversion* 20 (1): 106–114.

25 Blánquez, F., Platero, C.A., Rebollo, E. et al. (2015). On-line stator ground-fault location method for synchronous generators based on 100% stator low-frequency injection protection. *Electric Power Systems Research* 125: 34–44.

26 Wang, C., Liu, X., and Chen, Z. (2014). Incipient stator insulation fault detection of permanent magnet synchronous wind generators based on Hilbert–Huang transformation. *IEEE Transactions on Magnetics* 50 (11): 1–4.

27 Nandi, S. (2005). Detection of stator faults in induction machines using residual saturation harmonics. In: *IEEE International Conference on Electric Machines and Drives,* 27–29 September 2005. Nanjing, China.

28 Mirafzal, B., Povinelli, R.J., and Demerdash, N.A. (2006). Interturn fault diagnosis in induction motors using the pendulous oscillation phenomenon. *IEEE Transactions on Energy Conversion* 21 (4): 871–882.

29 Rehaoulia, H., Henao, H., and Capolino, G. (2007). Modeling of synchronous machines with magnetic saturation. *Electric Power Systems Research* 77 (5–6): 652–659.

30 Romeral, L., Urresty, J., Ruiz, J.R. et al. (2010). Modeling of surface-mounted permanent magnet synchronous motors with stator winding interturn faults. *IEEE Transactions on Industrial Electronics* 58 (5): 1576–1585.

31 Ma, H., Li, H., Wei, S. et al. (2008). Transient analysis of synchronous generator with stator winding faults based on starting process. In: *Third International Conference on Electric Utility Deregulation and Restructuring and Power Technologies,* 6–9 April 2008. Nanjing, China.

32 Bi, D., Wang, X., Wang, W. et al. (2005). Improved transient simulation of salient-pole synchronous generators with internal and ground faults in the stator winding. *IEEE Transactions on Energy Conversion* 20 (1): 128–134.

33 Tallam, R.M., Lee, S.B., Stone, G.C. et al. (2007). A survey of methods for detection of stator-related faults in induction machines. *IEEE Transactions on Industry Applications* 43 (4): 920–933.

34 Williamson, S. and Mirzoian, K. (1985). Analysis of cage induction motors with stator winding faults. *IEEE Transactions on Power Apparatus and Systems* 7: 1838–1842.

35 Grubic, S., Aller, J.M., Lu, B. et al. (2008). A survey on testing and monitoring methods for stator insulation systems of low-voltage induction machines focusing on turn insulation problems. *IEEE Transactions on Industrial Electronics* 55 (12): 4127–4136.

36 Tavner, P., Gaydon, B., and Ward, D. (1986). Monitoring generators and large motors. *IEE Proceedings B (Electric Power Applications)* 133 (3): 169–180.

37 Bowers, S.V., Piety, K.P., and Davis, W.A.K. (1993). Proactive motor monitoring through temperature shaft current and magnetic flux measurements. *CSI Technoloy Inc, Users Manual.*

38 Sdid, M. and Benbouzid, M.E.H. (2000). HG diagram based rotor parameters identification for induction motors thermal monitoring. *IEEE Transactions on Energy Conversion* 15 (1): 14–18.

39 Mehala, N. and Dahiya, R. (2007). Motor current signature analysis and its applications in induction motor fault diagnosis. *International Journal of Systems Applications, Engineering & Development* 2 (1): 29–35.

40 Tavner, P., Penman, J., and Sedding, H. (2008). *Condition Monitoring of Rotating Electrical Machines,* 2e. The Institution of Engineering and Technology.

41 Bonnett, A. (1976). The cause of winding failures in Three Phase Squirrel Cage Induction Motors. *IEEE PCIC Conference.* Philadelphia, USA

42 Bonnett, A.H. and Soukup, G.C. (1992). Cause and analysis of stator and rotor failures in three-phase squirrel-cage induction motors. *IEEE Transactions on Industry Applications* 28 (4): 921–937.

43 Lee, Y.-S., Nelson, J.K., Scarton, H.A. et al. (1994). An acoustic diagnostic technique for use with electric machine insulation. *IEEE Transactions on Dielectrics and Electrical Insulation* 1 (6): 1186–1193.

44 Singal, R., Williams, K., and Verma, S.P. (1987). Vibration behaviour of stators of electrical machines, Part II: Experimental study. *Journal of Sound and Vibration* 115 (1): 13–23.

45 Verma, S., Singal, R., and Williams, K. (1987). Vibration behaviour of stators of electrical machines, Part I: Theoretical study. *Journal of Sound and Vibration* 115 (1): 1–12.

46 Leonard, R. and Thomson, W. (1986). Vibration and stray flux monitoring for unbalanced supply and inter-turn winding fault diagnosis in induction motors. *British Journal of Non-Destructive Testing* 28 (4): 211–215.

47 Trutt, F.C., Sottile, J., and Kohler, J.L. (2001). Detection of AC machine winding deterioration using electrically excited vibrations. *IEEE Transactions on Industry Applications* 37 (1): 10–14.

48 Nassar, O. (1987). The use of partial discharge and impulse voltage testing in the evaluation of interturn insulation failure of large motors. *IEEE Transactions on Energy Conversion* 4: 615–621.

49 Tanaka, T. (1995). Partial discharge pulse distribution pattern analysis. *IEE Proceedings: Science, Measurement & Technology* 142 (1): 46–50.

50 Kemp, I. (1995). Partial discharge plant-monitoring technology: present and future developments. *IEE Proceedings: Science, Measurement & Technology* 142 (1): 4–10.

51 Stone, G.C. and Sedding, H.G. (1993). In-service evaluation of motor and generator stator windings using partial discharge tests. In: *Conference Record of Twenty-Eighth IEEE Industry Applications Conference*, 02-08 October 1993. Toronto, ON, Canada.

52 Stone, G.C., Sedding, H.G., and Costello, M.J. (1996). Application of partial discharge testing to motor and generator stator winding maintenance. *IEEE Transactions on Industry Applications* 32 (2): 459–464.

53 Tetrault, S.M., Stone, G.C., and Sedding, H.G. (1999). Monitoring partial discharges on 4-kV motor windings. *IEEE Transactions on Industry Applications* 35 (3): 682–688.

54 Penman, J. and Jiang, H. (1996). The detection of stator and rotor winding short circuits in synchronous generators by analysing excitation current harmonics. In: *International Conference on Opportunities and Advances in International Electric Power Generation*, 18–20 March 1996. Durham, UK.

55 Penman, J., Hadwick, J., and Stronach A. (1995). Protection strategy against the occurrence of faults in electrical machines. Supplemental Notes, University of Aberdeen, UK, A monograph.

56 Sadeghi, I., Ehya, H., Zarandi, R.N. et al. Condition monitoring of large electrical machine under partial discharge fault – A review. In: *2018 International Symposium on Power Electronics, Electrical Drives, Automation and Motion (SPEEDAM)*, 20–22 June 2018. Amalfi, Italy.

57 Thorsen, O. and Dalva, M. (1997). Condition monitoring methods, failure identification and analysis for high voltage motors in petrochemical industry. In: *Eighth International Conference on Electrical Machines and Drives*, 1-3 September 1997. Cambridge, UK.

58 Yaghobi, H., Ansari, K., and Rajabi-Mashadi, M. (2011). Analysis of magnetic flux linkage distribution in salient pole synchronous generator with different kinds of inter turn winding faults. *Iranian Journal of Electrical and Electronic Engineering, Iran University of Science and Technology* 7 (4): 260–272.

59 Kliman, G., Premerlani, W.J., Koegl, R.A. et al. (1996). A new approach to on-line turn fault detection in AC motors. In: *Thirty-First IAS Annual Meeting*, 6–10 October 1996. San Diego, CA, USA.

60 Bone, J. and Schwarz, K. (1973). Large AC motors. *Proceedings of the Institution of Electrical Engineers* 120 (10R): 1111–1132.

61 Hsu, J.S. and Stein (1994). J., Shaft signals of salient-pole synchronous machines for eccentricity and shorted-field-coil detections. *IEEE Transactions on Energy Conversion* 9 (3): 572–578.

62 Sottile, J., Trutt, F.C., and Leedy, A.W. (2006). Condition monitoring of brushless three-phase synchronous generators with stator winding or rotor circuit deterioration. *IEEE Transactions on Industry Applications* 42 (5): 1209–1215.

63 Shuting, W., Heming, L., Yonggang, L. et al. (2003). The diagnosis method of generator rotor winding inter-turn short circuit fault based on excitation current harmonics. In: *The Fifth International Conference on Power Electronics and Drive Systems, PEDS*, 17–20 November 2003. Singapore.

64 Mirabbasi, D., Seifossadat, G., and Heidari, M. (2009). Effect of unbalanced voltage on operation of induction motors and its detection. In: *2009 International Conference on Electrical and Electronics Engineering-ELECO*, 5-08 November 2009. Bursa, Turkey.

65 Quispe, E., Gonzalez, G., and Aguado, J. (2004). Influence of unbalanced and waveform voltage on the performance characteristics of three-phase induction motors. In: *International Conference on Renewable Energies and Power Quality*, 31 March–2 April 2004. Barcelona.

66 Hiendro, A. (2010). A quantities method of induction motor under unbalanced voltage conditions. *TELKOMNIKA Indonesian Journal of Electrical Engineering* 8 (2): 73–80.

67 Pillay, P. and Manyage (2001). M., Definitions of voltage unbalance. *IEEE Power Engineering Review* 21 (5): 50–51.

68 Jalilian, A. and Roshanfekr, R. (2009). Analysis of three-phase induction motor performance under different voltage unbalance conditions using simulation and experimental results. *Electric Power Components & Systems* 37 (3): 300–319.

69 Von Jouanne, A. and Banerjee, B. (2001). Assessment of voltage unbalance. *IEEE Transactions on Power Delivery* 16 (4): 782–790.

70 Cruz, S.M. and Cardoso, A.M. (2001). Stator winding fault diagnosis in three-phase synchronous and asynchronous motors, by the extended Park's vector approach. *IEEE Transactions on Industry Applications* 37 (5): 1227–1233.

71 Nandi, S., Toliyat, H.A., and Li, X. (2005). Condition monitoring and fault diagnosis of electrical motors – A review. *IEEE Transactions on Energy Conversion* 20 (4): 719–729.

72 Cash, M.A., Habetler, T.G., and Kliman, G.B. (1998). Insulation failure prediction in AC machines using line-neutral voltages. *IEEE Transactions on Industry Applications* 34 (6): 1234–1239.

73 Xiangheng, W., Weijian, W., and Shanming, W. (2000). Research on internal faults of generators and their protection schemes in Three Gorges Hydro Power Station. In: *IEEE Power Engineering Society Winter Meeting Conference*, 23–27 January 2000. Singapore.

74 Dallas, S.E., Safacas, A.N., and Kappatou, J.C. (2011). Interturn stator faults analysis of a 200-MVA hydrogenerator during transient operation using FEM. *IEEE Transactions on Energy Conversion* 26 (4): 1151–1160.

75 Dallas, S., Safacas, A., and Kappatou, J. (2012). A study on the transient operation of a salient pole synchronous generator during an inter-turn stator fault with different number of short-circuited turns using FEM. In: *International Symposium on Power Electronics Power Electronics, Electrical Drives, Automation and Motion*, 20–22 June 2012. Sorrento.

76 Vazquez, J. and Salmeron, P. (2003). Active power filter control using neural network technologies. *IEE Proceedings-Electric Power Applications* 150 (2): 139–145.

77 Hemmati, S., Shokri, S., and Saied, S. (2011). Modeling and simulation of internal short circuit faults in large hydro generators with wave windings. In: *International Conference on Power Engineering, Energy and Electrical Drives*, 11–13 May 2011. Malaga, Spain.

78 Rahnama, M. and Nazarzadeh, J. (2007). Synchronous machine modeling and analysis for internal faults detection. In: *IEEE International Electric Machines & Drives Conference*, 3–5 May 2007. Antalya, Turkey.

79 Kim, K.-H., Gu, B.-G., and Jung, I.-S. (2011). Online fault-detecting scheme of an inverter-fed permanent magnet synchronous motor under stator winding shorted turn and inverter switch open. *IET Electric Power Applications* 5 (6): 529–539.

80 Ebrahimi, B.M. and Faiz, J. (2010). Feature extraction for short-circuit fault detection in permanent-magnet synchronous motors using stator-current monitoring. *IEEE Transactions on Power Electronics* 25 (10): 2673–2682.

81 Kohler, J.L., Sottile, J., and Trutt, F.C. (2002). Condition monitoring of stator windings in induction motors. I. Experimental investigation of the effective negative-sequence impedance detector. *IEEE Transactions on Industry Applications* 38 (5): 1447–1453.

82 Sottile, J., Trutt, F.C., and Kohler, J.L. (2002). Condition monitoring of stator windings in induction motors. II. Experimental investigation of voltage mismatch detectors. *IEEE Transactions on Industry Applications* 38 (5): 1454–1459.

83 Khan, M., Ozgonenel, O., and Rahman, M.A. (2007). Diagnosis and protection of stator faults in synchronous generators using wavelet transform. In: *IEEE International Electric Machines & Drives Conference*, 3–5 May 2007. Antalya, Turkey.

84 Pennacchi, P. (2008). Computational model for calculating the dynamical behaviour of generators caused by unbalanced magnetic pull and experimental validation. *Journal of Sound and Vibration* 312 (1–2): 332–353.

85 Lipo, T.A. and Chang, K.C. (1986, 22). A new approach to flux and torque-sensing in induction machines. *IEEE Transactions on Industry Applications* 4: 731–737.

86 Byars, M. (1982). Detection of alternator rotor winding faults using an on-line magnetic field search coil monitoring unit, in *Proceedings of the 17th Universities Power Engineering Conference*.

87 Conolly, H. et al. (1985). Detection of interturn faults in generator rotor windings using airgap search coils. *Proceedings of the 2nd International Conference Electrical Machine Design and Applications*.

88 Neti, P. (2007). Stator fault analysis of synchronous machines. PhD Thesis, Department of Electrical and Computer Engineering, University of Victoria, Canada.

89 Penman, J., Sedding, H.G., Lloyd, B.A. et al. (1994). Detection and location of interturn short circuits in the stator windings of operating motors. *IEEE Transactions on Energy Conversion* 9 (4): 652–658.

90 Auckland, D., Pickup, I.E.D., Shuttleworth, R. et al. (1995). Novel approach to alternator field winding interturn fault detection. *IEE Proceedings-Generation, Transmission and Distribution* 142 (2): 97–102.

91 Streifel, R.J., Marks, R.J., El-Sharkawi, M.A. et al. (1996). Detection of shorted-turns in the field winding of turbine-generator rotors using novelty detectors-development and field test. *IEEE Transactions on Energy Conversion* 11 (2): 312–317.

92 Nadarajan, S., Bhangu, B., Panda, S.K. et al. (2015). Feasibility analysis of auxiliary winding for condition monitoring of wound field brushless synchronous generators. In: *41st Annual Conference of the IEEE Industrial Electronics Society (IECON)*, 9–12 November 2015. Yokohama, Japan.

93 Nadarajan, S., Panda, S.K., Bhangu, B. et al. (2014). Hybrid model for wound-rotor synchronous generator to detect and diagnose turn-to-turn short-circuit fault in stator windings. *IEEE Transactions on Industrial Electronics* 62 (3): 1888–1900.

94 Touma-Holmberg, M. and Hjarne, S. (2003). Suppression of slot discharges in a cable wound generator. *IEEE Transactions on Energy Conversion* 18 (3): 458–465.

95 Lin, X., Tian, Q., Gao, Y. et al. (2007). Studies on the internal fault simulations of a high-voltage cable-wound generator. *IEEE Transactions on Energy Conversion* 22 (2): 240–249.

96 Rahman, M. and Jeyasurya, B. (1988). A state-of-the-art review of transformer protection algorithms. *IEEE Transactions on Power Delivery* 3 (2): 534–544.

97 Mao, P.L. and Aggarwal, R.K. (2001). A novel approach to the classification of the transient phenomena in power transformers using combined wavelet transform and neural network. *IEEE Transactions on Power Delivery* 16 (4): 654–660.

98 Kim, C. and Russell, B.D. (1989). Classification of faults and switching events by inductive reasoning and expert system methodology. *IEEE Transactions on Power Delivery* 4 (3): 1631–1637.

99 Phadke, A. and Thorp, J. (1983). A new computer-based flux-restrained current-differential relay for power transformer protection. *IEEE Transactions on Power Apparatus and Systems* 11: 3624–3629.

100 Ma, H. and Pu, L. (2009). Fault diagnosis based on ANN for turn-to-turn short circuit of synchronous generator rotor windings. *Journal of Electromagnetic Analysis and Applications* 1 (3): 187–191.

101 Elez, A., Tomcic, B., and Petrinic, M. (2010). Detection of inter-coil short circuits in coils of salient pole synchronous generator field winding on the basis of analysis of magnetic field in the machine. In: *International Conference on Renewable Energies and Power Quality (ICREPQ)*, 23–25 March 2010. Granada, Spain.

102 Stone, G.C. et al. (2012). Using magnetic flux monitoring to detect synchronous machine rotor winding shorts. In: *Conference Record of 2012 Annual IEEE Pulp and Paper Industry Technical Conference (PPIC)*. IEEE.

103 Albright, D. (1971). Interturn short-circuit detector for turbine-generator rotor windings. *IEEE Transactions on Power Apparatus and Systems* 2: 478–483.

104 Groth, I.L. (2019). On-line magnetic flux monitoring and incipient fault detection in hydropower generators. MSc Thesis, NTNU, Norway.

105 Ehya, H., Nysveen, A., Groth, I.L. et al. (2020). Detailed magnetic field monitoring of short circuit defects of excitation winding in hydro-generator. In: *International Conference on Electrical Machines (ICEM)*, 23–26 August 2020. Gothenburg, Sweden.

106 Pyrhonen, J., Jokinen, T., and Hrabovcova, V. (2013). *Design of Rotating Electrical Machines*. Wiley.

107 Shuting, W., Yonggang, L., Heming, L. et al. (2006). The analysis of generator excitation current harmonics on stator and rotor winding fault. In: *IEEE International Symposium on Industrial Electronics*, 9–13 July 2006. Montreal, QC, Canada.

108 Zhou, Z., Wang, Y., Guo, W. et al. (2011). Fault identification of turbine generator rotor system based on spectrum monitor and analysis. In: *Asia-Pacific Power and Energy Engineering Conference*, 25–28 March 2011. Wuhan, China.

109 Hao, L., Sun, Y., Qiu, A. et al. (2011). Steady-state calculation and online monitoring of inter-turn short circuit of field windings in synchronous machines. *IEEE Transactions on Energy Conversion* 27 (1): 128–138.

110 Valavi, M., Jørstad, K.G., and Nysveen, A. (2018). Electromagnetic analysis and electrical signature-based detection of rotor inter-turn faults in salient-pole synchronous machine. *IEEE Transactions on Magnetics* 54 (9): 1–9.

111 D'Angelo, M.F. and Costa, P.P. (2001). Detection of shorted turns in the field winding of turbogenerators using the neural network mlp. In: *IEEE International Conference on Systems,*

Man and Cybernetics. e-Systems and e-Man for Cybernetics in Cyberspace, 7–10 October 2001. Tucson, AZ, USA.

112 Wang, L., Cheung, R., Ma, Z. et al. (2008). Finite-element analysis of unbalanced magnetic pull in a large hydro-generator under practical operations. *IEEE Transactions on Magnetics* 44 (6): 1558–1561.

113 Ehya, H. and Nysveen, A. (2021). Pattern recognition of inter-turn short circuit fault in a synchronous generator using magnetic flux. *IEEE Transactions on Industry Applications* 57 (4): 3573–3581.

114 Ehya, H., Nysveen, A., Nilssen, R., and Pattern recognition of inter-turn short circuit fault in wound field synchronous generator via stray flux monitoring (2020). *International Conference on Electrical Machines (ICEM)*, 23–26 August 2020. Gothenburg, Sweden.

115 Ehya, H., Nysveen, A., and Nilssen, R. (2021). Static, and dynamic eccentricity fault diagnosis of large salient pole synchronous generators by means of external magnetic field. *IEEE Transactions on Industry Applications* 15 (7): 890–902.

116 Cuevas, M., Romary, R., Lecointe, J.-P. et al. (2016). Non-invasive detection of rotor short-circuit fault in synchronous machines by analysis of stray magnetic field and frame vibrations. *IEEE Transactions on Magnetics* 52 (7): 1–4.

117 Ehya, H., Nysveen, A., and Skreien, T.N. (2021). Performance evaluation of signal processing tools used for fault detection of hydro-generators operating in noisy environments. *IEEE Transactions on Industry Applications* 57 (4): 3654–3665.

118 Antonino-Daviu, J.A., Pons-Llinares, J., and Lee, S.B. (2016). Advanced rotor fault diagnosis for medium-voltage induction motors via continuous transforms. *IEEE Transactions on Industry Applications* 52 (5): 4503–4509.

119 Pons-Llinares, J., Antonino-Daviu, J.A., Riera-Guasp, M. et al. (2014). Advanced induction motor rotor fault diagnosis via continuous and discrete time–frequency tools. *IEEE Transactions on Industrial Electronics* 62 (3): 1791–1802.

120 Ehya, H., Nysveen, A., and Antonino-Daviu, J.A. (2021). Advanced fault detection of synchronous generators using stray magnetic field. *IEEE Transactions on Industrial Electronics* 69 (11): 11675–11685.

121 Ehya, H., Nysveen, A., Antonino-Daviu, J.A. et al. (2021). Inter-turn short circuit fault identification of salient pole synchronous generators by descriptive paradigm. In: *Energy Conversion Congress & Exposition*, 10–14 October 2021. Vancouver, BC, Canada.

12

Electromagnetic Signature Analysis of Mechanical Faults

12.1 Introduction

Electrical generators experience a wide range of electromagnetic and mechanical tensions, as well as abnormal operating conditions. The origin of these stresses can be traced back to the external factors from the power grid side, environmental factors, and internal factors that arise within electrical generators. Excessive tension on a generator can cause early aging and severe damage to different parts. Early detection of faults in electrical generators is required to avoid unplanned generator outages that can result in huge economic losses. Faults in synchronous generators are divided into electrical faults and mechanical faults. The electrical faults were discussed in Chapter 11. The mechanical faults are described in the present chapter. The mechanical faults in synchronous generators include:

1. Eccentricity faults
2. Misalignment fault
3. Broken damper bar fault
4. Broken end ring fault
5. Stator core deformation
6. Stator core interlaminar fault
7. Stator core joint fault

Various methods have been developed to diagnose these faults in the early stages. Mechanical fault detection methods include both invasive and non-invasive approaches involving analysis of air gap magnetic fields, stator and rotor currents and voltages, shaft voltages and currents, vibration, and stray magnetic fields. Mechanical fault detection methods are discussed in the following three sections.

Electromagnetic Analysis and Condition Monitoring of Synchronous Generators, First Edition. Hossein Ehya and Jawad Faiz.
© 2023 The Institute of Electrical and Electronics Engineers, Inc. Published 2023 by John Wiley & Sons, Inc.

12.2 Eccentricity Faults

Eccentricity faults in electrical generators imply a non-uniformity of the distance between the stator and rotor. These faults are divided into static, dynamic, and mixed eccentricity faults. The mixed eccentricity fault is a combination of both static and dynamic eccentricity faults. The major difference between static and dynamic eccentricity faults is that the air gap length only varies with the position in an eccentricity fault, whereas it varies both in time and position in the case of a dynamic eccentricity fault. Up to 10% of the inherent eccentricity fault severity in electrical generators is unavoidable, even in brand new machines [1–7].

An eccentricity fault results in a non-uniform magnetic field in the air gap. One of the main consequences of an eccentricity fault is an unbalanced magnetic pull (UMP), which worsens the eccentricity fault situation. The existence of an eccentricity fault and an UMP in an electrical generator creates a positive feedback loop in control theory, whereby the eccentricity results in an UMP, which results in an increment in the severity of the eccentricity fault. This increase in severity then worsens the UMP. Ultimately, the rotor touches the stator and the generator stops operating. This is why early detection of an eccentricity fault is essential.

Numerous eccentricity fault detection methods, based on different types of signals, are available for synchronous generators. In this section, electromagnetic field signal methods are described, including those based on the stator voltage, stator current, rotor field current, air gap magnetic field, and stray magnetic field.

Notably, the applied eccentricity fault to the synchronous generator in this chapter is toward the positive direction of the *x*-axis, which results in a minimum air gap between the stator and rotor core. The right and left measuring points are assigned in both simulation and experimental results.

12.2.1 Invasive Detection Methods

12.2.1.1 Air Gap Magnetic Field
Static and dynamic eccentricity faults both lead to asymmetry of the magnetic field in the air gap [8]. Figure 12.1 shows a 14-pole synchronous generator air gap magnetic field with a 10% dynamic eccentricity fault severity. The dynamic eccentricity fault causes variation in the minimum air gap between the stator and rotor in both time and position. The peak amplitude of the air gap magnetic

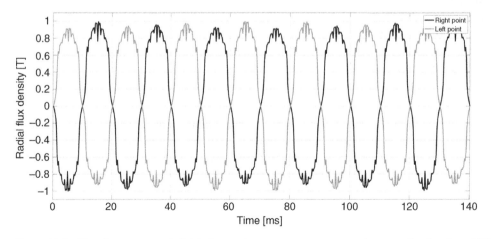

Figure 12.1 The air gap magnetic field of a synchronous generator operating at no-load under a 10% dynamic eccentricity fault – simulation result.

field fluctuates in the case of a dynamic eccentricity fault for both measuring points, indicating that the air gap magnetic field magnitude deviates from the average value when each rotor pole passes over the installed sensor. The deviation in the flux density under a 10% dynamic eccentricity fault severity is approximately 5% with respect to the average flux density of a healthy synchronous generator.

The trace showing the variation in the air gap magnetic field can be used to identify the dynamic eccentricity fault since the average flux density of each rotor pole varies in one mechanical revolution while the poles pass over the measuring sensor, and this results in a distinctive pattern for the fault detection. Figure 12.2 shows the flux density distribution of the synchronous generator with an eight-pole rotor under 25% dynamic eccentricity fault severity [9]. The concentration of the flux lines on the right side of the synchronous generator, while the rotor is closed to the stator core, is a clear indication of an eccentricity fault. The saturated stator slots in the right half of the synchronous generator also indicate that the flux density concentration toward the minimum air gap is caused by the dynamic eccentricity fault. Figure 12.2 shows the flux density distribution of the full-load synchronous generator, where the armature reaction causes the flux density lagging [9].

Figure 12.3 shows the impact of a 10% severity static eccentricity fault on the air gap magnetic field. This eccentricity fault causes variations in the air gap magnetic field in space. In addition, for the given static eccentricity fault severity, the air gap length is time independent, which means it is static. Therefore, the magnetic flux density increases on the measuring side, where the air gap length between the stator and rotor core is the minimum. Conversely, the air gap magnetic field is decreased by increasing the air gap length on the opposite side. According to Figure 12.4, the maximum peak amplitude of the magnetic field on the entire rotor pole for one mechanical revolution on the right side increases, whereas it decreases for the left measuring side. Detection of the static eccentricity fault based on analysis of the air gap magnetic field is very difficult, since the

Figure 12.2 The flux density distribution in the synchronous generator operating at full-load under a 25% dynamic eccentricity fault.

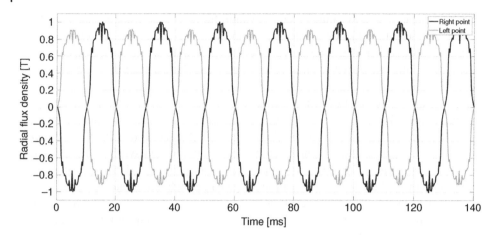

Figure 12.3 The air gap magnetic field of a synchronous generator operating at no-load under a 10% static eccentricity fault – simulation result.

location of the measuring point and the direction of the static eccentricity fault must be aligned, and this is not actually the case.

The proposed detection method is based on comparing the average flux density of each pole with the neighboring pole for a short circuit or dynamic eccentricity fault, but this is not applicable to static eccentricity faults because a static eccentricity fault affects all the poles. An alternative solution is to distribute the sensors inside the air gap and compare the average flux density of pairs of poles installed directly across from each other.

Figure 12.4 shows the polar diagram of the on-load synchronous generator with 5% static eccentricity fault severity. Comparison of the polar diagrams of the synchronous generator with a static eccentricity fault with those from a healthy generator indicates that the average magnetic flux density changes due to the fault. The average magnetic flux density of the right measuring point increases because the static eccentricity is directed toward the positive x-axis direction. The average magnetic flux density of the left measuring point also decreases compared with the healthy case. The average measured magnetic flux density is 0.669 T at the right point and 0.650 T at the left point, while the average magnetic flux density for a healthy generator is 0.655 T. Increasing the static eccentricity fault severity directly influences the average magnetic flux density: the average magnetic flux density of the synchronous generator under 10% static eccentricity fault severity measured is 0.679 T at the right point and 0.641 T at the left point, while the average magnetic flux density of the healthy generator is 0.655 T [9].

12.2.1.2 Frequency Analysis of the Air Gap Magnetic Field

The frequency spectrum analysis of the air gap magnetic field of 14-pole synchronous generators is shown in Figure 12.5, which shows a healthy generator and a generator with four different dynamic eccentricity fault severities. The frequency spectrum of the healthy generator contains only the fundamental frequency and its odd multiples. By contrast, the dynamic eccentricity fault results in the generation of several frequency components in the frequency spectrum. The frequency of the fault-related harmonics is equal to the mechanical frequency of the generator and its multiples. Figure 12.5 demonstrates that the amplitude of the fault-related harmonics increases with increasing dynamic fault severity.

A comparison with an inter-turn short circuit fault in the synchronous generator, which also integrates the same mechanical frequency components, reveals that the amplitude and the

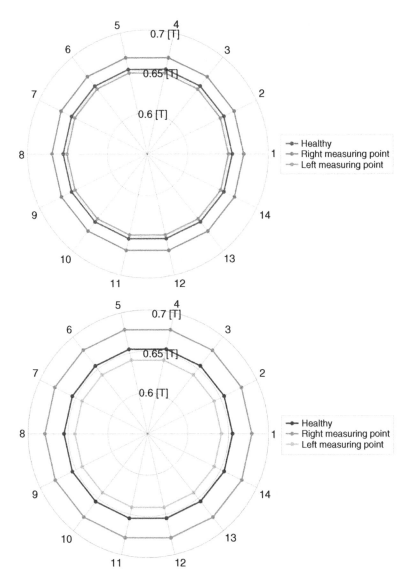

Figure 12.4 Polar diagram of a synchronous generator operating in an on-load condition under a 5% static eccentricity (top) and 10% static eccentricity fault (bottom).

distribution of the dynamic eccentricity fault-related harmonic components are entirely different. The inter-turn short circuit fault-related harmonics are dispersed along the entire frequency spectrum, whereas the harmonics of the fault-related components for dynamic eccentricity are concentrated around the fundamental component and its odd multiples. The largest fault-related harmonic components in the inter-turn short circuit fault are located at the lowest frequencies, whereas the largest amplitudes for the dynamic eccentricity fault are located in $k = 1$:

$$(1 \pm k/p)f_s \tag{12.1}$$

The impact of the dynamic eccentricity fault on the amplitude of the fault-related components depends on several factors, including the loading condition, saturation level, air gap length, generator topology, winding layout, rotor pole size, and distance between two adjacent poles. Therefore, the amplitude of the faulty components for the same severity of the dynamic eccentricity fault should not be expected to be the same in two synchronous generators with different topologies and power ratings.

Figure 12.5 indicates that detection of the dynamic eccentricity fault is possible using the frequency spectrum of the air gap magnetic field. The dynamic eccentricity fault results in the harmonic components being distributed with a frequency difference equal to the mechanical frequency around the main frequency component and its odd multiples. Figure 12.5 shows the impact of the loading of a synchronous generator with a dynamic eccentricity fault on the frequency spectrum of the air gap magnetic field. The loading clearly does not significantly change the fault-related harmonics behavior [9].

Figure 12.6 demonstrates the impact of a static eccentricity fault on the frequency components of the air gap magnetic field. The frequency spectra of 14-pole synchronous generators operating at no-load conditions under different static eccentricity fault severities are also shown in Figure 12.6. The applied static eccentricity fault in the positive *x*-axis direction indicates that an increase in

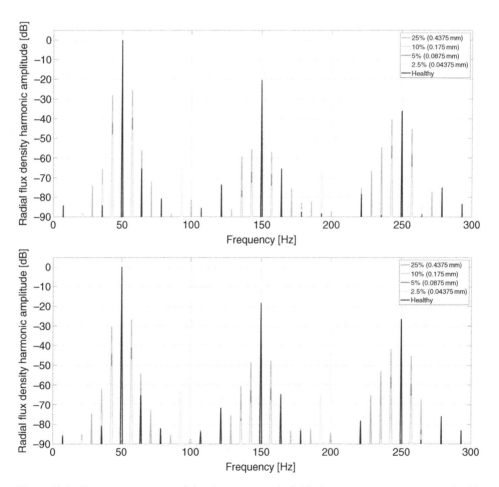

Figure 12.5 Frequency spectrum of the air gap magnetic field of a synchronous generator with 14 poles operating at no-load (top) and full resistive-inductive load under various degrees of dynamic eccentricity (bottom) – simulation results.

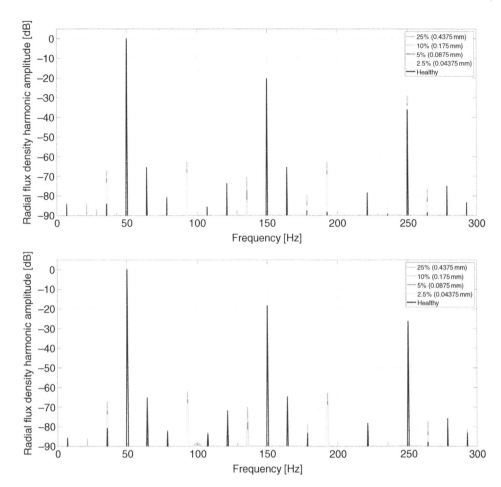

Figure 12.6 Frequency spectrum of the air gap magnetic field of a synchronous generator with 14 poles operating at no-load (top) and full resistive-inductive load under various degrees of static eccentricity – simulation results.

the fault severity results in an amplitude increment in the main frequency harmonic and its odd multiples. The frequency spectrum of the air gap magnetic field under a static eccentricity fault for both synchronous generators shows that the fundamental harmonic and its odd multiples are the major components of the frequency spectrum. However, some harmonics in the frequency spectrum are related to the generator topology. The static eccentricity fault does not change the mechanical frequency components of the air gap magnetic field; rather, it creates a slight variation in the fundamental frequency components and their odd multiples. However, some variations in the inter-harmonics are observed in the 14-pole synchronous generator. Therefore, the frequency spectrum of the air gap magnetic field in the faulty synchronous generator is not applicable for detection of the static eccentricity fault, because the fault only slightly changes the fundamental frequency component and its odd multiples. This variation in the fundamental frequency component and its odd multiples are not reliable indexes for static eccentricity fault detection for the following reasons [9–11]:

1. The load can change the amplitude of the fundamental frequency components and their odd multiples.

2. If the orientation of the static eccentricity is not aligned with the sensor location, its impact on the frequency components is negligible.
3. Although the 14-pole synchronous generator shows some inter-harmonics due to the static eccentricity fault, the highest amplitude for a 25% static eccentricity severity is −63 dB or 0.7 mT. Detecting this variation in the frequency spectrum of the air gap magnetic field is very difficult, and the possibility of false identification is high.

Figure 12.6 shows the load impact on the frequency spectrum of the air gap magnetic field. It indicates that the static eccentricity fault of the on-load synchronous generator does not change the frequency spectrum.

12.2.1.3 Spectral Kurtosis

Spectral kurtosis is a statistical tool that can be used to identify the non-Gaussian distribution and transient behavior of the time-domain signals in the frequency domain [12]. Spectral kurtosis is also employed to determine the location of the non-Gaussian and non-stationary components in the frequency domain [13, 14]. This tool can be used to determine the hidden pattern in the air gap magnetic field of the synchronous generator that might be caused by a fault. Figure 12.7 shows the kurtogram of the air gap magnetic field of a healthy no-load 100 kVA synchronous generator, as well as when under a 20% static eccentricity fault. The following two features, based on kurtosis analysis, are defined for static eccentricity fault detection:

1. Maximum kurtosis of the kurtogram.
2. The energy of the spectral kurtosis.

A comparison between the kurtogram of the healthy state and under a 20% static eccentricity fault of a synchronous generator shows that the kurtogram intensity changes between levels one and five. A correlation of the maximum kurtosis with the eccentricity fault severity also indicates its sensitivity to the fault severity. The maximum kurtosis is 2.1 for a 10% fault severity and 1.9 for the healthy case. Increasing the static eccentricity fault severity to 20% results in a maximum kurtosis of 3.

The energy of the kurtosis is introduced in reference [13] to differentiate the occurrence of the static eccentricity fault as follows:

$$E_{b_\varphi}^\kappa = \int_{-\infty}^{\infty} \left| b^\kappa(\varphi) \right|^2 dt \tag{12.2}$$

where $E_{b\phi}{}^K$ is the energy of the kurtosis and b_ϕ is the air gap flux density. The energy for a healthy case is 7.18, but it decreases to 6.68 and 5.46, respectively, for 10 and 20% static eccentricity fault severities. The eccentricity fault reduces the energy because the kurtosis is applied to the measured air gap magnetic field of the sensor with a large air gap between the stator core and rotor core.

12.2.2 Non-invasive Detection Methods

12.2.2.1 Inductance Variation Index

The asymmetry of the air gap magnetic field causes parameter changes in the synchronous generator. These parameters are the self-inductance and mutual inductance of the magnetic circuit of the generator. These inductances increase (according to the location of the minimum air gap) with increasing dynamic eccentricity fault severity, especially in the salient pole synchronous generator.

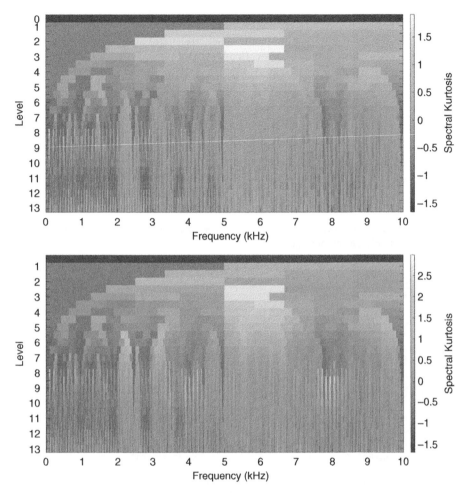

Figure 12.7 The kurtogram of the air gap magnetic field of the synchronous machine operating in the no-load in a healthy case (top) and under a 20% static eccentricity fault (bottom).

According to Figures 12.8 and 12.9, an 85% eccentricity fault in the salient pole synchronous generator increases the stator self-inductance and the mutual inductance between the stator and rotor compared with the healthy case. Although the eccentricity fault changes the parameters of the synchronous generator, the self-inductance and mutual inductance show high sensitivity to high severity eccentricity faults, indicating that the method based on the parameter change is not appropriate for early-stage detection of the fault.

12.2.2.2 Harmonics of the Stator Current

Eccentricity fault detection based on harmonic analysis of the stator current is applicable when a synchronous generator is connected to the power grid or the local load current passing through the stator winding. An eccentricity fault in a synchronous generator causes a non-uniform air gap; therefore, the induced voltage in the stator winding contains some harmonics. When the stator winding of a generator has a closed circuit, these voltages lead to currents with similar harmonics.

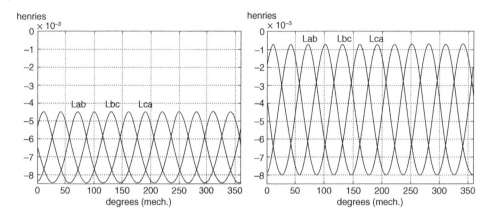

Figure 12.8 The self-inductance of the stator windings in a healthy case and under an 85% dynamic eccentricity fault in a salient pole synchronous generator.

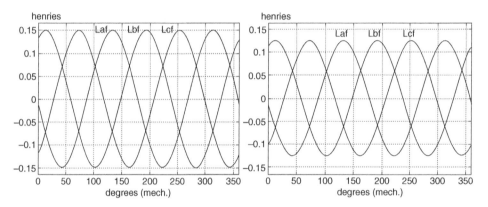

Figure 12.9 The mutual inductance between the stator and rotor windings in a healthy case and under an 85% dynamic eccentricity fault in a salient pole synchronous generator.

These current harmonics can be used to detect the eccentricity fault in salient pole synchronous generators [15, 16]. The shape of the rotor poles and the topology of the generator have a significant impact on both the current and voltage harmonics of the stator winding. In some cases, the amplitude of the current harmonics is decreased by increasing the fault severity in a salient pole synchronous generator.

Some current harmonics such as the 5th, 7th, 11th, 13th, 17th, and 19th, have been introduced to detect the eccentricity fault in a salient pole synchronous generator. Table 12.1 shows the amplitude variations of the aforementioned harmonics under a 50% dynamic eccentricity fault in a synchronous generator. Notably, in the analytical modeling of the generator, the slotting effect must be taken into account, as neglecting the slotting effect increases the amplitude of the current harmonics [15].

The 17th and 19th harmonics of the stator current are frequently practiced for eccentricity fault detection [16–18]. The 3rd harmonic of the stator current has been used in reference [19] as an indicator of a dynamic eccentricity fault in the salient pole synchronous generator. The 21st and 23rd harmonics are also evaluated for eccentricity fault detection. Figure 12.10 shows the current frequency spectrum in the healthy generator and under a 50% eccentricity fault, which results in

Table 12.1 Stator current harmonic amplitude rises under a 50% dynamic eccentricity fault.

Harmonic order	5th	7th	11th	13th	17th	19th
Stator current rising (%)	22.8	12.4	20.9	28.4	47.1	36.9

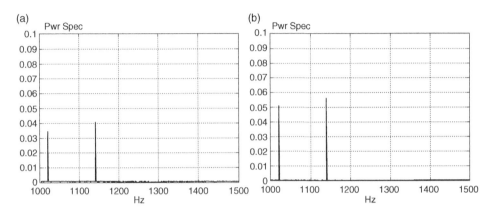

Figure 12.10 Frequency spectrum of the stator current of a synchronous generator with four salient poles in (a) a healthy case and (b) under a 50% dynamic eccentricity fault.

an increment of the 21st and 23rd harmonics [20]. Although the method shows the possibility of detecting eccentricity faults in a synchronous generator, fault diagnosis based on the stator current is not a reliable method. This is because the following factors can change the amplitude of the harmonics current:

1. Other types of faults, such as a short circuit fault in the stator or rotor field winding, also result in the variation of the same harmonics in the generator. However, their amplitudes differ, because the amplitude increment depends on the type of fault or the generator configuration.
2. In equipment equipped with power electronic components, the same harmonics that are injected into the power grid may interfere with the synchronous generator. This generates a false alarm if the fault detection method is solely based on the current analysis.

In a generator with a parallel-connected stator winding, the current must be measured in each branch separately because the resultant current of multiple parallel branches can change the harmonic contents of the faulty current. Figure 12.11 shows the current passing through the parallel branches of the synchronous generator in a healthy case, as well as under eccentricity and short circuit faults.

12.2.2.3 Harmonics of the Open-Circuit Voltage of the Stator Winding

The no-load voltage of the stator winding of a synchronous generator is a widely used signal to detect the eccentricity fault [20, 21]. This signal has superiority over the stator current because it is easily measured in both the no-load and on-load conditions. Figure 12.12 shows the induced voltage of a 675 MVA salient pole synchronous generator in a healthy state and under different types of fault. The waveform variation caused by the fault indicates that the fault does not generate low-order harmonics. Note that the low-order harmonics can significantly change the waveform

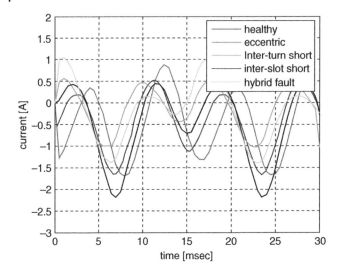

Figure 12.11 Current of one of the parallel branches of a synchronous generator in a healthy case and under different types of faults.

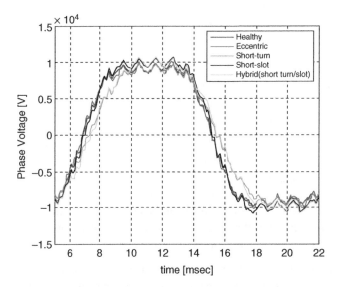

Figure 12.12 No-load voltage of a 675 MVA salient pole synchronous generator in a healthy state and under various types of faults.

shape [20]. Therefore, the high-frequency harmonics in reference [20] have been selected as indicators of an eccentricity fault for detection. One of these harmonics is the 21st harmonic, which changes markedly in the case of an eccentricity fault.

In addition, the eccentricity fault in the salient pole synchronous generator causes an increase in some harmonics, such as the 3rd, 5th, 7th, 11th, 13th, 17th, and 19th. As the fault severity increases, these harmonics increase in the phase voltage and line voltage [20, 22–24].

The topology of the synchronous generator, and specifically the winding layout, has a paramount impact on the amplitude of the harmonics in the induced voltage [25]. For example, the sensitivity

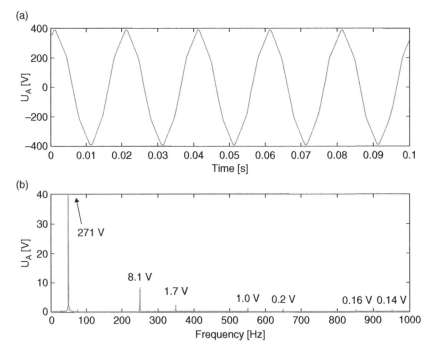

Figure 12.13 Phase voltage and frequency spectrum of a healthy synchronous generator.

of the no-load voltage spectrum to an eccentricity fault is much higher for the series-connected stator winding than for the parallel-connected case. Moreover, harmonics are also seen in the frequency spectrum of the healthy synchronous generator due to manufacturing problems, making the fault difficult to detect [22, 23]. Figures 12.13 and 12.14 show the phase and line voltages of the healthy synchronous generator and their corresponding frequency spectra. Figures 12.15 and 12.16 depict the impact of a 50% eccentricity fault on the phase and line voltages of a synchronous generator and their frequency content.

The sensitivity to the eccentricity fault is greater for the no-load terminal voltage of the stator of a synchronous generator than for the on-load case. The reason is that the no-load generator operates in the linear part of the magnetization characteristic, which shows a linear reaction to the fault. By contrast, an on-load generator operates in a knee part of the magnetization characteristic that exhibits low sensitivity to the occurrence and progression of the fault.

12.2.2.4 Analysis of the Space Vector Loci of the Electromotive Force

In addition to the changes in the frequency content of the electromotive force (EMF) under an eccentricity fault in the synchronous generator, changes also occur in the space vector of the EMF (see Figure 12.17). The eccentricity faults alter the radius of the EMF space vector, causing increases in this radius for both static eccentricity and dynamic eccentricity faults. The covered area under the healthy case must be compared with that under faulty conditions. The difference between these areas indicates the occurrence of the fault. The type of eccentricity fault is also examined by comparing the shape of the space vector, because that shape is elliptical under a static eccentricity fault but circular under a dynamic eccentricity fault [25]. The severity of the fault is determined by the difference in area between the healthy case and the faulty case.

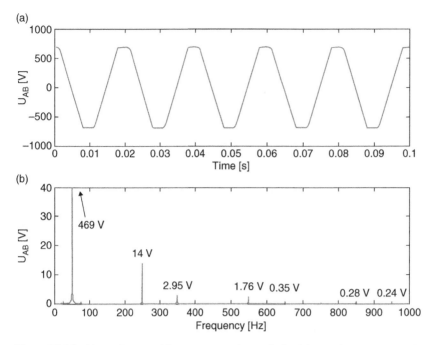

Figure 12.14 Line voltage and frequency spectrum of a healthy synchronous generator.

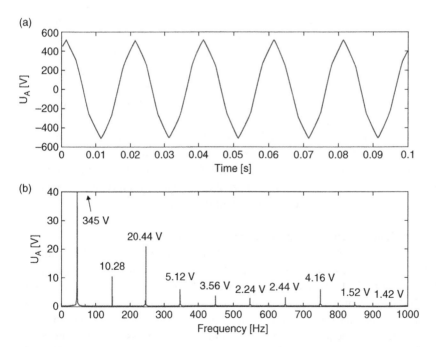

Figure 12.15 Phase voltage and frequency spectrum of a synchronous generator under a 50% dynamic eccentricity fault.

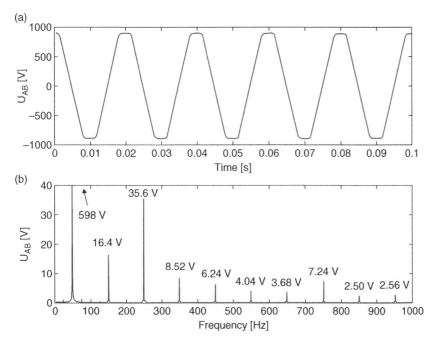

Figure 12.16 Line voltage and frequency spectrum of a synchronous generator under a 50% dynamic eccentricity fault.

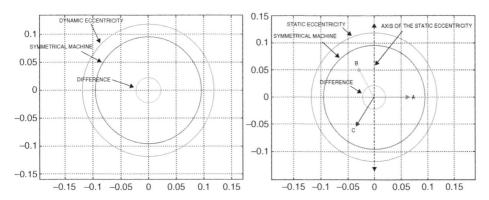

Figure 12.17 Variation in the space vector loci of the EMF of a synchronous generator operating in a healthy state and with static and dynamic eccentricity faults.

12.2.2.5 The Harmonic Component in the Current of the Rotor Field Winding

Analysis of a rotor current is another approach for eccentricity fault detection. In this case, the DC power is fed to the rotor field winding of the synchronous generator, thereby inducing a frequency component in the DC current of the rotor field winding that is twice that of the stator terminal frequency. However, the frequency harmonics for a static eccentricity fault are sensitive to both the short circuit fault and the supply unbalanced voltage. The dynamic eccentricity fault results in the following frequency component:

$$f_{de} = \frac{f_{source}}{p} \tag{12.3}$$

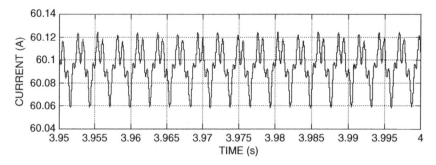

Figure 12.18 Waveform of rotor field current under an eccentricity fault.

where f_{source} is the frequency of the synchronous generator at the stator terminal and p is the number of pole pairs. In the case of static and dynamic eccentricity faults, both frequency components are expected to appear in the rotor current [24]. Figure 12.18 shows the impact of an eccentricity fault on the rotor field current of the synchronous generator [22, 26, 27].

12.2.2.6 Stator Split-Phase Current

The stator split-phase current is analyzed to detect the static and dynamic eccentricity faults in synchronous generators. This method is applicable if the synchronous generator has a parallel winding connection [26, 27]. In the case of an eccentricity fault in the synchronous generator, an additional flux component of $2(p \pm 1)$ appears in the air gap magnetic field distribution. However, the induced voltage in the stator winding due to the additional flux components cannot create any component in the stator phase current because of the mismatch of poles. The additional component is created due to the $2(p \pm 1)$ component in the current of a stator winding with a parallel winding layout. The circulating current related to the additional flux components results in an unbalanced current inside the stator winding that compensates for the UMP [27].

A current sensor or Rogowski coil can be utilized to measure the stator split-phase current in the parallel winding [26]. This method does not need the historical background of the generator because the amplitude of the circulating current in the parallel branch in the healthy generator is negligible, while the presence of a static or dynamic eccentricity fault results in the circulating current. Figure 12.19 shows the three-phase current of a healthy no-load synchronous generator. Figures 12.20 and 12.21 present the impact of static and dynamic eccentricity faults on the split-phase current of the generator. The phase angle and the amplitude of the split phase of each phase depend on the position of the coils inside the stator core.

The space vector representation of the stator split-phase current can also be used to determine the type of eccentricity fault in the synchronous generator because the direction of the space vector rotation for each time data series is different. This indicates that the shape type tendency of these vectors can be used as an indicator to identify the type of eccentricity fault [28].

12.2.2.7 Stator Voltage Subharmonics Index

The stator voltage and currents can be used to detect the eccentricity fault in the synchronous generator. The generated harmonics, such as the 7th, 11th, 13th, 17th, 19th, and 21st, in the stator current or voltage can detect the eccentricity fault, while these harmonics coexist in the frequency spectrum of the synchronous generator due to its specifications. This indicates the difficulty of

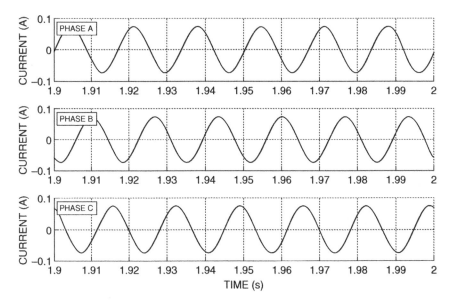

Figure 12.19 Three-phase stator current of a no-load synchronous generator.

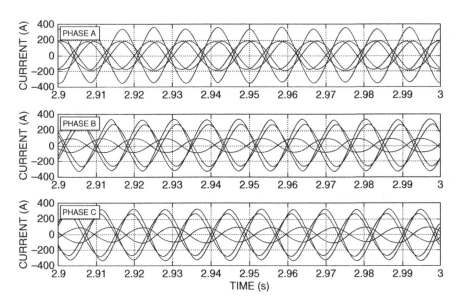

Figure 12.20 Split-phase current of stator phase windings of a 1950 kVA no-load synchronous generator under a 66% static eccentricity fault.

determining whether the harmonics in the spectrum are due to the fault or whether they exist because of the generator topology [15–29].

The fault in the synchronous generator yields an unbalanced magnetic field. This field, due to the dynamic eccentricity fault, creates subharmonics in the stator-induced voltage. The subharmonics can be determined using the stator frequency and the number of poles of the synchronous

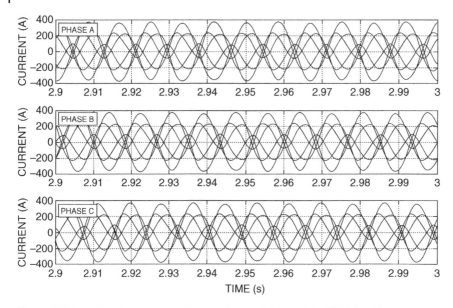

Figure 12.21 Split-phase current of stator phase windings of a 1950 kVA no-load synchronous generator under a 66% dynamic eccentricity fault.

generator, as follows:

$$f_{eccentricity} = \left(1 \pm \frac{k}{P}\right) f_s \tag{12.4}$$

where $f_{eccentricity}$ is the mechanical frequency of the synchronous generator that is invoked when the dynamic eccentricity fault happens. The main drawback of this method is the need for data from the healthy generator for comparison. Moreover, the frequency components of the induced voltage must be compared at the same loading condition. Figure 12.22 shows the frequency spectrum of the induced voltage in a four-pole no-load synchronous generator. The 10 and 20% dynamic eccentricity faults increase the amplitude of the subharmonics above that of the healthy frequency spectrum [8].

The eccentricity fault can be detected using the average amplitude of the increased harmonics and subharmonics in the stator terminal voltage. The total harmonic distortion of the generator voltage gives an insight into the health status of the generator [29]. A normalized index, which compares the total harmonic distortion of the healthy synchronous generator with a faulty case, is introduced as follows [29]:

$$\text{ETHDF} = \frac{\text{THD}_{Faulty} - \text{THD}_{Healthy}}{\text{THD}_{Healthy}} \times 100 \tag{12.5}$$

where ETHDF is the total harmonic distortion factor of the induced voltage in the stator winding and $\text{THD}_{Healthy}$ and THD_{Faulty} are the total harmonic distortions of the healthy and faulty cases. Since the index is based on the comparison between the healthy and the faulty cases, the measurement must be performed in the same load or no-load conditions.

12.2.2.8 Shaft Voltage

The magnetic flux inside a synchronous generator is symmetrical and is based on the number of its poles. Therefore, the sum of the magnetic flux in the shaft of the healthy generator is also symmetrical. In the faulty case, including both mechanical faults and electrical faults, the symmetry of the flux distribution in the generator is distorted, as shown in Figure 12.23. This indicates that the unbalanced magnetic field results in the passage of a flux through the shaft; therefore, its net flux is

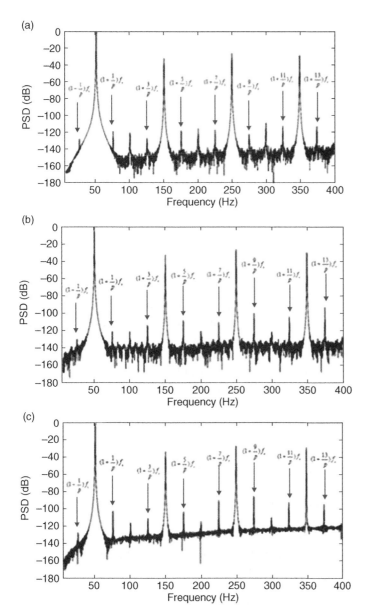

Figure 12.22 Frequency spectrum of no-load voltage of a four-pole no-load synchronous generator (a) at the healthy state, (b) under a 10% dynamic eccentricity fault, and (c) under a 30% dynamic eccentricity fault.

no longer zero. The variation in the non-zero flux induces a voltage in the shaft of the generator. In power plants, the bearings are isolated to avoid the closed path between the shaft and the ground. If a flux passes from the shaft to the grounding, a current flows in the shaft. An induced voltage in the generator shaft can range from a few mV to a few hundred V, depending on the level of asymmetry of the flux. If the shaft resistance, bearings, and connection resistance of the generator are low, the current passing through the generator shaft can cause severe damage to the bearings [30–32].

The induced voltage of the shaft is initially measured to reduce and eliminate it. However, further investigations have shown that this voltage can also be used to diagnose the operating condition of the generator and the presence of faults within it. Short circuits of the rotor winding, static or

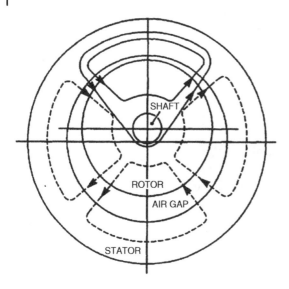

Figure 12.23 Magnetic flux distribution in a synchronous generator under eccentricity fault.

dynamic eccentricity, and misalignment, in which the flux symmetry is lost in the air gap, are among the faults that can be detected by this method [33]. Since the optimal design of the electric generator takes place at the knee part of the magnetization characteristic, the third order harmonic is usually seen in the shaft voltage. If the generator operates in the saturation region, then the principal component in the induced shaft voltage is reduced significantly.

Asymmetry in the air gap due to an eccentricity fault results in an induced voltage in the shaft of the machine. Increasing the fault severity in the machine increases the induced voltage in the shaft of the synchronous machine. Increasing the severity of the fault (either static or dynamic eccentricity faults) results in increases in the induced voltage increment in the shaft, whereas increases in the severity of the misalignment do not lead to an increased induced voltage.

The load has a significant impact on the induced voltage in the shaft of the faulty synchronous generator if the fault severity is high. Otherwise, the induced voltage remains the same in the shaft of a no-load or in a full-load faulty generator with a low severity fault. The reason is that the flux passing through the shaft is higher for an on-load generator than for the no-load case, but the net flux in the shaft due to the symmetry is close to zero [33].

The winding layout connection is also an important factor that alters the amplitude of the induced voltage in the generator. This voltage does not increase for a faulty synchronous generator with a parallel stator winding connection because the circulating current in the stator parallel winding tries to compensate for the non-uniform air gap distribution caused by the fault. Therefore, the type of winding connection must be considered when the detection algorithm is based on shaft voltage assessment.

Several factors influence the induced voltage in the shaft of a synchronous generator. In particular, the structural specifications of the stator core, including the number of stator segments, weld recess, keyway, and the manufacturing quality of the stator lamination sheets, have a marked impact on shaft-induced voltage [34]. The induced voltage in the shaft is reduced when the keyways and the weld recess are eliminated in the stator lamination sheets, as shown in Figure 12.24. The reason is that the flux distribution in the generator is symmetrical and the net flux in the shaft is almost zero, whereas the weld recess and keyways distort the magnetic field uniformity and increase the shaft-induced voltage [35–37]. Including both a weld recess and a keyway in the stator lamination core therefore increases the induced voltage, as shown in Figure 12.24.

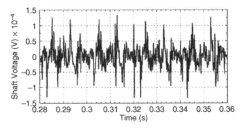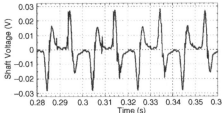

Figure 12.24 Induced voltage in the shaft of a synchronous generator without considering the keyway and weld recess (left) and by including both in the stator lamination sheet (right).

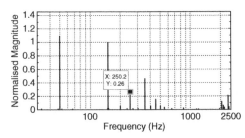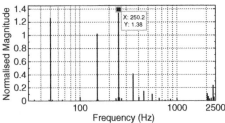

Figure 12.25 Frequency spectrum of shaft-induced voltage in a synchronous generator in (a) the healthy case (left) and under an eccentricity fault (right).

The experiment has been performed on a 20 kVA synchronous generator and the amplitude of the induced shaft voltage in the healthy generator is 79 mV [35]. Increasing the static eccentricity severity from 5 to 15% increases the shaft-induced voltage from 88.9 to 100.1 mV [34]. The presence of a non-uniform air gap increases the shaft-induced voltage and a frequency index is required to detect the fault. The 5th harmonic of the shaft-induced voltage is used to detect the eccentricity fault in the synchronous generator. The 5th harmonic has a high sensitivity to the eccentricity fault [35].

The 5th harmonic of the shaft-induced voltage is also used as an indicator of the fault and increases significantly with fault severity. Other harmonics can also be excited by the eccentricity fault due to the generator structure. Figure 12.25 shows the 5th, 6th, and 8th harmonics in a cylindrical pole generator excited by the fault [38]. Figure 12.25 shows the frequency spectrum of the shaft-induced voltage in a synchronous generator in the healthy state and under an eccentricity fault. The invoked harmonics in the frequency spectrum of the synchronous generator have a direct relationship with the stator core structure.

12.2.2.9 Stray Magnetic Field

The stray magnetic field offers a non-invasive approach that can provide informative details regarding the health status of a synchronous generator. In fact, the stray magnetic field is called the "mirror" of the air gap magnetic field [39]. The eccentricity fault changes the asymmetry and amplitude of the air gap magnetic field and consequently results in a variation in the self-inductance, mutual inductance, and linkage fluxes. Variations in the synchronous generator parameters result in variations in the stray magnetic field. Therefore, the stray magnetic field can be utilized to monitor the static and dynamic eccentricity faults in the synchronous generator.

Although the eccentricity fault influences the self-inductance and mutual inductance of the stator and rotor winding, the impact of a low severity eccentricity fault on the self-inductance and mutual inductance is very low. This indicates that various quantities, such as the stator voltage, stator current, rotor current, and rotor voltage, cannot be appropriate signals for the early-stage detection of synchronous generator faults [16]. The stray magnetic field has a high sensitivity to both static and dynamic eccentricity faults and can be a reliable candidate for health assessment [39]. The location of the stray magnetic field sensor must be chosen based on the fault type. The static and dynamic faults have a significant influence on the radial flux compared with the axial flux. However, the sensor can be mounted on the stator core to pick up both fluxes, where the main contributing flux is the radial flux.

The amplitude of the stray magnetic field is very low compared with the air gap magnetic field; however, it contains information similar to the air gap magnetic field. Figure 12.26 shows the radial magnetic field in the synchronous generator, including the air gap magnetic field, the magnetic field in the middle of the stator yoke, and the stray magnetic field at the stator backside. The stray magnetic field pattern is approximately the same as the air gap magnetic field, but the amplitude of the stray magnetic field is in the order of microTesla.

The stray magnetic field can be detected using a passive coil installed on the stator back core. The high sensitivity of the induced voltage in the search coil is due to the correlation of the mutual inductance of the stator and rotor windings with the search coil winding. Figure 12.27 shows the mutual inductance of the rotor winding with the search coil installed on the stator core side of the no-load synchronous generator in the healthy case and under a 20% static and dynamic eccentricity fault. The static eccentricity on the mutual inductance of the rotor winding and the search coil causes an amplitude reduction in the stray magnetic field because only the air gap length changes; the center of rotor rotation remains fixed. The dynamic eccentricity fault results in rotor whirling, where both the center of rotation and the length of the air gap vary. Therefore, the mutual inductance of the rotor winding and the search coil should represent this fluctuation. As shown in Figure 12.27, a dynamic eccentricity fault causes a fluctuation in the upper and lower envelope of the mutual inductance between the rotor and search coil windings [40].

Figure 12.28 displays the mutual inductance between the three phases of the stator windings (phase A, phase B, and phase C) and the search coil winding installed on the stator backside.

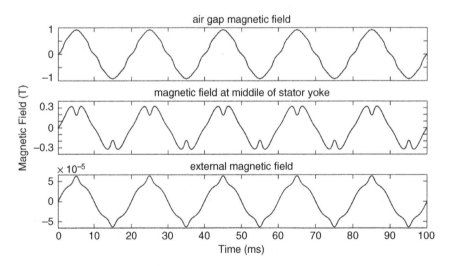

Figure 12.26 Radial magnetic field in different parts of a no-load synchronous generator, including an air gap magnetic field (top), a magnetic field at the middle of the stator yoke (middle), and a stray magnetic field (bottom).

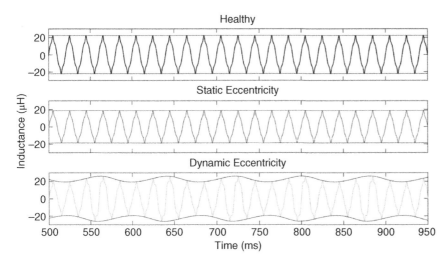

Figure 12.27 Mutual inductance between rotor field winding of a synchronous generator and search coil winding installed in the stator backside operating in a healthy state and under 20% static and dynamic eccentricity faults.

A difference is evident between the mutual inductance of all three phases in the healthy case, depending on the location of the distributed stator winding in the core. The pattern of the mutual inductance between the stator winding and the search coil under a static eccentricity fault is similar to that for the healthy mutual inductance, whereas the amplitude changes. The dynamic eccentricity fault results in both amplitude and pattern variations of the mutual inductance. This is because the upper envelope of the mutual inductance of phase A has high variations, while the lower envelope in phase B undergoes significant fluctuations. Furthermore, both the upper and lower envelope of the mutual inductance in phase C fluctuate due to the location and distribution of the stator winding and the search coil. A comparison of the mutual inductance of the rotor and the stator with the search coil reveals a higher influence of the stator mutual inductance than the rotor mutual inductance due to the short flux path from the stator winding to the search coil on the backside of the stator core.

The shape of the induced voltage in the sensor installed on the backside of the synchronous generator depends on the following factors:

1. Power rating of generator.
2. Number of slot/pole.
3. Shape of rotor pole.
4. Stator core thickness.
5. Winding layout.
6. Fault type.

Figure 12.29 shows the induced voltage of the sensor in an eight-pole 22 MVA synchronous generator. The induced voltages of the sensors are completely different in the two healthy synchronous generators. The impact of a 20% static eccentricity fault in the 22 MVA generator increases the induced voltage of the sensor; however, the location of the sensor is important. If the minimum air gap length due to the static eccentricity fault lies in the direction of the sensor, the induced voltage

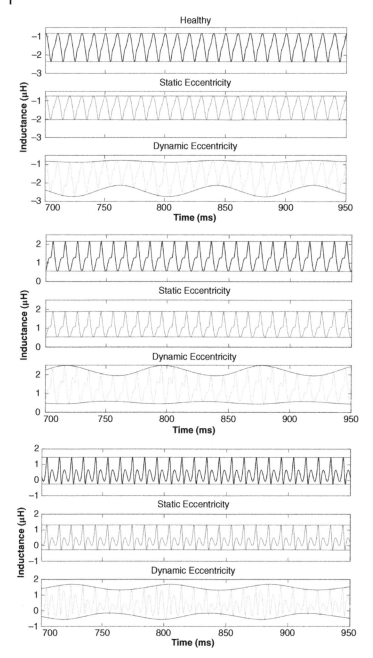

Figure 12.28 Mutual inductance between the stator phase winding of a synchronous generator and a search coil installed on the stator backside for a healthy case and under a 20% static and dynamic eccentricity fault. Phase A (top), phase B (middle), and phase C (bottom).

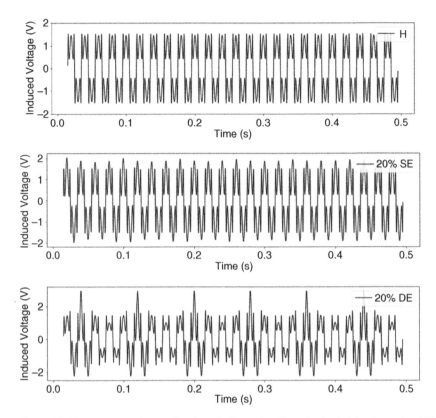

Figure 12.29 Induced voltage of an installed search coil on the backside of a no-load 22 MVA synchronous generator in a healthy case and under 20% static eccentricity and 20% dynamic eccentricity faults – simulation results.

increases, whereas the amplitude decreases at the opposite side. This dynamic eccentricity fault causes variations in the amplitude and pattern of the induced voltage of the sensor; this increases where the rotor pole becomes close to the stator core. Both the upper and lower envelopes of the induced voltage fluctuate under a dynamic eccentricity fault, as shown for both generators in Figure 12.29 [40].

One possible solution for detecting a static eccentricity fault is to analyze the stray magnetic field measured on the stator core backside because a static eccentricity fault results in a variation in the induced voltage of the sensor. Therefore, at least four sensors with 90 mechanical degrees of displacement are distributed, and this induced voltage is used. The static eccentricity orientation fault can also be detected by analyzing the induced voltage amplitude. Figure 12.30 shows the measured induced voltage of the sensor in a no-load 42 MVA synchronous generator. Four sensors are distributed on the stator backside; sensors S1, S3, and S2, S4 are located in front of each other. The induced voltage is approximately 1 V for the sensor in S1 and more than 2 V for the sensor in S3. The difference between the induced voltages in S2 and S4 is approximately 0.3 V. Clearly, a severe static eccentricity exists in the direction of S1 and S3 in the 42 MVA synchronous generator.

A similar approach for the differential air gap magnetic field can be applied to the stray magnetic field, where the induced voltage of two sensors installed opposite each other can be differentiated.

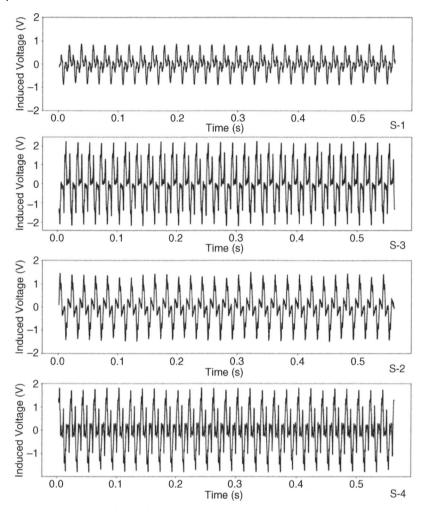

Figure 12.30 Induced voltages in the installed sensors distributed circumferentially on the backside of a no-load 42 MVA synchronous generator – field test results.

The differential stray magnetic field for the healthy synchronous generator is assumed to be zero and is non-zero in the case of a static or dynamic eccentricity fault. However, the amplitude of the differential stray magnetic field for the healthy synchronous generator is non-zero due to the location and topology of the generator. The mean value and the standard deviation of the differential stray magnetic field are introduced for the static and dynamic eccentricity fault detection in reference [39]. Figure 12.31 shows the variation in the applied standard deviation and mean value of the differential stray magnetic field for various static eccentricity and dynamic eccentricity fault severities. A linear trend is evident, where both the standard deviation and mean value of the differential stray magnetic field are increased by increasing the fault severity. For the same severity, the variation rate is greater for the standard deviation of the dynamic than the static eccentricity fault. In addition to the amplitude variation, new harmonics are introduced to the stray magnetic field [41].

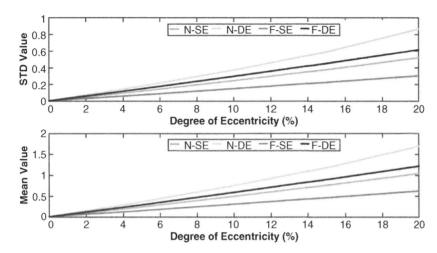

Figure 12.31 Variation in the standard deviation (STD) and mean value for static eccentricity versus dynamic eccentricity fault severity for the no-load 22 MVA synchronous generator at (N) and the full-load (F) case. (DE = dynamic eccentricity; SE = static eccentricity).

The impact of loading on the differential stray magnetic field is presented in the form of the standard deviation and the mean value. The rate of change is lower in the on-load synchronous generator than in the no-load case because a parallel connection of the stator minimizes the asymmetry in the air gap magnetic field by increasing the current in the winding where the air gap length is maximum due to the eccentricity fault.

The frequency spectrum of the induced voltage of the sensor of the synchronous generator under static eccentricity and dynamic eccentricity faults can be used to assess the frequency content variations. Figure 12.32 shows the impact of the static eccentricity and dynamic eccentricity faults on the frequency contents of the no-load synchronous generator. The impact of the static eccentricity on the frequency spectrum is not noticeable, which is similar to the frequency spectrum of the air gap magnetic field. The static eccentricity fault does not introduce subharmonics or inter-harmonics to the stray magnetic field; therefore, the variation in the harmonic components is almost the same for various severity static eccentricity faults.

The dynamic eccentricity fault results in a variation in the harmonic components of the induced voltage of the sensor. Figure 12.32 shows a comparison of the frequency components of the induced voltage of the sensor spectrum in the healthy case and in the case of a 5% dynamic eccentricity fault, where the amplitudes of the frequency components of 25, 75, and 125 Hz rise from −86.3, −86.4, and −82.5 dB in the healthy case to −68, −52, and 51.8 dB. The amplitude of the fault components is increased by increasing the dynamic eccentricity of the fault severity.

Although detection of the frequency spectrum of the induced voltage of the sensor using fast Fourier transform can provide information regarding the variation of the harmonic components caused by the eccentricity fault, the same harmonics are invoked by an inter-turn short circuit fault. Therefore, a signal processing tool that can provide a unique pattern for the eccentricity fault detection based on the discrete wavelet transform has been proposed in reference [39]. A mother wavelet from the Daubechies family has been used, and the signal was divided into eight sub-bands, where the highest sub-band contained a frequency between 5000 and 2500 Hz.

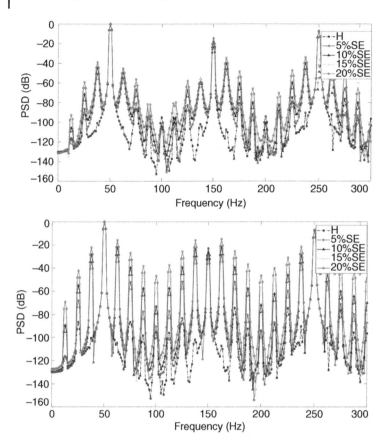

Figure 12.32 Frequency spectrum of the induced voltage of the sensor in an eight-pole 22 MVA no-load synchronous generator under various severities of static eccentricity (top) and dynamic eccentricity (bottom).

Figure 12.33 shows the application of a discrete wavelet transform to the differential induced voltage of the sensor in a 22 MVA no-load synchronous generator in the healthy state and under 20% static eccentricity and 20% dynamic eccentricity faults. The amplitude of the differential induced voltage of the sensor in the case of an eccentricity fault is 100 times higher than that of the healthy case, indicating that the differential induced voltage for the healthy case must be insignificant. The amplitude of the differential induced voltage of the sensor for the healthy case is almost zero; however, it still contains the frequency components. A detailed comparison of the differential induced voltage of the sensor of the healthy case and a 20% static eccentricity fault shows that the frequency contents and, consequently, the signal pattern for each sub-band must be similar. However, the amplitude of the sub-bands is increased under a static eccentricity fault. Conversely, the sub-bands pattern under a dynamic eccentricity fault change compared with the healthy case due to the variations in the nature of the dynamic eccentricity fault and the minimum air gap both in time and space. Figure 12.34 exhibits the on-load synchronous generator in the healthy case and under static eccentricity and dynamic eccentricity faults. Loading the generator reduces the amplitude and dampens the pattern due to the eccentricity fault; however, the variation in the sub-bands due to the eccentricity fault is sufficient for accurate fault detection.

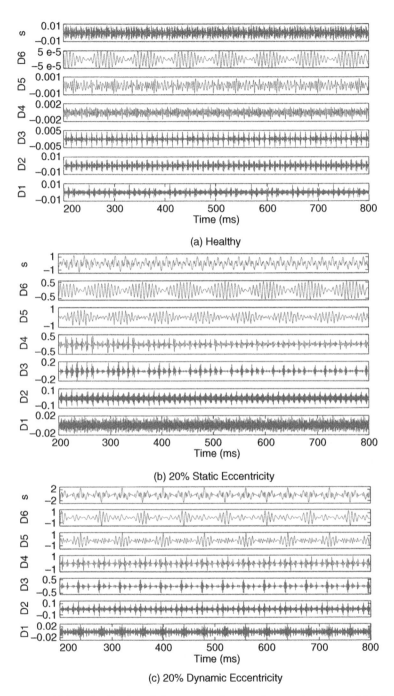

Figure 12.33 Application of discrete wavelet transform to the differential induced voltage of a sensor in the 22 MVA no-load synchronous generator (a) in a healthy case, (b) under a 20% static eccentricity fault, and (c) under a 20% dynamic eccentricity fault.

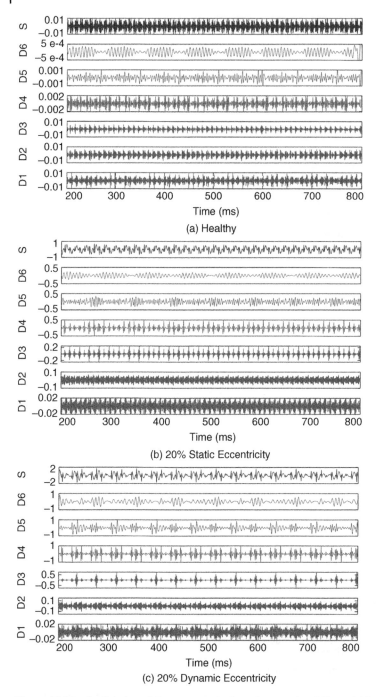

(a) Healthy

(b) 20% Static Eccentricity

(c) 20% Dynamic Eccentricity

Figure 12.34 Application of discrete wavelet transform to the differential induced voltage of a sensor in the 22 MVA on-load synchronous generator (a) in a healthy case, (b) under a 20% static eccentricity fault, and (c) under a 20% dynamic eccentricity fault.

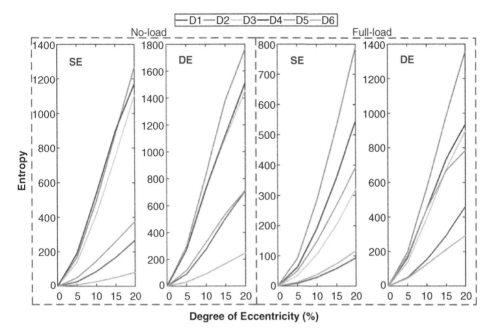

Figure 12.35 Entropy of wavelet sub-bands for static eccentricity and dynamic eccentricity faults with various severities in a 22 MVA full-load synchronous generator.

The entropy, in combination with the wavelet transform, has been used to detect the eccentricity fault in reference [39]. The entropy has been used to demonstrate the disorder level in the different sub-bands. Figure 12.35 shows the effects of entropy applied to the various wavelet sub-bands of the differential induced voltage of a sensor in a healthy case and under a static and dynamic eccentricity fault. The entropy of the healthy differential induced voltage of the sensor is zero, indicating a uniform air gap magnetic field of the synchronous generator. The static eccentricity and dynamic eccentricity fault with a small severity result in a wavelet entropy increment in the differential induced voltage of the sensor. However, for the same fault severity, the variation rate of the wavelet entropy is higher for a dynamic eccentricity fault than for a static eccentricity fault. Among the different sub-bands, the variation rate due to an eccentricity fault in the wavelet entropy of D2, D3, and D4 is higher than for the other sub-bands. The magnitude of the wavelet entropy of the differential induced voltage of a sensor in a healthy generator is zero; therefore, the detection method does not require a threshold. The threshold definition can differ due to the generator topology, its power rating, and the loading condition [39].

Fault diagnosis methods based on the differential induced voltage of the sensor can detect an early-stage fault in the synchronous generator; however, the differential induced voltage of the sensor of the large generator cannot be zero, even during healthy operation, as claimed [42]. Therefore, a time-frequency plot of the signal using a spectrogram of the short-time Fourier transform (STFT) was proposed in reference [42] that only requires the induced voltage of the sensor. Figure 12.36 shows an eccentricity fault detection algorithm based on an analysis of the induced voltage of the sensor using the STFT. An easy method to detect the static eccentricity fault is to analyze the induced voltage of a sensor installed on the stator backside. If the measured induced voltage of the sensors on each pair differs, this is an obvious indication of a static eccentricity fault. The next step is to deduce the pattern to determine the dynamic eccentricity fault.

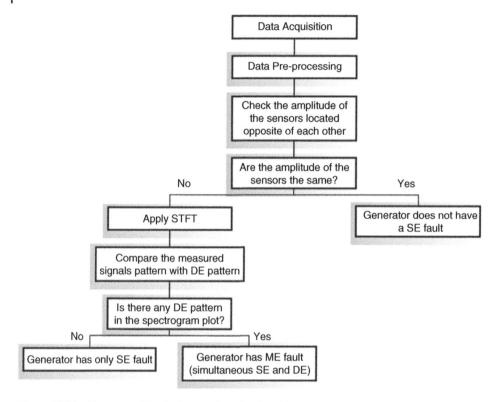

Figure 12.36 The eccentricity fault detection algorithm.

Figure 12.37 shows the time-frequency plot of the induced sensor voltage of the 22 MVA no-load synchronous generator in the healthy case and under a 20% static eccentricity, 10 or 20% dynamic eccentricity, and mixed eccentricity faults (10% static eccentricity and 10% dynamic eccentricity faults). The white window defined in the time-frequency plot shows one mechanical revolution of the machine. In other words, each white window contains all the rotor poles. For example, the width of the white window is 80 ms in the eight-pole 22 MVA synchronous generator. Three frequency bands exist in the time-frequency plot of the induced voltage of the sensor, which can be utilized for fault detection purposes. Two of the frequency bands are depicted in gray (35–50 Hz and 65–80 Hz) and one frequency band in dark gray (50–65 Hz).

None of the three frequency bands in the healthy synchronous generator contains any pattern and the color intensity is uniform. The static eccentricity fault does not introduce any new pattern into the time-frequency plot, but the color intensity of the time-frequency plot changes, as shown in Figure 12.37. In the case of the static eccentricity fault, the color intensity increment or decrement depends on the induced voltage of the sensor being analyzed. The color intensity increases if the minimum air gap between the stator and rotor is decreased due to the static eccentricity fault and the color intensity decreases if the minimum air gap increases.

The 10% dynamic eccentricity fault distorts the uniform pattern in the time-frequency plot of the induced sensor voltage. The dynamic eccentricity fault introduces a new pattern in all three frequency bands and the color intensity increases in all three bands. Several instances of this pattern are evident in the time-frequency plot, indicating the periodic nature of the dynamic eccentricity fault. The introduced pattern in the time-frequency plot becomes more evident with increases in

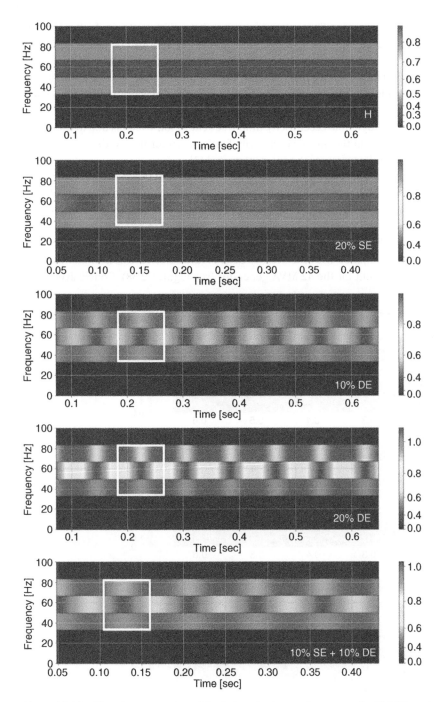

Figure 12.37 Time-frequency plot of the induced voltage of a sensor of a 22 MVA no-load synchronous generator in the healthy case and under static eccentricity, dynamic eccentricity, and mixed eccentricity faults – simulation result.

the dynamic eccentricity fault severity. The pattern of the 20% dynamic eccentricity fault is similar to that of the 10% dynamic eccentricity fault, but the intensity of the color band changes [42].

The occurrence of mixed eccentricity faults in a synchronous generator is unavoidable; therefore, the algorithm must be able to detect mixed faults. Figure 12.37 shows the co-existence of 10% static eccentricity and 10% dynamic eccentricity fault. The pattern of the mixed eccentricity fault is similar to that of a dynamic eccentricity fault; however, the color intensity of the frequency bands is different from that of the 10% dynamic eccentricity time-frequency plot, which is due to the static eccentricity fault. The color bar of the mixed eccentricity fault also confirms that the co-existence of both static and dynamic faults increases the color intensity [42].

The introduced algorithm is applied to 16-pole 42 MVA and 14-pole 100 kVA synchronous generators to verify the generalizability of the method. Figure 12.38 shows the time-frequency plot of the induced voltage of the sensor for the no-load 42 MVA generator in a healthy state and under 20% static eccentricity and 20% dynamic eccentricity faults. The white window length is 160 ms for the 16-pole generator. The color intensity of the time-frequency plot in the case of a static eccentricity fault increases compared with the healthy case. Figure 12.37 reveals a similar pattern for the dynamic eccentricity fault for the 42 MVA generator (see Figure 12.38). Conclusively, the

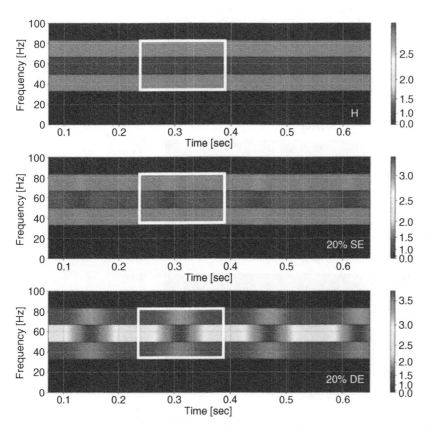

Figure 12.38 Time-frequency plot of the induced voltage of a sensor in a no-load 42 MVA synchronous generator in the healthy case and under static eccentricity and dynamic eccentricity faults – simulation results.

introduced algorithm in reference [42] is able to detect static, dynamic, and mixed eccentricity faults in the synchronous generator, regardless of their topology and power rating.

Although the introduced method based on STFT analysis in reference [42] does not require a differential induced voltage (which might result in a false identification of the fault compared with the discrete wavelet analysis), the pattern of the eccentricity fault is similar to the pattern intro- duced by the inter-turn short circuit fault of the rotor field winding. A method based on the analysis of the continuous wavelet transform introduced in reference [43] can provide a distinct pattern for these two fault types. Application of the continuous wavelet transform has been used to diagnose the inter-turn short circuit fault of a synchronous generator in reference [43].

The detection method based on analysis of the induced voltage of the sensor by continuous wavelet transform produces the time-frequency plot shown in Figure 12.39, which contains sev- eral stalks, each one representing one rotor pole (a stalk is indicated by the arrow in Figure 12.39). The widths of each stalk are equal to the time that the rotor needs to revolve once, divided by a num- ber of poles. A window is defined that contains stalks equal to the number of rotor poles. For an eight-pole synchronous generator, the width of the window is 80 ms, while for a 16-pole generator, the widths are equal to 140 and 160 ms. The required length of each stalk for fault detection is between a frequency of 20 and 70 Hz.

Figure 12.39 shows the time-frequency plot of an eight-pole no-load 22 MVA synchronous gen- erator in a healthy case and under 20% static eccentricity and 20% dynamic eccentricity faults. In the healthy synchronous generator, the stalks have a uniform length and color intensity. The static eccentricity fault does not change the harmonic contents of the time-frequency plot, but the length of the stalks changes. Based on the location of the installed sensor, the length of the stalks increases or decreases compared with the healthy case. The length of the stalk increases if the minimum air gap between the stator core and the rotor decreases, and vice versa.

The dynamic eccentricity fault changes the time-frequency pattern because the induced sen- sor voltage under the dynamic eccentricity contains harmonics due to the asymmetrical air gap. The dynamic eccentricity fault results in the following two distinct patterns in the time-frequency plot of the induced sensor voltage:

1. A fire-flame shaped pattern is observed, where the lengths of the stalks are shorter at the sides of the window than at its center.
2. A sine wave pattern is observed in the top and bottom envelope of the time-frequency plot.

As shown in Figure 12.39, the dynamic eccentricity fault pattern is repetitive, indicating the peri- odic nature of the dynamic eccentricity fault.

Figure 12.40 shows the time-frequency plot of the induced sensor voltage of the installed sensor in a 16-pole no-load 42 MVA synchronous generator in the healthy case and under 20% static eccen- tricity and 20% dynamic eccentricity faults. The time width of the 16-pole synchronous generator is 160 ms. The 20% static eccentricity fault results in a stalk length reduction, while the pattern is similar to that of the healthy generator. The dynamic eccentricity pattern is similar for both the 16-pole 42 MVA synchronous generator and the 22 MVA synchronous generator and contains a sine wave envelope in both the upper and lower parts, in addition to the fire-flame shaped pattern.

The introduced method, based on analysis of the induced voltage of the sensor using a continuous wavelet transform, is applied to the measured sensor data of the 22 MVA generator. The amplitudes of the four distributed sensors with 90 mechanical degrees of displacement are the same, indicating

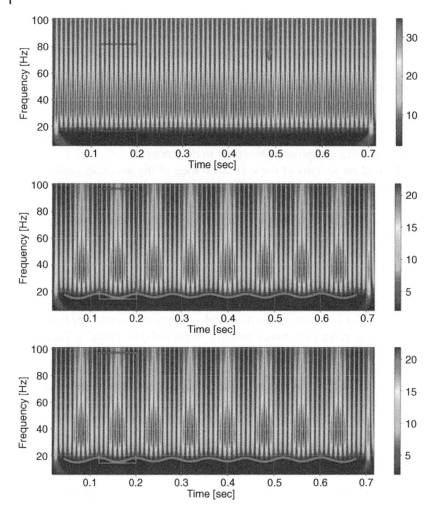

Figure 12.39 Time-frequency plot of the induced voltage of a sensor in an eight-pole 22 MVA synchronous generator in the healthy case (top row) and under 20% static eccentricity (middle row) and 20% dynamic eccentricity faults – simulation results. Based on [40].

that the synchronous generator does not suffer from a static eccentricity fault. Figure 12.41 shows the time-frequency plot of the induced voltage of the sensor of the no-load 22 MVA synchronous generator and the on-load 17 MW generator. Two indicators of the dynamic eccentricity faults are evident: the sine envelope and the fire-flame shaped patterns. Therefore, the synchronous generator operates under a dynamic eccentricity fault. In addition, a couple of arrows indicate the stalks in the time-frequency plot. The arrows represent the short circuit fault in the rotor field winding. Even in the case of the dynamic eccentricity fault, symmetry exists around the center of the fire flame in the window, whereas the length of the stalk is shorter in the synchronous generator with a short circuit fault in the rotor field winding than in the healthy state. The on-load operation of the synchronous generator indicates that the pattern is similar to the no-load case. Furthermore, the effect of an inter-turn short circuit fault in the time-frequency plot becomes more evident compared with the no-load operation. The reason is that the current passing through the rotor field winding

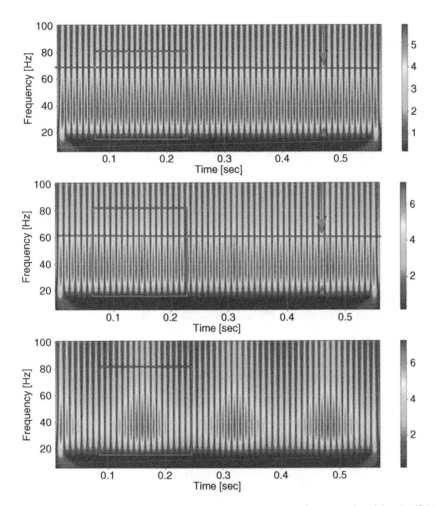

Figure 12.40 Time-frequency plot of the induced voltage of a sensor in a 16-pole 42 MVA synchronous generator in the healthy case (top row) and under 20% static eccentricity (middle row) and 20% dynamic eccentricity faults – simulation results.

increases in the on-load operation condition, resulting in a more evident impact on the asymmetric air gap magnetic field and, consequently, on the stray magnetic field on the stator backside.

Figure 12.42 shows the time-frequency plot of the induced voltage of the two installed sensors in front of each other on the backside of the 16-pole 42 MVA synchronous generator. The static eccentricity increases the stalk length on the side where the distance between the stator and rotor core shortens, whereas the stalk length decreases where the distance between the stator and rotor core is large. The static eccentricity fault impact is evident in the time-frequency plot of the no-load synchronous generator, as shown in Figure 12.42. Figure 12.43 shows the on-load synchronous generator and its impact on the time-frequency plot of the induced voltage of the sensor. The synchronous generator has a parallel winding connection that suppresses the impact of the eccentricity fault by increasing the current in the parallel winding. Therefore, the static eccentricity fault has less of an impact on the length reduction of the stalks of the on-load synchronous generator than during the no-load operation.

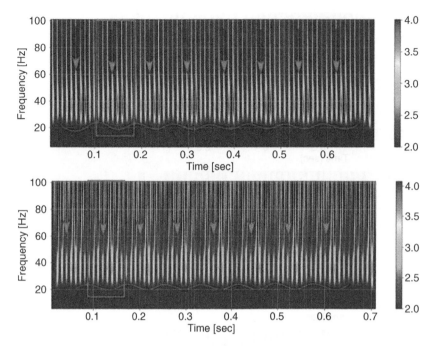

Figure 12.41 Time-frequency plot of the induced voltage of a sensor installed on the backside of a no-load 22 MVA synchronous generator (top) and an on-load 17 MW generator (bottom) – field test results.

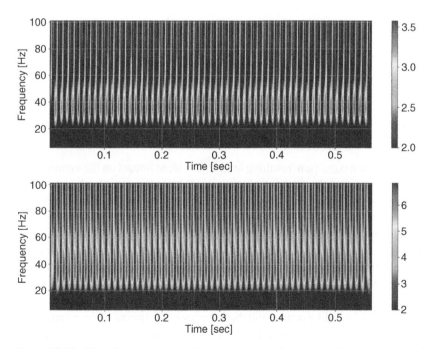

Figure 12.42 Time-frequency plot of induced voltage of a sensor in four sensors installed on the backside of a 42 MVA no-load synchronous generator - field test results.

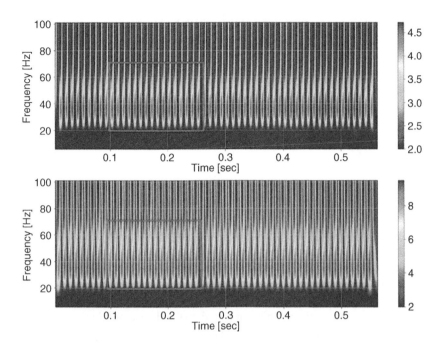

Figure 12.43 Time-frequency plot of the induced voltage of a sensor in four sensors installed on the backside of a 42 MVA on-load synchronous generator – field test results.

The detection method based on the application of the continuous wavelet transform and stray magnetic field has several advantages [13, 40–42]:

1. The method does not require healthy case data for the synchronous machine. Providing healthy state data for synchronous generators operating for decades in industry and power plants is impossible.
2. The method can detect the fault in its early stage, thereby avoiding serious damage to the generator.
3. The method can detect the fault type because it provides a distinctive pattern for each fault type.
4. The method can detect the no-load and on-load faults of the synchronous generator.
5. The method can detect the fault regardless of the power rating and the topology of the synchronous generator.

12.3 Stator Core Fault

12.3.1 Introduction

A stator core fault in a synchronous generator is not as common as electrical and mechanical faults, such as short circuits and eccentricity faults. It is also not a major concern in small electric machines, indicating that the stator core fault occurs mostly in large electric machines. The stator core maintenance and rebuilding process is costly and time consuming, which underscores the importance of stator core monitoring for large synchronous generators. A stator core fault in a 430-MVA synchronous generator reported in reference [41] resulted in a forced stoppage of the generator and replacement of the stator core and winding. The estimated repair cost for the generator was $17.5 M and the loss of revenue for the utility, which had to purchase power during the maintenance time, was $270 M [43]. Therefore, reliable condition monitoring of the stator core is essential.

The stator core of the synchronous generator is made of insulated steel laminations that allow magnetic flux penetration and reduce the eddy current losses. In large hydro-generators, the stator core is divided into several pieces and requires on-site manufacturing. The reason is that the large hydro-generators have a large diameter and are therefore difficult to transfer, so the stator core is divided. The following two faults are reported in the large synchronous generators:

1. Inter-laminar core failure.
2. Stator core joint failure.

The inter-laminar insulation is subject to deterioration and damage; therefore, over the long term, the inter-laminar insulation loses its properties. The lack of proper insulation between the laminations causes a high circulating eddy current compared with the normal operation of the generator. The loop of the circulating eddy current consists of the shorted laminations, the key bar, or the through bolt, as shown in Figure 12.44. The circulating current in the laminations results in a local temperature rise and a consequent insulation degradation of the inter-laminar region. The progression of the insulation degradation and the increased current eventually results in the burning and melting of the stator laminations. This can also damage the stator winding system. Several factors can result in the inter-laminar fault in the stator core [43]:

1. Mechanical damage to the stator inner surface during manufacturing or maintenance.
2. Foreign particles lodging between the inter-laminar areas.
3. Rubbing of the stator or rotor during assembly or maintenance.
4. Arcing due to a stator winding system failure.
5. Loose winding, lamination, and slot wedge vibration.

Several methods, which are mostly based on offline inspection of the stator core, are available to determine an inter-laminar core failure. The following is a brief description of these methods [20, 44–47]:

12.3.2 Core Loss

The inter-laminar fault detection method based on core loss is an appropriate method for small or medium core electric generators that can be transported to a maintenance service center. A faulty

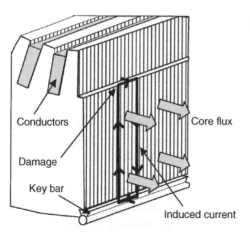

Figure 12.44 Current flow due to an inter-laminar fault [44]/with permission of IEEE.

Conductors

Damage

Key bar

Core flux

Induced current

core section requires more power than a healthy core, indicating that the comparison between the input power of the core can indicate the health status of the generator. However, this method only gives a general perspective of the core health status and cannot provide useful information regarding the location of the inter-laminar fault. Achieving a certain level of the magnetic field in the core requires constant power to magnetize the stator core. Defects in the lamination insulation create a circulating current in the stator core and increased power is then required to magnetize the stator core. The core loss method involves winding a series of turns (like a toroidal coil) around the stator core to provide a nominal magnetic field in the stator core. The required current and the voltage of the wattmeter input must be provided by a proper current and voltage transformer because the current and voltage for energizing the stator core are high. In addition, the stator core weight is required to calculate the W/kg. The stator core mass can be calculated using the inner and outer diameter of the core and the length of the stator stack. The density of the rolled steel utilized for the laminations is 7850 kg/m^3 [45].

The core loss of lamination steels is around 2–3 W/kg, while the core loss of the old manufactured lamination steel is 6 W/kg. Therefore, the measured value of the core loss should not exceed 10 W/kg; otherwise, this is a clear indication of an inter-laminar failure in the stator core. In a case where the trend of core loss is recorded, the value of the measured core loss should not exceed more than 5% compared with the previously measured core loss [45].

12.3.3 Rated Flux

The rated flux test is a loop test method to detect an inter-laminar fault in the stator core. In this method, the stator core is magnetized up to the nominal stator core magnetic field. The impact of flux in the stator core is then monitored using a thermal imaging camera or thermocouples mounted in the stator core. The rated flux test can be used for synchronous generators with different power ratings and topologies, and it can detect the fault severity and its location, even for a deep-seated core fault. However, the method requires a power supply for energizing the core to the rated flux density. This method can also aggravate the existing inter-laminar fault because the test must be performed for a few hours for deep-seated core faults.

A proper cable size and number of turns must be designed based on the core size [46]. The current passing through the cable in the axial direction of the stator core creates a magnetic field in the stator core. The magnetic field acts as a heat source in the stator core, so a thermal imaging camera is required to monitor the thermal distribution in the stator core. In the case of an inter-laminar fault, a hot spot, where its temperature is higher than the rest of the healthy core, is detected. The reason for the hot spot is the circular eddy current in the faulty core segment. Although a failure in the stator core surface can be quickly detected, the diagnosis of a deep defect in a large synchronous generator requires at least half an hour, because the healthy segments of the core behave as a heat sink for the faulty part. The recommended test time is 2 hours for a large synchronous generator, with a thermal measurement taken every 15 minutes. In addition, the core must be cooled down to the ambient temperature if an additional measurement is required.

Figure 12.45 illustrates the test setup required for the rated flux test. A commercialized power source is available that can adjust the current and voltage levels to magnetize the synchronous generators at the nominal level. However, in the case where a large synchronous generator, such as a hydro-generator, is examined, an autotransformer supplied by the internal supply of the power plant can be utilized. The power cables must be distributed in the circumferential of the stator cores of large synchronous generators to ensure a uniformly magnetized core, as shown in Figure 12.46.

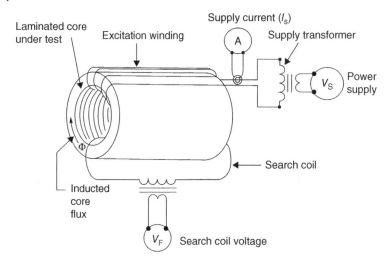

Figure 12.45 The rated flux test setup used for lamination degradation analysis.

Figure 12.46 Excitation windings are utilized to energize the stator core for the inter-laminar fault detection.

The main results of this test are the measured temperatures using thermocouples, a resistance temperature detector-probe (RTD), and the thermal imaging camera to indicate the temperature distribution in the stator core (see Figure 12.47). The thermal inspection must be performed from the time that the core is magnetized. A damaged inter-laminar section close to the surface can promptly indicate the hot spots, while a deep-seated fault takes a few minutes to an hour to show hot spots. Note that, due to the flux concentration in some parts of the stator core, a uniform temperature distribution should not be expected, even for a healthy stator core. However, if the thermal variation is above 5 °C in the large uniformly magnetized synchronous generator, this is an indication of an inter-laminar defect in the core. In addition, a local temperature rise above 20 °C is an indication of serious lamination insulation degradation [48].

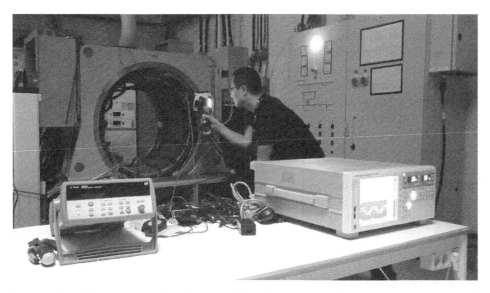

Figure 12.47   Detection of an inter-laminar fault in the stator core of a synchronous generator using a thermal camera.

12.3.4   EL-CID Method

The main drawback of the rated flux test method is the need for a custom-made power source and extra equipment, such as winding and measurement facilities, for inter-laminar fault detection in the large synchronous generator. The need for the custom-made high-power source to fully magnetize the stator core to reach a nominal flux density in the stator backside is the main problem of the rated flux density method. In the 1980s, an electromagnetic core imperfection detector (EL-CID) was invented to solve this problem. This EL-CID method can use a power source with a smaller capacity, since induction of only 3–4% of the rated flux is required in the stator core [48, 49]. In this method, an AC power source of 120 or 220 V can be used, even for large synchronous generators. In this method, the time of inspection is significantly shortened compared with the rated flux method.

In the EL-CID method, the axial current is assumed to flow through the damaged core segments, even when only part of the rated flux is applied. In this method, the axial current is detected based on its generated magnetic field rather than the temperature rise or loss. A special pickup, known as a Chattock coil, is used to determine the voltage, which is proportional to the axial current that flows through the damaged laminations. A solenoid coil with a double layer of wire is wound on a U-shaped form and placed on two stator teeth. The induced voltage in the coil due to the axial current is approximately proportional to the line integral of the magnetic field along its length. The output of the Chattock coil cannot be used directly to interpret the health status of the core laminations, since the coil-induced voltage is due to the axial current caused by the fault and the circumferential magnetic field that is used to magnetize the stator core. A solution to this problem arises from the fact that a 90° phase shift exists between the excitation winding flux and the faulty flux. Therefore, the induced voltage is fed to the signal processing system, where the induced voltage due to the circumferential magnetic field is eliminated by feeding a reference signal to the signal processing system. The remaining induced voltage is then due to the axial current of the faulty stator core. To avoid the danger due to the induced voltage in the stator winding caused by the excitation winding, all three stator windings must be grounded.

The amplitude of the measured induced voltage is converted to current in a healthy core. This value is around 100 mA when only 4% of the rated flux is applied. Therefore, a measured current in the stator teeth below this value indicates that the stator core laminations have good insulation. If an inter-laminar core fault is indicated by the EL-CID method, a further examination must be performed using a rated flux test to identify the severity and location of the fault.

The El-CID method is a significant improvement over the rated flux in terms of the power source capacity reduction. However, the signal-to-noise ratio decreases due to the reduced flux density in the stator core; therefore, data interpretation is difficult. A further difficulty is encountered when attempting to scan the stepped core end region of the stator using the Chattock coil because the coil alignment on a stepped core is difficult and the obtained signal has a high noise that requires multiple measurements. The iron core-based Chattock coil can be replaced with the air-core type to increase the signal-to-noise ratio, thereby providing a higher sensitivity. The main difficulty of the iron-core-based Chattock is that the air between the stator tooth iron and the probe iron generates fluctuations in the measured induced voltage. Since the stator tooth has a non-uniform surface, the variation in the air gap length as the probe slides over the tooth causes a fluctuation that obscures the results [43, 46, 47].

The stator core of the large hydro-generator assembled in the hydropower plants consists of several segments that are prone to movement for the following reasons:

1. Improper or loose connection of the stator core segments in the joint field.
2. Stator core expansion during on-load operation, and contraction due to stoppage of the generator.

Movement of the stator core segments results in variations in the air gap magnetic field and increases the vibration level in both the stator core and the core segment joints. Little research has focused on stator core joint failure in large synchronous generators. A solution for the detection of stator core joint faults is the use of an air gap sensor installed on the rotor pole, either at the top or at the bottom of the rotor pole. The synchronous generator is then rotated and the magnetic field of the stator is recorded. A polar diagram of the air gap magnetic field measured on the rotor side can provide a visual understanding of the stator core joints. If the polar diagram is circular, this indicates a healthy stator core, whereas any ovality or variation of the magnetic field in the polar diagram close to the core joints is an indication of a fault in the stator core joints.

12.4 Broken Damper Bar Fault

12.4.1 Introduction

Two sets of circuits are present in the rotor of a synchronous generator and are associated with the rotor field winding and the damper bars. The damper of a synchronous generator consists of a couple of copper bars distributed over the rotor pole shoe. Both ends of the damper bars are short-circuited by copper rings. Figure 12.48 shows a synchronous generator with damper bars in each pole that are short-circuited by the end rings. The number of damper bars and their distribution depend on the design of the synchronous generator [50].

A broken damper bar fault is less likely in a synchronous generator than in an induction motor because the damper bars are mostly active during the transient operation of the synchronous generator. The transient nodes of the synchronous generator include the start-up period, testing for fault existence in the power grid, and load variations. The damper bars in a synchronous generator

Figure 12.48 A salient pole synchronous generator with damper bars and end ring.

behave similarly to an induction motor and are active when the magnetic field of the rotor winding is not synchronized with the stator magnetic field. This indicates that when the stator and rotor magnetic fields are synchronized, the damper bars do not see any magnetic field variations, and the induced voltage and current are theoretically almost zero. However, the following conditions can result in current circulation in the damper bars during steady-state operation [22, 51–54]:

1. The air gap magnetic field pulsation due to the rotor and stator slotting effects.
2. The internal faults, such as eccentricity faults and short circuit faults.
3. The space harmonics in the air gap magnetic field due to the fractional slot winding.
4. Load variations.

The resistivity of the damper bars is also low and the low induced voltage can result in a significant circulating current in the damper bars. The stator short circuit fault in a synchronous generator is claimed in reference [55] to result in a non-uniform magnetic field in the air gap. The asymmetric air gap magnetic field induces a current in the damper bars; consequently, a larger current passes through the damper bars compared with the healthy damper bar current.

The following are reasons for broken damper bar faults in the synchronous generator [50]:

1. Improper connections between the damper bars and the end rings.
2. A high number of starts and stops in the synchronous generator, as this causes a huge current to pass through the damper bars and produces a local hot spot at the connection points.
3. Thermo-mechanical stress caused by the uneven distribution of the current in the damper bars, due to the location of the damper bars in the rotor pole shoes.

One broken damper bar fault can result in a multiple broken damper bar fault in a short time. The reason is that the current of the damaged bar is divided between the adjacent damper bars. Since the cross section of the damper bars is designed for a certain amount of current, the additional current due to the damaged damper bar results in additional losses, and the excessive temperature rise in the damper bars and the joint results in increasingly greater numbers of damaged damper bars. This section covers some approaches, based on analysis of the air gap magnetic field, stray magnetic field, and stator and rotor current, that deal with detection of broken damper bar faults.

12.4.2 Single-Phase Rotation Test

The most common way to detect a broken damper bar is by visual inspection. However, visual inspection is not appropriate for a large generator disassembly. An offline method applied to the induction motor can also be used to detect a broken rotor bar fault in a synchronous generator. In this method, a power supply with 10–20% of the rated voltage and the same frequency as the induction motor excites the stator winding, and the rotor is rotated in small steps. The current must be constant in the healthy induction motor and a broken damper bar fault results in current fluctuation. The single-phase test is applicable to synchronous generators. However, due to the rotor saliency, the stator winding current, even in the healthy case, does not have a constant value. The broken damper bar fault results in an increase in the current in the synchronous generator because of rises in the d-axis equivalent impedance with the broken damper bar. Figure 12.49 shows the result of the single-phase test for broken damper bar detection [56]. Although this method can detect a broken damper bar fault, rotating the rotor shaft, especially for a large synchronous generator, is almost impossible.

12.4.3 Air Gap Magnetic Field

Similar to the other types of faults, such as eccentricity or short circuit faults, a broken damper bar fault also leads to a variation in the air gap magnetic field, and this variation can be tracked during the transient period of synchronous generator operation. Note that the field winding of the line-started synchronous motors is short-circuited using a starting resistor up to 20 times the rotor field winding resistance. The three-phase supply of the stator winding creates a symmetric magnetic field in the air gap as the current passes through the windings. The induced voltages in the damper bars produce a magnetic field in the air gap that lags behind the stator magnetic field. Once the rotor reaches 95% of the synchronous speed, the rotor field winding produces a

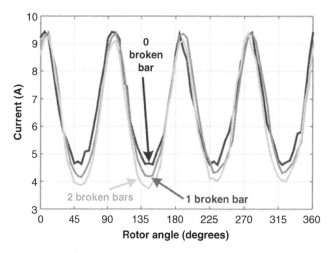

Figure 12.49 Current of a single-phase rotation test in a synchronous generator [57]/with permission of IEEE.

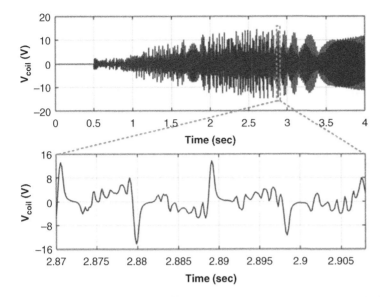

Figure 12.50 Induced voltage in a sensor installed on the stator tooth of a four-pole synchronous generator [57]/with permission of IEEE.

torque and the rotor and stator become synchronized. The rotor damper bar becomes inactive after synchronization. Therefore, a limited time window is available that can be utilized for a broken damper bar fault detection [56]. Figure 12.50 shows the induced voltage in the sensor installed on the stator tooth during the start-up transient [57].

The main frequency component in the induced sensor voltage of the air gap magnetic field is the supply frequency (f_s). In addition, a large fluctuation occurs in the air gap magnetic field due to the reluctance variation of the rotor pole saliency. Therefore, a rotor rotational frequency component is also induced in the induced sensor voltage, as follows:

$$f_r = \frac{(1-s)f_s}{p} \tag{12.6}$$

where s is the slip and p is the number of pole pairs. The f_s and the rotor rotational frequency are modulated, which induces the kf_r sidebands of the f_s in the induced sensor voltage, as follows:

$$f_{flux} = f_s \, k f_r \tag{12.7}$$

The f_r and f_{flux} components can be traced in the frequency spectrum of the induced sensor voltage of the air gap magnetic field. The frequency components of the f_r and f_{flux} are the slip-dependent components. The slip changes between 1 and 0 from the acceleration to the steady-state operation. Table 12.2 shows the variation in f_{flux} for different values of the slip and k. The f_{flux} shows the different changes in the different values of k. The air gap magnetic field contains the harmonic components with $k = \pm 4, \pm 8$, and ± 12 for the four-pole synchronous generator. The broken damper bar fault can change the frequency components in f_{flux} with $k = \pm 1$.

Figure 12.51 presents the time-frequency plot of the induced sensor voltage of the installed sensor on the stator tooth during start-up of a four-pole synchronous generator in a healthy case and under a 1 and 2 broken damper bar fault. The components of the rotor saliency in the f_{flux} are observed for $k = \pm 4, \pm 8$ in Figure 12.51. The broken damper bar fault increases the intensity of the region

Table 12.2 The variation of the f_{flux} component in the magnetic flux from acceleration until synchronous speed for various values of k (up to ±12).

k	0	+1	+2	+3	+4	+5	+6	+7	+8	+9	+10	+11	+12
$s=1$	60	60	60	60	60	60	60	60	60	60	60	60	60
$s=0.5$	60	75	90	105	120	135	150	165	180	195	210	225	240
$s=0$	60	90	120	150	180	210	240	270	300	330	360	390	420

k	0	−1	−2	−3	−4	−5	−6	−7	−8	−9	−10	−11	−12
$s=1$	60	60	60	60	60	60	60	60	60	60	60	60	60
$s=0.5$	60	45	30	15	0	−15	−30	−45	−60	−75	−90	−105	−120
$s=0$	60	30	0	−30	−60	−90	−120	−150	−180	−240	−240	−270	−300

surrounded by the saliency components in the f_{flux}. Increasing the fault severity results in a large increment of the frequency component of the f_{flux} with the $k=\pm1$, as shown in Figure 12.51.

The impact of a broken damper bar fault in a 430 kVA synchronous machine is considered in reference [58]. The machine starts by operating as a pump (asynchronous machine); therefore, a huge starting current passes through the damper bars. Although the air gap magnetic field changes during start-up due to varying the damper bar current, its shape under faulty conditions also differs from the healthy case. This magnetic field is not symmetrical and contains harmonics generated due to the fault. Therefore, analyzing the air gap magnetic field and comparing the harmonic contents of the healthy machine and the suspected faulty machine can indicate a broken damper bar fault. An indicator based on the start-up time of the synchronous machine is also introduced in reference [58]. Due to the broken damper bar fault, the starting time of the synchronous machine increases according to the number of broken damper bars. However, this method is not suitable for diagnosing a damper bar fault because many factors, such as faults occurring in mechanical parts of the machine or increases in the shaft load, can lead to a longer start-up time of a synchronous machine [58].

12.4.4 Stray Magnetic Field Monitoring

The method based on analysis of the air gap magnetic field can detect a broken damper bar fault during transient operation of the synchronous generator; however, installation of the search coil on the stator tooth is difficult and in some cases requires rotor removal. Therefore, methods based on application of stray magnetic field monitoring are proposed in references [51, 59, 60].

The faults in the rotor of the synchronous generator influence both the radial stray flux and the axial stray flux [61, 62]. During acceleration of the synchronous machine, a frequency component proportional to the rotor slip (sf_s) is induced in the rotor field winding and damper bars. As shown in reference [60], an axial leakage flux is generated in both the drive and non-drive ends of the machine due to the circulating current of the sf_s harmonic component in the rotor field winding and the damper bars. Since the magnetic field generated by the sf_s harmonic component is symmetrical in the healthy machine, the net axial magnetic field must be zero. Therefore, having a fault in the rotor field winding and the damper bars can distort the symmetric magnetic field and result in a

Figure 12.51 Time-frequency plot of the induced sensor voltage installed on the stator tooth of a four-pole synchronous generator operating (a) in the healthy condition, (b) under a 1 broken damper bar, and (c) a 2 broken damper bar fault [57].

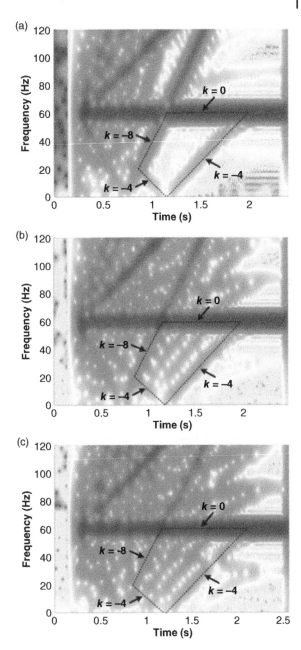

non-zero axial leakage flux generated by the sf_s harmonic component [59]. Figure 12.52 shows the induced voltage in the sensor installed in the axial direction of the synchronous generator around the rotor shaft in the healthy state and under a broken damper bar fault. The impact of one broken damper bar clearly results in an increase in the induced voltage of the sensor.

Figure 12.53 shows the time-frequency plot of the induced voltage of the sensor obtained using an STFT. Several harmonic components exist in the time-frequency plot of the induced axial voltage, where the main component is the fundamental frequency component (60 Hz).

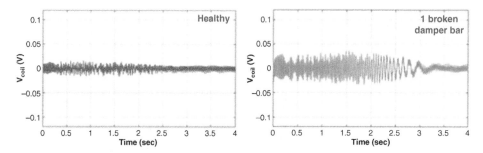

Figure 12.52 Induced voltage in a sensor installed in the axial direction of a four-pole synchronous generator during start-up [60].

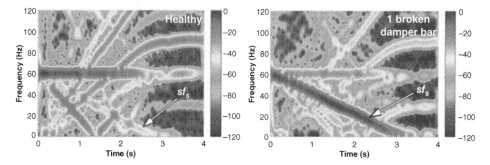

Figure 12.53 Time-frequency plot of the induced voltage in a sensor installed in the axial direction of a four-pole synchronous generator during start-up [60].

Moreover, several harmonic components are present in the induced axial voltage, in addition to the sf_s harmonic component. Since the magnetic field in the healthy generator is symmetric and the sf_s harmonic component must be theoretically zero, the sf_s harmonic component remains in the time-frequency plot of the healthy synchronous generator [59]. Having one broken damper bar fault increases the intensity of the sf_s harmonic component in the time-frequency plot; however, the high sf_s harmonic component changes due to the inter-turn short circuit fault in the rotor field winding. Therefore, diagnosis of the fault type is difficult. The impact of a broken damper bar fault on the radial flux is also investigated in reference [60], where the obtained result was the same as that of the axial leakage flux.

A method based on the analysis of the induced sensor voltage installed on the stator backside of the synchronous generator is proposed in reference [51]. In this method, a synchronous generator, coupled to the prime mover, rotates in a synchronous speed and then a trapezoidal-shaped DC current is fed to the rotor field winding. Figure 12.54 shows the waveform of the applied current to the rotor field winding. The input current consists of two parts: the transient parts (ramp-up and ramp-down) and the steady-state part. The maximum amplitude of the current is set based on the no-load rotor field current, and the length of the applied current can be adjusted based on the actual needs. Each rotor pole must pass over the installed sensor at least once; therefore, the minimum required time for each part is the time taken for one mechanical revolution of the rotor.

Although the current in the damper bars is assumed to be zero during the steady-state operation of the synchronous generator, a circulating current is always present in the rotor damper bars. The current amplitude and waveform pattern in the damper bars depend on their location in the rotor pole shoe, where the current is higher in the damper bars located at the pole edge than in

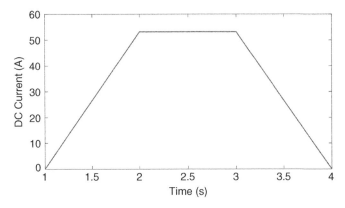

Figure 12.54 Current applied to a rotor field winding of 100 kVA no-load synchronous generator.

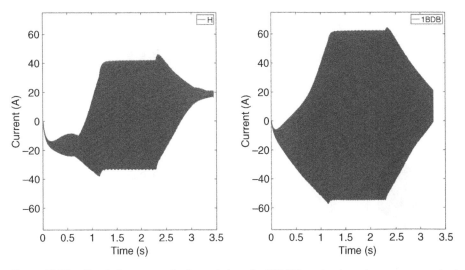

Figure 12.55 Circulating current in the end ring of a 100 kVA no-load synchronous generator in the healthy state and under one broken damper bar fault.

the middle of the rotor pole shoe. The current pattern in the damper bars and the end ring must follow the same pattern as a power source is fed to the rotor field winding. Figure 12.55 shows the circulating current in the end ring of a 100 kVA synchronous generator in the healthy state and under one broken damper bar fault. The current is higher in the end ring of the synchronous generator with one broken damper bar fault than in the healthy operation. Since the broken damper bar current is divided between the two adjacent damper bars, if the broken damper bar is the one at the rotor pole edge, the divided current passes through the end ring to reach the damper bar in the adjacent pole [51].

Figure 12.56 shows the induced current in the rotor damper bars due to the trapezoidal shape current of the rotor field winding in a healthy and under one broken damper bar fault. In Figure 12.56, the damper bar current of the two damper bars at the two separate rotor pole edges (damper bar 1 in pole 1 and damper bar 7 in pole 2, as shown in Figure 12.57). In addition, its neighboring damper bar (damper bar 2 in pole 1) is presented. The pattern and the amplitude of the current in the damper bars differ depending on the location of the damper bars. The current

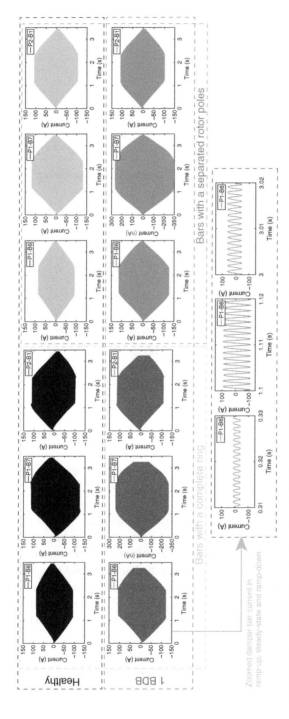

Figure 12.56 Induced current in rotor bars due to the trapezoidal shape of the field current in a healthy generator during a no-load operation (first row waveforms) and with one broken damper bar fault (pole 1, bar 7 [P1-B7]) (second row waveforms). The first three columns represent bars with a complete end ring and the next three columns represent bars with separated rotor poles. The first three columns show the current in damper bar 6 in pole 1 (P1-B6), damper bar 7 in pole 1 (P1-B7), and damper bar 1 in pole 2 (P2-B1).

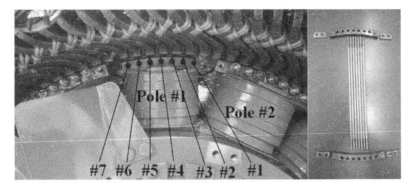

Figure 12.57 Pole with removed damper bars to simulate a broken damper bar fault in a 100 kVA synchronous generator.

amplitude is higher for the damper bars located at the rotor pole edges, due to the high reluctance path, than for the damper bars located in the middle of the rotor pole shoe [51].

Although the broken damper bar current is always assumed to be zero, a non-zero current always circulates in the faulty damper bar with the magnitude of few µA. The current of the adjacent damper bars of the faulty damper bar increases regardless of whether they are in the same pole. The amplitude of the maximum damper bar current in the adjacent damper bar in the same pole increases from 85 to 100 A, while the current of the adjacent damper bar in the neighboring pole increases from 100 to 106 A.

In the large synchronous generator, an inter-pole connection exists that connects the damper bars of each pole to each other to increase the subtransient reactance of the generator in the quadrature axis. A small subtransient reactance in the quadrature axis results in instability and vibration of the synchronous generator. Therefore, a connection between the poles is required to ensure synchronous generator stability. Figure 12.56 shows the damper bar current of the healthy and under one broken damper bar for a synchronous generator without an inter-pole connection in the rotor pole. The damper bar current in the synchronous generator without the inter-pole connection does not increase in the adjacent damper bar if the damper bars are not in the same pole.

The impact of current variation in the damper bars, both in the transient and steady-state operation of the synchronous generator, can be observed in the stray magnetic field on the stator backside. Figure 12.58 shows the induced voltage of the sensor and applied discrete wavelet transform to it in the 100 kVA synchronous generator. The shape of the induced voltage of the sensor is similar to the shape of the applied current to the rotor field winding. A comparison between the induced voltage of the sensor of the healthy and under one broken damper bar indicates that the broken damper bar changes the induced voltage of the sensor pattern for all three working regions of the synchronous generator, including ramp-up, steady-state, and ramp-down.

A discrete wavelet transform is applied to the induced voltage of the sensor. The Daubechies mother wavelet with eight sub-bands is utilized where only four sub-bands (D1–D4) show the highest sensitivity to the broken damper bar fault. The wavelet entropy is used to quantify the variation in the induced voltage of the sensor due to the broken damper bar fault. The magnitude of the wavelet entropy for sub-bands D4 and D3 under one BDB during the ramp-up period increases from 405 to 482 and 6610 to 7166, respectively. The amplitude of the wavelet entropy during the ramp-down period also increases for one broken damper bar fault in the D4 and D3 sub-bands, from

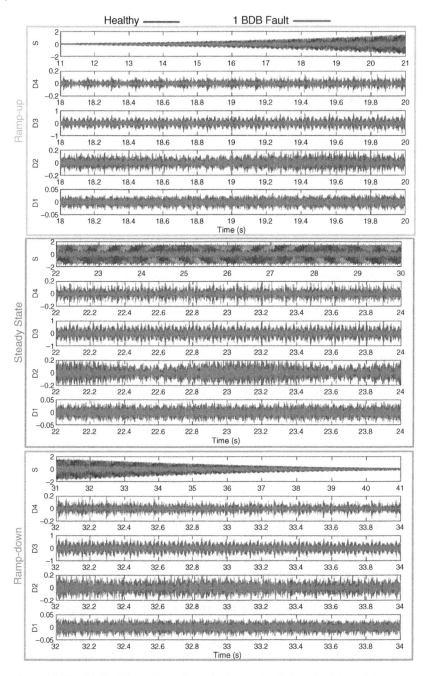

Figure 12.58 Applied discrete wavelet transform to the induced voltage of the sensor of a 100 kVA synchronous generator, where the trapezoidal shape current is applied to the rotor field winding.

382 to 607 and from 6397 to 8355, respectively. Entropy applied to the induced voltage of the sensor of the detailed sub-bands of D4 and D3 during steady-state operation shows marked variations. The amplitudes of the D4 and D3 sub-bands increase from 959 to 1181 and from 10 403 to 11 473, respectively [51].

The following criterion function, based on the analysis of the wavelet entropy of the faulty sub-band with the healthy sub-band, is proposed in reference [51]:

$$\text{Criterion function} = \frac{\mid \text{WEH}_{Di} - \text{WEF}_{Di} \mid}{\text{WEH}_{Di}} \times 100 \tag{12.8}$$

where WEH_{Di} and WEF_{Di} are the wavelet entropy of the corresponding wavelet sub-band in the healthy and faulty conditions. The results of the applied criterion function to the induced voltage of the sensor in the case of a broken damper bar fault in various locations of the rotor pole shoe and several faulty damper bars are presented in Table 12.3.

Since the location of the damper bars in the rotor pole shoe has a significant impact on the circulating current inside that bar, the criterion function values for broken damper bars distributed in the rotor pole shoe must be different. The criterion function is higher for the damaged damper bars located at the edges compared with the damper bars in the middle of the pole. The results indicate that the broken damper bar in the middle of the rotor pole has a smaller impact on the distortion of the air gap magnetic field and consequently on the stray magnetic field on the stator backside.

A comparison of the results of the various broken damper bar faults during ramp-up and ramp-down in Table 12.3 indicates that the amplitude of the criterion function during ramp-down is higher for the same location and the same number of broken damper bars. The inequality of the criterion function in the ramp-up and ramp-down is due to the saturation impact. The time constant is higher during ramp-up than ramp-down and requires a more magnetizing current to reach the operating point, which is close to the knee of the magnetization characteristic. By contrast, the working point during ramp-down is not close to the knee; therefore, decreasing the rotor magnetic field current causes the working point to locate in the linear part of the magnetization characteristic, which is a site more sensitive to a fault.

Detection of a middle broken damper bar fault in a synchronous generator with an odd number of damper bars in one rotor pole shoe is a very difficult task because the current is low and it has very low sensitivity to the broken damper bar fault. Therefore, the broken damper bar fault in the middle damper bar does not noticeably change the air gap magnetic field symmetry or, consequently, the

Table 12.3 Criterion function values for four wavelet sub-bands for different numbers of broken damper bar faults in a no-load synchronous generator (in %).

Cases	Ramp-up interval				Steady-state interval				Ramp-down interval			
	D-4	D-3	D-2	D-1	D-4	D-3	D-2	D-1	D-4	D-3	D-2	D-1
1BDB 1	15.97	7.75	3.37	7.4	18.8	9.3	10.1	7.9	37	23.4	17.1	15.9
1BDB 2	13	7	5.1	1.8	7.1	2.4	2.5	1.75	6	1.8	1.1	2.9
1BDB 4	0.6	0.4	2.9	1.8	5.1	3.1	3.5	3.5	11.36	7.6	6.7	8.7
2BDB 1&2	31.7	19.4	7.2	1.8	15.3	7.3	6.3	0.8	20.4	11.4	0.4	5.8
2BDB 1&7	25.1	12.1	4.2	5.5	13.3	4.1	2.0	6.1	20.9	9.4	0	4.3
3BDB 3&4&5	7.2	3.6	0.2	1.8	3.9	3.1	1.45	0	7.5	4.5	2.1	1.4
7BDB 1 to 7	31.5	18.1	12.1	5.5	29.7	15.1	18.8	18.4	44.1	28.4	24.1	23.1

Table 12.4 Values of the criterion function for a synchronous generator with the isolated poles under one broken damper bar fault.

Periods	Sub-band 4 (%)	Sub-band 3 (%)	Sub-band 2 (%)	Sub-band 1 (%)
Ramp-up	5.5	1.9	1.0	3.4
Steady-state	0.1	0.3	0.7	0.5
Ramp-down	4.0	5.9	5.5	3.3

stray magnetic field on the stator backside. However, detection of a middle broken damper bar is possible using the analysis of the criterion function during both steady-state and ramp-down operations.

The broken damper bar fault results in a division of the faulty bar current into its adjacent damper bars. Since the cross section of the damper bars is designed for a certain current density, the additional current results in increased loss and, consequently, the appearance of a local hot spot and vibration. The impact of local hot spots and vibration over the long term can damage the healthy damper bars. A single broken damper bar can clearly damage its neighboring damper bars. Table 12.3 shows that increasing the number of damaged broken damper bars increases the criterion function. However, the amplitude of the criterion function for three broken damper bars in the middle of the rotor pole shoe does not increase as much as for two broken damper bars in the pole edge, because the distortion of the magnetic field is less for the former condition.

In some synchronous generators, a separated damper bar is preferred due to the cost reduction and avoidance of deformation in the end ring due to the centrifugal force. The isolated winding circuit in the synchronous generator results in a subtransient salience (the difference between the subtransient reactance in the d-axis and q-axis). This, in turn, results in a significant subtransient short circuit current in the case of a phase-to-phase short circuit fault.

Table 12.4 presents the result of the application of the criterion function to detect a broken damper bar fault in an isolated synchronous generator. The results reveal the difficulty of using the proposed method for broken damper bar fault detection in the isolated damper bar synchronous generator because the impact of the broken damper bar current resulting in the asymmetric air gap decreases due to the impossibility of circulation of the faulty damper bar current in the end ring.

12.4.5 Stator Current

The stator current of the synchronous generators can be used to detect a broken damper bar fault [63]. A signature extracted from the harmonic components of the stator current is introduced as follows:

$$f_{BDB} = \frac{k.f}{p} \tag{12.9}$$

where f is the supply frequency of the machine, p is the number of pole pairs, and k is the integer. The harmonic components, such as 30, 90, 150, and 180 Hz, are generated based on Equation (12.9), when the machine is supplied by a grid frequency of 60 Hz. Figure 12.59 shows the frequency spectrum of the synchronous generator current in a healthy state and under a broken damper bar fault. The frequency components of 300 and 420 Hz already exist in the frequency spectrum and the broken damper bar fault does not change their amplitude. The broken damper bar fault and increasing the number of broken damper bars increase the harmonic component, as indicated in

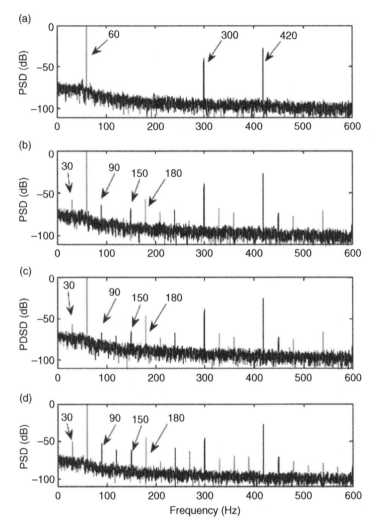

Figure 12.59 Frequency spectrum of the stator line current for (a) a balanced synchronous machine fed with a balanced power supply, (b) an unbalanced healthy synchronous machine fed with the balanced power supply, (c) an unbalanced healthy synchronous machine fed with the unbalanced power supply, and (d) a synchronous machine with three broken damper bars fed with a balanced power supply.

Equation (12.9). The sensitivity is higher for the 30 Hz harmonic component of the broken damper than for the other frequency components. Figure 12.59 shows the impact of three factors (an unbalanced power source, an unbalanced synchronous generator, and a broken damper bar fault) on the frequency spectrum of the stator current. No frequency components exist in the current spectrum when the generator is run off a balanced power source. The frequency components of 30, 90, 150, and 180 Hz appear in the frequency spectrum of the current when the synchronous generator becomes asymmetrical while being fed with a balanced source. If both the synchronous generator and the power supply become unbalanced, the same harmonic components that would appear during a broken damper bar fault now appear, but the amplitude of the 90 Hz component is lower.

The four harmonic components increase while both the generator and the power supply are balanced and the generator is under a broken damper bar fault. The 30 Hz harmonic component is sensitive to the broken damper bar fault if the generator and the power supply are balanced. Consequently, the index introduced for a broken damper bar fault must be used carefully, since both the generator and the power supply must be checked to determine whether they are balanced. Otherwise, the detection effort may lead to a false positive fault.

Since the behavior of the synchronous machine during start-up is similar to that of an induction machine, the same harmonics can be tracked during the transient operation of the synchronous machine. The sideband harmonics around the fundamental component of the stator current in the induction machine are used to detect the broken damper bar fault. A Hilbert Huang transform is used to investigate the impact of the broken damper bar fault on the sideband harmonics around the fundamental frequency component current. The harmonics components around the fundamental component are located in the second intrinsic mode functions (imf2), which include all harmonic components, principal components, and harmonics below 50 Hz. The left sideband harmonic component, as the main harmonic, decreases from 50 Hz to near zero due to a broken damper bar fault during the synchronous machine start-up and acceleration. The variation in the left sideband component from standstill until the synchronous machine reaches the synchronous speed forms a V-shaped curve under the broken damper bar fault. Therefore, no sign of a V-shape variation is found in the left sideband components of the stator current if the machine is healthy [64, 65].

Figure 12.60 shows the IMF2, 2D, and 3D time-frequency plot using HHT of the stator current of a 5 kVA no-load synchronous generator. In a healthy generator, the V-shaped curve is absent from the time-frequency plot, but it appears when a broken damper bar fault occurs. The rate of variation and the intensity of the V-shaped curve are increased by increasing the synchronous machine load [65].

These changes in the intensity of the V-shaped curve in the no-load and nominal load conditions are due to the increase in the synchronous machine start time, which is caused by increasing the moment inertia of the machine [66]. The energy of the IMF2 is used to introduce an index to quantify the fault in the synchronous machine. The results presented in Figure 12.61 show that a broken damper bar fault increases the energy of the IMF2 from 90.28 to 86.64 dB, while increasing the number of broken damper bar faults cause the energy of the IMF2 to reach 61.69 dB. The variation in the left sideband component of the stator current depends on the location, number, and shape of the damper bars in the synchronous machine, indicating the possibility of decreasing the amplitude of the left sideband harmonic in the case of a broken damper bar fault in some synchronous motors [67].

12.4.6 Rotor Field Winding Voltage

The air gap magnetic field of the synchronous generator during no-load operation only contains two magnetic fields: the rotor magnetic field and the damper bar magnetic field. The magnetic field of the air gap induces a voltage in both the stator winding and the rotor field winding. The broken damper bar fault changes the symmetry of the magnetic field contributed by the damper bars. Therefore, the asymmetry caused by the broken damper bar fault can be traced to the induced voltage in the rotor field winding. The induced voltage in the rotor field winding is calculated by obtaining the MMF of the damper bars, as follows [67]:

$$\text{MMF}_p(\alpha, t) = \sum_j \sum_\zeta \frac{1}{\zeta} \left(\frac{2 \sin \zeta p(\alpha_{k+1} - \alpha_k)}{\zeta p \pi} \cos \zeta \left(p - \frac{p(\alpha_{k+1} - \alpha_k)}{2} - \omega t \right) \right) I_j \qquad (12.10)$$

where ρ is the number of pole pairs and α is the angle of the damper bar with the reference point in the 2D plane in radians, as shown in Figure 12.62. I_j is damper bar current, ω is the angular

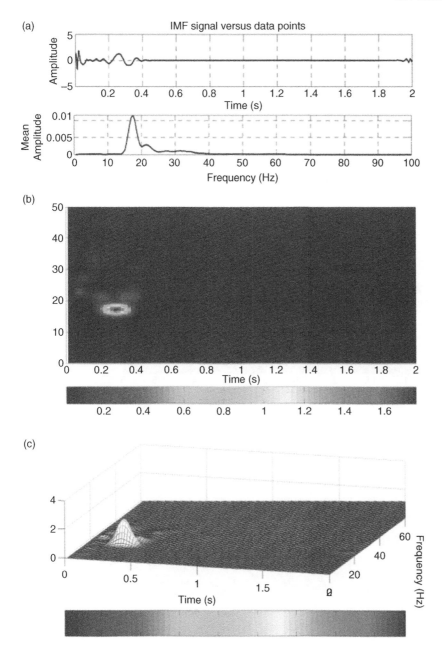

Figure 12.60 Application of HHT to a healthy synchronous machine current: (a) IMF 2, (b) HHT 2-D spectrum of the IMF 2, (c) HHT 3-D spectrum of IMF 2.

velocity, and ξ is the spatial harmonic number because the harmonics of the winding function are expressed by $\xi = 1 \pm 6n$, where n is an integer. The magnetic field for a damper bar with a length l is

$$B_p = \frac{2\mu_0}{\pi l p} \sum_j \sum_\zeta \frac{1}{\zeta} \left(\sin \zeta (p + p\alpha_k - \omega t) + \sin \zeta (-p + p\alpha_{k+1} + \omega t) \right) I_j \tag{12.11}$$

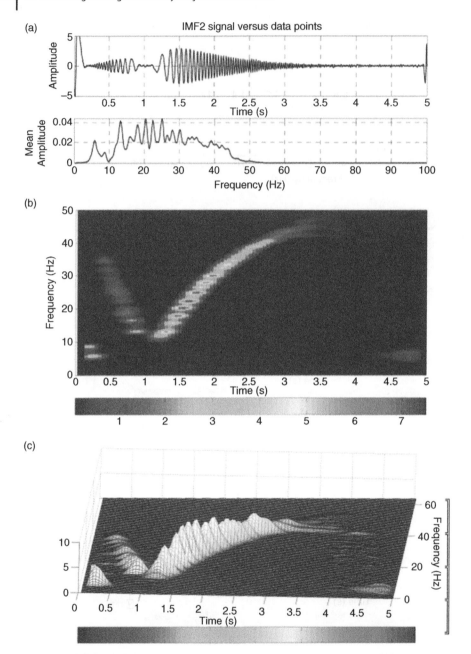

Figure 12.61 Application of HHT to a synchronous machine current under a broken damper bar fault: (a) IMF 2, (b) HHT 2-D spectrum of the IMF 2, (c) HHT 3-D spectrum of IMF 2.

The flux density in the no-load synchronous generator due to the magnetic field of the damper bar is

$$\varphi = \int_{\alpha_i}^{\alpha_j} B_p \, 2\pi \, r_r l \, d\alpha \tag{12.12}$$

Figure 12.62 Location of the damper bar in the rotor pole shoe of a synchronous generator.

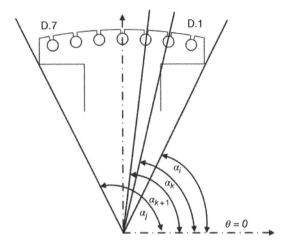

where α_i and α_j are positions of the rotor pole where the flux-linkage passes it and r_r is the outer radius of the rotor. Based on the Faraday law, the induced voltage in the rotor field winding is

$$e = \frac{2\mu_0 N\omega}{\pi p^2} \left\{ \sum_j \sum_\zeta \frac{1}{\zeta^2} (\sin\zeta(p + p\alpha_k - \omega t) + \sin\zeta(-p + p\alpha_{k+1} + \omega t))I_j \right\}_{\alpha_i}^{\alpha_j} \qquad (12.13)$$

where N is the number of turns in the rotor field winding. Equation (12.13) shows that a voltage is induced in the rotor field winding due to the damper bar magnetic field. Figure 12.63 shows the induced voltage in the rotor field winding in the no-load synchronous generator. The induced voltage in the rotor field winding is measured when the rotor rotates at the synchronous speed and the DC current increases from zero to the nominal no-load voltage measured at the stator terminals. The broken damper bar fault results in an asymmetric magnetic field that increases the induced voltage in the rotor field winding. The impact of the broken damper bar at the rotor pole edge is significant, since the current is higher in that damper bar than in the damper bars located in the middle of the rotor pole shoe [50].

One feature based on the application of time series data mining is proposed to quantify the broken damper bar fault. The radius of the gyration increases in the case of a broken damper bar fault compared with the healthy synchronous generator. Figure 12.64 shows the application of the time series data mining method to the induced voltage in the rotor field winding in the healthy case and under one broken damper bar fault. The radius of the gyration increases due to one broken damper bar fault, increasing from 0.1119 in the healthy case to 2.4307 under one broken damper bar fault.

The radius of the gyration depends on the following three factors:

1. Location of the broken damper bar.
2. Number of broken damper bars.
3. Symmetry of the broken damper bars.

When two damper bars at the motor edges are broken, the radius of gyration is less than that occurring when one damper bar at the edge is broken because the first fault is symmetrical and reduces the impact of the magnetic field asymmetry.

Figure 12.64 shows the various configurations of the broken damper fault in the synchronous generator. The radius of gyration increases twofold compared with the three broken damper bars

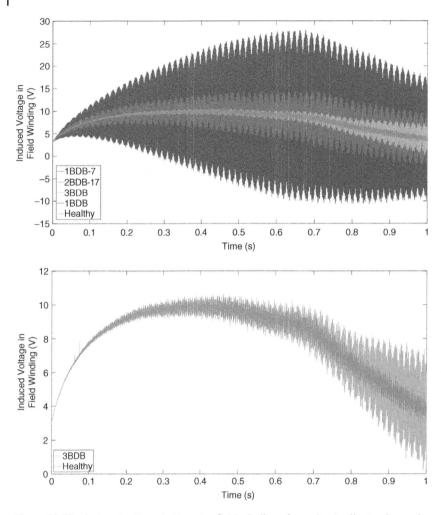

Figure 12.63 Induced voltage in the rotor field winding of a no-load salient pole synchronous generator in the healthy case and under a various number of broken damper bar faults.

in the middle of a rotor pole and when compared with case (a), where six damper bars are broken in two poles (three in each rotor pole) located opposite to each other. However, due to the low current density in the middle bars, the radius of gyration compared with one broken damper bar at the pole edge does not differ markedly. The radius of gyration is the same as for one broken damper bar in case (b), where two damper bars at the edge of two separate poles are broken because the flux density changes over the circumference of one of the poles (north or south). Having a broken damper bar in two poles opposite each other, as in case (c), results in an increase in the radius of gyration that is almost twice that arising from one broken damper bar at the edge. In case (d), the radius does not increase but partly decreases, which could be explained based on case (b). The results of different configurations of broken damper bars are shown in Table 12.5.

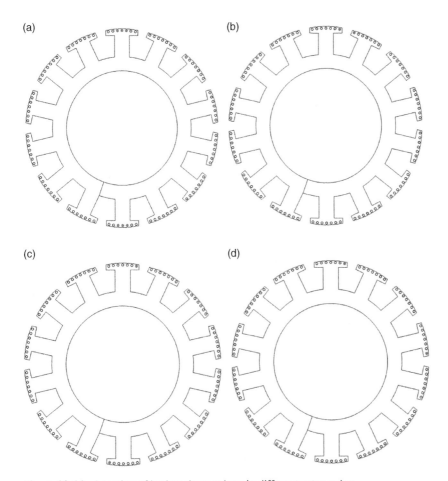

Figure 12.64 Location of broken damper bars in different rotor poles.

Table 12.5 Radius of gyration of the induced voltage in a rotor field winding in a healthy case and under various combinations of broken damper bar faults.

Cases	RG	Cases	RG
Healthy	0.1119	Healthy	0.1119
1BDB–No. 4	0.1121	3BDB (middle bars of 1 pole)	0.3424
1BDB–No. 6	0.4091	6BDB in two poles (case a)	0.6685
1BDB–No. 7	2.4307	2BDB–No. 7, 1 (case b)	2.3577
2BDB–No. 6, 7	4.0743	2BDB–No. 1, 7 (case c)	4.8344
2BDB–No. 1, 7	1.0442	2BDB–No. 1, 1 (case d)	1.3047a

12.5 Summary

In this chapter, mechanical faults in synchronous generators including an eccentricity fault, a stator core related fault, and a broken damper bar fault are explained. The detection method based on invasive and non-invasive methods is introduced. The root cause of the static eccentricity fault in hydropower plants and their detection method based on stary magnetic field analysis is introduced. Among various eccentricity faults, dynamic eccentricity is the most prevalent fault type in synchronous generators. Various methods and signals including voltage, current, air gap magnetic field, and stray magnetic field are proposed for dynamic eccentricity fault detection.

Although core-related faults are not the main concern in small synchronous machines, they can cause irreparable damage to the stator core of large synchronous machines. Therefore, a comprehensive inspection must be performed during periodic maintenance of the large machines to prevent core-related failures.

The broken damper bar fault is a common fault mostly in synchronous motors, but it can happen in pump storage generators due to high starts and stops. Detection of the broken damper bar is proposed mainly during transient operation of the synchronous machine. However, a broken damper bar fault can be detected during steady-state operation of the synchronous machines with fractional winding using a stary magnetic field analysis.

References

1 Ehya, H., Sadeghi, I., and Faiz, J. (2017). *Online condition monitoring of large synchronous generator under eccentricity fault. 2017 12th IEEE Conference on Industrial Electronics and Applications (ICIEA)*, 18–20 June 2017, Siem Reap, Cambodia.

2 Foggia, A., Torlay, J.-E., Corenwinder, C., et al. (1999). Circulating current analysis in the parallel-connected windings of synchronous generators under abnormal operating conditions. *Proceedings, IEEE International Electric Machines and Drives Conference. IEMDC'99*, 9–12 May 1999, Seattle, WA, USA.

3 Perers, R., Lundin, U., and Leijon, M. (2007). Saturation effects on unbalanced magnetic pull in a hydroelectric generator with an eccentric rotor. *IEEE Transactions on Magnetics* 43 (10): 3884–3890.

4 Burakov, A. and Arkkio, A. (2007). Comparison of the unbalanced magnetic pull mitigation by the parallel paths in the stator and rotor windings. *IEEE Transactions on Magnetics* 43 (12): 4083–4088.

5 Dorrell, D. and Smith, A. (1994). Calculation of UMP in induction motors with series or parallel winding connections. *IEEE Transactions on Energy Conversion* 9 (2): 304–310.

6 Wallin, M., Ranlof, M., and Lundin, U. (2011). Reduction of unbalanced magnetic pull in synchronous machines due to parallel circuits. *IEEE Transactions on Magnetics* 47 (12): 4827–4833.

7 Wallin, M., Bladh, J., and Lundin, U. (2013). Damper winding influence on unbalanced magnetic pull in salient pole generators with rotor eccentricity. *IEEE Transactions on Magnetics* 49 (9): 5158–5165.

8 Sadeghi, I., Ehya, H., and Faiz, J. (2017). Analytic method for eccentricity fault diagnosis in salient-pole synchronous generators. *International Conference on Optimization of Electrical and Electronic Equipment (OPTIM) & 2017 International Aegean Conference on Electrical Machines and Power Electronics (ACEMP)*, 25–27 May 2017, Brasov, Romania.

9 Groth, I.L. (2019). On-line magnetic flux monitoring and incipient fault detection in hydropower generators. MSc Thesis, Norwegian University of Science and Technology (NTNU), Norway.

10 Elez, A., Car, S., Tvorić, S. et al. (2016). Rotor cage and winding fault detection based on machine differential magnetic field measurement (DMFM). *IEEE Transactions on Industry Applications* 53 (3): 3156–3163.

11 Kokoko, O., Merkhouf, A., Tounzi A., et al. (2018). Detection of short circuits in the rotor field winding in large hydro generator. *XIII International Conference on Electrical Machines (ICEM)*, 3–6 September 2018, Alexandroupoli, Greece.

12 Dwyer, R. Detection of non-Gaussian signals by frequency domain kurtosis estimation, in *ICASSP'83, IEEE International Conference on Acoustics, Speech, and Signal Processing*, 14–16 April 1983, Boston, MA, USA.

13 Ehya, H., Nysveen, A., and Nilssen, R. (2020). A practical approach for static eccentricity fault diagnosis of hydro-generators. *International Conference on Electrical Machines (ICEM)*, 23–26 August 2020, Gothenburg, Sweden.

14 Antoni, J. (2006). The spectral kurtosis: A useful tool for characterising non-stationary signals. *Mechanical Systems and Signal Processing* 20 (2): 282–307.

15 Al-Nuaim, N.A. and Toliyat, H.A. (1998). A novel method for modeling dynamic air-gap eccentricity in synchronous machines based on modified winding function theory. *IEEE Transactions on Energy Conversion* 13 (2): 156–162.

16 Toliyat, H.A. and Al-Nuaim, N.A. (1999). Simulation and detection of dynamic air-gap eccentricity in salient-pole synchronous machines. *IEEE Transactions on Industry Applications* 35 (1): 86–93.

17 Ebrahimi, B.M., Etemadrezaei, M., and Faiz, J. (2011). Dynamic eccentricity fault diagnosis in round rotor synchronous motors. *Energy Conversion and Management* 52 (5): 2092–2097.

18 Faiz, J., Babaei, M., Nazarzadeh, J. et al. (2010). Time-stepping finite-element analysis of dynamic eccentricity fault in a three-phase salient pole synchronous generator. *Progress in Electromagnetics Research* 20: 263–284.

19 Babaei, M., Faiz, J., Ebrahimi, B.M. et al. (2011). A detailed analytical model of a salient-pole synchronous generator under dynamic eccentricity fault. *IEEE Transactions on Magnetics* 47 (4): 764–771.

20 Kiani, M., Lee, W.-J., Kenarangui, R., et al. (2007), Detection of rotor faults in synchronous generators. *IEEE International Symposium on Diagnostics for Electric Machines, Power Electronics and Drives*, 6–8 September 2007, Cracow, Poland.

21 Ehya, H., Faiz, J., and Abu-Elhaija, W. (2014). Detailed performance analysis of salient pole synchronous generator under dynamic eccentricity fault. *8th Jordanian International Electrical and Electronics Engineering Conference*, April 2014, Amman, Jordan.

22 Bruzzese, C. and Joksimovic, G. (2010). Harmonic signatures of static eccentricities in the stator voltages and in the rotor current of no-load salient-pole synchronous generators. *IEEE Transactions on Industrial Electronics* 58 (5): 1606–1624.

23 Joksimovic, G., Bruzzese, C., and Santini, E. (2010). Static eccentricity detection in synchronous generators by field current and stator voltage signature analysis – Part I: Theory. *The XIX International Conference on Electrical Machines – ICEM*, 6–8 September 2010, Rome, Italy.

24 Bruzzese, C., Joksimovic, G., and Santini, E. (2010) Static eccentricity detection in synchronous generators by field current and stator voltage signature analysis – Part II: Measurements. *The XIX International Conference on Electrical Machines – ICEM*, 6–8 September 2010, Rome, Italy.

25 Bruzzese, C., Giordani, A., and Santini, E. (2008). Static and dynamic rotor eccentricity on-line detection and discrimination in synchronous generators by no-load EMF space vector loci analysis. *2008 International Symposium on Power Electronics, Electrical Drives, Automation and Motion*, 11–13 June 2008, Ischia, Italy.

26 Bruzzese, C. (2013). Diagnosis of eccentric rotor in synchronous machines by analysis of split-phase currents – Part II: Experimental analysis. *IEEE Transactions on Industrial Electronics* 61 (8): 4206–4216.

27 Bruzzese, C. (2013). Diagnosis of eccentric rotor in synchronous machines by analysis of split-phase currents – Part I: Theoretical analysis. *IEEE Transactions on Industrial Electronics* 61 (8): 4193–4205.

28 Bruzzese, C., Mazzuca, T., and Torre, M. (2013). On-line monitoring of mechanical unbalance/misalignment troubles in ship alternators by direct measurement of split-phase currents. *IEEE Electric Ship Technologies Symposium (ESTS)*, 22–24 April 2013, Arlington, VA, USA.

29 Faiz, J., Ehya, H., and Ebrahimi, B.M. (2014). Dynamic eccentricity fault diagnosis in a salient-pole synchronous generator under non-linear loads. *IEEE International Magnetics Conference (Intermag 2014)*, May 4–8, 2014, Dresden, Germany.

30 D'Antona, G., Pennacchi, P., Pensieri, C., et al. (2012). Turboalternator shaft voltage measurements. *IEEE International Workshop on Applied Measurements for Power Systems (AMPS) Proceedings*, 26–28 September 2012, Aachen, Germany.

31 Costello, M.J. (1993). Shaft voltages and rotating machinery. *IEEE Transactions on Industry Applications* 29 (2): 419–426.

32 Nippes, P.I. (2004). Early warning of developing problems in rotating machinery as provided by monitoring shaft voltages and grounding currents. *IEEE Transactions on Energy Conversion* 19 (2): 340–345.

33 Hsu, J.S. and Stein (1994). J., Shaft signals of salient-pole synchronous machines for eccentricity and shorted-field-coil detections. *IEEE Transactions on Energy Conversion* 9 (3): 572–578.

34 de Canha, D., Willie, A.C., Meyer, A.S., et al. (2007). The use of an electromagnetic finite elements package to aid in understanding shaft voltages in a synchronous generator. *IEEE Power Engineering Society Conference and Exposition in Africa – Power Africa*, 16–20 July 2007, Johannesburg, South Africa.

35 de Canha, D., Willie, A.C., Meyer, A.S., et al. (2007) Methods for diagnosing static eccentricity in a synchronous 2 pole generator. *IEEE Lausanne Power Technology*, 1–5 July 2007, Lausanne, Switzerland.

36 Yucai, W., Yonggang, L., and Heming, L. (2012). Diagnosis of turbine generator typical faults by shaft voltage. *2012 IEEE Industry Applications Society Annual Meeting*, 7–11 October 2012, Las Vegas, NV, USA.

37 Torlay, J.-É., Corenwinder, C., Audoli, A., et al. (1999). Analysis of shaft voltages in large synchronous generators. *Proceedings, IEEE International Electric Machines and Drives Conference (IEMDC)*, 9–12 May 1999, Seattle, WA, USA.

38 Doorsamy, W., Abdallh, A.A., Cronje, W.A. et al. (2014). An experimental design for static eccentricity detection in synchronous machines using a Cramér–Rao lower bound technique. *IEEE Transactions on Energy Conversion* 30 (1): 254–261.

39 Ehya, H., Nysveen, A., Nilssen, R. et al. (2021). Static and dynamic eccentricity fault diagnosis of large salient pole synchronous generators by means of external magnetic field. *IET Electric Power Applications* 15 (7): 890–902.

40 Ehya, H., Nysveen, A., and Nilssen, R. (2022). Pattern recognition of inter-turn short circuit fault in wound field synchronous generator via stray flux monitoring. *International Conference on Electrical Machines (ICEM)*, September 5–8, 2022, Valencia, Spain.

41 Ehya, H., Nysveen, A., and Antonino-Daviu, J.A. (2021). Static, dynamic and mixed eccentricity faults detection of synchronous generators based on advanced pattern recognition algorithm. *13th IEEE International Symposium on Diagnostics for Electric Machines, Power Electronics and Drives (SDEMPED)*, 22–25 August 2021, Dallas, TX, USA.

42 Ehya, H., Nysveen, A., and Antonino-Daviu, J.A. (2021). Advanced fault detection of synchronous generators using stray magnetic field. *IEEE Transactions on Industrial Electronics* 69 (11): 11675–11685.

43 Lee, S.B., Kliman, G.B., Shah, M.R. et al. (2005). An advanced technique for detecting inter-laminar stator core faults in large electric machines. *IEEE Transactions on Industry Applications* 41 (5): 1185–1193.

44 Romary, R., Demian, C., Schlupp, P. et al. (2012). Offline and online methods for stator core fault detection in large generators. *IEEE Transactions on Industrial Electronics* 60 (9): 4084–4092.

45 Stone, G.C., Boulter, E.A., Culber, I. et al. (2004). *Electrical Insulation for Rotating Machines: Design, Evaluation, Aging, Testing, and Repair*. Wiley-IEEE Press.

46 Lee, S.B., Kliman, G., Shah, M. et al. (2006). Experimental study of inter-laminar core fault detection techniques based on low flux core excitation. *IEEE Transactions on Energy Conversion* 21 (1): 85–94.

47 Lee, S.B., Kliman, G., Shah, M., et al. (2003). An iron core probe based inter-laminar core fault detection technique for generator stator cores. *IEEE Power Engineering Society General Meeting*, 13–17 July 2003, Toronto, Ontario, Canada

48 Klempner, G. and Kerszenbaum, I. (2018). *Handbook of Large Turbo-Generator Operation and Maintenance*, 3e. Wiley-IEEE Press.

49 Sutton, J. (1994). Theory of electromagnetic testing of laminated stator cores. *Insight (Northampton)* 36 (4): 246–251.

50 Ehya, H., Nysveen, A., Nilssenet, R., et al. (2019). Time domain signature analysis of synchronous generator under broken damper bar fault. *45th Annual Conference of the IEEE Industrial Electronics Society, IECON*, 14–17 October 2019, Lisbon, Portugal.

51 Ehya, H. and Nysveen, A. (2022). Comprehensive broken damper bar fault detection of synchronous generators. *IEEE Transactions on Industrial Electronics* 66 (4): 4215–4224.

52 Traxler-Samek, G., Lugand, T., and Schwery, A. (2010). Additional losses in the damper winding of large hydrogenerators at open-circuit and load conditions. *IEEE Transactions on Industrial Electronics* 57 (1): 154–160.

53 Matsuki, J., Katagi, T., and Okada, T. (1992). Effect of slot ripples on damper windings of synchronous machines. *Proceedings of the IEEE International Symposium on Industrial Electronics*, 25–29 May 1992, Xi'an, China.

54 Ranlöf, M. and Lundin, U. (2010). The rotating field method applied to damper loss calculation in large hydrogenerators. *The XIX International Conference on Electrical Machines – ICEM, 6–8 September 2010*, Rome, Italy.

55 Dallas, S.E., Safacas, A.N., and Kappatou, J.C. (2011). Interturn stator faults analysis of a 200-MVA hydrogenerator during transient operation using FEM. *IEEE Transactions on Energy Conversion* 26 (4): 1151–1160.

56 Yun, J., Park, S.W., Yang, C. et al. (2019). Airgap search coil-based detection of damper bar failures in salient pole synchronous motors. *IEEE Transactions on Industry Applications* 55 (4): 3640–3648.

57 Park, Y., Lee, S.B., Yun, J. et al. (2020). Air gap flux-based detection and classification of damper bar and field winding faults in salient pole synchronous motors. *IEEE Transactions on Industry Applications* 56 (4): 3506–3515.

58 Karmaker, H.C. (2003). Broken damper bar detection studies using flux probe measurements and time-stepping finite element analysis for salient-pole synchronous machines. *4th IEEE International Symposium on Diagnostics for Electric Machines, Power Electronics and Drives (SDEMPED)*, 24–26 August 2003, Atlanta, GA, USA.

59 Shaikh, M.F., Park, J., and Lee, S.B. (2020). A non-intrusive leakage flux based method for detecting rotor faults in the starting transient of salient pole synchronous motors. *IEEE Transactions on Energy Conversion* 36 (2): 1262–1270.

60 Ceban, A., Pusca, R., and Romary, R. (2010). Eccentricity and broken rotor bars faults – Effects on the external axial field. *The XIX International Conference on Electrical Machines – ICEM*, 6–8 September 2010, Rome, Italy.

61 Erlicki, M.S., Porat, Y., and Alexandrovitz, A. (1971). Leakage field changes of an induction motor as indication of a nonsymmetric supply. *IEEE Transactions on Industry and General Applications* 6: 713–717.

62 Zidat, F., Lecointe, J.-P., Morganti, F. et al. (2010). Non-invasive sensors for monitoring the efficiency of AC electrical rotating machines. *Sensors* 10 (8): 7874–7895.

63 Neti, P., Dehkordi, A.B., and Gole, A. (2008). A new robust method to detect rotor faults in salient-pole synchronous machines using structural asymmetries. *IEEE Industry Applications Society Annual Meeting*, 5–9 October 2008, Edmonton, AB, Canada.

64 Antonino-Daviu, J., Roger-Folch J., Pons-Llinares J. et al. (2011). Application of the empirical mode decomposition to condition monitoring of damper bars in synchronous motors. *IEEE International Symposium on Industrial Electronics*, 27–30 June 2011, Gdansk, Poland.

65 Antonino-Daviu, J.A., Riera-Guasp, M., Pons-Llinares, J. et al. (2012). Toward condition monitoring of damper windings in synchronous motors via EMD analysis. *IEEE Transactions on Energy Conversion* 27 (2): 432–439.

66 Rahimian, M.M., Choi, S., and Butler-Purry, K. (2011). A novel analytical method for prediction of the broken bar fault signature amplitude in induction machine cage rotor and synchronous machine damper winding. *IEEE Energy Conversion Congress and Exposition*, 17–22 September 2011, Phoenix, AZ, USA.

67 Knight, A.M., Karmaker, H., and Weeber, K. (2002). Use of a permeance model to predict force harmonic components and damper winding effects in salient-pole synchronous machines. *IEEE Transactions on Energy Conversion* 17 (4): 478–484.

13

Vibration Monitoring

CHAPTER MENU

13.1 Introduction

The synchronous generator has been a prevalent electromechanical power conversion device for many years and is a key component of the power production sector. Various protection systems have played an important role in the evaluation of the health status of synchronous generators in power plants, but they protect the generators only against severe faults imposed by the power grid or severe faults, such as short circuit faults, that result in burnout of the whole phase winding. Synchronous generators are reliable and robust, with a life expectancy of a couple of decades; however, they may be vulnerable to different types of internal faults.

Faults in synchronous generators operating in power plants have a serious consequence, both in terms of economic loss and the safety of the operators. Therefore, early-stage detection of a fault in a synchronous generator is vital. Numerous methods developed in past years are available for condition monitoring of synchronous generators. One of the significant consequences of any fault type in synchronous machines is vibration, as this can be used to diagnose the fault. The main advantage of fault detection based on vibration measurement is its non-invasive nature, as a sensor can be attached to the generator frame without any requirement for stoppage of the generator. Therefore, condition monitoring in electric machines has a long history of application [1–4]. However, the only useful information based on vibration data is the increased level of vibration, which may be a symptom of a fault. Therefore, an in-depth study of vibration signals for fault detection in synchronous generators is required.

The objective of this chapter is to investigate condition monitoring based on the vibration signals in synchronous generators. General details about vibration in synchronous generators are presented. The detailed analysis of the air gap magnetic field and forces that create vibration in the machine body in a healthy and faulty condition are explained. The analytical findings developed in Chapter 6 are used to determine the frequency pattern of the vibration signal.

Electromagnetic Analysis and Condition Monitoring of Synchronous Generators, First Edition. Hossein Ehya and Jawad Faiz.
© 2023 The Institute of Electrical and Electronics Engineers, Inc. Published 2023 by John Wiley & Sons, Inc.

13.2   Condition Monitoring Using Vibration

Protective relays played a major role in the protection and reliable operation of the synchronous generator prior to the introduction of condition monitoring systems. Protection systems based on relays are retroactive, since they are designed to trip the generator from a grid network when a severe fault has occurred and reached its final limit. In other words, protective relays work when a fault has occurred, whereas the intention of a condition monitoring system is to avoid fault progression. Condition monitoring is an evaluation of the generator's health status throughout its serviceable lifetime [1].

Condition-based maintenance of a synchronous generator provides safe operation that ensures operator security, and it is a unique ability compared with breakdown maintenance and fixed time interval maintenance. Breakdown maintenance describes the situation in which an electric machine operates until it breaks and is then replaced. Fixed-interval maintenance involves occasional inspections and scheduled outages are performed. The possibility of fault development during fixed-interval maintenance exists, indicating that continuous condition monitoring would be a better strategy for avoiding problems. A predictive condition monitoring system can be developed to determine a fault in its early stage and alert the operating system to avoid fault progression. Scheduled stoppage based on fault conditions and production forecasts results in less maintenance down-time and milder consequences.

Most of the electrical or mechanical faults in synchronous machines affect the air gap magnetic field distribution and result in a distortion of that field and, consequently, a change in vibration behavior. Therefore, fault detection is possible by analyzing the vibration profile of a synchronous machine. Condition monitoring based on vibration signals is a well-established technique with the standard covering of electric machines [1]. A literature review on fault detection based on vibration signals shows that the majority of work on fault detection of electric machines based on vibration signals has been conducted on asynchronous machines, whereas synchronous generators have not been explored to a sufficient extent [1–29].

An accelerometer is a practical tool for measuring the vibration generated due to air gap magnetic field asymmetry, which is caused by a fault in electric machines. Both mechanical and electrical faults in the stator and rotor of electric machines result in a distorted magnetic field in the air gap that causes vibration of the stator core. Therefore, an accelerometer installed on a machine frame or on the stator backside can measure this vibration. Since the stator backside and frame are accessible for accelerometer attachment, even during machine operation, no machine shutdown is required.

13.3   Vibration in Salient-Pole Synchronous Generators

The radial force acting on the stator core due to irregularities in the air gap causes vibration in rotating electrical machines [30]. The net force in the air gap is decomposed into two components: tangential components that create useful electromagnetic torque required for power generation and radial components that generate the vibration. Tangential and radial forces are expressed as force per square meter and calculated according to Equations (13.1) and (13.2), based on the Maxwell stress tensor:

$$f_t = \frac{1}{\mu_0}(b_r b_t) \tag{13.1}$$

$$f_r = \frac{1}{2\mu_0}\left(b_r^2 - b_t^2\right) \tag{13.2}$$

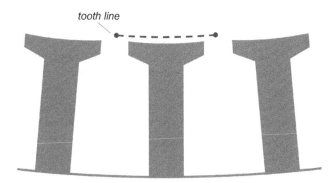

tooth line

Figure 13.1 Stator teeth and a stator tooth line used to calculate the force density on a tooth [2]/NTNU.

where μ_0 is the vacuum permeability and b_t and b_r are tangential and radial components of the air gap magnetic field, respectively. The dominant component of the air gap magnetic field is the radial component since the air-gap length in rotating electrical machines is small and the permeability of the stator and rotor iron core is higher than air. Therefore, the tangential magnetic field is neglected in the calculation of the radial air gap force [30–33]. A marked effect is caused by the tangential component of the air gap magnetic field on the radial force in large round-rotor permanent magnet machines [34, 35].

The radial force density from Equation (13.2) can be expressed as:

$$f_r(\phi, t) = \hat{f} \, \cos(k\omega_r t - m\phi) \tag{13.3}$$

where f_r is a radial force density wave that creates an attraction between the stator and rotor; $\hat{f}$ is the amplitude of the radial force; t and ϕ are time and angular positions of the radial force, respectively, and ω_r, k, and m are the angular velocity of the rotor, time-harmonic order, and spatial harmonic order, respectively. The radial force density rotates in the air gap, either in the same or opposite direction of the rotor, with an angular velocity of $k\omega_r/m$ [30]. The radial force density acts on the surface of the stator teeth that face the air gap and the net force density of the tooth propagates into the stator yoke and can cause deformations. The total force density on the tooth can be calculated by integrating a force density over the tooth line, L_t, and multiplying it by the stator stack length, L_s according to:

$$F_{tooth} = L_s \int_{L_t} f_r \, \mathrm{d}l \tag{13.4}$$

Figure 13.1 shows the stator tooth line [2].

13.4   Introduction to Utilized Terms in Vibration Analysis

The radial force density in the air gap is a function of both time and position, according to Equation (13.3). The radial force density wave is characterized by a time-harmonic order, k, that depends on an electrical frequency in the time domain and the spatial harmonic order, m, which shows its periodicity in the spatial domain. Time harmonics are important from a condition monitoring point of view, since they represent the frequency of the vibration and are linked to resonance. The spatial harmonics are important since their order determines the stator deformation caused by a force wave. A better understanding of both time and spatial harmonics provides a better understanding of the vibration characteristics in rotating machines.

13.4.1 Time Harmonics

An alternating current in the stator windings and the revolving rotor create time harmonics. The air gap magnetic field, which is a combination of the stator and rotor magnetic field, has a fundamental frequency equal to the synchronous speed. The radial force density peaks twice whenever the magnetic flux reaches the maxima and minima, indicating that the main harmonic of the radial force density has a frequency twice that of the main flux harmonic. In addition, the rotational speed of the rotor is similar to the air gap synchronous speed; therefore, the frequency of the main harmonic of the magnetic force density is twice the fundamental frequency.

In addition to the main time-harmonic frequency in the magnetic field, harmonics also exist with orders equal to odd multiples of the fundamental frequency and these generate natural time harmonics of higher orders in the magnetic force density. These time harmonics are generated due to the interactions between two harmonics, either from the stator and rotor together or one from each of them. In this context, the term "interaction" means that the frequencies and spatial order are added or subtracted. The radial magnetic force density is the square of the air gap magnetic field, according to Maxwell's stress tensor in Equation (13.2). This can be expressed as a summation of the cosine terms, which represent the various harmonics from the stator and the rotor magnetic field. When the air gap magnetic field is squared, all the time harmonics of the rotor and stator magnetic field are multiplied, indicating that the time harmonics contents of the air gap force density consist of numerous time harmonics. For instance, the air gap force density of a synchronous generator with a fundamental frequency of 50 Hz and natural air gap flux density harmonics of 50, 150, and 250 Hz, ... has time harmonics of 100, 200, 300 Hz, ... with $k = 2p, 4p, 6p, ...$

The time distribution of the air gap force density in a generator is determined by measuring the air gap magnetic field at a fixed location in the air gap and calculating the Maxwell stress tensor. Figure 13.2 shows a synchronous generator with eight salient poles and a point on the stator tooth on the right side that measures the air gap magnetic flux density. The magnetic field in the air gap, which is a combination of the rotor magnetic field and the stator magnetic field in a loaded case, is captured by a sensor attached to the stator tooth. Figure 13.3 shows the air gap magnetic field and the corresponding calculated air gap force density. The frequency of the signal can be evaluated by measuring how many times one period passes over a fixed sensor point. The number of periods per revolution determines the harmonic order of a signal. For instance, the harmonic order of an air gap flux density and air gap force density for a generator with eight poles is equal to four and eight, respectively.

Any fault types in a synchronous generator change the time variation of the air gap magnetic field; consequently, the force density is also affected, indicating that the spectral density of an air gap magnetic field and the air gap flux density will be enriched by unnatural time harmonics.

Figure 13.2 A salient pole synchronous generator with eight poles and a measuring point for acquiring the air gap magnetic field.

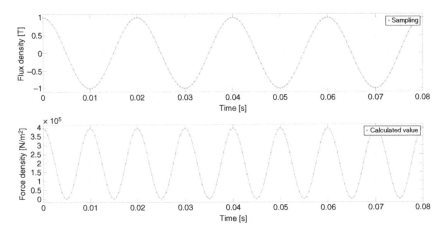

Figure 13.3 Time distribution of an air gap magnetic field (top) and air gap force density (bottom) of a synchronous generator with eight poles.

Subharmonics (harmonics between the zero and the fundamental frequency) and interharmonics (harmonics between the natural components) will occur in the frequency spectrum of both the air gap magnetic field and the air gap magnetic force density.

13.4.2 Spatial Harmonics

Spatial harmonics are time independent and arise due to geometric specifications, such as the rotor pole saliency, the stator and rotor slots, and the winding layout. The spatial order, m, of a magnetic force density wave describes the number of periods that are distributed around the air gap. The spatial order of a main spatial harmonic in the force density is equal to the number of poles. The main spatial harmonic is generated by the square of the main spatial flux density harmonic, which has an order equal to the number of pole pairs.

The stator slots create spatial harmonics in the flux density of the air gap according to:

$$m_{slot,flux} = Q_s \pm p \tag{13.5}$$

where Q_s is the number of stator slots and p is the number of pole pairs. The interaction between the main flux density and spatial harmonics due to the slot effect creates spatial force density harmonics:

$$m_{slot,force} = Q_s, \quad Q_s \pm 2p. \tag{13.6}$$

The low spatial harmonic order has a significant effect on stator core deformation, since it imposes a marked vibration. The lowest harmonic order and lowest spatial periodicity of a force are defined by the number of identical magnetic parts of an electric machine and can be determined based on the number of poles and the number of stator slots [36]. The greatest common divisor between the number of stator slots and the number of poles determines the lowest non-zero order taking place naturally during the no-load operation of the generator. This is an indication that some of the healthy synchronous generators, due to their specific design, might have low spatial harmonics during the no-load operation.

The distorted air gap magnetic field shows unnatural spatial harmonics due to faults and their interaction results in spatial harmonics with a force density waveform. For example, using two air gap flux density harmonics with an order that differs by one generates force density harmonics with a spatial order equal to one.

Comparison of the spatial distribution of the air gap magnetic force density with the time distribution is scarcely achievable in a power plant, since it requires measurement of points along the inner circumference of the stator core. As shown in Figure 13.4, measuring points are distributed along the inner surface of the stator at three mechanical degrees apart from each other. Therefore, a flux can be measured along the air gap at each point at one time instant and the Maxwell stress tensor can be calculated for each measuring point. Figure 13.5 represents both the air gap magnetic field and the air gap force density as a function of the angular position. The order of both the spatial air gap magnetic field and the spatial force density is determined based on the number of periods, which is equal to the number of pole pairs and the number of poles.

Measuring the spatial distribution of the synchronous generator is impractical, as it requires attachment of numerous sensors the same distance from each other to the stator teeth; this is especially difficult in slotted machines. Signal transmission using wires from the installed sensors to

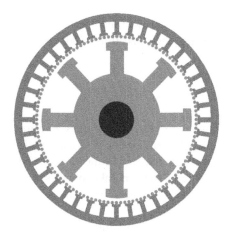

Figure 13.4 A salient pole synchronous generator with measuring points distributed along the inner surface of the stator core to acquire the spatial distribution of the air gap magnetic field.

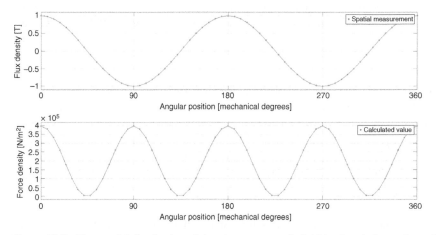

Figure 13.5 The spatial distribution of the air gap magnetic field (top) and air gap force density (bottom).

the output of the machine is also impractical and using wireless technology requires more space for sensors. Therefore, in reality, finding a spatial distribution of the synchronous generator is not feasible. However, a better understanding of vibration behavior depends on a better understanding of the spatial distribution of the machine and can be performed by exploring simulation data.

Finding a dominant vibration mode in a synchronous generator is possible if accelerometers are installed on the stator backside or machine frame to measure the vibration [37]. This is the so-called mode observation of an electric machine and provides the order of the spatial force density harmonic that has the greatest influence on vibration. However, mode observation does not provide a full spatial distribution of the force density.

The frequency of the force wave corresponds to the frequency of the vibration, while the amplitude of the produced vibration by a wave is proportional to the spatial harmonic order, indicating the importance of the coherence between the time and spatial harmonic orders. The number of time and spatial harmonics that exist in the air gap is limited to the generator specification and the current passing through the winding in a healthy operation. However, a fault changes the distribution of the air gap magnetic field and results in new harmonic components. Therefore, the relationship between the time harmonic order and spatial harmonic order is changed, since various interactions occur between the harmonics. Several waves with different time harmonic orders and the same spatial harmonic orders are possible, and vice versa, meaning that a particular spatial harmonic can affect the vibration level of the synchronous generator at more than one frequency [38, 39].

13.4.3 Mode Number and Deformation

The spatial harmonic order, m, also known as a mode number, is the most important parameter for studying vibration since it determines how the stator responds to the force density [30]. A low mode number is the most detrimental wave and is characterized by a few maxima and a long wavelength. The amplitude of the static deformation, Y_m, due to the mode number m is determined as follows:

$$Y_m = \frac{K_s \widehat{f}_m}{(m^2 - 1)^2}, \qquad m \geq 2 \tag{13.7}$$

where $\widehat{f}_m$ is the magnitude of the radial force density with a particular spatial harmonic order and K_s is the coefficient determined by the dimensions and structural properties of the stator core. According to Equation (13.7), the amplitude of the static deformation is inversely proportional to the m^4 for $m \geq 2$; consequently, low order mode numbers can cause a severe static deformation in the stator core and are important for vibration analysis. Moreover, inclusion of the force waves with mode numbers higher than eight is not useful, since these cannot significantly deform the stator. For a synchronous generator with a mode number equal to one, the magnitude of the static deformation is an estimate based on its specifications. The dominant vibration in the synchronous generator that causes a serious problem is generated by the lowest mode. The frequency of the oscillation of a force wave is determined based on the wave frequency, which is represented by a time harmonic order.

The static deformation of the stator must be evaluated using a numerical method since the analytical method cannot yield an accurate result. The magnitude of the deformation and the natural vibration frequencies of the stator are influenced by the structural support, which is not included in the analytical analysis, such as in Equation (13.7). The time characteristics of time harmonics that have a marked impact on the deformation are not included in the analytical equation. However, the

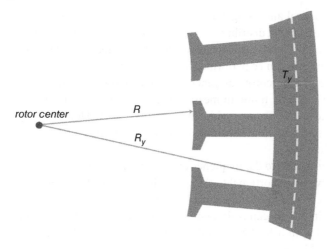

Figure 13.6 The stator parameter specification is utilized for estimation of the static deformation.

analytical analysis provides a better view of the importance of low-frequency modes for machine deformation.

The coefficient K_s, which is determined based on the machine dimensions and the structural properties, illustrates how the stator geometry impacts the vibration:

$$K_s = \frac{12RR_y^3}{ET_y^3} \tag{13.8}$$

where R_y is the average radius of the yoke, R is the inner radius of the stator, T_y is the thickness of the yoke, and E is the stiffness coefficient Young's modulus (see Figure 13.6). The salient pole synchronous generators in the hydropower section have a larger rotor diameter, which makes R and R_y larger in comparison to the thickness of the yoke. Therefore, the amplitude of the static deformation in synchronous generators is larger and special attention must be paid to vibration.

The number of attraction points in the stator core also depends on the mode number of the force density. Figure 13.7 shows how the radial force for each mode number can deform the stator core. The black arrow shows how the deformation is rotated around the circumference of the stator core and its speed is proportional to the wave speed. The arrow shows the place of maximum attraction and the broken line shows how the stator is deformed. The number of attraction points is increased

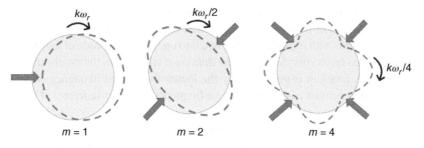

Figure 13.7 The original shape and the deformed shape of the stator core for different mode numbers.

for a mode number above one and its amplitude is also decreased. An asymmetrical movement also exists for a mode number equal to one and is called the unbalanced magnetic pull. A uniform attraction occurs between the stator and the rotor along the air gap for a mode number equal to zero, which is the so-called pulsating vibration.

13.4.4 Resonance

Resonance is an important feature of vibration that is not included in the analytical analysis (see Equation (13.7)). The electromechanical system tends to oscillate at specific natural frequencies based on the geometry configuration, material properties, and the supporting structure of the electric machine. The natural frequencies also depend on the vibration mode number. A resonance occurs if the electrical frequency of the excited mode number m_1 is at or sufficiently close to the natural vibration frequency that results in an amplified vibration magnitude.

Numerical modeling must be applied to complex mechanical systems, such as synchronous generators, in order to predict the natural frequencies. A simple analytical expression can clarify the relationship between the natural frequencies and the mode number. The natural frequency for a given mode ($m \geq 2$) is as follows:

$$f_{nat}^m \propto \frac{m(m^2 - 1)}{\sqrt{m^2 + 1}} \tag{13.9}$$

The natural frequency increases with the mode number, according to Equation (13.9). The natural frequency might intensify the vibration if a mode is low, thereby highlighting the importance of the low mode force waves.

13.5 Force and Vibration Analysis

This section includes the magnetic field, radial magnetic force, and total force analysis of the synchronous generator operating in no-load and full-load cases under an inter-turn short circuit fault and a static eccentricity fault. The analysis is performed in both the time and spatial domains, and the analysis is only limited to a frequency spectrum of the signals. The spectrum plot of the time domain quantities includes two x-axes, where the one located at the bottom of the figure is in hertz and the one on the top of the plot is labeled with the time-harmonic order. The y-axis is on a logarithmic scale to fit the harmonics with a large amplitude on the plot. The amplitude of the signal is converted into a logarithmic scale, as follows:

$$A_{dB} = 20 \log_{10} \left(\frac{A_{harmonic}}{A_{ref}} \right) \tag{13.10}$$

where $A_{harmonic}$ is the amplitude of the harmonic and A_{ref} is the reference value, which is 1. The y-axis range of the magnetic field is limited between 0 and -90 dB, which are equal to 1 T and 31.6 µT, and the y-axis range of the force density plots is between 100 and 20 dB, corresponding to 100 kN/m^2 and 10 N/m^2. The absolute values of the FFT plots are considered, since the direction of the harmonic rotation may result in a negative sign and is not of interest here.

The units utilized for vibration measurement are acceleration, velocity, or deformation. The impacts of vibration modes on the shape of stator deformation at different frequencies have also been studied and have shown how different modes may result in deformation. The form of the stator core might not change from the depicted deformed stator core, but the deformation profile must be the same.

13.5.1 Modal Analysis

Resonance phenomena intensify the vibration if an excited force with a particular mode has a certain frequency very close to the natural frequency of the machine. Table 13.1 shows the six natural frequencies (f_{nat}^m) of the first six vibration modes of a 100 kVA, 14 poles synchronous generator. According to Table 13.1, the mode numbers follow the order of the natural frequencies. The second mode during healthy operation of the synchronous generator is of interest, since the lowest spatial harmonic order for the 100 kVA synchronous generator is two and the main force harmonic has a frequency of 100 Hz with $k = 14$. This indicates that, for a force wave with $k = 14$ and $m = 2$ in the air gap, the second natural frequency is very close to causing resonance. If a force with the 1st order spatial harmonic and the first-order time harmonic at 7.14 Hz is excited due to the fault, it is close to the first natural frequency that would cause resonance.

Figure 13.8 shows the static deformation of the stator core at six natural frequencies. The maximum deformation is depicted in dark gray. In addition, the severity of the stator geometry deflection is enhanced to emphasize the profile of the stator core. The mode number is evident based on the number of attraction points in the deformed stator core, as shown in Figure 13.8. Note that the mode number zero is a stationary and pulsating mode and does not impose radial deformation.

13.5.2 Analysis of a Healthy Generator

13.5.2.1 Time-Domain Distributions of the Magnetic Field
The time-domain distribution of the radial flux density of the healthy synchronous generator operating in no-load and full-load conditions is presented in Figure 13.9. The armature reaction due to loading changes the radial flux waveform. The amplitude is slightly higher for the loaded case than for the no-load case for an inductive load with a power factor of 0.93. Some high-frequency components exist that cause a ripple, which is evident in the no-load case operation. The existing dips in the waveform are due to the damper slots, which are located in the rotor pole. The damper bar slots create a non-uniform permeability path for the flux that passes through the rotor pole. The natural frequencies at odd multiples of the fundamental frequency also cause distortions in the waveform around the zero crossing of the waveform.

Table 13.1 Six natural frequencies of a stator core.

Mode, m	f_{nat}^m (Hz)
1	−0.21
2	118
3	328
4	620
5	984
6	1400

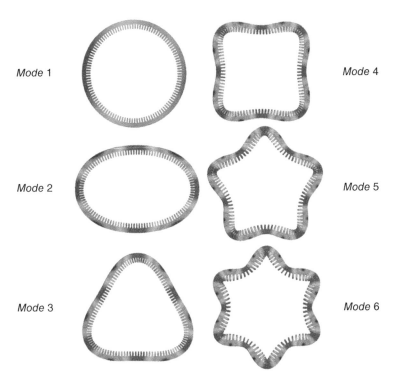

Mode 1

Mode 2

Mode 3

Mode 4

Mode 5

Mode 6

Figure 13.8 Deformation profiles of the six vibration modes at their natural frequencies.

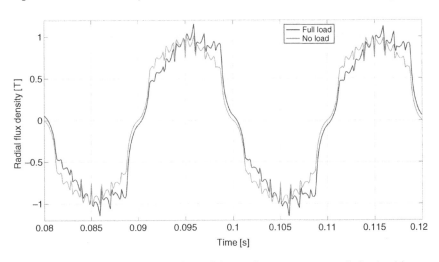

Figure 13.9 The air gap magnetic field of the synchronous generator during healthy operation in no-load and full-load cases.

Figure 13.10 presents the frequency spectrum of the healthy radial flux density in the air gap. The fundamental component caused by the seven rotor pole pairs corresponds to 50 Hz, given by the synchronous electrical frequency. It dominates the spectrum with a 10 times higher amplitude than the component at 150 Hz, when the logarithmic values are converted to Tesla. Natural harmonics occur at its odd multiples and a slight difference is evident between the no-load and full-load operation. The largest difference between the harmonics in no-load and full-load

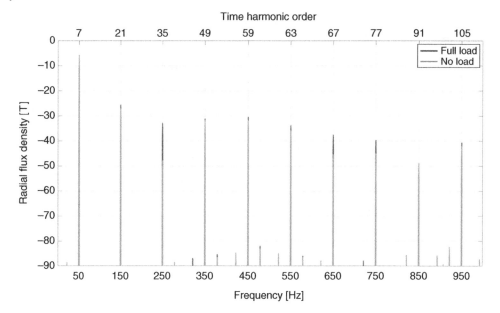

Figure 13.10 The frequency spectrum of the time distribution of the radial air gap flux density during healthy operation.

operation is seen at 250 Hz, where the logarithmic amplitude deviation corresponds to 18 mT. Some interharmonics occur with magnitudes below −80 dB, corresponding to 0.1 mT, which means that these harmonics are relatively small compared to the fundamental harmonics with an amplitude of 0.5 T.

Figure 13.11 shows the radial force density of the air gap magnetic field, which is a squared form of the air gap magnetic field depicted in Figure 13.9, in which negative periods are also changed into positive periods. The damper bar effect and load effect are more noticeable in the radial force density plot than in the air gap magnetic field plot due to the squaring of the air gap magnetic field. The lowest magnitude of the radial force wave is −32 N/m², which is due to the tangential flux density component that imposes a negative term in the Maxwell stress tensor.

The 14th harmonic depicted in Figure 13.12 is the main harmonic of the radial flux density and is twice the fundamental frequency. This coherence is directly deduced from Maxwell's stress tensor, as the squaring of the flux density causes the force density to peak when the flux is either at its positive or negative maximum. Every multiple of 100 Hz is representative of the natural harmonics caused by the interaction of natural flux harmonics. The difference between the no-load and full-load operation is also evident in the frequency spectrum of the radial flux density. Some interharmonics are present with magnitudes below 100 and 30 N/m², which are only 0.1 and 0.04% of the main component.

Figure 13.13 shows the radial flux density and the tangential flux density of the synchronous generator in a healthy condition in a no-load case. The maximum amplitude of the tangential flux density reaches up to 10% of the radial flux density, indicating that the amplitude of the tangential flux density is negligible compared with the radial flux density. This is why, in Equation (13.2), the term b_t is ignored.

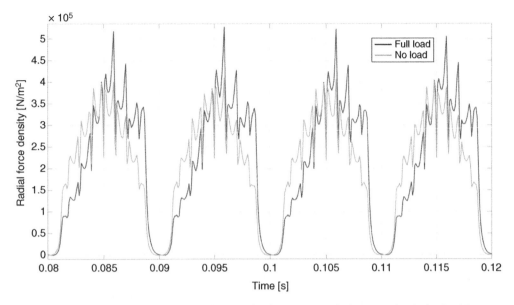

Figure 13.11 Time-domain distribution of the radial air gap magnetic force density during healthy operation in no-load and full-load cases.

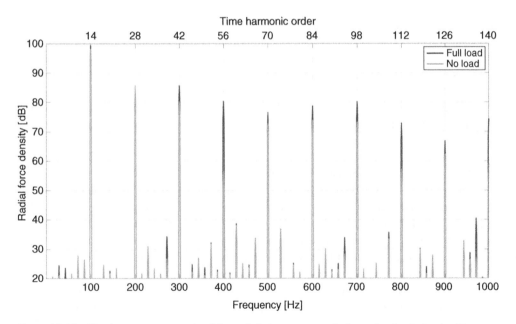

Figure 13.12 The frequency spectrum of the radial air gap magnetic force density during healthy operation in no-load and full-load cases.

According to Equation (13.2), the radial force density consists of two terms that arise due to the radial flux density and tangential flux density and can be rewritten as

$$f_r = \frac{1}{2\mu_0}\left(b_r^2 - b_t^2\right) = \frac{b_r^2}{2\mu_0} - \frac{b_t^2}{2\mu_0} = f_{rr} - f_{rt} \tag{13.11}$$

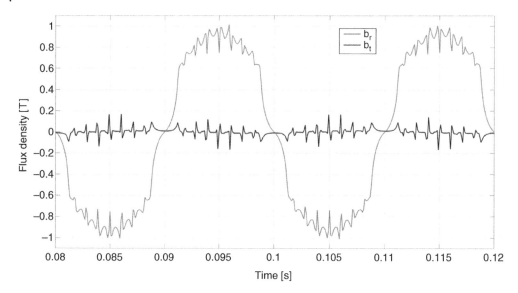

Figure 13.13 The radial and tangential components of the air gap flux density in the healthy condition under no-load operation.

where f_{rr} and f_{rt} are the force density caused by the radial and tangential flux density. The relationship between the radial force density and its two constituent components in the no-load healthy operation of the synchronous generator is shown in Figure 13.14. The contribution of the tangential force density to the radial force density is very small and the amplitude of the f_{rt} reaches as much as 3% of the f_{rr}. Therefore, the waveforms of f_r and f_{rr} are approximately the same and difficult to differentiate. Figure 13.15 shows the frequency spectrum of both components f_{rr} and f_{rt} that contribute to the radial force density. The contribution of the tangential component compared with the radial force component is negligible. Moreover, the frequency content of the f_{rt} is similar to the f_{rr}.

13.5.2.2 Spatial-Domain Distributions of the Magnetic Field

Figure 13.16 depicts the spatial distribution of the air gap magnetic field in a single time instant in a healthy condition under the no-load and full-load operation of a synchronous generator. The spatial distribution of the radial flux density is similar to the time distribution of the radial flux density since it includes seven positive and seven negative peaks, which correspond to the seven pole pairs. The stator slot effect and the load impact are more effective in the peak of the waveform in the spatial domain than in the time domain. The frequency spectrum of the spatial distribution of the radial flux is shown in Figure 13.17, where the fundamental component is number seven and the natural frequencies appear in its odd multiples in addition to its sidebands around natural frequencies. Two distinct harmonics exist that represent the slot harmonic with the order of 107 and 121 for a synchronous generator with 114 slots and seven pole pairs, according to the following:

$$m_{slots,flux} = Q_s \pm p = 114 \pm 7 \tag{13.12}$$

The spatial distribution of the radial force density at a single time instant, where the rotor is located at position $\omega_r t = 0$ degrees, in the no-load and full-load operation of the synchronous generator at a healthy condition is shown in Figure 13.18. The radial force density consists of

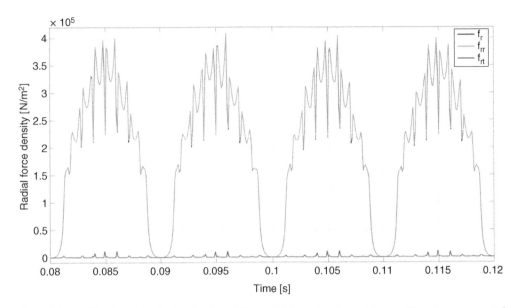

Figure 13.14 The time-domain distribution of the radial force density and its constituent components f_{rr} and f_{rt} in a healthy state and under no-load operation.

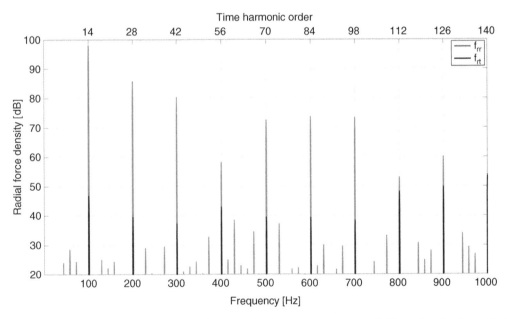

Figure 13.15 The frequency spectrum of the contributing components to radial force density, f_{rr} and f_{rt}, in a healthy and under no-load operation.

14 positive poles and the existing dips in all the peaks are due to the slotting effect. The lowest spatial order is discernible according to the pattern distribution of the force density, since the wave pattern between 0 to 180° is repeated from 180 to 360°. The second-order spatial harmonic is expected, as shown in Figure 13.19, since the greatest common divider between the number of poles and the

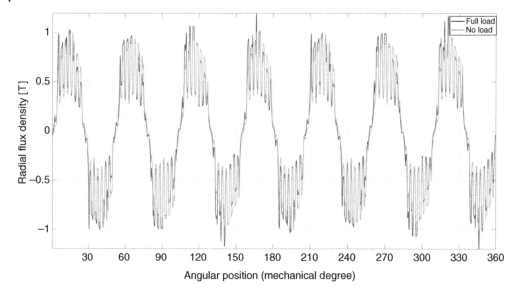

Figure 13.16 Spatial distribution of the radial flux density in the no-load and full-load operation of the synchronous generator in a healthy condition.

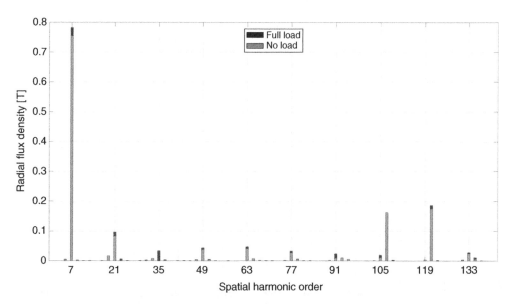

Figure 13.17 The frequency spectrum of the spatial distribution of radial flux density in a healthy condition during no-load and full-load operation.

number of slots for a generator with seven pole pairs and 114 slots is equal to two. The amplitude of the second-order harmonic is higher in a full-load operation than in a no-load operation due to the amplified flux density harmonics with orders that differ by two. This indicates that the amplified second-order harmonic during the loading condition compared with the no-load condition is due to the armature reaction, which results in an increase in the vibration level. Although several spatial slot harmonics are present in the radial force density waveform, they do not have an impact

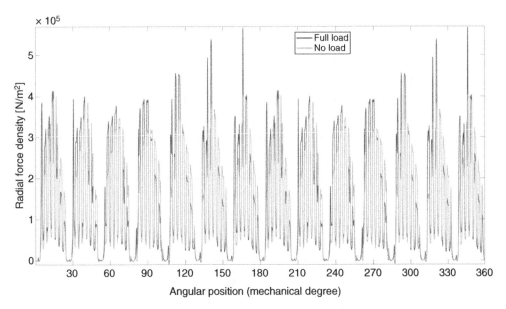

Figure 13.18 The spatial distribution of radial force density in a healthy condition during no-load and full load operation.

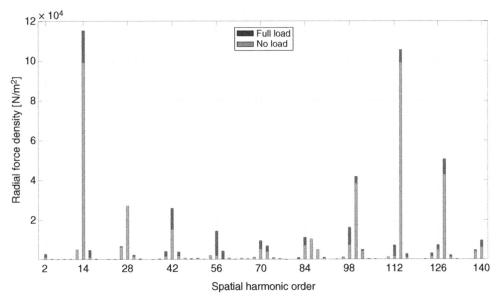

Figure 13.19 The frequency spectrum of the spatial distribution of radial force density in a healthy condition during no-load and full-load operation.

on the deformation of the stator core, since they are high-order spatial harmonics. The order of slot harmonics in the spatial distribution is 100, 114, and 128, according to Equation (13.12).

Although the radial force density of the air gap magnetic field provides detailed information about the acting force in the air gap, the total radial force that imposes vibration on the stator teeth is important. The total radial force is determined by integrating the radial force density of one stator

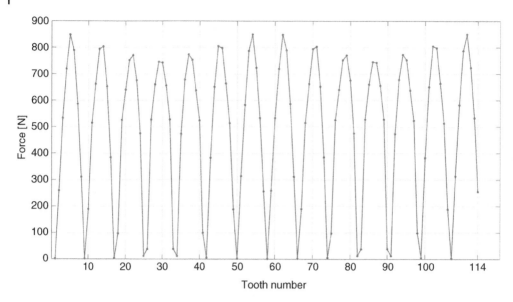

Figure 13.20 The total radial force density acting on each tooth in a healthy condition during no-load operation of the synchronous generator; each dot on the graph represents a tooth.

tooth over the entire interior circumference of the stator core and multiplying by its stack length. Figure 13.20 shows the total radial force density acting on each stator tooth. The maximum value of the force is 850 N and the second-order spatial harmonic is clear from the symmetry pattern in tooth number 58.

The spatial distribution of the radial flux density and tangential flux density during a healthy operation under a no-load condition is shown in Figure 13.21. The amplitude of the tangential component reaches 30% of the radial flux density component; however, the analysis of tangential flux

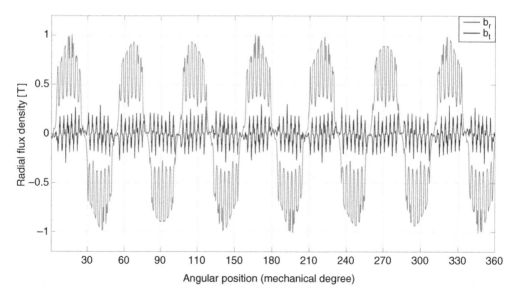

Figure 13.21 The spatial distribution of the radial component and the tangential component of the air gap flux density during healthy operation of the synchronous generator in a no-load case.

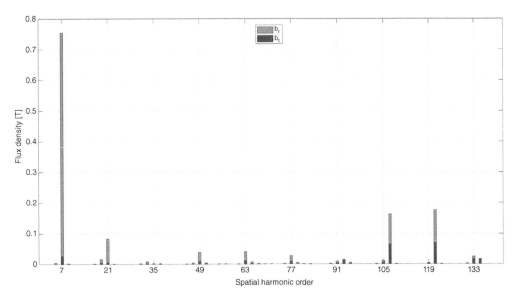

Figure 13.22 The frequency spectrum of the spatial distribution of radial flux density, b_r, and tangential flux density, b_t, in healthy operation of the synchronous generator at a load condition.

density components reveals that it contains high-frequency components rather than fundamental frequencies, as shown in Figure 13.22. A small variation appears in the seventh harmonic order, whereas the majority of variations are caused by the stator slot effect. In fact, the amplitude of the seventh harmonic order is 3.6% of the radial flux density, while the slot harmonics achieve up to 40% of the radial flux density.

The total force density, $f_r = f_{rr} - f_{rt}$, and its constituent components, radial force density and tangential force density, are depicted in Figure 13.23. Figure 13.24 also shows the frequency

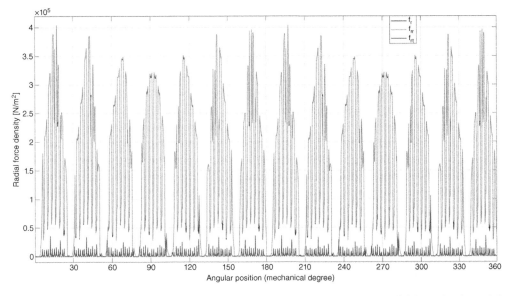

Figure 13.23 The spatial distribution of the radial force density caused by the radial force density and the tangential force density in the air gap of a healthy synchronous generator at no-load operation.

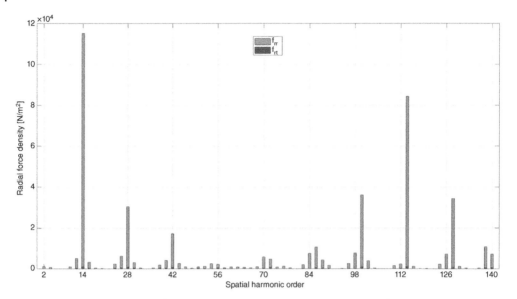

Figure 13.24 The frequency spectrum of the radial force density and tangential force density in the spatial domain during healthy operation of the synchronous generator at no-load operation.

spectrum of the total radial force density. The impact of the tangential radial flux is not considerable, although it contains the higher-order harmonics of the slotting effect. Since the air gap forces are produced by interactions between the flux density harmonics, the 14th order harmonic of f_{rt}, which is produced by the double product of the seventh order flux harmonic, $b_{t,7th} \cdot b_{t,7th}$, is relatively small and has an amplitude of 1% of f_{rr}. The same applies to the three slot harmonics of the force density, which are all caused by interactions between the slot harmonics and the seventh harmonic of the flux distribution. The largest harmonic f_{rt} was number 228, which was mainly caused by the two tangential slot harmonics $b_{t,107}$ $b_{t,121}$. Its amplitude was 9% of the corresponding harmonics f_{rr}. Notably, f_{rt} does not contain any harmonics that are not present in f_{rr}. The tangential radial flux density therefore has only a small impact on the radial force density in the spatial domain. The impact of loading and faults on the tangential radial force was also investigated and its effect on the total radial force density was also negligible, indicating the possibility of ignoring the tangential flux density in the calculation of the radial force density.

The low order harmonics, such as the second harmonic order, are important for examination of synchronous generator vibration behavior. The second harmonic order is excited by flux density harmonics with orders that differ by two. The interaction between two harmonics may result in positive or negative excitement of the second-order force density based on their sign, which is determined according to the direction of their rotation in the air gap. The frequency spectrum of the spatial flux density, as shown in Figure 13.17, indicates that a few harmonic components are present that have orders that differ by two and their amplitude is adequate to excite the second-order force harmonic to a noteworthy degree. The harmonic order above 140 may result in second-order harmonics, while their amplitude is not strong enough to result in vibration, indicating that the second-order force of spatial harmonics is achieved based on the following equation:

$$f_{rr,2nd} = \frac{1}{2\mu_o} \sum_{i,j} b_{r,ith} \, b_{r,jth}$$

for $i, j \in \{5, 7, 19, 21, 23, 33, 35, 37, 91, 93, 105, 107, 109, 119, 121, 123, 133, 135\}$,

$$j = i + 2 \tag{13.13}$$

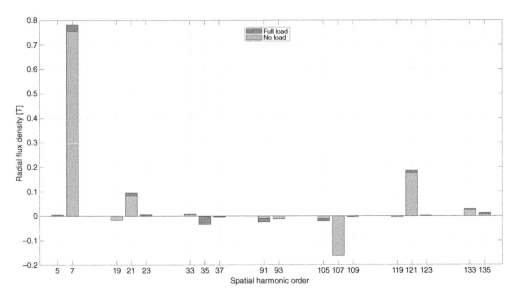

Figure 13.25 The most significant radial flux density components that produce the radial force density harmonics of second order during healthy operation.

Figure 13.25 presents the selected harmonics with their signs. Figure 13.26 elaborates how two adjacent harmonics with opposite signs contribute negatively to the generation of the second-order force harmonic density. The fundamental flux component and their sidebands, and the lowest order slot harmonics and their side-band components, excite the second-order force density. Some negative contributors, such as other slot harmonics with their left sidebands, are also evident. The selected flux density harmonics produce, in total, 3.0 kN/m^2 and 1.6 kN/m^2 of the second-order radial flux density during the full-load and no-load operation of the synchronous generator, respectively.

13.5.2.3 Mechanical Analysis

The stator core deformation of a healthy synchronous generator operating in the no-load and full-load conditions is shown in Figure 13.27, where the frequency spacing between each point is 7.14 Hz. Figure 13.27 shows two plots: the top one is the actual deformation values in millimeters, while the plot at the bottom is scaled logarithmically. The 100 Hz frequency component of the vibration profile is the dominant component and is caused by the radial force density in the air gap. The other natural multiple frequencies of 100 Hz, such as 200 Hz and 300 Hz, are also elevated. The deformation value at 100 Hz for a healthy machine is on the order of tens of nanometers, which is insufficient to cause any deformation in the stator core. However, the value of the deformation for each synchronous generator might differ and the evaluation of its health state must be performed based on its power rating.

The difference between the vibration at the no-load and full-load operation must be investigated in terms of the frequency contents of both the time harmonics and spatial harmonics distribution of the radial force density. The amplitude of the main component of the radial force is increased slightly, by approximately 10%, during full-load operation of the synchronous generator compared with a no-load condition, and this results in the augmented vibration at 100 Hz (see Figure 13.12). However, the amplitude of the frequency spectrum of the 200 Hz component of the radial force density in the time domain distribution is equal; the reason for the augmented vibration level must be

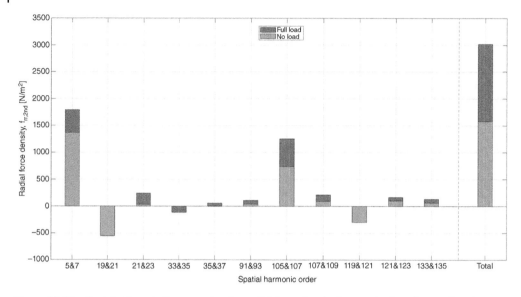

Figure 13.26 Contributions to the second-order radial force density harmonics from interactions between the most significant radial flux density harmonics during healthy operation.

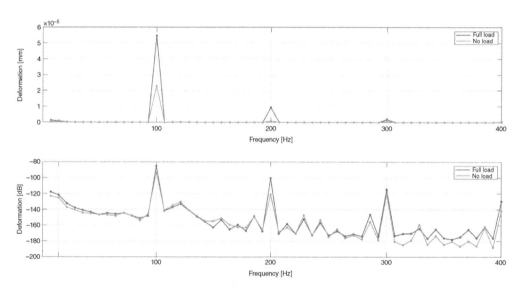

Figure 13.27 The frequency spectrum of stator yoke deformation during no-load and full-load operation of a healthy synchronous generator. Top: actual values, bottom: logarithmic values.

investigated in the spatial distribution of the force density. The lowest harmonic order, the second harmonic order, is doubled in a full-load condition, resulting in a larger deformation at frequencies dominated by the second mode (see Figure 13.19).

The deformation profile of the synchronous generator at six different frequencies in a healthy operation of the synchronous generator is shown in Figure 13.28. The shape of the deformed stator core for both the no-load and full-load cases is the same. In addition, the geometry deflection is increased to emphasize the deformed shape of the stator. The deformation at 100 Hz is dominated by the second mode, since the stator has two attraction points at the top and bottom of the

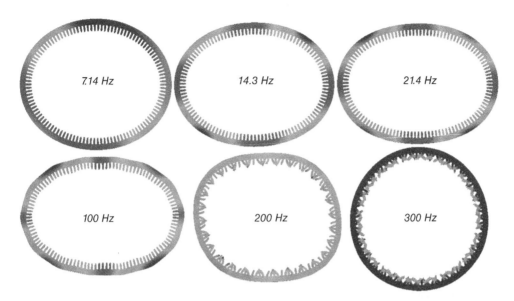

Figure 13.28 The stator deformation profiles of the synchronous generator at various frequencies during healthy operation.

core. However, the shape of the deformed stator core due to the second mode is not elliptical and contains 14 indents due to the 14th spatial harmonic. This means that the air gap consists of various 100 Hz force densities with different harmonic orders. The amplified 200 Hz component that causes deformation in the stator core during full-load operation is caused by the increased second-order spatial harmonics. The second vibration mode is absent from the 300 Hz frequency component and the twisted form of the stator teeth indicates the presence of more sophisticated frequency components in different modes. Thus, the small increment in 300 Hz vibration can be explained by the increase in the time-domain force density harmonic at 300 Hz at full-load operation. The analysis concludes that the second-order harmonic is the most significant factor for deformation of the stator core during healthy operation of the synchronous generator and that the increased vibration during full-load operation is due to the increased magnitude of the force wave with a spatial harmonic order of two at a frequency of 100 Hz.

The deformed stator at the three different rotor frequencies is depicted in Figure 13.28 for 7.14, 14.3, and 24.4 Hz. Vibration at the mentioned frequencies is insignificant, according to Figure 13.27, while a modal inspection is required to understand the reason for the low-frequency vibration increment in a full-load operation of a synchronous generator. The deformation at frequencies of 17.3 and 21.4 Hz are clearly caused by the second mode, while the ovality of the stator core in 7.14 Hz is also due to the second mode, but its severity is smaller, indicating that the deformation at these frequencies is due to the amplified second-order spatial harmonics during full-load operation. However, the insignificant increment is due to the low amplitude of these harmonics at the spatial domain.

Figure 13.29 shows the vibration spectrum of the stator core with two clear spikes at 118.3 and 327.7 Hz; these mentioned frequencies are in accordance with the natural frequencies of the second and third vibration modes. The deformation from no-load to full-load operation of the synchronous generator is almost the same. The deformation at two natural frequencies of the stator is increased by a factor of approximately 100 compared with cases in which the resonance impact was excluded; however, their amplitude does not exceed the 100 Hz component, indicating that the resonance

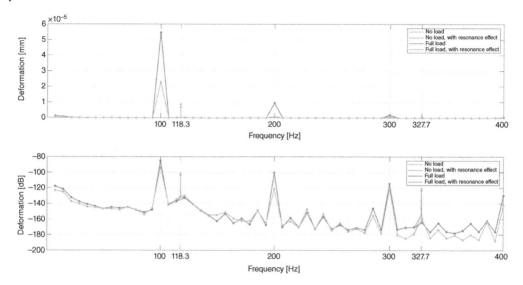

Figure 13.29 The frequency spectrum of the healthy stator core deformation in the no-load and full-load operation with specific calculations around natural frequencies. Top: actual values, bottom: logarithmic values.

effect is limited. This means that forces with a spatial harmonic order of two and three and a frequency close to 118.3 and 327.7 Hz, respectively, are not excited to a sufficient degree.

13.5.3 Analysis of a Synchronous Generator under an Interturn Short Circuit Fault

13.5.3.1 Time-Domain Distributions of the Magnetic Field

The impact of the interturn short circuit (ITSC) fault of a rotor field winding on a radial flux density of the synchronous generator is shown in Figure 13.30. The numbers of shorted turns are 1, 2, 3, 7, and 10 turns out of 35 turns in one pole. The faulty pole passes over the point in the air gap from 0.10 to 0.11 seconds. The impact of the fault severity is evident in Figure 13.30, which shows that increasing the number of shorted turns decreases the amplitude of the radial flux density due to a reduction in the magnetomotive force. The ITSC fault not only results in a reduction of the air gap magnetic field density in the faulty pole, but it also changes the amplitude of the magnetic field in the neighboring poles in a way that compensates for the reduced magnetic field by the faulty pole. The pattern of a reduced and increased magnetic field due to the ITSC fault is repeated in all the poles. Consequently, the short circuit fault in the rotor pole results in an air gap magnetic field asymmetry and impacts the frequency content of the radial force density.

Figure 13.31 shows the frequency spectrum of the radial flux density in a healthy case and under an ITSC fault. The natural frequency harmonics under the ITSC fault are reduced, which shows that the ITSC fault does not affect these components. However, the ITSC fault has a considerable impact on the interharmonics and subharmonics since their amplitude is increased significantly by increasing the fault severity, and even with one ITSC fault, the amplitudes of the sidebands are increased. The magnitude of the subharmonics and interharmonics in a low severity ITSC fault indicates that they are not very critical during generator operation, whereas by increasing the degree of severity, their amplitude becomes considerable and generator operation under the mentioned circumstance may result in winding deterioration. The lowest difference between the harmonics is one, and the interaction among them may excite in the first-order force density

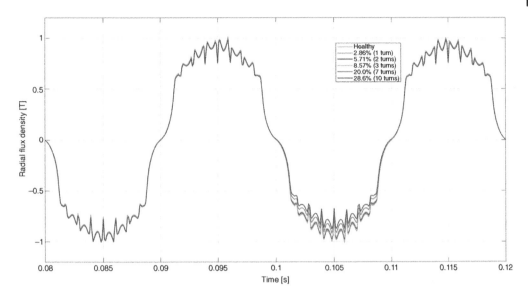

Figure 13.30 The time-domain distribution of the radial air gap flux density in a healthy state and under an inter-turn short circuit fault at no-load operation.

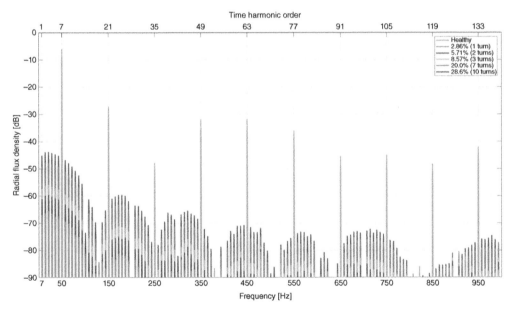

Figure 13.31 The frequency spectrum of radial air gap flux density in a healthy case and under an inter-turn short circuit fault at no-load operation.

harmonic, which is critical for the vibration of the generator. Note that the faulty generator under full-load operation has the same behavior as the no-load generator, except for the armature effect in the time distribution of the radial flux density; otherwise, the frequency contents and their amplitudes are the same.

The time-domain distribution of the radial force density of the air gap in a healthy state and under various severities of an ITSC fault is shown in Figure 13.32. The amplitude of the force produced

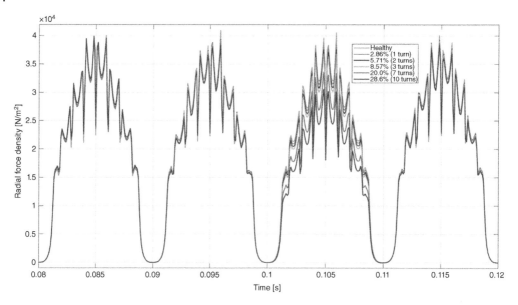

Figure 13.32 The radial air gap force density of a synchronous generator in a healthy state and under an inter-turn short circuit fault at no-load operation.

by the faulty pole is reduced compared with a healthy case due to the reduction in the radial flux density; consequently, the amplitude of the force density in the neighboring poles is slightly increased. The unbalanced magnetic pull is generated in the air gap due to the reduced radial force density in one of the rotor poles and the UMP is increased by increasing the fault severity.

The unnatural harmonics in the frequency spectrum of the radial force density of the air gap under the ITSC fault are substantially affected, as shown in Figure 13.33. The main harmonic

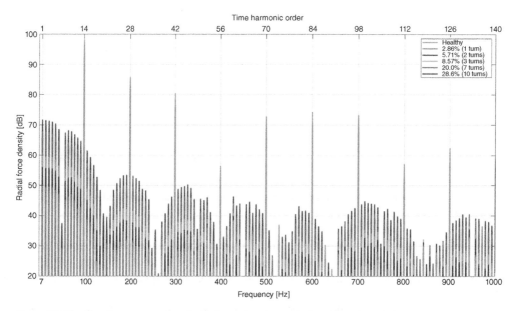

Figure 13.33 The frequency spectrum of the radial air gap force density in a healthy state and under an inter-turn short circuit fault at no-load operation.

component of order 14 is generated by the fundamental flux harmonic, while the low order subharmonics of the force density are the result of the interaction between the adjacent flux density harmonics. The amplitude of the first subharmonics with a 10 ITSC fault reaches 71.7 dB, which is 3.8 kN/m² or 4.7% of the main component. The net vibration due to a severe ITSC fault may damage the machine if the low-order time harmonics are combined with low-order spatial harmonics. The loading effect on the frequency spectrum of the radial force density indicates that the amplitude of the subharmonics and interharmonics are the same as the no-load frequency spectrum.

13.5.3.2 Spatial Domain Distributions of the Magnetic Field

The spatial distribution of the radial flux density of a healthy and under ITSC fault of a synchronous generator is shown in Figure 13.34. The flux pattern is distorted; consequently, the low-order spatial harmonics are excited in the frequency spectrum of the radial flux density, as shown in Figure 13.35. The difference between their order is one that results in the low-order spatial force harmonics, which are caused by their interactions. The main component reaches 0.76 T and all the natural harmonics are only marginally decreased and essentially are unaffected by the fault.

The spatial distribution of the radial force density under the ITSC fault condition is depicted in Figure 13.36. The amplitude of the force density in the case of 10 ITSC faults is reduced up to 23% and the symmetry around 180° is lost. The air gap has only one unique force density distribution, which indicates that the lowest spatial order is one. The same pattern can be seen in Figure 13.37, which shows the total force acting on each stator tooth. The low-order spatial harmonics are present in the frequency spectrum of the radial force density, which is due to the interaction between the flux density harmonics. As expected, the lowest order is one, while the subharmonics of higher orders are also excited by flux density harmonics with orders that differ by 2, 3, 4, and up to 13. The low-order time harmonics and the low-order spatial harmonics with relatively high amplitude may belong to the same force wave, which causes significant vibration and consequently imposes a large deformation on the stator core (see Figure 13.38).

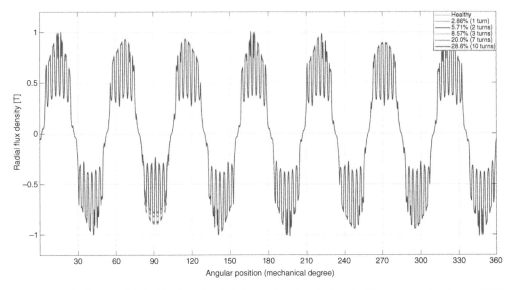

Figure 13.34 The spatial distribution of radial air gap flux density in a healthy state and under an ITSC fault of a synchronous generator at no-load operation.

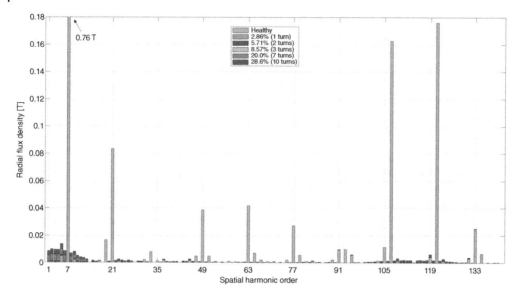

Figure 13.35 The frequency spectrum of the spatial distribution of radial air gap flux density in a healthy state and under an ITSC fault of a synchronous generator at no-load operation.

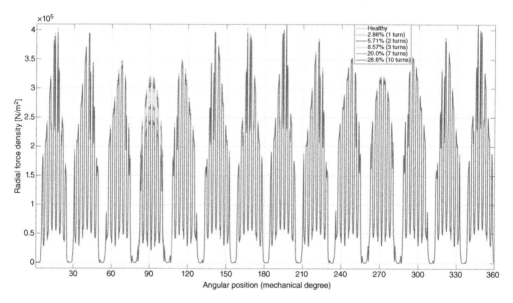

Figure 13.36 Spatial distribution of radial air gap force density in a healthy state and under an ITSC fault of a synchronous generator at no-load operation.

The ITSC fault changes the frequency spectrum of the radial flux density and the resulting low-order force density, leading to an increased level of vibration and a shift in the dominant vibration to the first order. The orders of the flux density harmonics that interact and produce the first-order force density harmonic differ by one, which means that several components contribute when the harmonic content of Figure 13.35 is examined. However, the amplitude of the two flux harmonics is increased due to the interaction between them, and when the magnitudes of

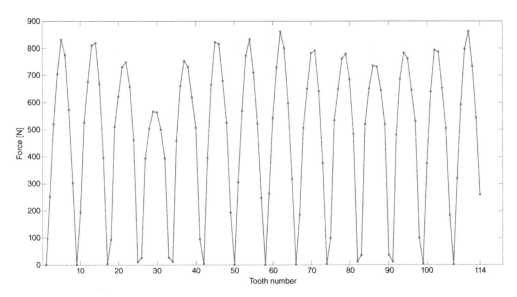

Figure 13.37 The total force acting on each tooth during no-load operation under an ITSC fault in the rotor field winding.

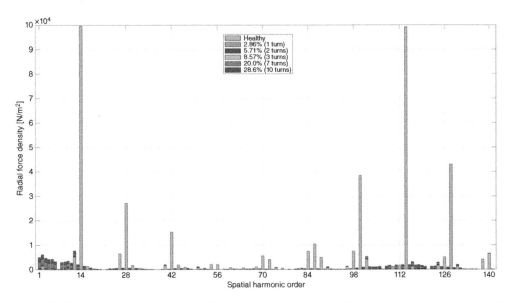

Figure 13.38 The frequency spectrum of the spatial distribution of radial air gap force density in a healthy state and under an ITSC fault of a synchronous generator at no-load operation.

the harmonics in Figure 13.35 are considered, a few components clearly have a considerable impact: the fundamental harmonic with both sidebands and the slot harmonics with their sidebands. The impact of a higher-order harmonic, such as 140 and higher, has a negligible impact; hence, the main first-order force density harmonic is

$$
\begin{aligned}
f_{rr,1st} = \frac{1}{2\mu_0}(&b_{r,6th}\, b_{r,7th} + b_{r,7th}\, b_{r,8th} + b_{r,106th}\, b_{r,107th} + b_{r,107th}\, b_{r,108th} \\
&+ b_{r,120th}\, b_{r,121st} + b_{r,121st}\, b_{r,122nd})
\end{aligned}
\tag{13.14}
$$

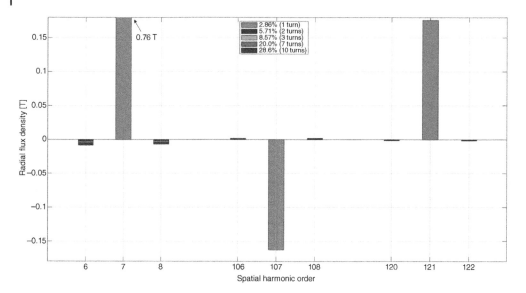

Figure 13.39 The most significant radial flux density components that produce the radial force density harmonics of the first order during no-load operation under an ITSC fault in a rotor field winding of a synchronous generator (component number 7 has an amplitude of 0.76 T for all five fault scenarios).

The sign of the contributed radial flux density harmonics, b_r, is also important for the resulting $f_{rr, 1st}$. Figure 13.39 presents the radial flux density harmonics obtained based on Equation (13.14) with their included signs. All the sidebands have an opposite sign to that of their adjacent natural harmonic component, resulting in a negative amplitude when they are multiplied with each other, as shown in Figure 13.40. These harmonics are completely negative and produce 5 kN/m^{2}of the first-order radial force density harmonic that is above the 4.9 kN/m^2 magnitude

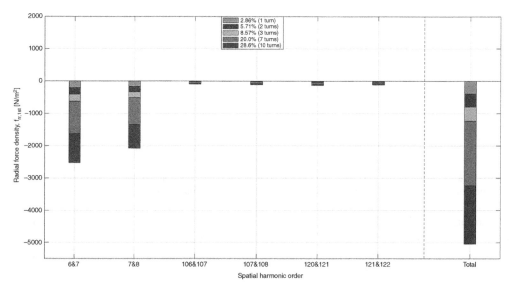

Figure 13.40 The contributions to the first-order radial force density harmonics from interactions between the most significant radial flux density harmonics during no-load operation under an ITSC fault in a rotor field winding of a synchronous generator.

of the first-order component shown in Figure 13.35. This indicates that six interactions among the subharmonics and interharmonics dominated by the fundamental flux component are the main contributors to the expected variation in a vibration of the generator during an ITSC. It also means that the negative 0.1 kN/m^2 of $f_{rr, 1st}$ is generated by the remaining radial flux density harmonics and the tangential field. The tangential flux density had an insignificant impact due to its magnitude, with harmonics accounting for approximately 0.1 kN/m^2 or 2% of 4.9 kN/m^2.

13.5.3.3 Mechanical Analysis

A significant augmentation of vibration occurs due to subharmonics, especially at a frequency of 7.14 Hz, in a synchronous generator under an ITSC fault during no-load operation, as shown in Figure 13.41. The rotor frequency vibration is increased approximately 6500 times from the healthy condition by having one ITSC fault in the rotor field winding, while it increases to 80 000 times by having a 10 ITSC fault. The deformation profiles for subharmonics with frequencies of 14.3, 21.4, and 35.7 Hz are also increased significantly. The frequency spectrum of the time domain radial force, as shown in Figure 13.33, contains subharmonics that are excited due to the ITSC fault and contribute to the low-frequency vibration. Moreover, the same sub-harmonics are excited in the spatial domain force density, which has a significant influence on the stator core deformation, as shown in Figure 13.38.

Figure 13.42 shows how six frequencies contribute to the deformation of the stator core. It indicates how the spatial-harmonic number, m, and the time-harmonic number, k, follow each other at low frequencies. The stator at the four lowest frequencies, which correspond to the time-harmonic numbers of 1, 2, 3, and 4, are merely deformed by the first, second, third, and fourth modes, respectively. The force density wave of 7.14 Hz in the air gap has a first mode number that illustrates why deformation at 7.14 Hz is the most severe, although the magnitude of the excited force is, to a certain extent, the same. The rotor frequency of 7.14 Hz is the highest frequency at which the equal values for the time-harmonic number and the spatial-harmonic number have an evident influence on the stator deformation. The deformed stators at 100 and 14.3 Hz are the same, and

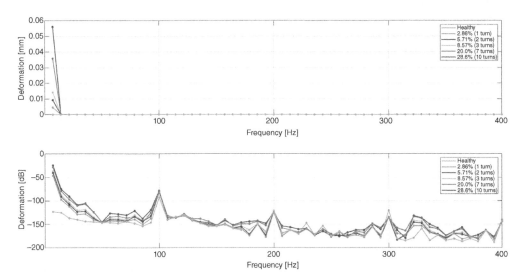

Figure 13.41 The frequency spectrum of stator yoke deformation during no-load operation of a synchronous generator under an ITSC fault. Top: actual values, bottom: logarithmic values.

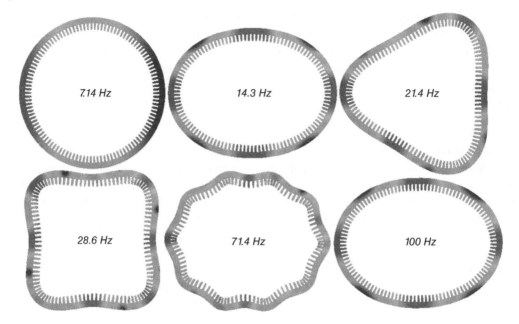

Figure 13.42 The deformation profiles of the stator core at different frequencies during no-load operation of the synchronous generator under 10 ITSC faults.

both are influenced by the second mode number. However, the 14 minor indents that exist in the healthy operation are ineffectual due to the amplified second-order harmonic. The lowest naturally occurring spatial harmonic order (i.e., the second mode in this synchronous generator) is the dominant vibration mode at the main frequency of 100 Hz during an ITSC fault, despite the excited first-order harmonic.

The ITSC fault does not affect the deformation profile between 100 and 200 Hz, even though deformation occurred mostly at two attraction points, indicating that the magnitude of the second spatial harmonic mainly affects force waves with a frequency of 14.3 and 100 Hz. The slight modification in the vibration between 100 and 200 Hz means that the air gap forces with the mentioned frequencies have a low impact on the vibration of the generator. The vibration level is higher during an ITSC fault for a frequency above 300 Hz.

The deformation profile of the stator core at a majority of frequencies above 300 Hz has three attraction points, which are due to the increased third-order spatial harmonic. The vibration close to the third natural frequency of the stator at 327.7 Hz is also escalated. The behavior of the synchronous generator for a loaded condition under an ITSC fault is the same as in the no-load condition, except that the vibration level is increased by 5% in a full-load case due to an increase in the forces.

The calculation resolution is enhanced to investigate the deformation profile around the natural frequencies during no-load operation under 10 ITSC faults (see Figure 13.43). The dark line overlaps the light line that represents the increased frequency resolution at 7.14 Hz. This overlap means that the vibration at this frequency is not influenced by a resonance effect, indicating that the highly excited force at 7.14 Hz and mode number one is sufficiently separated from the first natural frequency, which is 0.21 Hz. A spike occurs at 118.3 Hz, similar to the healthy case, while the frequency spectrum of the radial force density (Figure 13.33) indicates the appearance of excited harmonics at 114.3 and 121.4 Hz. Although the deformed shape of the stator at both 114.3

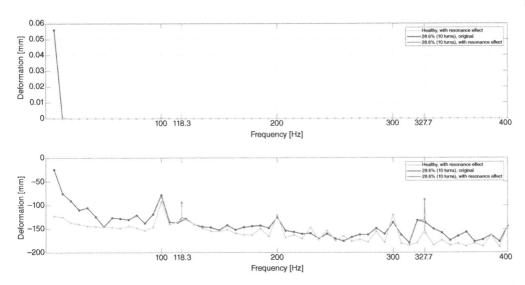

Figure 13.43 The frequency spectrum of stator core deformation during no-load operation under 10 ITSC faults with specific calculations around natural frequencies. Top: actual values, bottom: logarithmic values.

and 121.4 Hz is dominated by the second-order mode, the presence of a small frequency separation between the excited force and the natural frequency may provide a sufficiently safe margin to avoid resonance. The deformation at a frequency of 327.7 Hz under the ITSC fault is increased up to 40 times compared with the healthy case, indicating that the frequency of the excited harmonic is quite close to the natural frequency to cause resonance. In addition, excited spatial harmonics during an ITSC fault play an important role in the generation of the deformation at the natural frequency. For instance, the third-order spatial harmonic might be the reason for the spike observed at 327.7 Hz.

13.5.4 Analysis of a Synchronous Generator under Static Eccentricity

13.5.4.1 Time-Domain Distributions of a Magnetic Field

The behavior of the synchronous generator is studied under a static eccentricity (SE) fault for a better understanding of the radial force and corresponding vibration. The radial flux density distribution measured at two points in the synchronous generator operating in a no-load case under 30% SE fault is presented in Figure 13.44. The SE fault is applied to the generator by moving the stator core along the positive direction of the x-axis. The amplitude of the radial flux density (blue waveform) is increased by decreasing the distance between the rotor and stator core due to the SE fault, while the amplitude of the radial flux density on the opposite side is decreased due to a large air gap caused by the SE fault. The difference between the radial flux density at the two measuring points due to the 30% SE fault is 17%. The harmonics that have not appeared in the frequency spectrum of the healthy generator become evident due to the strengthened magnetic field caused by the SE fault.

The frequency spectrum of the two radial flux density measuring points is shown in Figure 13.45. A clear difference is evident between the frequency spectrum of the two measuring points of the radial flux density due to the different field strengths. Some harmonic sidebands are only visible on the right measuring point, indicating that the same sideband components exist for the left measuring point, but their amplitude is weak.

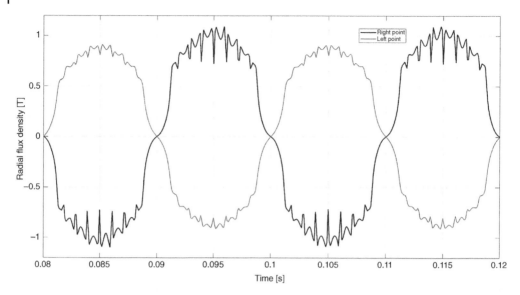

Figure 13.44 The time distribution of the radial flux density measured at the right and left sides of the synchronous generator in a no-load operation under a 30% SE fault.

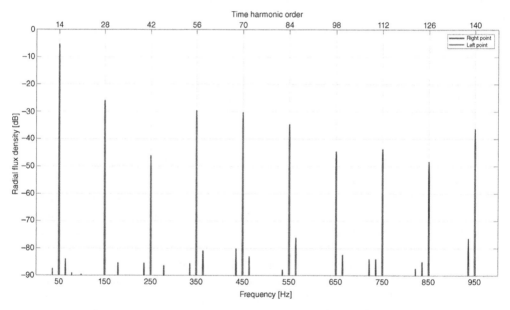

Figure 13.45 The frequency spectrum of the time distribution of the radial flux density measured at the right and left sides of the synchronous generator in a no-load operation under a 30% SE fault.

Increasing the severity of the SE fault results in severe asymmetry of the radial flux density in the air gap. Figure 13.46 compares the different radial flux density waveforms in a healthy state and under 10, 20, and 30% SE faults. The left measuring point is appointed for analysis purposes; hence, the increased degree of SE fault results in an air gap increment and, consequently, a reduced radial flux density. Apart from the magnitude changes under the SE fault, the shape, symmetry, and periodicity of the waveform are maintained under the SE fault.

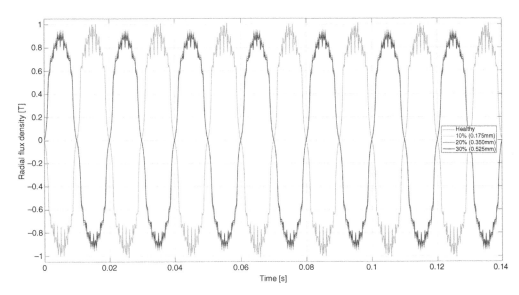

Figure 13.46 The time-domain distribution of the radial air gap flux density during no-load operation of a synchronous generator under different degrees of an SE fault.

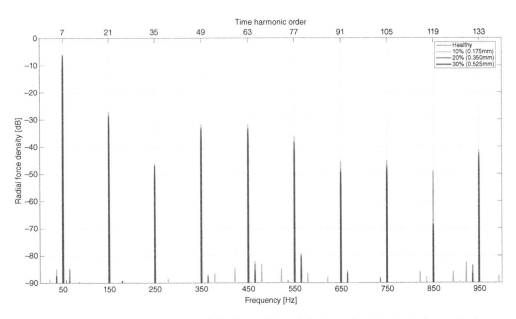

Figure 13.47 The frequency spectrum of the time of the radial air gap flux density during no-load operation of a synchronous generator under different degrees of an SE fault.

Figure 13.47 shows the frequency spectrum of the time distribution of the radial flux density. The natural harmonics under the SE fault are decreased and a few new sideband components appear, indicating that some difference exists between the distribution of the harmonics in the healthy and faulty cases. The largest sideband appears at 564 Hz with an amplitude around −80 dB corresponding to 0.1 mT, which is insignificant for fault detection purposes. The inter-harmonics between 350 and 650 Hz do not appear in the faulty case, unlike the healthy case.

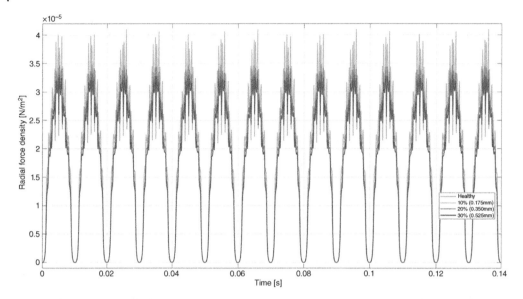

Figure 13.48 The time distribution of the radial force density during no-load operation of a synchronous generator under various degrees of an SE fault.

The disappearance of this harmonic is not due to the reduced field strength; rather, it is caused by the SE fault. The variation in the frequency spectrum of the radial flux density in the time domain from the left measuring point clearly cannot provide a clear signature for SE fault detection

The time distribution of the radial force density of a healthy case and under different degrees of SE fault is shown in Figure 13.48. The amplitude of the radial force density is decreased by increasing the SE fault severity for the left measuring point and vice versa for the right measuring point. The damper bar effect that causes the dip in the waveform is diminished due to the SE fault, but the shape of the waveform and the periodicity are unchanged. The frequency spectrum of the radial force density, as shown in Figure 13.49, reveals a decrease in the amplitude of the natural frequencies, whereas some sideband components are excited in a frequency range between 500 and 600 Hz. The loading condition shows the same results as the no-load case.

13.5.4.2 Spatial-Domain Distributions of the Magnetic Field

The spatial distribution of the radial flux density, compared with the time distribution of the radial flux density, also indicates some changes. Figure 13.50 shows the spatial distribution of the radial flux density in a healthy state and under various degrees of SE fault for the no-load operation of the synchronous generator. The magnitude of the radial flux density is changed in two intervals, between zero to 90° and from 270 to 360°, due to the displacement along the positive direction of the x-axis. The amplitude of the radial flux density in the mentioned interval is clearly reduced due to the increased air gap length. The frequency spectrum of the radial flux density in a spatial domain must differ from the time-domain distribution as an unbalanced distribution exists in its waveform. Figure 13.51 shows the frequency spectrum of the spatial domain of the radial flux density under various severities of SE fault. All the natural harmonics have sideband components whose order differs by one. The components adjacent to the fundamental harmonic and two slot harmonics are the most influenced by the SE fault. The SE fault excites the harmonics in the spatial domain that may interact with natural harmonic components and result in force

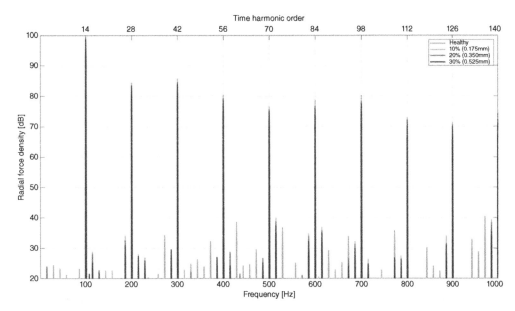

Figure 13.49 The frequency spectrum of the time distribution of the radial force density during no-load operation of a synchronous generator under various degrees of an SE fault.

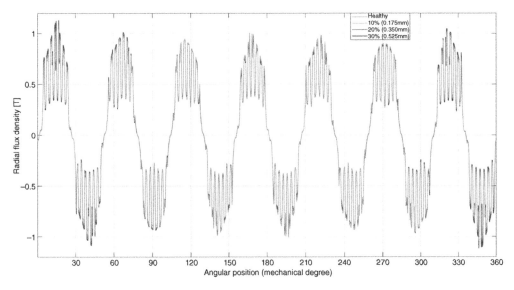

Figure 13.50 The spatial distribution of the radial flux density of the air gap during no-load operation of the synchronous generator in a healthy state and under different degrees of an SE fault.

density amplification. The seventh-order harmonic reaches 0.76 T for different degrees of SE fault, and comparison with ITSC fault reveals that its natural odd multiples have minor elevations rather than reductions. The analysis of the full-load condition also shows no changes compared with the no-load condition.

The radial force density wave during the SE fault differs from a healthy case as the magnetic field is no longer symmetric and the lowest spatial order has changed. Figure 13.52 shows the radial

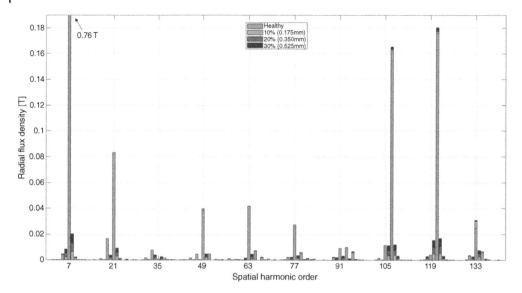

Figure 13.51 The frequency spectrum of the radial flux density of the air gap during no-load operation of the synchronous generator in a healthy state and under different degrees of an SE fault.

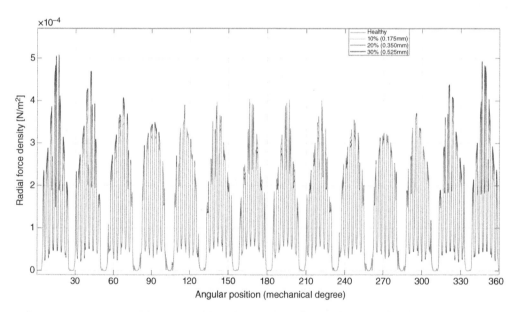

Figure 13.52 The spatial distribution of the radial flux density of the air gap during no-load operation of the synchronous generator in a healthy state and under different degrees of an SE fault.

force density of the synchronous generator in a healthy state and under various degrees of the SE fault. The second-order harmonic, which is evident in the healthy waveform, is not recognizable in the SE fault waveform, indicating that the lowest spatial order in a case of SE fault is one. However, the second-order spatial harmonic exists in the faulty waveform due to the synchronous generator topology, but it is not the lowest harmonic order. Rather, the lowest peak during SE fault occurs

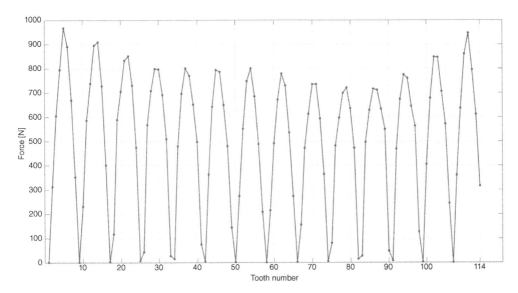

Figure 13.53 The total force acting on each tooth during no-load operation of the synchronous generator under a 30% SE fault.

at 270° due to the small increase around 180° caused by the second-order harmonic. This is more apparent in the spatial distribution of the total forces in Figure 13.53.

The first-order harmonic is the most important difference between the frequency spectrum of the healthy and faulty cases (Figure 13.54). The amplitude of the first-order harmonic is markedly excited and reaches 14% of the fundamental harmonics under a 30% SE fault. A severe deformation may happen if it belongs to one or more force waves with sufficient amplitude. However, the frequency spectrum of the time-domain distribution (see Figure 13.49) shows that the low-order time harmonics have a small magnitude that might influence the magnitude of the resulting low-mode vibration under the SE fault.

The first-order harmonic is produced under the SE fault in a synchronous generator and can be examined similarly to an ITSC fault. According to Figure 13.51, all the natural harmonic components have one or two sidebands with an order that differs by one. The most relevant harmonic components are included in the following equation:

$$f_{rr,1st} = \frac{1}{2\mu_o} \sum_{i,j} b_{r,ith}\, b_{r,jth}$$

for $i, j \in \{6, 7, 8, 20, 21, 22, 49, 50, 62, 63, 64, 76, 77, 78, 106, 107, 108, 120, 121, 122, 133, 134\}$,

$$j = i + 1 \tag{13.15}$$

Figure 13.55 shows these harmonic components with their included signs.

All the side-band components, except those between 64 and 78, have the same sign as their adjacent natural components and contribute to the generation of the $f_{rr,1st}$ (see Figure 13.55). Figure 13.56 shows the contributions of the interactions between the most significant radial flux density harmonics under the SE fault to the first-order radial force density harmonics. The fundamental harmonic component, in addition to its sidebands, and the slot harmonics, with their sidebands, are the main contributors to the force density component, while the flux harmonic

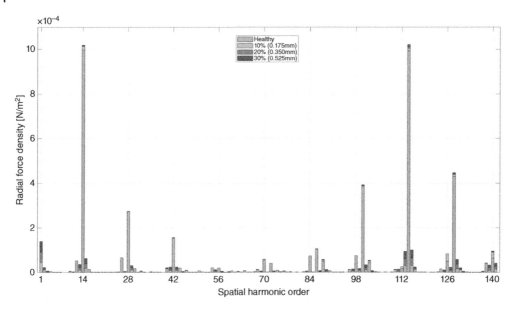

Figure 13.54 The frequency spectrum of the spatial distribution of the radial flux density of the air gap during no-load operation of the synchronous generator in a healthy state and under different degrees of an SE fault.

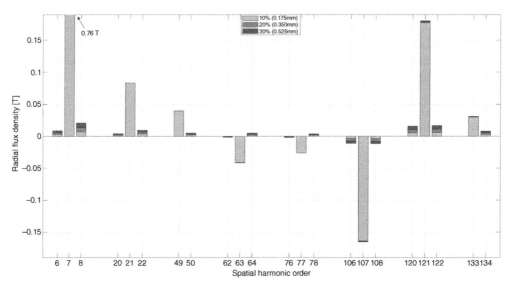

Figure 13.55 The most significant radial flux density components to produce the radial force density harmonics of the first order in no-load operation of the synchronous generator under an SE fault.

numbers 63, 64, 77, and 78 have negative signs and reduce the force. Summation of the chosen interactions between harmonics adds up to $f_{rr,\,1st} = 13.2\ \mathrm{kN/m^2}$, which is close to the lowest spatial harmonic $f_{r,\,1st} = 14.0\ \mathrm{kN/m^2}$. The remaining force ($0.8\ \mathrm{kN/m^2}$) is produced by the remaining radial flux density components with orders that differ by one and by all tangential flux density harmonics with orders that differ by one, indicating the small influence of the tangential field.

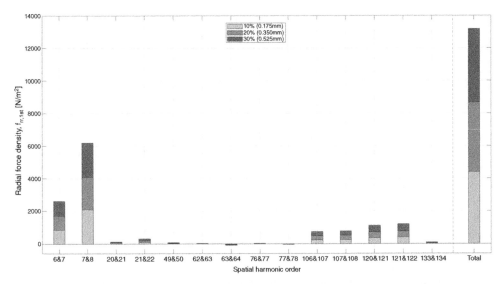

Figure 13.56 Contributions to the first-order radial force density harmonics from interactions between the most significant radial flux density harmonics during a no-load operation of the synchronous generator under an SE fault.

13.5.4.3 Mechanical Analysis

Figure 13.57 presents a comparison of the deformation profile of a synchronous generator in a healthy and under 30% SE fault during no-load operation. A clear rise is evident from 100 to 7.14 Hz. The rotor frequency vibration has increased by 131 times compared with a healthy case under 30% SE fault, and it is four times higher than the 100 Hz component. The frequency component of the 100 Hz is identical during healthy and faulty operation, while every frequency below

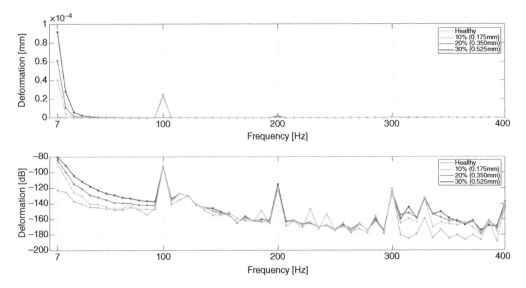

Figure 13.57 The frequency spectrum of stator yoke deformation during a no-load operation and static eccentricity. Top: actual values, bottom: logarithmic values.

100 Hz is influenced by the SE fault. The frequency spectrum of the time-domain force density (see Figure 13.49) contains no noteworthy harmonics below 100 Hz, indicating that the time-domain harmonics are not the reason for vibration increment during an SE fault, while the frequency spectrum of the spatial domain presents a highly excited first-order harmonic (Figure 13.54). The deformation of the stator core also proves that the spatial harmonic order dominates the deformation at five frequencies below 100 Hz.

The shapes of the stator core are depicted in Figure 13.58; they are completely circular and no higher-order modes appear with more than one attraction point. In addition, the same tendency was found in all of the 13 subharmonic frequencies. Conclusively, the excited first-order spatial harmonic is distributed among the force density waves at several frequencies. The vibration at rotor frequency, 7.14 Hz, has the highest elevation according to the vibration frequency spectrum, which indicates that the force density waves with $k = m = 1$ is influenced the most. The rotor frequency vibration is amplified under SE fault, while it is not as high as the ITSC fault, which is due to the low amplitude of the first time harmonic. Despite the low amplitude of the time domain harmonics, a vibration is still induced at a rotor frequency of 7.14 Hz due to the high effect of the excited first-order spatial harmonic.

The stator deformation at 100 Hz caused by the SE fault is similar to the healthy case, which is dominated by the second mode with a minor impact of the 14th mode. The color intensity of the deformed stator core on the left-hand side is different from that on the right-hand side, which is the indication of the first mode. However, the frequency spectrum of vibration at the mentioned frequency presents a marginal elevation during an SE fault, and it is considered unaffected by the excited first-order harmonic, indicating that the small increase in the second-order harmonic does not affect the 100 Hz component to a notable degree. However, it is probably the cause of the small elevation at 200 Hz.

The vibration is unaffected in a frequency range between 100 and 300 Hz due to an SE fault, while the amplitude of the vibration is increased for the frequency range of 300–400 Hz. The vibration

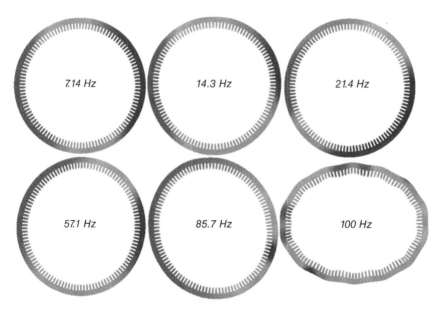

Figure 13.58 Deformation profiles of the stator core at various frequencies during no-load operation of the synchronous generator with static eccentricity.

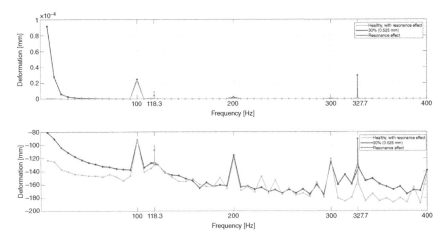

Figure 13.59 The frequency spectrum of stator yoke deformation in the no-load operation of the synchronous generator under a 30% SE fault with specific calculations around natural frequencies. Top: actual values, bottom: logarithmic values.

level for the mentioned range is approximately 2000–3000 times smaller than a 100 Hz component. Moreover, the third mode component is the reason for the deformed stator core at frequencies from 307.14 to 371.4 Hz. According to Figure 13.54, the third spatial harmonic is raised from 0 N/m² at the healthy operation of the synchronous generator to 550 N/m² under a 30% SE fault. The characteristics of the vibration in the full-load operation of the synchronous generator are similar to the no-load case.

The frequency resolution is increased to investigate the resonance phenomena in a synchronous generator under an SE fault (see Figure 13.59). The rotor frequency vibration is unchanged by the first natural frequency, 0.21 Hz, due to adequate frequency spacing. The vibration at the second natural frequency is influenced marginally less than for a healthy operation, while the spike at 327.7 Hz is 31 times higher during the SE fault and exceeds the 100 Hz peak. No excited time harmonics are evident around these frequencies and the reason for the large increase is somewhat unclear. However, the third vibration mode at the frequencies between 307.14 and 371.4 Hz and the minimum increment in the third spatial harmonic is the anticipated cause. Nevertheless, the resonance impact does not induce vibration beyond the one at rotor frequency and the generator would not experience any dramatic resonance between 7.14 and 400 Hz if operated with an SE fault.

13.5.5 Load Effect

A marked difference was found between the no-load and full-load operation in the spatial-domain harmonics during healthy operation. The second-order harmonic is approximately twice as high in the full-load compared with the no-load condition, due to the amplified main harmonic and slot harmonics in the flux density distribution. The dominance of the second vibration mode at frequencies of 100 and 200 Hz consequently yields an increment in the vibration, specifically with factors of 2.4 and 10.6, respectively. However, the amplitude of the vibration in a faulty operation is not significantly influenced by the load. The excited harmonics in both time and spatial domains are equally likely to be as large during full load as during no-load, yielding minor differences in the vibration frequency spectrum. For instance, the rotor frequency vibration during full load is less than 5% higher than no-load under the most severe scenario of both fault types. In addition, the power factor of the operating point must be considered for vibration analysis of the synchronous generator, since the amplitude of the vibration differs during a fully resistive load, inductive load, and capacitive load.

13.5.6 Comparison of Fault Impacts on the Magnetic and Vibration Signatures

Vibration is the most prevalent type of signal expected in either a healthy or faulty machine and is one of the first signals utilized for condition monitoring purposes. The vibration in the stator is the symptom of different kinds of electrical or mechanical faults that originate in the stator or the rotor, since they can influence the air gap magnetic field and impose force on the stator core. Vibration measurement is easy to access simply by installing a non-invasive sensor on the stator yoke, thereby allowing measurements to be taken, even during synchronous generator operation Figure 13.60.

Table 13.2 summarizes how the vibration is influenced in the synchronous generator from healthy to faulty operation. The lowest rotor frequency, f_r, and its six multiples, in addition to two frequencies where vibration naturally occurs, f_s, are used to evaluate the vibration behavior of the synchronous generator under different degrees of ITSC and SE fault. The vibration during faulty cases is expressed as a multiple of the healthy condition level. Figure 13.61 demonstrates the data from Table 13.2 on a logarithmically scaled y-axis. Vibration during the healthy operation of a synchronous generator with a relative value of 1 equals 0 dB and values between 0 and 1 are converted to a negative dB.

Analysis of Table 13.2 shows that the vibration at the rotor frequency f_r is strongly affected by the ITSC fault in the rotor field winding during no-load operation. For instance, having one ITSC fault results in a relatively dramatic increment in vibration. This increment in the rotor frequency amplitude can definitely be used to diagnose the ITSC fault. A similar pattern to that of the ITSC fault is achieved for the SE fault, where the rotor frequency can detect the occurrence of the SE fault. A 10% SE fault results in a vibration that is approximately 60 times stronger than in the healthy case, thereby providing a sufficiently distinguishable feature for SE detection. In addition to the rotor frequency, its multiples can be used to confirm the detection of both ITSC and SE faults.

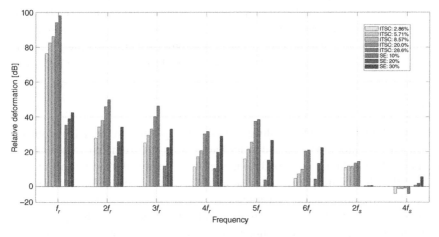

Figure 13.60 The logarithmic comparison of the deformation at certain frequencies during no-load operation of the synchronous generator under both ITSC and SE faults.

Table 13.2 The amplitude of the vibration frequencies caused by faults relative to the healthy case in the no-load operation of the synchronous generator (f_r is the rotor frequency and f_s is the synchronous frequency).

Frequency (13.Hz)	Healthy	ITSC in the rotor winding (%)					SE (%)		
		2.86	5.71	8.57	20.0	28.6	10	20	30
f_r (13.7.14)	1	6513.7	13 255.0	20 277.5	51 307.3	80 225.6	58.1	87.4	131.3
f_r (13.14.2)	1	24.7	51.0	78.8	194.5	311.2	7.6	19.6	50.6
f_r (13.21.4)	1	17.7	29.2	44.2	100.9	204.9	3.8	13.0	44.2
f_r (13.28.62)	1	3.6	7.1	10.5	32.0	53.9	3.2	9.5	27.7
f_r (13.35.7)	1	6.1	11.8	18.6	73.5	82.0	1.5	5.7	21.0
f_r (13.42.9)	1	1.7	2.3	3.1	10.4	11.4	1.6	4.6	12.9
f_s (13.100)	1	3.5	3.8	3.8	4.7	5.2	1.0	1.1	1.1
f_s (13.200)	1	0.6	0.6	0.9	0.9	0.9	1.1	1.2	1.9

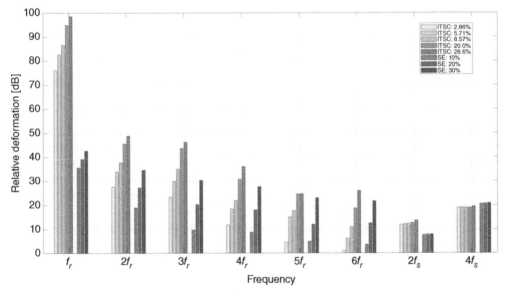

Figure 13.61 The logarithmic comparison of the deformation at certain frequencies in full-load operation of the synchronous generator under both ITSC and SE faults.

The behavior of the rotor frequency for fault detection purposes in a synchronous generator under SE and ITSC faults in a full-load condition is similar to the no-load cases. Table 13.3 summarizes the rotor frequency and its multiples used for ITSC and SE fault diagnosis in a full-load case. The characteristics of the two frequencies, $2f_s$ and $4f_s$, differ for both the ITSC and SE fault during the loaded condition compared with the no-load condition. Notably, their amplitudes are increased up to 10 times for both faults, indicating that, in a case of fault detection according to $4f_s$, the load situation must be considered. The amplitude of the vibration

Table 13.3 The amplitude of the vibration frequencies caused by faults in the full-load operation of the synchronous generator relative to the healthy case (f_r is the rotor frequency and f_s is the synchronous frequency).

Frequency, (13.Hz)	Healthy	ITSC in the rotor winding (%)					SE (%)		
		2.86	5.71	8.57	20.0	28.6	10	20	30
f_r (13.7.14)	1	6373.8	13 567.1	21 242.2	55 624.8	84 157.6	61.1	91.8	136.1
f_r (13.14.2)	1	23.9	49.9	76.7	191.4	278.6	9.0	23.1	54.3
f_r (13.21.4)	1	14.7	31.1	57.3	151.7	204.6	3.1	10.4	33.1
f_r (13.28.62)	1	3.9	8.4	12.5	34.7	17.0	2.7	8.0	24.5
f_r (13.35.7)	1	1.7	5.7	7.7	16.8	64.0	1.5	4.2	12.1
f_r (13.42.9)	1	1.1	2.0	3.5	9.5	17.0	1.5	4.2	12.1
f_s (13.100)	1	3.9	4.0	4.2	4.3	4.8	2.4	2.5	2.5
f_s (13.200)	1	8.8	8.9	8.9	8.9	9.3	10.6	10.8	11.0

at a frequency of $2f_s$ for the ITSC fault is similar in both the no-load and loaded cases, while the amplitude of the frequency for the SE fault in the loaded case is three times higher than the no-load case, indicating that the load situation must be considered if the frequency component is considered for fault detection.

A problem arises regarding the fault type detection of the synchronous generator based on the rotor frequency characteristic, since both ITSC and SE faults involve the same frequencies. However, 1 ITSC and 10 ITSC faults cause about 50 to 60 times higher vibrations compared with a 30% SE fault. Therefore, an ITSC fault with a low degree of severity is easily distinguishable from the high degree of the SE fault. In addition, the rotor frequency harmonics and amplitude increase evenly in the case of the SE fault, while the amplitude of the first rotor frequency is twice that of the second rotor frequency in the case of an ITSC fault. If vibration measurements are conducted in the no-load operation, increased 100 Hz vibration can be used to detect the ITSC in the generator and distinguish it from SE.

In general, fault detection is based on a comparison between the healthy and faulty signals of the electric machines operating under the same conditions. However, finding a power plant that has been operated for several years and has conducted a vibration measurement when the machine was healthy is almost impossible. Therefore, having a reference healthy signal for comparison is difficult to achieve. However, frequency spectrum analysis of healthy and faulty generators indicates that the amplitudes of some frequency harmonics, such as 107.14 and 192.9 Hz, do not change under the faulty condition, and their amplitudes remain the same as the rotor frequency components in the healthy case. This indicates that if the measurement from a healthy case is not available, these frequency components might be used as a reference value to substitute for the healthy case.

13.6 Summary

Vibration analysis is the most dominant approach for fault detection of electrical machines, especially in hydropower plants. This chapter explained the detailed modeling of the synchronous generator in finite element software in order to measure the appearance of vibration on the stator core. One of the main consequences of a fault in the generator is the unbalanced magnetic field in the air gap that gives rise to the unbalanced force that is the root cause of vibration. The finite element analysis of the air gap magnetic field and radial forces indicates that the ITSC fault excites the unnatural harmonics in the time domain and the spatial domain that gives rise to the low-frequency vibration. The impact of SE fault on the radial force only excites the spatial harmonics and, consequently, the vibration level compared with a short circuit fault is lower.

The finite element analysis also demonstrated that the lowest order spatial harmonics are the reason for the dominating vibration. The lowest-order spatial harmonics are initiated by the slot harmonics and can be reduced using magnetic wedges or semi-closed slots.

It was shown that the loading condition does not change the vibration behavior of the generator both during healthy and fault operation, indicating that the proposed signatures can be used for fault detection regardless of the load conditions.

References

1 Tavner, P., RanJim, L., Penman, L. et al. (2008). *Condition Monitoring of Rotating Electrical Machines*. UK: IET.

2 Rødal, G.L. (2020). Online condition monitoring of synchronous generators sing vibration signal. MSc Thesis, NTNU, Norway.

3 Ehya, H., Rødal, G.L., Nysveen, A. et al. (2022). Condition monitoring of wound field synchronous generator under inter-turn short circuit fault utilizing vibration signal. *23rd International Conference on Electrical Machines and Systems (ICEMS)*, pp. 24–27, November 2020, Hamamatsu, Japan.

4 Sadeghi, I., Ehyam, H., Faiz, J. et al. (2017). Online fault diagnosis of large electrical machines using vibration signal – A review. *International Conference on Optimization of Electrical and Electronic Equipment (13.OPTIM) and International Aegean Conference on Electrical Machines and Power Electronics (ACEMP)*, 25–27 May 2017, Brasov, Romania.

5 de Canha, D., Cronje, W.A., Meyer, A.S., et al. (2007). Methods for diagnosing static eccentricity in a synchronous 2 pole generator. *IEEE Power Technical Conference*, 1–5 July 2007, Lausanne, Switzerland.

6 Djurović, S., Vilchis-Rodriguez, D.S., and Smith, A.C. (2014). Investigation of wound rotor induction machine vibration signal under stator electrical fault conditions. *The Journal of Engineering* 5: 248–258.

7 Ebrahimi, B. and Faiz (2012). J., Magnetic field and vibration monitoring in permanent magnet synchronous motors under eccentricity fault. *IET Electric Power Applications* 6 (1): 35–45.

8 Sadeghi, I., Ehya, H., and Faiz, J. (2007). Eccentricity fault indices in large induction motors an overview, in *8th Power Electronics, Drive Systems & Technologies Conference (13.PEDSTC)*. 14–16 February 2017, Mashhad, Iran.

9 Ebrahimi, B.M., Faiz, J., and Roshtkhari, M.J. (2009). Static-, dynamic-, and mixed-eccentricity fault diagnoses in permanent-magnet synchronous motors. *IEEE Transactions on Industrial Electronics* 56 (11): 4727–4739.

10 Cuevas, M., Romary, R., Lecointe, J.P. et al. (2016). Non-invasive detection of rotor short-circuit fault in synchronous machines by analysis of stray magnetic field and frame vibrations. *IEEE Transactions on Magnetics* 52 (7): 1–4.

11 Nadarajan, S., Wang, R., Gupta, A.K., et al. (2015). Vibration signature analysis of stator winding fault diagnosis in brushless synchronous generators. *IEEE International Transportation Electrification Conference (ITEC)*, 27–29 August 2015, Chennai, India.

12 Cuevas, M., Romary, R., Lecointe, J.P. et al. (2018). Noninvasive detection of winding short-circuit faults in salient pole synchronous machine with squirrel-cage damper. *IEEE Transactions on Industry Applications* 54 (6): 5988–5997.

13 Sadeghi, I., Ehya, H., Faiz, J., et al. (2018). Online condition monitoring of large synchronous generator under short circuit fault – A review. *IEEE International Conference on Industrial Technology (ICIT)*, 20–22 February 2018, Lyon, France.

14 Ma, H. and Pu, L. (2009). Fault diagnosis based on ANN for turn-to-turn short circuit of synchronous generator rotor windings. *Journal of Electromagnetic Analysis and Applications* 01 (03): 187–191.

15 Gritli, Y., Di Tommaso, A.O., Miceli, R., et al. (2013) Vibration signature analysis for rotor broken bar diagnosis in double cage induction motor drives. *4th International Conference on Power Engineering, Energy and Electrical Drives*, 13–17 May 2013, Istanbul, Turkey.

16 Gritli, Y., Di Tommaso, A.O., Miceli, R., et al. (2013). Quantitative rotor broken bar evaluation in double squirrel cage induction machines under dynamic operating conditions, in *8th International Conference and Exhibition on Ecological Vehicles and Renewable Energies (EVER)*, 27–30 March 2013, Monte Carlo, Monaco.

17 Di Tommaso, A., Miceli, R., and Galluzzo, G.R. (2011). Monitoring and diagnosis of failures in squirrel-cage induction motors due to cracked or broken bars. *8th IEEE Symposium on Diagnostics for Electrical Machines, Power Electronics & Drives*, 5–8 September 2011, Bologna, Italy.

18 Concari, C., Franceschini, G., and Tassoni, C. (2008). Differential diagnosis based on multivariable monitoring to assess induction machine rotor conditions. *IEEE Transactions on Industrial Electronics* 55 (12): 4156–4166.

19 de Jesus Rangel-Magdaleno, J., Romero-Troncoso, R., Osornio-Rios, R.A. et al. (2009). Novel methodology for online half-broken-bar detection on induction motors. *IEEE Transactions on Instrumentation and Measurement* 58 (5): 1690–1698.

20 Ágostonak, K. (2015). Fault detection of the electrical motors based on vibration analysis. *Procedia Technology* 19: 547–553.

21 Martinez, J., Belahcen, A., and Muetze, A. (2017). Analysis of the vibration magnitude of an induction motor with different numbers of broken bars. *IEEE Transactions on Industry Applications* 53 (3): 2711–2720.

22 Nandi, S. and Toliyat, H.A. (1999). Condition monitoring and fault diagnosis of electrical machines-a review. *34th IEEE IAS Annual Meeting Conference Record*, 3–7 October 1999, Phoenix, AZ, USA.

23 Patil, S. and Gaikwad, J. (2013). Vibration analysis of electrical rotating machines using FFT: A method of predictive maintenance. *4th IEEE International Conference on Computing, Communications and Networking Technologies (ICCCNT)*, 4–6 July 2013, Iruchengode, India.

24 Plante, T., Nejadpak, A., and Yang, C.X. (2015). Faults detection and failures prediction using vibration analysis. *2015 IEEE AUTOTESTCON*, 2–5 November 2015, National Harbor, MD, USA.

25 Marcin, B. (2014). Vibration diagnostic method of permanent magnets generators-detecting of vibrations caused by unbalance. *9th IEEE International Conference on Ecological Vehicles and Renewable Energies (EVER)*, 25–27 March 2014, Monte-Carlo, Monaco.

26 Zhou, W., Habetler, T.G., and Harley, R.G. (2007). Bearing condition monitoring methods for electric machines: A general review. *IEEE International Symposium on Diagnostics for Electric Machines, Power Electronics and Drives*, 6–8 September 2007, Cracow, Poland.

27 Shrivastava, A. and Wadhwani, S. (2014). An approach for fault detection and diagnosis of rotating electrical machine using vibration signal analysis. *IEEE International Conference on Recent Advances and Innovations in Engineering (ICRAIE)*, 9–11 May 2014, Jaipur, India.

28 Kumar, T.C.A., Singh, G., and Naikan, V., (2016). Effectiveness of vibration and current monitoring in detecting broken rotor bar and bearing faults in an induction motor. *IEEE 6th International Conference on Power Systems (ICPS)*, 4–6 March 2016, New Delhi, India.

29 Sadeghi, I., Ehya, H., Zarandi, R.N., et al. (2018). Condition monitoring of large electrical machine under partial discharge fault – A review. IEEE *International Symposium on Power Electronics, Electrical Drives, Automation and Motion (SPEEDAM)*, 20–22 June 2018, Amalfi, Italy.

30 Gieras, J.F., Wang, C., and Lai, J.C. (2018). *Noise of Polyphase Electric Motors*. CRC Press, USA.

31 Traxler-Samek, G., Lugand, T., and Uemori, M. (2011). Vibrational forces in salient pole synchronous machines considering tooth ripple effects. *IEEE Transactions on Industrial Electronics* 59 (5): 2258–2266.

32 Zhu, Z.Q., Xia, Z.P., Wu, L.J. et al. (2010). Analytical modeling and finite-element computation of radial vibration force in fractional-slot permanent-magnet brushless machines. *IEEE Transactions on Industry Applications* 46 (5): 1908–1918.

33 Engevik, E.L., Valavi, M., and Nysveen, A. (2017). Influence of winding layout and airgap length on radial forces in large synchronous hydrogenerators, in *20th International Conference on Electrical Machines and Systems (ICEMS)*, 11–14 August 2017, Sydney, NSW, Australia.

34 Valavi, M., Nysveen, A., and Nilssen, R. (2014). Effects of loading and slot harmonic on radial magnetic forces in low-speed permanent magnet machine with concentrated windings. *IEEE Transactions on Magnetics* 51 (6): 1–10.

35 Valavi, M., Nysveen, A., Nilssen, R. et al. (2014). Slot harmonic effect on magnetic forces and vibration in low-speed permanent-magnet machine with concentrated windings. *IEEE Transactions on Industry Applications* 50 (5): 3304–3313.

36 Devillers, E., Hecquet, M., Cimetiere, X., et al. (2018). Experimental benchmark for magnetic noise and vibrations analysis in electrical machines. *XIII International Conference on Electrical Machines (ICEM)*, 3–6 September 2018, Alexandroupoli, Greece.

37 Valavi, M., Nysveen, A., Nilssen, R. et al. (2013). Influence of pole and slot combinations on magnetic forces and vibration in low-speed PM wind generators. *IEEE Transactions on Magnetics* 50 (5): Article Sequence Number: 8700111.

38 Jang, I.-S., Ham, S.H., Kim, W.-H. et al. (2014). Method for analyzing vibrations due to electromagnetic force in electric motors. *IEEETransactions on Magnetics* 50 (2): 297–300.

39 Jang, I.-S. and Kim (2020). W.-H., Study on electromagnetic vibration analysis process for PM motors. *IEEE Transactions on Applied Superconductivity* 30 (4) Article Sequence Number: 5202406: 1–6.

14

Application of Machine Learning in Fault Detection

14.1 Introduction

Machine learning as part of artificial intelligence (AI) is capable of tackling the solution to a problem without being specifically programmed [1, 2]. It can be classified into supervised, unsupervised, and reinforcement types [3]. A new data sample can be categorized into one of the predefined classes or some unknown values can be estimated from sample data by a trained model in the supervised learning method. The technique follows a certain pattern in an unstructured data set, which is supplied to a machine in the unsupervised learning case.

The idea of reinforcement learning is based on maximizing some rewards to achieve an appropriate action by the trained model. The supervised method is the most appropriate technique for condition monitoring. Training machine learning requires pre-processed data, a model type, and model parameters. The trained machine can then be used for new data sets. Applying machine learning techniques in electric machine fault diagnosis has huge advantages. This is because the system is atomized and no expert is needed for data interpretation. However, the machine needs to be tailored based on the existing data sets and applications.

Electromagnetic Analysis and Condition Monitoring of Synchronous Generators, First Edition. Hossein Ehya and Jawad Faiz.
© 2023 The Institute of Electrical and Electronics Engineers, Inc. Published 2023 by John Wiley & Sons, Inc.

14.2   Supervised Learning

In supervised learning, the similarities of the classes are determined by comparing classified data sets with new data sets, which are labeled during the machine training stage. The approach is called a classification model and the value is predicted by a regression model. A simple example to indicate the distinction between classification and regression is the power consumption in an electric power system. This is a classification job if the goal is to predict if the load may be over a specific limit. If the goal is to predict the absorbed power, it is a regression task. To perform either task, the models must be trained with tagged data.

Labeled training data are samples belonging to a specific target value or classes. Generally, supervised learning approaches are called classifiers or predictors. Some well-known supervised models, widely used in fault diagnosis, include a support vector machine (SVM), a K-nearest neighbor (KNN), an artificial neural network (ANN), and decision tree machines. The trained labeled data are always predicted with high accuracy by the trained model. However, it is very difficult to achieve a similar accuracy for new sample data sets. The main goal of applying the supervised learning method is to achieve a general model that can work with any new data sample sets. The following three steps are followed to generate a classifier:

1. Feature extraction and selection.
2. Data set balancing.
3. Data set training.
4. Validation.
5. Testing.

14.2.1   Feature Extraction and Selection

The main prerequisite for machine learning is data sets. To avoid a complex model, the number of data sets is kept limited. In addition, to prevent model degradation, the non-informative samples must be removed from the training data sets [3]. An extracted feature from signal processing tools is employed to achieve the required features for model training. The informative side-lobes that might create an informative feature can be used to generate a frequency spectrum using a frequency domain processor. The features employed as training inputs to a model might include:

1. The energy of the spectrum of a certain decomposition level in discrete wavelet transforms.
2. Relative energy of certain sub-bands.
3. Entropy wavelet decomposition.
4. Many other properties of a signal processing tool.

In addition to many signal-processing tools, some techniques generate features automatically from a sample time series. The time series feature extraction based on scalable hypothesis tests (TSFRESH) feature selection package is one example. Although feature extraction is vital, feature selection among the generated features is crucial because reliable and informative data selection leads to a more amenable model for visual representation, shortens the training time, decreases the storage requirement, and enhances the model performance [4]. Random forest is another method for feature selection and will also be described later in this chapter. Figure 14.1 presents a procedure for extraction and selection of features.

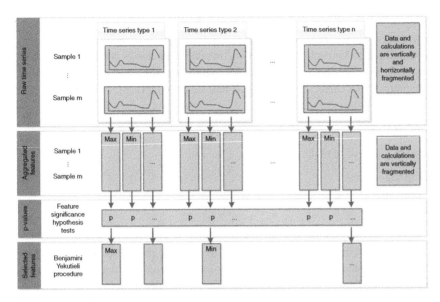

Figure 14.1 Feature extraction and selection procedure.

14.2.1.1 Time Series Feature Extraction Based on Scalable Hypothesis Tests (TSFRESH)

An algorithm has been proposed in reference [5] to generate the features from the time series data sets. This is called FeatuRe (FR) extraction and is based on Saleable Hypothesis tests (FRESH) that extract features from time series while the feature selection is implemented. The goal is to extract fast and accurate features beside feature selection, meaning that the method is highly parallelized.

If a feature matches the target predictions satisfactorily, it is recognized as relevant. This is performed using the statistical inference technique of hypothesis testing, which evaluates a p-value between the feature and the target and quantifies the probability that the feature is irrelevant for predicting the target. A feature is selected using a p-value above a predefined threshold for rejection. The number of features is reduced by employing the FRESH algorithm with a principal component analysis (PCA).

The FRESH algorithm has been integrated into an algorithmic feature generation package called TSFRESH [5]. TSFRESH can generate 794 time series features using 63 time series characterization techniques and then apply feature selection methods. The run time of TSFRESH scales linearly with the number of features extracted, the number of samples, and the number of different time series. However, for some of the more advanced features, such as calculation of the sample entropy, TSFRESH does not scale linearly with the length of the time series. Adjustment of the calculated features can drastically affect the run time of the algorithm. TSFRESH has worked well in extracting relevant features from ensembles of torque sensors in a robot, in determining the failure of a task execution, and as an industrial data set in steel production [1].

14.2.2 Data Set Balancing

Considering the available training data and classifier design, the data set may require balancing if one of the classes is over-represented (i.e., one class is more frequently observed) [6]. For example, in industry, the expectation is that more sample series containing healthy electrical machines than containing faulty machines will be captured. The design of the classifiers can also create imbalances in the data sets; for example, the utilization of "one-versus-all" classifiers results in the classifiers attempting to separate each class from the remaining classes. If five similar frequency classes are present in the data set (i.e., the data set is balanced at the outset), the "one-versus-all" technique leads to imbalanced training data for each classifier. This imbalance in classes is rectified by some common approaches, such as collecting more data, weighting the classes based on their frequencies, using evaluation metrics that correct for imbalance, and resampling [1, 6].

14.2.3 Training and Testing

Generally, three steps are followed to make a predictor:

1. Choosing the model/algorithm.
2. Initializing with random learning variables.
3. Defining a cost function.

The variables are those trained by a model, while the cost function indicates the impacts of each variable on the output. As shown in Figure 14.2, before the training procedure, the data set is split into training and testing sets. The models are then trained by applying a gradient descent to minimize the cost function. This involves introducing labeled samples from the training set to the predictor, comparing the model output to the sample label and adjusting the weights to nudge the output in the correct direction. This process ends when no more samples remain in the training set or when some early stopping criteria are met. An early stop may be desirable because the model can be over-fitted to the data. If the training is allowed to continue, it leads to worse predictions for samples outside the training set. The performance of the trained model is evaluated using a test set of samples not used in model training. The performance of the model is decided by the quality of the prediction of the labels of the test set [7].

The data sets are not completely uniform; therefore, the data splitting method affects the train/test procedure. One split can provide good test results, while another may give unreasonable test results. A proposed model may perform well on the test set, but it may generalize poorly. As seen in Figure 14.3, one solution is to use k-fold cross-validation (CV) [7], which takes in a data set and makes many splits, or folds. Each fold consists of a training set and a validation set. For each fold, the training set trains the model and the validation set measures the performance, which is the average performance across all the folds. This performance presumably can reflect the correct model performance on hidden data.

Generally, the train/test or CV procedure can be repeated for configuration of many candidate models and the researcher can then select the best one. However, if the model is selected based on the performance of the model on the test set, data set behavior is properly reflected in the model. Therefore, the estimated performance of the model is probably optimistic and reflects the "principle of optimism." This principle states that selection of a model based on the data that gave it birth

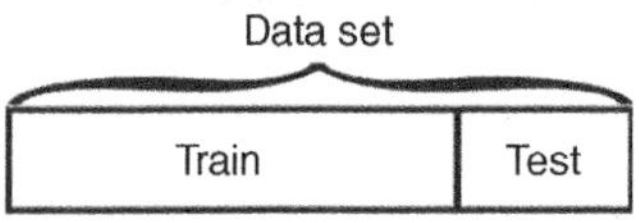

Figure 14.2 Train/test split of a data set.

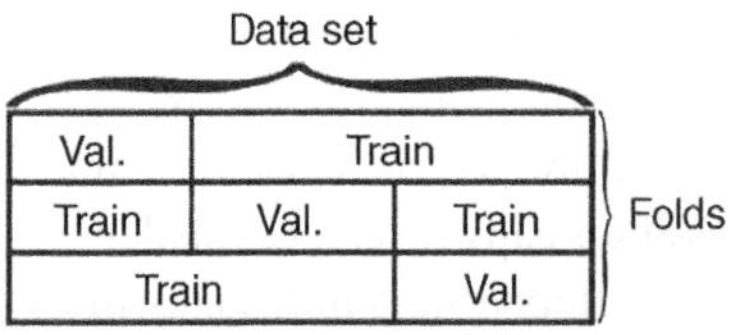

Figure 14.3 Threefold cross-validation. Each fold consists of a training and a validation set.

Figure 14.4 Cross-validation with a hold-out data set.

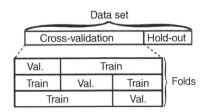

probably works better than almost any other data in practice [8]. The interesting point in testing a new model is its performance on new and hidden data; part of the data set is set aside to determine the model's performance. This is called a hold-out data set (Figure 14.4). The whole selection and tuning process of modeling must be performed without a hold-out data set; the latter set is merely utilized for performance evaluation of the final model(s).

14.2.4 Evaluation Metrics

The performance of classifiers is evaluated using a number of methods that lead to different results. The simplest technique is counting classifications and dividing them by the total number of samples, as shown below:

$$\text{Accuracy} = \frac{\text{TP} + \text{TN}}{\text{TP} + \text{FP} + \text{FN} + \text{TN}} \tag{14.1}$$

This is called the classifier accuracy and shows some aspects of the performance of the classifier. However, this accuracy has problems if it uses unbalanced data sets. A measured data set of an unbalanced electric machine contains samples that are 99% from healthy machines and 1% from faulty machines. Therefore, using a healthy machine, the classifier will always classify a sample with 99% accuracy. Clearly, this is a poor classifier because it cannot properly classify a single faulty machine. Adding other measurements often emphasizes the misclassified samples. Some interesting measures include the F-score and receiver operating characteristic area under the curve (ROC AUC), which operate by combining sensitivity, specificity, and precision.

A confusion matrix for a binary classifier is a useful tool for these measures. It classifies samples as true (belonging to the class) or false (not belonging to the class), as summarized in Table 14.1 [1]. The number of samples in the confusion matrix can be properly classified as a true positive (TP)/false negative (FN) class or improperly as a false positive (FP) class/true negative (TN) class.

The sensitivity is defined as follows:

$$\text{Sensitivity} = \frac{\text{TP}}{\text{TP} + \text{FN}} \tag{14.2}$$

Table 14.1 Confusion matrix [1]/Norwegian University of Science and Technology.

		Actual	
		True	**False**
Predicted	True	True positive (TP)	False positive (FP)
	False	False negative (FN)	True negative (TN)

The sensitivity determines how well the model picks up on the class (i.e. the probability of detecting the class). This is determined based on the number of properly classified samples of the class divided by all occurrences of the class.

Specificity is defined as follows:

$$\text{Specificity} = \frac{\text{TN}}{\text{TN} + \text{FP}} \tag{14.3}$$

The result gives an impression of the capacity of the model to properly classify false samples. The specificity is the number of TNs divided by the total number of actual false samples.

Precision is defined as follows:

$$\text{Precision} = \frac{\text{TP}}{\text{TP} + \text{FP}} \tag{14.4}$$

Precision is the ratio of TPs divided by the total number of true classified samples. A high precision means that the classifier provides a correct prediction when it returns a true classification. For unbalanced data sets, classifying all samples (as true or false) carries risks.

To weight the classification reliability and the chance of detecting the class [8], the F-score is defined as follows:

$$F_\beta - \text{score} = \frac{(\beta^2 + 1) \times \text{precision} \times \text{sensitivity}}{\beta^2 \times \text{precision} + \text{sensitivity}} \tag{14.5}$$

For $\beta = 1$, the F_1 score will be [9]

$$F_1 - \text{score} = 2\frac{\text{precision} \times \text{sensitivity}}{\text{precision} + \text{sensitivity}} \tag{14.6}$$

To balance the likely risks, the F-score is particularly good for unbalanced classes. For more balanced data sets, the ROC AUC is a better metric. The weighted F_1 score: (1) estimates the F_1 score of each class (i.e. faulty or healthy for electrical machines), (2) multiplies each score by the prevalence of each class, (3) adds the adjusted scores, and (4) divides the sum by the total number of samples.

The weighted F_1 function of classes a and b with n_a and n_b samples, respectively, is as follows:

$$F_{1,weighted} = \frac{F_{1,a}\, n_a + F_{1,b}\, n_b}{n_a + n_b} \tag{14.7}$$

A classifier does not often return 1 for true and 0 for false; it does return a number between 0 and 1. The threshold set classifies the sample. For the threshold equal to 0.5, the value $\geq$ classifies the sample as true and vice versa. For every threshold between 0 and 1, the sensitivity plotted against (1-specificity) is the receiver operating characteristic curve (ROC). Figure 14.5 shows the AUC as the area under the ROC.

14.3 Ensemble Learners

Weak learners may have poor performance in generating a good strong learner performance. By combining many of these weak learners, ensemble learners are obtained. Generally, a few methods can do this, with the key ones being bagging, boosting, and stacking [1].

Bagging (bootstrap aggregating) may be applied with any learning algorithm, but is most commonly used with decision tree algorithms. The strategy is to generate several bootstrap data sets from the training set and then train the model with each bootstrapped data set. A bootstrapped data set of size N is created by drawing N samples, with replacements from the original data set.

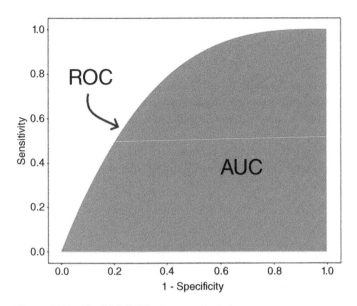

Figure 14.5 The ROC AUC is the area shaded gray.

All the models are combined and their classifications are aggregated so that each model classifies a sample within the ensemble. The final classification is based on the mean or majority vote of the constituent models.

Bagging returns a majority vote or predicted mean from several weak learners, making this process similar to boosting. However, the model generation differs, as bagging utilizes a fully random process, whereas boosting creates models consecutively to improve on the last trained model. This can be achieved by first training a single model on the data set. That model is then placed into the ensemble and the ensemble makes predictions on the training data set. The samples that are poorly classified provide more emphasis for training the next model to go into the ensemble and this eventually improves the ensemble where it performs worst. This continues until the ensemble approaches the desirable size. The samples can also be weighted to indicate an importance in a cost function or for oversampling into a bootstrapped set. In that case, the ensemble focuses on the hard-to-classify samples. In aggregating the prediction, the ensemble often weights its constituent models based on their performance.

Training a meta-learner is called stacking. The learner is trained to interpret the outputs of several other models and makes predictions based on those of several other learners. In this scenario, the base learners provide predictions to the meta-learner. Each base learner is first fitted to the training set; then its prediction is utilized as the training set for the meta-learner. The base learner can be any machine learning model that returns predictions. The benefit of stacking is that different models can be recruited as base learners, so the weaknesses of one model may be remedied by another.

14.4 Logistic Regression

Logistic regression estimates the probability that a sample belongs to a particular class [3]. This estimate is obtained by fitting a logistic function to samples in a two-class training set $X = \{x_n, d_n\}_{n=1}^{N}$ of N samples. Each sample $x_n = \left(x_n^1, \ldots, x_n^p\right)$ is a vector having p features with a class $d = 0, 1$.

The logistic function is defined as follows:

$$p(x_{N+1}) = \frac{e^{\beta_0 + \beta_1\, x_{N+1}^1 + \ldots + \beta_p\, x_{N+1}^p}}{1 + e^{\beta_0 + \beta_1\, x_{N+1}^1 + \ldots + \beta_p\, x_{N+1}^p}} \tag{14.8}$$

Note that the logistic function output range for all inputs is between 0 and 1.

The maximum likelihood method is typically employed to estimate the regression coefficients $\beta_0, \beta_1, \ldots, \beta_p$. The result is a set of the most likely regression coefficients using the training set. A value on the interval of 0 to 1 is returned if an unknown sample x_{N+1} is introduced into the function. This indicates the probability of the sample belonging to class $d = 1$. Since a logistic function asymptotically approaches 0 and 1, a decision threshold θ is introduced. Above this threshold, a sample is classified as belonging to class 1:

$$p(x_{N+1}) > \theta \Rightarrow d_{N+1} = 1$$
$$p(x_{N+1}) < \theta \Rightarrow d_{N+1} = 0 \tag{14.9}$$

Reducing the value of θ causes the classifier to become more or less conservative. In the fault diagnosis of electric machines, the classifier may return a probability of an existing fault in the machine. In the case where conducting maintenance of the equipment comes with a high cost: a high θ could be justified, as only the samples with a very high probability of having a fault will be classified. If no fault is detected, θ could be lowered.

14.5   K-Nearest Neighbors

The class or value of a labeled data set can be predicted by a supervised learning algorithm, K-Nearest Neighbors (KNNs). This algorithm compares a sample to a labeled data set by estimating the distance from the sample to be classified to the samples in the training set. The class of the sample is then obtained as the most frequent class of its KNN. In using KNN for regression, the sample is set to the mean value of its KNN.

KNN is not a generalized model, as it does not really involve training. Rather, it compares samples to the training set. Figure 14.6 presents an illustration of the KNN. The sample x is to be classified into one of the three classes in the training set with $k = 5$, where its distance to any sample in the training set is estimated [1]. Here, there are three "O"'s, and two "–"'s among the five nearest neighbors. Sample x is therefore classified as belonging to the category "O." Local topography may disturb classification, so the KNN suffers from noise. This is realized from the illustration, as changing to $k = 3$ would also change the classification. KNN works well with uneven class borders, which could be considered its strength. However, because of outliers in the training set, the KNN can easily misclassify samples, which is considered its main weakness. Condensed Nearest Neighbors (CNNs) have been advanced as an improvement, and this also accelerates the operation of the algorithm. CNN chooses its prototypes from the data that best represent each class in the training set.

14.6   Support Vector Machine

The Support Vector Machine (SVM) is a supervised learning algorithm that classifies samples by placing them into a Euclidean space subdivided by hyperplanes. Each subspace corresponds to a specific class and the sample is classified according to its location in the sample space. This is done

Figure 14.6 An illustration of KNN.

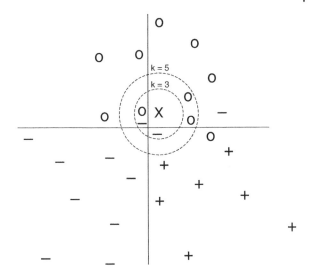

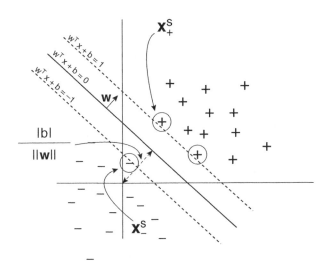

Figure 14.7 An illustration of an SVM distinguishing between two classes. The hyper-plane is the bold black line and the margins are illustrated by the dotted lines. The support vectors are circled.

by finding an optimal hyperplane in the data space that best divides the different classes and maximizes the margins between the classes. For example, a two-class sample set $X = \{x_n, d_n\}_{n=1}^N$ of N can be separated linearly and classed into $d = \pm 1$. Finding the hyperplane with the maximum margin separating the two classes can correctly classify an unknown sample, as shown in Figure 14.7. The hyperplane is defined as follows:

$$w^T x + b = 0 \tag{14.10}$$

where the parameters w and b fully define it. They are the parameters that the SVM seeks to optimize by finding the support vectors, $x_{\pm}^s$, at the frontiers between the two classes that give the widest margins. The following equation evaluates new samples for classification:

$$f(x_{N+1}) = w^T x_{N+1} + b$$

$$f(x_{N+1}) > 0 \Rightarrow d_{N+1} = 1$$
$$f(x_{N+1}) < 0 \Rightarrow d_{N+1} = -1 \tag{14.11}$$

The machine described so far functions as a linear classifier; therefore, it only properly classifies linearly separable sets. Kernels are introduced to deal with nonlinear cases by "adding" a dimension by performing a nonlinear operation on the samples. The following equations are examples of kernels:

$$K(x_i, x_j) = \left(x_i^T x_j + 1\right)^p \tag{14.12}$$

$$K(x_i, x_j) = \tanh\left(\beta_0 \, x_i^T \, x_j + \beta_0\right) \tag{14.13}$$

The kernels in Equations (14.12) and (14.13) are polynomial and hyperbolic tangent kernels, respectively. The parameters p, β_0, and β_1 are the tunable parameters in these kernels. The computation time and required memory are reduced because the kernels evaluate the distance between the samples in the new dimension, but do not actually map the samples into it.

14.7 Decision Tree Learning

A decision tree is a hierarchical structure consisting of many nodes branching out from a root node and ending in leaves. The most common decision tree applied in machine learning is a binary decision tree. In this learning model, some information is evaluated as true or false in each node and the decision guides the next branch to be followed. This true/false evaluation continues until a leaf is reached. The final decisions of the decision tree are expressed by the leaves. The number of nodes in the longest branch (maximum number of decisions) defines the depth of the tree. Figure 14.8 presents a decision tree and shows that the evaluated features may be continuous, such as temperature, or Boolean, such as the presence of rain.

An optimal decision tree is fitted to the training data in decision tree learning. This is carried out by: (1) evaluation of the predictive power of each feature using an impurity measure and (2) selection of the feature with the lowest impurity as a node.

Many impurity measures, such as information gain, gain ratio, gini-index, and variance, can be used in decision tree learning. A very large decision tree may grow to establish perfect predictions on the training data, as there is effectively a branch for every sample. However, this produces a

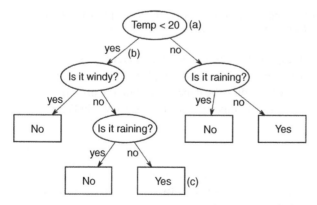

Figure 14.8 A decision tree determining whether a person should go outside: (a) the root node, (b) branches, and leaves (c).

decision tree that generalizes poorly because it struggles to classify new samples and is over-fitted. Over-fitting is combated by the introduction of the random forest, described next.

14.8 Random Forest

A random tree grows similarly to a normal decision tree, but it uses bagging to combine a number of trees previously generated from bootstrapped data sets into a forest. The forest generalization capability is further advanced by selection of a random subset of the features employed to train each tree. This leads to full-size trees of a forest that is less prone to over-fitting. In the final classification, each tree has an equal vote. The performance can be further improved by introducing boosting.

Random forests can also be used for feature selection. In this case, each tree, t, assigns an importance to each feature in its feature subset. It does this by assessing the misclassification rate on its out-of-bag (OOB) samples, OOB_t. These samples were not included in the bootstrapped data set and therefore were not used to build the tree. This is the baseline of the performance of the tree, shown by $errOOB_t$. This method then considers the feature column of feature $j(X^j)$ in the OOB samples and randomly permutes the values in the column. The misclassification rate is again assessed on this randomly permuted OOB sample set (OOB_t^j). The $errOOB_t^j$ is then the new misclassification rate. If scrambling the features does not change the $errOOB_t^j$, then feature j is deemed less important. However, if $errOOB_t^j$ decreases, then feature j is important. This process is repeated for every feature X_j and every tree t in the forest of T trees. The feature importance of each feature, $VI(X^j)$, is the average importance across the entire forest, derived as [10]:

$$VI(X^j) = \frac{1}{T} \sum_{t=1}^{T} \left(errOOB_t^j - errOOB_t \right) \tag{14.14}$$

14.9 Boosted Trees

Boosted decision tree models are very powerful classifiers. The first practical boosting algorithm was adaptive boosting (AdaBoost), created in 1995, and is still useful today. It is a forest of stumps [11], which are trees with just a root node and two leaves. The stumps are made sequentially and weighted based on their predictive performance. The samples are initially given equal weight, and their weights increase if the samples are misclassified by the last generated stump. The weights of the stumps are calculated as follows:

$$\text{Stump weight} = \frac{1}{2} \log \left(\frac{1 - \text{total error}}{\text{total error}} \right) \tag{14.15}$$

where the total error is the sum of the weights of the misclassified samples.

The weights of each sample can be adjusted as follows:

$$\text{New sample weight} = \text{Old sample weight } e^{a \cdot StumpWeight} \tag{14.16}$$

The scalar a is either 1 or -1. If the sample was misclassified by the stump, then $a = 1$ and if it was correctly classified, then $a = -1$. When every sample has been adjusted, all the sample weights are normalized.

When the probability of being drawn is proportional to the sample's weight, the bootstrapped data set has an equal size with the original. The new data set weights are set as equal and a new stump is trained, with extra emphasis given by the hard-to-classify samples because these are now more numerous in the data set. The process continues until the forest is complete. AdaBoost has now been superseded by modern alternatives, as boosted trees outperform boosted stumps [12]. XGBoost, CatBoost, and LightGBM are popular modern boosting algorithms and use what is known as gradient boosting.

14.10 Gradient Boost Decision Trees

XGBoost was first used in 2015 and has become extremely popular, as it does very well in Kaggle competitions, achieving over half the winning implementations [13]. Kaggle is an international online machine learning platform where machine learning experts compete to implement the best machine learning algorithms with a given data set. XGBoost is a rapid and powerful algorithm that increases tree depth and is heavily optimized using parallel processing [13]. By contrast, Cat-Boost and LightGBM are still considered frontier algorithms, but in some respects, they are likely to outperform XGBoost. LightGBM is an extremely optimized boosting algorithm especially suited to extremely large data sets, as it shows a 20-fold increase in speed with little loss of accuracy [14]. CatBoost is another algorithm designed to reduce target leakage and thereby increase generalization in the models. CatBoost has outperformed both XGBoost and LightGBM in several popular machine learning tasks [13].

14.11 Artificial Neural Network

The Artificial Neural Network (ANN) is a machine learning approach inspired by biology. It consists of individual neurons organized into layers that are interconnected with other layers from the input to the output layer. As shown in Figure 14.9, any neuron will have several inputs x_k, weighted with weights w_k and fixed input x_0 [1]. The x_0 is weighted by the weight b, called the bias. The contributions of the weighted inputs and the bias (sum v) are passed to the activation function φ, resulting in the output $y = \varphi(v)$. Before initiating a learning algorithm, the hyper-parameters, consisting of the number of layers, the number of neurons in each layer, and φ, are selected by the researcher. The weights and biases of each neuron are fitted in the training process.

14.11.1 Perceptron

The simplest single-layer ANN model is the perceptron, shown in Figure 14.10. It was first proposed by Rosenblatt in 1958 [15] and contains only the output layer. It can classify some problems; however, it is only capable of correct classification of linearly separable classes.

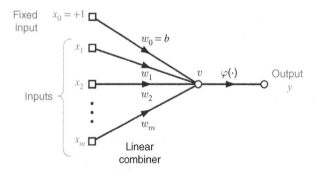

Figure 14.9 An artificial neuron.

Figure 14.10 A single layer perceptron consisting of (a) inputs, (b) neurons, and (c) outputs.

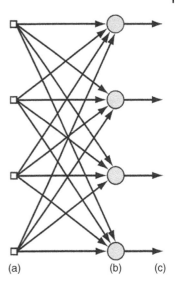

(a)   (b)   (c)

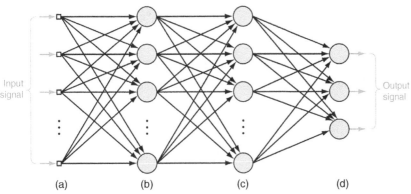

(a)   (b)   (c)   (d)

Figure 14.11 A fully connected three-layer perceptron consisting of inputs (a), the first hidden layer (b), the second hidden layer (c), and outputs (d).

14.11.2 Multi-Layer Perceptron

A multi-layer perceptron (MLP) is able to classify non-linear problems. The addition of one or more hidden layers between the input and output makes the MLP even more complex. This leads to a universal approximator that can approximate any continuous function. While the output layer is constrained by the problem, any hidden layer can have as many or as few neurons as the problem requires. Figure 14.11 shows a three-layer MLP.

14.11.3 Activation Function

The activation function used in a neuron influences the performance of the model and the training computational load. Some activation functions are the Heaviside, ReLu, and hyperbolic tangent equations:

$$\phi(v) = \max\left(0, \frac{v}{|v|}\right) \tag{14.17}$$

$$\phi(v) = \max(0, v) \tag{14.18}$$

$$\phi(v) = \tanh(v) \tag{14.19}$$

The activation function adds non-linearity to aid in decision-making. The activation function must be at differentiable least piecewise to be able to train the model by error back-propagation. The ReLU is the most popular activation function as it reduces the numerical burden imposed by continuous functions, such as the hyperbolic tangent.

14.11.4 Training

MLPs are usually trained using error back-propagation. The output o_j depends on neurons in every previous layer and the input is as follows:

$$o_j = y_j^{(L)} = \phi_j \left(\sum_i w_{ji} \phi_i \left(\sum_k w_{ik} \phi_k \left(\ldots \phi_r \left(\sum_m w_{rm} x_m \right) \right) \right) \right) \tag{14.20}$$

where L is the number of layers in the MLP and w_{ji} is the weight on the connection from neuron i to neuron j.

The error e_j, of the output, compared to the target value d_j, is given by

$$e_j = d_j - y_j^{(L)} \tag{14.21}$$

The error energy ε, summed over the entire output, is

$$\varepsilon = \frac{1}{2} \sum_j e_j^2 \tag{14.22}$$

where ε is a linear combination of differentiable functions. Therefore, a derivative of ε exists for every weight w_{ji}. The weights $w_{ji}[k]$ of epoch k can be adjusted for epoch $k + 1$ by gradient descent, as follows:

$$w_{ji}[k + 1] = w_{ji}[k] - \eta \frac{\delta \varepsilon[k]}{\delta w_{ji}[k]} \tag{14.23}$$

The learning rate η is set for each layer and usually decreases near the output. This continues until $\varepsilon[k]$ approaches a constant value.

14.12 Other Artificial Neural Networks

Many other ANNs have been proposed, such as convolutional neural networks that are suited for classifying images, auto-encoders that can be used for compression, anomaly detection, and generative models, and radial base function networks that substitute the weights of ANNs for vector coordinates.

14.13 Real Case Application

This section explains a fundamental problem of short circuit fault detection in the salient pole synchronous generator based on machine learning algorithms' application. The following subsections describe the procedures of data processing, feature extraction, feature reduction, and the performance of various classifiers on the experimental data.

14.13.1 Data Pre-processing

The data are processed in a manner analogous to processing in a production environment sample. In a production deployment of a fault detection system, the measurement series is required to be windowed with the classification run on a sliding window of the last electrical periods for fault diagnosis similar to real-time events. Incipient faults are not critical; therefore, a long window of several mechanical periods is possible. A minimum viable window length of one mechanical period is required to ensure passing of any fault by the sensors. An excessively long window length is prohibitive and can provide little new information, while slowing down feature extraction. Conversely, the window length must be long enough to remediate end effects in the signal processing tools. For periodic signals, the end effects can be alleviated by analyzing a concatenated series. The typical electrical machine has seven pole pairs; therefore, it captures seven electrical periods, which is equal to one complete mechanical period [7].

The measurement series before windowing is called the original sample series (OSS) and the series after windowing is the reduced sample series (RSS). Each RSS has extracted features that can be used as a sample in the data set and to train the classifiers. Non-fault-related variations in each RSS are reduced by cutting the OSS at the rising zero crossing to obtain integer electrical periods in each RSS.

The data scarcity for synchronous generators creates a need to generate several RSS from each OSS. The system is stationary in its steady-state mode and the RSS cut from the same OSS is identical. By skipping one electrical period after each captured RSS length of the integer mechanical periods, the faults pass each sensor one electrical period earlier in each contiguous RSS. As shown in Figure 14.12, this is done to provide training samples with faults in every position possible. An RSS length of seven electrical periods (or one mechanical period) was chosen to create the maximum possible number of RSS [7].

A visual inspection of all OSS indicates a continuous measurement series without sudden jumps in values that would indicate sensor malfunction or data corruption. Furthermore, no missing values are observed in the data set and the quality of the data is excellent. However, the provided data set to examine the application of machine learning for fault detection in this chapter is slightly imbalanced, as it contains 65.9% faulty case samples. In total, 24 experiments were conducted, each of which was sampled with two sensors and two sampling frequencies. The samples will later be split into train/test sets based on their OSS. Measurements taken from the same machine state with the same sensor are assigned to the same OSS irrespective of their sampling frequency. If measurements are from different machine states or different sensors, they are assigned to a different OSS. This is done because the OSS that the data belong to will be used later when splitting the data set into training/validation/hold-out data sets.

14.13.2 Feature Extraction

Raw time series are very sensitive to small perturbations. Therefore, they are not directly utilized as tabular training data and features are extracted from each RSS. These features are used as a basis for

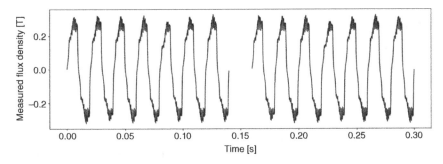

Figure 14.12 Two consecutive RSS cut from the same OSS. They each represent seven electrical periods, with a one electrical period between the two. Note the smaller negative peak occurring in periods four and three in the first and second RSS, respectively. Due to one period shift between each RSS, the fault indication appears one position earlier.

feature selection and, ultimately, as training data. The proposed feature extraction methods were fast Fourier transform (FFT), discrete wavelet transform energies (DWTE), and TSFRESH feature extraction. In total, 475 distinct features were extracted.

14.13.2.1 Fast Fourier Transform

The frequency content of each RSS was extracted by FFT. Previous studies showed that the faulty signal had a marked increase in the harmonic frequency components at intervals of f_m = 50/7 Hz, the mechanical frequency of the generator, outside the odd multiples of fundamental frequency compared to the healthy case. The frequency components of integer multiples of f_m up to 500 Hz were extracted as features, as follows:

$$f_{k,extracted} = k f_m = k \frac{2f_{sync}}{p}, k = 0, 1, 3, \ldots \tag{14.24}$$

14.13.2.2 DWT Wavelet Energies

Wavelet energies are good indicators of inter-turn short circuits (ITSC) and can be included as features. A 12-level-decomposition Haar wavelet DWT was taken of each RSS and instantaneous, Teager, hierarchical, and relative wavelet energies were computed for each decomposition level.

An issue with DWT is its end effects, which are worsened substantially in each decomposition level as the length of the data series that is transformed is effectively halved in each decomposition level with the Haar wavelet. The adverse effects are diminished as the length of the data series increases because the portions affected by the end effects are proportionally smaller. Therefore, before taking the discrete wavelet transform, each RSS was concatenated to four times its length. This exploits the assumption that the behavior of a generator is stationary and relieves the issue of end effects. In addition, before running the DWT, every 10 kHz RSS was up-sampled to 50 kHz. Up-sampling ensures that each DWT decomposition level contains the same frequencies for every RSS, even if sampling was originally conducted at different sampling rates. The frequencies within each of the 12 decomposition levels are shown in Table 14.2 [7].

14.13.2.3 TSFRESH

Features have been comprehensively extracted using TSFRESH but without including the TSFRESH FFT features. The reason is that the FFTs with informative frequency bins were already evaluated, as detailed above. Notably, TSFRESH has no way to select the frequencies of interest in a frequency spectrum. Before the feature extraction, the 50 kHz measurement series were downsampled by a factor of 5–10 kHz because many features extracted by TSFRESH are sensitive to the length of the sample series analyzed. This also shortens the computation time to 4.7 s and is able to create all the features required for a single RSS.

14.13.3 Exploratory Data Analysis

A feature's correlation with the target value and its variance indicate its usefulness in classifications. In addition, if the features are strongly correlated with each other, some features are likely

Table 14.2 Frequencies in each decomposition level of a 12-level DWT.

Level	A12	D12	D11	D10	D9	D8	D7	D6	D5	D4	D3	D2	D1
Frequency (Hz)	0-6	6-12	12-21	21-48	48-97	97-195	195-390	390-781	781-1562	1562-3125	3125-6250	6250-12500	12500-25000

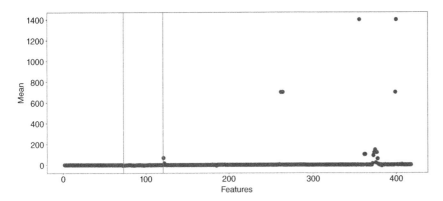

Figure 14.13 Calculated mean values across all samples for each feature. A few features have far larger means than the others. The plot is divided by vertical lines into three portions. The first portion from the left contains the FFT-derived features, the second from the left is the DWT energy feature portion, and the last is the collection of TSFRESH generated features.

to be redundant. These are some of the things looked for in exploratory data analysis (EDA), as performed in this section. The output of the feature extraction was formatted before the EDA. Note that any invariant features are removed before the EDA, which reduces the number of features from 475 to 417.

The qualitative inspection of the features indicates that most features have a mean close to zero, with some means far in excess of this. The same is true for the standard deviation. Some features have much greater variability than the norm. Figures 14.13 and 14.14 present the feature means and standard deviations. The results indicate that the features need to be standardized to work with some techniques. Standardization is a requirement for many techniques and learners, including PCA, KNN, and SVM with radial bias [3]. Exactly which specific feature's mean and standard deviation deviate from the rest is not of interest, since the existence of any in the set necessitates standardization.

The Pearson correlation of each feature used to target values to determine how correlated the feature is with the number of ITSCs applied to the poles showed largely uncorrelated features with a few exceptions. DWT energy features have many correlated features and some TSFRESH generated

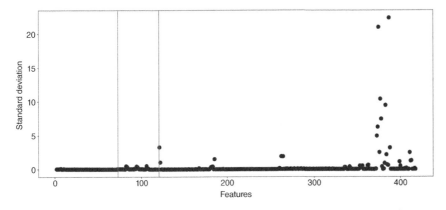

Figure 14.14 Standard deviations across all samples for each feature. The plot is divided by vertical lines into three portions. The first portion from the left contains the FFT-derived features, the second from the left is the DWT energy feature portion, and the last is the collection of TSFRESH generated features.

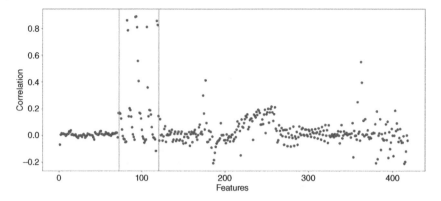

Figure 14.15 An overview of feature correlations. The plot is divided by vertical lines into three portions. The first portion from the left contains the FFT-derived features, the second from the left is the DWT energy feature portion, and the last is the collection of TSFRESH generated features.

features are strongly correlated. The FFT-generated features are largely uncorrelated. Figure 14.15 shows an overview of the features' correlation with the number of ITSCs. Both negative and positive correlations are useful for classification, but correlation only shows linear relationships. Other non-linear relationships may be present that are not revealed by this test. Table 14.3 presents the features with an absolute Pearson correlation above 0.3 [1].

These correlations reveal several relevant features. If these relevant features are strongly correlated among themselves, many may be redundant. An efficient visual method is to construct a

Table 14.3 Features most correlated with the number of ITSCs (DL = decomposition level) [1]/Norwegian University of Science and Technology.

Feature	Correlation
TWE, DL level 9	0.890734
TWE, DL level 8	0.886363
IWE, DL level 10	0.861306
RWE, DL level 10	0.856935
RWE, DL level 11	0.826443
HWE, DL level 10	0.811841
TWE, DL level 10	0.810356
IWE, DL level 11	0.786659
TWE, DL level 11	0.556887
TSFRESH longest strike above mean	0.549258
TSFRESH approximate entropy $(m = 2, r = 0.7)$	0.412482
TWE, DL 12	0.407188
TSFRESH longest strike below mean	0.39544
HWE, DL 11	0.359168
TSFRESH approximate entropy $(m = 2, r = 0.1)$	0.300201

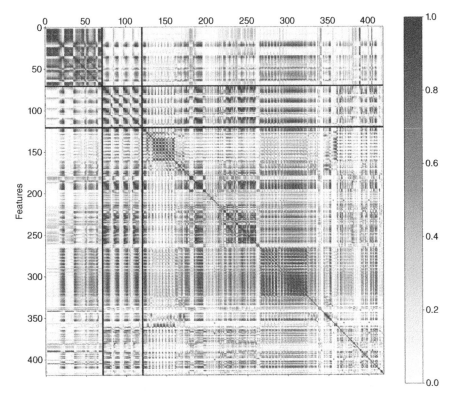

Figure 14.16 Feature correlation matrix. A darker color indicates a higher correlation between the features. Solid lines separate FFT features (left/top), wavelet energy features (middle), and TSFRESH features (right/bottom).

correlation matrix, which displays each feature's correlation to every other feature. Figure 14.16 shows the correlation matrix of the features. The features' correlations with themselves is indicated by a center diagonal line, which is necessarily 1. On either side of the diagonal are mirror images of the inter-feature correlations. As shown, both FFT and DWT features are strongly correlated among themselves, while TSFRESH exhibits this to a lesser degree. This large correlation between the features warrants further study of feature selection and reduction methods.

Since a high degree of correlation is evident between the features, a PCA would give the variation level of the samples. Fewer principal components are required to capture a given portion of the original variance, which shows that the data set contains many low variance features or high inter-feature correlations. A PCA of the data set is used to span 95% of the variance within the data set, leading to 31 principal components. Of the 31 principal components, 85% of the variance is contained within the 10 first components. This indicates that many features are uninformative or are strongly correlated with each other, coinciding with the results from the correlation analysis.

Direct plots of high-dimensional data sets are unsuitable. Instead, the PCA high-variance principal components can be plotted to visualize the data set and obtain some intuition about its distribution. Figure 14.17 presents a plot of the data set along the two first principal components and shows 16 distinct clusters, and 24 healthy and faulty clusters that are counted separately, although faulty and healthy sample distributions overlap in most parts. No clear decision boundary is evident for

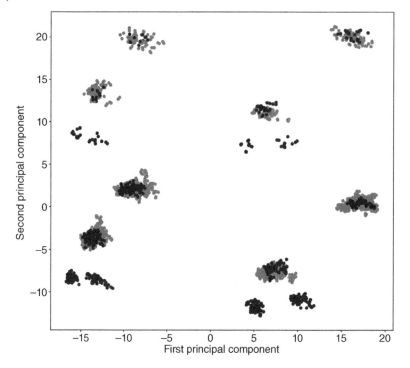

Figure 14.17 Plot of samples along the first and second principal component. Each point represents one sample; gray samples represent faulty machines and dark gray samples represent healthy machines.

discrimination of faulty from healthy. Consequently, SVM and KNN classifiers may perform poorly on the data set.

14.13.4 Feature Selection

Random forest feature selection and TSFRESH methods have been selected to apply to the feature data set. A hold-out data set is first extracted from the data set before selecting features. The reason is to prevent any target leakage resulting from selecting features based on the entire data set and inadvertently providing the classifiers with features selected for their hold-out data set relevance. Normally, samples originating from the same OSS are very similar and are split to ensure that no samples are split among the hold-out and remainder data sets. The hold-out set contains 15% of the total samples that are used to assess the final classifier performance [7].

14.13.4.1 Random Forest Feature Selection
The random forest feature is selected based on a forest of 1000 decision tree estimators and is trained on the training set using a splitting criterion called Gini impurity. In the training process, every feature is assigned an importance based on its impurity. All features that are greater than the mean importance are selected; the rest are discarded. This leads to a reduction of features from 417 to 81.

14.13.4.2 Time Series Feature Extraction Based on Scalable Hypothesis Tests (TSFRESH)
A subset of features deemed relevant is extracted using the feature extraction module included in TSFRESH. Considering the correlations discovered during the EDA, TSFRESH is configured to assume dependent features. False discovery rates in the interval 0.001, 0.01, 0.05, and 0.1 are tested,

Table 14.4 Three data sets taken into machine learning.

Set	Selection method	Numerical features
A	None	417
B	Random forest	81
C	TEFRESH	301

which leads to a similar number of features. The false discovery rate settled upon was 0.05, the rate used in reference [16]. This reduced the features from 417 to 301.

14.13.4.3 Summary of Feature Selection

Table 14.4 summarizes three versions of the feature data set, called feature data sets A, B, and C. By comparing the performance of classifiers trained on the different collections of features, some insight was gained into which features are most useful for classifying the fault and which feature selection algorithms are most useful with this data. This knowledge could be used in a final version of the fault detection system to ensure selective computation of only the most useful features.

14.13.5 Fault Detection

Details are shared of a classifier development intended to detect the ITSC presence using previously generated data sets. All classifiers are implemented from Scikit-Learn version 0.23.1, with the exception of XGBoost. The following four phases are followed to conduct this:

1. Feature data set selection.
2. Hyper-parameter optimization of single machine learning models.
3. Stacking classifiers evaluation.
4. Selection of a final classifier and its evaluation on a hold-out data set.

The feature data set selection is accomplished by evaluation of the training results of a host of different classifiers on each data set. The following classifiers are chosen [7]:

1. Logistic regression
2. Logistic regression with PCA
3. KNN
4. KNN with PCA
5. Radial basis function SVM
6. Radial basis function SVM with PCA
7. Linear SVM
8. Linear SVM with PCA
9. XGBoost
10. MLP
11. Stacking classifier

Logistic regression, SVM, and KNN are implemented with and without a PCA, with a view to determine the effectiveness of PCA in this application. The interpretability of the model as the key strength of decision trees is reduced by PCA; therefore, PCA is not combined with XGBoost. The stacking classifier combines other models, with the exception of KNN and KNN with PCA.

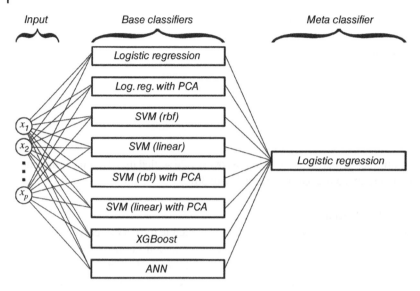

Figure 14.18 Implemented stacking classifier. The outputs of all the base classifiers are combined via a logistic regression model to make the final classification.

The KNN models have a long prediction time and poor performance compared to the other classifiers; therefore, they are excluded. Figure 14.18 shows the stacking classifier architecture.

14.13.5.1 Initial Hyper-parameter Choices

Table 14.5 shows the hyper-parameter settings of each classifier. The classifiers not shown in Table 14.5 use default parameters. These hyper-parameters have not been optimized; only reasonable rule-of-thumb values are applied. Note that the PCA was identically executed in all four applications with the same setting as those used in the EDA.

Table 14.5 A summary of hyper-parameters employed for comparison of feature data sets and classifiers. The table is not exhaustive, but it includes the most important hyper-parameters. The hyper-parameters not included were kept as defaults for their respective software libraries [1]/Norwegian University of Science and Technology.

Classifier	Hyper-parameters	Setting
KNN	K	20
	Weight	Uniform
SVM (rfb)	Kernel	Radial basis function
	Gamma	$\dfrac{1}{\text{Number of features}}$
PCA	Spanned variance	95%
XGBoost	Eta	0.3
	Max depth	6
Neural net	Number of hidden layers	2
	Number of neurons first layer	200
	Number of neurons second layer	100
	Number of neurons third layer	14
	Activation function	ReLU

14.13.5.2 Metrics

Sensitivity and precision are chosen as metrics due to their lower susceptibility to imbalanced classes. They are useful metrics for gauging the probability of a fault detection and the confidence of the correctness of the detection. Avoiding false alarms is as important to a power plant operator as being alerted of every possible fault.

14.13.5.3 Cross-Validation

CV heuristic selection methods are vulnerable to random chance. This is used to address this vulnerability. Previously, during the feature selection process, the data sets were split into a hold-out test data set and a remainder data set. Since a single train/test cycle result can largely depend on the split of the samples, the classifiers were evaluated by their average performance across a fivefold CV. This generates five folds of CV-train and CV-validation sets drawn from the remainder data set of the initial split. The folds are identical across all classifiers and feature data sets.

The samples in the remaining data are split by the CV based on the OSS they belonged to. Care must be taken to ensure that no samples from the same OSS are in both the training and validation sets, as RSS from the same OSS were deemed to be too similar.

The objective is to train a classifier to detect faults, but not to identify from which OSS the samples are drawn. To check this assertion, the classifiers were evaluated once by CV with random splitting. This leads to classifiers with near perfect accuracy, indicating that the OSS-dependent split was necessary.

14.13.5.4 Standardization

Logistic Regression, KNN, and SVM are sensitive to the variance of the samples, which is realized by applying standardization. Each CV split was standardized to zero-mean and unity variance. Estimates of the mean and variance of every feature were obtained from the CV-train set. The CV-train means and variances are employed to standardize both the CV-test and CV-validation sets by the CV-train means and variances.

14.13.5.5 Results

To gather performance metrics, this procedure was repeated for every classifier on every feature data set. Table 14.6 presents the results. This method of model fitting was used for every classifier evaluation at later stages of the classifier development.

14.13.6 Feature Selection and Reduction Performance

To select the best feature data set, three things are considered. First, the feature data set with the best average performance across the different classifiers is superior. Any aid to classification performance should be pursued. Second, the one with more consistent scores (i.e. with smaller variance in the results from CV) is preferred. Third, if the performance is similar among all the sets, the feature data set with the fewest features is superior for two reasons: a smaller number of features reduces training and prediction time and fewer features reduce the risk of over-fitting to the data.

A rough summary of the classifier performances on each feature data set is presented in Figure 14.19. Clearly, the choice of data set does not greatly affect the classifier performance and the variance of the results is large. However, feature data set C, the TSFRESH feature selection data set, slightly outperforms the rest on every averaged metric. Data set C is thus preferred and is utilized from this point onward.

Table 14.6 A summary of Logistic Regression results, KNN, SVM (radial base function kernel), SVM (linear), Logistic Regression with PCA, KNN with PCA, SVM (radial base function kernel) with PCA, SVM (linear) with PCA, XGBoost, MLP, and stack classifiers trained in each data set. Average scores across all classifiers for each data set are also included [1]/Norwegian University of Science and Technology.

Data set	Classifier	Sensitivity	Precision	ROC AUC
A	Logistic Regression	0.8853	0.7722	0.6774
	Logistic Regression with PCA	0.8622	0.8131	0.7179
	KNN	0.8269	0.6747	0.5288
	KNN with PCA	0.8201	0.6775	0.5335
	SVM (rbf)	0.8492	0.7050	0.5834
	SVM (rbf) with PCA	0.8538	0.6312	0.4453
	SVM (linear)	0.8859	0.7612	0.6705
	SVM (linear) with PCA	0.8576	0.8176	0.7175
	XGBoost	0.8518	0.7766	0.6788
	Multi-layer Perception	0.8833	0.7390	0.6420
	Stack	0.8652	0.8191	0.7179
	Average classifier score	0.8583	0.7443	0.6285
B	Logistic Regression	0.8675	0.7772	0.6721
	Logistic Regression with PCA	0.8140	0.7394	0.6279
	KNN	0.8074	0.7237	0.6053
	KNN with PCA	0.8392	0.7207	0.6075
	SVM (rbf)	0.8117	0.7029	0.5728
	SVM (rbf) with PCA	0.8149	0.6453	0.4761
	SVM (linear)	0.8790	0.7925	0.6905
	SVM (linear) with PCA	0.7878	0.7392	0.6155
	XGBoost	0.8407	0.7193	0.5938
	Multi-layer Perceptron	0.8702	0.7322	0.6233
	Stack	0.8712	0.7981	0.6955
	Average classifier score	0.8367	0.7355	0.6164
C	Logistic Regression	0.8966	0.7998	0.7092
	Logistic Regression with PCA	0.8663	0.8082	0.7271
	KNN	0.8282	0.6743	0.5287
	KNN with PCA	0.8222	0.6756	0.5306
	SVM (rbf)	0.8531	0.7226	0.6062
	SVM (rbf) with PCA	0.8492	0.6327	0.4477
	SVM (linear)	0.8972	0.8106	0.7389
	SVM (linear) with PCA	0.8615	0.8162	0.7289
	XGBoost	0.8313	0.7816	0.6747
	Multi-layer Perceptron	0.8859	0.7643	0.6869
	Stack	0.8714	0.8485	0.7627
	Average classifier score	0.8603	0.7577	0.6492

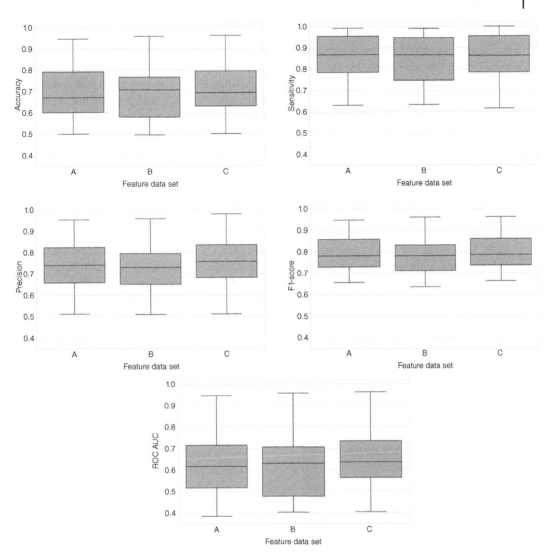

Figure 14.19 Performances across all classifiers on each feature data set are shown in box-and-whisker plots. The boxes extend from the upper to the lower quartile of the distribution, the center line in each box denotes the median score, and the whiskers envelope the greatest and lowest scores.

For feature reduction or PCA application, the application of PCA resulted in a performance drop in every classifier in nearly every metric. The radial basis function SVM with PCA consistently had an AUC lower than 0.5 ROC, indicating that it performed worse than chance because PCA was abandoned. The use of PCA might still have been justified on the grounds of training reduction and time prediction if more features were available or an extremely large number of samples, but no such considerations were necessary.

14.13.7 Hyper-parameter Optimization and Selection

Since a performance classifier depends heavily on its hyper-parameters, all the candidate classifiers were optimized before selecting among them. The optimization procedure was a fivefold cross-validating grid search. In this procedure, a hyper-parameter grid is defined that contains a range of values for each hyper-parameter to be optimized. The grid search algorithm then executes a CV of the classifier for every possible combination of these hyper-parameters. For each hyper-parameter combination, a mean CV performance is calculated and the hyper-parameter combination yielding the best performance on the chosen performance metric is selected.

Depending on the complexity of the classifier and the number of hyper-parameters to be optimized, the grid search can span thousands of hyper-parameter combinations. The data set contains a few samples, so each hyper-parameter combination has a short training time, which allows investigation of a large hyper-parameter grid. The performance metric used was the F_1-score because it combines sensitivity and precision.

Table 14.7 shows the hyper-parameter grids tested. The complexity of the classifiers increases by a higher size of the search grid [1]. KNN, SVM, and Logistic Regression have 175, 50, and 175 different hyper-parameter combinations, respectively, while XGBoost boasts 2304 different combinations. This is because XGBoost, as an ensemble classifier, requires a large set of hyper-parameter variations to search thorough a grid. Finally, 256 different combinations of hyper-parameters were tested for the MLP.

Table 14.8 presents the hyper-parameter sets with the highest mean performances across fivefold CVs for each classifier. Table 14.9 summarizes the scores of these classifiers across several metrics. Of the optimized classifiers, the XGBoost and KNN were outperformed by the others. The accuracy of the KNN was 64.0% in an unbalanced data set of a 65.9% majority class. This performance was worse than that of a dummy classifier that classifies randomly or always classifies samples as the majority class. Furthermore, with $k = 1$, the KNN is entirely non-generalizing, implying that the algorithm is not well suited for this problem at all, since this was the best result from a grid search of k-values from 1 to 351.

14.13.8 Stacking Classifiers

Since a stacking classifier improved the performance during the feature data set selection, the same approach is made again using the optimized classifiers. Four stacking classifiers with different meta-classifiers (Logistic Regression, MLP, gradient boosting forest, and a random forest classifier) were chosen. The gradient boosting forest classifier was chosen over XGBoost as a meta-classifier because of its greater compatibility with Sci-kit Learn's stacking framework. Since XGBoost is also a variant of the gradient boosting forest, it should return similar results at the expense of computing power. The stacks all include the optimized Logistic Regression, SVM, MLP, and XGBoost classifiers as base classifiers. KNN has a poor performance and slow prediction time; therefore, Table 14.10 shows the results [1].

Of the stacking classifiers, the Logistic Regression stacking classifier outperformed the others by a large margin. Figure 14.20 presents the best stacking classifier.

A somewhat surprising result surfaces by comparing the performance of the best stacking classifier with that of the best non-ensemble classifier. The logistic regression classifier alone, on average, slightly outperforms the stacking classifier that it is a part of across the CV folds. This may reflect

Table 14.7 Hyper-parameter search grids for Logistic Regression, KNN, SVM, and XGBoost classifiers. Note that 11 and 12 are Lasso and ridge regresssion, respectively. The *rbf* and *linear* kerners correspond to radial basis function and linear SVMs. Regarding hidden layer sizes: in a configuration of (a, b, c), the depth of the MLP is determined by how many numbers there are – in this case, three hidden layers deep. Each of these layers have a, b, c neurons each in order of increasing distance from the input layer [1]/Norwegian University of Science and Technology.

Classifier	Hyperparameter	Values	Description
Log. Reg.	C	10^k, $k = -10, -9.5, ..., 10$	Inverse of regularization strength
	penalty	"l1," "l2"	Penalization norm
KNN	n_neighbors	1, 3, 5, ..., 351	Number of nearest neighbors
SVM	C	10^k, $k = -1, 0, 1, 2, 3$	Inverse of regularization strength
	gamma	10^k, $k = 0, -1, -2, -3, -4$	Inverse of regularization strength
	kernel	"rbf," "linear"	Kernel type
XGBoost	learning_rate	0.01, 0.2, 0.3, 0.5	Learning rate
	n_estimators	100, 400, 700, 1000	Number of trees in ensemble
	max_depth	3, 10, 15, 25	Maximum tree depth
	col_sample_by tree	0.8, 1	Per tree column subsampling ratio
	subsample	0.6, 0.8, 1	Sample subsampling ratio
	reg_alpha	0.7, 1, 1.3	L1 regularization term on weights
	reg_lambda	0, 0.5, 1	L2 regularization term on weights
MLP	activation	"identity," "logistic," "tanh," "relu"	The activation function
	batch_size	200, 133, 66, 32	Size of minibatches
	max_iter	200, 500, 1000, 1200	The maximum number of epochs
	hidden layer_ sizes	(50,25,3), (100,50,7), (200 100,14), (300 150,21)	Size and number of hidden layers

the fact that the logistic regression algorithm was optimized by way of a grid search on the very same CV splits, while the stacking classifier has not been similarly optimized.

14.13.9 Final Classifier

Stacking classifiers often generalize better than single classifiers, which is considered advantageous. They also usually outperform their base classifiers. However, the hyper-parameters of the

Table 14.8 The best hyper-parameters found from the grid search [1]/Norwegian University of Science and Technology.

Classifier	Hyperparameter	Value
Logistic Regression	C	$10^{8.5}$
	penalty	12
KNN	n_neighbors	1
SVM	C	10
	gamma	1
	kernel	linear
XGBoost	colsample_bytree	0.800
	learning_rate	0.500
	max_depth	3
	n_estimators	100
	reg_alpha	1.300
	reg_lambda	0
	subsample	1
Multi-layer Perceptron	activation	identity
	batch_size	200
	hidden_layer_sizes	(50, 25, 3)
	max_iter	200

Table 14.9 The accuracy, sensitivity, precision, F_1-score, and ROC AUC of the best models found in the hyperparameter grid search [1]/Norwegian University of Science and Technology.

Classifier	Accuracy	Sensitivity	Precision	F_1-score	ROC AUC
Logistic Regression	0.7986	0.8740	0.8376	0.8506	0.7606
KNN	0.6395	0.8350	0.6990	0.7501	0.5723
SVM	0.7940	0.8854	0.8247	0.8501	0.7500
XGBoost	0.7438	0.8576	0.7846	0.8142	0.6927
Multi-layer perceptron	0.7958	0.9022	0.8170	0.8542	0.7340

meta-classifier have not been optimized on the training set, whereas they have been optimized for the simple logistic regression classifier. To gauge their performance on unseen samples, both are trained on the entire training set and tested on the hold-out data set. Table 14.11 presents the results [1]. On the hold-out set, the stacking classifier outperforms the simple logistic regression classifier. The stacking classifier could likely be further improved by running a grid search for the optimal hyper-parameters of the logistic regression meta-classifier, but this involved far too long a computation time to complete without a large and time-consuming refactoring of the code base.

Coefficients of the logistic regression stacking classifier weigh each of its base classifiers. Since all base classifiers return predictions in the same interval (0 to 1), the absolute value of the coefficients

Table 14.10 The results from the stacking classifier comparison [1]/Norwegian University of Science and Technology.

Meta-classifier	Accuracy	Sensitivity	Precision	F_1-score	ROC AUC
Logistic Regression	0.7840	0.8701	0.8260	0.8432	0.7472
Multi-layer Perceptron	0.7479	0.8057	0.8276	0.8107	0.7267
Gradient boosting forest	0.7663	0.8268	0.8304	0.8255	0.7314
Random Forest	0.7704	0.8216	0.8388	0.8265	0.7440

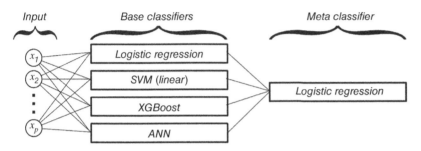

Figure 14.20 A stacking classifier with Logistic Regression as its meta-classifier.

Table 14.11 The results of the best of the single and stacking classifiers on the hold-out data samples [1]/Norwegian University of Science and Technology.

Classifier	Accuracy	Sensitivity	Precision	F_1-score	ROC AUC
Logistic Regression	0.7569	0.6961	0.9435	0.8011	0.7986
Logistic Reg. stack	0.8448	0.8456	0.9274	0.8846	0.8443

Table 14.12 The base-classifier coefficients of the logistic regression classifier used as meta-classifier in the stacking classifier. The models are ranked in order of importance to the final prediction [1]/Norwegian University of Science and Technology.

Rank	Base-classifier	Coefficient
1	SVM	3.464
2	XGBoost	1.365
3	Logistic regression	−1/012
4	Multi-layer Perceptron	1.036

is correlated with how large an emphasis is placed on each particular base-classifier. Table 14.12 shows the coefficients of each base-classifier [7].

14.13.9.1 Feature Usefulness

Extraction of feature importance is possible from both logistic regression and XGBoost classifiers. In the logistic regression classifiers, the importance corresponds to the weights associated with the

Table 14.13 The 20 most useful features for the optimized XGBoost classifier [1]/Norwegian University of Science and Technology.

Rank	Feature	Description
1	DWT_ _RWE10	RWE, decomposition level 10
2	time_reversal_asymmetry_statistic (lag = 1)	TSFRESH
3	partial_autocorrelation (lag = 7)	TSFRESH, partial autocorrelation with a lag of 7
4	change_quantiles(f_agg = "mean," isabs = True,qh = 0.4, ql = 0.0)	TSFRESH, mean absolute change of all measurements between 0.0 and 0.4
5	DWT_ _RWE11	RWE, decomposition level 11
6	time_reversal_asymmetry_statistic (lag = 2)	TSFRESH
7	DWT_ _HWE1	HWE, decomposition level 1
8	c3(lag = 1)	TSFRESH, measure of nonlinearity in time series
9	FFT 121.4 Hz	FFT, frequency magnitude at 121.4 Hz
10	autocorrelation(lag = 3)	TSFRESH, autocorrelation with a lag of 3
11	FFT 450.0 Hz	FFT, frequency magnitude at 450.0 Hz
12	DWT_ _TWE10	TWE, decomposition level 10
13	autocorrelation(lag = 6)	TSFRESH, autocorrelation with a lag of 6
14	autocorrelation(lag = 3)	TSFRESH, autocorrelation with a lag of 2
15	DWT_ _IWE1	IWE, decomposition level 1
16	DWT_ _IWE10	IWE, decomposition level 10
17	DWT_ _TWE11	TWE, decomposition level 11
18	approximate_entropy(m = 2, r = 0.5)	TSFRESH, measure of regularity in time series
19	skewness	TSFRESH, skewness of the sample
20	c3(lag = 1)	TSFRESH, measure of nonlinearity in time series

features, and in XGBoost, it is the average gain across all splits of the features used. Table 14.13 shows the most important features for the XGBoost and logistic regression classifiers.

14.13.9.2 Fault Severity Assessment

An attempt has also been made to determine fault severity using some of the above-mentioned methods. This is done by generating several one-versus-all classifiers that try to determine the fault severity. This means that the final classification is really made by a collection of several binary classifiers. Each classifier checks whether the sample does belong to a certain class; for example, the no-fault class. The final classification chooses the class with the associated binary classifier with the highest confidence that the sample belongs to its class.

The objective of these classifiers is to assess the fault severity of a sample by classifying it as one of several defined fault severities. The exceptions are XGBoost, KNN, and MLP classifiers. They are capable of multi-class classification without employing the one-versus-all technique. Since so

Table 14.14 The severity degrees of the classifier defined by the number of ITSCs. The rightmost column contains the number of experiments conducted in that state [1]/Norwegian University of Science and Technology.

Severity case	ITSCs	Coefficient
No fault	0	16
Low severity	1–6	12
Moderate severity	7–10	8
High severity	Above 10	12

few experiments were run, only the no-fault condition has more than four experimental cases. It was immediately clear that it would be fruitless to attempt to obtain the exact fault severity due to data constraints. In effect, four samples of each case were present in the data set, not counting different sampling rates as separate samples. In an attempt to remedy this, the classifiers were made to classify samples into either no fault, low severity, moderate severity, or high severity. These degrees of severity were defined to split the available samples as evenly as possible between the severity classes while still being informative, as shown in Table 14.14 [1].

The resulting classifiers gave either terrible predictions that were near random guesses when the RSS were split according their OSS, or they gave close to perfect predictions when the RSS were split randomly.

14.13.10 Data Management and Pre-processing

The RSS length was chosen to maximize the number of samples from each OSS as a way to exploit the following assumptions [7]:

1. The captured signal is stationary.
2. The signal could be cut easily in zero-passing to capture integer electrical periods.

The reason is that the air gap magnetic field is periodic. This may not be the case if the same method is applied to signals originating from other sensors mounted on the machine. These sensors include vibration and stray magnetic field sensors.

Two Hall-effect sensors for concurrent measurements were mounted on the generator. The measurements of these sensors are considered independent and assigned to separate OSSs. The motivation is to maximize the information extracted from the source data. The sensors measure effectively independently of the same machine state, because their idiosyncrasies and differences in mounting would affect the measurements. It was expected that there would be improbably high detection rates if they were too similar; however, this was not observed.

In evaluating the final classifiers on the hold-out set, the classifiers' performance varied largely between folds of the same CV run and a leap in classification performance occurred. This highlights some issues with small data sets. By drawing few OSSs, the data splitting set drastically changes the data in the training set. This is also the reason for the performance change in the final evaluation. The previously well-performed logistic regression classifier lost its performance when introduced to more data, while the logistic regression stacking classifier improved its performance by relying primarily on its linear SVM base-classifier. Determination of the best performing classifier depends very much on parts of the trained data set.

The best classifier for the generated data set was the logistic regression stacking classifier. However, it is likely that this changed by the introduction of more data. Since the performance varies between CV-folds, splitting the data set in the same way for every CV for every classifier and feature data set is important to ensure direct comparability between the results.

14.13.11 Feature Extraction and Importance

Some features calculated by TSFRESH are very computationally demanding. The computation time of TSFRESH on a system with an Intel Core i5-6200U CPU for generating all features of a single RSS was 4.7 s. This affords a monitoring system of comparable computational power to gauge the condition of the machine roughly once every five seconds. Nevertheless, including all the features increases the chance that useful features were not overlooked [1].

The four most computationally intensive generated features among the TSFRESH are approximate entropy, change quantiles, entropy, and aggregated linear trend lines [5]. Note that these features are more than two orders of magnitude more computationally intensive than the average TSFRESH feature [5]. They were included among the most useful features for both the optimized logistic regression and the optimized XGBoost classifiers. Taking into account that the DWT wavelet energy and FFT features have insignificant computation time compared to TSFRESH generated features, there is little room to shorten the time of feature extraction. However, a condition monitoring system once every 5 or 10 s is sufficient for fault diagnosis of a machine because of the non-critical nature of incipient faults.

The classifiers included both linear and nonlinear models and therefore their range was intentionally broad. The best performers after the grid search were the linear models, including logistic regression, SVM with a linear kernel, and MLP with the identity activation function. Note that although MLPs are generally nonlinear models, in the case of the linear activation function they are linear. This is in accordance with the EDA, which shows a great deal of linear correlation between the target and several features.

The most useful features are the high decomposition level DWT RWE features, aggregate linear trend features, approximate entropy features, and change quantile features. The TSFRESH-generated features perform similarly to those of the wavelet energy and FFT features. Correlation with the target value is a strong indication that features are useful in classification and could thus be employed to screen a large number of potential features without having to train classifiers. However, screening in this manner would come at the risk of missing nonlinear relationships, as mentioned above.

14.13.12 Feature Selection and Target Leakage

The feature selection algorithms for the small size data set were run on the entire data set, where the hold-out data set is excluded. This likely affects the classifier CV results run on the feature selected data sets. Since the algorithms of feature selection selects the best features, based on all samples appearing in each CV-fold, some target leakage occurs. This leads to optimistic CV results. This does not, however, influence the results from the hold-out data set because it was set aside before feature selection [7].

An alternate approach for eliminating the aforementioned target leakage is to delete the hold-out data set and run a feature selection step as part of each CV fold. This can eliminate target leakage and allow feature selection on a data set of similar size; however, it has another drawback. Without a hold-out data set, the final evaluation of the classifiers would be carried out on the same samples used to choose among the classifiers and their hyper-parameters. This introduces another target leakage source, and the final evaluation of the classifiers performance on unseen samples is unrealistically optimistic.

14.13.13 Classifier Selection

As a result of the grid search, the classifier performance increases markedly. However, a general trend among the optimized classifiers is the observation that less complicated classifiers performed better. For instance, the optimized hyper-parameters of MLP have the lowest number of neurons, even in the search with the identity function as an activation function. The hyper-parameters with the highest performance are on the lower-complexity extremes of their associated search grids, with the exception of the optimized logistic regression classifier. Another grid search with less complicated hyper-parameters could likely lead to better results.

14.13.14 Performance

The performance of the classifiers developed in this chapter has been, to some standards, unimpressive, but this is to be expected when the source data set is reasonably diverse and small. Maximizing the information extracted with a small data set is a challenging matter. As mentioned before, this involves correct handling of splitting and avoiding target leakage. When samples are randomly split, a close to 100% accuracy is an indication of mishandled data management, as exemplified when non-optimized classifiers achieve nearly perfect predictions.

14.13.15 Real-World Validity

The data sets used here were obtained from a 100 kVA synchronous generator at the NTNU laboratory, an experimental generator constructed to resemble hydropower generators but with its own unique geometry. Its sensors were mounted in a certain way and position. The generator was tested in a noisy electromagnetic environment laboratory. These discrepancies with a real production hydropower generator make any classifier trained on the measurements taken from this test generator useless on machines other than the NTNU setup. In general, this is nearly true for all machine learning models; therefore, a new model must be made for each machine [1].

The generator was driven by an induction motor. Turbines were not used because they introduce vibrations into the generator that could affect the features obtained here. Some of these problems are hydraulic imbalance [17], cavitation [18], and runner blade damage [17]. These vibrations can affect the air gap magnetic field and introduce torque variations to the shaft. While cavitation-induced vibrations are usually wide band, vibration induced by high-frequency noise, hydraulic imbalance, and runner blade damage occur at multiples of the mechanical frequency of the rotor. Since the turbine mechanical frequency and the generator necessarily share a common frequency, the state of the turbine would affect the frequency content of features based on the generator rotor mechanical frequency. The FFT feature set generated here would be particularly susceptible to this.

14.13.16 Real-World Applicability

Two prerequisites determine the applicability of a fault classification system. One of them has been created here for use in a production machine. The machine is outfitted with an air gap Hall-effect sensor. Measurements in several different load and fault conditions of the machine were taken prior to developing the classifier. The sensor itself is small relative to the air gap of the machine and is easily installed. However, running this machine with induced faults to construct the required data set is a highly unlikely scenario. Therefore, this approach is not practical, with a few possible exceptions. If the machine is manufactured based on the standard and then modeled, then it will be possible to gather faulty machine data with sensors in one or a few machines, and use them for training other machines of the same type. Imperfections in generators can arise during production and/or assembly. Differences in the connected turbines could also interfere with the classification.

The degree to which this would be a problem is speculative. Many generated features depend on machine geometrical specifications, such as the number of poles. This would be an expensive data-gathering method because of the generator cost and the small production runs. The next section presents some recommendations for providing more generalized classifiers.

14.13.17 Anomaly Detection

Anomaly detection is needed to develop a healthy machine model operating state, and then to develop a faulty machine model. This can feasibly be applied to the machine because the training data are simply the data from the healthy machine operating under different load conditions, so the data are much more easily obtained than fault condition measurements. The auto-encoder is a machine learning model that would be suitable for this.

The auto-encoder is a multi-layered neural net with smaller intervening layers where the input and output layers are of the same size. Since the intervening layers are smaller than the input layer, the auto-encoder is forced to discard some input information. It is trained using back-propagation to reconstruct the input in the output layer until it does so with a low error rate. A normal state is then encoded into the neurons of the model. Anomalous inputs would thus be reconstructed poorly and incur a high reconstruction error, indicating an anomaly. The least informative model is the simplest model to produce. It does not indicate the condition but only shows the presence of an anomalous operating condition. This model could use features similar to the ones generated in this chapter, but it should have a great deal of healthy training data spanning every acceptable operating condition to decrease the FP rate [1].

14.13.18 Simulated Data Generation

One of the greatest hindrances in developing classifiers for fault detection for the machines is the lack of labeled fault data. This is addressed using a simulation that includes the generator along with the mounted sensors. The FE simulation could be calibrated against measurements of the machine; therefore, the simulation provides sample series that agree with the sample series of healthy machine operation. To create a faulty measurement series, the same simulation can be modified to include induced faults. Sample series from simulated faulty and healthy machine operation would then be utilized to train the fault classifier. This is more easily implemented in industry because very few labeled data exist for fault conditions in the machines. The method is also very non-invasive and requires only the mounted sensor in the generator. The FP performance model could be assessed by predictions on a healthy data set gathered from the machine, but a major challenge is the lack of faulty testing data from the real machine to assess its efficacy in detecting true faults.

Table 14.15 The 20 most useful features for the optimized logistic regression classifier. ALTL is an abbreviation of aggregated linear trend line.

Rank	Feature	Description
1	DWT_ _RWE11	RWE, decomposition level 11
2	agg_linear_trend(chunk_len = 50, f_agg = 'max')	TSFRESH, intercept of maximum value ALTL with chunk length 50
3	DWT_ _RWE12	RWE, decomposition level 12
4	agg_linear_trend(chunk_len = 50, f_agg = 'min')	TSFRESH, standard deviation of minimum value ALTL with chunk length 50
5	approximate_entropy($m = 2, r = 0.3$)	TSFRESH, measure of regularity in time series
6	c3(lag = 1)	TSFRESH, measure of nonlinearity in time series
7	DWT_ _RWE10	RWE, decomposition level 10
8	agg_autocorrelation(f_agg = "var," maxlag = 40)	TSFRESH, variance of autocorrelations for lags up to 40
9	FFT 450.0 Hz	FFT, frequency magnitude at 450.0 Hz
10	FFT 314.3 Hz	FFT, frequency magnitude at 314.3 Hz
11	c3(lag = 2)	TSFRESH, measure of nonlinearity in time series
12	FFT 442.9 Hz	FFT, frequency magnitude at 442.9 Hz
13	binned_entropy(max_bins = 10)	TSFRESH, entropy of the sample magnitude distribution
14	DWT_ _HWE7	HWE, decomposition level 7
15	FFT 371.4 Hz	FFT, frequency magnitude at 371.4 Hz
16	change_quantiles(f_agg = "var,"isabs = False, qh = 0.6,ql = 0.0)	TSFRESH, variance of consecutive changes in measurements between 0.0 and 0.6
17	FFT 350.0 Hz	FFT, frequency magnitude at 350.0 Hz
18	FFT 100.0 Hz	FFT, frequency magnitude at 100.0 Hz
19	FFT 178.6 Hz	FFT, frequency magnitude at 178.6 Hz
20	time_reversal_asymmetry_statistic(lag = 2)	TSFRESH

14.14 Summary

This chapter shows how signal processing and machine learning tools can be used to diagnose the ITSCs of rotor field windings. A fault detection system was introduced to diagnose the ITSC faults. This was done based on measurements from a single Hall-effect sensor mounted on a stator tooth inside the air gap of a salient-pole synchronous generator. This procedure follows three stages: data pre-processing, feature extraction and selection, and classifier development. The objectives were specifically to find the most useful features for the best performance of machine learning models and to determine whether a single air gap magnetic field sensor is sufficient for reliable fault detection or if more sensors are required.

The power spectral densities of integer multiples of the mechanical frequency of the generator were the features extracted by FFT, DWT energies, and the entire TSFRESH feature extraction suite, excluding their FFT features. Tables 14.13 and 14.15 show that the most useful features were the RWE features and some TSFRESH features. The performance of TSFRESH generated features that paralleled DWT features and surpassed the FFT features, indicating that automatic feature extraction is useful.

For fault detection on this data set, linear machine learning models, especially the logistic regression and linear SVM classifiers, are the most suitable. However, KNN is not appropriate and provides results worse than random chance. When the classifiers are stacked, the performance is decreased somewhat on averaged CV. However, when it is tested on the hold-out data set, it is better generalized.

The best classifier is an ensemble stacking classifier, with logistic regression as the meta-classifier taking inputs from logistic regression, XGBoost, linear SVM, and MLP classifiers as base classifiers. The ITSC fault classification using machine learning on the air gap magnetic field measured by a single sensor leads to good results. The accuracy, sensitivity, and precision of applying logistic regression stacking classifiers were 0.8448, 0.8456, and 0.9274, respectively. This means that the classifier correctly classified 84.48% of all the samples in the hold-out data set and that 84.56% of the faulty samples present were correctly classified as such.

Of the samples that were classified as faulty, 92.74% were correctly classified. Since a large portion of faults go undetected, this fault detection system should not be relied upon as the only detection system. However, if the system alerts the operator to a fault, it should warrant investigation, since it is likely to be correct. This is predicated on a similar performance on out-of-set samples. Assuming similar performance from this classifier on novel samples is naive due to its limited training data. The robustness of the classifier could likely be improved by generating a more diversified data set.

References

1 Skreien, T.-N. (2020). Application of signal processing and machine learning tools in fault detection of synchronous generators, MSc Thesis, NTNU, Norway.

2 Koza, J.R., Bennett, F.-H., Andre, D. et al. (1996). *Automated design of both the topology and sizing of analog electrical circuits using genetic programming, Artificial Intelligence in Design'96*, 151–170. Dordrecht: Springer.

3 James, G., Witten, D., Hastie, T. et al. (2013). *An Introduction to Statistical Learning*. New York, NY: Springer.

4 Guyon, I. and Elisseeff, A. (2003). An introduction to variable and feature selection. *Journal of Machine Learning Research* 3: 1157–1182.

5 Christ, M., Braun, N., Neuffer, J. et al. (2018). Time series feature extraction on basis of scalable hypothesis tests (tsfresh – a python package). *Neurocomputing* 307: 72–77.

6 He, H. and Garcia, E.A. (2009). Learning from imbalanced data. *IEEE Transactions on Knowledge and Data Engineering* 21 (9): 1263–1284.

7 Ehya, H., Skreien, T., and Nysveen, A. (2022). Intelligent data-driven diagnosis of incipient inter-turn short circuit fault in field winding of salient pole synchronous generators. *IEEE Transactions on Industrial Informatics* 5 (5): 3286–3294.

8 Picard, R.R. and Cook, R.D. (1984). Cross-validation of regression models. *Journal of the American Statistical Association* 79 (387): 575–583.

9 Lewis, D.D., Schapire, R.E., Callan, J.P. et al. (1996). Training algorithms for linear text classifiers. In: *Proceedings of the 19th Annual International ACM SIGIR Conference on Research and Development in Information RetrievalAssociation for Computing Machinery*, 298–306. Zurich Switzerland, August 18–22, 1996, New York: USA.

10 Genuer, R., Poggi, J.-M., and Tuleau-Malot, C. (2010). Variable selection using random forests. *Pattern Recognition Letters* 31 (14): 2225–2236.

11 Freund, Y. and Schapire, R.E. (1997). A decision-theoretic generalization of on-line learning and an application to boosting. *Journal of Computer and System Sciences* 55 (1): 119–139.

12 Caruana, R. and Niculescu-Mizil, A. (2006). An empirical comparison of supervised learning algorithms. In: *Proceedings of the 23rd International Conference on Machine Learning*, 161–168.

13 Chen, T. and Guestrin, C. (2016). Xgboost: A scalable tree boosting system. In: *Proceedings of the 22nd ACM Sigkdd International Conference on Knowledge Discovery and Data Mining, Association for Computing Machinery*, 785–794. New York, NY, USA.

14 Ke, G., Meng, Q., Finley, T. et al. (2017). Lightgbm: A highly efficient gradient boosting decision tree. In: *Advances in Neural Information Processing Systems, 31st Conference on Neural Information Processing Systems (NIPS)*, 3146–3154. Long Beach, CA, USA.

15 Rosenblatt, F. (1958). The perceptron: A probabilistic model for information storage and organization in the brain. *Psychological Review* 65 (6): 386.

16 Benjamini, Y. and Hochberg, Y. (1995). Controlling the false discovery rate: A practical and powerful approach to multiple testing. *Journal of the Royal Statistical Society: Series B (Methodological)* 57 (1): 289–300.

17 Mohanta, R.K., Chelliah, T.R., Allamsetty, S. et al. (2017). Sources of vibration and their treatment in hydro power stations – A review. *Engineering Science and Technology, an International Journal* 20 (2): 637–648.

18 Kumar, P. and Saini, R. (2010). Study of cavitation in hydro turbines – A review. *Renewable and Sustainable Energy Reviews* 14 (1): 374–383.

15

Insulation Defect Monitoring

15.1 Introduction

Repair, maintenance, and condition monitoring have vital roles to play in the proper performance of electrical machines [1–3]. One of the most fundamental factors affecting the proper performance of electrical machines is their insulation system; indeed, the useful lifespan of electrical machines is often determined by their insulation system life. This is the reason why condition monitoring of the insulation of rotating electrical machines is essential. The insulation system is often under the influence of conditions due to partial discharge (PD), which profoundly affects the lifespan of the insulation [3].

PD consists of small sparks that occur in electrical machine insulation. Today, online and offline techniques are widely used to diagnose faults arising from insulation problems in large motors and power station generators. PD techniques are used in large high voltage (equal to or above 3 kV) electrical machines. Various methods are available for condition monitoring of large machines to diagnose faults; for example, IEC and IEEE standards offer and classify implementation methods [4–9]. Four types of discharge occur in electrical equipment [10]:

- Corona discharge.
- Partial discharge.
- Spark discharge.
- Arc discharge.

In electrical machines with robust insulation systems, the corona discharge occurs very rarely on AC voltages above 4 kV. If the electrical machine insulation is porous, the PD depends on

Electromagnetic Analysis and Condition Monitoring of Synchronous Generators, First Edition. Hossein Ehya and Jawad Faiz.
© 2023 The Institute of Electrical and Electronics Engineers, Inc. Published 2023 by John Wiley & Sons, Inc.

temperature and electric field intensity. In that case, a substantial field presence, the high porosity in the machine insulation, and dependency on the mechanical and electrical conditions create a situation where the probability of insulation defects is high. During a PD, pulse occurrence on the surface or inside electrical insulation causes high-energy electrons and/or accelerated ions to collide with the insulation material, leading to chemical reactions and finally defects in the insulation [11].

This chapter presents the history of insulation of rotating electrical machines, different types of PDs, and PD measurement sensors.

15.2 History and Advantages of Using Partial Discharge Techniques

PD occurs in electrical equipment mainly operating at high voltage. The insulation materials used in electrical equipment are pure organic materials. However, continuous PD damages the insulation. Basically, electrical machines operating at voltages of 3 kV and above are exposed to the PD, but some insulation, such as mica and fiberglass, are robust against the PD. Generally, the PD occurring in the machine causes damage to the machine insulation system and substantial economic losses to the operating power system. An increasing number of PDs in a new machine indicates existing problems that arose during its production process, whereas PDs in old machines indicate the declines in the machine caused by temperature rises, existing contamination in the end windings, and/or loosing of windings inside the stator slots. A fault due to these factors can damage the winding, making re-winding of the whole winding necessary [11].

The methods used for the PD diagnosis in electrical machines can be divided into online and offline categories. Winding fault diagnosis using the offline PD approach dates back to the 1930s, whereas the first online fault diagnosis was reported in 1950 [11]. Some general standards for diagnosis and measurement of the PD, such as IEC 60270 and the USA Materials and Measurement Association standard (ASTM D1868), have been presented. The first specialized standard for evaluating PD in electrical machines was IEC 1434, and its updated versions are IEC 60034-27 and IEC 60034-27-2. Initially, the measured current and voltage in the machine were used as an index for the PD, as electrical current and voltage have no information about electrical discharge. Therefore, pColun, which shows the number of electrons and ions, was used as an index for measuring insulation damage detected with a low-pass filter below 1 MHz. However, the high-frequency high-pass filter is now employed to measure electric charge. Measurement of the PD using high-frequency components in high-voltage electrical equipment typically follows the IEC TS 62478 standard, which has been recommended for PD measured using acoustic and electromagnetic methods.

15.3 Electrical Machine Fault Generation Factors

Approximately 56% of the faults in generators are caused by an insulation system fault. Insulation problems in generators are divided into several classes, including internal PD, winding contamination, loose buses in the slot and/or in the overhang, over-voltage, corona protection, thermal cycling, and/or overload. Figure 15.1 shows different factors causing an insulation fault in a hydro-generator [12]. The defects of the insulation system due to PDs play a significant role

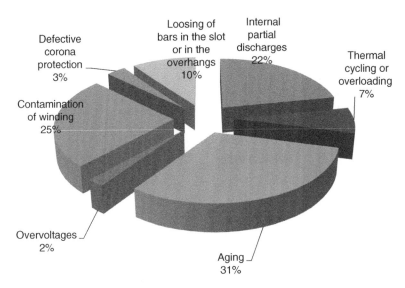

Figure 15.1 Different factors causing damage to hydro-generator insulation systems [12].

in damaging rotating generators and motors. The factors causing PDs in the rotating machines consist of:

1. Thermal stress.
2. Electrical stress.
3. Aging.
4. Mechanical stress.

15.4 Rotating Machine Insulation System

In rotating machines, the rotor and stator consist of the core, bars, and insulation system. The insulation system has a significant influence on the proper performance of the rotating machines so knowledge of the insulation system is essential for condition monitoring of rotating electric machines and in almost all HV equipment.

15.4.1 Rotor Insulation System

The voltage applied to the rotor of generators is up to 500 V (DC), which is medium compared to the insulation voltage level. This voltage is distributed between the turns such that a small voltage of about 10 V is present in each turn. This level of voltage in normal operation is tolerable for insulation, whereas transient cases (such as impulse voltages generated in the excitation system) can lead to five times the normal voltage, resulting in insulation degradation [13, 14]. The voltage impulses successively influence the rotor insulation. Alternatively, static excitation systems used to generate DC voltage affect the rotor insulation, which further decreases the lifespan of the rotor insulation system due to PD. Since insulation between the turns is the thinnest part of the insulation under mechanical stress, these voltage impulses can cause a turn-to-turn short circuit (SC) in the rotor. This process is similar to the aging of the stator winding insulation supplied by the drive. Stator faults also affect the rotor due to the induced voltage along the air gap. Disturbances

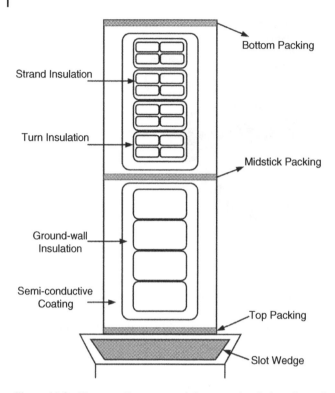

Figure 15.2 The overall structure of the stator insulation of rotating machines [6]/with permission of IEEE.

in the power systems, such as circuit breaker opening, maloperation of equipment, and lightning, are transferred to the rotor and can increase the voltage level along the insulation by up to five times, leading to a turn-to-turn SC in the rotor. The possible solution for diagnosis of this fault is the use of air gap flux monitoring [14].

15.4.2 Stator Insulation System

The stator insulation system of rotating electrical machines consists of mica, resin, and polyethylene. Figure 15.2 shows the stator insulation structure while Figure 15.3 presents the stator insulation system components, which consist of:

A. Conductors.
B. Turn insulation.
C. Stack consolidation materials.
D. Main wall insulation (mica tape).
E. Sealing tapes.
F. Corona protection (slot semi-conducting and field grading tapes).
G. Bracing materials.
H. Impregnating resin.
I. Slot wedges.

Essentially, the PD in an HV stator relates to the current of the generator conductors. Generally, the PD process in the stator winding occurs as follows [11]:

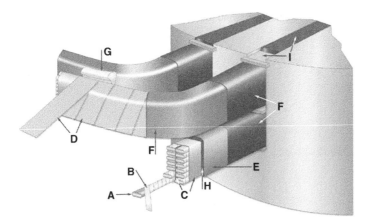

Figure 15.3 Different parts of the stator insulation of rotating machines.

- Discharge in the conductor bending point in the generator slot.
- Discharge in the slot due to very high resistive losses of slot coating.
- Internal discharge due to porosity and/or insulation delamination.
- Surface discharge due to contamination or humidity in the winding.

All the cases mentioned above provide the conditions causing PD due to the electrical HV.

15.5 PD Types in Rotating Machines

PDs can occur in the stator winding insulation system due to specific manufacturing technologies, defects during production, normal in-service aging, defects during service time, and/or abnormal aging due to overloading. Electric machine design methods, the nature of the materials used in the manufacturing process, and operational conditions can seriously affect the severity, location, and trend of destruction of the PD. Generally, a PD happens in electrical machines with a voltage level higher than 3 kV (rotating machines with no drive, such as power station synchronous generators). Of course, inverter-fed induction motors with voltage levels below 3 kV also experience the PD phenomenon and IEC TS 61934:2011, IEC60034-18-41:2014, and IEC60034-18-42:2017 standards have been published for these electric machines.

Different sources of PD can be identified and detected by investigating PD characteristics. Different types of PD in rotating electrical machines can be classified as described in the following subsections [15].

15.5.1 Internal Discharge

This type of discharging in machines occurs in the following three forms.

15.5.1.1 Internal Void
Although product processes are designed to minimize internal voids, some of these voids inevitably occur in the impregnated mica tape insulation system generally used in HV and rotating machinery. In fact, mica in the insulation system prevents the extension of the PD to a full breakdown. When

Figure 15.4 Void and mica insulation delamination [4].

0.30 mm

0.22 mm   0.40 mm

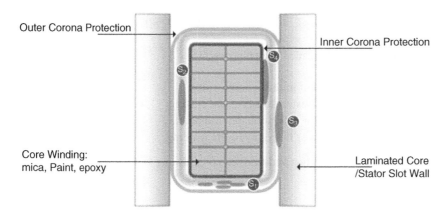

Outer Corona Protection

Inner Corona Protection

Core Winding:
mica, Paint, epoxy

Laminated Core
/Stator Slot Wall

S1: Micro voids
S2: Delamination of tape layers
S3: Slot discharge
S4: Delamination of winding

Figure 15.5 PD inside slot and insulation.

the internal voids are small, the rotating machine can be safely utilized (Figure 15.4 and S1 in Figure 15.5).

15.5.1.2 Internal Delamination

Internal delamination in the main insulation occurs due to imperfect insulation systems during the manufacturing and/or high mechanical or thermal stresses applied during utilization. The voids may extend over a large area and cause PD with rather high energy, thereby significantly damaging the insulation system. Notably, delamination reduces the thermal conduction of the insulation and may accelerate the aging and/or lead to thermal instability. Therefore, when a PD process is evaluated, delamination needs precise attention (Figure 15.5).

15.5.1.3 Delamination Between the Conductor and Insulation

Delamination between the copper conductor and the main insulation is usually caused by thermal cycling. It is a serious and dangerous condition that happens in multi-turn coils (S4 in Figure 15.5).

15.5.1.4 Electrical Treeing

Electrical treeing in electric machines is an aging process that happens in the epoxy around the mica barriers. An erosion channel that propagates in the mica may result in the complete breakdown of the main insulation system. Many factors can cause electrical treeing, including gas-filled voids and delamination in the insulation.

15.5.2 Slot Discharges

When the conductive slot coating is damaged, slot discharge occurs in HV electrical machines. This may be due to bar displacement and/or slot exit. High oscillations produce high-energy discharge, which may cause further deficiencies in the main insulation and, finally, the breakdown of the insulation. Although the time between the start of this phenomenon and final insulation breakdown is unknown, it may be short. Therefore, a reliable diagnosis in the initial stages is needed to prevent the process (S3 in Figure 15.5).

15.5.3 Discharges in the End Winding

Discharge at the end winding may occur in different locations due to the high electric field intensity.

15.5.3.1 Surface Discharge

Surface discharge at the end winding of an electric machine happens when the magnitude of the electric field intensity exceeds the breakdown field of the surrounding gas. If the stress control coating of the end winding loses its properties due to low-quality materials, contamination, porosity (careless application of materials), and/or thermal effects, then surface discharges occur and can gradually cause erosion of insulation materials. This type of discharge is typically a very slow breakdown mechanism. This type of discharge can also occur in phase-to-ground faults.

15.5.3.2 Conductive Particles

Conductive particles, and particularly small particles, arising from winding contamination can lead to local PDs and can create a pinhole in the insulation.

15.5.3.3 Phase-to-Phase Discharge

PD is a possibility if no adequate clearance exists between the phases or at the overhang support system. The intensity of the PD, in this case, depends on the design specification of the electric machine and the discharge can be internal and/or surface discharge. Therefore, the time that elapses between the detection of the phenomenon and complete breakdown is uncertain. Phase-to-phase discharge in an electric machine may result in a phase-to-phase short circuit fault.

15.5.4 Arcing and Sparking

Arcing and sparking occur due to current interruption caused by the stator magnetic field and are not due to a locally enhanced electric field. The energy and temperature created by the arcing and sparking are stronger than those of a PD, indicating that the deterioration process of insulation

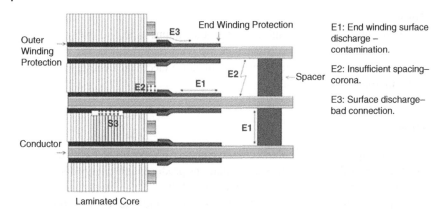

Figure 15.6 Partial discharge around the slot.

material happens much faster. Sparking and arcing may be detected by the PD system since they create a transient impulse.

15.5.4.1 Arcing at Broken Conductors
Mechanical stress may result in conductor breakage that causes occasional contact and may result in arcing.

15.5.4.2 Vibration Sparking
The magnetic field in the stator core causes an axial parasitic current circulation along the slot coating of a bar. When the bar vibrates, these parasitic currents may generate a spark at the contact point with the iron core, known as vibration sparking [16, 17]. The magnitude of the current is high if the resistivity of the conductive coating is low. A high parasitic current may result in deterioration of the ground wall insulation.

In addition to the cases mentioned above, corona discharge (E2 in Figure 15.6) may also occur within the machine. This discharge is one of the critical PDs, but its impact has not been investigated. However, it is considered a type of interference in PD measurement. Another type of discharge can occur between two bars in the overhang part of a winding or between a bar and a stator core finger press (left side of Figure 15.4) in a process called gap discharge.

15.6 Risk Assessment of Different Partial Discharge Faults

The relatively non-destructive and predictable nature of the PD has made it an attractive area of research for assessing the condition and measuring insulation systems. In recent decades, different techniques have been considered for PD assessment and much research has focused on understanding the nature of the PD phenomenon and developing measurement methods and devices for detecting, measuring, locating, and assessing PDs. Assessment of the faults that can give rise to PDs is now also recognized as essential. Table 15.1 shows the risk level of introduced faults based on the experiments conducted according to the IEC 60034-27 standards.

Table 15.1 Risk assessment of PD [15].

PD source	Risk	Frequency range (MHz)	Failure time (year)	Description
Internal void	Low	50	Over 30	This type of fault mostly originates in the manufacturing process, which does not show generator aging. In normal PD conditions, the PD due to internal voids does not significantly affect the aging
Internal delamination	High	50	Under 10	The reason for this type of fault is very high heat and/or high mechanical forces. Each factor causes breaking off of large parts of the insulation system
Spacing between conductor and insulation	High	100	Under 3	This type of fault is produced due to very high heat and/or strong mechanical forces. Each of these factors can cause breaks in a large part of the insulation system
Slot discharge	High	100	Under 2	Typically, these discharges appear only during utilization of the device. Electromechanical forces and vibrations lead to a spark, which can be measured as a slot discharge. This defect can be detected only in the case of large, destructive, and high discharges
Surface discharges at end winding	Normal	250	Over 10	Since surface discharge typically only occurs on an insulation surface, it does not lead to considerable aging. However, the presence of other factors, such as high concentrations of ozone and/or surface contamination, can accelerate insulation aging
Conducting particles	Normal	500	Over 20	Since these particles exist only on the insulation surface, they cannot lead to significant aging. Normally, the presence of other factors, such as high concentrations of ozone and/or surface contamination can accelerate insulation aging

15.7 Frequency Characteristics of Current Pulses

The PD pulses are rapid pulses with a very short time duration. According to the definition in the IEC 60270 standard, the period of these pulses is shorter than $1\,\mu s$ [18]. A critical parameter of these pulses is their rising time, which specifies the frequency width of the pulses. For electrical discharges in generators, the rising time of the pulses can be even 1 ns, which means these pulses can have frequency characteristics up to GHz [19].

These characteristics concern the discharge pulse, and this signal must travel a path to reach the sensor for electrical measurement. This path has a capacitive-inductive model, which generally has a low-pass frequency characteristic. This signal misses a large number of the frequency characteristics. In measurements using electromagnetic methods, a longer range of the frequency band can be used.

15.8 Measurement of PD Signals

Electrical pulses generated by PD in shielded HV equipment are detectable only in their terminals. Therefore, finding a discrimination method for external current pulses and internal current pulses (caused by internal insulation) is an exciting research topic. Analysis of PD involves the three-capacitor model, which models the void as different spherical, elliptical, and cylindrical shapes, instead of considering a real case study.

The measurable external charge, also called the apparent charge, is evaluated by integration of the measured PD current in the tested cell using the following equation [18, 20].

$$q_a = \int_0^\infty i(t)\, dt = I(0) \tag{15.1}$$

where $i(t)$ is the instantaneous single-pulse current of the PD in tested device terminals and I is the Fourier transform. I is a measurable discharged electric charge equal to the integral of the PD current or zero frequency currents in the terminals. Two actions can be followed, based on the standard for the discharged electric charge measurement. An external path is generated for passing external current and performing the calibration process. Note that the parameters of the external path do not affect the measurement results. The measurement circuit according to the IEC standard is shown in Figure 15.7.

Completion of the measurement process first requires calibration before applying the voltage. The calibration procedure is shown in Figure 15.8. No voltage is applied to the circuit, and the only calibrator is connected to the terminals of the tested device (C_a). The applied load is specified by

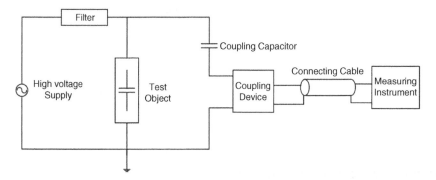

Figure 15.7 Measurement of PD based on IEC 60270 standard.

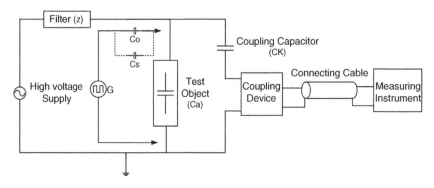

Figure 15.8 Calibration procedure based on the IEC 60270 standard.

the calibrator, so the ratio of the charge measured by the measurement device and the applied load is defined as the calibration coefficient.

Two points must be noted about the calibration circuit. The first is that this type of calibration is again based on the three-capacitor model. The step voltage source (G) shows an SC between the two terminals of the void (C_c), and its series capacitor is representative of the same capacitor (C_0). The second point is that the calibration coefficient, as well as the sensitivity level of the measurement, is defined by the C_k/C_a ratio. This means that the larger ratio leads to a calibration coefficient closer to 1 and enhances the measurement sensitivity. According to Equation (15.1), measurement of the discharged charge requires the use of the current integral (zero frequency current). However, this measurement is impossible because the noise in this frequency is very high in practice. For this reason, this measurement is conducted over higher frequencies. The following standard numerical values are defined as the permissible frequencies for measurement:

30 kHz $< f_1 <$ 100 kHz	Low cut-off frequency
$f_2 <$ 500 kHz	High cut-off frequency
100 kHz $< \Delta f = f_2 - f_1 <$ 400 kHz	Bandwidth

The basis of the standard is the use of the three-capacitor model to justify the introduced methods. This measurement method and calibration introduced by the standard are therefore only valid when the test device is modeled by a three-capacitor system. This modeling is also valid only over low frequencies, as raising the frequency will further increase the error of the three-capacitor model. Hence, the range of high frequency is defined by the standard ($f_2 <$ 500 kHz). In practice, the noise level and interferences over these frequency ranges are high and measuring the PD off-line and under these conditions is challenging. This is why the standard has recommended an alternative method called narrow-band measurement, which is defined as follows:

50 kHz $< f_0 <$ 1 MHz	Central frequency
9 kHz $< \Delta f <$ 30 kHz	Bandwidth

In this standard-based measurement method, the measurement bandwidth should first be severely limited to suppress various existing interferences. Second, the central measurement frequency should be raised to 1 MHz to avoid different interfaces. Note that this method enables measurement of the PD in a space with great interfaces; otherwise, this method has a more significant error than the previous measurement method.

The necessity of using statistics for analysis of the PD pulses arises because the electric discharge phenomenon and, similarly, the PD, is a random phenomenon. PD may occur over several successive cycles of sinusoidal high voltage in a particular phase winding with different amplitudes and/or some cycles essentially do not occur. Therefore, observing the PD pulses in one cycle or a number of cycles cannot be a precise criterion for diagnosing and distinguishing the type and intensity of these pulses. Sampling the PD pulses as precisely as possible over a rather long time and storing and analyzing the amplitudes and phases of these pulses is required. Analyzing the frequency distribution diagram of different quantities, such as the number and mean of the pulse amplitudes for each of the phases of a 50 Hz cycle, and utilizing the statistical quantities describing these distributions, seems essential.

The main quantities appearing from the detector system and by analog-to-digital (A/D) converters are digitally transmitted to a computer. By processing these quantities as raw data, other quantities can be obtained.

In addition to measuring the level of the discharged charges based on the standard, measuring the phase in which a PD has occurred is also possible. Knowing the amplitude and phase of the PD, the phase resolved partial discharge (PRPD) pattern can be determined. To determine the phase of PD pulses based on the applied voltage in each cycle, every cycle can be segregated into many numbers of divisions. Every unit of this division is called a phase window. The number of phase windows in every cycle is determined based on the device time resolution. In the PRPD data format, in addition to the PD phase and apparent charges, the number of pulses, with their phase and apparent charge, is also stored as the main quantity (Figure 15.9). The most prevalent representation of the PRPD is a ϕ-q-n graphical diagram, which can be plotted in two-dimensional or three-dimensional formats (Figure 15.10).

Statistical quantities extracted from the main quantities are generally obtained by collecting data in several periods. Some of these quantities are as follows:

A pattern of n–ϕ: The number of PD pulses occurring in every phase window versus the phase of an applied voltage plotted as a 2D diagram, as shown in Figure 15.11

A pattern of q–ϕ: In this pattern, the peak of the apparent charge of PD pulses in every phase window is plotted as a 2D diagram (Figure 15.12).

A pattern of n–q: In this 2D pattern, the number of occurring pulses having an apparent charge with similar absolute value is presented (Figure 15.13)

The recording method for these data has a historical meaning. In the past, recording devices had only the capability of recording the amplitude of the discharge (Figure 15.14), which is why the analysis of PD amplitude is considered to be the only PD analysis method. The discharge phase value has also been added to this measurement technique, leading to more PD data and further analysis of PRPD patterns (Figure 15.15). The tremendous advances in the electronic area in recent years now mean that PD signal recording can now be done quickly and can include all the details. This is why the role of PD current signals in an analysis has recently become significant.

Numerical factors, such as the total PD activity (NQN) and the peak PD magnitude (Q_{max}) associated with the diagrams, as mentioned earlier, are used to analyze the data obtained by measuring the PD and diagnosing the insulation faults. As shown in Figure 15.16, the NQN factor shows that the normalized factor NQN is the area under the diagram pulse per second per amplitude window (PHA). Q_{max} is the highest PD amplitude with an iteration rate of at least 10 times per second [21].

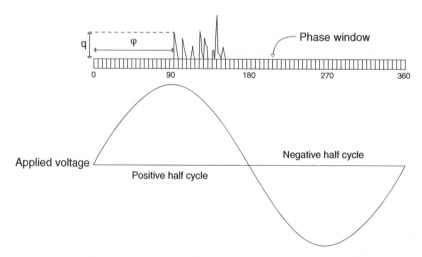

Figure 15.9 Phase window and related apparent charge.

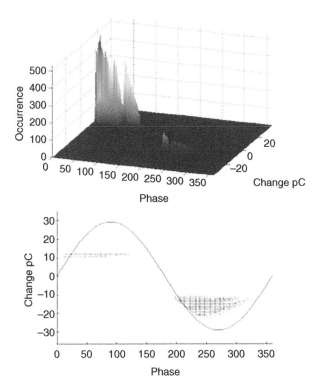

Figure 15.10 Graphical depiction of the ϕ-q-n pattern.

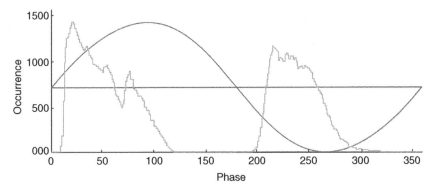

Figure 15.11 The n-ϕ pattern.

The NQN factor is as follows:

$$\text{NQN} = \frac{FS}{GN}\left[\frac{\log_{10}}{2} + \sum_{i=2}^{N-1}\log_{10}P_i + \frac{\log_{10}P_N}{2}\right] \tag{15.2}$$

where P_i is the number of pulses per second in magnitude window i, N is the number of magnitude windows, G is the gain of the PD partial discharge detector (arithmetic not decibels), and FS is the maximum magnitude window in mV at unity gain.

In addition to PRPD characteristics, other characteristics have been used for identification of the discharging source. The pulse sequence analysis (PSA) is the best-known method next to PRPD.

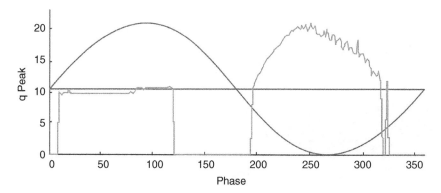

Figure 15.12 The *q-φ* diagram.

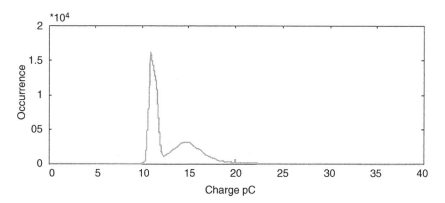

Figure 15.13 The *q-n* pattern.

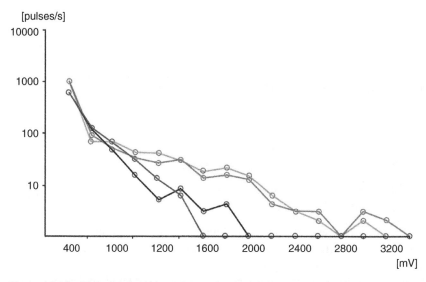

Figure 15.14 Diagram of the analysis amplitude of the pulse.

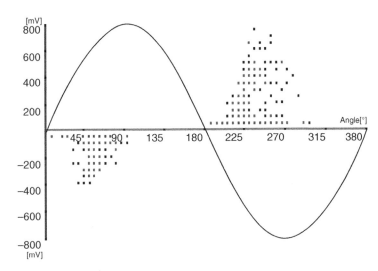

Figure 15.15 Diagram for analysis of the phase of the pulse.

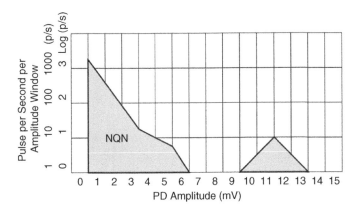

Figure 15.16 NQN factor.

PSA has apparent differences in characteristics compared to PRPD (Figure 15.17). In this analysis, the discharged charge is not used and the only sequence of PD occurrence is taken into account. To obtain this characteristic, the voltage difference between the two sequences of PD is calculated and the 2D characteristic of the present voltage difference and the previous voltage difference is plotted. This characteristic is useful only for recognition of the source type. Detection of the PD severity requires that the discharge be evaluated by other characteristics as well.

In addition to the PRPD and PSA characteristics, other characteristics can also be used for the analysis of the PD signals. They include:

- Histograms of the voltage difference between the two sequence PDs (Figure 15.18a).
- Histograms of the time difference between the two sequence PDs (Figure 15.18b).
- The voltage at the time of discharge versus the discharged charge (Figure 15.18c).
- The voltage difference between the two sequence PDs versus the voltage amplitude.
- Time difference between two sequence PDs versus the previous time difference.
- Time difference between two sequences PDs versus the discharged charge.
- The voltage difference between the two sequence PDs versus the discharged charge.

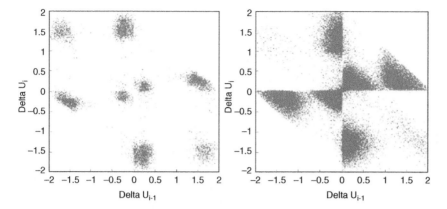

Figure 15.17 PSA characteristics of two types of defects.

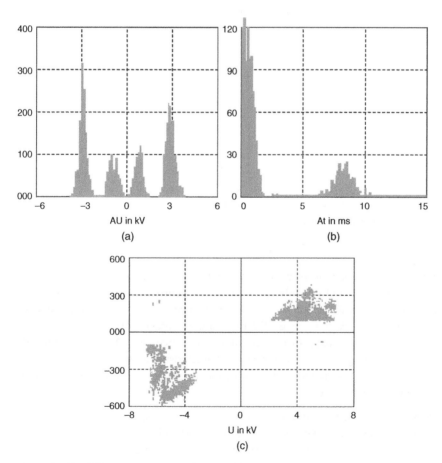

Figure 15.18 Different discharge characteristics.

15.9 Online Measurements of PD in Rotating Electrical Machines

The PD signals have a broad frequency spectrum that can be up to GHz. The basic concepts and general idea concerning the standard measurement of PD are given in the IEC 60270 standard and were investigated in the previous section. This section now introduces some methods for online measurement of the PD based on the IEC 60034-27 and IEC 60034-27-2 standards and IEEE Std 1129-2014 and IEEE Std 1434-2014 concerning the PD measurement in rotating electrical machines.

The main advantage of online measurement is that this measurement is performed when the rotating electrical machine operates under stresses, such as thermal, electrical, environmental, and mechanical stress. If the measurement is appropriately conducted, this method can provide the highest probability of evaluation of the capability of a rotating machine to continue reliable operation. The PD online measurement has the following advantages [15, 17]:

- Appropriate distribution of voltage along the winding.
  The possibility of obtaining a PD result more than the pessimistic limit is reduced (particularly in equipment sensitive to voltages higher than the rated value).
- Measurement is performed at the operating temperature.
  The second advantage is also significant due to the dependency of the PD on the temperature of the rotating machinery and also other insulation systems.
- Mechanical stress.
  Mechanical stress also influences the PD behavior and may generate a vibration arc.

Various methods can diagnose PD in large rotating electrical machines due to their many effects on the machine. These methods can be classified into the following five general classes:

1. Electrical (electromagnetic radiation, and induced current pulse).
2. Mechanical (acoustic).
3. Optical (visible and UV light).
4. Chemical (ozone gas).
5. Visual inspection.

Different techniques have been employed using different sensors to detect PD. Equipment including special imaging equipment for perceiving the UV components and directional ultrasonic microphones for detecting the acoustic noise of the PD phenomenon are some of these types of equipment. However, low sensitivity and limitation in measuring the surface PDs restrict the application of these methods in electrical machines that have a complex insulation system and possible internal PDs.

By neglecting the sensor type for PD measurement, the output data format of these techniques differs. They include the frequency spectrum of the PD pulse, the PD pulse waveform in the time domain, and/or statistical behavior of the PD pulse during several measurement periods presented in different frames. Today, the latter data format associated with the extension of measurement methods and using capacitive couplers for detecting PD pulse current is common in commercial systems. In the next section, various measurement types will be proposed.

15.9.1 Electrical Measurement of Partial Discharge

Generally, electrical measurement methods can be classified based on the frequency range of measurement [17, 22, 23]. Figure 15.19 shows the techniques and frequency bands that utilize PD measurement. Table 15.2 summarizes different frequency ranges for the electrical measurement of

f$_{lM}$ low cut-off frequency of PD measuring system   f$_{uPDt}$ upper cut-off frequency of PD signal at terminal (sensor)
f$_{uM}$ upper cut-off frequency of PD measuring system   f$_{uPDo}$ upper cut-off frequency of PD signal at source

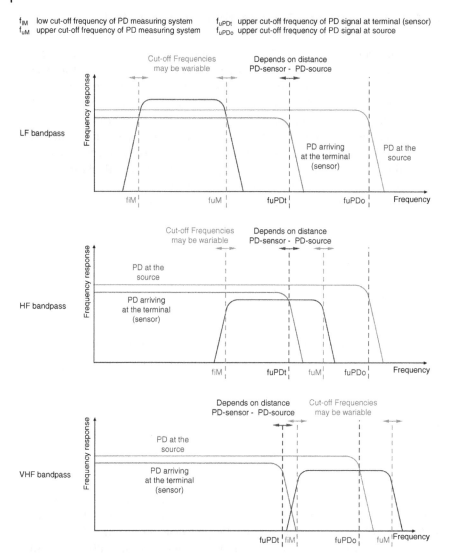

Figure 15.19 Different techniques for electrical measurement of PD according to IEC 60034-27-2 standard.

Table 15.2 Different frequency ranges (in MHz) to measure the PD electrically.

LF (low frequency)	0–3
HF (high frequency)	3–30
VHF (very high frequency)	30–300
UHF (ultra high frequency)	300–3000

the PD. Various sensors are used for electrical measurement and each sensor has a good response over a specific frequency range. Different types of sensors and their applications are discussed in the following section.

15.9.1.1 Capacitive Coupling Method

Capacitive couplers are the most common sensors for PD measurement in large generators and motors. These capacitors are built in different sizes and can be fixed in the terminals of generators during the test permanently. According to the IEC 60270 standard, the use of these sensors is the only possible way to measure PD in all HV equipment. Normally, the standard measurement method is not applied due to the presence of considerable low-frequency noise in the online measurement. However, high-frequency measurement is done using these sensors. This is why the use of smaller capacitors is more appropriate for high-frequency measurement fixed permanently in the terminal of generators and motors.

The PD can be measured over different frequency bands using a capacitive sensor. The frequency band ultimately depends on the coupling capacitor fixed in the generator terminals. This measurement circuit consists of an HV capacitor and LV equipment (impedance) in series. A series connection of these two elements creates a high-pass filter for the PD current signals. Therefore, by increasing the HV capacitance, the cut-off frequency of the high-pass filter is reduced. The measurement circuit can measure a broader range of the PD frequency, which enhances the measurement sensitivity. Referring to Figure 15.19, if the measuring impedance is a resistance, the circuit has a high-pass frequency response, where its low cut-off frequency is determined as follows:

$$f_{LOW} = \frac{1}{2\pi RC} \tag{15.3}$$

Referring to Figure 15.20 and Equation (15.3), measuring a wider band of the PD signal is possible, which enhances the measurement sensitivity. The only problem is that increasing the capacitance will require a larger capacitor, which will restrict its fixation in different places. This is the reason why the capacitances permanently installed on the generators are low and that is why these capacitances can only be used to measure high frequencies.

The typical size of a capacitance is 80 pF with a usable frequency range of 40–350 MHz [21]. This means that these sensors are used in the VHF frequency range. Increasing the capacitance allows measurement of low frequencies while enhancing the measurement sensitivity. Application of the IEC 60270 standard online measurement also requires a capacitance greater than 1 nF. Figure 15.21 shows a typical sensor fixed within a generator terminal.

15.9.1.2 Implement Capacitive Coupler Method

In this method, a signal containing the PD data is the impedance response of the detector (Zm) to the PD current in a closed circuit provided by the coupling capacitor (Ck) (Figure 15.22), which is a voltage signal. In the European systems, to present the output data format with apparent charge, assuming a fixed amplitude of the PD pulse frequency spectrum, the signal peak voltage is taken to correspond to the apparent charge. Therefore, determination of the amplitude of the voltage in the input of the device and the output apparent charge is required and is done by calibrator equipment that applies a definite voltage with a specific load to extract this ratio. In the Canadian commercial systems, the format of the output data is the same as the voltage signal and is not converted to an apparent charge. Instead, it is only amplified with the definite gain, and then divided by that gain using signal processing. The conclusion is that a calibrator is not needed.

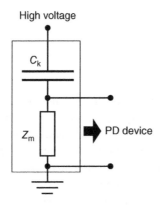

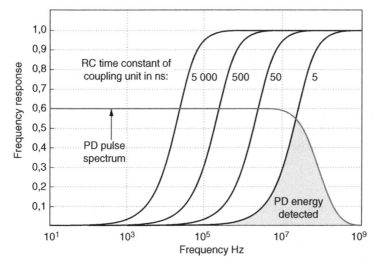

Figure 15.20 The frequency response of a series capacitor coupled with resistive measuring impedance for different capacitances [15].

Figure 15.21 Coupling capacitor fixed in the terminal of a generator for online PD measurement.

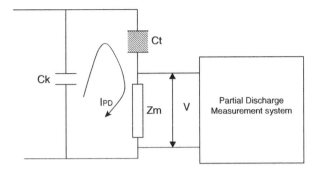

Figure 15.22 Schematic circuit for PD measurement.

Inserting the detector impedance in series under the test insulation system enhances the sensitivity. However, in practical cases, the detector impedance and measurement device must be in series with capacitive coupling to increase protection.

Temporary Capacitive Couplers A power company introduced an online PD test in about 1951. This test was an adjusted version of radio frequency (RF) monitoring, which had already been designed in reference [24] to measure the slot PD in a hydro-generator. This contrasted with the RF monitoring method, which diagnosed the PD from the generator neutral. The test introduced by the power company was performed using three capacitive couplers, which temporarily connected to the terminals of a generator. From 1951 until the end of the 1960s, measurements and applications of this test were sporadic because the credibility of the tests was not decisively verified. This is why a long time was needed to obtain the signal classification method and to relate the test results with the real conditions of winding. In the late 1960s, the test was completed, as described in the following section. It was the most important measuring and diagnostic tool for determining insulation failures in North America [25].

Utilizing a sensor. The used sensors were capacitors consisting of loops of 27.6 kV polyethylene with PVC-covered power cables, which are fixed on to the terminal of a machine or other appropriate location. Usually, the HV terminals of the PTs were chosen for this fixation [26]. The couplers had 375 pF capacitance and when the generator was operating in a normal condition, a portable couple was fixed in the terminal for each phase of the generator. Note that due to the energized busbars, precautionary measures, such as live-line, on distribution lines must be followed [25].

Test method. The couplers were connected to the generator terminal by connecting the LV side of the coupler to a five-pole RC filter with a passing band between 30 kHz and 30 MHz. The output of the filter consists of high-frequency signals related to the discharges in the stator winding, as shown by an appropriate analog oscilloscope. A power signal frequency was also taken from the generator terminal and simultaneously presented by an oscilloscope to allow easy detection of the PD pulses. Figure 15.23 shows the employed filter and Figure 15.24 exhibits the patterns recorded by this method for the generators with thermoplastic and thermoset insulation [26]. Diagnosed faults included loose wedges, delaminated ground-wall insulation, and contaminated end windings. Pattern interpretation and separation of noise pulses from the PD need the participation of experienced and expert personnel.

A key parameter for expressing the stator insulation condition in hydro- and turbo-generators is the observable peak of positive and negative PD in the rated operating voltage and power of the generator. In addition to this parameter, the behavior of PD over time was also investigated. Increments in the peak of PD up to two or three times over many years were also defined to show the worsening

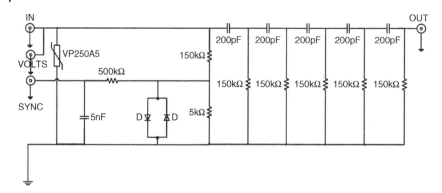

Figure 15.23 Low-pass filter to pass PD pulses with a cut-off frequency around 30 kHz [26].

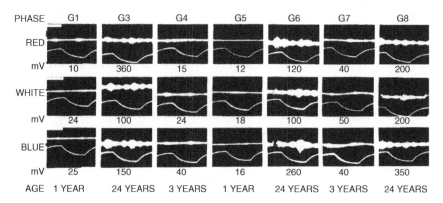

Figure 15.24 PD pattern taken from different generators with thermoplastic and thermoset insulation 1–24 years in age [26].

condition of the insulation. Figure 15.25 shows the PD variations of two generators, along with the time and impact of maintenance upon the PD reduction. In this method, the stator insulation condition observed on the oscilloscope was mainly related to the largest discharge. Recognition of the noise and PD inside the generator was also too complicated and required an experienced operator. In addition, no similar view was shared among different experts; therefore, the possibility of interpreting noise as the PD, or vice versa, was always present. Another problem in this method arises from the nature of the PD pulses. Since a PD pulse is very rapid, detecting its peak was very difficult in the past, but detecting the location of the PD occurrence against the power frequency cycle is very important. Another significant problem with this test, in practice, was that only the highest value of the pulses was registered. Recording and analyzing other information, such as the number of pulses and their manner of distribution, was not possible, while vital information about the nature and range of insulation deterioration required full determination of the PD.

Permanent Capacitive Coupler Application problems related to temporarily placed capacitive couplers prompted the introduction of a method for PD measurement. The most important problem, as mentioned above, was noise and a resulting incorrect interpretation of the measurement. The Partial Discharge Analyzer (PDA) system was offered in the mid-1970s by a power company. In this method, HV capacitive couplers were permanently connected to the generator winding. This was a

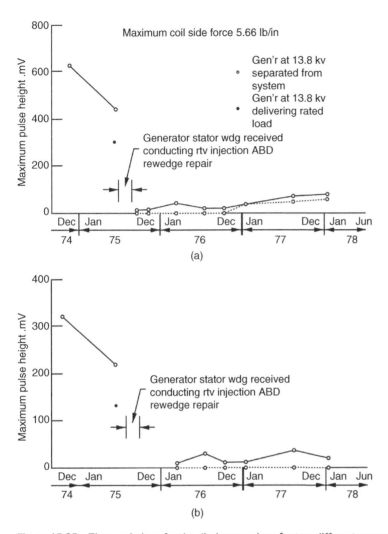

Figure 15.25 Time variations for the discharge values for two different generators [26].

portable device for measuring PD activity through permanent capacitive couplers during the operation of the generator. The application of permanent capacitive couplers and its related problems
have been investigated in references [27–35]. The capacitive coupler is the basis of all well-known
commercial systems used in this field today.

Utilizing sensors. The initial couplers utilized in this method were first built from HV cables
with 80 pF capacitance. In recent years, and with signs of progress in this field, mica epoxy capacitance couplers have been introduced that are significantly smaller than the cable couplers and
have no limitations to placement. The coupler's other limitations have been removed by introducing mica epoxy capacitance couplers, which allows the use of capacitors with a higher capacitance.
The frequency bandwidth of capacitances of 80, 500, and 1000 pF have been compared in references
[31] and [35]. Increasing the capacitance and the frequency bandwidth increased the sensitivity for
PD diagnosis (Figure 15.26).

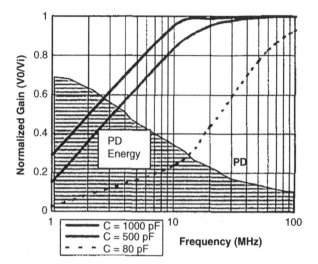

Figure 15.26 The frequency characteristics of capacitance couplers with different capacitances [35].

Figure 15.27 Two different forms for HFCT sensors.

15.9.1.3 Current Transformer

RF and other high-frequency current transformers (CTs) can be used to monitor a motor, large and small generators, and/or any other equipment that may be exposed to the PD activity. A transformer with a large bandwidth can be used to measure a PD. These CTs are generally built with a Ferrite core and are covered by a metallic cover. The frequency characteristics and output signal level are defined for most CTs in the industry. As shown in Figure 15.27, these CTs are built in two different forms [36, 37].

Generally, CTs are fixed in the LV, earth, and neutral of the monitored equipment. A common place for fixing a radio frequency current transformer (RFCT) is the generator neutral. The CT tank must be earthed. High-frequency current transformers (HFCTs) and Rogowski coils are the two usual sensors employed in this measurement. These sensors are built over the high-frequency range and can be used in the HF and VHF frequency range. The HFCT has a very high sensitivity compared to the Rogowski coil and is widely used for PD measurement. The HFCT sensors are built in two forms. One has a separable core, which is a benefit as it allows operation without taking the device from the network. The other form requires cable disassembly. Figure 15.28 shows a typical fixation of these sensors in three phases [38].

One of the first methods for the PD measurement in rotating machines was the use of HFCT in the neutral of the machine. This method is often called an RF monitoring method and has been successfully applied to several machines [39]. The method has been used for the diagnosis of secondary conductors with an arc in large turbo-generator winding [40]. Later, RF monitoring was used for diagnosis of some faults, such as stator core loss, end winding corona, and crack in the stator

Figure 15.28 Fixation of HFCT sensors [38].

conductors [41, 42]. In addition to fault diagnosis, this method has also been used to estimate the remaining life of the insulation [43]. In the next section, the problems related to the measurement methods are briefly reviewed.

Used Coupler Type In this method, CT is mostly used as a coupler and is introduced as an RFCT, an HFCT, and a low-frequency current transformer (LFCT), according to the frequency range [29, 42, 44, 45]. Typically, in the measurements, the CT is fixed in the neutral of the winding, as shown in Figure 15.29. The best connection location for these CTs to the neutral is near the neutral conductor connection by earthing (resistor, reactor, or transformer) [42].

Other combinations of using HFCT for PD diagnosis within machines have been investigated in references [29] and [44]. In this method, the HFCTs with different bandwidths over the 0.1–2 MHz frequency range have been used to achieve the optimal signal-to-noise ratio in the PD measurement in different positions of the generator (neutral or stator terminals). The wide bandwidth provides the possibility of extracting more data from the PD. To prevent the AM frequency interference, the low limit of the frequency can be adjusted to be higher than 20 MHz for the HF PD measurements. Figure 15.30 shows a number of possible and non-troublesome positions for online and live-line measurement of the PDs on the generator.

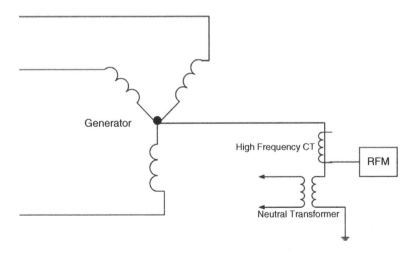

Figure 15.29 Schematic diagram of the coupler fixing method in RF monitoring [45].

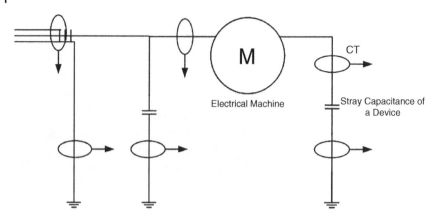

Figure 15.30 Possible positions for fixing measurement CT [29].

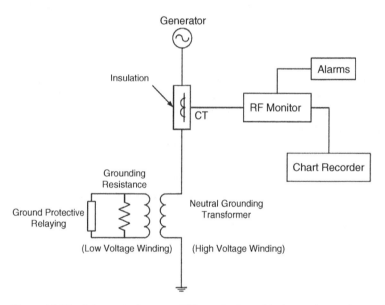

Figure 15.31 Schematic diagram of RF monitoring with the static coupler method [46].

Considering the low coupling power of CT for reflecting the RF currents passed the generator neutral, the electrostatic coupler has been introduced for RF monitoring and can conduct the HF pulses with a better gain to the RF device [46, 47]. At this end, a high-pass RC filter is used, which shows that this filter can operate as a sensitive and effective coupler. Figure 15.31 shows the procedure of RF monitoring with the static coupler.

Interpretation of Measurement Results The results in the frequency domain [40–42] and the time domain (fixed frequency) [40] are now analyzed. After the coupler receives the PD signals and existing noise in the environment, the output is analyzed with a spectrum analyzer. The PD is separated from the noise through frequency components and by use of the spectrum analyzer as a narrow-band filter. Generally, both frequency and pattern recognition characteristics are used for noise rejection [45], but the noise in turbo-generators is too high and usually varies from one test to another [45]. Therefore, the user who extracts and separates the PD pulses from the noise

Figure 15.32 Mean RF spectrum measured in neutral for 500–750 MVA generators [40].

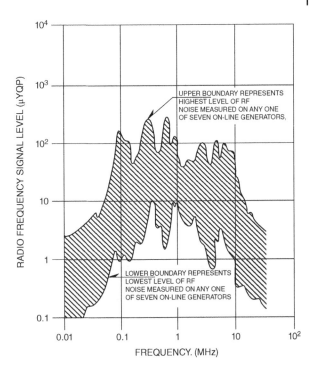

must be highly skilled. In addition, the PD pulses occur in the line side of the winding and are severely weakened as they approach the neutral; this itself decreases the signal-to-noise ratio [39]. The measurements include quantities such as the peak, quasi-peak, effective slide back peak (SBP), and mean of the field intensity [42].

Data interpretation in the frequency domain is obtained based on the frequency spectrum from the measurement. For example, in reference [40], the initial measurement was first carried out on seven machines, and the RF spectra of these machines in the range of 100 kHz up to 150 MHz were obtained. These measurements have been done in 500–750 MVA generators and the required data are provided for determining the necessary conditions of the arc diagnosis in this range of the generators. These measurements and studies resulted in a combined spectrum, as shown in Figure 15.32, which presents two upper and lower borders for generator RF responses within the mentioned limits. If the generator RF response is placed inside the hatching region, it is healthy and if it crosses this border, particularly the upper border, it probably has a broken strand.

Exceeding the permissible region shown in Figure 15.32 is equivalent to an internal arc in the winding of the generator and the results typically show the capability of the RF method in this field.

15.9.1.4 Antenna Monitoring Method

Antenna sensors are employed to measure the electromagnetic radiation of the PD pulses. The PD signals have a wide frequency band and radiated electromagnetic wave and the frequencies of the pulse range up to 2 GHz.

The main problem encountered in the online measurement of PD is that the existing interferential in LF and HF frequency ranges are rather high, so measurement in these frequency ranges is a problem in the online case. However, these interferentials in the higher frequency ranges are lower, so the UHF range is a very appropriate range for measurement by the antenna. This results

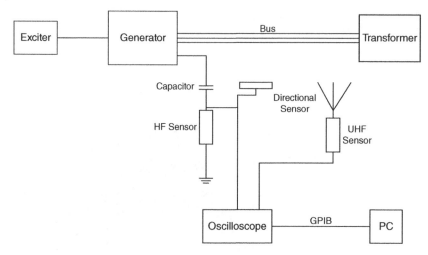

Figure 15.33 Installation of an on-site detector system [47].

in the robustness of the sensors against interferentials and so they are very suitable for online measurement. The application of these sensors in GISs is widespread, and despite some problems, their use in transformers has also increased in recent years. The use of this measurement technique in generators and all rotating electrical machines runs into serious problems. The most important problem is that the central part of these machines is made of metals. Conductors cause the reflection of electromagnetic waves, which prevents good propagation of the electromagnetic waves in space. Nevertheless, despite these issues, sensors can be used in different ways to measure PD in rotating machines. The sensors can be fixed outside or inside the generator stator.

Fixing an Antenna outside the Generator On-site measurement is done in the bus-bar room of the power station. The installation procedure of the detector system is shown in Figure 15.33.

Utilizing sensors. The sensors used in this method are described in the following sections.

- **Directional sensor**

The directional sensor consists of two busbar couplers and one pre-amplifier and is capable of distinguishing the generator's internal PD signals from the signals coming from the power system side (transformer). Since the sensor is designed for the detection of signals coming from inside the generator, the sensors are fixed along with busbars and the received signals are amplified and then used as the input to the oscilloscope.

- **HF sensor**

The HF sensor consists of a CT and a pre-amplifier with a 20 dB gain. It is fixed in a shielded measurement box. The bandwidth of the HF sensor is in the range of 0.1–1 MHz. For signal coupling, an HV 1000 pF capacitance is used in the generator neutral conductor. The HF sensor detects the pulse of the current flowing through the neutral wire. The apparent PD level can be calibrated by a step voltage pulse calibrator. The HF sensor signals consist of the PD signals and a broad level of noise. The signals are sent to the oscilloscope via a shielded 50 Ω coaxial cable.

- **UHF sensor**

This sensor is portable and can be installed in any location. It can be placed in two locations: first in the bus-bar room and then beside the exciter. In on-site detection, the UHF envelope signals with a low sampling ratio are measured. The HHF envelop signals are sent to an oscilloscope through a 50 Ω shielded coaxial cable.

Data collection and their process. A digital oscilloscope is a data collection tool. The most extended data storage length is 1 Mbyte and the maximum sampling rate is 200 MHz. The first channel of the oscilloscope shows the HF signal, the second channel presents the directional sensor signal, and the third channel shows the UHF sensor signal. All collected data are transferred to a PC via a GPIB card and the oscilloscope is controlled by a processing software. This software consists of a recorder part and adata analyzer part, which shows the data related to the PD pulse in the time domain.

Defects and virtues:

- A UHF sensor and directional sensor discriminate the PD inside the generator and the PD signals from the transformer and power system.
- The UHF and HF sensors can typically operate using the time windows in a de-noising mode and also calibrate the PD level.
- The UHF sensor frequency range is between 500 and 1500 MHz, so the UHF sensor cannot detect a corona discharge with a frequency around 300 MHz.

Fixing the Antenna on the Generator Frame Figure 15.34 shows the measurement system used for receiving the radiated waves from the PD [48, 49]. In this measurement system, microwaves are received by two double ridge guide horn antennas fixed on the generator frame. The signal then passes an amplifier and enters a declining converter. The output signal of the declining converter is an input for a computer via an A/D converter with a sampling frequency of 1 MHz. The other triggering signal starts the A/D converter when the applied voltage passes zero. The central frequency of the internal generator of the declining converter in periods of 100 MHz varies from 2 to 3 GHz. Combining the input signal and generated signal by the generator converter leads to the following output signals:

$$f_{out} = |f_{in} - f_{local}| \tag{15.4}$$

where f_{out} is the central frequency of the output signal, f_{in} is the central frequency of the input signal, and f_{local} is the central frequency of the generated signal by an internal generator of the converter. The central frequency of the input signal is transformed as a declining mode without varying the frequency spectrum distribution.

Figure 15.35 presents an alternative measurement system [50] that uses a patch antenna in the place of the horn antenna. This system is newer than the previous one, but the operation principle of both systems is similar. Figure 15.36 shows the establishment of a complete measurement unit.

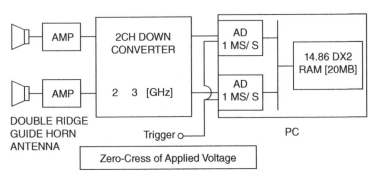

Figure 15.34 PD measurement system [48].

Figure 15.35 Internal schema of the rotating machine and locations of fixing antenna [50].

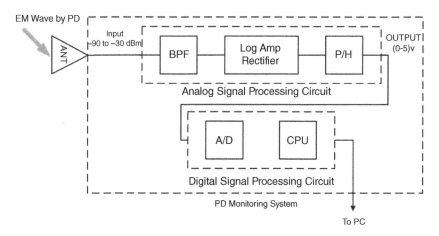

Figure 15.36 Structure of PD monitoring system [50].

Defects and virtues. The voltage received by the antennas is very well correlated with Q_{max} over the above picocoulomb range [50]. However, the major weakness of this method is that it cannot detect the PDs associated with the low apparent charge.

Fixing Antenna inside Generator (Stator) Extreme internal noises are produced in large turbo-generators; therefore, permanent capacitive couplers used for measurements will receive data that contains some noise. This problem has been addressed by online PD measurement in gas turbines and steam turbines since 1978 with a new sensor called the stator slot coupler (SSC) [51]. Applications of this sensor have been described in references [28], [29], [31], [39], and [51–53].

Theory and principles of stator slot coupler operation. The SSC is a directional electromagnetic coupler. This technology has been well-known in microwave research. The SSC consists of a ground plate and a sensing line with coaxial output cables on its two sides. Figure 15.37 shows the structure of an SSC. When an electromagnetic wave propagates as a PD pulse along the SSC and

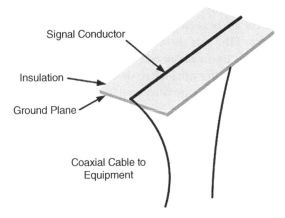

Figure 15.37 Simplified structure of an SSC [39].

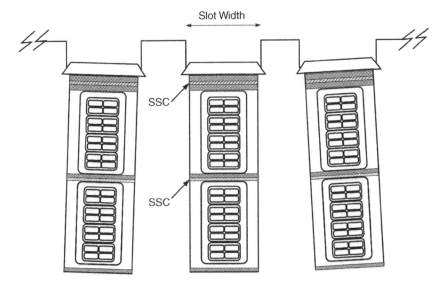

Figure 15.38 SSC fixing location inside slot [53].

near the sensing line, some pulses appear in two-output terminals of the SSC. The impedance of the SSC characteristic is 50 Ω, which has been considered to adopt two-output cables. This is made possible by a device that collects and processes the SSC data.

The SSC must not be considered a purely capacitive or inductive coupler. This tool is known as an antenna with distributed parameters. Referring to Figure 15.38, the SSC is normally placed in the slot and under the wedge. Of course, it is installed for a number of generators on wedges, but the SSC is fixed under the wedge.

Defects and virtues of SSC. The advantage of the SSC sensor is that it deletes the internal noises of the large turbo-generators. The bandwidth of the SSC (10 MHz up to1 GHz) allows identification of the PD pulses with a short rising time and duration period from the noise pulses, which have a longer rising time and duration period compared with the PD. In addition, the SSC is fixed inside the winding and near the points where the PD is most likely to occur; therefore, the impact of the external noises, such as bus-bar noises, is low. The SSC offers a much better signal-to-noise ratio compared with the terminal capacitive couplers. Meanwhile, considering the two-output feature of the SSC, separating and detecting the PD sources are possible, as the PD inside slot is easily separated and detected from the end-winding surface PDs. This also allows the detection of some phenomena, such as cross-coupling and end-winding PDs. In addition to these virtues, since the SSC has no electrical contact with the generator windings, an earth fault or short circuit is unlikely to occur in relation to the use of this sensor, whereas the use of capacitive couplers in the terminal can lead to very high probabilities of these occurring.

The disadvantage of this sensor is the necessity of pulling out the rotor to fix the sensor. This requires at least three days and the generator body is likely to need to be drilled to take out the cables. If the generator is hydrogen cooled, the holes must be appropriately seamed. One of the most significant drawbacks of the SSC, which is a common defect of HF sensors, is that the SSC detection region is limited to an area around the sensor. Therefore, the SSC diagnoses only the PDs around the sensor and PDs in the slot in which the sensor has been fixed. The SSCs are typically fixed in the slots that contain bars with the highest electrical stress; therefore, the occurrence of the PD is more likely. However, the SSC can only detect the PDs occurring mainly due to lost coils, aging insulation, and contamination of the end-winding in the slot. Should a PD occur in an adjacent slot, the SSC would be unable to detect it [28].

RTDs as Sensors   To overcome the problems associated with the SSC, a resistive temperature detector (RTD) is often used in HV machines as the PD system sensor. The resistive element of the RTD is sensitive to temperature variations and is generally used to measure temperature. These resistors are made from platinum, nickel, and copper. The sensors usually are wound elements and/or layer elements.

The wound elements used in electrical machines are wound on a ceramic core and self-inductance is eliminated with a double-twist method. These sensors are shielded by steel, glass, and/or ceramic to protect them mechanically. For each phase, two of these sensors are typically placed between two coils inside a slot. The sensor terminal wires are seamed with epoxy glues and/or ceramic glues. The wire terminals are insulated with Teflon or fiberglass and connected to the distribution box and the measurement circuit. These thermometers are often placed on the slot's bottom and/or inner folds of bars for measuring core and stator winding temperatures. The thickness of these sensors is about 4–5 mm and the length and number of their applications depend on the bar dimensions and the number of slots in the machine.

Since these sensors have an appropriate cross section, they can be used as a PD sensor. This sensor is used like coupling capacitance and/or an antenna. In this method, the end of the sensor is connected to a wide-band filter and then the received signals, after amplification, are transferred to a computer using an A/D converter to record and analyze the signals [25]. The use of this sensor as the PD receiver has been described in many references [54–60]. Of course, some criticisms have been raised regarding this method [61, 62]. The last successful results and responses to the criticisms in reference [62] were published in 2007 [63]. The use of the RTD as a PD sensor led to an efficient system for online measurement, based on the suggestion in references [54] and [55]. The RTD sensor sensitivity and results of some tests, after damping the pulses inside the windings, have been described, as mentioned above. Online measurement of the PD using the RTD wires as the PD sensor can lead to PD detection, where the wires are connected to the earthed terminals of the RTD by RFCT [56–58]. A similar method has been introduced in reference [59], where pickup coils with a performance similar to the RFCT were used instead of the RFCT.

Description of a typical PD measurement system using RTD. Figure 15.39 shows a schematic diagram of a PD monitoring system. The PD and other signals are inputs of the digital

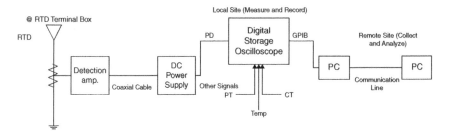

Figure 15.39 Schematic diagram of a PD monitoring system based on RTD.

storage oscilloscope (DSO). In every channel, the DSO has an A/D converter with a sampling rate of 1 G/s and is a peak holding function. Digital signals are converted by the DSO and sent to the personal computer (PC) by a GP/IP cable and data are stored in its disk. The PC used here has a 32 Mbyte memory and 1 GB internal hard disk and receives or sends the control messages from the DSO at a constant speed (normally one hour). The digital data contain PD and noise.

To improve the reliability of the monitoring, a notebook type PC is used, because its power supply is supported by a battery and, in the case of an electric power cut-off, its operation will not be disturbed. By creating a network, the stored data in the local PC can be sent to a remote PC using a telephone line where the data are processed and displayed in the last PC. The data processing software has been generated by macro language. This program converts the data to a set of phase angles (ϕ) and charge quantities (q) and deletes the Thyristor-related noise pulses.

Table 15.3 summarizes the specifications of a typical generator and the measured signals. The PD signal is in the input of channel 1 and the PT signal (generator voltage monitor) is in the input of channel 2. The PT is isolated from the DSO and its voltage is transformed into 10 V by an insolating transformer. To record the coil temperature, the RTD temperature is converted into a DC voltage using the temperature converter. In some generators, the CT signal (generator output current monitor) is also stored. The sequence of the signal recording is first the PD signal, second the PT signal, third the temperature, and fourth the CT signal. In the case of collective monitoring, the second generator signals of the fifth section are stored. For example, the fifth data will be the PD input of the second generator.

Output Data Format. Figure 15.40 shows a model of processed data recorded by an oscillogram and converted to a ϕ-q-Φ graph. The central unit of the PD signal is in mV and is converted to the

Table 15.3 Specifications of generators and measured signals.

Generators	Operation commencement (replacement)	Rated voltage [kV]	Rated output [MVA]	Resin	Signals[a]
A(G)	1956 (1970)	13.2	93	Epoxy	PD, PT, Te
B(G)	1944	11	9	Polyester	PD, PT, Te
C(G)	1976	13.2	53	Epoxy	PD, PT, Te, CT
D(G)	1961	13.2	50	Epoxy	PD, PT[b], Te
E(G&M)	1969	13.2	100	Epoxy	PD, PT, Te, CT
F(G&M)	1981	18	290	Epoxy	PD, PT, Te, CT

a) PT: line voltage, Te: coil temperature, CT: output current.
b) PT is only used as a synchronizing signal.

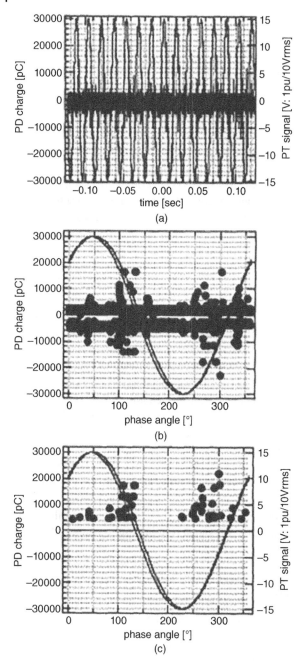

Figure 15.40 Output data format: (a) main oscillograph of the PD and PT pulse, (b) scattering distribution between the PD pulse peak and phase angle of the PT voltage, and (c) the Φ-q-n graph of a PD pulse by removing the thyristor pulses.

PD custom unit. The calibration test of the load quantity has also been done. This test consists of two parallel stages. The first stage of the test consists of detection of the PD pulse and a load calibration pulse using the traditional method determined via a coupling capacitor connected to the generator output terminal. The second stage of the test is detection of the PD pulse by the RTD terminal. Applying the results of the tests, mV is converted to PC.

Figure 15.41 shows variations of the maximum charge (qm) and the RTD temperature measured every hour. These data were approximately measured in generator C for 17 months. This figure

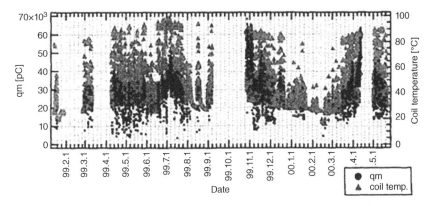

Figure 15.41 An example of long-term variations in q_m and coil temperature.

shows that the highest value of qm is 6000 pc, but qm varies seasonally and rises proportionally with generator C temperature. Figure 15.42 presents qm as a function of the RTD temperature in generator C. This figure shows that qm depends on the coil temperature.

Figure 15.42 indicates a relationship between the CT signal and the q_m in generator C. The coil temperature varies over a long time where the output current rises. This is the reason that, as shown in Figure 15.43, the q_m has a wide scattering range. The proposed generator has a larger q_m than other generators. According to the diagnostic criteria of deterioration and damage, considering q_m,

Figure 15.42 Maximum charge quantity as a function of the coil temperature.

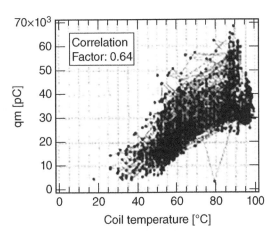

Figure 15.43 Maximum load as a function of the output current.

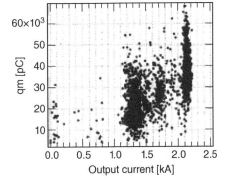

the coil isolator has severely deteriorated. Therefore, mechanical stress and/or thermal stress has led to the extension of insulated internal voids, which increases the q_m. If the coil temperature is considered to be the PD variation factor, this phenomenon will clearly be a diagnostic index.

Advantages and disadvantages. Some concerns have been raised regarding the use of the RTD as a PD sensor:

1. Use of the PD measurement and diagnosis systems under the frequency of 10 MHz. In this range of frequency, the windings of the machine consist of a complicated network of inductive and capacitive components. When this network is excited by a rapid PD pulse, many types of resonance frequencies appear. Therefore, any winding, even those that apparently have the same power and capacity, offer different responses. This means that making conclusions regarding the winding conditions based on the detected pulses is very complicated [61].

2. Cross-coupling problem. This phenomenon occurs mostly due to unshielded RTD terminals, as the diagnosis of a pure PD with an RTD is very tricky. Of course, as claimed in reference [55], this problem has been solved using unique methods.

3. The type of RTD of machines is generally undefined. The length and size of the RTDs in different machines are not consistent, which causes differences in the effective capacity, as well as differences in the RTD sensitivity in various machines [55].

4. Basically, in many cases, the RTD wires (and not the RTD itself) sense the majority of the PD signal and the RTDs with shielded wires (most machines have this system) have much lower sensitivity and exhibit a minimal value of the PD. In addition, by varying the length of non-shielded wires of the RTD and their location displacement, the RTD sensitivity rises or falls. As a result, the PD value detected by the RTD depends on the fixing conditions and its wires [61, 62].

5. Calibration must be done offline by injecting simulated PD pulses into different points of winding and measuring the sensor response. This requires opening up the machine completely to access the end windings and is, in fact, very time consuming and more costly than fixing sensors in the terminal of the machine.

6. Performing the test and interpretation of the results need a very high level of knowledge and skill. The presumption is that the lack of progress in using the PD monitoring and measuring systems introduced in the 1950s and continuing to the –1980s has been due to the complexity, application specialty, and interpretation of the results. A similar view has been mentioned in reference [62] in relation to the difficulty in interpreting and explaining the results. The introduction of some patterns has made the interpretation of the results similar to that of other measurement methods and with no more difficulty [63].

Coupling System Fixed on Rotor This method is capable of locating the slot discharge of an operating generator. The signals are received slot by slot and processed to give details on the size and location of the PD. The system is also capable of detecting differences between different types of PDs by diagnosing the selectable frequency used in the processing system. This system was introduced only in reference [31] and no reports followed regarding its application. In this measurement system, a specially designed antenna has been used. Figure 15.44 shows the fixing procedure and the PD measurement circuit. Other details of this measurement system have been given in references [31] and [64].

Sensor Applications A group of sensors, known as frame sensors, can measure the impulse electromagnetic fields related to the PD or induced impulse currents by the PD in the external earth circuits of the machine. As an example, a sensor can be fixed between the generator terminal box

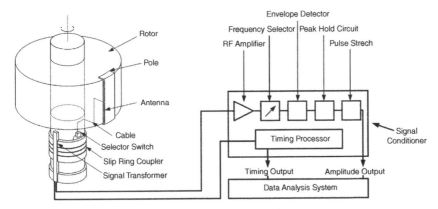

Figure 15.44 Schematic diagram of a coupling system fixed on the rotor [31].

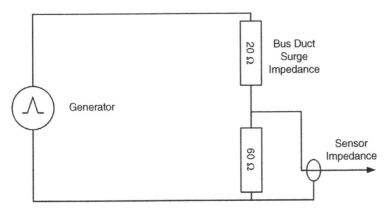

Figure 15.45 An equivalent circuit of a frame sensor of a generator [55].

and phase isolated bus-duct shield; the latter is always earthed in the distance from the generator frame, which can be assumed for the HF pulses of the isolated generator. In this case, the connected sensor between the generator frame and bus-duct shield short circuits all the HF currents through the sensor. Figure 15.45 shows the equivalent circuit of this method [55].

Application of these sensors must be restricted to initial detection of the insulation conditions in machines that have not been equipped with permanent sensors. The application of these sensors for measurement and evaluation of the results needs a relevant expert. More details in this area and some results related to the measurement analysis have been given in reference [55].

15.9.2 Acoustic Measurement of PD

The PD generates acoustic noise due to the temperature rise in the location of a gas discharge. These waves have a frequency of up to 150 kHz. Acoustic noises have the LF characteristic and the maximum acoustic noises can be 40 kHz and measured by a low-pass filter at the acoustic sensor output; this filter has a minimum impact on the measurement. The advantages of this method are the measurement without disconnecting the machine from the mains and the easy measurement of the signals because of the LF characteristic. The drawback of this method is its sensitivity; for example, if a PD occurs inside the slot and/or its walls, the PD cannot be diagnosed by this method [65–67].

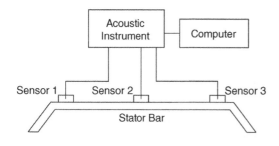

Figure 15.46 Measurement of a PD acoustic signal and its use in PD localization [66].

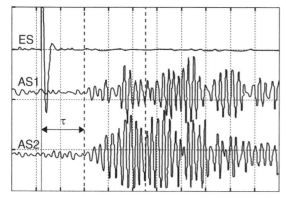

Figure 15.47 A typical measured acoustic signal in the generator.

One application of this method is locating the PD. To do so, at least four acoustic piezoelectric sensors must be fixed on the body of the machine. Since these sensors are placed with distance, the PD pulses reach different sensors with a time interval, and by considering this time interval, the location of the PD can be estimated. Figure 15.46 shows the location of fixed piezoelectric sensors for PD measurement using the acoustic signal. Figure 15.47 presents a typical measured acoustic signal for PD diagnosis.

15.9.3 Chemical Measurement of PD

Detection of ozone gas is the only chemical method used to diagnose a PD in generators. In the air-cooled machines, a chemical reaction occurs between the PD and its surrounding air. One of the elements of this chemical reaction is ozone gas. Ozone gas has a distinct smell and increasing the number of PDs increases the amount of this gas. A PD inside the slot between bars and/or between a bar and the body does not generate a large amount of ozone gas usable for condition monitoring. Several methods are available to study the ozone gas association, which can be under the influence of different environmental factors, such as temperature, humidity, and air displacement. This method of PD diagnosis can also be used online [68].

15.9.4 Visual Inspection and Optical Measurement of PD

This method can be carried out by a blackout test of the winding. In this case, the coil and/or bar are tested under a fully dark environment and by studying the photo, the PD occurrence in the machine is detected. Since this method operates based on the level of the emitted light from the PD, it cannot be applied to fault detection if the level of the emitted light is low or not visible (such as discharge in the hole). Therefore, this method is appropriate merely for detecting surface discharges, as other defects cannot be diagnosed by this method [69].

This method can be used in the production process by applying a phase-to-phase voltage to the windings. In addition, if the PD is diagnosed by this method, the defect location and even the discharge level can be estimated based on the observed light density.

A substitution method for the blackout is to observe ultraviolet (UV) light instead of the PD in a dark environment [13]. The main advantage of this method is that it does not need a dark area, as the defect and its location can be detected under a low light. The safety of this method is high because it does not need the HV.

15.10 Summary

This chapter presented the historical background of electrical machine insulation and explains in detail the insulation system of the stator and rotor of the synchronous generators. Different types of partial discharges (PDs) and their risk assessment were explained. The online developed PD measurement approaches are explained, including electrical, acoustic, chemical, and visual methods. The required sensors and practical applications of the methods have been discussed.

References

1 Tavner, P. (2008). *Condition Monitoring of Rotating Electrical Machines*, vol. 56. UK: IET.

2 IEEE Std 1129-2014 (2014). *IEEE Guide for Online Monitoring of Large Synchronous Generators (10 MVA and Above)*, 1–62. IEEE.

3 Stone, G.C. (2012). A perspective on online partial discharge monitoring for assessment of the condition of rotating machine stator winding insulation. *IEEE Electrical Insulation Magazine* 28 (5): 8–13.

4 Hoof, M. and Lanz, S. (1999). PD diagnostics on rotating machines. Possibilities and limitations. In: *Proceedings, Electrical Insulation Conference and Electrical Manufacturing & Coil Winding Conference*, 28 October 1999, 195–200. Cincinnati, OH, USA.

5 Hudon, C. and Belec, M. (2005). Partial discharge signal interpretation for generator diagnostics. *IEEE Transactions on Dielectrics and Electrical Insulation* 12 (2): 297–319.

6 (2002). Annual Report. In: *Conference on Electrical Insulation and Dielectric Phenomena* (Cat. No.02 CH37372). Cancun, Quintana Roo, Mexico: IEEE Dielectrics and Insulation Society.

7 Pascoli, G., Kral, C., Pirker, F. et al. Experiences with online partial discharge diagnoses on turbogenerators. In: *4th IEEE International Symposium on Diagnostics for Electric Machines, Power Electronics and Drives*, SDEMPED, 24–26 August 2003, 20–24. Atlanta, GA, USA.

8 Farahani, M., Borsi, H., Gockenbach, E. et al. (2005). Partial discharge and dissipation factor behavior of model insulating systems for high voltage rotating machines under different stresses. *IEEE Electrical Insulation Magazine* 21 (5): 5–19.

9 Luo, Y., Li, Z., and Wang, H. (2017). A review of online partial discharge measurement of large generators. *Energies* 10 (11): 1964–1996.

10 Stone, G.C., Stranges, M.K.W., and Dunn, D.G. (2016). Common questions on partial discharge testing: A review of recent developments in IEEE and IEC standards for offline and online testing of motor and generator stator windings. *IEEE Industry Applications Magazine* 22 (1): 14–19.

11 Brutsch, R., Tari, M., Frohlich, K. et al. (2008). Insulation failure mechanisms of power generators. *IEEE Electrical Insulation Magazine* 24 (4): 17–25.

12 IEEE Motor Reliability Working Group (1986). Report of large motor reliability survey of industrial and commercial installations. *IEEE Transactions on Industry Applications* IA-21 (4): 853–864.

13 Stone, H., Boulter, G.C., Culbert, E.A. et al. (2004). *Electrical Insulation for Rotating Machines: Design, Evaluation, Aging, Testing, and Repair*. USA: Wiley-IEEE Press.

14 T. S. IEC, 60034–27: 2006, Off-Line Partial Discharge Measurement for Stator Winding and Insulating Rotating Electric Machine.

15 Stone, G.C. and Maughan, C. (2008). Vibration sparking and slot discharge in stator windings. In: *IEEE International Symposium on Electrical Insulation* (ISEI), 9–12 June 2008, 148–152. Vancouver, BC, Canada.

16 T. S. IEC 60034-27-4-2018, On-Line Partial Dischcharge Measurement for Stator Winding and Insulating Rotating Electric Machine, Geneva, Switzerland.

17 IEC 60270 (2000). *High Voltage Test Techniques – Partial Discharge Measurement*, 1–99.

18 Bartnikas, R. (2002). Partial discharges. Their mechanism, detection, and measurement. *IEEE Transactions on Dielectrics and Electrical Insulation* 9 (5): 763–808.

19 Lemke, E., Berlijn, E., Gulski, E. et al. (2008). *Guide for Partial Discharge Measurements in Compliance to IEC 60270*, vol. 366. CIGRE Technical Bochure.

20 (2014). IEEE Guide for the Measurement of Partial Discharges in AC Electric Machinery. In: *IEEE Std 1434–2014 (Revision of IEEE Std 1434-2000)*, 1–89, 4 December 2014, https://doi.org/10.1109/IEEESTD.2014.6973042. USA.

21 Stone, G. (2000). Importance of bandwidth in PD measurement in operating motors and generators. *IEEE Transactions on Dielectrics and Electrical Insulation* 7 (1): 6–11.

22 Strehl, T., Muhr, M., Tenbohlen, S. et al. (2010). *Guidelines for Unconventional Partial Discharge Measurements*, no. 253, WG D1.33. CIGRE.

23 Johnson, S. (1951). Slot discharge detection between coil surfaces and the core of high voltage stator windings. *AIEE Transactions* 70: 1993–1997.

24 Sadeghi, I., Ehya, H., Nasiri, R. et al. Condition monitoring of large electrical machine under partial discharge fault – A review. In: *24th International Symposium on Power Electronics, Electrical Drives, Automation, and Motion*, SPEEDAM 2018, 20–22 June 2018. Amalfi, Italy.

25 Kurtz, M. and Lyles, J.F. (1979). Generator insulation diagnostic testing. *IEEE Transactions on Power Apparatus and Systems* PAS-98 (5): 1596–1603.

26 Stone, G.C. (2005). Partial discharge diagnostics and electrical equipment insulation condition assessment. *IEEE Transactions on Dielectrics and Electrical Insulation* 12 (5): 891–903.

27 IEEE Std 1434-2000 (2000). *IEEE Guide to the Measurement of Partial Discharges in Rotating Machinery*. USA: IEEE.

28 Campbell, S.R., Stone, G.C., Sedding, H.G. et al. (1994). Practical online partial discharge tests for turbine generators and motors. *IEEE Transactions on Energy Conversion* 9 (2): 281–287.

29 Stone, G.C. (1991). Partial discharge – Part VII: Practical techniques for measuring PD in operating equipment. *IEEE Electrical Insulation Magazine* 7 (4): 9–19.

30 Kurtz, M., Stone, G.C., Freeman, D. et al. (1980). *Diagnostic Testing of Generator Insulation Without Service Interruption*. Paris, France: CIGRE, Paper 11-09, 27 August–4 September.

31 Zhu, H., Green, V., Sasic, M. et al. (1999). Increased sensitivity of capacitive couplers for in-service PD measurement in rotating machines. *IEEE Transactions on Energy Conversion* 14 (4): 1184–1192.

32 Kurtz, M., Lyies, J.F., Stone, G.C. et al. (1984). Application of partial discharge testing to hydro generator maintenance. *IEEE Transactions on Power Apparatus and Systems* PAS-103 (8): 195–200.

33 Stone, G.C. and Sedding, H.G. (1995). In-service evaluation of motor and generator stator windings using partial discharge tests. *IEEE Transactions on Industry Applications* 31 (2): 299–303.

34 Zhu, H., Green, V., Sasic, M., and Halliburton, S. Development of high sensitivity capacitors for PD measurement in operating motors and generators. In: *Proceedings of the 6th International Conference on Properties and Applications of Dielectric Materials,* June 21–26, 2000, 571–574. Xi'an, China: Xi'an Jiaotong University.

35 Renforth, L.A., Armstrong, R., Clark, D. et al. (2014). High-voltage rotating machines: A new technique for remote partial discharge monitoring of the stator insulation condition. *IEEE Industry Applications Magazine* 20 (6): 79–89.

36 Renforth, L.A., Hamer, P.S., Clark, D. et al. (2015). Continuous remote online partial discharge monitoring of HV Ex/ATEX Motors in the oil and gas Industry. *IEEE Transactions on Industry Applications* 51 (2): 1326–1332.

37 Renforth, L., Hamer, P.S., Clark, D. et al. (2013). Continuous, remote on-line partial discharge (OLPD) monitoring of HV EX/ATEX motors in the oil and gas industry. In: *Industry Applications Society 60th Annual Petroleum and Chemical Industry Conference,* 23–25 September 2013. Chicago, IL, USA.

38 Sedding, H.G., Campbell, S.R., Stone, G.C. et al. (1991). A new sensor for detecting partial discharges in operating turbine generators. *IEEE Transactions on Energy Conversion* 6 (4): 700–706.

39 Emery, F.T. and Harrold, R.T. (1980). Online incipient arc detection in large turbine generator stator windings. *IEEE Transactions on Power Apparatus and Systems* PAS-99 (6): 2232–2240.

40 Timperley, E.J. (1983). Incipient fault identification through neutral RF monitoring of large rotating machines. *IEEE Transactions on Power Apparatus and Systems* PAS-102 (3): 693–698.

41 Timperley, J.E. and Chambers, E.K. (1992). Locating defect in large rotating machines and associated systems through EMI diagnostics, CIGRE-Paper 11-311, Session.

42 Lo, P., Magerkurth, W., and Phillipson, J.T. (1992). Stator insulation monitoring of large high voltage motors and generators using on-line r.f. techniques in the petrochemical environment. In: *Procceedings of the Petroleum and Chemical Industry Conference,* 177–183. San Antonio, TX, USA: Industry Application Society.

43 Zhou, Y., Dix, G.I., and Quaife, P.W. (1996). Insulation condition monitoring and testing for large electrical machines. In: *Conference Record of the 1996 IEEE International Symposium on Electrical Insulation,* 239–242. Montreal, Quebec, Canada, June 16–19, 1996: IEEE.

44 Zhou, Y., Gardiner, A.I., Mathieson, G.A., and Qin, Y. New methods of partial discharge measurement for the assessment and monitoring of insulation in large machines. In: *Proceedings of the Electrical Insulation Conference,* 16–19 June 1996, 111–114. Quebec, Canada.

45 Mohammed, O.A. and Mundulas, J. (1989). Improvements in RF monitoring system on generators Improvements. *IEEE Transactions on Energy Conversion* 4 (2): 237–243.

46 Li, X., Li, C.R., Ding, L. et al. Generator PD measurement by using UHF and HF sensors. In: *Conference Record of the 2004 IEEE International Symposium on Electrical Insulation,* 22–25. Indianapolis, IN, USA, 19–22 September 2004.

47 Kawada, M., Zen-Ichiro Kawasaki, Z.-I., Matsuura, K. et al. (1997). Use of spatial phase difference method to detect microwaves associated with stator coil partial discharge in generator. *Electrical Engineering in Japan* 121 (2): 16–26.

48 Muto, H., Kaneda, Y., Aoki, H. et al. (2008). On-line PD monitoring system for rotating machines using narrow-band detection of EM wave in GHz range. In: *International Conference on Condition Monitoring and Diagnosis*. Beijing, China, April 21–24, 2008: IEEE.

49 Kawada, M., Kawasaki, Z.-I., Matsuura, K. et al. (1998). Detection of partial discharge in operating turbine generator using GHz-band spatial phase difference method. In: *Proceedings of International Symposium on Electrical Insulating Materials. 1998 Asian International Conference on Dielectrics and Electrical Insulation. 30th Symposium on Electrical Insulating Materials*, 30 September 1998. Toyohashi, Japan: IEEE.

50 Stone, G.C. and Sedding, H.G. New technology for partial discharge testing of operating generators and motors. In: *Proceedings of Electrical Insulation on Electrical Manufacturing and Coil Winding Conference*, 4–7 October 1993, 667–672. Chicago, IL, USA.

51 Allahbakhshi, M. and Akbari, A. (2011). A method for discriminating original pulses in online partial discharge measurement. In: *Measurement*, vol. 44, No. 1, 148–158. Elsevier.

52 Evagorou, D., Kyprianou, A., Lewin, P.L. et al. (2010). Feature extraction of partial discharge signals using the wavelet packet transform and classification with a probabilistic neural network. In: *IET Science, Measurement and Technology*, vol. 4, No. 3, 177–192. UK: The Institution of Engineering and Technology.

53 Blokhintsev, I., Golovkov, M., Golubev, A. et al. (1999). Field experiences with the measurement of partial discharges on rotating equipment. *IEEE Transactions on Energy Conversion* 14 (4): 930–938.

54 Stone, G.C. and Campbell, S.R. (2001). Discussion of field experiences on the measurement of partial discharges on rotating equipment. *IEEE Transactions on Energy Conversion* 16 (4): 380–381.

55 Itoh, K., Kaneda, Y., Kitamura, S. et al. On-line partial discharge measurement of turbine generators with new noise rejection techniques on pulse-by-pulse basis. In: *Conference Record of the EEE International Symposium on Electrical Insulation*, 197–200. Montreal, Quebec, Canada, June 16–19,1996.

56 Itoh, K., Kaneda, Y., Kitamura, S. et al. (1996). New noise rejection techniques on pulse-by-pulse basis for on-line partial discharge measurement of turbine generators. *IEEE Transactions on Energy Conversion* 11 (3): 585–594.

57 Kanegami, M., Okamoto, T., Noda, T. et al. (2001). Partial discharge monitoring system with use of resistance temperature detector laid in stator coil slot of hydro power generator. In: *Proceedings of the International Symposium on Electrical Insulation Materials (ISEIM 2001)*, 657–660.

58 Huang, C., Yu, W., and Wei-Wei (2002). Development of PD on-line monitoring system for large turbine generators. In: *Conference Record of the 2002 IEEE International Symposium on Electrical Insulation*, 23–26. Boston, MA, USA, April 7–10, 2002.

59 Campbell, S.R. and Stone, G.C. Investigations into the sse of temperature detectors as stator winding partial discharge detectors. In: *Conference Record of the 2006 IEEE International Symposium on Electrical Insulation*, 369–375. Toronto, ON, Canada.

60 Tsurimoto, T., Itoh, K., Kaneda, Y. et al. (1999). Development of partial discharge monitor for turbine generators. In: *Proceedngs of the Electrical Insulation on Electrical Manufacturing and Coil Winding Conference*, 185–189. Cincinnati, OH, USA.

61 Kane, C., Golubev, A., Blokhintsev, I., and Patterson, C. (2007). Our response – Use of resistive temperature detectors as partial discharge sensors in rotating equipment. In: *Electrical Insulation Conference and Electrical Manufacturing Exposition*. Nashville, TN, USA: IEEE.

62 Naghashan, M.R., Seyfi, H., and Grayili, A. (2004). Measurement of RTD characteristic in partial discharge detection of a high power electrical machine. In: *19th International Conference of Electrical Engineering*. Tehran, Iran: PSC.

63 Technical Report, Different methods of capacitive couplers installment and their noise rejection methods in partial discharge measurement (2008). Niroo Research Institute of Iran, Tehran, Iran.

64 Ahmadi, S., Naghashan, M.R., and Shadmand, M. (2012). Partial discharge detection during electrical aging of generator bar using the acoustic technique. In: *Electrical Insulation (ISEI), Conference Record of the 2012 IEEE International Symposium*, 576–578. Tehran, Iran.

65 Bozzo, R. and Guastavino, F. (1995). PD detection and localization by means of acoustic measurements on hydro generator stator bars. *IEEE Transactions on Dielectrics and Electrical Insulation* 2 (4): 660–666.

66 Zhu, H., Kung, D., Cowell, M., and Cherukupalli, S. (2010). Acoustic monitoring of stator winding delaminations during thermal cycling testing. *IEEE Transactions on Dielectrics and Electrical Insulation* 17 (5): 1405–1410.

67 Lépine, L., Lessard-Déziel, D., Bélec, M. et al. (2008). Using ozone measurements to diagnose partial discharge in generators. *Hydro Review* 27 (7).

68 Stranges, M.K.W., Haq, S.U., and Dunn, D.G. (2014). Black-out test versus UV camera for Corona inspection of HV motor stator endwindings. *IEEE Transactions on Industry Applications* 50 (5): 3135–3140.

69 Rosolem, J.B., Floridia, C., and Sanz, J.P.M. (2010). Field and laboratory demonstration of a fiber-optic/RF partial discharges monitoring system for robust applications. *IEEE Transactions on Energy Conversion* 25 (3): 884–890.

16

Noise Rejection Methods and Data Interpretation

16.1 Introduction

Despite the advantages of an online electrical partial discharge (PD) measurement method, application of these methods has some difficulties, including:

- Electrical interferences.
- Huge data.
- Data interpretation.

Solving these three problems allows the use of several advantages of online PD measurement, such as condition-based maintenance (CBM) and predictive maintenance (PdM). Two continuous and periodic measurement methods are applied to attain the benefits of CBM. In the continuous method, the measurement system is installed on the machine and the data measurement and analysis are performed continuously. In the periodic method, the measurement is done over definite time ranges. The load level in this measurement must be identical over the measurement ranges to enable comparison of the measured results over different ranges [1].

The first problem with an online PD measurement is the existing electrical interferences, which create a severe problem in data interpretation. Various methods are available to solve this problem. The second problem is the considerable amount of data associated with online measurement. This type of measurement can be done over a long time; therefore, if the data are being stored, the amount of data increases enormously. This is the reason why algorithms and methods presented at different stages must be able to apply online to PD data. This problem can be avoided by taking into account available algorithms and methods used for denoising and data interpretation. For example, denoising by applying the wavelet method requires huge computations, so that technique cannot be applied online for denoising, for the reason that the method has no application in denoising online

Electromagnetic Analysis and Condition Monitoring of Synchronous Generators, First Edition. Hossein Ehya and Jawad Faiz.
© 2023 The Institute of Electrical and Electronics Engineers, Inc. Published 2023 by John Wiley & Sons, Inc.

PD signals [2]. The third problem is the interpretation of the PD signals. Data interpretation is a fundamental problem in condition monitoring of a generator, even after solving the noise problem in the measurement. Several methods have been applied to interpret the data at different times. Of course, these methods depend mostly on the available facilities at the application time.

This data interpretation was initially conducted by analyzing the amplitude of the current pulses, because that was the only measurable parameter possible with the available facilities at that time. The phase in which the PD appeared was then added to this analysis, because more important data were available for the data interpretation. In recent years, the current signal analysis has been used to interpret the PD signals due to greater advances in electronics, sampling, and signal recording. The method of using these three techniques for the PD interpretation is given in this chapter.

16.2 Noise Rejection in Online Measurement

One of the most critical steps in electrical measurement of the PD is choosing from among the different electrical denoising techniques available for the measurement. Before introducing various methods for denoising, the following terms are defined.

Sensitivity: The sensitivity of measurement equipment is almost always defined by the ratio of the real PD and the PD energy, which is detected by the measurement device. According to the IEC 60270 Standard, this sensitivity can be expressed by the pC value, where the lowest pC measured by the device or sensor is expressed as the sensitivity of that device or sensor.

Noise and signal-to-noise ratio: Electronic measurement devices generate basic noise. This noise depends mostly on heat and increases with the temperature rise of the electronic devices used for PD measurement. The signal-to-noise ratio (SNR) depends on the PD measurement bandwidth and rises as the bandwidth of the SNR is increased.

Disturbance: Noise and disturbance are fundamentally different. Disturbance is generated periodically or at random. For example, converters and/or PD that occur outside the proposed equipment are disturbance generation sources [3].

16.3 Noise Sources in Generators

Noise and disturbances in generators arise from different sources, which can be classified as follows (Figure 16.1):

1. **Excitation system noise**: Electronic pulses in the range of several kHz can generate many problems in measuring low-frequency PD.
2. **Ground noise:** Considering the high common earth system of the power device, this part can also affect many noises in measurement.
3. **HV busbar noise:** The generator output is connected to the step-up transformers. The PDs occur in these transformers and other equipment can enter the PD measurement system of the generator.
4. **Internal noise:** The possible arcs occurring within the generator, such as busbar arcs, are not necessarily a PD, but the measurement system can detect it as a PD. Examples are the arcs in the generator busbars. Arcs are generated in the generator brushes earthing the turbine shaft and arcs from slip ring brushes are considered as internal noise.

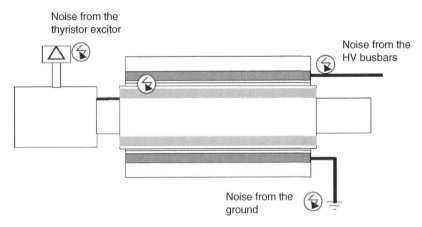

Noise from the
thyristor excitor

Noise from the
HV busbars

Noise from the
ground

Figure 16.1 Different sources of noise in generators.

16.4 Different Methods for Denoising

Based on a variety of PD sources, several methods have been suggested in recent years for practical denoising. Some of these denoising methods, such as wavelet transform, have not yet been used in practice. The methods described in the following sections are those that have been utilized in practice by standard companies and/or developed by active companies in the PD measurement field.

16.4.1 Restricting the Frequency Range

The main idea underlying this method is that the disturbance frequency characteristics, such as the PD frequency characteristic, are not extensive and have a limited frequency band. The central frequency measurement methods with minimum disturbances use narrow band. As shown in Figure 16.2, considerable noise/interference occurs in the frequency under 1 MHz and by moving it to higher frequencies, the PD can be measured without the noise.

Figure 16.2 Choosing the frequency range for PD measurements [3].

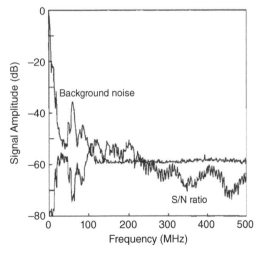

Table 16.1 Desirable VHF and UHF frequency range.

Frequency range	MHz
VHF	30–300
UHF	300–3000

Noise is a serious problem when performing standard measurements over the low-frequency range. Using measurement methods over the high-frequency range can eliminate many noises, such as noise and disturbances from the thyristor of the excitation system. Table 16.1 shows the desirable very-high frequency (VHF) and ultra-high frequency (UHF) ranges in measurement methods for noise rejection.

The lower frequency band in the VHF and UHF measurement methods is where most noises and disturbances fall, which is the reason why these two measurement methods are very convenient from the denoising point of view. PD can be measured using several sensors. For measurements over the VHF range, the capacitive couplers fixed on the terminals of the generator are employed. In addition, antennas, particularly SSC antennas fixed inside the generator, can measure the PD in the UHF range.

16.4.2 Pulse Shape Analysis

One efficient and applicable approach to noise separation is the time domain analysis of the PD signal. Pulse shape analysis could provide a unique characteristic of the time domain signal, such as rise time or decay time, to predict the PD. In this approach, a signal detected by a PD sensor and located close to the machine would have a shorter rise time compared with the rise time of noise signals. Digital circuits could be used to recognize the type of signal as a stator PD or noise, according to the rise time of a signal. The rise time of a detected pulse depends on the location of the PD or noise sources with respect to the PD sensor. In other words, the farther the PD or noise sources are from the located sensors, the longer the rise time of the signal will be [4]. If the rise time of a detected pulse of 80 pF in an installed capacitive coupler is longer than 6 ns, the detected pulse is classified as noise. It is a PD pulse if the rise time of the detected pulse is less than 5 ns.

16.4.3 Noise Rejection by Propagation Time

This method is used only if capacitive coupler sensors are employed. The propagated PD pulse passes through the tested device and cable. Therefore, pulses from different places of the tested device and cable reach the PD measurement equipment with a time delay. The propagated pulse direction is determined by two capacitive couplers fixed in two different positions. Consequently, the external PD signals and the external disturbances can be separated from the PD signals that occurred inside the tested device (Figure 16.3) [4–6]. In this method, two permanent sensors for each phase or each pair of the parallel paths of the winding are used; the capacitors must be insulated in specific locations depending on the machine type.

For instance, in turbo-generators, capacitors are fixed in a directional form (Figure 16.4) and provide conditions whereby external noises reach the device at the same time. The noises are

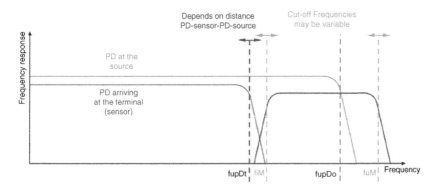

Figure 16.3 VHF measurement of PD.

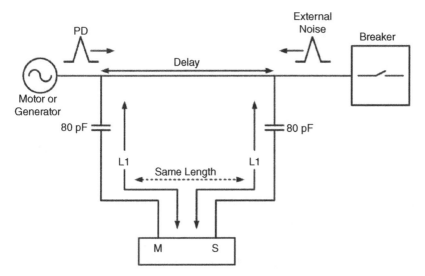

Figure 16.4 Time-domain disturbance separation based on the arrival time of the pulse.

subtracted, giving a zero result, and they are eliminated. This denoising method is very convenient for eliminating the noise sources that enter the HV busbar of the transformer.

16.4.4 Residue of Two Channel Signals

This denoising method proposed by the IEC 60270 standard is usable for a case in which capacitive coupler sensors are utilized (Figure 16.5). Since the disturbances and the PD signals pass through the tested equipment as well as through the coupling capacitor, these signals can be measured in both loops. The difference is that the external disturbance signal in both loops has the same polarity, but the PD signal has different polarities in the two places. Therefore, the external interferences are eliminated by the difference between the two signals, whereas the PD signal is amplified [5].

16.4.5 Gating

In this method, the measurement device must have at least two channels: the first one is the main channel for measurement and the second channel is for gating. This method has been proposed

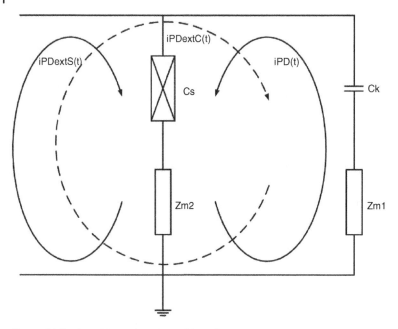

Figure 16.5 Denoising using the residue of two channel signals.

by the IEC standard. The basis of this method is that if the gating channel detects a signal, the PD signal either will or will not be measured. The denoising method using gating can be performed in the following two forms:

- The gating channel is used for noise detection.
- The gating channel is employed for PD signal detection.

If the gating channel is used as a noise detection source, the relevant sensor of this channel is located where there is external noise and there are no PD signals. Therefore, both PD signals and noise and interferences appear in the first channel, but only noise appears in the second channel, with no PD signals or only a weak amplitude.

In this gating method, when the signal amplitude in the second channel is higher than the defined threshold, the data recording in the first channel is stopped for a short period. Therefore, the signals due to interferences in the first channel are eliminated and the equipment records only the PD signals. Figure 16.6 shows the denoising procedure using the gating method when the second channel is used for detecting the interferences. The parts denoted by the blue window have been eliminated and the remaining pulses are considered to be PD pulses.

If the UHF sensor is placed in the excitation side and/or in the side with the power network and transformer, it is used as the gating channel for eliminating the interference due to the excitation system and/or transformer. Figure 16.7 shows a typical gating using UHF sensors.

Another gating method involves using the second channel to record the PD. In this case, this sensor only detects the PD signals and shows no impact of the interferences. The measurement in the first channel begins when the signal in the second channel exceeds the defined threshold. For the second channel, the VHF and UHF measurement is used, but particularly the UHF, which has a high capability for denoising. For example, in Figure 16.7, the UHF sensor has been located such that only the PD pulses due to the generator are detected. In the following section, this signal

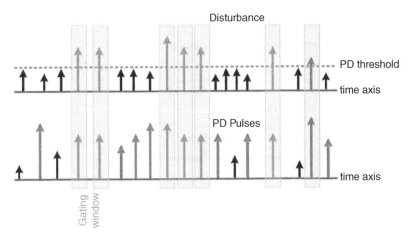

Figure 16.6 Denoising by the gating method.

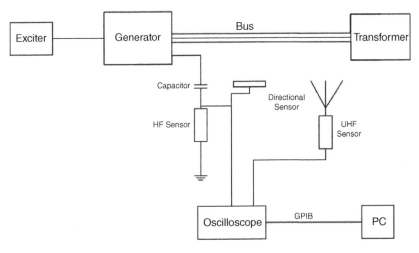

Figure 16.7 Using UHF sensors as gating channels [7] / with permission of IEEE.

is used for gating the capacitor coupler measurement, which is an HF measurement. As seen in Figure 16.8, in this gating method, measurement in the main channel is performed when there is a pulse in the gating channel.

16.4.6 Three-Phase Amplitude Relation Diagram (3PARD)

This method is applicable to three-phase systems and can detect interferences, such as a corona outside the equipment. This method does not need other equipment. It is sufficient to measure the PD in three phases simultaneously. This method operates such that the discharged charge in any phase is mapped on to a 3D diagram. In this case, the external discharges for any three-phase system have almost the same values and are placed in the middle of the 3D diagram. The basic principle of noise rejection based on a three-phase amplitude relation diagram (3PARD) is shown in Figure 16.9.

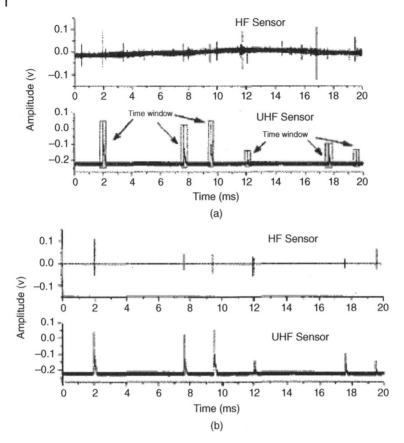

Figure 16.8 Gating of a capacitive coupler measurement using a UHF sensor.

If an internal discharge occurs in a specific phase, the discharge level in that phase is high and the data related to the discharge charges are on the same axis. In addition to diagnosing the electric interferences, this method can discriminate different PD sources. Figure 16.10 presents a typical use of this method for grouping different faults and noise. Although the proposed method works well to detect PD in transformers and cables, its use in electric machines entails some difficulties. This method assumes that noise occurs in three phases at the same time, whereas PD is concentrated on a phase axis that is not valid for electric machines. This is because all three windings are co-mingled in the stator, while this is not the case for transformers or cables.

16.4.7 Current Signal Features

A general procedure for separating the PD pulses from the noise pulses is as follows [5]:

- A suitable number of pulses is stored. A suitable number of pulses is stored.
- For each stored pulse, the features generating a difference between the PD pulses and noise are extracted. With these features, this can be done online and without requiring the total stored pulses.

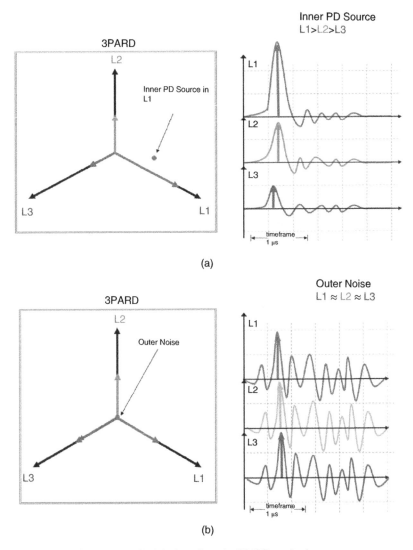

Figure 16.9 Denoising principle based on the 3PARD method.

- The pulses with similar features are placed in one group. This can be done automatically using clustering methods.
- The phase-resolved partial discharge (PRPD) pattern is obtained for each group.
- The noise is separated from the obtained patterns. This can diagnose, either automatically or manually, the source and/or sources of the PD by an expert.

The process shown in Figure 16.11 is performed automatically or manually. However, in a monitoring system, this must be done automatically. At present, this process is manually performed by experts to analyze the PD test results.

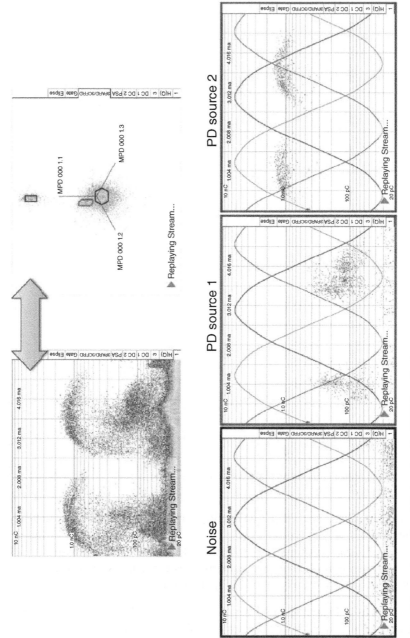

Figure 16.10 Denoising and grouping defects using the 3PARD method.

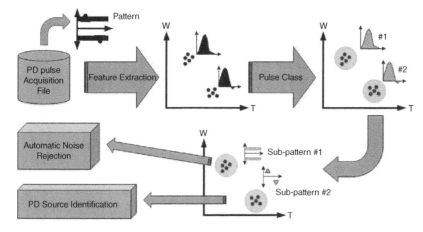

Figure 16.11 Discrimination between PD signals and noise using the clustering method [8].

Several features can be extracted from the PD current signal [4, 9]. However, two sets of features that are widely used are the time-frequency (TF) map and the three-center frequency relation diagram (3CFRD).

In the TF method, general features are extracted from the PD pulse signal, as follows:

$$F\{p(t)\} = P(f) = \int_{-\infty}^{+\infty} p(t)e^{-j2\pi ft}dt \tag{16.1}$$

$$t_0 = \frac{\sum_{i=0}^{N} t_i p(t_i)^2}{\sum_{i=0}^{N} p(t_i)^2} \tag{16.2}$$

$$T^2 = \frac{\sum_{i=0}^{N} (t_i - t_0)^2 p(t_i)^2}{\sum_{i=0}^{N} p(t_i)^2} \tag{16.3}$$

$$F^2 = \frac{\sum_{i=0}^{N/2} f_i^2 |P(f_i)|^2}{\sum_{i=0}^{N/2} |P(f_i)|^2} \tag{16.4}$$

where time (T) is calculated from the time-dependent signal and frequency (F) is evaluated from the frequency-dependent signal [10]. This method has been investigated by a power company. Figure 16.12 indicates how this method is used to separate the PD signals from the noise caused by six pulses. Figure 16.13 shows the method of separating the base noise from the PD signals using the TF method.

Another method involves the use of the current signal feature by the 3CFRD method, which is mostly based on the TF approach. The basis of this method is that the PD and noise signals have different frequency feature interference (Figure 16.14). This method has been employed by a power company. In this method, three features are extracted for every pulse. This means that the current signal passes three filters with different limited bandwidths and central frequencies. These three features are placed in 3D coordinates to be able to discriminate the PD from noise signals. Figure 16.15 presents how this method works. Of course, the features of this method are not general and are defined as before. The central frequencies of these three filters must be chosen by the user, which makes it difficult to monitor the condition. Figure 16.16 shows how to separate different noises from the PD signals using the 3CFRD method.

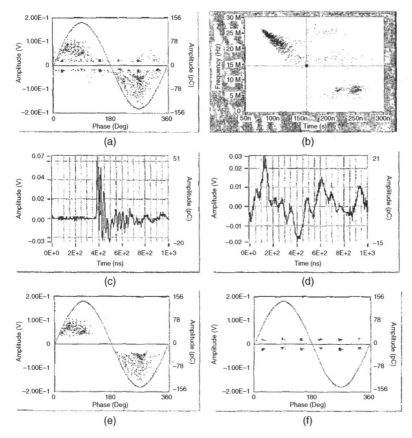

Figure 16.12 Rejecting noises caused by six pulses using the TF method: (a) PD characteristic with noise, (b) TF characteristic of signals, (c) PD of current signal, (d) six-pulse current signal (noise), (e) separated characteristic of PD, and (f) separated noise characteristic [10] / with permission of IEEE.

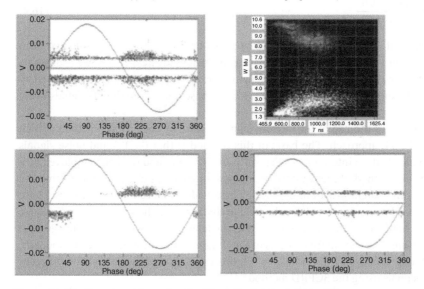

Figure 16.13 Base denoising using the TF method. Adapted from[4].

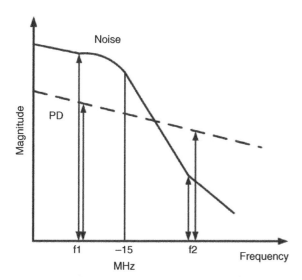

Figure 16.14 Typical relation between a spectrum of the frequency of PD and noise pulses. Modified from [11].

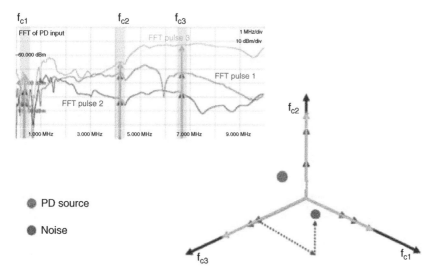

Figure 16.15 Performance of the 3CFRD method for denoising.

16.4.8 Noise Rejection Using Fourier Transform

A signal can be separated into its sinusoidal components by Fourier transform. If a significant difference is observed between the frequency spectrum of the PD pulse and the noise in the measured signal, this difference can be the basis for the elimination of noise. In other words, a signal transformation to the frequency domain distinguishes the frequency components of the PD pulse and sinusoidal noise, so that by removing the noise-related components and re-transformation of the signal to the time domain, a signal free of noise is achieved.

A PD pulse with its frequency spectrum is shown in Figure 16.17, which covers a wide range of frequencies; in other words, the PD contains a frequency enriched spectrum [13, 14]. Among different types of noises, only the narrow-band communication noise in the frequency

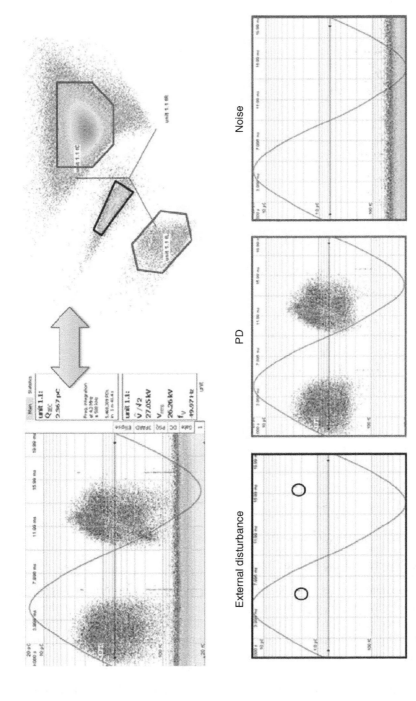

Figure 16.16 Separating different noises from the PD signals using the 3CFRD method [12] / with permission of IEEE.

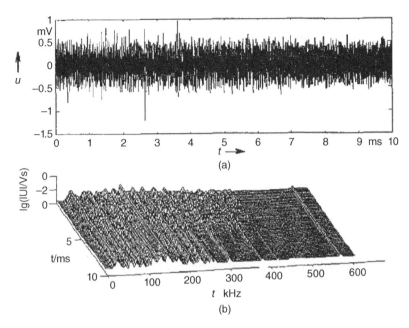

Figure 16.17 Obtained PD signal by on-site measurement: (a) time domain and (b) STFT plot. Adapted from [13].

domain has a feature distinct from the PD pulse. The communication noises in the frequency domain exhibit a relatively peaky form, which gathers around the central frequencies (carrier frequencies).

Figure 16.17a presents a measured signal and Figure 16.17b shows its short-time Fourier transform (STFT). The narrow-band noises are striped lines independent of time with fixed frequency and the PD signals with perpendicular wide-band structures. As shown in Figure 16.17, there are many PD pulses between 0 and 5 ms, and the region between 5 and 10 ms is free of the pulse. The filter can reject the noise, which is performed by specifying a predefined threshold. Other types of noise similar to the PD pulse have an enriched frequency spectrum and it is not possible to distinguish them from the PD in the frequency domain. Therefore, the Fourier transform is only suitable for eliminating sinusoidal noises from the PD pulses.

16.4.8.1 Principles of Noise Rejection Using Fourier Transform

The basis for eliminating the narrow-band noises using Fourier transform is peak shaving in the frequency domain. This means that, following the Fourier transform application on the main signal, the frequency peaks in the amplitude-frequency (narrow-band noises) are reduced. The Fourier transform technique for removing noise is rarely used today and to the best of the authors' knowledge, no practical system is commercially available. The available algorithms consist of the following stages [15]:

1. Determining the Fourier transforms of the noisy signal.
2. Obtaining an amplitude versus frequency diagram.
3. Detecting frequency peaks depending on the narrow-band noises, which can be done in two ways:
   - **Constant threshold method**: In this method, a threshold value obtained by trial-and-error methods experimentally is chosen and amplitudes higher than this threshold are considered to be the peak.

- **Constant peak number method**: This method is based on the frequency peak number depending on the limited narrow-band noises that are already predictable. From the experimental point of view, this method is efficient and justifiable because communication systems have a few carrier frequencies to prevent wave interferences. Therefore, a sufficiently large threshold value is initially chosen and the constant threshold method is applied based on this assumption of the algorithm. The obtained peaks are then counted. If this number is less than the pre-defined number, the threshold value is reduced by one step and the threshold method is applied again. This recursive algorithm continues until the detected number of peaks exceeds the pre-defined number.

4. Peak shaving of the detected cases in the second item is possible in two ways:
   - **Hard peak shaving**: In this method, the peaks are mapped to the zero number.
   - **Soft peak shaving**: The PD signal has an enriched frequency spectrum. Therefore, applying the soft peak shaving method inevitably changes the PD signal, because the existing frequency components in the PD signal, which are equal to the noise frequency, tend to zero. To avoid this, the following two methods are suggested:
     a. **Local peak shaving**: After detecting the central frequencies related to the narrow-band noises, the mean value of the complex components around this frequency is substituted as the initial signal complex value in the frequency domain. In other words, at frequency fx, a peak amplitude is detected; in this case, the following value must be replaced by the sinusoidal component with frequency fx:

$$Z_x = \left(\sum_{i=fx-dfx}^{fx+df} Z_i \right) / n \tag{16.5}$$

   where n is the number of samples that are averaged, Z is the complex Fourier value at frequencies x or i, and the choice of the number of samples is achievable by the trial-and-error method and considering the sampling frequency and distances between the peaks. Considering the distributed PD spectrum density, this algorithm is justifiable because, in the amplitude–frequency diagram, the amplitude of the sudden variations with frequency is eliminated, leading to a signal Fourier transform more similar to the PD.

     b. **Proportional peak shaving:** Local peak shaving largely prevents the variation in the PD signal. A more precise method, but that is more time consuming, involves obtaining the first frequency spectrum of an expected sample from the PD signal (this sample spectrum can be extracted from a part of the noisy signal known to contain a PD pulse). The values at frequencies depending on the narrow-band noises are then substituted by values obtained from this spectrum. The drawback of this method is that the application of the algorithm is more complicated and time-consuming. No significant difference is noted between the proportional peak shaving and local peak shaving methods.

5. Transferring the obtained signal to the time domain: The obtained signal, after passing the above algorithm, is considered free of the narrow-band noise.

16.4.9 Denoising Using Wavelet Transform

Wavelet transform can provide very appropriate data for the frequency components of the signal over any frequency range by considering the precise decomposition of a signal mixed with noise over different frequency ranges. Wavelet transform can eliminate wide-band noise with the

Figure 16.18 Decomposition and reconstruction of a signal using the wavelet transform method.

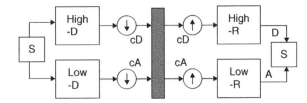

help of a threshold for separating the main signal from the noise over different ranges. The continuous wavelet transform generates many numbers of data, while the discrete wavelet transform produces 2^j numbers of coefficients related to the scale and $K\ 2^j$ coefficients related to displacement to achieve higher calculation speeds. A single-level discrete wavelet generates a series of approximate coefficients (CA) and a series of detail coefficients (CD) passing through a low-pass filter and high-pass filter, respectively (Figure 16.18) [16]. The approximated and detailed data can be recovered by the high-pass filter and the low-pass filter. Multi-level recombination with the maximum level of j gives a series of approximated and detailed coefficients corresponding to a few frequency sub-band signals [14, 17].

Noise rejection based on wavelet transform has a simple principle. First, the main signal is divided into approximated and detailed components, based on a number of defined levels. This step is performed by selecting a suitable mother wavelet function according to the noise characteristics. The noise and PD components are then identified at various levels by visual and frequency characteristics. Finally, the coefficients related to the noise in different levels are eliminated and the PD-dependent coefficients with acceptable threshold values remain. Signal reconstruction based on modified coefficients will give a noise-free signal. The signal decomposition method and extraction of the main signal based on the wavelet transform are shown in Figure 16.19. In summary, the decomposition process using wavelet transform involves the following steps:

- Mother wavelet function selection.
- Find the approximate and detailed signal at each level.
- Remove noise-related coefficients.
- Signal reconstruction (noise-free signal).

Figure 16.20 illustrates one of the proposed algorithms for the wavelet transform. The following factors have a significant impact on noise rejection and must be chosen carefully:

- Proper selection of the mother wavelet function.
- Selection of the number of levels.
- Selection of threshold parameters.

16.5 Data Interpretation

Different methods have been proposed for the measurement of PD. Interpretation of the data related to each method has its particular complexities, where the variety of the methods makes it difficult to interpret the data. However, it is necessary to interpret these data and identify the PD. Different types of PD do not have the same level of risk; some of the discharges in the generator do not constitute a problem and the generator can continue to operate. Data interpretation can be separated into three parts: low range, VHF/UHF range, and based on artificial intelligence (AI).

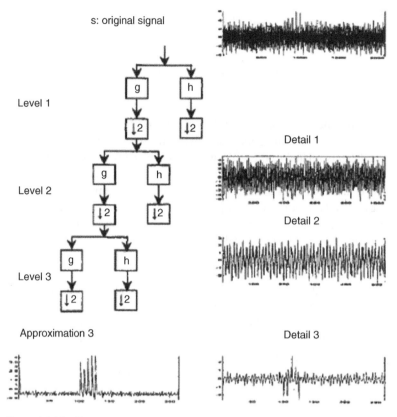

Figure 16.19 Signal three-level decomposition with wavelet transform.

Note that a true calibration cannot be achieved for stator windings. Discharges occurring some distance away from the calibration point will already have attenuated with propagation.

16.5.1 Data Interpretation in the Low-Frequency Range

Considering the measurement and simplicity of the data analysis based on the low-frequency method compared with other measurement methods, excellent information is available on the data interpretation in this method. Initially, the data analysis for this measurement method is carried out using the analysis of the PD signal amplitude and then is performed using the PRPD characteristics. The defect type is detected based on the obtained characteristics [19, 20]. In Figure 16.21, the right-hand figure is the distribution of the amplitude of the PD pulses and the left-hand figure is the PRPD characteristic [12]. For different sources of PD, these characteristics are unique and can be used for recognition of the PD source.

Following the PD measurement (the measurement of amplitude and/or amplitude as well as phase angle), the obtained characteristic is compared with the characteristics of Figure 16.21 and a figure more similar to that type of PD is chosen as the PD source. As seen in Figure 16.21, this can be conducted with higher precision by the PRPD characteristic because it contains sufficient data. The PRPD characteristic gives considerably more data compared to the amplitude analysis; therefore, using this characteristic for data interpretation is preferred to the amplitude analysis method.

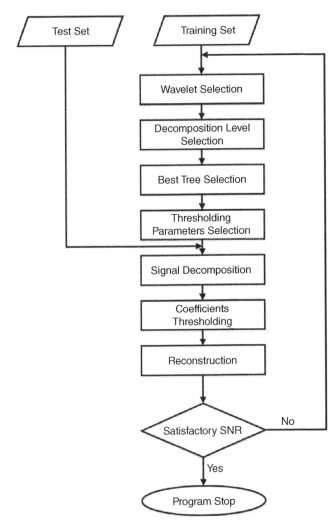

Figure 16.20 A suggested algorithm for denoising based on the wavelet approach [18]/ The Institution of Engineering and Technology.

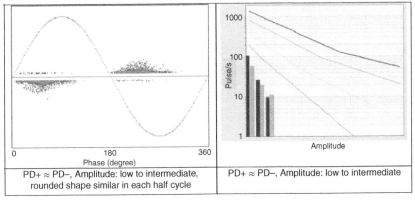

(a) Internal discharge

Figure 16.21 Different PD characteristics [12] / with permission of IEEE: (a) internal discharge, (b) slot discharge, (c) gap discharge, (d) surface discharge, and (e) discharge due to delamination.

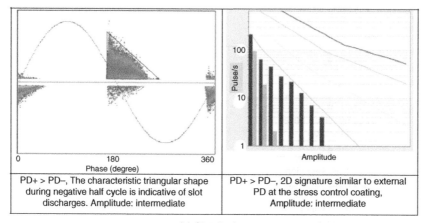

PD+ > PD−, The characteristic triangular shape during negative half cycle is indicative of slot discharges. Amplitude: intermediate	PD+ > PD−, 2D signature similar to external PD at the stress control coating, Amplitude: intermediate

(b) Slot discharge

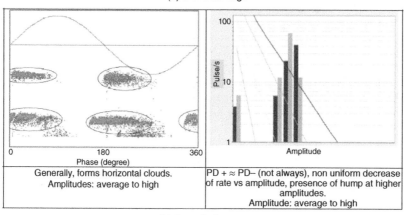

Generally, forms horizontal clouds. Amplitudes: average to high	PD + ≈ PD− (not always), non uniform decrease of rate vs amplitude, presence of hump at higher amplitudes. Amplitude: average to high

(c) Gap discharge

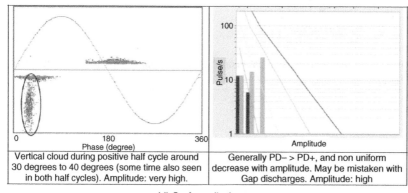

Vertical cloud during positive half cycle around 30 degrees to 40 degrees (some time also seen in both half cycles). Amplitude: very high.	Generally PD− > PD+, and non uniform decrease with amplitude. May be mistaken with Gap discharges. Amplitude: high

(d) Surface discharge

Figure 16.21 *(Continued)*

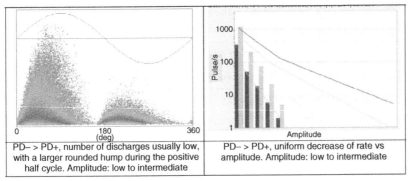

| PD– > PD+, number of discharges usually low, with a larger rounded hump during the positive half cycle. Amplitude: low to intermediate | PD– > PD+, uniform decrease of rate vs amplitude. Amplitude: low to intermediate |

(e) Discharge due to delamination

Figure 16.21 (*Continued*)

16.5.2 Data Interpretation in VHF and UHF Measurements

VHF and UHF measurements, particularly the online VHF measurements using coupler capacitors, are more common than the low-frequency measurements. Since the PD measurement in these methods has no calibration capability, the data interpretation is considerably difficult. In the low-frequency measurement, the calibration can be performed; therefore, the measured output is the electric charge in pC and the PD severity can be detected. However, in other methods with no calibration capability, the PD signal in mV cannot interpret the PD severity [21].

To be able to obtain more data from the measured signals using these data, the data for measurement methods, encompassing more than 400 000 PD tests, have already been collected for different generators. The summary of this database for the two measurement methods is given in Table 16.2. The results of the VHF measurements using an 80 pF capacitive coupler sensor (in mV) have also been collected. This table classifies the results for voltage level and insulation type (air or hydrogen). For example, in generators with voltage levels between 16 and 18 kV and hydrogen pressure between 206 and 345 kPa, 25% of the PD have a level less than 24 mV, 50% less than 43 mV, and 75% less than 85 mV.

Table 16.3 summarizes the results similar to Table 16.2, except that an SSC sensor has been used in Table 16.3. According to the inspections of the tested generators and more than 10 years

Table 16.2 Cumulative probability of PD occurrence in a turbo-generator using an 80 pF capacitive coupler sensor (in mV) [22].

Cumulative probability range (%)	Operating value					
	13–15 kV	16–18 kV		>19 kV		
	Air	Air	206–345 kPAgH$_2$	Air	206–345 kPAgH$_2$	>345 kPAgH$_2$
<25	55	41	24	42	23	9
<50	124	72	43	85	55	28
<75	265	157	85	165	108	77
<90	529	310	194	504	161	255*
<95	778	579	307	750	206	732*

Table 16.3 Cumulative probability of PD occurrence in a turbo-generator using an SSC sensor (in mV) [22].

Cumulative probability range (%)	Operating Value						
	11–15 kV	16–18 kV		>19–22 kV		23–27 kV	
	Air	Air	206–345 kPAgH$_2$	206–345 kPAgH$_2$	>345 kPAgH$_2$	206–345 kPAgH$_2$	>345 kPAgH$_2$
<25	0	0	0	0	0	0	0
<50	15	1	5	7	3	8	3
<75	37	13	15	19	11	31	9
<90	87	77	22	41	25	66	18
<95	119	122	34	57	39	100	29

of experiments on this type of measurement, if the PD level exceeds 90% of the generator's PD level, it will be a serious warning about the generator insulation condition [23]. For example, in the generators with voltage levels between 16 and 18 kV and hydrogen pressures between 206 and 345 kPa, the surface PD of 194 mV for measurement using a capacitive coupler sensor of 80 pF and the PD level of 22 mV for measurement with an SSC sensor indicate severe problems in the generator insulation condition.

In addition to the interpretation of numerical values in the online measurement using these tables, identifying the type of PD source is also necessary. This has already been implemented for the low-frequency offline measurement. The PRPD characteristics related to the online measurement over high frequency (2–20 MHz) are then given in reference [1]. Figure 16.22 shows that these online characteristics are very similar to the offline case.

16.5.3 Data Interpretation Based on Artificial Intelligence

AI provides an unmanned recognition of PDs. PD pattern recognition based on AI techniques is divided into supervised and unsupervised methods. The supervised methods use labeled datasets to train the algorithm, while the unsupervised methods do not provide information, or labels, on the nature of the data. The main intention of supervised learning is to discriminate a single PD source or type based on the labeled data extracted using phase-resolved or pulse sequence PD patterns. The expert system method [24] was an early example of a supervised AI method. Different kinds of neural networks, such as backpropagation [25], cascade [26], counter propagation [27], radial basis function [28], extension [29], ensemble [30], and probabilistic [31] neural networks, as well as support vector machine (SVM) [32, 33], are commonly used to determine the PD type or source detection based on supervised learning.

The classification of multiple PD sources is achieved using unsupervised machine learning tools, which are used for extracting features based on PD pulse shapes [34, 35]. Note that most studies for unsupervised and supervised studies are performed under laboratory conditions, with artificial defects. A few studies were performed on high-voltage machines in operation, such as described in reference [36], which were performed on eight 11 kV machines, and reference [37], which were performed on a 6-kV machine. Further independent studies on machines in operation are needed to validate the various AI data interpretation approaches.

The extracted feature sets from laboratory tests or field tests are fed into a neural network. The number of neurons and the number of layers in a neural network depend on the number of labeled features used. The number of outputs also depends on the number of predefined classes [38]. Neural networks have some disadvantages, as listed below:

1. Require a large number of datasets for training.
2. Sensitive to the training algorithm.
3. Sensitive to weight functions and local minimums.

SVM is also used for PD data interpretation. The SVM method works satisfactorily to classify two categories, although the performance for more than two categories is not satisfactory. A combination of different methods, such as least square, fuzzy logic, and particle swarm optimization, with SVM leads to acceptable results [32, 33, 39–42].

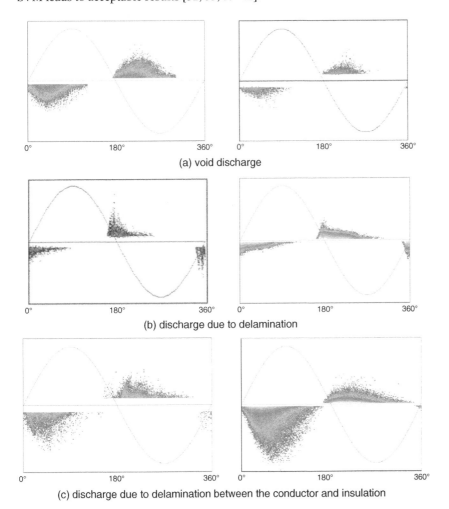

(a) void discharge

(b) discharge due to delamination

(c) discharge due to delamination between the conductor and insulation

Figure 16.22 PRPD due to a different type of discharges (right – offline measurement, left – online measurement) [1]: from top, (a) void discharge, (b) discharge due to delamination, and (c) discharge due to delamination between the conductor and insulation, and (a) slot discharge, (b) corona discharge, (c) surface discharge, and (d) gap discharge.

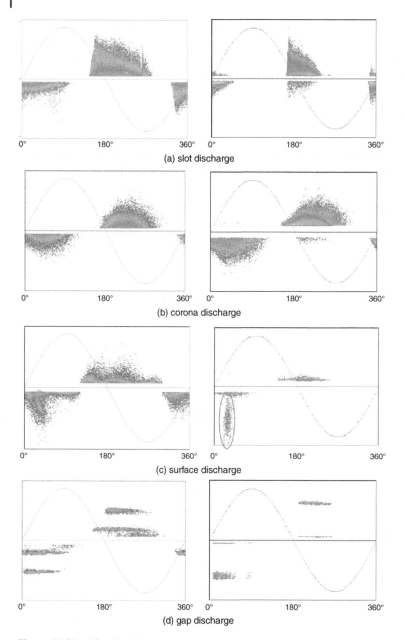

(a) slot discharge

(b) corona discharge

(c) surface discharge

(d) gap discharge

Figure 16.22 (*Continued*)

Although PD data interpretation using supervised learning, without the subsequent use of a human expert, is a great achievement, the sources of PD in large electric machines are diverse. Consequently, PD source detection based on unsupervised machine learning tools, such as k-means, and its combination with other methods, such as fuzzy network, have attracted research attention [43]. In reference [44], k-means are applied to various PD sources to separate them into clusters, where k defines the number of clusters. The input data to k-means can be statistical data derived from PRPD, pulse shape data, or extracted features using FFT or wavelet transforms.

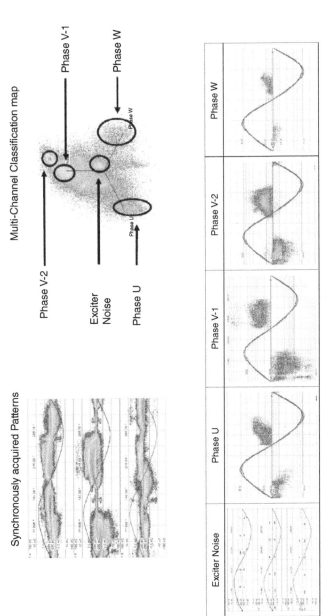

Figure 16.23 Separating PD different sources using the 3PARD method [12] / with permission of IEEE.

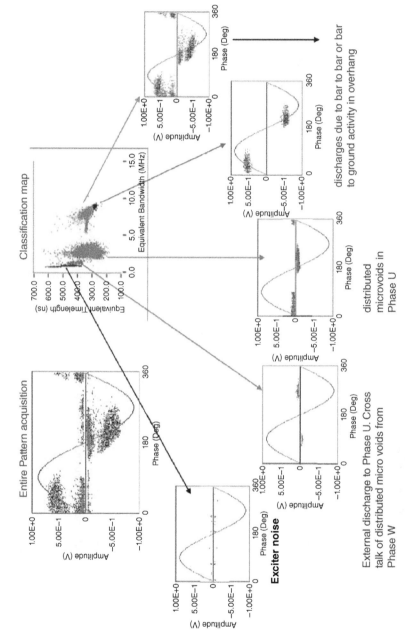

Figure 16.24 Classification of different sources of PD using the TF method [12] / with permission of IEEE.

Although *k*-means is a simple approach, optimization of its primary centroids is critical, and it is also sensitive to the local minimum. Prior knowledge is also required to estimate the number of clusters that PD sources fall into. The main advantage of using AI techniques is that the competence of a highly specialized human expert is not needed for data interpretation. However, many challenges impede their implementation, as listed below:

1. A large volume of tagged PD data is required for reliable supervised machine learning to determine the single source of PD. In the industry, the number of PD sources is not limited to one. Providing adequate labeled data for multiple online PD source detection based on supervised machine learning is time-consuming and difficult.
2. Multiple PD source classification based on unsupervised machine learning requires a priori knowledge, which is difficult to provide.
3. Few studies have been performed on operating machines or with validation of proposed algorithms by separate identification of the real PD sources in the machine.

16.6 Separating PD Sources

In the previous discussions, the assumption was made that only one type of PD occurs. When several types of PD occur simultaneously, analyzing the results is difficult. Therefore, separating the different sources of the faults is necessary and then the results can be interpreted. In the following section, different methods of separating the PD sources methods are discussed.

The first solution is to use the 3PARD method. This method, as previously used for noise detection, can also be used to separate PD sources. Figure 16.23 shows how to use this method to separate PD sources. This method is not very good at separating PD sources and is more applicable to noise separation, especially external noise, such as corona.

Two other methods for using the current features to separate the PD sources have already been used for diagnosing and separating the noise. Figure 16.24 shows how the PD sources are separated using the TF method [45]. Another method for separating the PD sources is the use of 3CFRD. Figure 16.25 shows its principles. In this method, the PD signal is passed to three band-pass filters with different central frequencies, and the peak filtered signal is considered the feature of the

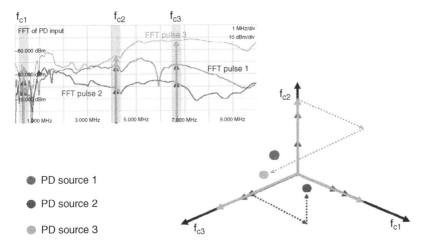

Figure 16.25 Principle of the 3CFRD method for separating different sources of PD.

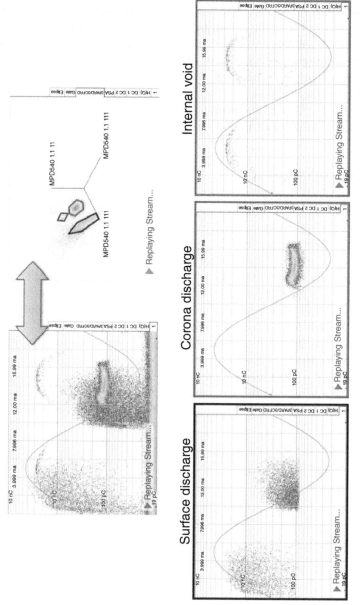

Figure 16.26 Separating different sources of PD using the 3CFRD technique.

signal. Consequently, three features are extracted for every PD pulse. These three features are plotted in Figure 16.26, which shows that the PD signals coordinate with different frequencies form separate sets. The central frequency for these three band-pass filters is selected so that it can show the greatest difference between different PD sources. Figure 16.26 shows how to use this method to isolate different sources of PD using the 3CFRD method.

16.7 Summary

To ensure a robust and reliable PD diagnostic system, the type and number of sensors, fault detection type, PD source separation method, noise-rejection system, and data interpretation method must be selected. The number of sensors that give adequate data must be sufficient to facilitate data interpretation. If no sensors have already been fixed on the installed motor or generator for PD detection, retrofitting of capacitive sensors is most commonly applied today.

Regarding SSC sensors, a limited number of sensors have been installed in the stator, but they may not cover all discharges occurring in the machine. Thus, the use of capacitive couplers may be preferable. If LF or HF measurement is applied, several noise-rejection approaches or source separation methods can be used, such as TF mapping or the 3PARD/3CFRD method. If VHF and UHF measurements are utilized, the methods themselves are able to eliminate noise, so an additional method for noise rejection is often unnecessary. To detect the main cause of the apparent PD signals, the PRPD characteristic is often used, along with a comparison to existing databases. However, this should be performed with care, as the PRPD analysis does not always reveal the actual failure process. During online measurements, the variation in PD sensor signals with data from other sensors detecting the load, winding temperature, and humidity should be assessed. Using the trend of essential parameters of the PD is a precise method for data interpretation. Moreover, the threshold level also provides an accurate method wherein the level of threshold relies on the electric machine parameters and measurement method.

The approaches for noise cancellation, PD source segregation, and data interpretation provide adequate information from the PD regarding the insulation condition. An application of the online PD detection based on improved approaches is increasing in a large range of industries since it could reduce economic losses.

References

1 IEC TS 60034-27-2: 2017 (2017). On-line partial discharge measurements on the stator winding insulation of rotating electrical machines

2 Carvalho, A.T., Lima, A.C.S., Cunha, C.F.F.C. et al. (2015). Identification of partial discharges immersed in noise in large hydro-generators based on improved wavelet selection methods. *Measurement* 75: 122–133.

3 Boggs, S.A. and Stone, G.C. (1982). Fundamental limitations in the measurement of Corona and partial discharge. *IEEE Transactions on Electrical Insulation* EI-17 (2): 143–150.

4 IEC,TS 60034-27-1: 2017 (2017). Off-line partial discharge measurements on the stator winding insulation of rotating electrical machine.

5 (2014). *IEEE Guide for the Measurement of Partial Discharges in AC Electric Machinery*, IEEE Std 1434-2014 (Revision of IEEE Std 1434-2000), 1–89.

6 Stone, G.C. and Warren, V. (2006). Objective methods to interpret partial-discharge data on rotating-machine stator windings. *IEEE Transactions on Industry Applications* 42 (1): 195–200.

7 Li, X., Li, C.R., Ding, L. (2004). et al., Generator PD measurement by using UHF and HF sensors. *IEEE International Symposium on Electrical Insulation*, 19–22 September 2004, Indianapolis, IN, USA, pp. 22–25.

8 Cavallini, A., Montanari, G.C., Fabiani, D., et al. (2009). Advanced technique for partial discharge detection and analysis in power cables. *International Conference on Condition Monitoring and Diagnostic Engineering Management of Power Station/Substation Equipment*, pp. 1–4, 2009.

9 Campbell, S.R., Stone, G.C., and Sedding, H.G. (1992). Application of pulse width analysis to partial discharge detection. *IEEE International Symposium on Electrical Insulation*, 7–10 June 1992, Baltimore, MD, USA, pp. 345–348.

10 Cavallini, A., Contin, A., Montanari, G.C. et al. (2003). Advanced PD inference in on-field measurements. I. Noise rejection. *IEEE Transactions on Dielectrics and Electrical Insulation* 10 (2): 216–224.

11 Itoh, K., Kaneda, Y., Kitamura, S. et al. (1996). New noise rejection techniques on pulse-by-pulse basis for on-line partial discharge measurement of turbine generators. *IEEE Transactions on Energy Conversion* 11 (3): 585–594.

12 (2014). *IEEE Guide for the Measurement of Partial Discharges in AC Electric Machinery*, IEEE Std 1434-2014 (Revision of IEEE Std 1434-2000), 1–89. USA.

13 Kopf, U. and Feser, K. (1995). Rejection of narrow-band noise and repetitive pulses in on-site PD measurements. *IEEE Transactions on Dielectrics and Electrical Insulation* 2 (6): 1180–1191.

14 Xiaoning, W., Zhu, D., Fuqi, L., et al. (2003). Analysis and rejection of noises from partial discharge (PD) on-site testing environment. *Proceedings of the 7th International Conference on Properties and Applications of Dielectric Materials*, Nagoya, June 1–5, 2003.

15 Özkan, A.O. and Kalenderli, O. (2010). Noise reduction on partial discharge data with wavelet analysis and appropriate thresholding. *IEEE International Conference on High Voltage Engineering and Application (ICHVE)*, 11–14 October 2010, New Orleans, LA, USA.

16 Wang, P, Lewin, P.L., Tian, Y., et al. (2004). Application of wavelet-based de-noising to online measurement of partial discharges. *International Conference on Solid Dielectrics*, July 5–9, 2004, Toulouse, France.

17 Ma, X., Zhou, C., and Kemp, I.J. (2002). Interpretation of wavelet analysis and its application in partial discharge detection. *IEEE Transactions on Dielectrics and Electrical Insulation* 9 (3): 446–457.

18 Chang, C.S., Jin, J., Kumar, S. et al. (2005). De noising of partial discharge signals in wavelet packets domain. *IEE Proceedings -Science Measurement and Technology* 152 (3): 129–140.

19 Li, S. and Chow, J.M.Y. (2007). Partial discharge measurements on hydro generator stator windings case studies. *IEEE Electrical Insulation Magazine* 23 (3): 5–15.

20 Bélec, M., Hudon, C., and Nguyen, D.N. (2006) Statistical analysis of partial discharge data. *Conference Record of the 2006 IEEE International Symposium on Electrical Insulation*, 11–14 June 2006, Toronto, ON, Canada, pp. 122–125.

21 Stone, G.C. and Sedding, H.G. (2013). Comparison of UHF antenna and VHF capacitor PD detection measurements from turbine generator stator windings. In: *IEEE International Conference on Solid Dielectrics (ICSD)*, 30 June–4 July 2013, 63–66. Bologna, Italy.

22 Warren, V. (2014). How much PD is too much PD. *University of Texas First Conference on On-Line Monitoring of Electric Assets*.

23 Stone, G.C., Chan, C., and Sedding, H.G. (2015). Relative ability of UHF antenna and VHF capacitor methods to detect partial discharge in turbine generator stator windings. *IEEE Transactions on Dielectrics and Electrical Insulation* 22 (6): 3069–3078.

24 Kranz, H. (2005). PD pulse sequence analysis and its relevance for on-site PD defect identification and evaluation. *IEEE Transactios on Dielectrics and Electrical Insulation* 12 (2): 276–284.

25 Jin, J., Chang, C., Hoshino, T. et al. (2006). Classification of partial discharge events in gas-insulated substations using wavelet packet transform and neural network approaches. *IEE Proceedings-Science, Measurement and Technology* 153 (2): 55–63.

26 Salama, M.M. and Bartnikas, R. (2002). Determination of neural-network topology for partial discharge pulse pattern recognition. *IEEE Transactions on Neural Networks* 13 (2): 446–456.

27 Hoof, M., Freisleben, B., and Patsch, R. (1997). PD source identification with novel dis-charge parameters using counter propagation neural networks. *IEEE Transactions on Dielectrics and Electrical Insulation* 4 (1): 17–32.

28 Shim, I., Soraghan, J.J., and Siew, W. (2001). Detection of PD utilizing digital signal processing methods. Part 3: Open-loop noise reduction. *IEEE Electrical Insulation Magazine* 17 (1): 6–13.

29 Chen, H.C., Gu, F.C., and Wang, M.H. (2012). A novel extension neural network-based partial discharge pattern recognition method for high-voltage power apparatus. *Expert Systems with Applications* 39 (3): 3423–3431.

30 Masud, A.A., Stewart, B., and McMeekin, S. (2014). Application of an ensemble neural net-work for classifying partial discharge patterns. *Electric Power Systems Research* 110: 154–162.

31 Karthikeyan, B., Gopal, S., and Venkatesh, S. (2008). Partial discharge pattern classification using composite versions of probabilistic neural network inference engine. *Expert Systems with Applications* 34 (3): 1938–1947.

32 de Oliveira Mota, H., da Rocha, L.C.D., de Moura. Salles, T.C. et al. (2011). Partial discharge signal denoising with spatially adaptive wavelet thresholding and support vector machines. *Electric Power Systems Research* 81 (2): 644–659.

33 Hao, L. and Lewin, L.P. (2010). Partial discharge source discrimination using a support vector machine. *IEEE Transactions on Dielectrics and Electrical Insulation* 17 (1): 189–197.

34 Sahoo, N.C., Salama, M.M.A., and Bartnikas, R. (2005). Trends in partial discharge pat-tern classification: a survey. *IEEE Transactions on Dielectrics and Electrical Insulation* 12 (2): 248–264.

35 Raymond, W.J.K., Illias, H.A., Mokhlis, H. et al. (2015). Partial discharge classifications: review of recent progress. *Measurement* 68: 164–181.

36 Gopinath, S., Sathiyasekar, K., Padmanaban, S. et al. (2019). Insulation condition assessment of high-voltage rotating machines using hybrid techniques. *IET Generation, Transmission and Distribution* 13 (2): 171–180.

37 Cavallini, A., Conti, M., Contin, A. et al. (2003). Advanced PD inference in on-field measure-ments. II. Identification of defects in solid insulation systems. *IEEE Transactions on Dielectrics and Electrical Insulation* 10 (3): 528–538.

38 Masud, A.A., Ardila, J.A., Albarracin, R. et al. (2017). Comparison of the performance of artifi-cial neural networks and fuzzy logic for recognizing different partial discharge sources. *Energies* 10 (7): 1060.

39 Wenrong, S., Junhao, L., and Peng, Y. (2008). Digital detection, grouping, and classification of partial discharge signals at DC voltage. *IEEE Transactions on Dielectrics and Electrical Insulation* 15 (6): 1663–1674.

40 Warren, V. (1998). How much PD is too much PD?, in *Iris Rotating Machinery Conference*.

41 Robles, G., Parrado-Hernández, E., Ardila-Rey, J. et al. (2016). Multiple partial discharge source discrimination with multiclass support vector machines. *Expert Systems with Applications* 55: 417–428.

42 Poyhonen, S., Conti, M., Cavallini, A. et al. (2004). Insulation defect localization through partial discharge measurements and numerical classification. *IEEE International Symposium on Industrial Electronics* 1: 417–422.

43 Likas, A., Vlassis, N., and Verbeek, J.J. (2003). The global k-means clustering algorithm. *Pattern Recognition* 36 (2): 451–461162.

44 Chatpattananan, V., Pattanadech, N., and Yutthagowith, P. (2006). Partial discharge classification on high voltage equipment with k-means. In: *IEEE 8th International Conference on Properties and Applications of Dielectric Materials*, 26–30 June 2006, 191–194. Bali, Indonesia.

45 Cavallini, A., Conti, A., Contin, M.A., and Montanari, G.C. (2003). Advanced PD inference in on-field measurements. II. Identification of defects in solid insulation systems. *IEEE Transactions on Dielectrics and Electrical Insulation* 10 (3): 528–538.

Index

Electromagnetic Analysis and Condition Monitoring of Synchronous Generators, First Edition. Hossein Ehya and Jawad Faiz.
© 2023 The Institute of Electrical and Electronics Engineers, Inc. Published 2023 by John Wiley & Sons, Inc.

IEEE Press Series on Power and Energy Systems

Series Editor: Ganesh Kumar Venayagamoorthy, Clemson University, Clemson, South Carolina, USA.

The mission of the IEEE Press Series on Power and Energy Systems is to publish leading-edge books that cover a broad spectrum of current and forward-looking technologies in the fast- moving area of power and energy systems including smart grid, renewable energy systems, electric vehicles and related areas. Our target audience includes power and energy systems professionals from academia, industry and government who are interested in enhancing their knowledge and perspectives in their areas of interest.

Printed and bound by CPI Group (UK) Ltd, Croydon, CR0 4YY

16/04/2025

14658605-0004